Space Sciences Series of ISSI

Volume 49

For further volumes:
www.springer.com/series/6592

Maurizio Falanga • Tomaso Belloni •
Piergiorgio Casella • Marat Gilfanov •
Peter Jonker • Andrew King
Editors

The Physics of Accretion onto Black Holes

Previously published in *Space Science Reviews* Volume 183,
Issues 1–4, 2014

Springer

Editors
Maurizio Falanga
International Space Science Institute
Bern, Switzerland

Marat Gilfanov
Max Planck Institute for Astrophysics
Garching, Germany

Tomaso Belloni
The National Institute of Astrophysics
Merate, Italy

Peter Jonker
Netherlands Institute for Space Research
Utrecht, The Netherlands

Piergiorgio Casella
The National Institute of Astrophysics
Rome, Italy

Andrew King
University of Leicester
Leicester, UK

ISSN 1385-7525 Space Sciences Series of ISSI
ISBN 978-1-4939-2226-0 ISBN 978-1-4939-2227-7 (eBook)
DOI 10.1007/978-1-4939-2227-7
Springer New York Heidelberg Dordrecht London

Library of Congress Control Number: 2014952460

Cover illustration: Artist impression of an accreting black hole by H. Flinterman (Studio WW15) & SRON, Netherlands Institute for Space Research.

Printed on acid-free paper

Springer is part of Springer Science+Business Media (www.springer.com)

Contents

About the Editors

Professor Maurizio Falanga is the Science Program Manager at the International Space Science Institute. His research areas include high-energy astrophysics, observations and numerical simulations.

Professor Tomaso Belloni is a senior scientist at INAF's Brera Astronomical Observatory. He is an expert in X-ray astronomy, accreting sources and time series analysis.

Dr. Piergiorgio Casella is an astronomer at INAF's Rome Astronomical Observatory. He is an expert in high-energy astrophysics, accretion and jet physics.

Professor Marat Gilfanov is a senior scientist at the Max Planck Institute for Astrophysics. His research areas include theoretical and observational astrophysics, black holes and galaxies.

Professor Peter Jonker is an astronomer at SRON. He is an expert in high-energy astrophysics and observations.

Professor Andrew King is a senior scientist at the University of Leicester. He is an expert in theoretical astrophysics, accretion, black holes and galaxies.

DOI 10.1007/978-1-4939-2227-7_1
Reprinted from *Space Science Reviews* Journal, DOI 10.1007/s11214-014-0044-8

Foreword

Maurizio Falanga · Luigi Stella

Published online: 16 April 2014

Black holes were predicted long before the beginning of the space age; they were perceived as by-products of mathematical theories, existed only in the imagination of a few scientists. The idea of "dark stars" (they were dubbed "black holes" only in 1968) can be traced back to the late 18th century, when John Michell (English philosopher and geologist) and some years later to Pierre-Simon Laplace (French mathematician and astronomer) speculated that, if a planet or a star were dense enough, their escape velocity would equal the speed of light. Light particles (photons) leaving the surface of such a world, would rise, stop, and then fall back down like projectiles do. This "Newtonian" view of black holes, while conceptually interesting, is not an adequate description of what happens to light near a massive dense body.

By the end of the 19th century strong evidence had been found that the speed of light is a universal constant, which remains the same in any reference frame. By exploiting the constancy of the speed of light and the principle of relativity (stating that the laws of physics should remain the same in any *inertial* reference frame) in application to Maxwell's equations of electromagnetism, Albert Einstein developed in 1905 a new theory (the Theory of Special Relativity) that led to a deep revision of the concepts of space and time: contrary to simple intuition, space and time intervals do not remain unchanged for observers in motion with respect to one another.

After a decade of attempts, Einstein succeeded in formulating a theory gravity (and electromagnetism) obeying the general principle of Relativity, i.e. that physical laws are the same in *all* reference frames (inertial or non-inertial). Einstein's basic concept was to drop Newton's idea of a force (the gravitational force) that is responsible for the attraction of masses. In place of that he was guided by of what he defined the "happiest thought of his life": that the effects of gravity are cancelled in a body that accelerates because it falls

M. Falanga (✉)
International Space Science Institute (ISSI), Hallerstrasse 6, 3012 Bern, Switzerland
e-mail: mfalanga@issibern.ch

L. Stella
INAF—Osservatorio Astronomico di Roma, Via Frascati, 33, Monteporzio Catone, Rome 00040, Italy

freely, no matter what the body is. Stated differently, acceleration can mimic gravity and, vice-versa, gravity can mimic acceleration: that is the "Equivalence principle". By using the mathematical instrument of differential geometry, Einstein formulated a new theory, the Theory of General Relativity, in which gravitation and motion of both matter and light result from the geometric properties of space-time (rather than Newton's attraction force). In turn the geometry of space-time, the matter and light in particular, determine its varying curvature.

Some solutions of Einstein's equations of General Relativity predict that a sufficiently compact mass will curve space-time such much that nothing, not even light, can escape from inside a critical surface, the so-called event horizon: that provided the modern foundation of the black hole concept. According to the so-called "No Hair Theorem" stationary black holes are completely characterized by only three observables: mass, angular momentum and electric charge (the latter being irrelevant in astrophysical black holes). All other information (for which "hair" is a metaphor) about the matter which formed the black hole (or fell into it after formation) is lost behind the event horizon, and remains permanently inaccessible to external observers.

Since the early seventies, a wealth of observational evidence has been found for the presence of black holes in the universe. While nothing can escape from the event horizon, the regions in its immediate surroundings (that is, say, tens of times the horizon radius) can become very luminous and launch jets at speeds close to the speed of light. This happens when matter flows towards the black hole and releases up $\sim$40 % of its rest mass energy in the process. This "accretion" energy, might be supplemented by the extraction of part of the black hole's rotational energy.

Stellar mass black holes (4–15 solar masses) in binary systems are being discovered in increasing numbers in the Milky Way and nearby galaxies. Super-massive black holes, from millions to billion solar masses, exist in the centre of most if not all galaxies; the radiation they release when they are active deeply influences the evolution of their hosts. Though relatively quiescent, the $\sim$4 million solar mass black hole in the centre of our Milky Way is one of the best studied and offers very good prospects for direct imaging of the "shadow" caused by light bending in the vicinity of the event horizon.

Black holes are ideal laboratories for studying both physical properties of accretion onto compact objects and probing the effects of General Relativity in the strong field regime. These extreme phenomena are inaccessible to laboratory experiments. Through observations at high energies (mainly in the X-rays) and multiwavelength programs spanning the widest range of the electromagnetic spectrum, from the radio to TeV energies), our knowledge of astrophysical black holes has advanced considerably over the last two decades. Diagnostics have emerged which can directly probe the dynamics of matter motion very close to the black hole, where the strong field general relativistic effects become important. At the same time, considerable progress has been made developing advanced models and understanding the physics of accretion onto compact objects. Yet, a number of key issues remain poorly understood. For instance the interpretation and decomposition of the energy spectra of accreting black holes are still much debated. Similarly, different competing models are being investigated which explain at least part of the variability properties of black.

This book is presents a collection of reviews of astrophysical black hole. The first section of the contains very valuable introductory material about the history of the first observed black hole. The second section describes the physical models for the accretion flow around black holes of all masses, where the third and fourth sections describe the accretion on black holes from stellar mass to supermassive and its fundamental parameters. The fifth section is devoted to the accretion-jet interplay, while the last section reports an overview and outlook of black hole research.

It is our honour to warmly compliment the conveners and organizers of the Workshop; they conducted the whole workshop with great enthusiasm and dedication. We thank all those who participated in the workshop it is them who made it successful. This excellent book represents also an important outcome of the workshop: congratulations to all.

DOI 10.1007/978-1-4939-2227-7_2
Reprinted from *Space Science Reviews* Journal, DOI 10.1007/s11214-013-0011-9

Searching for Black Holes in Space

The Key Role of X-Ray Observations

Ken Pounds

Received: 15 February 2013 / Accepted: 19 July 2013 / Published online: 31 August 2013

Abstract Although General Relativity had provided the physical basis of black holes, evidence for their existence had to await the Space Era when X-ray observations first directed the attention of astronomers to the unusual binary stars Cygnus X-1 and A0620-00. Subsequently, a number of faint Ariel 5 and Uhuru X-ray sources, mainly at high Galactic latitude, were found to lie close to bright Seyfert galaxies, suggesting the nuclear activity in AGN might also be driven by accretion in the strong gravity of a black hole. Detection of rapid X-ray variability with EXOSAT later confirmed that the accreting object in an AGN is almost certainly a supermassive black hole.

Keywords Black holes · X-ray astronomy · Uhuru · Ariel 5 · EXOSAT · GINGA

1 Introduction

The first recorded suggestion that there may be stars too massive for light to escape from their surface appears to have been made in 1783 by John Michell (McCormack 1968; Hockey 2007), a rector working at a church near Leeds in Yorkshire. Michell, who had studied geology at Cambridge, but chose a better-remunerated position in the church, also proposed that binary stars must have a common origin (rather than by chance encounter) and hence seems a worthy first reference in this historical review. However, a physical basis for the implied 'light bending' had to await developments in general relativity by Einstein and Schwarzschild more than a century later.

A key to the subsequent discovery of black holes, a term proposed by John Wheeler in 1965, is the extremely small (Schwarzschild) radius of the event horizon, with a correspondingly large potential energy release from accreting matter prior to falling into the hole. A further consequence of the small emission region is that the radiation temperature will be high, indicating that much of the luminosity will lie in the far UV or X-ray wavebands.

K. Pounds (✉)
Department of Physics and Astronomy, University of Leicester, Leicester, UK
e-mail: kap@le.ac.uk

Such simple considerations make it clear, with hindsight, why the first evidence that black holes actually exist had to await the Space Era, when the first opportunity arose to send X-ray instruments above the Earth's atmosphere. In the event, although a number of remarkably bright non-solar X-ray sources were discovered from brief sounding rockets flights during the 1960s, convincing evidence for the existence of stellar black holes had to await the launch of the first dedicated satellites, in particular Uhuru and Ariel 5, launched in 1970 and 1974.

Theoretical developments through the 1960s were slowed by the uncertain nature of most cosmic X-ray sources, though the binary star identification of Sco X-1 was a strong stimulus to accretion disc theory (Shakura and Sunyaev 1973). Accretion onto a massive object in the nucleus of an active galaxy (AGN) was considered by Salpeter (1964) and by Zel'dovich and Novikov (1964), and developed with the benefit of more efficient disc accretion by Lynden-Bell (1969). However, it was only after the identification of Cygnus X-1 as a strong black hole candidate that a comparable stellar model was developed by Pringle and Rees (1972) and by Shakura and Sunyaev (1973).

It is hoped that this—necessarily—personal account will provide a useful review of one of the most significant developments in astrophysics during the Space Era.

2 The Origins of X-Ray Astronomy

Prior to the historic discovery (Giacconi et al. 1962) of a remarkably bright X-ray source in the constellation of Scorpius, in June 1962, most astronomers considered that observations in the ultraviolet and gamma ray bands offered the best promise for exploiting the exciting potential of space research. X-ray observations were expected to focus on the study of active stars, with fluxes scaled from that of the solar corona, the only known X-ray source at that time. Predictions ranged up to a thousand times the Sun's X-ray luminosity, but seemed beyond the reach of detection with then-current technology.

In a reflection of the contemporary thinking the recently formed US National Aeronautics and Space Agency (NASA) was planning a series of Orbiting Astronomical Observatories, with the first missions devoted to UV astronomy, although a proposal in 1961 from the UCL and Leicester groups to make simultaneous X-ray observations of the primary UV targets was accepted, and eventually flown on OAO-3 (Copernicus) eleven years later. In the USA, Riccardo Giacconi at American Science and Engineering (ASE) and Bruno Rossi of MIT had, still earlier, published the design of a more ambitious imaging X-ray telescope, with nested mirrors for increased throughput (Giacconi and Rossi 1960), while Rossi made what was to prove a visionary statement in declaring that 'Nature so often leaves the most daring imagination of man far behind'.

The ASE/MIT Aerobee 150 sounding rocket flight from the White Sands Missile Range in June 1962, finding in Sco X-1 a cosmic X-ray source a million time more luminous than the Sun (and brighter than the non-flaring corona at a few keV), began a transformation that laid the foundations for a revolution in High Energy Astrophysics.

Further sounding rocket observations quickly followed, with the US Naval Laboratory group—responsible for still earlier, but unrewarded night-time flights (Friedman 1959)—confirming Sco X-1 and finding a further source in Taurus (Bowyer et al. 1964a). The AS&E group (Gursky et al. 1963), and a team at Lockheed (Fisher et al. 1966) continued with launches from the White Sands missile range, while the British Skylark rocket was used to explore the southern sky from Woomera in South Australia from 1967 (Harries et al. 1967; Cooke et al. 1967), finding several sources in Centaurus and the Galactic centre region, exhibiting both thermal and non-thermal X-ray spectra (Cooke and Pounds 1971). With a hint of what lay ahead, repeated Skylark observations found the X-ray flux from Cen X-3 (first seen by the Lawrence Livermore Group; Chodil et al. 1967) varied by an order of magnitude, while Cen X-2 briefly outshone Sco X-1, before apparently disappearing.

Most sources remained of unknown nature, a notable exception being Tau X-1 which the NRL group had identified with extended X-ray emission from the Crab Nebula supernova remnant in a classic use of the Moon as an occulting disc (Bowyer et al. 1964b). Development of the modulation collimator (Oda et al. 1965), with an important refinement by Gursky, then yielded an arc minute position for Sco X-1 (Gursky et al. 1966), enabling its optical identification with a 13th magnitude blue star (Sandage et al. 1966). By the end of the decade some 20–30 cosmic X-ray sources had been reliably detected, the uncertainty being partly due to many cosmic X-ray sources being highly variable or transient. However, the majority remained unidentified, a major problem being their poorly determined positions on the sky.

The first small orbiting satellites dedicated to observations of cosmic X-ray sources began a transformation which quickly led to X-ray astronomy becoming a major branch of astrophysics. Uhuru led the way in December 1970, being launched into a low equatorial orbit and carrying a large proportional counter array to undertake a deep all-sky survey. Within a few months the extended observations made possible from orbit had shown that many X-ray sources were variable. This led to the important finding that Cen X-3 and Her X-1—and by implication probably many of the most luminous galactic sources—were binary star systems. Moreover, the rapid and periodic X-ray variations of Cen X-3 and Her X-1 showed the X-ray component to be most likely a neutron star.

A second remarkable Uhuru discovery—of extended X-ray emission from galaxy clusters quickly followed—while the number of known X-ray sources increased by an order of magnitude. The 3U catalogue (Giacconi et al. 1974), listing 161 sources, was a key milestone in the development of X-ray astronomy, and the major scientific impact of Uhuru is well recorded in Giacconi and Gursky (1974). The identification of Cygnus X-1 and its recognition as the first strong candidate to contain a black hole is described in more detail in Sect. 3.

Other Uhuru-class satellites followed, with Ariel 5 (UK), SAS-3 (USA) and Hakucho (Japan) dedicated to X-ray observations and OSO-7 (USA) and ANS (Netherlands) being solar and UV astronomy missions carrying secondary X-ray instrumentation. For astronomers in the UK, Ariel 5 brought an ideal opportunity to play a part in the rapid advances taking place. Like Uhuru, Ariel 5 (Fig. 1) was launched on a Scout rocket into a circular near-Earth orbit from a disused oil platform off the coast of Kenya. It carried 6 experiments, the most important in the context of this paper being a Sky Survey Instrument (SSI), similar to that on Uhuru, and an All Sky Monitor (from Goddard Space Fight Center). The Ariel 5 orbit was a good choice, not only in minimising background due to cosmic rays and trapped radiation, but in allowing regular data dumps from the small on-board data recorder. A direct ground and satellite link to the UK provided 6 orbits of 'Quick Look' data from the SSI within an hour of ground station contact. The remaining 'bulk' data were received within 24 hours, an immediacy that contributed substantially to the excitement of the mission operations, while also ensuring a rapid response to new discoveries.

One such discovery was particularly well-timed, with the SSI detecting a previously unseen source in the constellation Monoceros, just 2 days before the start of the first European Astronomy Society meeting, held in Leicester, where new X-ray results from the Ariel 5 and SAS-3 missions dominated the programme. The new X-ray source was to become a further strong black hole candidate, as reported in Sect. 4. In all, some 27 soft X-ray transients were

Fig. 1 Ariel 5 spacecraft. The Sky Survey Instrument (SSI) detector array is seen on the *upper left*

discovered as the Ariel 5 mission continued until 1980, though none were as powerful as A0620-00, the majority being neutron star binaries. A second enduring result from the Ariel 5 SSI was in establishing powerful X-ray emission as a characteristic property of Seyfert galaxies. The challenge of correctly identifying many previously unidentified sources, individually located to a few tenths of a sq. deg, was possible only because Seyfert nuclei are also unusually bright in the optical band. While the initial classification was made on a statistical basis, it held up well as new data emerged to establish Seyfert galaxies as a major class of extragalactic X-ray source (Sect. 5).

3 Cygnus X-1

Cygnus X-1 was first detected (Fig. 2) in an NRL Aerobee launch from White Sands in New Mexico in 1964 (Bowyer et al. 1965). Subsequent observations, including a balloon-borne telescope (Overbeck and Tananbaum 1968), showed an unusually hard power law spectrum, not unlike the Crab Nebula, but the lack of an obvious radio or optical counterpart (within the large position uncertainty typical of those early detections) limited detailed investigation.

The unusual nature of Cygnus X-1 became clearer from extended Uhuru observations (Fig. 3) which showed large amplitude fluctuations in the X-ray intensity on timescales down to 100 ms (Schreier et al. 1971), implying a correspondingly small emission region. An improved X-ray source location from Uhuru and an MIT rocket flight (Rappaport et al. 1971) led to the discovery of a weak transient radio source by Braes and Miley (1971) from

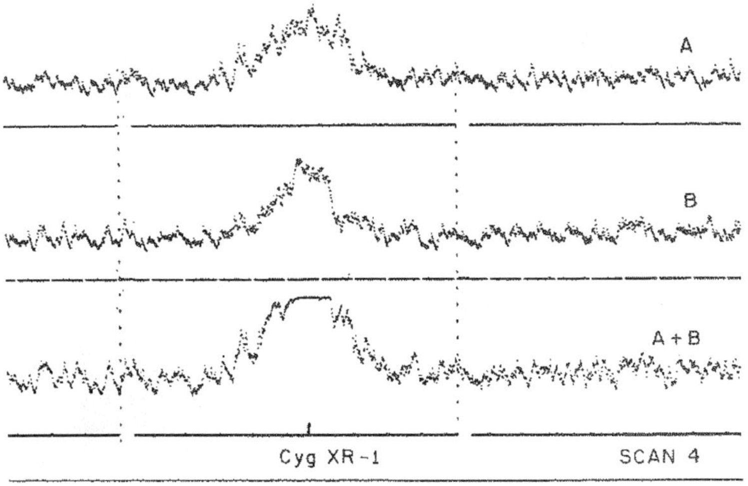

Fig. 2 Telemetry traces from the NRL rocket-borne Geiger counters which first detected Cygnus X-1 in 1964 (Bowyer et al. 1965)

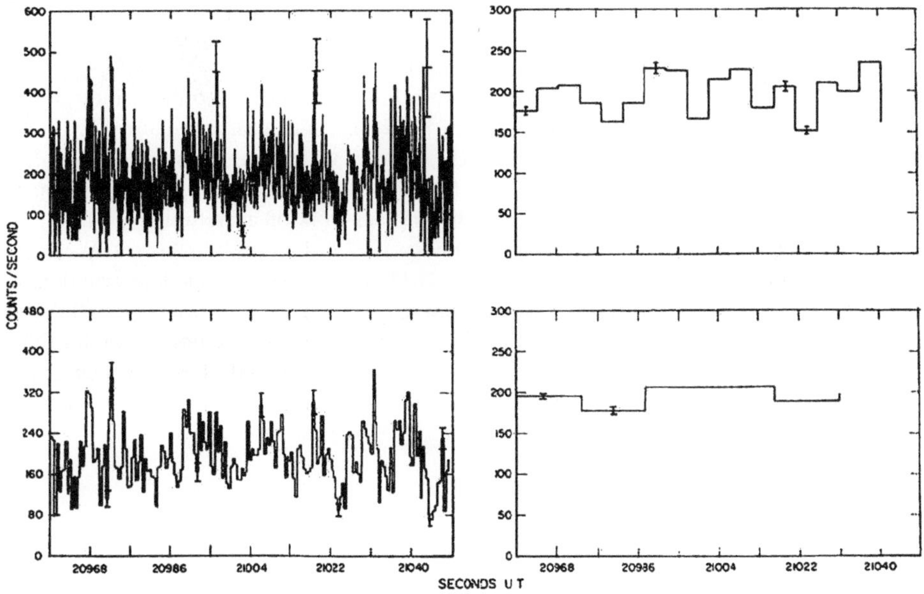

Fig. 3 Uhuru observation of rapid non-periodic variability in Cygnus X-1 binned at 0.096 s, 0.48 s, 4.8 s and 14.4 s. Typical 1σ error bars are shown (Schreier et al. 1971)

Leiden Observatory, and independently by Hjellming and Wade (1971) at the NRAO, which subsequent Uhuru analysis showed to coincide with a change in the X-ray appearance. The greatly improved source position (Fig. 4) finally allowed the optical identification of 'Cyg X-1—a Spectroscopic Binary with a Heavy companion' by Webster and Murdin (1972), at the Royal Greenwich Observatory, and Bolton (1972), at the David Dunlap Observatory in Toronto.

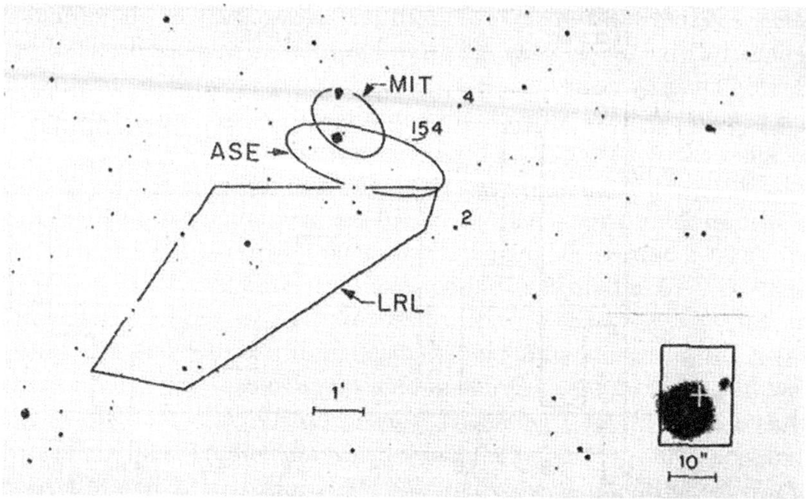

Fig. 4 X-ray observations of Cygnus X-1. HDE 226868 is the bright star in the overlap of the ASE and MIT error boxes. The *insert* shows the more precise coincidence of HDE 226868 with a transient radio source

The 'heavy' companion was HDE 226868, a supergiant star at some 2 kpc distance. Notwithstanding a likely mass of ~30 $M_\odot$, optical spectroscopy showed Doppler shifts of amplitude ~65 $km\,s^{-1}$ which, together with a binary period of 5.6 days, indicated the unseen (X-ray) companion to have a mass of ~15 $M_\odot$, much larger than the maximum value expected for a neutron star. Cygnus X-1 became widely—though not universally—acclaimed as the first stellar black hole (e.g. Shipman 1975), and understanding the nature of Cygnus X-1 was a primary motivation of theoretical work on accretion discs during that period (see Pringle 1977 for a contemporary review).

VLBA observations in 2009-10 (Reid et al. 2011) finally resolved the long-standing uncertainty in the distance of Cygnus X-1, obtaining a parallax distance of 1.86 ± 0.12 kpc. Re-analysis of extensive X-ray and optical data using the new value has allowed a refinement of the black hole mass of 14.8 ± 1.0 $M_\odot$ (Orosz et al. 2011), and shown that Cygnus X-1 contains a near-extreme Kerr black hole with a spin parameter $a_* > 0.95$, corresponding to a spin rate of ~800 s^{-1} (Gou et al. 2011).

4 A0620-00

A0620-00 (=V616 Mon) was much the brightest of several X-ray transients discovered by the Ariel 5 satellite during an extended scan of the Galactic plane in autumn 1975. Within 3 days of its first sighting it was brighter than the Crab Nebula, while 2 days later it outshone Scorpius X-1 (Fig. 5), becoming—for a few weeks—the brightest cosmic X-ray source ever seen (Elvis et al. 1975), a record to be held for 30 years. Well before peaking at a flux level ~3 times that of Sco X-1, the new source was being monitored by Ariel 5, SAS-3 and other space- and ground-based telescopes around the world. The X-ray emission then faded over several months, being continually monitored by the Goddard All Sky Monitor on Ariel 5 (Kaluzienski et al. 1975).

Despite being only 30 degrees from the Sun, the optical counterpart of A0620-00, subsequently designated V616 Mon, was rapidly located through its nova-like behaviour (Boley

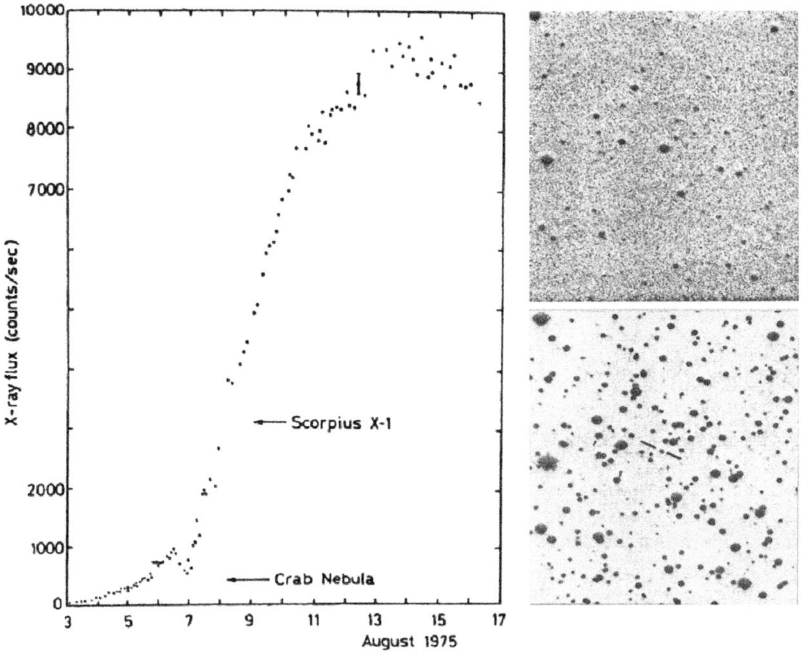

Fig. 5 (*left*) Soft X-ray transient source A0620-00 (Nova Mon) detected in an Ariel 5 Galactic Plane survey. (*right*) Comparison of a short UK Schmidt telescope exposure taken during outburst (*top*) with the corresponding Palomar Sky Survey red plate (*lower*), from which Boley and Wolfson (1975) identified A0620-00 with the K5V star indicated

and Wolfson 1975), the accurate stellar position then allowing identification on the Palomar Observatory Sky Survey charts (Ward et al. 1975) with a $m_B \sim 20$ star (Fig. 5). Ward et al. suggested the faint counterpart was a low mass, solar type star. A radio counterpart was identified with the Mk 2 telescope at Jodrell Bank (Davis et al. 1975), albeit delayed for a week as the telescope was in use by a guest observer.

A more detailed optical study, when the nova light had faded, confirmed the companion as a K5V star in a 7.8 hr period binary (McClintock et al. 1983). The small separation of the binary explained why Roche lobe overflow created an accretion disc around the compact companion, with a thermal-viscous disc instability causing the X-ray and optical outburst.

Spectroscopic observation of the binary companion in quiescence revealed the presence of narrow absorption lines, showing an extremely large amplitude radial velocity (Fig. 6) which allowed McClintock and Remillard (1986) to derive a precise value for the binary period, and a Doppler-corrected spectrum of the companion. The outcome was a 3σ lower limit of 3.2 $M_\odot$ for the mass of the compact X-ray source, *independent* of distance and mass of the companion star. With reasonable assumptions regarding the K dwarf star, the lower limit increased to $\sim$7.3 $M_\odot$, well above the maximum mass of a neutron star.

The low mass of the optical counterpart to A0620-00 was critical in enabling the mass of the X-ray star to be determined with greater certainty than for Cygnus X-1, where the large and (at the time) uncertain mass of the companion supergiant led to residual doubts regarding the nature of the X-ray source (e.g. Trimble et al. 1973).

The return of A0620-00 has been forecast for circa 2033, based on the discovery from Harvard plates that V616 Mon previously flared up in 1917, although that return date is

Fig. 6 Radial velocity curve of the companion star to A0620-00. The extreme amplitude and low companion mass provided strong evidence of a black hole (from Charles and Seward 1995)

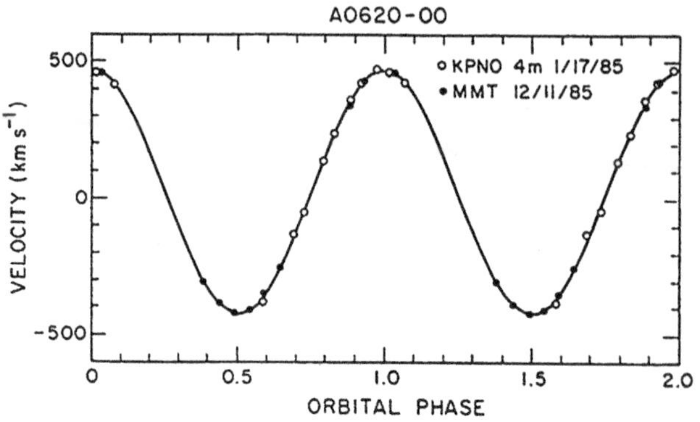

subject to limitations in modelling the thermal-viscous limit cycle in the accretion disc. Meanwhile, optical studies in quiescence remain of considerable interest, with the detection of rapid optical flaring, with rise times of 30 seconds or less, and a power-density spectrum that may be characteristic of black hole binaries in their low state (Hynes et al. 2003). Hynes et al. concluded that the flares are associated with the accretion flow rather than with an active companion, though it remains unclear whether they originate in the outer disc, or are driven by events in the inner region.

In another important step, using a SAS-3 spectrum taken in 1975 to estimate the radius of the innermost orbit, Gou et al. (2010) have shown the black hole in A0620-00 to be spinning quite slowly, with a spin parameter $a_* = 0.12 \pm 0.19$, suggesting the radio jet seen in both flaring and quiescent states is probably disc driven.

Finally, a recent determination of the inclination of A0620-00 by Cantrell et al. (2010), has allowed the black hole mass to be further refined to 6.6 ± 0.25 $M_\odot$.

5 Supermassive Black Holes in AGN

A second important contribution from the Ariel 5 Sky Survey to the search for black holes was in establishing powerful X-ray emission as a characteristic property of Seyfert galaxies, alongside the bright optical nucleus and broad permitted lines.

Prior to the launch of Ariel 5 the majority of extragalactic X-ray source identifications were with rich galaxy clusters. Only NGC 4151 and 3C 273 had been identified uniquely with AGN in the 3U catalogue (Kellogg 1974). 60 Uhuru sources at latitude greater than 20 degrees remained unknown and it was suspected that the majority of these unidentified high Galactic latitude sources (UHGLSs) might represent a new form of 'X-ray Galaxy' (Giacconi 1973). The difficulty of identifying many sources at high Galactic latitude, in both Uhuru and Ariel 5 catalogues (Fig. 7), was again due to their positions being known only to a few tenths of a sq. deg.

The Seyfert galaxy NGC3783 was the first new Ariel 5 identification (Cooke et al. 1976), being the brightest object in the 2A1135-373 error box (Fig. 8). Optical spectra obtained with the 3.8m Anglo Australian Telescope (AAT) revealed the presence of [FeX] and other high ionisation lines, further strengthening the X-ray association. Nine further coincidences of bright Seyferts with Ariel 5 sources quickly followed, enabling a report at the 1976 Relativistic Astrophysics meeting in Boston (Pounds 1977) that those 10 Seyfert identifications (Fig. 9), together with 21 new rich galaxy cluster/X-ray identifications, had essentially solved the mystery of the 'UHGLSs'. How a combination of modest gains in source location

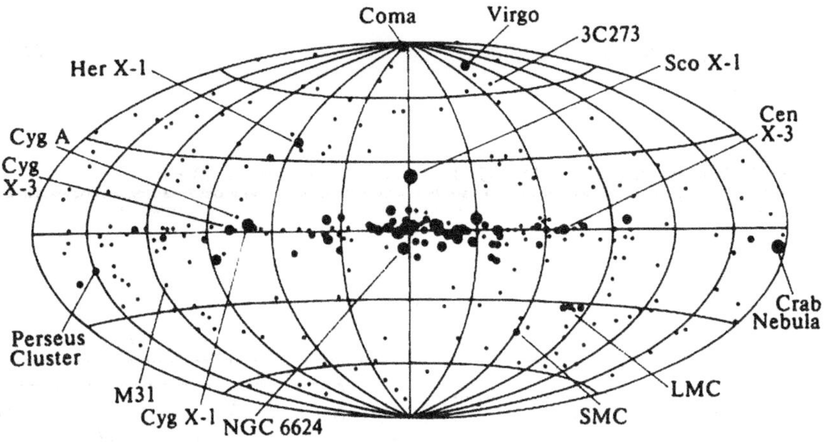

Fig. 7 Map of 297 X-ray sources in the Ariel 5 3A catalogue plotted in Galactic coordinates, with source diameter proportional to the log of the X-ray flux. Many of the faint sources at high Galactic latitude were identified with Seyfert galaxies and galaxy clusters

and in source detection sensitivity allowed the SSI to achieve that breakthrough is recalled by a key participant at the time (Elvis 2012).

A subsequent search of Ariel 5 X-ray error boxes led to the discovery of several previously unknown Seyfert galaxies (Ward et al. 1978), while other Seyfert identifications were found in more precise SAS-3 error boxes (Schnopper et al. 1977, 1978) and from analysis of the final 4U source catalogue (Tananbaum et al. 1978).

Establishing powerful X-ray emission as a characteristic property of Seyfert galaxies was further strengthened when Elvis et al. (1978) showed, for a sample of 15 Seyfert galaxies, that the X-ray luminosity was correlated with the infrared and optical luminosity and with the width of the broad emission lines, but not with the radio flux, strongly suggesting a common origin in the innermost $\lesssim$0.1 pc.

Although it was by then widely believed that the X-ray emission was produced by accretion onto a supermassive object (see Rees et al. 1981 for a review), confirmation that it was the signature of a supermassive black hole required more information, in particular on the compactness of the emission region. An initial search for tell-tale rapid X-ray variability was disappointing, with a sample of 38 bright AGN observed by HEAO-2 finding no variability on a timescale less than 12 hours (Tennant and Mushotzky 1983).

That crucial next step came from uniquely long uninterrupted observations afforded by the highly eccentric orbit of the ESA satellite EXOSAT (Pallavicini and White 1988). Operational from 1983-86, it was only in the final months when ESA was persuaded to approve observations lasting for a full ~4 day spacecraft orbit. Several bright AGN were accepted as prime targets, since Ariel 5 had found evidence for variability on timescales of days (Marshall et al. 1981). The outcome was remarkable (Fig. 10), finding the X-ray emission from several Seyfert galaxies to vary, with large amplitude (i.e. coherently), on hours or less (Pounds and McHardy 1988). Light travel time arguments showed the X-ray sources to be very compact, with the clear implication that Seyfert X-ray sources were massive analogues of accreting stellar-mass black holes.

Further evidence supporting that conclusion followed with the discovery of 'X-ray reflection' from observations of several Seyfert galaxies with the Japanese GINGA satellite (Pounds et al. 1990; Nandra and Pounds 1994), offering a probe of the optically

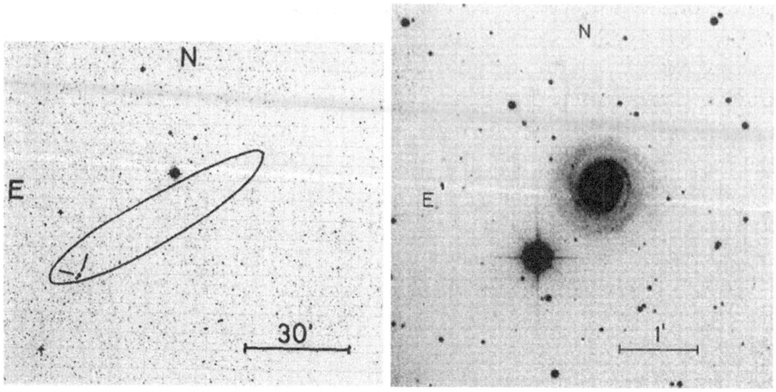

Fig. 8 (*left*) 90 % confidence error box of 2A 1135-373 superimposed on a UK Schmidt Telescope plate. The Seyfert galaxy NGC 3783 is indicated. (*right*) Blue filter 3.8m AAT plate showing the Seyfert galaxy NGC 3783, the first of a new class of X-ray emitters identified with the Ariel 5 SSI

Fig. 9 Luminosity-redshift plot of 13 Seyfert galaxies (10 new) identified in the 2A catalogue, together with the quasar 3C 273

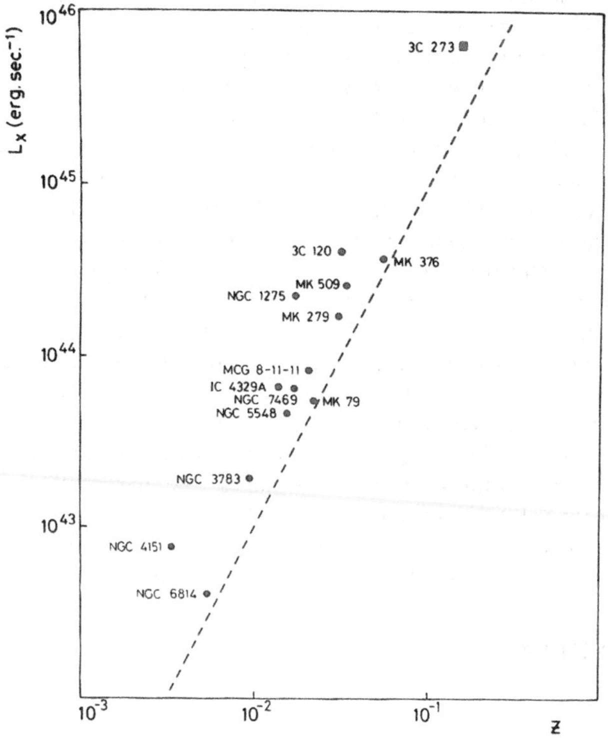

thick matter in the vicinity of the putative black hole. The succeeding Japanese satellite, ASCA, obtained higher resolution spectra which showed the characteristic fluorescent Fe K line associated with reflection to have a broad red wing (Tanaka et al. 1995; Nandra et al. 1997), being widely interpreted as a gravitational redshift, and inspiring a major and continuing research effort to explore the effects of strong gravity in the inner parts of the accretion disc (Fabian et al. 2000).

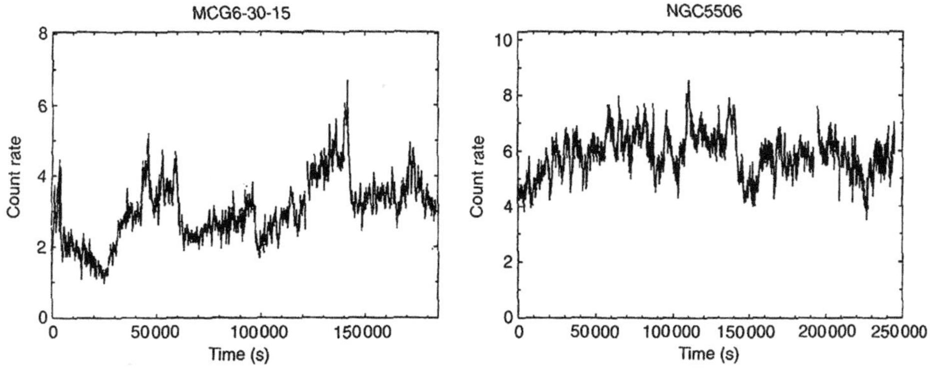

Fig. 10 X-ray light curves of two Seyfert galaxies with rapid, high amplitude variability providing a compelling argument for a supermassive black hole

6 ULXs and Intermediate Mass Black Holes

The discovery of several ultra-luminous X-ray sources (ULX) in nearby galaxies (Fabbiano 1989) has suggested the possible existence of a third class of black hole, with a mass intermediate between the now well-established categories of stellar and supermassive black holes. With X-ray luminosities, assuming isotropic emission, up to $\sim 10^{41}$ erg s^{-1}, the implied black hole mass accreting at the Eddington limit would be $\sim 10^3$ M$_\odot$. The existence of such intermediate mass black holes remains unclear, however, with a strong possibility being that the X-ray flux of a ULX is enhanced by being beamed in our direction (King et al. 2001).

SS433 is an example of a stellar mass black hole with X-ray emission known to be beamed away from the line of sight. In the context of this review it is interesting to note that SS433 remained largely ignored until flagged up by the adjacent Ariel 5 source A1909+04 (Seward et al. 1976). Follow-up observations at the AAT by Clark and Murdin (1978) found a strong emission line spectrum, with several lines they could not identify. Margon et al. (1979) made further observations at the Lick Observatory, finding the unidentified lines to shift in wavelength from night to night. Over the following months it emerged that SS433 is a truly remarkable object, with a pair of jets travelling at relativistic speeds and precessing about a common axis with a 164 day period (Fabian and Rees 1979). While SS433 stands out as an example of extreme super-Eddington accretion in a black hole binary, it can also be seen as a ULX viewed side-on.

7 Summary and Update

Establishing that black holes exist with masses in the range ~5–15 M$_\odot$ and $\sim 10^6$–10^9 M$_\odot$ has been one of the major scientific returns from space observations. The high luminosity and powerful outflows from black holes, when amply fed, offer exciting prospects for studying the physics of accretion processes that play such a wide role in astronomy, while also exploring matter in regions of strong gravity.

Looking back over 50 years of X-ray astronomy the discovery of A0620-00 still stands out, only the Magnetar SGR 1806 (Palmer et al. 2005) having exceeded its peak X-ray flux. Given the limited sensitivity and infrequent deployment of all-sky X-ray detectors, similar dramatic events may have been missed. The GINGA transient GS2000+25 (Tsunemi et al. 1989) appears to have been similar to A0620-00, but more distant and a factor 6 less bright. Just weeks before this Workshop, on 16 September 2012, the SWIFT GRB monitor triggered on a source in Scorpius which became as bright as the Crab nebula 2 days later, with early indications of being a further 'once in a mission event' (Neil Gehrels in NASA PR). However, the conclusion must be that such events are indeed quite rare.

Ozel et al. (2010) review the mass distribution of stellar-mass black hole candidates, including 23 confirmed black hole binaries. Dynamical data show 17 of those to have low mass optical companions with a narrow distribution of black hole masses, of $7.8 \pm 1.2\ M_{\odot}$, for the best determined systems. A significant absence of black hole masses between $5M_{\odot}$ and the neutron star mass limit of $2M_{\odot}$ was a surprise, which Ozel et al. speculate may be due to a sudden fall in the energy of the supernova explosion with increasing progenitor mass.

Binary periods for the low mass transient systems range from $\sim$4 hr to $\sim$6 d (an exception being GRS 1915+105), and $\sim$32 hr to $\sim$5.6 d for the persistent high mass sources, comfortably meeting the requirement anticipated in the Introduction for copius mass transfer by Roche lobe overflow. Such powerful X-ray sources will be the exception, of course, with the vast majority of Galactic black holes being limited to inefficient Bondi (radial) accretion from the local ISM, and much too faint to be detected.

Since launch in 2004 the NASA Swift satellite has been detecting $\sim$100 Gamma Ray Bursts (GRB) each year (Gehrels and Meszaros 2012), with the few-arc-sec positions obtained by rapidly detecting the X-ray afterglow allowing many bursts to be optically identified. Many GRB are found to be associated with supernovae in galaxies at high redshift. With a likely origin in the collapse of a massive star (MacFadyen and Woosley 1999), the remarkable inference is that Swift is observing the birth of stellar mass black holes from across the Universe, at a rate—allowing for beaming and other factors—of $\sim 10^5$ a year.

One unusually long event observed by Swift in March 2011 has been interpreted as the capture of a star by a SMBH (Burrows et al. 2011; Levan et al. 2011), prefacing an important new way of studying timescales in the accretion disc of an AGN. Follow-up X-ray observations with Suzaku and XMM-Newton detected Quasi Period Oscillations (QPO), with a transient periodicity at 3.5 minutes, evidently relating to processes very close to the innermost stable circular orbit (Reis et al. 2012). That transient event adds to the first reliable QPO detection in an AGN (Gierlinski et al. 1998), providing a further example of the similarities in disc physics shared by accreting black holes in X-ray binaries and AGN.

The interpretation of flickering in the emission from Sgr A as the disruption of passing asteroids (Zubovas et al. 2012) provides a further indication that the feeding of SMBH will range from short sub-Eddington events to less frequent minor and major mergers. However, mergers seem more likely to have led to the heavily dust-obscured galaxies (DOGS) found in the WISE sky survey (NASA PR 29/8/12); NuStar observations should confirm whether these are super-Eddington AGN.

The study of supermassive black holes has been a major beneficiary of three powerful X-ray observatories, Chandra, XMM-Newton and Suzaku, launched since 1999. Long exposures on HST deep fields have shown, via coincident X-ray emission, that powerful X-ray emission—and implicitly the presence of supermassive black holes—extends to AGN at high redshift. The high angular resolution of Chandra resolved two point-like X-ray sources near the centre of NGC 6420 (Fig. 11), believed to be a pair of merging galaxies (Komossa et al. 2003). The eventual merger of the SMBHs will be a spectactular event, with a substantial fraction of the combined rest mass released over a few hours, mainly as a QPO gravity wave outburst.

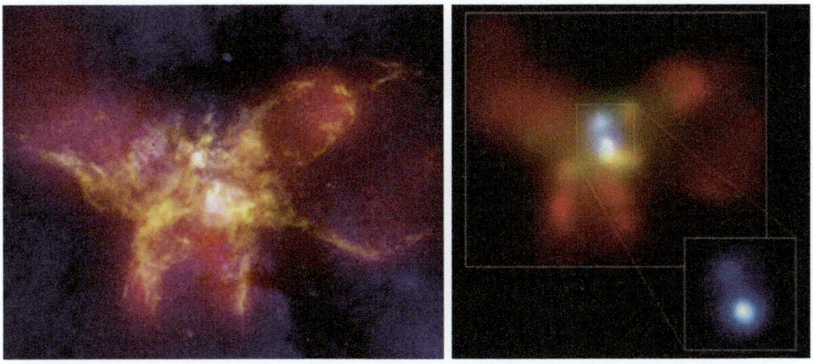

Fig. 11 Optical and Chandra images of NGC 6420, believed to be a pair of merging galaxies. The two bright point-like X-ray sources in the *right hand panel* are probably SMBH at the respective nuclei

Finally, as discussed elsewhere at this Workshop, highly ionised outflows with velocities ~0.1–0.2c were first detected in X-ray spectra of non-BAL AGN a decade ago (Pounds et al. 2003; Reeves et al. 2003). Subsequent observations (Cappi 2006) and searches of the XMM-Newton (Tombesi et al. 2010, 2011) and Suzaku (Gofford et al. 2013) data archives have shown such ultra-fast outflows to be a common feature of luminous AGN, indicating that powerful black hole winds are likely to have a wider importance in galaxy feedback models.

Acknowledgements The outstanding efforts of many colleagues and former students enabled the X-ray Astronomy Group at Leicester to play a key role in establishing the existence of black holes in both stellar systems and in the nuclei of Active Galaxies. That opportunity was, of course, only possible by building on the pioneering achievements of other scientists and engineers who have established X-ray astronomy as a vibrant discipline in Space Science. Grateful thanks are also due to Jeff McClintock, Paul Gorenstein and Martin Ward who provided valuable comments on the draft manuscript.

References

F. Boley, R. Wolfson, IAU Circ., 2819 (1975)
C.T. Bolton, Nature **235**, 271 (1972)
S. Bowyer, E.T. Byram, T.A. Chubb, H. Friedman, Nature **201**, 1307 (1964a)
S. Bowyer, E.T. Byram, T.A. Chubb, H. Friedman, Science **146**, 912 (1964b)
S. Bowyer et al., Science **147**, 394 (1965)
L. Braes, G.K. Miley, Nature **246**, 232 (1971)
D. Burrows et al., Nature **476**, 421 (2011)
A.G. Cantrell et al., Astrophys. J. **710**, 1127 (2010)
M. Cappi, Astron. Nachr. **327**, 1012 (2006)
P.A. Charles, F.D. Seward, *Exploring the X-Ray Universe* (Cambridge University Press, Cambridge, 1995)
G. Chodil, H. Mark, R. Rodrigues, F. Seward, C.D. Swift, W.A. Hiltner, G. Wallerstein, E.J. Mannery, Phys. Rev. Lett. **19**, 681 (1967)
D.H. Clark, P. Murdin, Nature **276**, 44 (1978)
B.A. Cooke, K.A. Pounds, E.A. Stewardson, D.J. Adams, Astrophys. J. **150**, 189 (1967)
B.A. Cooke, K.A. Pounds, Nature **229**, 144 (1971)
B.A. Cooke, M. Elvis, M.J. Ward, R.A.E. Fosbury, M.V. Penston, T. Maccacaro, Mon. Not. R. Astron. Soc. **177**, 21P (1976)
R.J. Davis, M.R. Edwards, I. Morison, R.E. Spencer, Nature **257**, 659 (1975)
M. Elvis, C.G. Page, K.A. Pounds, M.J. Ricketts, M.J.L. Turner, Nature **257**, 656 (1975)
M. Elvis, T. Maccacaro, A.S. Wilson, M.J. Ward, M.V. Penston, R.A.E. Fosbury, G.C. Perola, Mon. Not. R. Astron. Soc. **183**, 129 (1978)

M. Elvis, in *Fifty Years of Quasars*, ed. by J.W. Sulentic, P. Marziani, M. D'Onofrio. Astrophysics and Space Science Library, vol. 386 (2012), pp. 41–45
A.C. Fabian, M.J. Rees, Mon. Not. R. Astron. Soc. **187**, 13 (1979)
A.C. Fabian, K. Iwasawa, C.S. Reynolds, A.J. Young, Publ. Astron. Soc. Pac. **112**, 1145 (2000)
G. Fabbiano, Annu. Rev. Astron. Astrophys. **27**, 87 (1989)
P.C. Fisher, H.M. Johnson, W.C. Jordan, A.J. Mayerott, L.W. Acton, Astrophys. J. **143**, 203 (1966)
H. Friedman, Proc. Inst. Radio Eng. **47**, 278 (1959)
N. Gehrels, P. Meszaros, Science **337**, 932 (2012)
R. Giacconi, B. Rossi, J. Geophys. Res. **65**, 773 (1960)
R. Giacconi, H. Gursky, F. Paolini, B. Rossi, Phys. Rev. Lett. **9**, 439 (1962)
R. Giacconi, Phys. Today (1973)
R. Giacconi, H. Gursky, S. Murray, E. Kellogg, E. Schreier, T. Matilsky, D. Koch, H. Tananbaum, Astrophys. J. Suppl. Ser. **27**, 37 (1974)
R. Giacconi, H. Gursky (eds.). *Astrophysics and Space Science*, vol. 43 (Reidel, Dordrecht, 1974)
M. Gierlinski, M. Middleton, M.J. Ward, C. Done, Nature **455**, 369 (1998)
J. Gofford, J.N. Reeves, F. Tombesi, V. Braito, T.J. Turner, L. Miller, M. Cappi, Mon. Not. R. Astron. Soc. **430**, 60 (2013)
L. Gou, J.E. McClintock, J.F. Steiner, R. Narayan, A.G. Cantrell, C.D. Bailyn, J.A. Orosz, Astrophys. J. **718**, L122 (2010)
L. Gou et al., Astrophys. J. **742**, 85 (2011)
H. Gursky, R. Giacconi, F. Paolini, B. Rossi, Phys. Rev. Lett. **11**, 530 (1963)
H. Gursky et al., Astrophys. J. Lett. **146**, 310 (1966)
J. Harries, K.G. McCracken, R.J. Francey, A.G. Fenton, Nature **215**, 38 (1967)
R.M. Hjellming, C.M. Wade, Nature **134**, 238 (1971)
T. Hockey (ed.), *Biographical Encyclopedia of Astronomers*, vol. 2 (Springer, Berlin, 2007), pp. 778–779
R.I. Hynes, P.A. Charles, J. Casares, C.A. Haswell, C. Zurita, T. Shahbaz, Mon. Not. R. Astron. Soc. **340**, 447 (2003)
L. Kaluzienski et al., Astrophys. J. **201**, L21 (1975)
E. Kellogg, *Astrophysics and Space Science Library*, vol. 43 (Reidel, Dordrecht, 1974), pp. 321–357
A.R. King, M.B. Davies, M.J. Ward, G. Fabbiano, M. Elvis, Astrophys. J. **552**, 109 (2001)
S. Komossa, V. Burwitz, G. Hasinger, P. Predehl, J.S. Kaastra, Y. Ikebe, Astrophys. J. **582**, L15 (2003)
A. Levan et al., Science **333**, 199 (2011)
D. Lynden-Bell, Nature **233**, 690 (1969)
A.I. MacFadyen, S.E. Woosley, Astrophys. J. **524**, 262 (1999)
B. Margon, H.C. Ford, J.I. Katz, K.B. Kwitter, R.K. Ulrich, R.P. Stone, A. Klemola, Astrophys. J. **230**, 41 (1979)
N. Marshall, R.S. Warwick, K.A. Pounds, Mon. Not. R. Astron. Soc. **194**, 987 (1981)
J.E. McClintock, L.D. Petro, R.A. Remillard, G.R. Ricker, Astrophys. J. Lett. **266**, L27 (1983)
J.E. McClintock, R.A. Remillard, Astrophys. J. **308**, 110 (1986)
R. McCormack, Br. J. Hist. Sci. **4**, 126 (1968)
K. Nandra, K.A. Pounds, Mon. Not. R. Astron. Soc. **268**, 405 (1994)
K. Nandra, I.M. George, R.F. Mushotzky, T.J. Turner, T. Yaqoob, Astrophys. J. **477**, 602 (1997)
M. Oda et al., Nature **554**, 207 (1965)
J. Overbeck, H. Tananbaum, Astrophys. J. **153**, 899 (1968)
J. Orosz et al., Astrophys. J. **742**, 84 (2011)
F. Ozel, D. Psaltis, R. Narayan, J.E. McClintock, Astrophys. J. **725**, 1918 (2010)
R. Pallavicini, N.E. White (eds.), *Mem. Soc. Astron. Ital.*, vol 59 (1988)
D.M. Palmer et al., Nature **434**, 1107 (2005)
K.A. Pounds, in *8th Texas Symposium on Relativistic Astrophysics*. Ann. New York Acad. Sci., vol. 302 (1977), p. 361
K.A. Pounds, I.M. McHardy, in *Physics of Neutron Stars and Black Holes*, ed. by Y. Tanaka (University Academic Press, Tokyo, 1988), pp. 285–299
K.A. Pounds, K. Nandra, G.C. Stewart, I.M. George, A.C. Fabian, Nature **344**, 132 (1990)
K.A. Pounds, J.N. Reeves, A.R. King, K.L. Page, P.T. O'Brien, M.J.L. Turner, Mon. Not. R. Astron. Soc. **345**, 705 (2003)
J.E. Pringle, M.J. Rees, Astron. Astrophys. **21**, 1 (1972)
J.E. Pringle, Ann. N.Y. Acad. Sci. **302**, 6 (1977)
S. Rappaport, W. Zaumen, R. Doxsey, Astrophys. J. **168**, L17 (1971)
M.J. Rees, M.C. Begelman, R.C. Blandford, in *10th Texas Symposium on Relativistic Astrophysics*, ed. by R. Ramaty, F.E. Jones (Academic Science, New York, 1981)
J.N. Reeves, P.T. O'Brien, M.J. Ward, Astrophys. J. **593**, 65 (2003)

M. Reid et al., Astrophys. J. **742**, 83 (2011)
R.C. Reis, J.M. Miller, M.T. Reynolds, K. Gültekin, D. Maitra, A.L. King, T.E. Strohmayer, Science **337**, 949 (2012)
E.E. Salpeter, Astrophys. J. **140**, 796 (1964)
A. Sandage et al., Astrophys. J. **146**, 314 (1966)
H.W. Schnopper, A. Epstein, J.P. Delvaille, W. Tucker, R. Doxsey, G. Jernigan, Astrophys. J. **215**, L7 (1977)
H.W. Schnopper et al., Astrophys. J. **222**, L91 (1978)
E. Schreier, H. Gursky, E. Kellogg, H. Tananbaum, R. Giacconi, Astrophys. J. **170**, 21 (1971)
F.D. Seward, C.G. Page, M.J.L. Turner, K.A. Pounds, Mon. Not. R. Astron. Soc. **175**, 39P (1976)
N.I. Shakura, R.A. Sunyaev, Astron. Astrophys. **24**, 337 (1973)
H.L. Shipman, Astrophys. J. **16**, 9 (1975)
Y. Tanaka et al., Nature **375**, 659 (1995)
H. Tananbaum, G. Peters, W. Forman, R. Giacconi, C. Jones, Y. Avni, Astrophys. J. **223**, 74 (1978)
A.F. Tennant, R.F. Mushotzky, Astrophys. J. **264**, 92 (1983)
F. Tombesi, M. Cappi, J.N. Reeves, G.C. Palumbo, T. Yaqoob, V. Braito, M. Dadina, Astron. Astrophys. **521**, A57 (2010)
F. Tombesi, M. Cappi, J.N. Reeves, G.C. Palumbo, V. Braito, M. Dadina, Astrophys. J. **742**, 44 (2011)
V. Trimble, W.K. Rose, J. Weber, Mon. Not. R. Astron. Soc. **162**, 1P (1973)
H. Tsunemi, S. Kitamoto, S. Okamura, D. Roussel-Dupre, Astrophys. J. **337**, L81 (1989)
M.J. Ward, M.V. Penston, C.A. Murray, E.D. Clements, Nature **257**, 659 (1975)
M.J. Ward, A.S. Wilson, M.V. Penston, M. Elvis, T. Maccacaro, K.P. Tritton, Astrophys. J. **223**, 788 (1978)
L. Webster, P. Murdin, Nature **235**, 37 (1972)
Y.B. Zel'dovich, I.D. Novikov, Dokl. Akad. Nauk SSSR **155**, 1033 (1964)
K. Zubovas, S. Nayakshin, S. Markoff, Mon. Not. R. Astron. Soc. **421**, 1315 (2012)

DOI 10.1007/978-1-4939-2227-7_3
Reprinted from *Space Science Reviews* Journal, DOI 10.1007/s11214-013-9985-6

General Overview of Black Hole Accretion Theory

Omer Blaes

Received: 16 January 2013 / Accepted: 17 April 2013 / Published online: 14 May 2013

Abstract I provide a broad overview of the basic theoretical paradigms of black hole accretion flows. Models that make contact with observations continue to be mostly based on the four decade old alpha stress prescription of Shakura and Sunyaev (1973), and I discuss the properties of both radiatively efficient and inefficient models, including their local properties, their expected stability to secular perturbations, and how they might be tied together in global flow geometries. The alpha stress is a prescription for turbulence, for which the only existing plausible candidate is that which develops from the magnetorotational instability (MRI). I therefore also review what is currently known about the local properties of such turbulence, and the physical issues that have been elucidated and that remain uncertain that are relevant for the various alpha-based black hole accretion flow models.

Keywords Accretion · Accretion disks · Black hole physics · Instabilities · MHD

1 Introduction

Accretion is the very process that allows black hole sources to emit electromagnetic radiation and other forms of energy. Because black holes are so small in size compared to the spatial scale of their sources of fueling, and because centrifugal forces on matter of given angular momentum increase more rapidly ($\propto R^{-3}$) than gravity ($\propto R^{-2}$) as one moves inward in radius R, accretion is generally believed to be a process involving rotationally supported flows. Matter in such a flow must lose angular momentum in order to move inward and release gravitational binding energy. It is the nature of the angular momentum loss mechanism, and the process whereby gravitational binding energy is converted into observable forms of energy, that are the two central questions of black hole accretion theory. At least three mechanisms have been proposed for angular momentum extraction:

(1) *External stresses associated with large scale magnetic fields in a magnetohydrodynamic (MHD) outflow*. This mechanism (Blandford and Payne 1982) may be relevant in low

O. Blaes (✉)
Department of Physics, University of California, Santa Barbara, CA 93106, USA
e-mail: blaes@physics.ucsb.edu

luminosity sources where accretion power may be largely converted into mechanical power in outflows. It may also be relevant in resolving the fueling and self-gravity problems in the outer accretion flows in active galactic nuclei (Goodman 2003). Whether and how large scale magnetic fields can be created remains an open question, however.

(2) *Magnetorotational (MRI) turbulence*. Such turbulence is generic for plasmas that are sufficiently electrically conducting and not too strongly magnetized (Balbus and Hawley 1991, 1992, 1998; Hawley and Balbus 1991). Because turbulence is inherently dissipative, this process is almost certainly relevant for sources whose power output is dominated by thermal radiative emission mechanisms.

(3) *Nonaxisymmetric waves and shocks*. Nonaxisymmetric (e.g. spiral) waves can transport angular momentum outward through the flow. Such waves can also transport energy away from the region where gravitational binding energy is released, depositing it elsewhere. Waves are almost certainly relevant in disks around supermassive black hole binaries, and also in the outer, self-gravitating parts of disks in active galactic nuclei. They probably also play a role in the outer parts of black hole X-ray binary disks due to tidal excitation by the companion star. Nonaxisymmetric shocks can also play an important role in the inner regions of accretion flows whose angular momenta are misaligned with the black hole spin axis (Fragile and Blaes 2008).

Among these options, only the second—MRI turbulence—is a mechanism that *might* be describable by the classical alpha prescription of Shakura and Sunyaev (1973), at least in some aspects (Balbus and Papaloizou 1999). The angular momentum transporting stress $w_{R\phi}$ in the turbulence is given by local space and time averages of correlated fluctuations in radial (R) and azimuthal (ϕ) fluctuations of velocity $\mathbf{v}$ (the Reynolds stress) and magnetic field $\mathbf{B}$ (the Maxwell stress),

$$w_{R\phi} = \left\langle \rho v_R \delta v_\phi - \frac{B_R B_\phi}{4\pi} \right\rangle, \tag{1}$$

where ρ here is the mass density and δv_ϕ is the local deviation of the azimuthal velocity component from the mean background shear flow. Maxwell stresses are generally larger in magnitude than the Reynolds stresses by factors of at least several. I say that the total stress *might* be describable by the classical alpha prescription because these stresses appear to be mostly local in the sense that simulations show that radial correlations in stress drop rapidly on scales larger than the local disk scale height. However, as I discuss in Sect. 3.1 below, there remain correlations on larger radial scales.

This article provides a broad overview of alpha-based models of black hole accretion flows, focusing on structure, dynamics, and thermodynamics. These models continue to dominate theoretical efforts to explain observations, but a slow revolution is occurring as simulations of MRI turbulence, both local and global, continue to become more powerful and to incorporate more and more of the relevant physics. This article will also discuss what has been learned recently from local, shearing box simulations of MRI turbulence as this pertains directly to some of the alpha-based modeling. A review of global simulations can be found in Chap. 2.4. Spectral modeling of accretion flows is discussed in Chap. 2.3. I will also mainly focus on *accretion* rather than the formation of jets and outflows here, though jets and outflows are clearly important (both observationally and theoretically, in certain flow states). See Chap. 5.3 on jet launching mechanisms.

2 Hydrodynamic Disk Models with the Alpha Prescription

Decades of theory and models of black hole accretion flows have critically relied on the alpha prescription for a local stress introduced by Shakura and Sunyaev (1973). There are

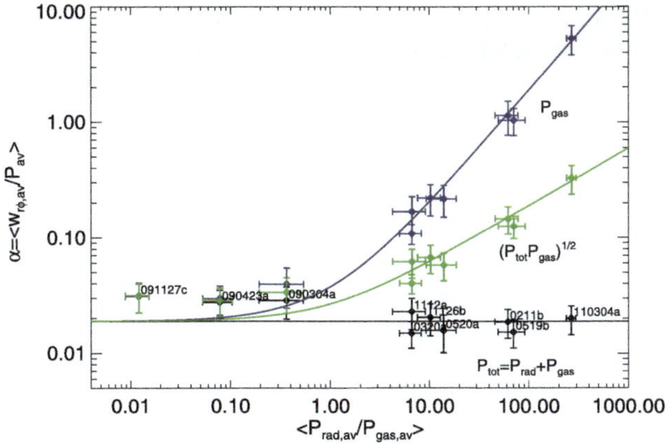

Fig. 1 Time-averaged values of the ratio of spatially averaged stress to various measures of spatially averaged thermal pressure, i.e. the Shakura and Sunyaev (1973) parameter alpha, as a function of the time-averaged ratio of spatially averaged radiation pressure to spatially averaged gas pressure, for a number of radiation MHD, vertically stratified, shearing box simulations of MRI turbulence. *Black*, *green* and *blue points* are the results for thermal pressures defined to be the total (radiation plus gas) pressure, the geometric mean of the total and gas pressures, and the gas pressure alone, respectively. *Horizontal* and *vertical error bars* on all points indicate one standard deviation in the respective time-averages. The *horizontal black line* is the average alpha value of the total pressure prescription (black) points, while the *green* and *blue curves* are what would result if the total pressure prescription were correct, but one nevertheless insisted on defining alpha in terms of the other thermal pressure definitions used in the green and blue points, respectively. The stress prescription that is most consistent with the simulation data is one in which the total thermal pressure is used, though it is perhaps noteworthy that the alpha values in the gas pressure dominated simulations are consistently higher than the alpha values in the radiation pressure dominated simulations (Updated from Hirose et al. 2009)

numerous variants of this prescription which produce order unity changes in the definition of α, and one of the most common is

$$w_{R\phi} = \alpha P, \tag{2}$$

where P is the thermal pressure. Most models have assumed that this is the *total* thermal pressure (gas plus radiation), but prescriptions in which the stress is taken to be proportional to just the gas pressure alone (e.g. Sakimoto and Coroniti 1981) or the geometric mean of the gas and total thermal pressures (e.g. Taam and Lin 1984) have also been suggested. However, as illustrated in Fig. 1, recent radiation MHD simulations of MRI turbulence find that the stress scales best with total thermal pressure, at least on long time scales (Ohsuga et al. 2009; Hirose et al. 2009).

The alpha prescription (2) is usually used to solve for the radial structure of vertically-integrated geometrically thin or slim accretion disks, in which case it enters the equations through the vertically-integrated stress:

$$W_{R\phi} = \int_{-\infty}^{\infty} w_{R\phi} dz \sim 2H\alpha P, \tag{3}$$

where P is now some vertically averaged thermal pressure, of order the midplane pressure, and H is the vertical half-thickness of the disk. This is consistent with MRI simulations, but the prescription is also occasionally used even more locally by assuming that the vertical profiles of stress and dissipation at a given radius are proportional to the local vertical profile of thermal pressure. As we discuss below in Sect. 3.2, this is not consistent with vertically stratified simulations of MRI turbulence, which generally have vertical profiles of stress that are broader than the thermal pressure profile. Alpha defined locally would therefore increase rapidly outward from the disk midplane.

2.1 Local Thermal Equilibria and Secular Instabilities

Virtually all (non-simulation-based) models of black hole accretion flows are based on vertically integrated hydrodynamic equations. These models often neglect the possibility of significant losses of mass, angular momentum, and energy in outflows and jets, though some models do attempt to include them with various prescriptions, particularly in advection-dominated flows which we will come to shortly. As discussed in Chap. 5.1, neglect of outflows is likely to be a bad approximation in some sources and accretion states. Nevertheless, if we adopt this assumption for simplicity, then for stationary flows, the conservation laws of mass, radial momentum, angular momentum, and internal energy can be written as

$$\dot{M} = 2\pi R\Sigma v, \tag{4}$$

$$\rho v\frac{dv}{dR} = \rho\left(\Omega^2 - \Omega_{\rm K}^2\right)R - \frac{dP}{dR}, \tag{5}$$

$$\dot{M}\frac{d\ell}{dR} = \frac{d}{dR}\left(2\pi R^2 W_{R\phi}\right), \tag{6}$$

and

$$Q_{\rm adv} \equiv \frac{-\dot{M}}{4\pi R}\left[\frac{dU}{dR} + P\frac{d}{dR}\left(\frac{1}{\rho}\right)\right] = Q^+ - Q^-, \tag{7}$$

where we have neglected general relativity for the purposes of physical transparency. Here $\rho \sim \Sigma/(2H)$ is a vertically-averaged density, Σ is the surface mass density, Ω is the fluid angular velocity which may differ from the test particle (Kepler) angular velocity $\Omega_{\rm K}$, $\ell = \Omega R^2$ is the fluid specific angular momentum, v is the inward radial drift speed, U is a vertical average of the internal energy per unit mass, Q^- is the radiative cooling rate per unit surface area on each face of the disk, $Q^+ = -(1/2)W_{R\phi}Rd\Omega/dR$ is half the turbulent dissipation rate per unit surface area, and $Q_{\rm adv}$ is half the inward radial advection of heat per unit surface area. Vertical hydrostatic equilibrium implies that the vertical half-thickness of the disk is $H \sim (P/\rho)^{1/2}/\Omega_{\rm K}$.

Once one adopts the alpha prescription (2), together with an equation of state and opacities and/or optically thin cooling functions, it is possible to solve these equations with assumed boundary conditions to derive the radial profiles of vertically-averaged fluid variables in the flow. Such models generally invoke a regularity condition at an inner sonic point and/or a no-torque inner boundary condition at, for example, the innermost stable circular orbit, although magnetohydrodynamic stresses can be important once one enters the plunging region near the black hole (Gammie 1999; Krolik 1999). (See Chap. 2.4 and, e.g., Penna et al. 2010 and Noble et al. 2010 for recent simulation work on this issue for geometrically thin disks.) Another approach is to consider radii much larger than the gravitational radius $R_{\rm g} \equiv GM/c^2$ and invoke self-similarity by assuming a constant ratio of advective cooling over turbulent dissipation $Q_{\rm adv}/Q^+$ (Narayan and Yi 1994).

For a fixed black hole mass, models that are stationary generally depend on a number of chosen parameters, the most important being the accretion rate $\dot{M}$ which is everywhere constant through the flow (remember, we are neglecting outflows here). A number of possible equilibria have been discovered in this way, and the primary method of choosing which ones are physically realizable in nature has been to check if they are stable to secular perturbations. The growth rates of such instabilities are related to one of two characteristic time-scales. The first is the thermal time, defined as the characteristic heating time

$$t_{\text{th}} \sim \frac{U\Sigma}{2Q^+} \sim \frac{1}{\alpha\Omega}. \tag{8}$$

Thermal instabilities, in which a local patch of the flow undergoes runaway heating or cooling, generally grow on this time scale. The second time scale is the inflow time, i.e. the time it would take for turbulent stresses to cause a fluid element to drift inward over a distance comparable to its current radius,

$$t_{\text{inflow}} \sim \frac{R}{v} \sim \frac{\Sigma\Omega R^2}{2Hw_{r\phi}} \sim \frac{\Omega}{\alpha\Omega_{\text{K}}^2}\left(\frac{R}{H}\right)^2. \tag{9}$$

"*Viscous*" or *inflow* instabilities tend to grow on this time scale, where I have enclosed the term "viscous" in quotation marks here (only) to emphasize that it is turbulent stresses, not microscopic viscosity, that play a role here. In geometrically thin accretion disks, where dynamical equilibrium on the inflow time requires the angular velocity and specific angular momentum of fluid orbits to be a function of radius and not of time, one can write down a time-dependent diffusion equation for the evolution of the surface mass density in the flow (Lynden-Bell and Pringle 1974; Lightman and Eardley 1974),

$$\frac{\partial\Sigma}{\partial t} = \frac{1}{R}\frac{\partial}{\partial R}\left[\frac{1}{\ell'}\frac{\partial}{\partial R}\left(R^2 W_{R\phi}\right)\right], \tag{10}$$

where $\ell' = (d/dR)(R^2\Omega)$ is the radial derivative of specific angular momentum. In this case, instabilities happen if $W_{R\phi}$ is inversely related to the surface mass density, as this equation then corresponds to a diffusion equation with a negative diffusion coefficient. Perturbations in surface mass density would then tend to grow, rather than be smoothed out, by this *anti*-diffusion.

Following theoretical work on dwarf nova outbursts in cataclysmic variables (e.g. Smak 1984), it has proved convenient to depict *local* thermal equilibrium ($Q^+ = Q^- + Q_{\text{adv}}$) solutions at one particular radius in a diagram of accretion rate vs. local surface mass density. Figure 2 (Chen et al. 1995) depicts the topology of the space of such solutions at a radius of $10GM/c^2$ around a ten solar mass Schwarzschild black hole. Each curve corresponds to a different chosen value of α, as labelled. The locations of these curves in this graph can shift considerably, depending on the particular physics being included in the models, but the topological structure is robust. Naively, we expect equilibrium curves with negative slopes in this diagram to be viscously unstable, as the vertically integrated stress $W_{R\phi}$ is proportional to the dissipation rate per unit area Q^+ which in turn is proportional to the accretion rate $\dot{M}$, because, after all, it is the inflow of matter which is the source of accretion power. The negative sloped curve portions in the right of this diagram near Eddington accretion rates are radiation dominated, geometrically thin disks, and are viscously unstable by this criterion (Lightman and Eardley 1974). They are also thermally unstable (Shibazaki and Hōshi 1975; Shakura and Sunyaev 1976), as can be seen because Q^+ exceeds (is less than) $Q^- + Q_{\text{adv}}$ above (below) these curve portions. Hence a perturbation upward (downward) from this curve will cause runaway heating (cooling), moving it away from the equilibrium curve. Similarly, the middle bold line in the lower portion of this diagram, which corresponds to a hot, optically thin flow where turbulent dissipation is balanced by radiative cooling is viscously stable but thermally unstable. This solution was first discovered by Shapiro et al. (1976).

Fig. 2 Thermal equilibrium curves of particular accretion flow models around a ten solar mass black hole at a particular radius $10GM/c^2$ on the accretion rate (scaled with $\dot{M}_E$, the Eddington luminosity divided by c^2) vs. surface mass density (Σ) plane, from Chen et al. (1995). Each curve is labelled by the value of the Shakura and Sunyaev (1973) stress parameter α chosen in the model. The locations of these curves in this plane can change considerably depending on the physics being incorporated and how it is treated in these models—see Chen et al. (1995) for more details. The topology of the curves is, however, reasonably robust

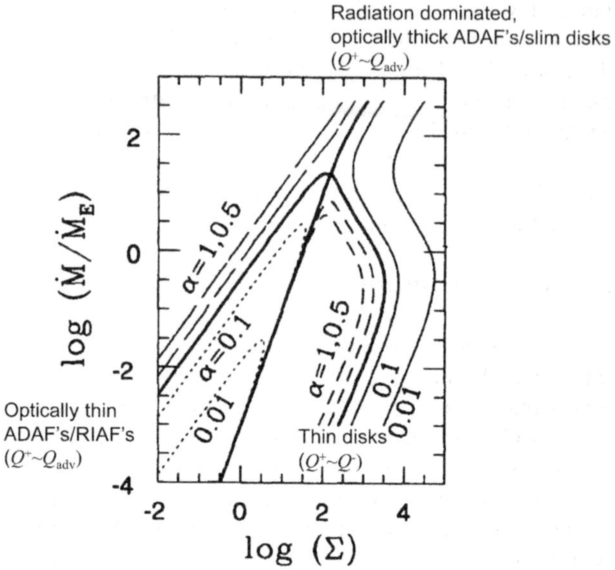

This leaves three regions of the diagram which appear to correspond to thermally and viscously stable solutions. The lower right set of curves correspond to the gas pressure dominated regime of the original geometrically thin, optically thick accretion disks of Shakura and Sunyaev (1973). At higher accretion rates, radiation pressure starts to dominate these geometrically thin solutions and the curves bend over to the unstable negative slopes. At still higher accretion rates, the inflow time becomes shorter than the cooling time, so that heat is advected inward. This advection is stabilizing (Abramowicz et al. 1988) and, because the scale height of the disk can become quite large, these flows have been dubbed "slim disks".

Note that for larger radii, the optically thick, geometrically thin disk curves would also bend toward unstable negative slopes and then back toward positive stable slopes as one *lowers* the accretion rate and passes through the transition where ionized hydrogen becomes neutral. This unstable branch is responsible for the transient outburst behavior observed in many black hole and neutron star X-ray binaries, as well as dwarf novae in accreting white dwarf systems. We will have little to say about this instability here, as it generally occurs in the outer, less luminous portion of the disk in accreting black hole systems, but it is crucial for explaining the phenomenology of black hole transients. See Lasota (2001) for a good review of the theory. Note, however, that these outbursts provide some of the only constraints on the levels of turbulent stress in accretion disks, with $\alpha \simeq 0.1$–0.3 in the outburst phase. This is significantly higher than that measured in local shearing box simulations with no net vertical magnetic flux, as illustrated in Fig. 1 above. It may be that the character of the turbulence changes when one is so close to the regime of hydrogen ionization, or it may be that external magnetic flux is necessary to explain the observations. This is a significant unsolved problem (e.g. King et al. 2007; Kotko and Lasota 2012).

In addition to these two optically thick accretion disk solutions, a third set of stable solutions exists which is optically thin and which only exists at low accretion rates, provided α is not too high. Here the cooling time significantly exceeds the infall time, so that advection again balances turbulent dissipation, and this apparently also stabilizes the flow (Ichimaru 1977; Narayan and Yi 1994, 1995a, 1995b; Abramowicz et al. 1995; Chen et al. 1995). Such flows are known as Advection Dominated Accretion Flows (ADAF's, a term which can also be applied to the optically thick, radiation dominated slim disks) or radiatively inefficient accretion flows (RIAF's).

While the above arguments suggest that these three sets of solutions are thermally and viscously stable, attempts to rigorously demonstrate this involve subtle issues, particularly in the two advection dominated solutions, and the situation is in fact not entirely clear. Advection dominated solutions are fairly geometrically thick ($H \sim R$), so the thermal and inflow time scales are comparable, and there is no longer a clean separation of thermal and viscous instabilities. Thermal pressure is also not dynamically negligible, so that thermal perturbations can alter the specific angular momentum distribution and the surface mass density in the disk, even without mass diffusion due to turbulent stresses. Attempts to tackle this problem have been made (e.g. Kato et al. 1996, 1997; Wu and Li 1996, 1997; Yamasaki 1997), but they involve considerations of turbulent heat diffusion and turbulent bulk viscosity, for which there is currently very little understanding in the context of MRI turbulence. One hopes that simulations will shed light on these issues, and currently global MRI simulations in the RIAF regime appear to be consistent with thermal and viscous stability (see Chap. 2.4). As we discuss further below, simulations of the optically thick solutions, which are far more challenging, currently cast doubt on alpha prescription stability analyses, even in the supposedly unstable geometrically thin, optically thick radiation pressure dominated solution.

It should also be noted that, in addition to the thermal and viscous instabilities, other, shorter time scale instabilities exist that are driven by the thermodynamics of the alpha stress prescription, particularly the excitation of acoustic modes (e.g. Blumenthal et al. 1984; Kato et al. 1988; Chen and Taam 1993), and these might be relevant for explaining high frequency variability in black hole sources. Again, however, it is far from clear that the time-dependent thermodynamics of the alpha prescription accurately represents the time-dependent thermodynamics of MRI turbulence. On the other hand, *dynamical* excitation of nonaxisymmetric acoustic waves clearly occurs in MRI turbulence (Heinemann and Papaloizou 2009a, 2009b).

Each of the three solutions is expected to have a distinct relationship between overall radiative luminosity and accretion rate. Geometrically thin, radiatively efficient accretion disks are expected to have a luminosity which varies linearly with accretion rate, as for a given black hole spin, all the released binding energy is equivalent to an approximately fixed fraction of the rest mass energy of the accreted material. Advective models have reduced radiative efficiency. For the optically thin RIAF solutions, the luminosity is approximately proportional to the square of the accretion rate (Narayan and Yi 1995b) and therefore the radiative efficiency drops as the accretion rate is reduced. Observational tests of this predicted relationship are discussed in Chaps. 3.2 and 5.7. In the high luminosity regime of slim disks, the radiative output is expected to approximately saturate to the Eddington limit as photons become trapped (Begelman and Meier 1982), and this is expected to remain true even if outflows are driven (e.g. Poutanen et al. 2007), although much of this luminosity will be emitted anistropically toward the rotation axis (e.g. Ohsuga and Mineshige 2011; Chap. 5.3).

The three basic flow paradigms have been used over the years to explain the observed variety of black hole accretion sources. Something like the radiation pressure dominated advective slim disks are probably relevant for luminous quasars and QSO's ("quasi-stellar objects"), narrow line Seyfert 1's, ultraluminous X-ray sources, SS433, and perhaps some of the intermediate/steep power law states of black hole X-ray binaries. Geometrically thin, optically thick, radiatively efficient accretion disks extending down close to the central black holes are probably most relevant for the high/soft state of black hole X-ray binaries and perhaps for some QSO's. Optically thin RIAF's are almost certainly relevant for low luminosity active galactic nuclei (AGN), jet-dominated nonthermal AGN such as M87, the Galactic Center source Sgr A*, the inner regions of some broad line Seyfert 1's, and the inner regions of the low/hard state of black hole X-ray binaries. Many of the other chapters in this book address how well these models work in explaining the observed properties of these sources.

2.2 Tying Local Models Into Global Models: The Overall Geometry of the Flow

Figure 2 shows that, provided alpha is not too high, the RIAF solutions at any particular radius always terminate above a critical accretion rate where radiative cooling becomes comparable to turbulent dissipation (Abramowicz et al. 1995; Narayan and Yi 1995b). This critical accretion rate generally decreases with radius at large radii, so that provided the accretion rate is not too low, the accretion flow will generally be in the radiatively efficient, geometrically thin disk state at large radii. However, at smaller radii, provided the accretion rate is not too high ($\dot{M} \lesssim 10\alpha^2 \dot{M}_{\rm E}$, where $\dot{M}_{\rm E}$ is the Eddington luminosity divided by c^2; Narayan and Yi 1995b), the accretion flow can exist in one of two distinct thermal equilibria: the optically thin RIAF solution, and the radiatively efficient geometrically thin disk solution. Narayan and Yi (1995b) argue that evaporation from the surface of the thin disk will tend to drive the accretion flow into the RIAF regime whenever it is possible. One would then be left with a geometry which consists of an outer geometrically thin disk extending down toward a transition radius inside of which the flow adopts the RIAF solution. As the external accretion rate increases, the transition radius moves inward. This then provides an explanation for transitions between hard and soft states in black hole X-ray binaries (Esin et al. 1997, Fig. 3). As discussed in Chap. 2.5, it is now well-established that the external accretion rate is not the only parameter that controls state transitions: hard to soft transitions generally occur at higher accretion rates than soft to hard transitions. Nevertheless, this geometry of an outer thin disk and an inner RIAF has become a popular model for hard states of black hole X-ray binaries, and of certain classes of active galactic nuclei.

RIAF's are not the only way to produce hard X-rays, however. As illustrated in the top most panel of Fig. 3, it is possible that a corona containing hot or energetic nonthermal electrons exists above and below the geometrically thin disk. This could be locally generated by, for example, flares associated with buoyant magnetic field lines in a manner analogous to the production of the solar corona (Galeev et al. 1979; Haardt and Maraschi 1991). Alpha disk models in which some fraction of the locally generated accretion power is dissipated in an external corona have been developed by Svensson and Zdziarski (1994). Just as in the sun, the actual geometry of the corona could be quite complicated, with multiple coronal patches.

Starting with the work of Meyer et al. (2000a, 2000b) and Różańska and Czerny (2000), many attempts have also been made to build alpha-based models of outer thin disks, inner RIAF's, and coronae that themselves are treated as accreting RIAF's but which can sandwich portions of the thin disk. Each of the different flow regions exchange energy and mass through thermal conduction, evaporation and condensation, and irradiation. As shown in Fig. 4, one can even form inner condensed pieces of radiatively efficient thin disks embedded inside the RIAF/corona flow in these models (e.g. Meyer et al. 2007; Liu et al. 2011). While such flow geometries may well occur in nature, using the same alpha prescription everywhere, especially at high latitudes off the midplane, might be problematic, as discussed briefly in Sect. 3.2 below. One hopes that thermodynamically consistent global simulations of MRI turbulence may shed light on how transitions between thin disks, coronae, and RIAF's actually occur.

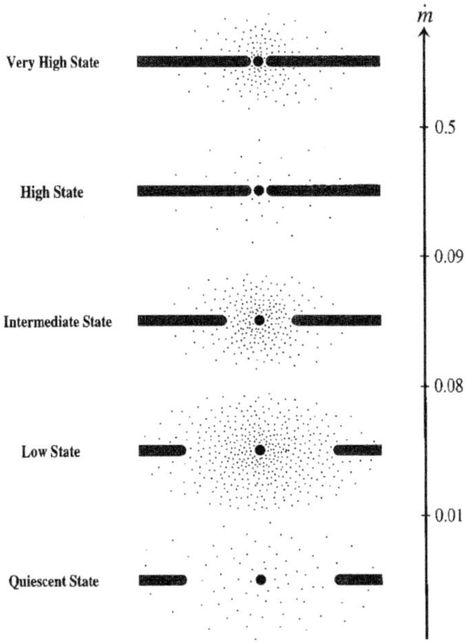

Fig. 3 Suggested flow geometries involving outer geometrically thin/optically thick disks and inner optically thin RIAF's for various observed states in black hole X-ray binaries, from Esin et al. (1997). Here $\dot{m}$ is the accretion rate scaled by ten times the Eddington luminosity divided by c^2

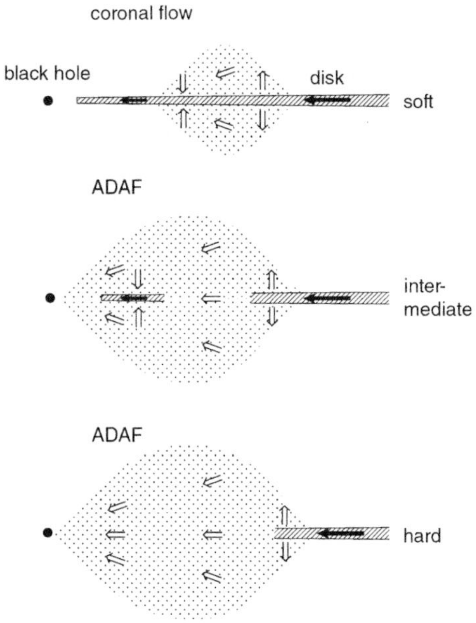

Fig. 4 Possible accretion flow geometries in evaporation/condensation models, from Meyer et al. (2007)

2.3 Other Variants

In addition to the three basic local models of optically thin RIAF's, radiatively efficient geometrically thin disks, and radiation pressure dominated advective slim disks, numerous other models have been proposed over the years. Some of these are variations on the three basic models, while others involve more significant departures from the physics included in these models.

Because much of the dissipated accretion power is not radiated away, both the optically thin and optically thick advection dominated solutions involve fluid which is only weakly bound to the black hole, i.e. with internal energy comparable in magnitude to the orbital binding energy. Outflows are therefore very likely to occur in these regimes (Narayan and Yi 1994; Blandford and Begelman 1999), and one way of incorporating them (with little physics beyond the invocation of self-similarity) is to simply assume that the accretion rate varies as a power law with radius $\dot{M} \propto R^p$ (Blandford and Begelman 1999). Assuming accretion velocities scale with the free-fall speed $\propto R^{-1/2}$, as self-similarity would require, such solutions have a radial density profile $\rho \propto R^{p-3/2}$, with $p = 0$ corresponding to the standard ADAF. Such solutions have been dubbed Advection-Dominated Inflow-Outflow Solutions (ADIOS) by Blandford and Begelman (1999). Outflows have been commonly observed in global simulations of these flow regimes (Chaps 2.4 and 5.3), though exactly how much mass is lost compared to how much is accreted remains theoretically uncertain. Very recent simulations by Narayan et al. (2012) in the low luminosity RIAF regime find that outflows are not as powerful as previously thought. Time-dependent self-similar ADAF solutions by Ogilvie (1999), that differ from the stationary self-similar solution of Narayan and Yi (1994), appear to be consistent with this result. At high luminosities, outflows can also be driven directly by radiation pressure, particularly with the high atomic opacities expected for gas around supermassive black holes in active galactic nuclei (Murray et al. 1995; Proga et al. 2000).

Another consequence of not dissipating accretion power in advection dominated flows is that the entropy of the plasma increases inward, implying that the flow will be, at least hydrodynamically, convectively unstable. It has been suggested that large scale convection in the flow can transport angular momentum *inward*, and stationary solutions in which only small accretion rates occur as material circulates again and again in convective eddies have been proposed which have radial density profiles $\rho \propto R^{-1/2}$ (Convection Dominated Accretion Flows or CDAF's; Narayan et al. 2000; Quataert and Gruzinov 2000). The physical consistency of such solutions is controversial (Balbus and Hawley 2002; Narayan et al. 2002). The recent RIAF simulations by Narayan et al. (2012) also do not find obvious signs of convection.

Yet another variant on the optically thin RIAF model is the luminous hot accretion flow (LHAF) model of Yuan (2001). As we discussed above, unless alpha is large, RIAF's have a maximum accretion rate above which the solutions with advective cooling do not exist. This creates a problem for using them to explain hard state sources that are observed to exist at high luminosities. The LHAF model solves this problem by positing that advective cooling is replaced by advective *heating*. The heating here is due to compressional work, and is non-dissipative. LHAF's therefore essentially balance compressional work plus turbulent dissipation with radiative cooling. However, such models appear to be thermally unstable given their local equilibrium curves on the accretion rate vs. surface mass density plane, although the growth rates might be long enough to not significantly affect the flow (Yuan 2003).

Models involving strong and/or large-scale magnetic fields in the flow have also been proposed, for example flows in which magnetic pressure dominates thermal pressure (Pariev et al. 2003; Meier 2005; Machida et al. 2006; Begelman and Pringle 2007), and flows in which angular momentum transport is dominated by large-scale MHD outflows (e.g. Ferreira and Pelletier 1995). It is even possible for enough magnetic field to be advected inward in the accretion flow that it becomes strong enough to disrupt the flow, allowing accretion to only proceed through magnetic Rayleigh-Taylor interchange motions (so-called Magnetically Arrested Disks; Narayan et al. 2003, see also Chap. 2.4). Such flows are beyond the scope of this particular review, which focuses on accretion driven by turbulence, although they may very well be important in nature. It is now time for us to turn to what is known about the local properties of MRI turbulence in accretion flows.

3 Going Beyond Alpha: MRI Turbulence

Since the discovery of its relevance to the physics of accretion flows (Balbus and Hawley 1991; Hawley and Balbus 1991), tremendous theoretical effort has been expended to try and understand the properties of MRI turbulence in various regimes of relevance to astrophysics. Studying MRI turbulence has allowed us to pose questions that simply cannot be asked within the alpha prescription, and has led to considerable new physical insight into how the accretion process works, as well as sharpened the true physical uncertainties. It is, after all, only by building models based on real physics, rather than precriptions that sweep undertainties into a single parameter, that real scientific progress can be made. However, it has to be admitted that the ultimate goal of replacing alpha-based modeling of accretion powered sources with observationally falsifiable models based on the actual physics of the turbulence has *not* yet been achieved. In this section I will review some of the fundamental physics issues that have been understood, or at least revealed, by *local* simulations of MRI turbulence. Global MRI simulations of black hole accretion flows have also provided considerable insight, and are in fact closest to realizing the goal of providing observationally testable models of the accretion flow onto the Galactic Center black hole source Sgr $A^{\star}$. These global simulations are reviewed in Chap. 2.4.

3.1 Shearing Box Simulations of MRI Turbulence

It is in the very nature of fully nonlinear, strong turbulence that energy released or injected from large spatial scales passes quickly down to microscopic dissipation scales through a turbulent cascade. In our case gravitational binding energy is released through the MRI which grows by tapping directly into the free energy associated with the differential rotation inherent in the accretion flow. The microscopic dissipation scales are associated with the true viscosity and resistivity of the plasma. The actual physical dissipation scales relevant to black hole accretion flows are extremely small compared to the energy release scales of the MRI (presumably of order the disk thickness), but numericists have nevertheless hoped that by putting in enough grid zones into their simulations, that some convergence can be achieved in describing the properties of MRI turbulence. That hope is best achieved in local shearing box simulations of the turbulence, where all the computing power is devoted to resolving scales within the turbulent cascade, and not on the larger scale dynamics associated with the overall flow geometry (as important as these larger scales are to ultimately understanding observed sources).

The geometry and properties of the shearing box are very nicely described by Hawley et al. (1995). Essentially a small, perfectly rectangular Cartesian box is placed in the rotating shear flow, and corotates with the background flow at the center of the box. The curvature in the background flow streamlines are entirely neglected, but the effects of rotation are nevertheless included through Coriolis forces as well as centrifugal forces that are combined with the gravitational force through an effective potential. Boundary conditions are such that the flow is assumed to be perfectly periodic in the azimuthal direction, but *shearing* periodic in the radial direction: one imagines many identical shearing boxes sliding past each other according to the background differential rotation. If the box is placed in the midplane of the flow, one sometimes neglects the vertical gravity and adopts periodic boundary conditions in the vertical direction (*unstratified shearing boxes*), but one can also include vertical gravity (*stratified shearing boxes*) and adopt outflow boundary conditions, or, for computational convenience, retain vertical periodic boundary conditions (a stack of accretion disk pancakes!).

The symmetries of the standard shearing box mean that there is *no* net accretion of mass through the box, and therefore in fact *no* release of gravitational binding energy. All the energy associated with the turbulence in fact arises from the net work done by the turbulent stresses on the shearing walls of the box.

Any initial net vertical magnetic field must be conserved in a shearing box simulation, and this is also true of net azimuthal magnetic field if the shearing box is unstratified (such field can leak out of the vertical boundaries of stratified boxes if outflow boundary conditions are employed). Shearing box simulations can therefore have the net magnetic flux through the box as a fixed external parameter. Unstratified simulations with no external flux, and no explicit treatment of viscosity and resistivity which would resolve dissipation on scales larger than the grid scale, actually produce MRI turbulent stresses that monotonically decrease with increasing numerical resolution (Pessah et al. 2007; Fromang and Papaloizou 2007)! However, this is a singular situation, as including explicit viscosity and resistivity in the MHD equations (albeit at far larger values than are relevant for black hole accretion flows) or a net magnetic flux does lead to converged levels of stress as numerical resolution is increased. This stress increases with the amount of external magnetic flux (Hawley et al. 1995; Pessah et al. 2007) and also increases with the dimensionless ratio of kinematic viscosity to Ohmic resistivity, known as the magnetic Prandtl number (Lesur and Longaretti 2007; Fromang et al. 2007; Simon and Hawley 2009). The result that stress increases with net magnetic flux has also been confirmed in localized regions of global simulations (Sorathia et al. 2010).

Adding in vertical gravity in stratified simulations enables convergence of the turbulent stresses even without net magnetic flux and explicit viscosity and resistivity (Shi et al. 2010; Davis et al. 2010). A major difference between unstratified and stratified shearing box simulations is that gravity allows for magnetic buoyancy, and this is clearly playing a role as alternating signs of azimuthal field continually develop in a dynamo within the MRI turbulence and rise outward (Brandenburg et al. 1995). For weak magnetic fields, this produces a quasiperiodic pattern of field reversals, as shown in Fig. 5. (When time-reversed, the pattern here resembles the butterfly diagram of latitudinal distributions of sunspots over the course of the solar cycle. By analogy, this behavior is occasionally referred to as the MRI butterfly diagram.) Moreover, stratified shearing box simulations with vertical outflow boundary conditions and net vertical magnetic flux can actually locally produce magnetocentrifugally driven outflows (Suzuki and Inutsuka 2009; Fromang et al. 2012; Lesur et al. 2012; Bai and Stone 2012). Increasing the net vertical magnetic flux in such simulations can also destroy the periodicity of the MRI butterfly dynamo, and ultimately suppress it (Bai and Stone 2012).

The whole premise of the alpha prescription is that the stresses are inherently local: (vertically-averaged) stress just depends on (vertically-averaged) thermal pressure. This appears to be mostly confirmed by shearing box simulations which have wide radial extents, in that spatial correlations between stresses at different locations decrease rapidly on radial scales larger than the disk scale height. However, there remain $\sim$20 percent correlations in the Maxwell stress on larger radial scales, indicating that the turbulence may not be entirely

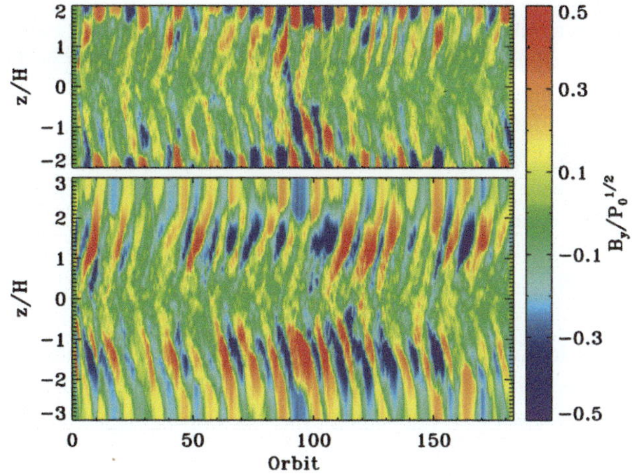

Fig. 5 Horizontally-averaged azimuthal component of the magnetic field as a function of height z and time in two vertically stratified shearing box simulations whose only difference is the height of the box, from Davis et al. (2010). There exists a dynamo in such vertically stratified simulations that causes quasiperiodic azimuthal field reversals

local (Simon et al. 2012). The butterfly dynamo cycles have also been observed in global simulations and have significant radial coherence on scales much larger than a disk scale height (O'Neill et al. 2011).

3.2 Aspects of the Vertical Structure Revealed by MRI Simulations

Provided any external magnetic flux is not too high, stratified shearing boxes generally result in a structure that is dominated by thermal pressure in the midplane and magnetic pressure and tension forces in the outer layers (Stone et al. 1996; Miller and Stone 2000; Hirose et al. 2006). MRI turbulence is generally confined to the weakly magnetized regions near the midplane, while Parker instability dynamics dominates the outer regions (Blaes et al. 2007). This basic structure of weakly magnetized midplane regions and more strongly magnetized high altitude regions is also generally observed in global simulations (e.g. Hawley and Balbus 2002; Penna et al. 2010; Sorathia et al. 2010), although here the strongly magnetized regions at high altitude can involve significant radial flows and circulation, which cannot happen in a shearing box. The fact that the magnetic pressure profile is broader than the thermal pressure profile, and that Maxwell stresses generally dominate Reynolds stresses in the turbulence, implies that an average of the ratio of stress to thermal pressure (alpha!) generally increases outwards: alpha should never be treated as a local quantity, but instead it is, at best, a representation of the ratio of vertically-averaged stress to pressure. Disk atmospheres are generally supported by magnetic fields, not thermal pressure, so that if atmosphere models of thermal and reflection spectra rely critically on vertical hydrostatic equilibrium between gravity and thermal pressure gradients, they may not be accurate. In addition, the fact that MRI turbulence is generally confined to the weakly magnetized midplane regions suggests that models of accreting coronal flows discussed in Sect. 2.2 above are probably not well described by a simple alpha prescription.

The magnetically-dominated outer layers are very suggestive of the locally generated magnetized corona discussed above in Sect. 2.2. Indeed, very tall stratified shearing box simulations by Miller and Stone (2000) found that approximately a quarter of the magnetic energy generated in the turbulent midplane regions was carried out into the corona. These simulations assumed an isothermal equation of state, however, and vertically stratified simulations that capture turbulent dissipation as heat and incorporate diffusive radiation transfer have generally found that the fraction of accretion power that is dissipated in the magnetized corona outside the photosphere is very small (Hirose et al. 2006, 2009; Krolik et al. 2007). On the other hand, the magnetic buoyancy exhibited in the butterfly diagram illustrated in Fig. 5 can be very energetically important in transporting significant amounts of thermal energy outward in the form of trapped photons in the radiation pressure dominated regime (Blaes et al. 2011). It should also be emphasized that the existing radiation MHD simulations have not included a net vertical magnetic flux, which might in principle enhance the coronal energetics.

3.3 Physics Issues in the RIAF Regime

Low luminosity RIAF models continue to be plagued by significant uncertainties in the microphysics of the plasma, whether they are globally simulated with MRI turbulence or modeled with an alpha prescription. Because the accreting plasma retains a significant fraction of its binding energy as internal energy, temperatures must approach virial temperatures: $kT \sim GMm_\mathrm{p}/R = (R_\mathrm{g}/R)m_\mathrm{p}c^2$, where m_p is the proton mass. This corresponds to $\sim 10^{12}$ K at ten gravitational radii. Optically thin cooling by electrons at such temperatures is very fast, unless the accretion rate and density is extremely low, so in order to not radiate away all the heat on an inflow time, the electron temperature T_e must be significantly less than the ion temperature T_i. This in turn implies that the electrons should not receive the vast majority of the turbulent dissipation of accretion power (otherwise they would radiate it away), and the ions must be at least partially thermally decoupled from the electrons on the inflow time. Coulomb collisions alone will be insufficient to thermally couple ions and electrons provided the accretion rate is not too high (Rees et al. 1982; Narayan and Yi 1995b). Plasma instabilities may in principle exist that couple the species more rapidly, but so far this has not been demonstrated (Begelman and Chiueh 1988; Park et al. 2010).

Even the MRI behaves differently in the collisionless regime (Quataert et al. 2002; Sharma et al. 2003; Balbus 2004; Islam and Balbus 2005), especially in giving rise to anisotropic pressure tensors that themselves can give rise to significant angular momentum transport (Sharma et al. 2006), and this is not captured in simulations that assume MHD. How the turbulent dissipation is ultimately channeled into heating of the ions and electrons (or energization—the ion and electron distribution functions need not be thermal) is another significant uncertainty. Local simulations find that direct heating by the anisotropic pressure tensor can account for 50 percent of the heating by the turbulence, and that the ratio of electron to ion heating is $\sim 0.3(T_\mathrm{e}/T_\mathrm{i})^{1/2}$ (Sharma et al. 2007). It is also just now becoming possible to do fully kinetic simulations of collisionless MRI turbulence, at least locally (Riquelme et al. 2012), so that there is hope for further resolving some of these issues in the not so distant future.

All these effects remain to be included in global simulations of the RIAF regime, discussed in Chap. 2.4. Currently these generally assume regular MHD, and adopt prescriptions for treating the electron and ion distribution functions, such as assuming thermal distributions with a constant ion to electron temperature ratio.

3.4 Attempts to Simulate the Radiation Pressure Thermal Instability

For some years now, it has been possible to do vertically stratified shearing box simulations that explore the thermodynamics of MRI turbulence. These simulations capture grid scale losses of magnetic and kinetic energy, and incorporate radiation transport and cooling through flux limited diffusion. It has also been possible to do global simulations using flux limited diffusion under axisymmetry—see Chap. 5.3. Recently, even more accurate radiation transport algorithms have been successfully developed (e.g. Davis et al. 2012; Jiang et al. 2012; Sadowski et al. 2012; Takahashi et al. 2012).

Vertically stratified shearing box simulations with optically thick cooling and which incorporate the dynamics of radiation pressure enable the exploration of the thermal instability predicted by alpha disk modeling on the negative slope branch of the thermal equilibrium curves on the right hand side of Fig. 2. The origin of this thermal instability is very easy to understand. Assuming that radiation diffusion dominates the vertical heat transport (which is generally found in simulations), the local cooling rate per unit area is of order the radiation energy density at the midplane aT^4 times the speed of light over the vertical optical depth. Because the opacity is dominated by electron scattering in these high temperature regimes, this then implies that $Q^- \propto T^4/\Sigma$. On the other hand, the heating rate per unit area is the vertically integrated stress times the rate of strain, $Q^+ \sim H w_{R\phi} R|d\Omega/dR|$. Because the disk is supported vertically by radiation pressure, and the vertical gravity increases linearly with height above the midplane, the disk half thickness is simply proportional to the surface radiation flux, i.e. $H \propto Q^- \propto T^4/\Sigma$. Hence a standard alpha prescription, in which $w_{R\phi} = \alpha P = \alpha a T^4/3$, will mean that the heating rate $Q^+ \propto T^8/\Sigma$. Because the inflow time is much longer than the thermal time for geometrically thin disks, the surface mass density Σ cannot vary significantly on the thermal time scale, and the heating rate therefore depends much more sensitively on temperature than the cooling rate. Hence a perturbative increase (decrease) in temperature would lead to runaway heating (cooling).

As shown in Fig. 1, simulations are consistent with the standard alpha prescription that the time and space averaged stress scales with total thermal pressure, which is mostly radiation pressure in this regime. Nevertheless, at least some simulations have been able to establish thermal equilibria between heating and cooling in the radiation pressure dominated regime that last for many thermal times (Turner 2004; Hirose et al. 2009). On the other hand, recent simulations by Jiang et al. (2013b) find that such equilibria, even if established, always eventually suffer runaway heating or cooling. It is not clear what is producing the differences between the different simulations, which are run using different codes.

Note that, in contrast to the hydrogen ionization driven thermal/viscous instability that is responsible for dwarf nova and outbursts in X-ray binaries, there is no clear observational evidence for the putative radiation pressure driven thermal instability predicted by alpha disk theory. It could be that the intrinsically stochastic nature of the turbulent dissipation, which is not captured in the alpha prescription, acts as a stabilizing influence (Janiuk and Misra 2012). But in addition, it is clear that the alpha prescription breaks down on short time scales. As shown in Fig. 6, fluctuations in thermal energy *lag* fluctuations in turbulent energy by approximately a thermal time. This is easy to understand physically: it is the dissipation of turbulence that heats the plasma and produces thermal pressure, and thermal energy will therefore only respond to fluctuations in turbulent energy on time scales of order the heating time $\sim(\alpha\Omega)^{-1}$. There appears to be no direct feedback from pressure to stress on the thermal time scale, and the alpha prescription is therefore only established on longer time scales, although exactly *why* the prescription is established remains mysterious. More thought needs to be applied to understanding the relationship between stress and pressure, and to understanding what is going on to explain the different results seen in different simulation codes. Recently, Lin et al. (2012) have taken a useful first step in developing analytic, time-dependent ordinary differential equations that successfully reproduce a number of the observed features in these simulations.

Note that while the thermal instability has been studied using local simulations of MRI turbulence, the inflow or viscous instability has not as the boxes that have been used are

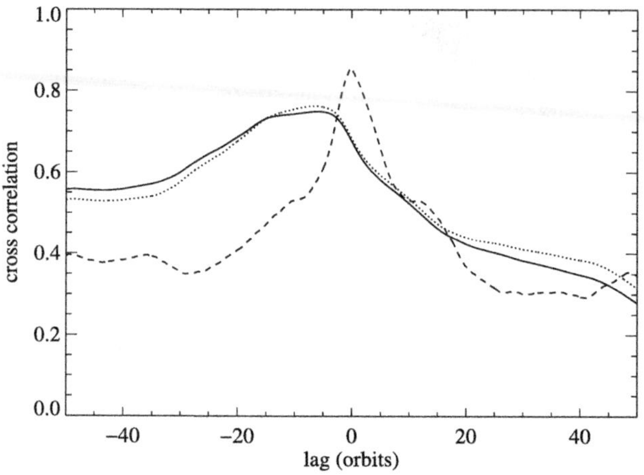

Fig. 6 Cross correlation coefficient for various forms of volume integrated energy as a function of time difference with respect to variations in volume integrated magnetic energy, for a radiation dominated stratified shearing box simulation. Negative values on the horizontal axis mean that the energy lags behind magnetic energy. The *dashed line* shows the turbulent kinetic energy, which is highly correlated with magnetic energy at zero lag. This is because both energies are aspects of the same MRI turbulence! The *solid* and *dotted curves* show radiation and gas internal energies, respectively. Both of these are very similar as they are both thermal energies, and both are again correlated with magnetic energy, but with a significant time lag of order 5–15 orbital periods. This is comparable to the thermal time in this simulation (From Hirose et al. 2009)

radially too narrow to allow for significant variations in surface mass density. Very wide radial boxes, or global simulations, will be necessary to explore this physics.

3.5 Other Issues in the Radiation Dominated Regime

Radiation dominated plasmas have a number of other interesting properties that are likely relevant to black hole accretion flows in the high luminosity regime. One in particular is that the sound speed in the fluid $\simeq [4aT^4/(9\rho)]^{1/2}$ is determined by radiation pressure, not gas pressure, and when the former exceeds the latter, it is possible to be in a regime where turbulent motions are subsonic and yet supersonic with respect to the sound speed from gas pressure alone. But photons generally diffuse through the plasma, and if they do so rapidly, then even fluid motions that are subsonic with respect to the radiation pressure sound speed, but supersonic with respect to the gas sound speed, can be highly compressible, because photon diffusion reduces the photon pressure response. Large density fluctuations can therefore be produced, and this has been observed in unstratified shearing box simulations of radiation dominated MRI turbulence (Turner et al. 2003; Jiang et al. 2013a).

Radiation damping of temperature fluctuations in radiation pressure dominated MRI turbulence can be a significant source of dissipation which, like the pressure anisotropies in collisionless MRI turbulence discussed above in Sect. 3.3, can be resolved in numerical simulations. Such fluctuations can be compressible in nature as we just mentioned (Agol and Krolik 1998), or due to nonlinear isobaric fluctuations associated with regions of high magnetic pressure (Blaes et al. 2011). Some tens of percent of the total dissipation has been observed to occur through radiation damping in shearing box simulations (Turner 2004; Blaes et al. 2011). This radiation damping can also increase the bulk viscosity and therefore the magnetic Prandtl number, and can increase the Maxwell stress in the turbulence (Jiang et al. 2013a).

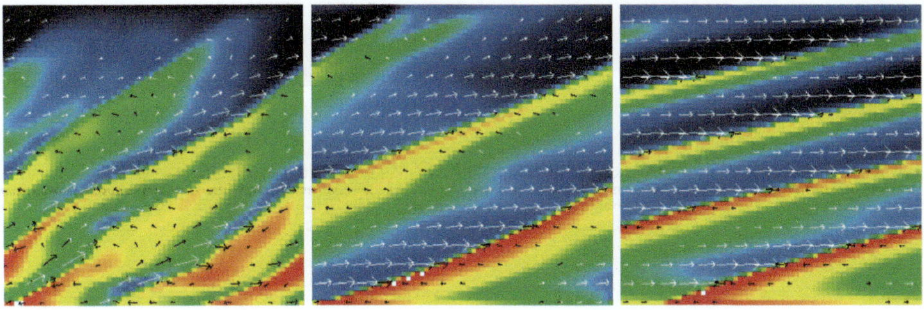

Fig. 7 2D simulations of shock trains produced by the photon bubble instability in radiation pressure supported media with initially uniform magnetic fields of increasing strength from left to right. *Arrows* show fluid velocity and *colors* show density on a logarithmic scale, with warm colors being high and cold colors being low. The weight of the dense shocked fluid causes the weaker magnetic fields in the left hand figure to buckle (From Turner et al. 2005)

The fact that MRI turbulence can be supersonic with respect to the sound speed in the gas alone in the radiation dominated regime implies that, in principle, turbulent speeds can exceed mean thermal speeds not only of the ions in the plasma, but also the electrons. If turbulent motions are limited by the radiation sound speed then this may start to happen at radiation to gas pressure ratios in excess of the proton to electron mass ratio, and may happen at even lower ratios in the photosphere regions which tend to be dominated by magnetic pressure, not thermal pressure. Differences in bulk turbulent velocities on the scale of a photon mean free path that exceed in magnitude characteristic electron thermal speeds will mean that bulk Comptonization by the turbulence itself will dominate thermal Comptonization, and this may provide an alternative means of producing a Comptonized high energy spectrum in radiation dominated luminous states of black hole sources (Socrates et al. 2004; Socrates 2010).

Advection of heat is a key ingredient to the radiation pressure dominated slim disk solutions discussed in Sect. 2.1 above. However, instabilities might produce inhomogeneities in the flow that allow photons to escape more readily through underdense channels, rather than be advected inward. The most well-explored of such instabilities are magnetically-mediated "photon bubble" instabilities (Arons 1992; Gammie 1998). On short length scales where photons diffuse rapidly, such instabilities amount to radiatively amplified magnetosonic modes (Blaes and Socrates 2003) that develop into highly inhomogeneous trains of shocks (Begelman 2001; Turner et al. 2005, see Fig. 7). In principle such inhomogeneities could allow locally super-Eddington fluxes to escape from the disk atmosphere without driving an outflow (Begelman 2002). Simulating photon bubbles in the presence of MRI turbulence has proved computationally challenging in the radiation pressure dominated regime due partly to the small length scales (of order the *gas* pressure scale height) that must be resolved, and partly by the fact that Parker instabilities in the magnetically dominated surface layers also produce significant inhomogeneity. Models of slim disks with porous outer layers *and* winds have recently been developed by Dotan and Shaviv (2011). It has also been suggested that accretion flows in the radiation dominated regime might be highly inhomogeneous structures that are not well-described by any of the standard accretion flow models discussed in Sect. 2.1 (Dexter and Agol 2011).

4 Summary

The primary reason why the alpha prescription continues to be a mainstay of black hole accretion theory is that it enables models to be built that couple the dynamics of the flow (outward angular momentum transport through the plasma by a stress described phenomenologically by alpha) to the thermodynamics (dissipation of accretion power described phenomenologically by the same alpha stress prescription times the shear rate). Once this is all combined with radiative cooling processes, one has models that can be used to generate spectra, time variability, etc. that can be compared to observations. While decades of theoretical work within this research paradigm have yielded valuable insights (e.g. scalings of luminosity and temperature with accretion rate, the importance of advection of heat), one can only carry this program so far without addressing the fundamental physics that the alpha prescription hides.

(Fewer) decades of research have now been spent on understanding the properties of MRI turbulence, but until recently, this has focused mostly on the dynamics of the turbulence, not the thermodynamics, and it is the latter which is ultimately required to connect to observations, which are, after all, detecting the photons emitted by the source. We still have a lot of unanswered fundamental questions. For example, what is the true nature of the thin disk/RIAF transition radius, upon which hangs so much phenomenology of black hole X-ray binaries (hard/soft state transitions, band-limited noise, low frequency quasi-periodic oscillations, ...)? The thermal/viscous instabilities have guided theoretical effort to exclude uphysical equilibria, but are these instabilities always real, and how do they manifest? (Only those driven by ionization/recombination are observationally known to exist, e.g., in dwarf novae.) What determines how accretion power is partitioned into various forms? These are all questions of thermodynamics, not just dynamics.

As discussed in Chap. 2.4, global MRI simulations of low luminosity RIAF's have been most successful in connecting to observations, largely because, at least in some regimes, the radiative cooling does not dramatically affect the dynamics of the flow, and therefore can be calculated after the fact by post-processing the simulation. Moreover, because the flow is optically thin at most photon frequencies, the simulation hardware has actually developed to the point where radiative cooling can be fully incorporated in the actual dynamical simulation itself (Dibi et al. 2012). However, as discussed in Sect. 3.3 above, even here there remain issues in the microphysics that still need to be understood theoretically.

Optically thick radiation transport is computationally more expensive, and also more important for the dynamics in the high luminosity black hole accretion regimes where radiation pressure plays a critical role. As discussed in Sect. 3.4 above, local shearing box simulations have shed some light on how MRI turbulence works in this regime, and in what ways the alpha prescription does, and does not, describe the physics. Unfortunately, this still leaves major unanswered questions as to the global structure of the flow. Two dimensional (axisymmetric) global simulations have also been done which have confirmed the existence of discrete flow states (Ohsuga et al. 2009; Ohsuga and Mineshige 2011, see also Chaps. 2.4 and 5.3), but these cannot be run over long time scales as MRI turbulence cannot be sustained in axisymmetry. With the ongoing increase of computer power, combined with the development of new radiation transport algorithms, it should be possible to do global 3D simulations of accretion flows in optically thick regimes too. This will better enable us to understand the origins of state transitions in black hole X-ray binaries, transitions between radiatively efficient thin disks and RIAF's, and other fundamental problems in black hole accretion flows. Achieving this goal should complete the shift to a more physics-based research paradigm.

Acknowledgements The author is grateful to ISSI and to the organizers of this workshop for enabling such a productive set of scientific interactions, and thanks a number of the participants, particularly Chris Fragile, Tom Maccarone, Shin Mineshige, Ken Ohsuga, Juri Poutanen, and Chris Reynolds, for enlightening him on various issues of relevance to this paper. The author is also grateful to the referees for suggestions that significantly improved the manuscript. The author's research is supported by the US National Science Foundation under grant AST-0707624.

References

M.A. Abramowicz, B. Czerny, J.P. Lasota, E. Szuszkiewicz, Astrophys. J. **332**, 646–658 (1988)
M.A. Abramowicz, X. Chen, S. Kato, J.P. Lasota, O. Regev, Astrophys. J. **438**, L37–L39 (1995)
E. Agol, J. Krolik, Astrophys. J. **507**, 304–315 (1998)
J. Arons, Astrophys. J. **388**, 561–578 (1992)
X.-N. Bai, J.M. Stone, Astrophys. J. (2012). doi:10.1088/0004-637X/767/1/30
S.A. Balbus, Astrophys. J. **616**, 857–864 (2004)
S.A. Balbus, J.F. Hawley, Astrophys. J. **376**, 214–222 (1991)
S.A. Balbus, J.F. Hawley, Astrophys. J. **400**, 610–621 (1992)
S.A. Balbus, J.F. Hawley, Rev. Mod. Phys. **70**, 1–53 (1998)
S.A. Balbus, J.F. Hawley, Astrophys. J. **573**, 749–753 (2002)
S.A. Balbus, J.C.B. Papaloizou, Astrophys. J. **521**, 650–658 (1999)
M.C. Begelman, Astrophys. J. **551**, 897–906 (2001)
M.C. Begelman, Astrophys. J. **568**, L97–L100 (2002)
M.C. Begelman, T. Chiueh, Astrophys. J. **332**, 872–890 (1988)
M.C. Begelman, D.L. Meier, Astrophys. J. **253**, 873–896 (1982)
M.C. Begelman, J.E. Pringle, Mon. Not. R. Astron. Soc. **375**, 1070–1076 (2007)
O. Blaes, A. Socrates, Astrophys. J. **596**, 509–537 (2003)
O. Blaes, S. Hirose, J.H. Krolik, Astrophys. J. **664**, 1057–1071 (2007)
O. Blaes, J.H. Krolik, S. Hirose, N. Shabaltas, Astrophys. J. **733**, 110 (2011), 24 pp.
R.D. Blandford, M.C. Begelman, Mon. Not. R. Astron. Soc. **303**, L1–L5 (1999)
R.D. Blandford, D.G. Payne, Mon. Not. R. Astron. Soc. **199**, 883–903 (1982)
G.R. Blumenthal, L.T. Yang, D.N.C. Lin, Astrophys. J. **287**, 774–784 (1984)
A. Brandenburg, Å. Nordlund, R.F. Stein, U. Torkelsson, Astrophys. J. **446**, 741–754 (1995)
X. Chen, R.E. Taam, Astrophys. J. **412**, 254–266 (1993)
X. Chen, M.A. Abramowicz, J.-P. Lasota, R. Narayan, I. Yi, Astrophys. J. **443**, L61–L64 (1995)
S.W. Davis, J.M. Stone, Y.-F. Jiang, Astrophys. J. Suppl. Ser. **199**, 9, 19 pp. (2012)
S.W. Davis, J.M. Stone, M.E. Pessah, Astrophys. J. **713**, 52–65 (2010)
J. Dexter, E. Agol, Astrophys. J. **727**, L24 (2011), 5 pp.
S. Dibi, S. Drappeau, P.C. Fragile, S. Markoff, J. Dexter, Mon. Not. R. Astron. Soc. **426**, 1928–1939 (2012)
C. Dotan, N.J. Shaviv, Mon. Not. R. Astron. Soc. **413**, 1623–1632 (2011)
A.A. Esin, J.E. McClintock, R. Narayan, Astrophys. J. **489**, 865–889 (1997)
J. Ferreira, G. Pelletier, Astron. Astrophys. **295**, 807–832 (1995)
P.C. Fragile, O.M. Blaes, Astrophys. J. **687**, 757–766 (2008)
S. Fromang, J. Papaloizou, Astron. Astrophys. **476**, 1113–1122 (2007)
S. Fromang, J. Papaloizou, G. Lesur, T. Heinemann, Astron. Astrophys. **476**, 1123–1132 (2007)
S. Fromang, H.N. Latter, G. Lesur, G.I. Ogilvie, Astron. Astrophys (2012). doi:10.1051/0004-6361/201220016
A.A. Galeev, R. Rosner, G.S. Vaiana, Astrophys. J. **229**, 318–326 (1979)
C.F. Gammie, Mon. Not. R. Astron. Soc. **297**, 929–935 (1998)
C.F. Gammie, Astrophys. J. **522**, L57–L60 (1999)
J. Goodman, Mon. Not. R. Astron. Soc. **339**, 937–948 (2003)
F. Haardt, L. Maraschi, Astrophys. J. **380**, L51–L54 (1991)
J.F. Hawley, S.A. Balbus, Astrophys. J. **376**, 223–233 (1991)
J.F. Hawley, S.A. Balbus, Astrophys. J. **573**, 738–748 (2002)
J.F. Hawley, C.F. Gammie, S.A. Balbus, Astrophys. J. **440**, 742–763 (1995)
T. Heinemann, J.C.B. Papaloizou, Mon. Not. R. Astron. Soc. **397**, 52–63 (2009a)
T. Heinemann, J.C.B. Papaloizou, Mon. Not. R. Astron. Soc. **397**, 64–74 (2009b)
S. Hirose, O. Blaes, J.H. Krolik, Astrophys. J. **704**, 781–788 (2009)
S. Hirose, J.H. Krolik, O. Blaes, Astrophys. J. **691**, 16–31 (2009)
S. Hirose, J.H. Krolik, J.M. Stone, Astrophys. J. **640**, 901–917 (2006)

S. Ichimaru, Astrophys. J. **214**, 840–855 (1977)
T. Islam, S. Balbus, Astrophys. J. **633**, 328–333 (2005)
A. Janiuk, R. Misra, Astron. Astrophys. **540**, A114 (2012), 6 pp.
Y.-F. Jiang, J.M. Stone, S.W. Davis, Astrophys. J. Suppl. Ser. **199**, 14 (2012), 29 pp.
Y.-F. Jiang, J.M. Stone, S.W. Davis, Astrophys. J. **767**, 148 (2013a), 14 pp.
Y.-F. Jiang, J.M. Stone, S.W. Davis, Astrophys. J. (2013b, submitted)
S. Kato, F. Honma, R. Matsumoto, Mon. Not. R. Astron. Soc. **231**, 37–48 (1988)
S. Kato, M.A. Abramowicz, X. Chen, Publ. Astron. Soc. Jpn. **48**, 67–75 (1996)
S. Kato, T. Yamasaki, M.A. Abramowicz, X. Chen, Publ. Astron. Soc. Jpn. **49**, 221–225 (1997)
A.R. King, J.E. Pringle, M. Livio, Mon. Not. R. Astron. Soc. **376**, 1740–1746 (2007)
I. Kotko, J.-P. Lasota, Astron. Astrophys. **545**, A115 (2012), 9 pp.
J.H. Krolik, Astrophys. J. **515**, L73–L76 (1999)
J.H. Krolik, S. Hirose, O. Blaes, Astrophys. J. **664**, 1045–1056 (2007)
J.-P. Lasota, New Astron. Rev. **45**, 449–508 (2001)
G. Lesur, P.-Y. Longaretti, Mon. Not. R. Astron. Soc. **378**, 1471–1480 (2007)
G. Lesur, J. Ferreira, G.I. Ogilvie, Astron. Astrophys. (2012). doi:10.1051/0004-6361/201220395
A.P. Lightman, D.M. Eardley, Astrophys. J. **187**, L1–L3 (1974)
D.-B. Lin, W.-M. Gu, T. Liu, M.-Y. Sun, J.-F. Lu, Astrophys. J. **761**, 29 (2012), 5 pp.
B.F. Liu, C. Done, R.E. Taam, Astrophys. J. **726**, 10 (2011), 5 pp.
D. Lynden-Bell, J.E. Pringle, Mon. Not. R. Astron. Soc. **168**, 603–637 (1974)
M. Machida, K.E. Nakamura, R. Matsumoto, Publ. Astron. Soc. Jpn. **58**, 193–202 (2006)
D.L. Meier, Astrophys. Space Sci. **300**, 55–65 (2005)
F. Meyer, B.F. Liu, E. Meyer-Hofmeister, Astron. Astrophys. **354**, L67–L70 (2000a)
F. Meyer, B.F. Liu, E. Meyer-Hofmeister, Astron. Astrophys. **361**, 175–188 (2000b)
F. Meyer, B.F. Liu, E. Meyer-Hofmeister, Astron. Astrophys. **463**, 1–9 (2007)
K.A. Miller, J.M. Stone, Astrophys. J. **534**, 398–419 (2000)
N. Murray, J. Chiang, S.A. Grossman, G.M. Voit, Astrophys. J. **451**, 498–509 (1995)
R. Narayan, I. Yi, Astrophys. J. **428**, L13–L16 (1994)
R. Narayan, I. Yi, Astrophys. J. **444**, 231–243 (1995a)
R. Narayan, I. Yi, Astrophys. J. **452**, 710–735 (1995b)
R. Narayan, I.V. Igumenshchev, M.A. Abramowicz, Astrophys. J. **539**, 798–808 (2000)
R. Narayan, E. Quataert, I.V. Igumenshchev, M.A. Abramowicz, Astrophys. J. **577**, 295–301 (2002)
R. Narayan, I.V. Igumenshchev, M.A. Abramowicz, Publ. Astron. Soc. Jpn. **55**, L69–L72 (2003)
R. Narayan, A. Sadowski, R.F. Penna, A.K. Kulkarni, Mon. Not. R. Astron. Soc. **426**, 3241–3259 (2012)
S.C. Noble, J.H. Krolik, J.F. Hawley, Astrophys. J. **711**, 959–973 (2010)
G.I. Ogilvie, Mon. Not. R. Astron. Soc. **306**, L9–L13 (1999)
K. Ohsuga, S. Mineshige, Astrophys. J. **736**, 2 (2011), 18 pp.
K. Ohsuga, S. Mineshige, M. Mori, Y. Kato, Publ. Astron. Soc. Jpn. **61**, L7–L11 (2009)
S.M. O'Neill, C.S. Reynolds, M.C. Miller, K.A. Sorathia, Astrophys. J. **736**, 107 (2011), 7 pp.
V.I. Pariev, E.G. Blackman, S.A. Boldyrev, Astron. Astrophys. **407**, 403–421 (2003)
J. Park, C. Ren, E.G. Blackman, X. Kong, Phys. Plasmas **17**, 022901 (2010)
R.F. Penna, J.C. McKinney, R. Narayan, A. Tchekhovskoy, R. Shafee, J.E. McClintock, Mon. Not. R. Astron. Soc. **408**, 752–782 (2010)
M.E. Pessah, C.-K. Chan, D. Psaltis, Astrophys. J. **668**, L51–L54 (2007)
J. Poutanen, G. Lipunova, S. Fabrika, A.G. Butkevich, P. Abolmasov, Mon. Not. R. Astron. Soc. **377**, 1187–1194 (2007)
D. Proga, J.M. Stone, T.R. Kallman, Astrophys. J. **543**, 686–696 (2000)
E. Quataert, W. Dorland, G.W. Hammett, Astrophys. J. **577**, 524–533 (2002)
E. Quataert, A. Gruzinov, Astrophys. J. **539**, 809–814 (2000)
M.J. Rees, M.C. Begelman, R.D. Blandford, E.S. Phinney, Nature **295**, 17–21 (1982)
M.A. Riquelme, E. Quataert, P. Sharma, A. Spitkovsky, Astrophys. J. **755**, 50 (2012), 20 pp.
A. Różańska, B. Czerny, Astron. Astrophys. **360**, 1170–1186 (2000)
A. Sadowski, R. Narayan, A. Tchekhovskoy, Y. Zhu, Mon. Not. R. Astron. Soc. (2012). doi:10.1093/mnras/sts632
P.J. Sakimoto, F.V. Coroniti, Astrophys. J. **247**, 19–31 (1981)
N.I. Shakura, R.A. Sunyaev, Astron. Astrophys. **24**, 337–355 (1973)
N.I. Shakura, R.A. Sunyaev, Mon. Not. R. Astron. Soc. **175**, 613–632 (1976)
S.L. Shapiro, A.P. Lightman, D.M. Eardley, Astrophys. J. **204**, 187–199 (1976)
P. Sharma, G.W. Hammett, E. Quataert, Astrophys. J. **596**, 1121–1130 (2003)
P. Sharma, G.W. Hammett, E. Quataert, J.M. Stone, Astrophys. J. **637**, 952–967 (2006)
P. Sharma, E. Quataert, G.W. Hammett, J.M. Stone, Astrophys. J. **667**, 714–723 (2007)

J. Shi, J.H. Krolik, S. Hirose, Astrophys. J. **708**, 1716–1727 (2010)
N. Shibazaki, R. Hōshi, Prog. Theor. Phys. **54**, 706–718 (1975)
J.B. Simon, K. Beckwith, P.J. Armitage, Mon. Not. R. Astron. Soc. **422**, 2685–2700 (2012)
J.B. Simon, J.F. Hawley, Astrophys. J. **707**, 833–843 (2009)
J. Smak, Publ. Astron. Soc. Pac. **96**, 5–18 (1984)
A. Socrates, Astrophys. J. **719**, 784–789 (2010)
A. Socrates, S.W. Davis, O. Blaes, Astrophys. J. **601**, 405–413 (2004)
K.A. Sorathia, C.S. Reynolds, P.J. Armitage, Astrophys. J. **712**, 1241–1247 (2010)
J.M. Stone, J.F. Hawley, C.F. Gammie, S.A. Balbus, Astrophys. J. **463**, 656–673 (1996)
T.K. Suzuki, S.-I. Inutsuka, Astrophys. J. **691**, L49–L54 (2009)
R. Svensson, A.A. Zdziarski, Astrophys. J. **436**, 599–606 (1994)
R.E. Taam, D.N.C. Lin, Astrophys. J. **287**, 761–768 (1984)
H.R. Takahashi, K. Ohsuga, Y. Sekiguchi, T. Inoue, K. Tomida, Astrophys. J. (2012). doi:10.1088/0004-637X/764/2/122
N.J. Turner, Astrophys. J. **605**, L45–L48 (2004)
N.J. Turner, J.M. Stone, J.H. Krolik, T. Sano, Astrophys. J. **593**, 992–1006 (2003)
N.J. Turner, O.M. Blaes, A. Socrates, M.C. Begelman, S.W. Davis, Astrophys. J. **624**, 267–288 (2005)
X.-B. Wu, Mon. Not. R. Astron. Soc. **292**, 113–119 (1997)
X.-B. Wu, Q.-B. Li, Astrophys. J. **469**, 776–783 (1996)
T. Yamasaki, Publ. Astron. Soc. Jpn. **49**, 227–233 (1997)
F. Yuan, Mon. Not. R. Astron. Soc. **324**, 119–127 (2001)
F. Yuan, Astrophys. J. **594**, L99–L102 (2003)

DOI 10.1007/978-1-4939-2227-7_4
Reprinted from *Space Science Reviews* Journal, DOI 10.1007/s11214-014-0076-0

Fast Variability from Black-Hole Binaries

Tomaso M. Belloni · Luigi Stella

Received: 26 May 2014 / Accepted: 23 July 2014 / Published online: 2 August 2014

Abstract Currently available information on fast variability of the X-ray emission from accreting collapsed objects constitutes a complex phenomenology which is difficult to interpret. We review the current observational standpoint for black-hole binaries and survey models that have been proposed to interpret it. Despite the complex structure of the accretion flow, key observational diagnostics have been identified which can provide direct access to the dynamics of matter motions in the close vicinity of black holes and thus to the some of fundamental properties of curved spacetimes, where strong-field general relativistic effects can be observed.

Keywords Accretion · Accretion disks · Black hole physics · X-rays: binaries

1 Introduction

The first power density spectra (PDS) from the black-hole binary (BHB) Cyg X-1 were obtained with the Uhuru satellite and it became clear that no obvious periodicity was present, but the data were consistent with a noise process (Terrell 1972). For two decades, the dearth of known sources and the small effective area of X-ray instruments allowed only the study of the PDS of Cyg X-1, which features strong continuous noise (10–40 % fractional rms), with a few breaks which allow to identify characteristic time scales (Nolan et al. 1981; Belloni and Hasinger 1990). Only with the Ginga satellite, which included a large-area instrument together with an all-sky monitor to detect transient sources, quasi-periodic oscillations (QPOs) at frequencies between 0.01 and 10 Hz were found in a few systems (see Tanaka and Lewin 1996 for a review). The launch of the Rossi X-ray Timing Explorer

T.M. Belloni (✉)
INAF—Osservatorio Astronomico di Brera, Via E. Bianchi 46, 23807 Merate, Italy
e-mail: tomaso.belloni@brera.inaf.it

L. Stella
INAF—Osservatorio Astronomico di Roma, Via di Frascati, 33, 00040 Monte Porzio Catone, Italy
e-mail: luigi.stella@oa-roma.inaf.it

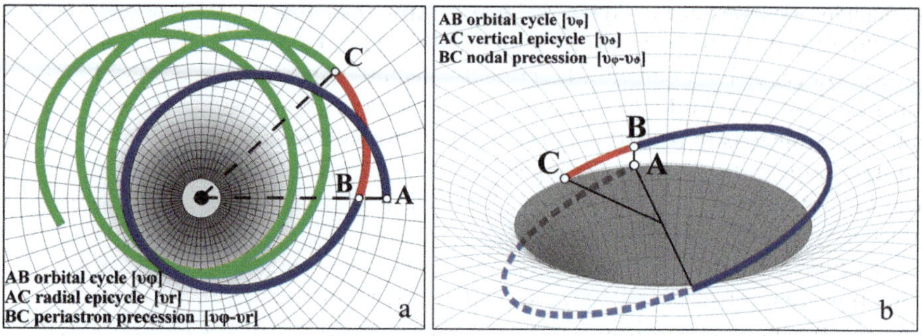

Fig. 1 Example of an eccentric and tilted orbit around a Kerr black, as seen face on (*left panel*), and from a ∼60 deg inclination angle (*right panel*). Cycles are represented for each of the three different fundamental frequencies of motion: azimuthal (aka orbital), and radial epicyclic and vertical epicyclic. The way in which the orbit undergoes periastron and nodal precession is also shown. Embedding diagrams are plotted to help visualize the perspective

(RXTE) increased dramatically the number of known sources, their coverage and the detection sensitivity, allowing classification of these low-frequency QPOs into different flavours (see Motta et al. 2011), as well as the discovery of weak signals at higher frequencies (see e.g. Belloni et al. 2012 and references therein). Here, we briefly summarise the basic knowledge deriving from this observational body and discuss its relevance for accretion theory and general relativity (GR).

2 Accretion and General Relativity

Early in the development of models of accretion disks around black holes it was realised that disk inhomogeneities orbiting in the innermost regions of the disk, where the bulk of the energy is released, can give rise to fast X-ray variability. If the lifetime and radial drift timescale of such inhomogeneities is longer than their orbital period, a quasi periodic signal ensues that can provide key information on the properties of matter motion and light deflection in the strong field regime of general relativity. For instance, back in 1972 Sunyeav discussed the possibility to identify black holes and "discriminate between the cases where a Schwarzschild or Kerr metric prevails", based on the fastest QPO signals arising from a disk at its innermost stable circular orbit (ISCO) (Sunyaev 1972; see also Shakura and Sunyaev 1973; Novikov and Thorne 1973). Stoeger (1980) went even further and calculated the characteristics of quasi-periodic signals produced by matter spiralling inwards in the unstable region between the inner disk boundary (considered to be at the ISCO radius) and the black hole horizon. These theoretical works as well as others were published well before the first detections of QPOs in accreting black holes and neutron stars and provided in many cases the framework within which QPOs were interpreted and models further developed.

Based on the discovery of multiple, simultaneous QPOs whose frequency varies in some cases in a correlated fashion (especially neutron star systems, see van der Klis 2006 for a review) different scenarios were proposed in which different quasi-periodic phenomena can take place at the same time. Amongst these are models that involve the fundamental frequencies of bound motion around a collapsed object, namely the azimuthal (aka orbital) frequency ν_ϕ, the radial epicyclic frequency ν_r and the vertical epicyclic frequency ν_θ (for a sketch of the relevant motions see Fig. 1). The expressions for these frequencies as a

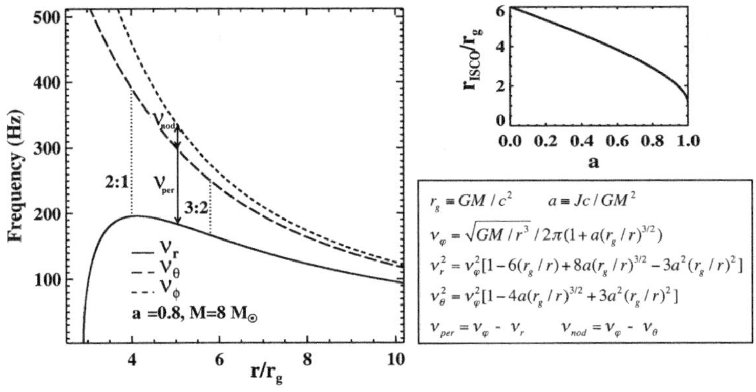

Fig. 2 *Left panel*: Fundamental frequencies of motion for an infinitesimally eccentric and tilted orbit around a Kerr black hole with $M = 8\ M_\odot$ and $a = 0.8$, as a function of radius. In contrast to the other two frequencies, the radial epicyclic frequency decreases with radius close to the compact object, reaching zero at the ISCO radius. The *dotted lines* show the radii at which the two epicyclic frequencies are in a 2:1 and 3:2 ratio. The *double arrowed segments* show the periastron and nodal precession frequencies at a radius of 5M. *Right top panel*: ISCO radius as a function of the Kerr spin parameter a for corotating orbits. *Right bottom panel*: formulae for the fundamental frequencies in the Kerr metric

function of the radius r of slightly eccentric, slightly tilted orbits close to the equatorial plane of a Kerr black hole are given in Fig. 2, where $r_g = GM/c^2$ is the gravitational radius, $a = Jc/GM^2$ the spin parameter and J the specific angular momentum of the black hole. The behaviour of the fundamental frequencies of matter motion in the strong field gravity regime that characterizes the inner region of accretion flows towards black holes (see Fig. 2 for an example), besides being considerably different from that of a point mass in Newtonian gravity (for which all three frequencies are identical), cannot be approximated by weak field expansions of General Relativity (GR). For instance, the radial epicyclic frequency increases from zero at the ISCO, to a broad maximum at a $\sim$40 % larger radii and then decreases as $r^{-3/2}$: this behavior is characteristic of the strong gravitational field close to a collapsed object. Therefore if QPOs can be associated unambiguously to fundamental frequencies of motion (or combinations or functions of them), they hold the potential to verify some key predictions of GR in the strong field regime, and measure the mass and spin of black holes. We note that the timescale corresponding to these frequencies are usually faster than viscous and thermal timescales of the disk (e.g. Pringle 1981).

3 Fast Variability and Source States

The low-frequency (0.01–30 Hz) fast variability from BHBs can take different forms and can range from being barely detectable to very strong. It is clear that what is observed is strongly connected to the state of the source, i.e. to the shape of the energy spectrum. For a definition of states, see Belloni (2010) and Belloni et al. (2011). We divide here the discussion in two parts: the Low Hard State (LHS) and High Soft State (HSS) with their noise components, and the intermediate states with their QPOs.

3.1 Hard and Soft States: Noise

For a full description of the energy spectra that characterise the LHS and HSS see Gilfanov (2010). In the LHS, the X-ray emission is very noisy, with fractional rms values as high

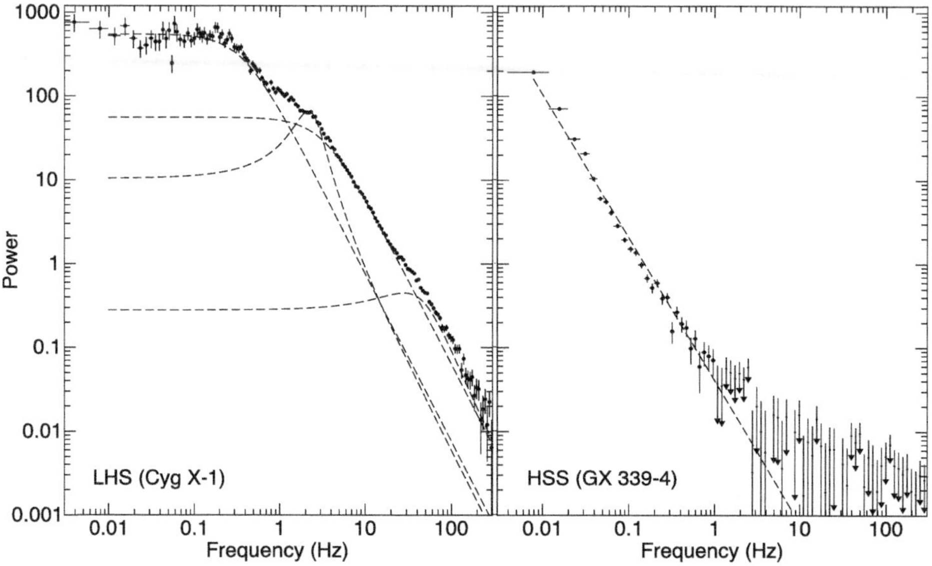

Fig. 3 *Left*: Representative PDS from the LHS obtained with the RossiXTE satellite (Cyg X-1). The *dashed lines* indicate separate Lorentzian components used for the fit. *Right*: PDS from the HSS (GX 339-4). The *dashed line* indicate a power law fit

as 40–50 %. The PDS shows that this variability is in form of band-limited noise, with usually no peaked components or QPOs (see Fig. 3, left panel). The shape of the PDS is not compatible with a simple model and is usually fitted with a combination of Lorentzian components (Belloni et al. 2002). These components provide characteristic frequencies (see Belloni et al. 2002 for a detailed description), but since they are very broad the determination of a characteristic frequency is not unique as in the case of a narrow peak and are therefore model dependent. Four main components have been identified: a low-frequency one providing the flat-top part (L_b), a peaked (sometimes QPO-like) component (L_q) and two broad Lorentzians at higher frequencies (L_ℓ and L_u). In a black-hole transient, the LHS is observed at the start and end of the outburst. In the early phases, as source flux (and accretion rate) increases, the total rms decreases slightly (see e.g. Muñoz-Darias et al. 2011), all characteristic frequencies increase and the energy spectrum steepens. The reverse takes place at the end of the outburst. In Cyg X-1, a persistent source, there is only one of these components whose frequency does not appear to vary (Pottschmidt et al. 2003).

It is difficult to associate the observed characteristic frequencies to physical frequencies either from the accretion flow or related to General Relativity. This is possible only when considering the intermediate states, to which the observed correlations extend (see below). One interesting correlation is the one that links the break frequency of the lowest component (the flat-top break ν_b) and the flat top fractional rms value (Belloni and Hasinger 1990; Belloni et al. 2002).

Both the linear relation observed over a broad range of frequencies between rms variability and source intensity, and the log-normal distribution of count rates indicate that the strong variability observed in the LHS is not consistent with being a linear process (see Heil et al. 2012 and references therein). Propagating fluctuation models have been proposed to incorporate these observables (Lyubarskii 1997).

The LHS noise is slightly stronger at lower energies. The fractional rms as a function of energy (the "rms spectrum") is flat or decreases by a few % from 2 to 20 keV (see Belloni et al. 2011 and references therein). The rms spectrum has been interpreted in the framework of Comptonization models (Gierlinski and Zdziarski 2005). Recent studies with XMM-Newton have shown that the soft thermal disk emission present below 1 keV leads the hard emission at time scales longer than one second, while it lags at shorter time scales, consistent with propagating models modified by disk heating at short time scales (Uttley et al. 2011). Finally, in the LHS the variability at high energies is observed to lag that at lower frequencies (see e.g. Nowak et al. 1999), consistent with a Comptonization origin for the emission (Gilfanov 2010).

In the HSS, variability is limited to a few % and the typical PDS features a power law component (see Fig. 3, right panel). Weak QPOs have been detected in some sources (see e.g. Motta et al. 2012), identified as high-frequency extension of type C QPOs (see below). The energy spectrum is dominated by the thermal disk component (see Gilfanov 2010), which is not much variable (but see Uttley and Casella, this issue). It is possible that most or all of the observed variability is associated to the faint hard component which is usually observed (see Grove et al. 1998).

3.2 Intermediate States: Quasi-Periodic Oscillations

In most cases, considerable time during an outburst is spent in the LHS and HSS (Fender and Belloni 2012). It is however during the transition between these states that the most prominent features in the fast time variability are observed. Two separate intermediate states have been identified, called Soft Intermediate State (SIMS) and Hard Intermediate State (HIMS). The boundary between these to states is defined by sharp changes in the variability.

- The HIMS is when the most common type of LFQPO is observed, called 'type C QPO' (see Wijnands et al. 1999; Motta et al. 2011 and references therein). The most important characteristics of this oscillation is that it is observed to vary over a relatively large range of frequencies, roughly from 0.01 to 30 Hz (see Fig. 4 for an example). These oscillations are observed in a large number of sources (Belloni 2010; Remillard and McClintock 2006) and were the first oscillations discovered from BHBs. Their typical quality factor Q (defined as the centroid frequency divided by the FWHM of the peak in the PDS) is around 10 (see e.g. Casella et al. 2005; Rao et al. 2010) and their fractional rms variability of the order of 3–15 %. Often it appears with one or two overtones and at times with a sub-harmonic (see Fig. 4). It is always associated to a band limited noise and its frequency is anti correlated with the total broad-band fractional rms variability (see lower left panel in Fig. 5, from Motta et al. 2011). The total fractional rms (QPO peaks plus noise components) of observations with type C QPOs is between 10 and 30 % (Muñoz-Darias et al. 2011).

  The energy spectrum of type C QPOs is hard, like that of all BHB QPOs with the rms vs. energy increasing and flattening above 10 keV (see Casella et al. 2004), with some evidence of a decrease at higher energies (Rodriguez et al. 2004). Despite the fact that the QPO is present (and more intense) at high energies where the accretion disk does not contribute, the centroid frequency is correlated with the soft flux from the accretion disk (see Markwardt et al. 1999; Motta et al. 2011). Type C QPOs are often also observed in the LHS, at lower frequencies. In those cases, the frequency evolution from LHS to HIMS is continuous.

  The large span in centroid frequency of type C QPOs is important to identify their origin. Wijnands and van der Klis (1999) discovered a strong correlation between their

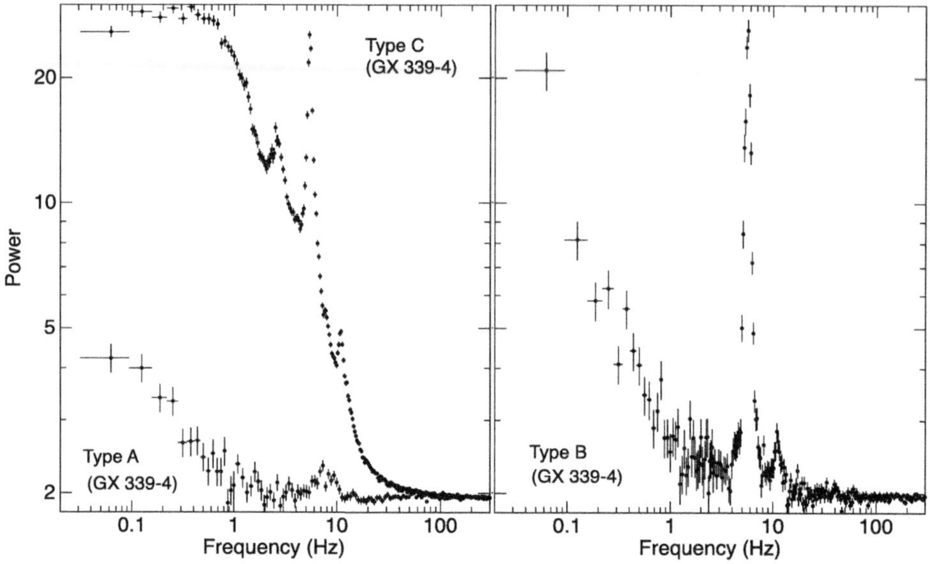

Fig. 4 Representative PDS if with a type C and type A QPO (*left*), and with a type B QPO (*right*). All three PDS are from RossiXTE observations of GX 339-4 (see Motta et al. 2011)

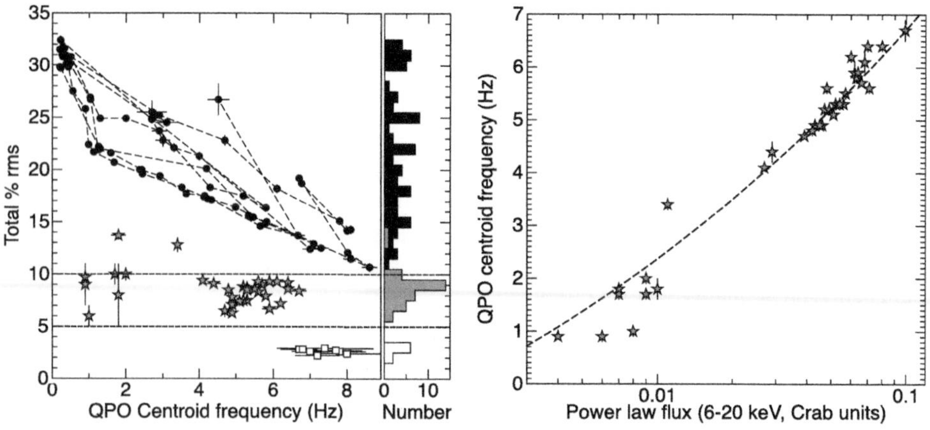

Fig. 5 *Left*: total fractional rms vs. QPO centroid frequency for all RXTE QPOs of GX 339-4, with the rms marginal distribution on the right side. The three types of QPOs (type A: *white squares*; type B: *grey stars*; type C: *black circles*) are clearly separated. *Right*: correlation between centroid frequency and power law flux (<6 keV) for type B QPOs in the same GX 339-4 data (both panels adapted from Motta et al. 2011)

frequency and the break frequency of the main noise component (around 1 Hz in the left panel of Fig. 3). This correlation is also valid for neutron star binaries when one considers the so-called Horizontal Branch Oscillations (HBO), quasi-periodic peaks which have been associated to type C QPO (van der Klis 2006; Casella et al. 2005).

- The SIMS, a short-lived state characterised by a softer energy spectrum than the HIMS, is defined in terms of sharp changes of properties of time variability. The band-limited noise is replaced by a weaker power-law noise component, and two alternative different types

Table 1 Narrow HFQPO features from BHBs. The columns are: source name, number of detected peaks, flag for simultaneity for the multiple peaks, approximate centroid frequencies. Question marks are discussed in the text

Source	N_{peaks}	Simultaneity	Frequency (Hz)
GRO J1655-40	2	Y	~300, ~400
XTE J1550-564	2	N	~180, ~280
XTE J1650-500	1	–	~250
H 1743-322	2	N	~160, ~240
IGR 17091-3624	1	–	~66
GRS 1915+105	4?	Y	~27, ~34, ~41, ~67

of QPO, called type A and type B, replace the type C QPO. These three types of QPO have been shown not to be the same signal, based on the fast transition between them (see e.g. Nespoli et al. 2003; Casella et al. 2004; Belloni et al. 2005) and in particular on the simultaneous detection of a type B and a type C (Motta et al. 2012). An example of each type of QPO can be seen in Fig. 4. Type A QPOs have centroid frequencies 6.5–8 Hz, are broad (quality factor $Q \sim 1$–3) and weak (fractional rms less than 5 %). Type B QPOs have centroid frequencies 0.8–6.4 Hz, are narrow ($Q > 6$) and have a 5–10 % fractional rms, in addition to appearing often together with an overtone and a sub-harmonic. In the left panel of Fig. 5 one can see that the three types of QPO are well separated as a function of the *total* integrated fractional rms in the PDS (noise+QPOs, see also Muñoz-Darias et al. 2011). The centroid frequency of type B QPOs shows a strong correlation with the flux from the hard component, as measured above 6 keV (right panel in Fig. 5).

Not much is known about type A QPOs, given their relative weakness and large FWHM, which prevent their detection in most cases. They are associated to a softer spectrum than type B QPOs. Both QPOs are observed in the SIMS, which is a short lived and transient state, possibly associated to the ejection of relativistic jets (see Fender et al. 2009 and references therein). They are much rarer than type C QPOs and span a smaller range in frequency.

A correspondence between the three flavours of LFQPO in BHBs and the three types of QPOs in neutron star binaries (normal branch oscillations: NBO; horizontal branch oscillations: HBO and flaring branch oscillations: FBO) has been suggested (Casella et al. 2005; van der Klis 2006). The similarities between type C QPOs and HBOs are very strong (van der Klis 2006) and the existence of a general correlation such as that described above strengthens this hypothesis (Wijnands and van der Klis 1999).

4 Variability at High-Frequencies

The large collecting area of the RXTE/PCA, which allowed the discovery of kHz QPOs in neutron star binaries (see van der Klis 2006), also opened a window onto high-frequency phenomena in BHBs. The first observations of the very bright system GRS 1915+105 led to the discovery of a transient oscillation at ~67 Hz (Morgan et al. 1997). Since then, sixteen years of RXTE observations have yielded only a handful of detections in other sources, although GRS 1915+105 seems to be an exception (see Belloni et al. 2012; Altamirano and Belloni 2012; Belloni and Altamirano 2013a, 2013b, and references therein).

Some of the reported detections were at low statistical significance, while others involved components too broad to be defined as QPO and difficult to assess due to the presence of counting noise. The remaining ones belong to six sources (see Table 1). Their properties can be summarized as follows:

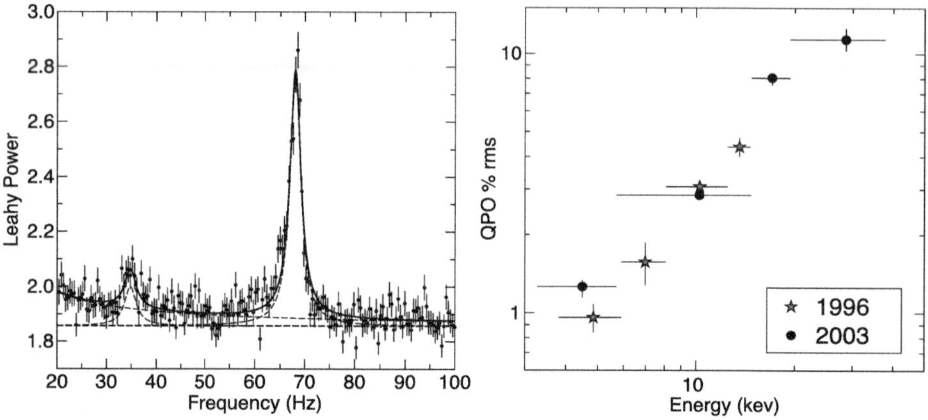

Fig. 6 *Left*: two HFQPO at 34 Hz and 68 Hz in a PDS of RXTE data of GRS 1915+105 (Belloni and Altamirano 2013b). *Right*: fractional rms as a function of energy for the 67 Hz QPO in GRS 1915+105 at two different epochs. From Belloni and Altamirano (2013a)

- They are observed only in observations at high flux/accretion rate. This is obviously at least partly due to a selection effect, but not all high-flux observations lead to the detection of a HFQPO, all else being equal, indicating that the properties of these oscillations can vary substantially even when all other observables do not change.
- They are observed as single or double peaks. From Table 1, two sources have shown a single peak: XTE J1650-500 (Homan et al. 2003) and IGR J17091-3624 (Altamirano and Belloni 2012). For the others, GRS J1655-40 showed two clear simultaneous peaks (Strohmayer et al. 2001; Belloni et al. 2012; Motta et al. 2014a). In XTE J1550-564, the two detected peaks (Remillard et al. 2002; Belloni et al. 2012) have been detected simultaneously, but the lower one with a 2.3σ significance after taking into account number of trials (Miller et al. 2001). Méndez et al. (2013), on the basis of their phase lags, suggested that the two detected peaks might be the same physical signal at two different frequencies. For H 1743-322, there is a weak detection of a second simultaneous peak (Homan et al. 2005), but two significant peaks have been detected, although not simultaneously (Remillard et al. 2006).
- While type C QPOs at low frequencies are rather common and define the presence of the HIMS, not only HFQPOs are very rarely detected with RXTE, but their detection is almost completely anticorrelated with the presence of type C QPOs. The only exception to this exclusion rule is GRO J1655-40 (see below). Often, HFQPOs are associated to type B QPOs or to the source being in the "anomalous state" (Belloni 2010; Belloni et al. 2012).
- A different case is that of GRS 1915+105, the first source where a HFQPO was discovered (Morgan et al. 1997). From the large archive of RXTE observations of this peculiar system (see Fender and Belloni 2004 for a review), a systematic analysis led to the detection of 51 HFQPOs, 48 of which at a centroid frequency between 63 and 71 Hz (Belloni and Altamirano 2013a, 2013b, see left panel of Fig. 7). This indicate that this range of frequencies is in some way special for this system. All detections corresponded to a very limited range in spectral parameters, as measured through hardness ratios. With a special time selection, Belloni et al. (2001) found a single 27 Hz QPO. Strohmayer (2001), averaging observations, discovered a 41 Hz HFQPO simultaneous to the 67 Hz one. Finally, Belloni and Altamirano (2013a, 2013b) discovered a 34 Hz peak simultaneous with a

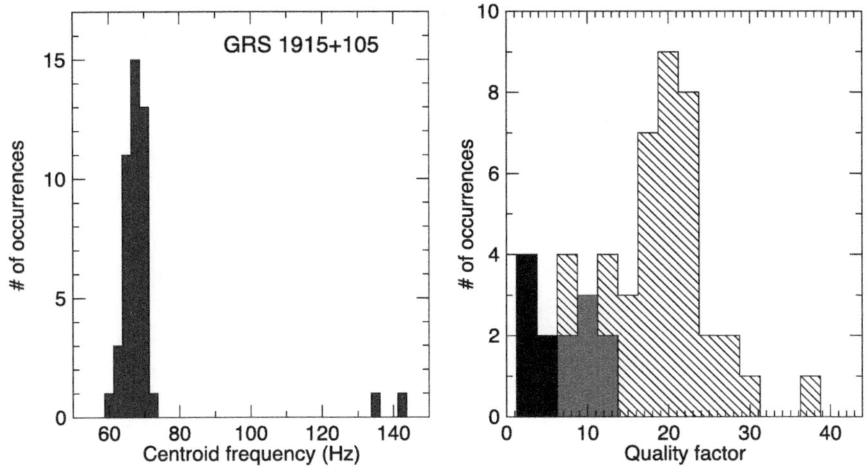

Fig. 7 *Left*: distribution of centroid frequencies of the HFQPOs detected in the full RXTE archive of observations of GRS 1915+105 (adapted from Belloni and Altamirano 2013a). *Right*: distribution of quality factors for the HFQPOs of GRS 1915+105 (*dashed histogram*, from Belloni and Altamirano 2013a), simultaneous pairs of HFQPOs in other sources (*black*: upper peak, *tray*: lower peak; from Belloni et al. 2012; Remillard et al. 2006 for H 1743-322). The single HFQPO in XTE J1650-500 (Homan et al. 2003) and the 34 Hz QPO in GRS 1915+105 (Belloni and Altamirano 2013b) have a Q of ~5

68 Hz one (Fig. 6, left panel). This could be the same as the 41 Hz one, in which case only one frequency would have changed, or a sub-harmonic of the upper one. The most recent HFQPO discovered, in IGR J17091-3624, is consistent with the average frequency of the 67 Hz QPO in GRS 1915+105 (Altamirano and Belloni 2012).

- Typical fractional rms for HFQPOs are 0.5–6 % (Belloni et al. 2012; Belloni and Altamirano 2013a, 2013b). The fractional rms increases steeply with energy, in the case of GRS 1915+105 reaching more than 19 % at 20–40 keV (see right panel in Fig. 6; Morgan et al. 1997; Belloni and Altamirano 2013a, 2013b). Quality factors (Q, the ratio between centroid frequency and FWHM of the QPO peak) are around 5 for the lower peak and 10 for the upper. In GRS 1915+105, a typical Q of ~20 is observed, but values as low as 5 and as high as 30 are observed (see Fig. 7, right panel).
- Time lags of HFQPOs have been studied for four sources (Méndez et al. 2013). The lag spectra of the 67 Hz QPO in GRS 1915+105 and IGR J17091-3624 and of the 450 Hz QPO in GRO J1655-40 are hard (hard photons variations lag soft photons variations), while those of the 35 Hz QPO in GRS 1915+105 are soft. The 300 Hz QPO in GRO J1655-40 and both HFQPOs in XTE J1550-564 are consistent with zero. The fact that the lag spectrum of both QPOs in the latter source are the same suggests that the 180 Hz and 280 Hz QPOs in this source might represent the same physical oscillation observed at two different frequencies.
- For three sources, GRO J1655-40, XTE J1550-564 and XTE J1743-322, the two observed frequencies are close to being in a 3:2 ratio (Strohmayer et al. 2001; Remillard et al. 2002, 2006), which has led to a specific model for their physical origin (see Sect. 5.2). For GRS 1915+105 the 67 Hz and 41 Hz QPOs, observed simultaneously, are roughly in 5:3 ratio. The 27 Hz would correspond to 2 in this sequence.

As for LFQPOs, it is interesting to compare the properties of HFQPOs with those of high-frequency (kHz) QPOs in neutron star binaries (see van der Klis 2006 for a review). kHz

QPOs also appear often in pairs and at frequencies generally higher than those of HFQPOs. However, they are much stronger signals, with a large number of detections in all bright systems, they are observed to change their frequency while keeping a roughly (but not exactly) similar frequency difference, and they correlate with other properties of the systems, including a possible relation with the neutron star spin frequency (but see Méndez and Belloni 2007). The phenomenology for NS kHz QPOs is complex and rather well defined, thanks to the large number of available detections.

Psaltis et al. (1999a; PBK) found that the known correlation existing in NS systems between HBO oscillations and the lower of the two kHz QPOs could be extended to lower frequencies if the kHz QPO frequency was replaced by the characteristic frequency of one of the broad components found in the PDS at low luminosities. This correlation was found to be valid also for black hole systems, who show the same high-frequency broad component simultaneous to a type C QPO (see the left panel of Fig. 3). This constitutes a strong link between the two classes of sources and suggests that the frequencies are associated to the same physical processes (see Sect. 5).

5 Interpretation

Early modeling of the fast variability properties of black hole X-ray binaries concentrated on exponential shot noise and other aperiodic processes, which can interpret the continuum power spectral components that were known at the time (Terrell 1972; Nolan et al. 1981; Belloni and Hasinger 1990). Then in the mid-eighties 5–50 Hz QPOs with *rms* amplitudes of several percents were discovered in the X-ray flux of neutron star low mass X-ray binaries. Being these QPOs tens to hundreds times slower than the orbital frequencies of matter in the innermost disk, interpretations were devised in which relatively low frequency signals could be generated from those regions (e.g. Alpar and Shaham 1985; Lamb et al. 1985). It was only with the RXTE discovery of the twin ∼1 kHz QPOs from neutron star systems that signals in the range of dynamical frequencies of the innermost disk could finally be studied (Strohmayer 1996; van der Klis et al. 1996).

Most QPO models were introduced in an effort to interpret the properties of neutron star QPOs. In particular a generic model soon emerged in which the faster kHz signal arises from the orbital (i.e. azimuthal) motion of matter in the inner disk region around the neutron star, as Sunyaev (1972) and others had suggested in the early seventies. The same idea was applied to the faster QPOs from black holes, soon after their discovery (Morgan et al. 1997). The phenomenology of QPOs in black hole systems is not as rich as that of neutron star binaries; however black holes provide a *cleaner* environment to probe matter motion in the very strong gravitational field regime, as they do not possess a solid surface, a stably-anchored magnetic field, nor a boundary layer at the disk inner edge.

Several models that were originally developed to interpret neutron star QPOs are not applicable to black hole QPOs, as they require the presence of a star surface and/or (offset) magnetic field. Among these are beat frequency models (Lamb et al. 1985; Miller et al. 1998). Other models are applicable to both neutron star and black hole QPOs and build on the analogy between the QPO modes and continuum power spectral components that are observed in both types of systems, especially the QPO pair at higher frequencies. Among these are the relativistic procession model and the epicyclic resonance model. We emphasize that both models are aimed at explaining the observed QPO frequencies based on the idea that QPOs are produced at a given radius in the disk: they are thus *local* models. We discuss them briefly below and emphasise that they provide also an alternative way of estimating black

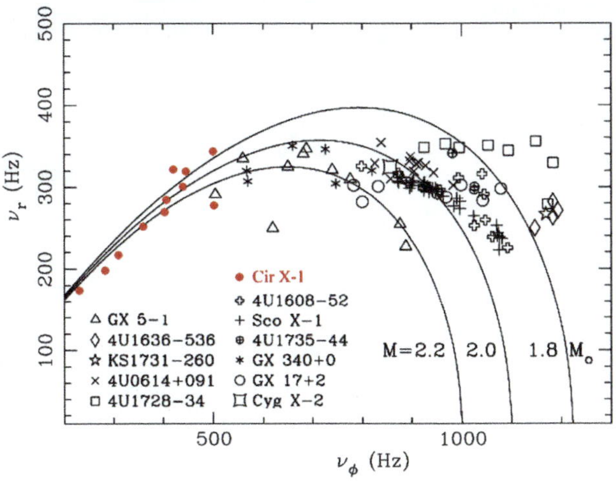

Fig. 8 kHz QPO frequency difference, versus upper kHz QPO frequency for a number of LMXRBs. Error bars are not plotted. Since in the RPM the upper kHz QPO corresponds to ν_ϕ and the lower kHz QPO to ν_{per}, their difference $\nu_\phi - \nu_{per}$ represents the radial precession frequency: therefore the *curves* give the ν_r vs. ν_ϕ of matter in nearly circular orbit around a non-rotating neutron star, of mass 2.2, 2.0 and 1.8 $M_\odot$ (a Schwarzschild spacetime is used here). The *filled red circles* refer to Cir X-1; these data provide a confirmation of the low frequency behaviour of kHz QPO predicted by the RPM (after Stella and Vietri 1999; Boutloukos et al. 2006)

spin (methods based on X-ray spectrum continuum-fitting and reflection/Fe-line modelling are described elsewhere in this volume).

5.1 The Relativistic Precession Model

The Relativistic Precession Model (RPM) was originally aimed at interpreting the twin kHz QPOs as well as a low-frequency QPO mode (the HBOs) of neutron star of low mass X-ray binaries of both the Atoll and Z-classes (Stella and Vietri 1998, 1999). In the RPM the higher and lower frequency kHz QPOs are identified respectively with the azimuthal frequency ν_ϕ and the periastron precession frequency ν_{per} of matter orbiting at a given radius in innermost disk regions. In terms of the fundamental frequencies of motion, the latter frequency is $\nu_{per} = \nu_\phi - \nu_r$, such that the difference frequency of the kHz QPOs, corresponds to the radial epicyclic frequency ν_r. Figure 8 shows different curves for the way in which ν_r is expected to vary as a function of increasing ν_ϕ, while the radius at which the QPOs are produced decreases. The space-time around neutron stars is approximated here by the equations in Fig. 2 with $a = 0$, i.e. a Schwarzschild spacetime. (The effects induced by neutron star rotation on the $\nu_r - \nu_\phi$ relation are small, though non-negligible, see Stella and Vietri 1999.) Plotted curves are for selected values of the neutron star mass. The measured values of the difference frequency of the twin kHz QPO vs. the frequency of the upper kHz QPOs for 11 neutron star LMXRBs is also plotted. It is apparent that for masses of ~2 $M_\odot$, the simple model above is in semi-quantitative agreement with measured values, including the decrease of the difference frequency for increasing QPO frequency, which was seen in Sco X-1, 4U 1608-52, 4U 1735-44 and 4U 1728-34. Note that most measured points lie near the broad peak region of the radial epicyclic frequency. The behaviour of ν_r at low frequencies is confirmed by the QPO separation in Cir X-1 which was found to increase with QPO frequency (Boutloukos et al. 2006).

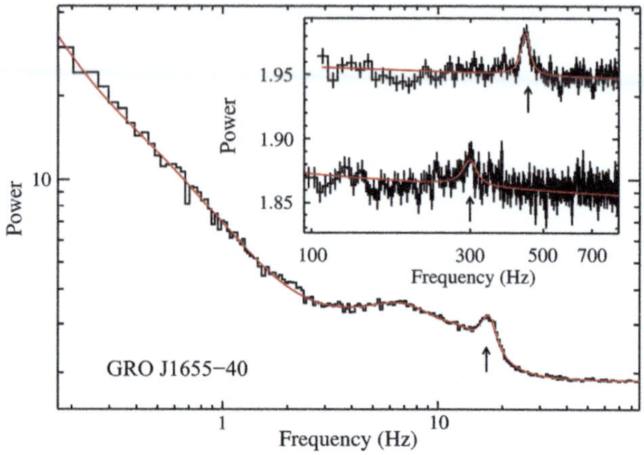

Fig. 9 Power spectrum of GRO J1655-40 displaying three simultaneous QPO peaks (marked by *arrows*): the type C at ∼18 Hz, upper and lower high frequency QPO at ∼300 and ∼450 Hz, respectively (after Motta et al. 2014a)

In the RPM the HBO frequency is related to the nodal precession frequency, ν_{nod}, at the same radius where the signals at ν_ϕ and ν_{per} are produced. ν_{nod} is given by $\nu_\phi - \nu_\theta$, the difference between the azimuthal frequency and the vertical epicyclic frequency. In the weak field limit, the relativistic nodal precession, scales as the square of the azimuthal frequency, as in the well-known Lense-Thirring effect. This dependence is in agreement with that observed in a number of neutron star systems (Stella and Vietri 1998; Ford et al. 1998; Psaltis et al. 1999b).

The RPM reproduces accurately also the PBK correlation, without resorting to additional assumptions. The measured QPO and peaked-noise frequencies giving rise to the PBK correlation is given in Fig. 1B of Stella et al. (1999), together with the curves obtained for ν_{nod} and ν_ϕ as a function of ν_{per} for some selected cases of rotating neutron star models. The RPM dependence of ν_{nod} on ν_{per} matches nicely the observed correlation over ∼3 decades in frequency and encompasses both neutron star and black hole systems. Indeed this was the context in which the RPM was first applied to black hole QPOs and broad noise components.

Figure 1A in Stella et al. (1999) shows frequency measurements for black hole systems, together with ν_{nod} and ν_ϕ as a function of ν_{per} for selected values of the mass and spin parameter. ν_{nod} depends weakly on M and strongly on a; the opposite holds for ν_ϕ. In principle, black hole mass and angular momentum can be thus be measured if three QPO frequencies (type C QPO, together with two high frequency QPOs) are detected simultaneously in a system. For the range of masses encompassed by black holes in X-ray binaries (roughly 5–10 $M_\odot$) the application of the RPM to black hole QPOs indicates relatively small values of $a/M \sim 0.1$–0.3.

Only in one instance a type C low-frequency QPO was detected simultaneously with two high-frequency QPOs in a black hole system, GRO J1655-40 (see Fig. 9). This made the first complete application of the RPM to a black hole possible (Motta et al. 2014a). By fitting the three measured QPO frequencies, precise values of the black hole mass ($5.31 \pm 0.07\ M_\odot$, consistent with the mass measured from optical/NIR spectro-photometric observations) and spin ($a = 0.290 \pm 0.003$) were obtained through the sole use of X-ray timing. Not only the quasi periodic oscillations, but also the broad band noise components of GRO J1655-40 and their variations match accurately the predictions of the RPM: this is shown in Fig. 10. The relative width of the QPO peaks was found consistent with being due to the jitter of the radius where QPOs are generated. In the case of XTE J1550-564 only type C and lower high-

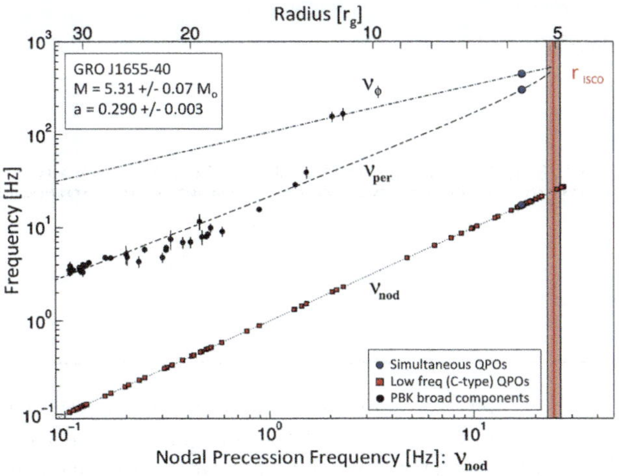

Fig. 10 Nodal precession frequency (*dotted line*), periastron precession frequency (*dashed line*) and orbital frequency (*dot-dashed line*) as a function of the nodal precession frequency around a Kerr black hole as predicted by the RPM. The lines are drawn for the mass and spin values that provide fit best fit the three simultaneous QPOs observed in GRO J1655-40 (*blue points* in the plot). The corresponding radii are given in the top x-axis. The *black circles* are the characteristic frequencies of the source's broad power spectrum components; they all lie close to the low-frequency extrapolation of the frequencies predicted by the RPM, as derived on the three simultaneous points only. The *squares*, giving the frequency of type C QPOs plotted against itself, illustrate the range of frequencies (and thus radii) over which these QPOs are seen. The *red line* marks the ISCO radius and nodal frequency; and the *red vertical band* indicates its corresponding 3σ uncertainty (after Motta et al. 2014a)

frequency QPOs were observed simultaneously (Fig. 11). By using the black hole mass as determined from optical/NIR measurements, the RPM yields a value of the spin parameter of $a = 0.34 \pm 0.01$ (Motta et al. 2014b).

5.2 Epicyclic Resonance Models

Another family of *local* models exploits the fact that at some specific radii in the accretion disk, the radial ν_r and vertical ν_θ epicyclic frequencies attain simple integer ratios and thus may resonate (Kluzniak and Abramowicz 2001, 2002; Abramowicz and Kluzniak 2001). Parametric and forced resonances have been discussed, especially those with small integer ratios 2:1, 3:1, 3:2. Examples of two such resonances are shown in Fig. 2. Resonances between ν_r and the azimuthal frequency ν_ϕ have also been considered (Török et al. 2005). These Epicyclic Resonance Models (ERMs) successfully explain black hole QPOs with frequency ratio consistent with 2:3 or 1:2 (see Sect. 3.2). As a given resonance condition is verified only at a fixed radius in the disk, the QPO frequencies are expected to remain constant, or jump from one resonance to another. In the application to neutron star QPOs various commensurabilities of epicyclic frequencies with other frequencies (e.g. with the spin frequency) have also been considered, which can help interpret kHz QPOs in the framework of ERMs. This included also the development of some *tunable* versions of the model, which can produce fairly continuum variations of the kHz QPO frequencies around the resonance condition. Simple applications of ERMs to black hole high frequency QPO pairs indicate large values of the spin parameter ($a > 0.9$) in the case of the 3:2 parametric resonance, and somewhat smaller values ($a \sim 0.3$–0.6) for forced resonances (Török et al. 2005). In ERMs low frequency QPOs (type C and HBOs in particular) remain to be interpreted.

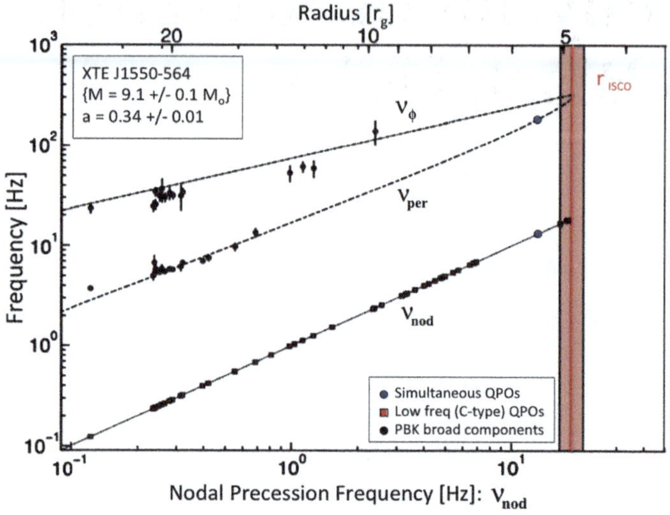

Fig. 11 Same as Fig. 10, but the points here are from XTE J1550-564. Note that in this case only two simultaneous QPOs (*blue points* in the plot) were observed, a type C QPO and the lower high frequency QPO; the best fit was obtained by using the black hole mass derived from optical spectrophotometric measurements (after Motta et al. 2014b)

5.3 Disk Oscillation Models

Different types of disk oscillation modes have been considered as mechanisms that can give rise to QPOs and/or rapid noise-like variability in accreting black hole systems.

Inertial g-modes can be trapped in the resonant cavity that is created by the broad peak of the radial epicyclic frequency (see Fig. 2). They give rise to the same global oscillations over a range of disk radii, the fundamental frequency of which is close to the maximum value of ν_r (Nowak and Wagoner 1991, 1992; Nowak et al. 1991). Global corrugation modes are expected to occur in disks around spinning compact objects because of the nodal precession induced by frame dragging (Silbergleit et al. 2001; Wagoner et al. 2001). Frame dragging can drive also warping modes close to the disk inner boundary at the ISCO (Markovic and Lamb 1998; Armitage and Natarajan 1999). In disk p-modes the restoring force arises from pressure gradients; the fundamental mode in a small region close to ISCO attains higher frequencies than the local value of ν_r (Wagoner et al. 2001). In a series of papers Kato studied trapped non-axisymmetric g-mode and vertical p-mode oscillations that are resonantly excited in the inner disk regions; mode frequencies for both magnetic and non-magnetic disks are compared with QPO frequencies from black holes as well as neutron star systems (Kato 2012a, 2012b, 2012c and references therein; see also Li et al. 2003). The global stability of non-axisymmetric trapped p-modes was analyzed by Lai and Tsang (2009) who showed that modes with frequencies $\sim$(0.5–0.7)$m\nu_r(ISCO)$ (with $m = 1, 2, 3\ldots$ is the azimuthal wavenumber) can grow as a result of corotational instability. Some of their models have the largest growth rate for $m = 2$ and 3, which might explain the 3:2 frequency ratio of some black hole high frequency QPOs.

5.4 Other Models

A variety of other models has been proposed which involve at least one fundamental frequency of motion in the strong-field regime. Ingram et al. (2009) consider a model in which the Lense-Thirring effect drives solid body-like nodal precession of a radially extended region of the hot inner flow. With the aid of numerical simulation they estimate the inner radius of the precessing region, work out the precession frequency and apply it to the low frequency QPOs of black holes (see also Ingram and Done 2012 and references therein). Petri (2008) studied the oscillations caused by the parametric resonance induced by the interaction between a spiral density wave in the accretion disk, excited close to the ISCO, and vertical epicyclic oscillations.

Tagger and Varniere (2006) developed the so-called accretion–ejection instability model for discs threaded by large-scale and intense poloidal magnetic fields. This B-field provides a strong coupling at corotation between spiral density waves and Rossby waves, leading to amplification and injection of energy to a hot disk corona.

There are also models in which the oscillation frequency is not related to the fundamental frequencies of motion. Among these is the disk-jet oscillation model, in which black hole-threated field lines interact with the inner disk, extracting rotational energy and powering jet ejection through the Blandford-Znajek mechanism; oscillations at $1/4$, $1/2$, and $3/4$ the rotation frequency at the black hole horizon are especially prominent in the general relativistic magnetohydrodinamical (MHD) simulations of magnetically-choked accretion flows by McKinney et al. (2012), the latter two being in a 3:2 frequency ratio.

5.5 QPOs in Accretion Disk Simulations

Searches for diskseismic modes were carried out in both hydrodynamical and MHD disk models. In the hydrodynamical simulations of Reynolds and Miller (2009) and O'Neill et al. (2009) trapped g-modes and inner p-modes in a narrow range of frequencies close to the maximum radial epicyclic frequency were found. On the contrary in the MHD case, turbulence created by the magneto-rotational instability (MRI) did not excite disk seismic modes, epicyclic frequencies, nor resonances with 3:2 frequency ratio at a level that could be revealed by simulations. On the other hand MRI turbulence was found to give rise to local hydrodynamic waves.

In the GR MHD disk simulations of Wellons et al. (2014) fluctuations induced by MRI-driven MHD turbulence were followed over a fairly long time and synthetic light curves of the disk emission as seen by a distant observed were calculated, by taking into account relativistic effects in the disk. The corresponding power spectra show a power-law like continuum and broad feature at high frequency, whose amplitude increases with observer inclination. Such a high-frequency feature is a product of the relativistic Doppler effect in the innermost disk regions and encodes information on the black hole mass and spin.

Despite the progress achieved in recent years, much remains to be learned from the study of disk variability and QPOs in disk simulations (see e.g. Blaes 2014; Fragile 2014; Henisey et al. 2009; Dolence et al. 2012; Romanova et al. 2013). The inclusion of full radiative transfer in 3D disk simulations will be especially important for the generation synthetic light curve simulations that can be directly compared to the observations. Steps in this direction have been undertaken (Jiang et al. 2013; Sadowski et al. 2014).

5.6 Conclusion

The past years (1996–2012) have seen an impressive increase of phenomenology of fast variations from BHBs, mostly due to the RXTE satellite. The available observational picture is very complex, but clear patterns have emerged and theoretical models to interpret them are now available and have been tested. Despite the amount of information available however, the number of detections of high-frequency oscillations from BHBs is very limited, due to their faintness and elusiveness. This is in contrast to the strong signals observed from neutron-star binaries, where the same mechanisms for QPO production could be at work. As the RXTE mission has been terminated, the observational picture can be improved only with new missions. The ASTROSAT satellite, expected to be launched in 2015, will provide an invaluable contribution by re-opening the window onto fast timing variability. Its sensitivity below 10 keV is similar to that of RXTE, but above that energy the effective area of its LAXPC instrument is considerably larger, offering a good opportunity to detect high-frequency oscillations. In addition, capitalizing on the RXTE heritage, it will be possible to concentrate observations in the outburst intervals when these oscillations have been detected, maximizing the probability of increasing the number of detections.

Most of what is currently known (and briefly surveyed in the present review) on the fast variability of BHBs relies upon X-ray timing information with little energy resolution, like that afforded by proportional counters (e.g. RXTE's). The potential of energy-resolved studies capable of addressing spectral variability and delays on dynamical timescales is beginning to emerge from observations of supermassive black holes in active galactic nuclei with CCD-type energy resolution (Iwasawa et al. 2004; Zoghbi et al. 2012). Relativistically shifted and broadened Fe Kα lines from the innermost disk regions and their fast variability in combination with the fast variability of the X-ray continuum spectral components appear to be very promising in this context. In particular tomography and reverberation studies at X-ray energies in stellar mass accreting black holes will be able to achieve outstanding sensitivity to the dynamics of matter motion in the strong field regime (Uttley et al. 2014). To this aim very large area instrumentation ($\sim$5–10 m^2 in the classical 1–10 keV energy range) with CCD-like spectral resolution, such as that of the proposed mission LOFT (see: http://sci.esa.int/loft/53447-loft-yellow-book/#), will be necessary.

Acknowledgements We thank Sara Motta for insightful discussions. TMB acknowledges support from PRIN INAF 2012-6 "Accreting X-ray binaries: understanding physics through periodic and aperiodic variability". LS acknowledges support from PRIN-INAF-2011 "A deep insight into strong gravity around black-holes: new theoretical and observational approaches".

References

M.A. Abramowicz, W. Kluzniak, Astron. Astrophys. **374**, L19 (2001)
M.A. Alpar, J. Shaham, Nature **316**, 239 (1985)
D. Altamirano, T. Belloni, Astrophys. J. **747**, L4 (2012)
P.J. Armitage, P. Natarajan, Astrophys. J. **525**, 909 (1999)
T. Belloni, in *The Jet Paradigm: from Microquasars to Quasars*, ed. by T.M. Belloni. Lecture Notes for Physics, vol. 794 (Springer, Berlin/Heidelberg, 2010), p. 53
T. Belloni, G. Hasinger, Astron. Astrophys. **227**, L33 (1990)
T. Belloni, D. Altamirano, Mon. Not. R. Astron. Soc. **432**, 10 (2013a)
T. Belloni, D. Altamirano, Mon. Not. R. Astron. Soc. **432**, 19 (2013b)
T. Belloni, M. Méndez, C. Sánchez-Fernández, Astron. Astrophys. **372**, 551 (2001)
T. Belloni, D. Psaltis, M. van der Klis, Astrophys. J. **572**, 392 (2002)
T. Belloni, J. Homan, P. Casella et al., Astron. Astrophys. **440**, 207 (2005)
T. Belloni, S. Motta, T. Muñoz-Darias, Bull. Astron. Soc. India **39**, 409 (2011)
T. Belloni, A. Sanna, M. Méndez, Mon. Not. R. Astron. Soc. **426**, 1701 (2012)
O. Blaes, Space Sci. Rev. (2014, this issue)
S. Boutloukos et al., Astrophys. J. **653**, 1435 (2006)
P. Casella, T. Belloni, J. Homan et al., Astron. Astrophys. **426**, 587 (2004)
P. Casella, T. Belloni, L. Stella, Astrophys. J. **629**, 403 (2005)
J.C. Dolence et al., Astrophys. J. **746**, 10 (2012)
R.P. Fender, T. Belloni, Annu. Rev. Astron. Astrophys. **42**, 317 (2004)

R.P. Fender, T. Belloni, Science **337**, 540 (2012)
R.P. Fender, J. Homan, T. Belloni, Mon. Not. R. Astron. Soc. **396**, 1370 (2009)
E.C. Ford et al., Astrophys. J. **508**, L155 (1998)
P.C. Fragile, Space Sci. Rev. (2014, this issue)
M. Gierlinski, A.A. Zdziarski, Mon. Not. R. Astron. Soc. **363**, 1349 (2005)
M. Gilfanov, in *The Jet Paradigm: from Microquasars to Quasars*, ed. by T.M. Belloni. Lecture Notes for Physics, vol. 794 (Springer, Berlin/Heidelberg, 2010), p. 17
J.E. Grove et al., Astrophys. J. **500**, 899 (1998)
L.M. Heil, S. Vaughan, P. Uttley, Mon. Not. R. Astron. Soc. **422**, 2620 (2012)
K.B. Henisey, O.M. Blaes, P.C. Fragile, B.T. Ferreira, Astrophys. J. **706**, 705 (2009)
J. Homan, M. Klein-Wolt, S. Rossi et al., Astrophys. J. **586**, 1262 (2003)
J. Homan, J.M. Miller, R. Wijnands et al., Astrophys. J. **623**, 383 (2005)
A. Ingram, C. Done, P.C. Fragile, Mon. Not. R. Astron. Soc. **397**, L101 (2009)
A. Ingram, C. Done, Mon. Not. R. Astron. Soc. **419**, 2369 (2012)
K. Iwasawa et al., Mon. Not. R. Astron. Soc. **347**, 411 (2004)
Y.F. Jiang, S.W. Davis, J.M. Stone, Astrophys. J. **778**, 65 (2013)
S. Kato, Publ. Astron. Soc. Jpn. **64**, 62 (2012a)
S. Kato, Publ. Astron. Soc. Jpn. **64**, 129 (2012b)
S. Kato, Publ. Astron. Soc. Jpn. **64**, 139 (2012c)
W. Kluzniak, M.A. Abramowicz, astro-ph/0105057 (2001)
W. Kluzniak, M.A. Abramowicz, astro-ph/0203314 (2002)
F.K. Lamb et al., Nature **317**, 681 (1985)
D. Lai, D. Tsang, Publ. Astron. Soc. Jpn. **393**, 979L (2009)
L.X. Li, J. Goodman, R. Narayan, Astrophys. J. **593**, 980 (2003)
Y.E. Lyubarskii, Mon. Not. R. Astron. Soc. **292**, 679 (1997)
D. Markovic, F.K. Lamb, Astrophys. J. **507**, 316 (1998)
C.B. Markwardt, J.H. Swank, R.E. Taam, Astrophys. J. **513**, L37 (1999)
J.C. McKinney et al., Mon. Not. R. Astron. Soc. **423** (2012)
M. Méndez, T. Belloni, Mon. Not. R. Astron. Soc. **381**, 790 (2007)
M. Méndez, D. Altamirano, T. Belloni et al., Mon. Not. R. Astron. Soc. **435**, 2132 (2013)
M.C. Miller, F.K. Lamb, D. Psaltis, Astrophys. J. **508**, 791 (1998)
J.M. Miller, R. Wijnands, J. Homan et al., Astrophys. J. **563**, 928 (2001)
E.H. Morgan, R.A. Remillard, J. Greiner, Astrophys. J. **482**, 993 (1997)
S. Motta, T. Muñoz-Darias, P. Casella et al., Mon. Not. R. Astron. Soc. **418**, 2292 (2011)
S. Motta, J. Homan, T. Muñoz-Darias et al., Mon. Not. R. Astron. Soc. **427**, 595 (2012)
S. Motta, T. Belloni, L. Stella et al., Mon. Not. R. Astron. Soc. **437**, 2554 (2014a)
S. Motta, T. Munoz-Darias, A. Sanna et al., Mon. Not. R. Astron. Soc. **439**, d5 (2014b)
T. Muñoz-Darias, S. Motta, T. Belloni, Mon. Not. R. Astron. Soc. **410**, 679 (2011)
E. Nespoli, T. Belloni, J. Homan et al., Astron. Astrophys. **412**, 235 (2003)
P.L. Nolan, D.E. Gruber, J.L. Matteson et al., Astrophys. J. **246**, 494 (1981)
I.D. Novikov, K.S. Thorne, in *Black Holes, Les Houches*, ed. by C. De Witt, B.S. De Witt (Gordon & Breach, New York, 1973), p. 343
M.A. Nowak, R.V. Wagoner, Astrophys. J. **378**, 656 (1991)
M.A. Nowak, R.V. Wagoner, Astrophys. J. **393**, 697 (1992)
M.A. Nowak, R.V. Wagoner, M.C. Begelman, D.E. Lehr, Astrophys. J. **477**, L91 (1991)
M.A. Nowak, J. Wilms, B.A. Vaughan, Astrophys. J. **515**, 726 (1999)
S.M. O'Neill, C.S. Reynolds, M.C. Miller, Astrophys. J. **693**, 1100 (2009)
J. Petri, Astrophys. Space Sci. **318**, 181 (2008)
K. Pottschmidt, J. Wilms, M.A. Nowak, Astrophys. J. **407**, 1039 (2003)
J.E. Pringle, Annu. Rev. Astron. Astrophys. **19**, 137 (1981)
F. Rao et al., Astrophys. J. **714**, 1065 (2010)
D. Psaltis, T. Belloni, M. van der Klis, Astrophys. J. **520**, 262 (1999a)
D. Psaltis et al., Astrophys. J. **520**, 763 (1999b)
R.A. Remillard, J.E. McClintock, Annu. Rev. Astron. Astrophys. **44**, 49 (2006)
R.A. Remillard, G.J. Sobczak, M.P. Muno et al., Astrophys. J. **564**, 962 (2002)
R.A. Remillard, J.E. McClintock, J.A. Orosoz et al., Astrophys. J. **637**, 1002 (2006)
C.S. Reynolds, M.C. Miller, Astrophys. J. **692**, 869 (2009)
J. Rodriguez, S. Corbel, D. Hannikainen, Astrophys. J. **615**, 416 (2004)
M.M. Romanova, G.V. Ustyugova, A.V. Koldoba, R.V.E. Lovelace, Mon. Not. R. Astron. Soc. **430**, 699 (2013)
A. Sadowski et al., Mon. Not. R. Astron. Soc. **493**, 503 (2014)

N.I. Shakura, R.A. Sunyaev, Astron. Astrophys. **24**, 337 (1973)
A.S. Silbergleit, R.V. Wagoner, M. Ortega-Rodríguez, Astrophys. J. **548**, 335 (2001)
L. Stella, M. Vietri, Astrophys. J. **492**, L59 (1998)
L. Stella, M. Vietri, Phys. Rev. Lett. **503**, 350 (1999)
L. Stella, M. Vietri, S. Morsink, Astrophys. J. **524**, L63 (1999)
W.R. Stoeger, Mon. Not. R. Astron. Soc. **190**, 715 (1980)
T.E. Strohmayer, Astrophys. J. **469**, L9 (1996)
T.E. Strohmayer, Astrophys. J. **554**, L169 (2001)
T.E. Strohmayer et al., Astrophys. J. **552**, L49 (2001)
R.A. Sunyaev, Astr. Zh. **49**, 1153 (1972) (Soviet Astr. **16**, 941 (1973))
M. Tagger, P. Varniere, Astrophys. J. **652**, 1457 (2006)
Y. Tanaka, W.H.G. Lewin, in *X-ray Binaries*, ed. by W. Lewin, J. van Paradijs, E.P.J. van den Heuvel. Cambridge Astrophysics Series, vol. 26 (Cambridge University Press, Cambridge, 1996), p. 126
N.J. Terrell, Astrophys. J. **174**, L35 (1972)
G. Török, M.A. Abramowicz, W. Kluzniak, Z. Stuchlık, Astron. Astrophys. **436**, 1 (2005)
P. Uttley et al., Mon. Not. R. Astron. Soc. **414**, L60 (2011)
P. Uttley et al., Annu. Rev. Astron. Astrophys. (2014, in press). arXiv:1405.6575
M. van der Klis, in *Compact Stellar X-ray Sources*, ed. by W. Lewin, M. van der Klis. Cambridge Astrophysics Series, vol. 39 (Cambridge University Press, Cambridge, 2006), p. 39
M. van der Klis et al., Astrophys. J. **469**, L1 (1996)
R.V. Wagoner, A.S. Silbergleit, M. Ortega-Rodríguez, Astrophys. J. **559**, 25 (2001)
S. Wellons et al., Astrophys. J. **785**, 142 (2014)
R. Wijnands, M. van der Klis, Astrophys. J. **514**, 939 (1999)
R. Wijnands, J. Homan, M. van der Klis, Astrophys. J. **526**, L33 (1999)
A. Zoghbi et al., Mon. Not. R. Astron. Soc. **422**, 129 (2012)

DOI 10.1007/978-1-4939-2227-7_5
Reprinted from *Space Science Reviews* Journal, DOI 10.1007/s11214-013-0033-3

Modelling Spectral and Timing Properties of Accreting Black Holes: The Hybrid Hot Flow Paradigm

Juri Poutanen · Alexandra Veledina

Received: 29 October 2013 / Accepted: 10 December 2013 / Published online: 1 January 2014

Abstract The general picture that emerged by the end of 1990s from a large set of optical and X-ray, spectral and timing data was that the X-rays are produced in the innermost hot part of the accretion flow, while the optical/infrared (OIR) emission is mainly produced by the irradiated outer thin accretion disc. Recent multiwavelength observations of Galactic black hole transients show that the situation is not so simple. Fast variability in the OIR band, OIR excesses above the thermal emission and a complicated interplay between the X-ray and the OIR light curves imply that the OIR emitting region is much more compact. One of the popular hypotheses is that the jet contributes to the OIR emission and even is responsible for the bulk of the X-rays. However, this scenario is largely ad hoc and is in contradiction with many previously established facts. Alternatively, the hot accretion flow, known to be consistent with the X-ray spectral and timing data, is also a viable candidate to produce the OIR radiation. The hot-flow scenario naturally explains the power-law like OIR spectra, fast OIR variability and its complex relation to the X-rays if the hot flow contains non-thermal electrons (even in energetically negligible quantities), which are required by the presence of the MeV tail in Cyg X-1. The presence of non-thermal electrons also lowers the equilibrium electron temperature in the hot flow model to $\lesssim 100$ keV, making it more consistent with observations. Here we argue that any viable model should simultaneously explain a large set of spectral and timing data and show that the hybrid (thermal/non-thermal) hot flow model satisfies most of the constraints.

Keywords Accretion, accretion discs · Black hole physics · Radiation mechanisms: non-thermal · X-rays: binaries

J. Poutanen (✉) · A. Veledina
Astronomy Division, Department of Physics, PO Box 3000, 90014 University of Oulu, Finland
e-mail: juri.poutanen@gmail.com

A. Veledina
e-mail: alexandra.veledina@oulu.fi

1 Introduction

Models for accretion onto a black hole (BH) have been discussed now for more than 40 years. During the last 10–15 years we have seen a dramatic increase in the amount of information on the BH X-ray binaries (BHBs). Spectral details (iron lines and Compton reflection), spectral transitions, and variability on various time scales has been studied in unprecedented details with the new generation X-ray telescopes such as *Rossi X-ray Timing Explorer (RXTE)* and *XMM-Newton*. Excellent recent reviews are devoted to these advances (Zdziarski and Gierliński 2004; Remillard and McClintock 2006; Done et al. 2007; Done 2013; Gilfanov 2010).

In addition to the X-ray data, we have seen an explosion of information coming from other wavelengths: radio, sub-mm, infrared, optical, UV, MeV and nowadays even from the GeV range. What is even more spectacular is that the properties of the BHs at these other wavebands are correlated with the X-ray flux and X-ray states. Among the most impressive achievements we find the discovery of correlated fast variability in the optical/infrared (OIR) band and in the X-rays (Kanbach et al. 2001; Durant et al. 2008; Gandhi et al. 2008) with some hints actually coming already 30 years ago (Motch et al. 1983). This got theoreticians to scratch their heads and invent new models that often were in disagreement with previously established theories and contradicted many other available data.

Here we discuss some of the recent discoveries. We would like to note that the time for theoretical (phenomenological) models based purely on spectral properties are long gone. In order to be considered seriously, any model has to address many observed facts together.

This review consists of two parts. In the first one, we discuss the most recent reincarnation of the hot flow model, which now also considers the role of the non-thermal particles. In the second part, we discuss recent observational advances. We review the spectral data in various energy bands concentrating on the X-rays and the OIR. Then we discuss the observed temporal properties and correlated variability in different energy bands, as well as more complicated temporal-spectral statistics such as Fourier resolved spectra. In this review we will concentrate on the hard state and interpret the observations in terms of the hot flow model.

2 Hot Flow Models

2.1 Comptonization Models for the X-Ray Emission

At high accretion rates exceeding typically 10 % of the Eddington value, BHs are in the "soft state" and have thermally looking spectra peaking in the standard 2–10 keV X-ray band, which are consistent with the thin α-disc model (Shakura and Sunyaev 1973; Novikov and Thorne 1973). These thermally-dominated spectra presumably depend only on the accretion rate, the BH mass and spin and the inclination. They potentially can be used to determine the BH spin if the distance to the source and e.g. the BH mass and inclination are known (see McClintock et al. 2013). However, often strong power-law tails are seen (see Fig. 2). This tails are interpreted as a signature of non-thermal "corona" atop of the standard Shakura-Sunyaev disc (Fig. 1b).

At the lower accretion rate, BHs are often found in the "hard state" and their spectra do not even remotely look thermal, but are close to the power-law in the X-ray band with a sharp cutoff at about 100 keV (e.g. Gierlinski et al. 1997; Zdziarski et al. 1998). These are well described by the thermal Comptonization model (see Zdziarski et al. 1997; Poutanen

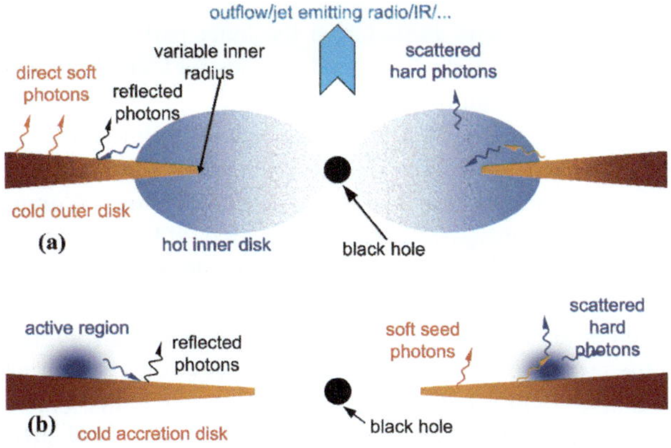

Fig. 1 (**a**) A schematic representation of the likely geometry in the hard state, consisting of a hot inner accretion flow surrounded by optically-thick cold accretion disc. The hot flow constitutes the base of the jet (with the counter-jet omitted from the figure for clarity). The disc is truncated far away from the minimum stable orbit, but it may overlap with the hot flow. The soft photons emitted by the disc (and possibly internally produced synchrotron photons) are Compton upscattered in the hot flow, and emission from the hot flow is partly Compton-reflected from the disc. (**b**) The likely geometry in the soft state consisting of flares/active regions above an optically-thick accretion disc extending close to the minimum stable orbit. The soft photons emitted by the disc are Compton upscattered in the flares by non-thermal electrons producing power-law spectra extending to γ-rays. Emission from the flares is partly Compton-reflected from the disc. From Zdziarski and Gierliński (2004)

1998, for reviews). The nature of the hard state emission and origin of the hot electrons has been already discussed in the 1970s (Shapiro et al. 1976; Ichimaru 1977) and is commonly associated with either a hot inner flow close to the BH (Esin et al. 1997, 1998; Poutanen et al. 1997; Narayan et al. 1998; Yuan and Zdziarski 2004) or a corona above the accretion disc (Galeev et al. 1979; Haardt and Maraschi 1993; Haardt et al. 1994; Stern et al. 1995b; Poutanen and Svensson 1996; Beloborodov 1999b, see Fig. 1a for a possible geometry).

Knowing the slope of the hard state spectra (with photon index $\Gamma = 1.6$–1.8) we can easily estimate the ratio of the total emitted power L to the soft seed photon luminosity L_s entering Comptonization region, i.e. the amplification factor $A = L/L_s$. Beloborodov (1999a) found an approximate relation between Γ and A for the Comptonized spectra:

$$\Gamma = \frac{7}{3}(A-1)^{-\delta}, \tag{1}$$

where $\delta = 1/6$ for BHBs and $\delta = 1/10$ for AGNs and the typical seed photon temperatures of 0.2 keV for BHBs and 5 eV for AGNs were assumed. If indeed the disc photons are being Comptonized, we get $A \approx 10$ for BHBs in their hard state. This fact puts serious constraints on the geometry of the emission region (Poutanen 1998; Beloborodov 1999a) and immediately rules out simple slab-corona models which predict much smaller amplification $A \lesssim 2$ and softer spectra because of the efficient X-ray reprocessing in the cold disc (Stern et al. 1995b). Assuming that coronal plasma has a mildly relativistic velocity away from the cold disc (Beloborodov 1999b; Malzac et al. 2001) one can in principle reconcile that model with the observed slopes as well as with the correlated changes of the spectral hardness and the amount of Compton reflection from the disc, but still one would have troubles explaining their correlations with the iron line width and characteristic variability frequencies (see Sect. 3.1).

If the accretion flow geometry is such that the inner part is occupied by the hot flow and the outer is the standard cold disc, the seed photons for Comptonization might be internal to the hot flow or come from the outer disc. For the truncation radius of the cold disc of more than $30R_S$ (where $R_S = 2GM/c^2$ is the Schwarzschild radius), most of the disc photons go directly to the observer and therefore the disc should be very prominent in the total spectrum. Furthermore, the luminosity in disc photons being Comptonized in the region of major gravitational energy release ($< 10R_S$) would be only about 1 % of the total luminosity resulting in an amplification factor of a hundred and a very hard Comptonization spectrum (see Eq. (1)). Neither is observed. An overlap of the inner hot flow with the cold disc (Poutanen et al. 1997; see Fig. 1a) was proposed as a solution to this, but does not really solve the problem, because most of the energy is dissipated within $10R_S$. Thus the disc should come to radii below $10R_S$. Another solution is that cold clouds embedded into the flow reprocess hard photons (Celotti et al. 1992; Krolik 1998; Zdziarski et al. 1998; Poutanen 1998) increasing thus the number of seed soft photons.

Alternatively, the hot flow itself can generate enough soft photons by synchrotron emission of the same hot electrons that emit the X-rays, if the electron temperature T_e is sufficiently high to overcome the self-absorption problem. The spectrum from the hot optically thin advection-dominated accretion flows (ADAF) indeed is produced mostly by synchrotron self-Compton (SSC) mechanism (Narayan and Yi 1994, 1995; Narayan et al. 1998; Yuan and Narayan 2014). Detailed radiative transfer calculations accounting for non-local Compton effect coupled with dynamics (also in the Kerr metric) predict, however, T_e exceeding the observed values of 50–100 keV by at least a factor of 2 (see Fig. 3 in Yuan et al. 2007, Fig. 1d in Xie et al. 2010 and Figs. 5 and 6 in Niedźwiecki et al. 2012). Most of the problems get solved if the electrons have a reasonably strong non-thermal tail. In this case, synchrotron emission becomes much more efficient (Wardziński and Zdziarski 2001) increasing the cooling, softening the spectrum and lowering T_e to the values which agree with observations. This is the essence of the hybrid Comptonization (or rather hybrid SSC) models developed for bright accreting BHs (Poutanen and Coppi 1998; Coppi 1999; Poutanen and Vurm 2009; Malzac and Belmont 2009; Veledina et al. 2011b, 2013a) that are described below in more details. Non-thermal electrons can also play a role in low-luminosity systems such as Sgr A* (Mahadevan 1998; Özel et al. 2000; Yuan et al. 2003). However, in these conditions the equilibrium electron temperature is very high ($\sim$ MeV) and the optical depth is very low, so that thermal synchrotron radiation is very effective. The role of the non-thermal electrons is then reduced to production of tails at higher and lower frequencies around the dominating thermal synchrotron peak.

2.2 Hybrid Comptonization Model

Arguments in favour of the presence of non-thermal particles in the accretion flow come from the significant detection of the MeV tails in the hard state spectra of Cyg X-1 (see McConnell et al. 1994, 2002; Ling et al. 1997; Jourdain et al. 2012a; Zdziarski et al. 2012, see Fig. 2) and marginally in GX 339–4 (Droulans et al. 2010). Also in the soft state, power-law tails extending to hundreds of keV and up to possibly 10 MeV are present (Grove et al. 1998; Gierliński et al. 1999; Zdziarski et al. 2001; McConnell et al. 2002, see Fig. 2) and are well described by non-thermal/hybrid Comptonization (Poutanen 1998; Poutanen and Coppi 1998; Gierliński et al. 1999; Coppi 1999). What is the nature of the non-thermal particles is an open question.

On the theoretical ground, it is expected that the electrons get some of the energy via Coulomb collisions with hot, nearly virial protons (as assumed in ADAF models) or diffu-

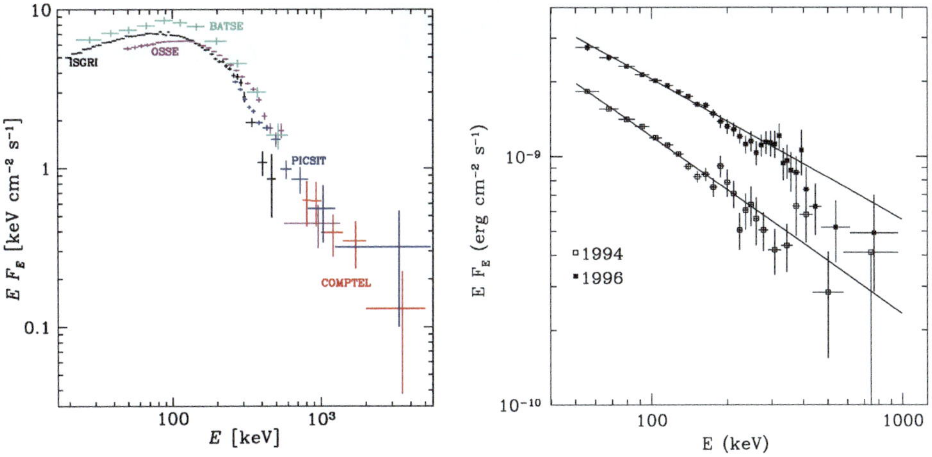

Fig. 2 Hard (*left panel*) and soft (*right*) state spectra of Cyg X-1. From Zdziarski et al. (2012) and Gierliński et al. (1999)

sive acceleration by MHD turbulence, resulting in "thermal heating", i.e. energy is transferred to the thermal population of electrons. However, some fraction of the energy can be transferred to them in the form of injection of relativistic electrons by shock acceleration (Fragile and Blaes 2008; Das et al. 2009; Henisey et al. 2012), magnetic reconnection (Ding et al. 2010; Riquelme et al. 2012; Hoshino 2013), or electron-position pair production by decay of pions born in proton-proton collisions (Mahadevan 1998). Because the microphysics of electron acceleration and heating in the hot flow is not well established from first principles, one can use a phenomenological prescription, where some fraction of the total power is given to the electrons as heating and the rest of the energy is given by non-thermal injection of power-law electrons. Such hybrid thermal/non-thermal models are reviewed by Coppi (1999). Models with the least number of free parameters are either purely thermal or purely non-thermal. Because the first option clearly contradicts the data, we consider in the following the second non-thermal option. Of course, the assumption that the electrons receive 100 % of their energy in the form of non-thermal injection is not realistic. Fortunately, the results are not very sensitive to the actual value of non-thermal injection fraction as long as it exceeds 10 % (see Appendix B1 in Veledina et al. 2013a). The reason is that even for pure non-thermal injection the steady-state electron distribution is thermal at low energies due to thermalisation via Coulomb collisions between electrons as well as via synchrotron self-absorption (Ghisellini et al. 1988, 1998; Nayakshin and Melia 1998; Vurm and Poutanen 2009). At higher energies a tail develops, whose shape is determined by the injection and the competition between various cooling/thermalisation mechanisms.

The most advanced hybrid models solve simultaneously for the momentum distribution of all considered particles, electrons and positrons (sometimes also protons), as well as the photons. This either can be done via Monte-Carlo simulations (e.g. Stern et al. 1995a), or by solving coupled kinetic equations (Coppi 1992, 1999; Belmont et al. 2008; Vurm and Poutanen 2009). The processes that need to be accounted for under the conditions of the hot flows are Compton scattering, synchrotron emission and absorption, pair production, Coulomb collisions (between leptons as well as with protons), and bremsstrahlung. The radiative transfer can be easily handled exactly with Monte-Carlo approach, while usually with the kinetic approach an escape probability formalism in a single-zone approximation is

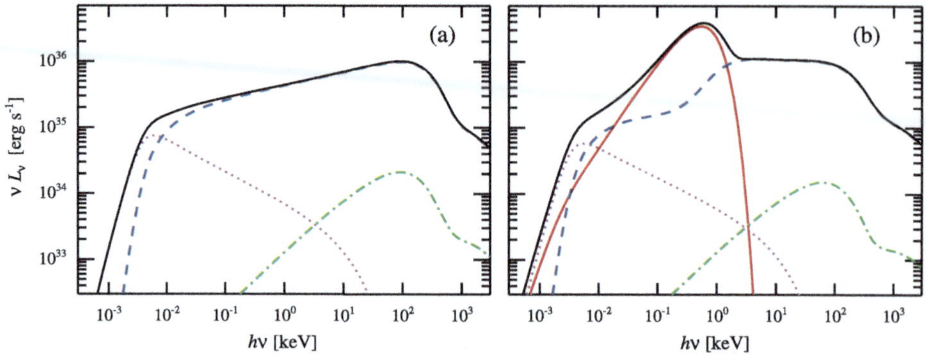

Fig. 3 Spectrum from the one-zone hot flow. (**a**) Spectral decomposition of the hybrid SSC model. The pink dotted line is the non-thermal synchrotron, the blue dashed is its Comptonization spectrum, and the *green dot-dashed curve* is the bremsstrahlung component. (**b**) Same as (**a**), but with the disc photons (*red triple-dot-dashed curve*) dominating the seed photons distribution. Now below 1 keV Comptonization spectrum is produced by SSC, while at higher energies Comptonization of the cold disc photons dominates

used. A single-zone approximation works reasonably well if most of the escaping radiation at all wavelengths is dominated by some narrow range of radii. For the X-ray production this is fine, because most of the energy is dissipated in the accretion flow spread from say $3R_S$ to $10R_S$. On the other hand, this approximation fails in the OIR. Here the outer zones of the hot flow can dominate the energy output in those bands as the inner zones are opaque for that radiation. Because the radiation from the inner zones can affect the energy balance and the escaping radiation from the outer zones, a multi-zone treatment with the radiative transfer is required (Veledina et al. 2013a).

The simplest model is described by the size, the magnetic field strength B, Thomson optical depth τ, the total injected power L, the spectrum of the injected electrons and the spectrum and luminosity of the external (blackbody/cold accretion disc) photons. For an extended multi-zone flow one can assume that the electron energy injection rate as well as B and τ have power-law distributions with radius, $B(R) \propto R^{-\beta}$, $\tau(R) \propto R^{-\theta}$ and thus the additional parameters, e.g., β and θ have to be introduced. Many parameters can be directly determined from observations or taken from theoretical accretion disc models. Now let us describe the main properties of such hybrid models.

2.3 Basic Properties of Hybrid Accretion Flows

If the truncation radius of the cold disc is significantly larger than the region of the major energy dissipation (i.e. $R_{tr} > 30R_S$), then the X-ray spectrum is dominated by the radiation from the innermost zone of the hot flow. Here locally generated non-thermal synchrotron photons are Comptonized by the thermal electron population (see Fig. 3a). Here we note that the internally generated synchrotron photons are much more efficient in cooling the plasma than the external disc photons. The first obvious difference comes from the geometry: all synchrotron photons are injected within the hot flow and have a chance to be Comptonized, while in the disc case only a small fraction gets to the hot flow. The second difference comes from the fact that the synchrotron photons have much smaller energies than the cold disc photons. Therefore, in order to produce the same spectral slope of the Comptonization continuum (with nearly the same total power), the synchrotron luminosity can be smaller by a factor a several than the disc luminosity intercepted by the hot flow. In other words, the amplification factor given by Eq. (1) is now closer to 50 than 10 (because we need to use $\delta \approx 1/10$).

At a few per cent of the Eddington luminosity corresponding to the bright hard state, the low-energy electrons are thermalised by Coulomb collisions and synchrotron self-absorption to the typical electron temperatures T_e of about 100 keV (Poutanen and Vurm 2009; Malzac and Belmont 2009; Veledina et al. 2011b). The hybrid model produces surprisingly stable spectra with photon index $\Gamma \sim 1.6$–1.8 largely independent of the model parameters (Fig. 3a). The high-energy electron tail can be approximated by a power-law, which is softer than the injected distribution due to the cooling. The observed MeV tail is produced by Compton up scattering of the 100 keV photons by these non-thermal electrons. The outer zones of the hot flow have softer spectra because of the additional cooling by the cold disc photons and because of more transparent conditions for the synchrotron radiation (see Fig. 3b). The overall X-ray spectrum is thus concave.

The OIR spectrum consists of two components: the multi-colour (possibly irradiated) cold accretion disc and the synchrotron radiation from the hot flow. Similarly to the inhomogeneous synchrotron models developed for extragalactic jets (Marscher 1977; Blandford and Königl 1979), the non-thermal synchrotron spectrum of the hot flow is a power-law $F_\nu \propto \nu^\alpha$ with the index (Veledina et al. 2013a):

$$\alpha_{\rm OIR} = \frac{5\theta + \beta(2p+3) - 2p - 8}{\beta(p+2) + 2\theta}, \tag{2}$$

where p and θ are the indices of the equilibrium distribution of electrons, $n_e(R,\gamma) \propto R^{-\theta}\gamma^{-p}$, at Lorentz factor γ emitting at the self-absorption frequency. Typically, spectral indices are $\alpha_{\rm OIR} \sim 0 \pm 0.5$. The turn-over at longer wavelengths is determined by the extent of the hot flow, while the transition to the optically thin synchrotron emission is hidden by the Comptonization spectrum (Fig. 3a).

The disc spectrum can be split also into two components: the inner warmer standard disc heated by viscous forces and the outer cooler disc heated by the X-rays. In the OIR one expects the dominance of the irradiated disc, which has the radial temperature dependence $T_{\rm irr} \propto R^{-3/7}$ (Cunningham 1976). Presence of the irradiated disc can be reflected in the optical echoes (Hynes et al. 1998; O'Brien et al. 2002), the X-ray time-lags (see Poutanen 2002, and references therein) and in the optical/X-ray cross-correlation function (Hynes et al. 2009a; Veledina et al. 2011a). Its signatures are also seen in the spectrum (e.g., Hynes et al. 2002; Gierliński et al. 2009). For typical parameters of LMXBs with the disc size of 10^{11} cm and the X-ray luminosity of 10^{37} erg s^{-1} the temperature of the outer disc is about $T_{\rm irr} \sim 20\,000$ K. The relative role of the components varies with the wavelength. The disc spectrum is hard in the OIR band, while the hot flow produces an excess emission dominating below ~1 eV (see Fig. 4a and Fig. 6).

At smaller accretion rates below a few percent of the Eddington value, the flow becomes more transparent to the synchrotron photons leading to their increasing role in cooling and resulting in slightly *softer* X-ray spectra (Veledina et al. 2011b). At higher accretion rates associated with the transition to the soft state (see Fig. 4), the outer zones of the hot flow gradually collapse, so that the disc truncation radius $R_{\rm tr}$ decreases (Poutanen et al. 1997; Esin et al. 1997). This leads to the rising role of the disc as a source of seed photons, which increases Compton cooling, leads to spectral softening and causes changes in the electron distribution from mostly thermal to nearly non-thermal (Poutanen and Coppi 1998; Poutanen and Vurm 2009; Malzac and Belmont 2009; Veledina et al. 2011b). This transition is accompanied by the increase in the reflection amplitude that scales with the solid angle at which the cold disc is seen from the hot flow. The OIR hot flow luminosity drops first

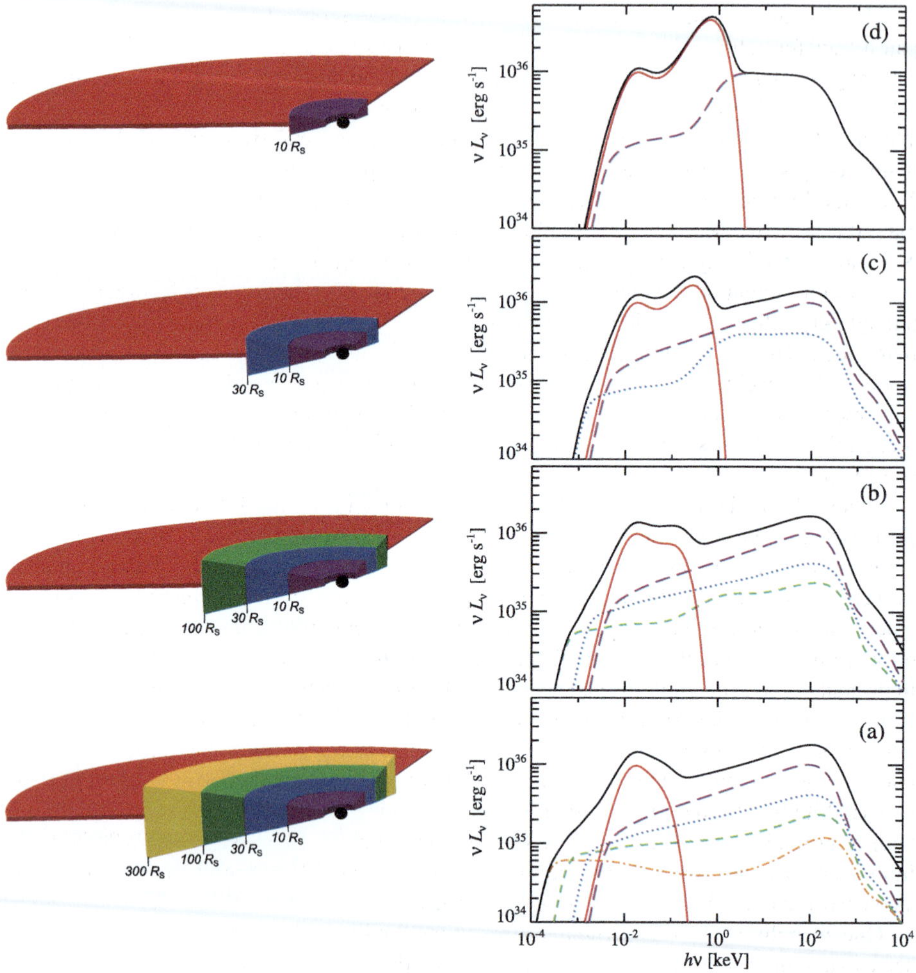

Fig. 4 *Left panels*: Schematic picture of the evolution of the hot flow size during spectral state transitions. In the low-luminosity hard state, the geometrically thick inner hot flow is large dominating the radiative energy output. With increasing accretion rate (from lower to upper panels) the outer zones of the hot flow gradually collapse swallowed by the cold thin accretion disc. For illustrative purposes, the hot flow is split into four zones with outer radii $10R_S$ (*violet*, zone 1), $30R_S$ (*blue*, zone 2), $100R_S$ (*green*, zone 3) and $300R_S$ (*yellow*, zone 4). *Red* outer component represents the truncated cold accretion disc. *Right panels*: corresponding spectral evolution at the state transition. Contribution of different zones are marked with different lines: zone 1 (*violet long-dashed*), zone 2 (*blue dotted*), zone 3 (*green short-dashed*), zone 4 (*orange dot-dashed*) and outer cold irradiated disc (*red three-dot-dashed*). At the *lower panel*, the hot flow spectrum was calculated not accounting for the seed photons from the disc. An IR excess is clearly visible above the irradiated disc spectrum. Collapse of the hot flow leads to dramatic changes in the OIR hot flow synchrotron spectrum. Changes in the X-ray spectral shape are insignificant until the truncation radius becomes as small as $10R_S$. The Comptonization spectrum from the hot-flow zone closest to the cold disc consists of two separate continua produced by Comptonization of the synchrotron and the cold disc photons, with the latter being dominant source of seed photons in this zone (see also Fig. 3b). For simplicity, in this illustration the total luminosity is kept constant. Adapted from Veledina et al. (2013a)

at longer wavelengths, where the outer zones radiate (see Fig. 4). Note that the X-ray spectrum changes much later, when the truncation radius comes closer to the zone of the main energy dissipation of about $10R_S$. At this moment, the Comptonization spectrum of the remaining hot flow consists of two segments: the hybrid SSC dominates at lower energies, while Comptonization of the disc photons takes over at energies above the cold disc peak (see Fig. 3b). Similarly curved spectra are expected for the hot flow zones closest to the cold disc (Fig. 4). The corresponding time delay between sharp luminosity changes at different wavelengths scale with the timescale of state transition and, depending on the separation of the wavelengths and accretion parameters, can be as short as hours (e.g., if one observes in different optical filters), as long as a few days (e.g., IR and UV) or weeks (e.g. IR and X-rays). The opposite evolution should be observed in the soft-to-hard spectral transition when the accretion rate drops after the outburst peak. Here first the X-ray spectral transition starts and at a timescale of a week or so the emission in the OIR peaks, when the size of the hot flow becomes large enough for the OIR synchrotron photons to escape.

At a high accretion rate, the accretion disc extends to the last stable orbit and the source switches to the soft state. In this case, no inner hot flow exists, but a non-thermal magnetically powered corona still could be present (Fig. 1b). Its presence is supported by the existence of the X-ray/γ-ray power-law tails. The non-thermal synchrotron from the corona may be present in the OIR band, but at a much lower level, because the electron cooling is dominated now by Comptonization of the disc photons. Actually, such a non-thermal corona atop the cold disc (in addition to the hot inner flow) may be present also in the hard state, but its emission scaled with the cold disc luminosity is weak.

3 Observational Properties

3.1 X-ray/γ-Ray Spectra

Let us first briefly summarise what we know about the spectral properties of BHBs. Further details can be found in reviews by Poutanen (1998), Zdziarski and Gierliński (2004), Done et al. (2007), and Done (2013). In the hard-state, the spectra constitute a power-law in the X-ray band with a rather stable spectral slope (with photon index $\Gamma \sim 1.6-1.9$ and ubiquitous sharp cut-off at around 100 keV (Gierlinski et al. 1997; Zdziarski et al. 1998; Ibragimov et al. 2005; see Fig. 10a in Zdziarski and Gierliński 2004). The shape of the spectra allows us to conclude that they are produced by (nearly) thermal Comptonization (e.g. Poutanen 1998; Zdziarski and Gierliński 2004, see Fig. 5). When the high-quality data above 100 keV were available (e.g. with OSSE/*CGRO* or IBIS/*INTEGRAL* or HXD-GSO/*Suzaku*) the electron temperature (measured with the accuracy of about 10 %)[1] was always lying in the interval 50–120 keV (e.g. Gierlinski et al. 1997; Zdziarski et al. 1998; Poutanen 1998; Aref'ev et al. 2004; Makishima et al. 2008), with temperature increasing with decreasing luminosity. The hard-state accreting BHBs also show weak but distinctive MeV tails (McConnell et al. 1994, 2002; Ling et al. 1997; Droulans et al. 2010; Jourdain et al. 2012a; Zdziarski et al. 2012, see

[1]Note that the electron temperature can be measured only if high-quality data are available above 100 keV and accurate Comptonization models such as COMPPS (Poutanen and Svensson 1996) or EQPAIR (Coppi 1999) are used for fitting. Using the exponentially cut-off power-law for the fits and identifying the e-folding energy with the electron temperature is dangerous, because that model does not correctly describe the shape of Comptonization continuum (see e.g. Fig. 5 in Zdziarski et al. 2003). This can result in over-estimation of T_e by a factor of 3–6.

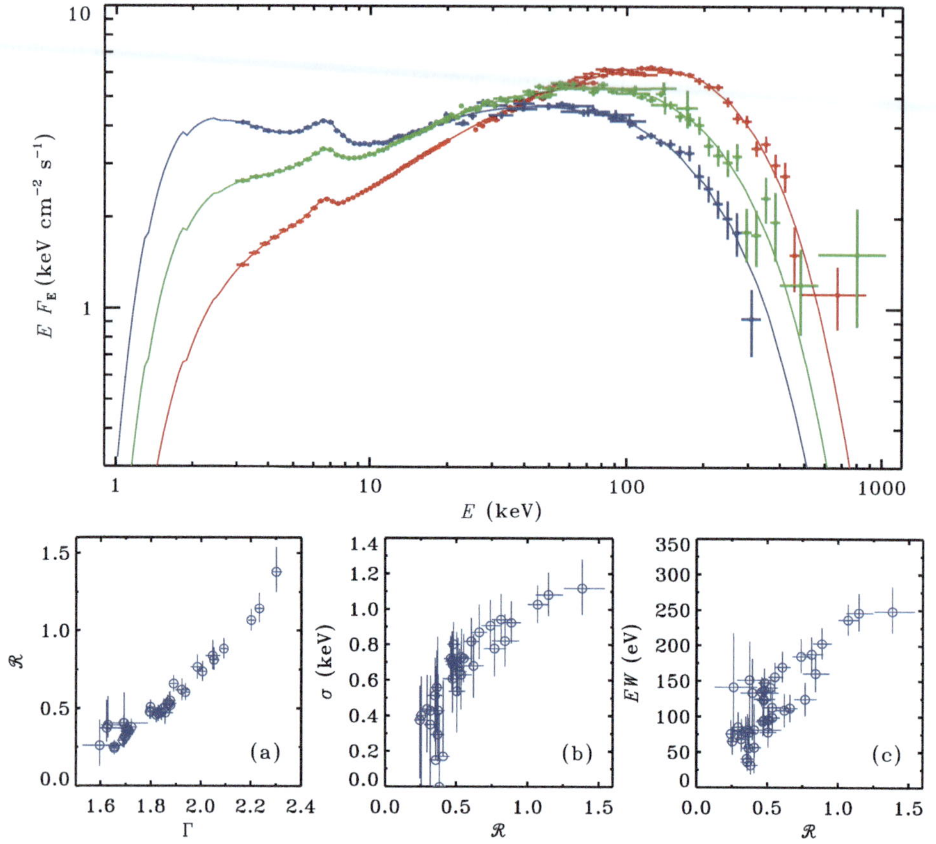

Fig. 5 Spectral variations of Cyg X-1 in its hard/intermediate state as shown in the top panel. The lower panels show correlation between the amplitude of Compton reflection $\mathcal{R}$, photon spectral index Γ, width σ and the equivalent width of the iron line. Adapted from Ibragimov et al. (2005)

Fig. 2). Such spectra are consistent with the hybrid hot-flow model. During the transition to the soft state, T_e is reduced. However, because of the growing importance of non-thermal tail, the spectral cutoff energy may actually increase (Poutanen and Vurm 2009).

The fact that the spectra are stable in the hard state with variable luminosity and never have $\Gamma < 1.6$ argues in favour of the hybrid SSC as the main emission mechanism. If the outer cold disc were the seed photons provider, one expects harder and strongly variable spectra when the truncation radius is changing in the soft-to-hard transition. Moreover, the best studied BHs, Cyg X-1 and GX 339–4, clearly have a concave spectrum that can be fitted with two Comptonization continua (Frontera et al. 2001; Ibragimov et al. 2005; Makishima et al. 2008; Shidatsu et al. 2011; Yamada et al. 2013). The inhomogeneous hot flow model naturally explains such spectra by the radial dependence of the slope of Comptonization spectrum. The spectral curvature can also appear if a non-thermal corona (similar to that producing power-law tail in the soft state) exists above the cold disc during the hard state too (Ibragimov et al. 2005).

In addition to the smooth continuum, in both states a Compton reflection feature and the fluorescent iron line at 6.4 keV originating from cool opaque matter (likely the cool accretion disc) are often detected. The strength of Compton reflection is correlated with the

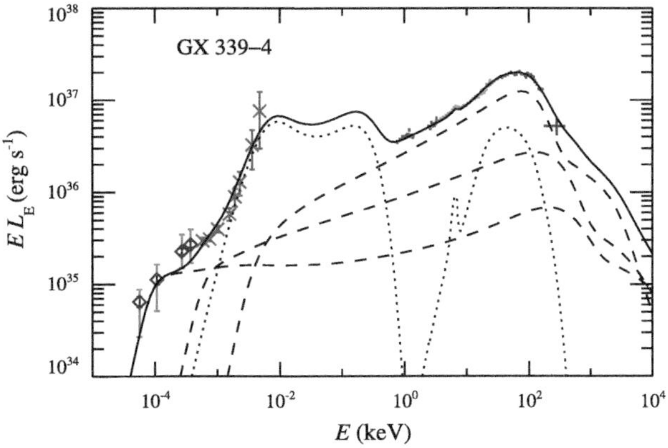

Fig. 6 Broad-band spectrum (corrected for absorption) of GX 339–4 in its hard state around March 5, 2010 from the mid-IR to the hard X-rays (Cadolle Bel et al. 2011). The *dashed lines* show contribution of different zones of the hot accretion flow (Veledina et al. 2013a), the *dotted lines* represent the spectra of the irradiated and standard discs as well as Compton reflection. From Veledina et al., in prep.

X-ray slope (Zdziarski et al. 1999, 2003), with the width of the iron line as well as with the quasi-periodic oscillation (QPO) frequency (Gilfanov et al. 1999; Revnivtsev et al. 2001; Ibragimov et al. 2005; Gilfanov 2010, see lower panels in Fig. 5). During the outbursts of BH transients, in the hard state the iron line width correlates well with the luminosity (Kolehmainen et al. 2014). These data are consistent with the hot-flow paradigm where all correlation are basically controlled by the cold disc truncation radius.

At luminosities above a few per cent of Eddington, BHBs show a strong correlation between spectral index and luminosity. At lower luminosities the trend is reversed, i.e. spectra become softer with decreasing luminosity (Wu and Gu 2008; Sobolewska et al. 2011). Similarly, an indication of the reverse trend was detected in low-luminosity AGNs (Constantin et al. 2009; Gu and Cao 2009). This was interpreted as a change of the source of seed photons for Comptonization from the disc photons dominating at higher luminosities to the synchrotron at lower luminosities. The whole spectral index—luminosity dependence is well explained by the hybrid hot flow model (see Figs. 7 and 12 in Veledina et al. 2011b). At very small luminosities, the flow becomes more transparent for the synchrotron radiation increasing the photon input and softening the Comptonization spectra.

3.2 Broad-Band Spectra and Infrared Flares

Numerous multiwavelength campaigns were conducted over the past decade. Broadband radio to X-ray spectral energy distributions (SEDs) for many BHBs were constructed (e.g. Hynes et al. 2000; McClintock et al. 2001; Chaty et al. 2003; Cadolle Bel et al. 2007, 2011; Durant et al. 2009). The OIR emission is normally dominated by the (irradiated) disc, but the IR excesses are observed in a number of sources: XTE J1859+226 (Hynes et al. 2002), GX 339–4 (Gandhi et al. 2011; Shidatsu et al. 2011; Buxton et al. 2012; Dinçer et al. 2012; Rahoui et al. 2012), A0620–00 (Gallo et al. 2007), SWIFT J1753.5–0127 (Chiang et al. 2010), V404 Cyg (Hynes et al. 2009b), XTE J1550–564 (Jain et al. 2001; Russell et al. 2011). In some cases the OIR spectrum can be described by a pure power-law $F_\nu \propto \nu^\alpha$ with index α close to zero (e.g. $\alpha_{\rm OIR} = -0.15$ in XTE J1118+480, see Esin et al. 2001; Chaty et al. 2003). Sometimes the OIR excess spectrum is rather soft with $\alpha \approx -0.7$ (see Kalemci et al. 2013, for a recent overview). The IR excesses were previously explained by the jet (Hynes et al. 2002; Gallo et al. 2007) or the dust heated by the secondary star (Muno and Mauerhan 2006). Veledina et al. (2013a) recently argued that the OIR excess emission may also be produced by synchrotron radiation from the hot flow.

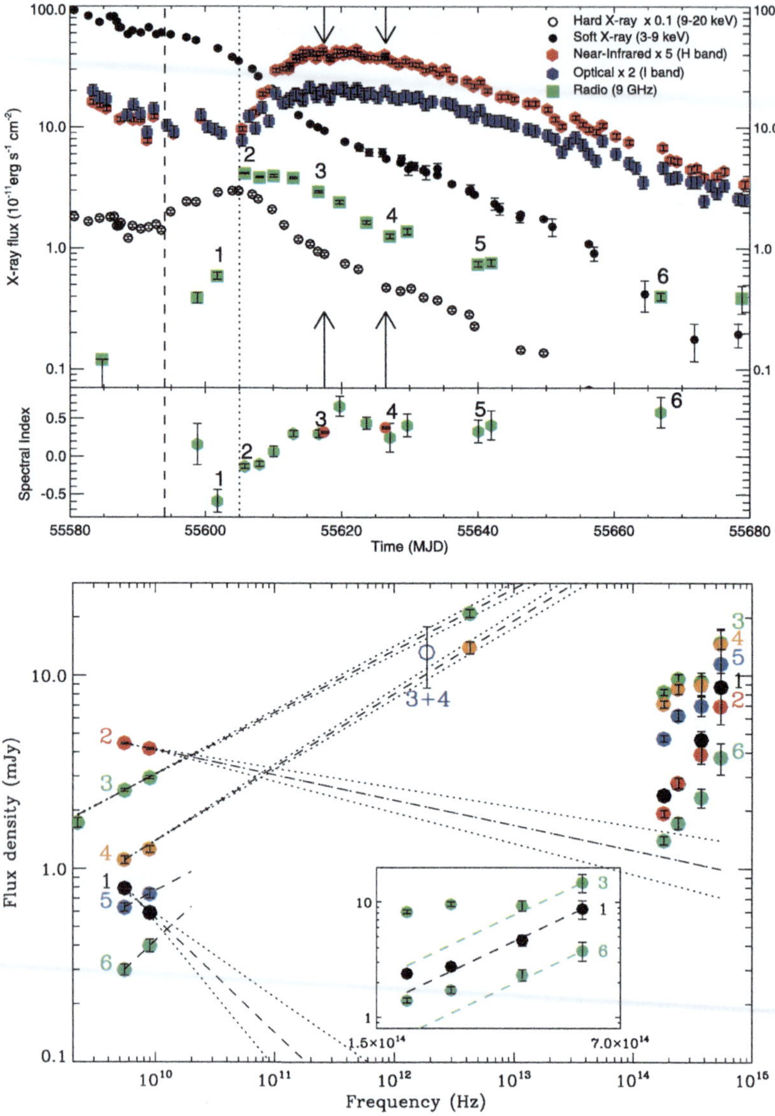

Fig. 7 *Upper panel*: The lightcurves of GX 339–4 during the decay of the 2010–2011 outburst in radio (*green*), NIR (*red*), optical (*blue*), soft X-ray (*filled black circles*) and hard X-ray (*open black circles*). The *bottom sub-panel* shows the evolution of the radio spectral index. The *dashed vertical line* indicates the start of the soft-to-hard state transition, while the *dotted line* marks the start of the OIR flare and the end of the transition (see Fig. 1 in Dinçer et al. 2012). The numbers refer to the individual observations displayed in the *lower plot*. *Lower panel*: Evolution of the radio to OIR spectra of GX 339–4. The *inset* zooms on the OIR spectra, with the *dashed lines* corresponding to power-law with $\alpha = 1.5$ (for an irradiated disc). From Corbel et al. (2013)

Here we only discuss a few representative examples of the recent studies. The broad-band spectrum of GX 339–4 in its hard state (Cadolle Bel et al. 2011, 2013) is shown in Fig. 6. The X-ray spectrum peaking at ~100 keV is well described by thermal Comptonization. The irradiated disc presumably dominates in the UV band. There is a clear excess in the mid-

and near-IR (Gandhi et al. 2011), but the spectrum becomes harder at longer wavelengths. These data are well explained by the non-thermal synchrotron emission from the hot flow (Veledina et al. 2013a) of about $500R_S$.

Excellent data covering both radio and OIR bands have been collected during the 2010–2011 outburst of GX 339–4 (Cadolle Bel et al. 2011; Dinçer et al. 2012; Corbel et al. 2013, see Fig. 7). A week after the start of the transition to the hard state (marked by the vertical dashed line) the OIR spectrum has a clear soft excess above the reprocessing thermal emission (point 1), while the radio jet optically thin emission can contribute at most 1 % to the OIR. Few days later, when the transition was completed (point 2), the fluxes in the H and I-band show a sharp increase and an obvious IR excess in the OIR spectra is visible, while the radio emission is still soft with $\alpha \sim -0.1$. On a week time scale the radio spectrum transits to the harder, optically thick state with $\alpha \sim 0.5$ (point 3) corresponding to the synchrotron emission from an inhomogeneous source analogously to the extragalactic jets (Blandford and Königl 1979). The radio spectrum stays hard and the IR excess is visible during the following decay. The high flux in the *Herschel* far-IR band lies exactly on the extrapolation of the radio spectra. Because after the break, the jet spectrum must be optically thin and soft, while the OIR spectra are flat (even after subtracting thermal component, as discussed by Dinçer et al. 2012; Buxton et al. 2012; Corbel et al. 2013), the jet does not contribute significantly to the OIR bands. An important conclusion from these data is that there is a rather strong, evolving component in the OIR band which cannot be produced by the jet or reprocessing in the accretion disc at any stage of the outburst. Its appearance in the hard state is, however, consistent with the hot flow interpretation.

The IR excess similar to that seen in GX 339–4 in the hard state appears also in XTE J1550–564 (Jain et al. 2001), 4U 1543–47 (Buxton and Bailyn 2004) and XTE J1752–223 (Russell et al. 2012). The properties of the excess can be studied using the colour-magnitude diagram (see Fig. 8c). During the 2000 outburst of XTE J1550–564 the data in the soft state, soft-to-hard transition and at very low-luminosity hard state can be adequately described by the reddened irradiated disc emission. Both at the rising (filled symbols) and decaying (open symbols) phase of the outburst, we see "flares" during which the spectrum becomes redder. One can interpret these flares as the appearance of the additional red component. Fitting the fluxes at the decaying stage with an exponential plus a constant, one can subtract the contribution from the irradiated disc and obtain a spectrum of the flare component only. We see that the second flare starts with the spectral index $\alpha \sim +0.7$, then becomes softer with $\alpha \sim -0.2$, and then harder again (see green arrows in Fig. 8). (The final points have large errors, because of the uncertainties in the subtraction of the disc.) What is important that the flare starts in I before H, so that the index measured between filters I and H is even larger, $\alpha_{IH} \approx 1.0$. This behaviour rules out immediately the interpretation of the flare in terms of optically thin jet emission.[2] Instead the data are consistent with the inhomogeneous hot accretion flow model of Veledina et al. (2013a). Such indices were also observed in the flare spectrum of XTE J1752–223 (Russell et al. 2012) and GX 339–4 (Dinçer et al. 2012; Buxton et al. 2012). This would mean that all these OIR flares are produced by the hot

[2]Colour-magnitude diagram for the 2000 outburst of XTE J1550–564 was constructed by Russell et al. (2011), who also related the observed colours to the intrinsic spectral indices and claimed very soft spectrum of the flare. Unfortunately, all their formulae are wrong for various reasons and the actual intrinsic spectra are much harder. Furthermore, the exponential fits to the OIR light curves to evaluate the flare spectrum were also flawed, as their fits overestimate the disc contribution in the V and I filters just before the flare (see Fig. 2 in Russell et al. 2010) resulting in over-subtraction of the flux in those filters and in a much too soft spectrum of the flare (compare our $\alpha \sim +0.7$ at the start of the flare with their $\alpha \sim -1.6$).

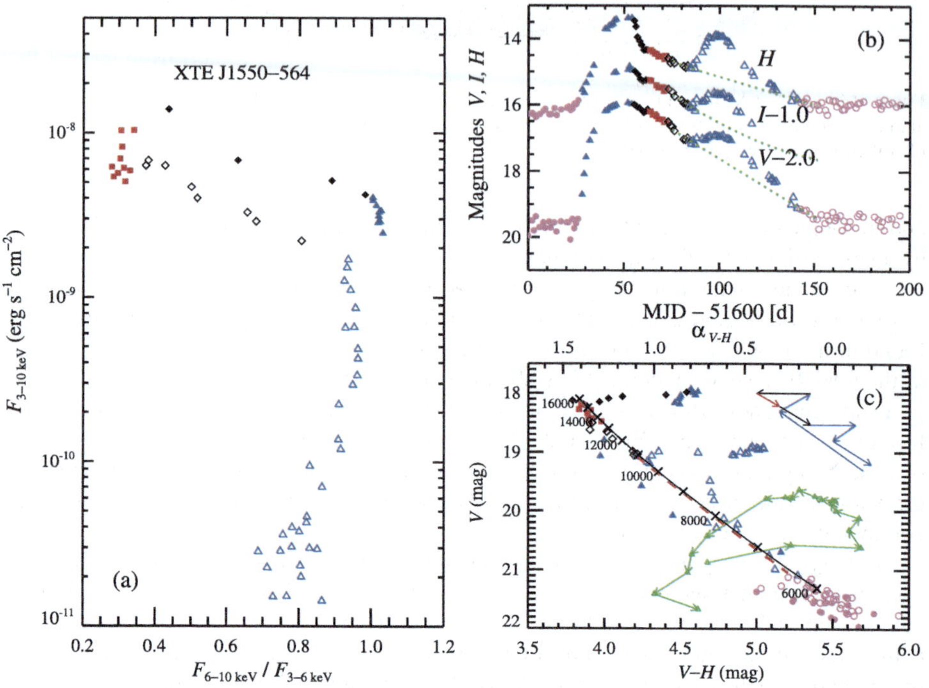

Fig. 8 (**a**) X-rays flux—hardness diagram for XTE J1550–564 during the 2000 outburst. Different colours indicate various stages of the outburst based on their X-ray hardness. The *blue colour* indicate the hard state, the *black colour* is the transition and the *red colour* the soft state. *Open* and *closed symbols* correspond to the rising and decaying outburst stages, respectively. (**b**) The *light curves* in V, I and H-filters of XTE J1550–564 (Jain et al. 2001). The *pink circles* correspond to the quiescent state. The *green dotted curves* show the best-fit model for the decaying disc component. (**c**) The observed V vs $V-H$ colour-magnitude diagram. The *black solid line* represent the relation expected for the black body disc of a characteristic radius 2.7×10^{11} cm inclined at $i = 75^{\circ}$ at distance of 4.38 kpc (Orosz et al. 2011) of different temperatures (marked next to the line). The model magnitudes were reddened following the extinction law of Fitzpatrick (1999) with $A_V = 5.0$. Much smaller A_V (that would lead to softer spectrum) is not possible, because the disc temperature would be below hydrogen ionisation temperature needed for the outburst to start. The *blue-black-red arrows* illustrate schematically the time evolution of the source during the outburst. The *green arrows* show the path followed by the flare after MJD 51680 (see the text). The upper x-axes show the intrinsic spectral index $\alpha_{VH} = 4.63 - 0.84\,(V - H)$ computed from the observed colour. We see that the flare component is never softer than $\alpha = -0.2$. From Poutanen et al., in preparation

flow but not the jet. The observed sharp colour change during the flares is related to the collapse/recover of a zone in the hot flow that is responsible for the H-band emission.

The delay of the IR flare peak by about 10 days from the start of the soft-to-hard transition is naturally expected, because the X-ray transition corresponds to the start of the retraction of the cold disc ($R_{\mathrm{tr}} \sim 10 R_{\mathrm{S}}$), while the IR flare peaks when the hot flow is large enough ($R_{\mathrm{tr}} \gtrsim 100 R_{\mathrm{S}}$) allowing the IR photons to escape (see Fig. 4). At the rising phase of the outburst, just a few day before the hard-to-soft spectral transition, a dip has been observed in the UV light curve of GX 339–4 (Yan and Yu 2012). In Swift J1910.2–0546 a dip first appears in the IR, then optical and finally in the UV (N. Degenaar, priv. comm.). The timing of the dips is consistent with the collapse of the hot flow with increasing accretion rate (see Sect. 2.3).

It is worth noticing that the hard-to-soft and the soft-to-hard spectral transitions occur at different X-ray luminosities (e.g. Zdziarski et al. 2004). This hysteresis is most probably related to the fact that at the same luminosity the cold disc is further away from the central source on the rising phase of the outburst, than on the decline. The hysteresis should be then also reflected in the OIR spectra, namely the fast colour change should occur at a higher X-ray luminosity on the rising phase, than on the decline, as indeed observed.

Some BHBs, however, do show signatures of the jet emission in the OIR band. The most obvious examples is microquasar GRS 1915+105, whose radio light-curve was found to be very similar to the IR one with a few hours delay (Fender et al. 1997) favouring a common origin. It, however, accretes at a nearly Eddington rate and is hardly representative. The OIR spectrum of 4U 1543–47 and MAXI J1836–194 in the hard state is rather soft with $\alpha \sim -0.7$ (Kalemci et al. 2005; Russell et al. 2013), which is consistent with the optically thin synchrotron emission from the jet. Furthermore, the rms spectrum of the IR variability of XTE J1118+480 during the 2005 outburst is close to a power-law with $\alpha \sim -0.8$ (Hynes et al. 2006), implying probably the jet origin. And finally, GX 339–4 in the hard state demonstrated strong correlated IR and X-ray variability with the IR lagging by 0.1 s, which could be interpreted as a signature of propagation delays between the X-ray producing accretion disc and the jet (Casella et al. 2010). It is well possible that three components (the irradiated disc, the hot flow and the jet) contribute to the OIR band and their contribution can vary not only from source to source, but also in the same source from the outburst to the outburst.

3.3 X-Ray Variability

In addition to the spectral properties, the variability in X-rays and longer wavelengths puts strong constraints on the models. In the hard state, the typical power-density spectra (PDS) that describe the X-ray variability can roughly be represented as a doubly-broken power-law with indices 0, -1 and -2 from low to high frequencies. A more accurate description of the PDS is achieved by representing it with the Lorentzians (e.g. Nowak 2000; Axelsson et al. 2005). The main source of the short-term variability in BHs is believed to be fluctuations in the mass accretion rate, propagating through the accretion flow (Lyubarskii 1997).

Often QPOs are observed in the range $\sim$0.1–10 Hz (see reviews by Remillard and McClintock 2006; Done et al. 2007). Their frequencies show correlation with the X-ray flux, amplitude of Compton reflection and anti-correlation with the hardness ratio (see Figs. 5 and 9). The origin of QPOs is often associated with the precession of orbits around the BH due to a misalignment of the BH and the orbital spins, known as Lense-Thirring precession (e.g., Stella and Vietri 1998), or with oscillation modes of the accretion flow itself (e.g., Wagoner et al. 2001). The problem with the Lense-Thirring precession models is that the frequency for test masses is a strong function of radius and the BH spin (see e.g. Schnittman et al. 2006). It is not clear why any specific radius gets selected to produce a QPO. If that radius is defined by the truncation of the cold disc, why the QPOs are then observed in the Comptonization spectrum? However, if the flow is hot and thick, it will precess as a solid body (Fragile et al. 2007) with the frequency mostly depending on its size (which is a function of the accretion rate) and weakly on the BH spin and the flow height-to-radius ratio. In such a case, the precession frequencies lie in the observed range and the model explains well their correlations with other quantities (Ingram et al. 2009; Ingram and Done 2011).

Another way of looking at the variability properties is through the autocorrelation function (ACF, which is related to the PDS via Fourier transform). The ACF becomes narrower at higher X-ray energies (Maccarone et al. 2000, see Fig. 10), which is equivalent to the

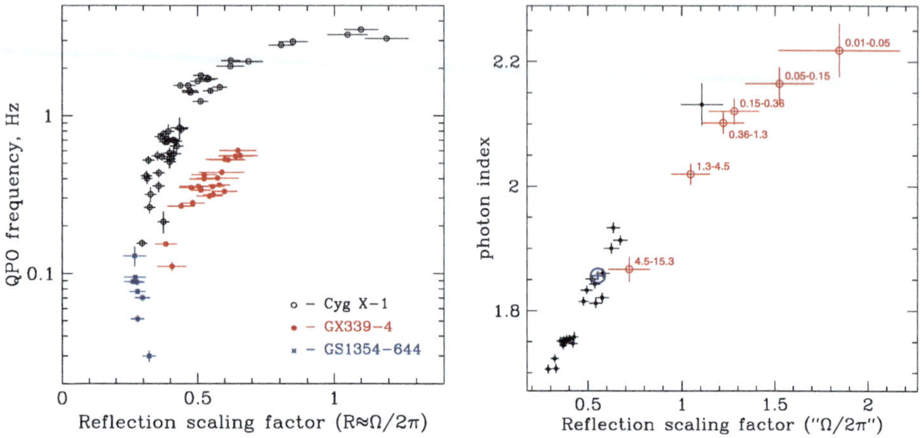

Fig. 9 *Left*: QPO frequency vs reflection correlation. From Gilfanov (2010). *Right*: Photon index–reflection correlation in the average spectra (same as Fig. 5a) and in the Fourier frequency resolved spectra (FFRS) at different frequencies. From Gilfanov et al. (1999)

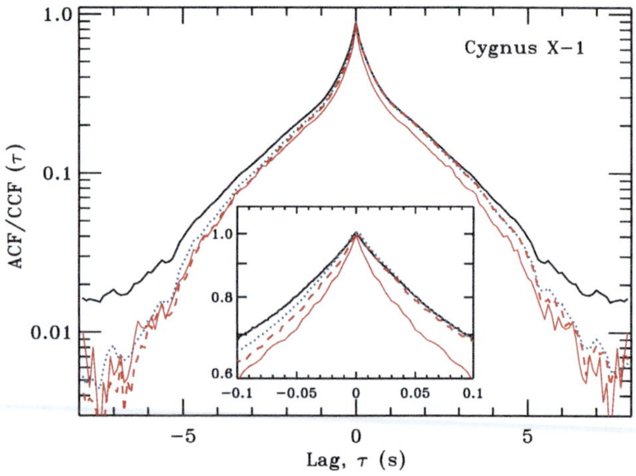

Fig. 10 The ACF and CCFs of Cygnus X-1 observed in December 1997. *Solid curves* show the ACFs for the 2–5 keV energy band (*black solid curves*) and the 24–40 keV band (*red solid curves*). The *blue dotted curve* shows the CCF between the 8–13 keV and the 2–5 keV bands, and the *red dashed curve* represents the CCF for the 24–40 keV vs the 2–5 keV bands. The positive lag corresponds here to the hard photons lagging the soft ones. The CCFs are asymmetric, but the peak do not show any shift from zero lag. From Maccarone et al. (2000)

excess variability at higher frequencies in the PDS at those energies. Moreover, the light curves at different energies are well correlated with each other, but the harder X-rays are delayed with respect to the soft X-rays (Miyamoto and Kitamoto 1989; Nowak et al. 1999a, 1999b, see Fig. 11). This effect is also reflected in the asymmetries of the cross-correlation function (CCF) between the hard and the soft X-ray energy bands (Priedhorsky et al. 1979; Nolan et al. 1981; Maccarone et al. 2000, see Fig. 10).

In order to have a better understanding for these asymmetries, it is useful to look at the time lags Δt between the light-curves at these X-ray energies as a function of the Fourier frequency f. For the hard state of Cygnus X-1 they are shown in Fig. 11a. As a function of energy E, the lags relative to energy E_0 follow the logarithmic law $\propto \ln(E/E_0)$. The rather large lags (exceeding 0.1 s) were first interpreted as produced by Comptonization in a large Compton cloud (Kazanas et al. 1997). For harder photons more scatterings is required, thus they spend more time in the medium before escape and therefore are delayed. Such an interpretation not only causes a problem with the energetics of the cloud (requiring large

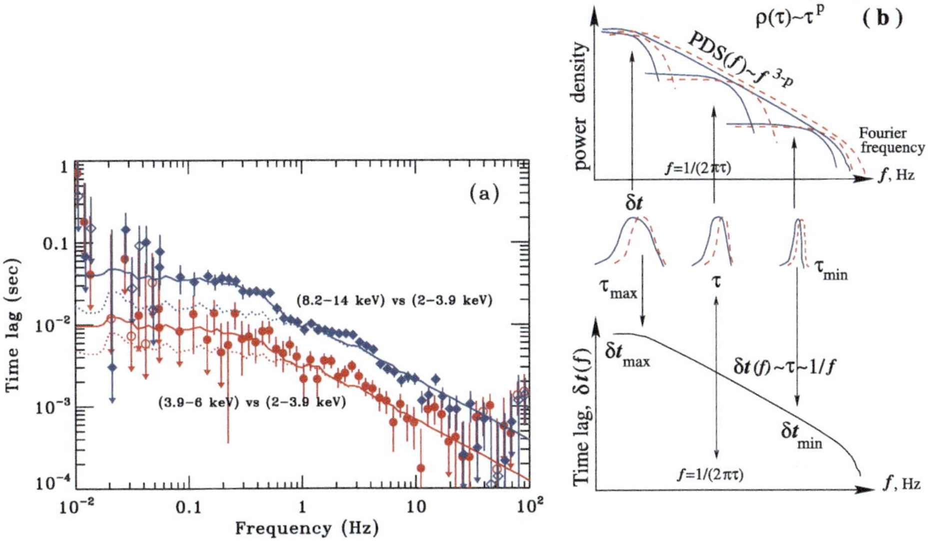

Fig. 11 (**a**) Time lags in Cyg X-1. The data points are from Nowak et al. (1999b). The *dotted curves* represent the model of Poutanen and Fabian (1999), where lags are produces by spectral evolution during flares. The *solid curves* have a contribution at low frequencies from the lags caused by reflection from the outer disc (Poutanen 2002). (**b**) A schematic picture how spectral evolution model produces time lags. The *red dotted curves* represent the flare *light curve* at high photon energies, while the *blue solid curves* represent those at low energies. Spectral hardening during flares produces hard time lags. Here p is the index of the power-law probability distribution of flare duration τ. From Poutanen (2001)

energy release at distances $> 10^4 R_S$), but also contradicts the energy dependence of the ACF width (Maccarone et al. 2000). On the other hand, the large lags, their frequency-dependence f^{-1} and the logarithmic energy dependence can be naturally explained by spectral pivoting of a power-law-like spectrum if the characteristic time-scale of the evolution scales with the duration of shots τ dominating variability at frequency $f \approx 1/(2\pi\tau)$ (see Poutanen and Fabian 1999; Poutanen 2001; Kotov et al. 2001; Körding and Falcke 2004, and Fig. 11b). Among the first physical models explaining the spectral evolution was the flaring magnetic corona model of Poutanen and Fabian (1999). The observed linear relation between the flux and the rms (Uttley and McHardy 2001), however, argues against the independent shots (flares) as the source of variability. In the propagation model of Lyubarskii (1997), spectral evolution can arise when the accretion rate fluctuations propagate towards the BH into the zone with the harder spectra (Miyamoto and Kitamoto 1989; Kotov et al. 2001; Arévalo and Uttley 2006).[3] This is consistent with the multi-zone hot flow model, because a lower flux of soft seed photon from the cold accretion disc as well as a stronger synchrotron self-absorption in the inner part of the flow produce harder Comptonization spectrum (Veledina et al. 2013a).

Additional contribution to the time lags are possible when the intrinsic X-rays from the hot flow are reflected from the distant matter, e.g. outer cold accretion disc (Kotov et al. 2001; Poutanen 2002). Due to the fluorescence and the energy dependence of the photoelectric opacity the contribution of reflection to the total spectrum is energy dependent. Thus the Fe line at 6.4 keV and the Compton reflection bump above 10 keV should show excess

[3]For the time lag production, this model is, however, mathematically identical to the flare evolution model.

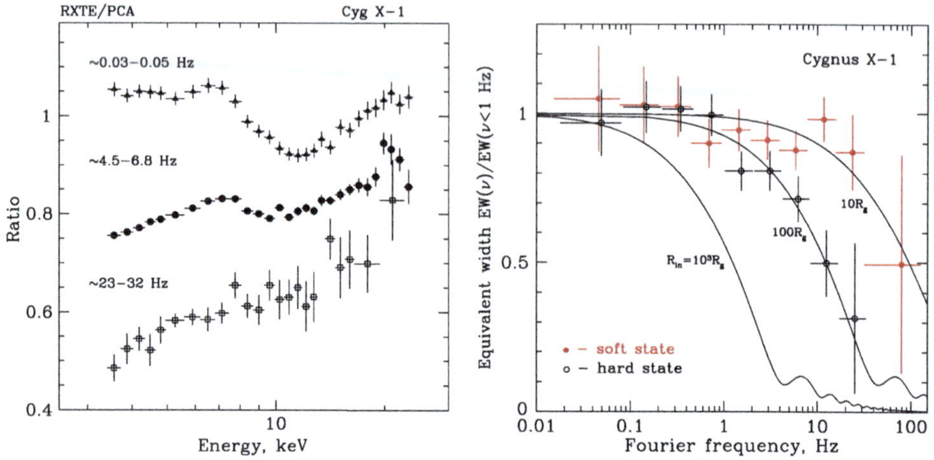

Fig. 12 *Left*: FFRS of Cyg X-1. From Revnivtsev et al. (1999). *Right*: equivalent width of the iron line as a function of Fourier frequency as measured from the FFRS. From Gilfanov et al. (2000)

lags. Surprisingly, the deficit of lags (or "anti-lags") relative to the logarithmic dependence has been observed at these energies (Kotov et al. 2001). Clearly, this deficit cannot be caused by a simple reverberation and the light travel time effect. Instead it could be explained within the hot flow paradigm as follows. The outer parts of the flow which are close to the cold disc have softer spectra and larger reflection amplitude, while the inner flow produces harder spectrum with low or no reflection. Thus in the propagating fluctuation model, the reflection will be *leading* the hard spectrum causing negative delays at photon energies where it contributes (Kotov et al. 2001). Delays due to reprocessing in the inner part of the cold disc are seen at high f in GX 339–4 (Uttley et al. 2011), while at low f the disc photons lead the Comptonized photons, which is consistent with the propagation of fluctuations from the disc to the hot flow.

Further support to the truncated cold disc—hot inner flow scenario comes from the Fourier-frequency-resolved spectra (FFRS; Revnivtsev et al. 1999; Gilfanov et al. 2000), which are softer and have larger reflection amplitude at low Fourier frequencies (see Fig. 9, right and Fig. 12, left). This implies that soft X-rays are mostly produced in the outer zones of the hot flow, closer to the cold reflecting medium, while the hard X-rays are produced in the inner zones that vary at high Fourier frequencies. The reduction of the equivalent width of the 6.4 keV Fe line in the FFRS above 1 Hz (Fig. 12, right) suggests that the cold disc truncation radius in the hard state is about 100 R_S (Revnivtsev et al. 1999; Gilfanov et al. 2000). Even larger truncation radius of the cold disc (300–700 R_S) was measured in the low-extinction BH transient XTE J1118+480 (Chaty et al. 2003; Yuan et al. 2005), where the peak is clearly seen in the UV. In the soft state instead the truncation radius is small ($< 10\ R_S$).

3.4 Optical (IR and UV) Variability and Its Relations to the X-Rays

Data on the fast variability are now available not only in the X-rays, but also at lower energies. The first simultaneous observations in the optical and X-rays were carried out already 30 years ago by Motch et al. (1983) for GX 339–4. Although no confident conclusion could be reached because of the short duration of observations, the optical/X-ray CCF revealed

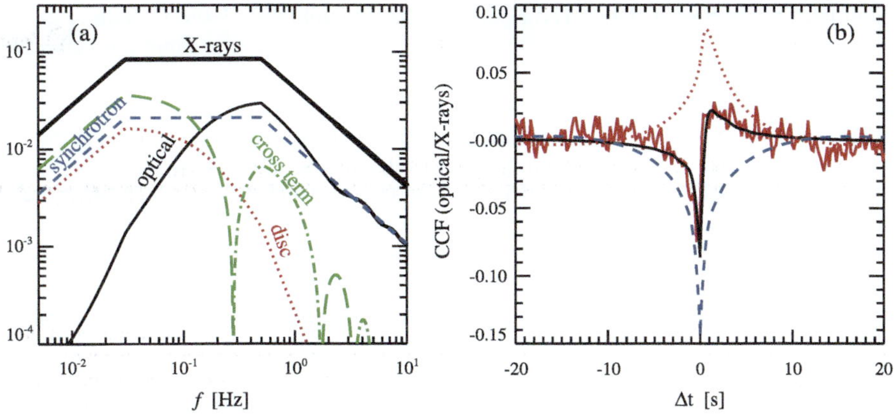

Fig. 13 Two-component model for the optical/X-ray CCF. (**a**) The PDSs of the X-ray (upper double-broken line) and optical light curves (*thin solid line*). Three terms contributing to the optical PDS are also shown: synchrotron (*blue dashed*), reprocessing in the disc (*red dotted*), and the cross term (*green dot-dashed*—positive contribution; *long-dashed*—negative contribution). (**b**) The optical/X-ray CCFs observed from Swift J1753.5–0127 in 2008 (*red noisy curve*, see Fig. A3 in Durant et al. 2011). The model CCF is shown by the *black solid curve*. The smooth lines represent the contributions of the synchrotron emission (*blue dashed*) and the reprocessed emission (*red dotted*). Adapted from Veledina et al. (2011a)

a complicated structure with a precognition dip (i.e. anti-correlation) at negative lags corresponding to optical leading the X-rays and a peak (i.e. correlation) at positive lags (see Fig. 13b). Recently, similar CCFs were obtained from the much longer duration simultaneous observations in three BHBs: XTE J1118+480 (Kanbach et al. 2001; Hynes et al. 2003; Malzac et al. 2003), Swift J1753.5–0127 (Durant et al. 2008, 2009, 2011; Hynes et al. 2009a) and GX 339–4 (Gandhi et al. 2008, 2010). These data provide an important information on the interrelation between various components and give clues to their physical origin.

The observed CCF shape cannot be explained by a simple reprocessing model (Kanbach et al. 2001; Hynes et al. 2003). However, if the optical emission consists of two components, e.g., one coming from the non-thermal synchrotron in the hot flow and another from reprocessed X-ray emission, the complex shape can be reproduced (Veledina et al. 2011a, see Fig. 13b). Increase of the mass accretion rate obviously causes an increase in the X-ray luminosity, but at the same time the optical synchrotron from the hot flow may drop because of higher self-absorption. A higher accretion rate can also lead to a decrease of the truncation radius, collapse of the hot flow at large radii, and the suppression of the OIR emission. Both scenarios leads to anti-correlation between optical and X-rays and to the negative contribution to the CCF, with the shape resembling that of the X-ray ACF. On the other hand, the second, reprocessed component correlates with the X-rays, but is delayed and smeared, giving rise to a positive CCF peaking at positive lags (optical delay). The combined CCF has a complicated shape consistent with the data (Fig. 13b). The PDS of the optical in this model consists of three components: the synchrotron (which has nearly identical shape to the X-ray PDS), the irradiated disc (which has less power at high frequencies because of smearing) and the cross-term of variable sign. The total optical variability is strongly reduced at low frequencies where the synchrotron and the disc vary out of phase (Veledina et al. 2011a, see Fig. 13a). These PDS shapes are very similar to that observed in GX 339–4 (see Fig. 9 in Gandhi et al. 2010).

Further clues on the origin of the optical emission come from the QPOs seen in the light curves of a number of low-mass BHBs (Motch et al. 1983, 1985; Imamura et al. 1990;

Steiman-Cameron et al. 1997; Hynes et al. 2003; Durant et al. 2009; Gandhi et al. 2010). The optical, UV and X-ray QPOs in XTE J1118+480 all have the same frequency, which evolves during the two months of observations (Hynes et al. 2003). There is also a clear connection between the optical and X-ray light curves in the 2007 data on Swift J1753.5–0127 (see Durant et al. 2008; A. Veledina et al. in prep.) seen as a modulation at the optical QPO frequency in the optical/X-ray CCF. Similarly, GX 339–4 shows oscillations at the same frequency in the optical and X-rays (Motch et al. 1983; Gandhi et al. 2010, Fig. 21). The question now arises how it is possible that the X-rays and OIR/UV vary at the same frequency and are phase-connected? This fact finds a simple explanation in the hybrid hot flow model because the hot flow can precess as a solid body (see Sect. 3.3) and therefore the long wavelength emission produced in the outer part of that flow is related to the X-rays produced in the inner part of the same flow (Veledina et al. 2013b).

Recently, periodic eclipses in the optical light curve of BHB Swift J1357.2–093313 were observed (Corral-Santana et al. 2013). If the period is related to the Keplerian frequency of the obscuring region, the sharpness of the eclipses implies that the size of the optical emission region is below 20 000 km (i.e. $< 700 R_S$ for a $10 M_\odot$ BH). These constraints are easily satisfied in the hot flow scenario.

3.5 Polarisation

Polarisation degree and polarisation angle provide two more observational constraints on the emission models. The only indication of the X-ray polarisation from BHB goes back to the OSO-8 satellite (Weisskopf et al. 1977), which measured 3.1 ± 1.7 % linear polarisation from Cyg X-1 at 2.6 keV. Such a polarisation can be produced by Compton scattering if the geometry of the X-ray emitting region is a flattened disc-like structure ($H/R \sim 0.2$ according to the calculations of Lightman and Shapiro 1976), consistent with the hot flow scenario.

Recently, strong linear polarisation ($\Pi = 67 \pm 30$ %) in the soft γ-rays above 400 keV was detected in Cyg X-1 with the IBIS instrument onboard *INTEGRAL* (Laurent et al. 2011). Similar polarisation ($\Pi = 76 \pm 15$ %) was also observed with the SPI spectrometer (Jourdain et al. 2012b). The polarisation angle of 40°–42° is about 60° away from the radio jet axis at $\approx -20°$ (Jourdain et al. 2012b; Zdziarski et al. 2012). Such a large polarisation degree in the MeV range is extremely difficult to get in any scenario. Synchrotron jet emission from non-thermal electrons in a highly ordered magnetic field can have a large polarisation degree (up to ~70 per cent) in the optically thin part of the spectrum, and indeed a high polarisation in the radio and the optical bands reaching 30–50 per cent is observed from extragalactic relativistic jets (Impey et al. 1991; Wills et al. 1992; Lister 2001; Marscher et al. 2002; Ikejiri et al. 2011). However, this scenario also needs a very hard electron spectrum as well as an extreme fine-tuning to reproduce the spectral cutoff at a few MeV (Zdziarski et al. 2012). In the hot-flow scenario, the MeV photons are produced by non-thermal Compton scattering of the 100 keV photons by electrons with Lorentz factors $\gamma \sim 2$–4. These electrons cannot be isotropic, because no significant polarisation is expected in that case (Poutanen 1994). This then implies that they must have nearly one-dimensional motion, e.g. along the large-scale magnetic field lines threading the flow. The 60° offset of the polarisation vector relative to the jet axis then implies the inclined field lines. If the measured high polarisation degree is indeed real, this would put strong constraints on the physics of particle acceleration in the hot flow and the magnetic field geometry.

In the OIR bands, polarisation is very small and does not exceed a few per cent (Schultz et al. 2004; Shahbaz et al. 2008; Russell and Fender 2008; Chaty et al. 2011). In the hot-flow scenario, polarisation degree of the optically thin synchrotron radiation in the OIR band is expected to be essentially zero (independently of the magnetic field geometry) because the Faraday rotation angle exceeds 10^5 rad (Veledina et al. 2013a). In the optically thick regime, the intrinsic polarisation (parallel to the magnetic field lines) is not more than about 10 per cent even for the ordered magnetic field (Pacholczyk and Swihart 1967; Ginzburg and Syrovatskii 1969). Faraday rotation in the disc atmosphere can still essentially depolarise that emission. The reprocessed emission from the outer disc can be slightly polarised because magnetic field there is smaller. If the jet were responsible for the OIR emission, one would expect instead a much higher polarisation because its radiation is optically thin, not consistent with the data. The observed small polarisation can also be produced by dust/electron scattering in the source vicinity or by the interstellar dust.

4 Summary

The purely thermal hot-flow model was shown to be consistent with many X-ray characteristics. However, that model fails to account for the MeV tails and a number of OIR properties. Addition of a small, energetically-negligible non-thermal component to the electron distribution dramatically changes the prediction of the model. The hybrid hot-flow model is now successful in explaining the following facts:

1. stability of spectra with photon index $\Gamma\sim 1.6$–1.8 and the cutoff at ~ 100 keV in the hard state (Poutanen and Vurm 2009; Malzac and Belmont 2009),
2. concave X-ray spectrum (Kotov et al. 2001; Veledina et al. 2013a),
3. low level of the X-ray and OIR polarisation (Veledina et al. 2013a),
4. presence of the MeV tail in the hard state (Poutanen and Vurm 2009; Malzac and Belmont 2009),
5. softening of the X-ray spectrum with decreasing luminosity below $\sim 10^{-2} L_{\rm Edd}$ (Veledina et al. 2011b),
6. weakness of the cold accretion disc component in the hard state,
7. correlation between the spectral index, the reflection amplitude, the width of the iron line and the frequency of the quasi-periodic oscillations,
8. hard X-ray lags with logarithmic energy dependence (Kotov et al. 2001),
9. non-thermal OIR excesses and flat OIR spectra (Veledina et al. 2013a),
10. OIR colours of the flares in the hard state (Poutanen et al., in prep.),
11. strong correlation between OIR and X-ray emission and a complicated CCF shape (Veledina et al. 2011a),
12. quasi-periodic oscillations at the same frequency in the X-ray and optical bands (Veledina et al. 2013b).

The model does not explain the radio points and the soft IR spectra. The jet is obviously a better model for those data. We, however, struggle to find any other observational fact that could be in conflict with the hybrid hot flow—truncated cold disc scenario.

Recently, the jet paradigm became popular and it was claimed that the jets are responsible not only for the radio emission from the BHBs, but also the OIR and even the X-ray emission. Unfortunately, that model is in contradiction with dozens of observed facts (see Veledina et al. 2013a, and references therein), which are usually ignored by the model proponents. When new data appear, they often are rather puzzling and difficult to understand within the available paradigms. However, it would be beneficial for the community when introducing brand new models to check also whether those models satisfy other observational constraints.

In spite of a serious progress in understanding of the viscosity in accretion discs around BHs, there are still many open questions. If non-thermal particles are present in the hot flow, it is now time to understand what is their nature. How are they accelerated: in shocks or in magnetic reconnections events, or maybe via diffusive acceleration by MHD turbulence? How are they related to the magneto-rotational instability that presumably drives the accretion? We hope that the observational advances will soon be reflected in the advance of the theory.

Acknowledgements The work was partially supported by the Academy of Finland grant 268740 (JP) and the Finnish Doctoral Program in Astronomy and Space Physics (AV). We thank Tomaso Belloni, Andrzej Zdziarski, and Feng Yuan for valuable comments and David Russell for the data on XTE J1550–564.

References

V.A. Aref'ev, M.G. Revnivtsev, A.A. Lutovinov, R.A. Sunyaev, Astron. Lett. **30**, 669–674 (2004)
P. Arévalo, P. Uttley, Mon. Not. R. Astron. Soc. **367**, 801–814 (2006)
M. Axelsson, L. Borgonovo, S. Larsson, Astron. Astrophys. **438**, 999–1012 (2005)
R. Belmont, J. Malzac, A. Marcowith, Astron. Astrophys. **491**, 617–631 (2008)
A.M. Beloborodov, in *High Energy Processes in Accreting Black Holes*, ed. by J. Poutanen, R. Svensson. ASP Conf. Ser., vol. 161 (1999a), pp. 295–314
A.M. Beloborodov, Astrophys. J. Lett. **510**, L123–L126 (1999b)
R.D. Blandford, A. Königl, Astrophys. J. **232**, 34–48 (1979)
M.M. Buxton, C.D. Bailyn, Astrophys. J. **615**, 880–886 (2004)
M.M. Buxton, C.D. Bailyn, H.L. Capelo et al., Astron. J. **143**, 130 (2012)
M. Cadolle Bel, M. Ribó, J. Rodriguez et al., Astrophys. J. **659**, 549–560 (2007)
M. Cadolle Bel, J. Rodriguez, P. D'Avanzo et al., Astron. Astrophys. **534**, 119 (2011)
M. Cadolle Bel, S. Corbel, A. Veledina et al., in *IAU Symp*, ed. by C.M. Zhang, T. Belloni, M. Méndez, S.N. Zhang. IAU Symposium, vol. 290 (2013), pp. 17–20
P. Casella, T.J. Maccarone, K. O'Brien et al., Mon. Not. R. Astron. Soc. **404**, 21–25 (2010)
A. Celotti, A.C. Fabian, M.J. Rees, Mon. Not. R. Astron. Soc. **255**, 419–422 (1992)
S. Chaty, C.A. Haswell, J. Malzac et al., Mon. Not. R. Astron. Soc. **346**, 689–703 (2003)
S. Chaty, G. Dubus, A. Raichoor, Astron. Astrophys. **529**, 3 (2011)
C.Y. Chiang, C. Done, M. Still, O. Godet, Mon. Not. R. Astron. Soc. **403**, 1102–1112 (2010)
A. Constantin, P. Green, T. Aldcroft et al., Astrophys. J. **705**, 1336–1355 (2009)
P.S. Coppi, Mon. Not. R. Astron. Soc. **258**, 657–683 (1992)
P.S. Coppi, in *High Energy Processes in Accreting Black Holes*, ed. by J. Poutanen, R. Svensson. ASP Conf. Ser., vol. 161 (1999), pp. 375–403
S. Corbel, H. Aussel, J.W. Broderick et al., Mon. Not. R. Astron. Soc. **431**, 107–111 (2013)
J.M. Corral-Santana, J. Casares, T. Muñoz-Darias et al., Science **339**, 1048–1051 (2013)
C. Cunningham, Astrophys. J. **208**, 534–549 (1976)
S. Das, P.A. Becker, T. Le, Astrophys. J. **702**, 649–659 (2009)
T. Dinçer, E. Kalemci, M.M. Buxton et al., Astrophys. J. **753**, 55 (2012)
J. Ding, F. Yuan, E. Liang, Astrophys. J. **708**, 1545–1550 (2010)
C. Done, in *Accretion Processes in Astrophysics*, ed. by I. González Martínez-País, T. Shahbaz, J. Casares Velázquez (Cambridge University Press, Cambridge, 2013). arXiv:1008.2287
C. Done, M. Gierliński, A. Kubota, Astron. Astrophys. Rev. **15**, 1–66 (2007)
R. Droulans, R. Belmont, J. Malzac, E. Jourdain, Astrophys. J. **717**, 1022–1036 (2010)
M. Durant, P. Gandhi, T. Shahbaz et al., Astrophys. J. Lett. **682**, L45–L48 (2008)
M. Durant, P. Gandhi, T. Shahbaz et al., Mon. Not. R. Astron. Soc. **392**, 309–324 (2009)
M. Durant, T. Shahbaz, P. Gandhi et al., Mon. Not. R. Astron. Soc. **410**, 2329–2338 (2011)
A.A. Esin, J.E. McClintock, R. Narayan, Astrophys. J. **489**, 865–889 (1997)
A.A. Esin, R. Narayan, W. Cui et al., Astrophys. J. **505**, 854–868 (1998)
A.A. Esin, J.E. McClintock, J.J. Drake et al., Astrophys. J. **555**, 483–488 (2001)
R.P. Fender, G.G. Pooley, C. Brocksopp, S.J. Newell, Mon. Not. R. Astron. Soc. **290**, 65–69 (1997)
E.L. Fitzpatrick, Publ. Astron. Soc. Pac. **111**, 63–75 (1999)
P.C. Fragile, O.M. Blaes, Astrophys. J. **687**, 757–766 (2008)
P.C. Fragile, O.M. Blaes, P. Anninos, J.D. Salmonson, Astrophys. J. **668**, 417–429 (2007)

F. Frontera, E. Palazzi, A.A. Zdziarski et al., Astrophys. J. **546**, 1027–1037 (2001)
A.A. Galeev, R. Rosner, G.S. Vaiana, Astrophys. J. **229**, 318–326 (1979)
E. Gallo, S. Migliari, S. Markoff et al., Astrophys. J. **670**, 600–609 (2007)
P. Gandhi, K. Makishima, M. Durant et al., Mon. Not. R. Astron. Soc. **390**, L29–L33 (2008)
P. Gandhi, V.S. Dhillon, M. Durant et al., Mon. Not. R. Astron. Soc. **407**, 2166–2192 (2010)
P. Gandhi, A.W. Blain, D.M. Russell et al., Astrophys. J. Lett. **740**, L13 (2011)
G. Ghisellini, P.W. Guilbert, R. Svensson, Astrophys. J. Lett. **334**, L5–L8 (1988)
G. Ghisellini, F. Haardt, R. Svensson, Mon. Not. R. Astron. Soc. **297**, 348–354 (1998)
M. Gierlinski, A.A. Zdziarski, C. Done et al., Mon. Not. R. Astron. Soc. **288**, 958–964 (1997)
M. Gierliński, A.A. Zdziarski, J. Poutanen et al., Mon. Not. R. Astron. Soc. **309**, 496–512 (1999)
M. Gierliński, C. Done, K. Page, Mon. Not. R. Astron. Soc. **392**, 1106–1114 (2009)
M. Gilfanov, in *The Jet Paradigm*, ed. by T. Belloni. Lecture Notes in Physics, vol. 794 (2010), pp. 17–51
M. Gilfanov, E. Churazov, M. Revnivtsev, Astron. Astrophys. **352**, 182–188 (1999)
M. Gilfanov, E. Churazov, M. Revnivtsev, Mon. Not. R. Astron. Soc. **316**, 923–928 (2000)
V.L. Ginzburg, S.I. Syrovatskii, Annu. Rev. Astron. Astrophys. **7**, 375–420 (1969)
J.E. Grove, W.N. Johnson, R.A. Kroeger et al., Astrophys. J. **500**, 899–908 (1998)
M. Gu, X. Cao, Mon. Not. R. Astron. Soc. **399**, 349–356 (2009)
F. Haardt, L. Maraschi, Astrophys. J. **413**, 507–517 (1993)
F. Haardt, L. Maraschi, G. Ghisellini, Astrophys. J. Lett. **432**, L95–L99 (1994)
K.B. Henisey, O.M. Blaes, P.C. Fragile, Astrophys. J. **761**, 18 (2012)
M. Hoshino, Astrophys. J. **773**, 118 (2013)
R.I. Hynes, K. O'Brien, K. Horne et al., Mon. Not. R. Astron. Soc. **299**, 37–41 (1998)
R.I. Hynes, C.W. Mauche, C.A. Haswell et al., Astrophys. J. Lett. **539**, L37–L40 (2000)
R.I. Hynes, C.A. Haswell, S. Chaty et al., Mon. Not. R. Astron. Soc. **331**, 169–179 (2002)
R.I. Hynes, C.A. Haswell, W. Cui et al., Mon. Not. R. Astron. Soc. **345**, 292–310 (2003)
R.I. Hynes, E.L. Robinson, K.J. Pearson et al., Astrophys. J. **651**, 401–407 (2006)
R.I. Hynes, C.K. Bradley, M. Rupen et al., Mon. Not. R. Astron. Soc. **399**, 2239–2248 (2009b)
R.I. Hynes, K. O'Brien, F. Mullally, T. Ashcraft, Mon. Not. R. Astron. Soc. **399**, 281–286 (2009a)
A. Ibragimov, J. Poutanen, M. Gilfanov et al., Mon. Not. R. Astron. Soc. **362**, 1435–1450 (2005)
S. Ichimaru, Astrophys. J. **214**, 840–855 (1977)
Y. Ikejiri, M. Uemura, M. Sasada et al., Publ. Astron. Soc. Jpn. **63**, 639–675 (2011)
J.N. Imamura, J. Kristian, J. Middleditch, T.Y. Steiman-Cameron, Astrophys. J. **365**, 312–316 (1990)
C.D. Impey, C.R. Lawrence, S. Tapia, Astrophys. J. **375**, 46–68 (1991)
A. Ingram, C. Done, Mon. Not. R. Astron. Soc. **415**, 2323–2335 (2011)
A. Ingram, C. Done, P.C. Fragile, Mon. Not. R. Astron. Soc. **397**, 101–105 (2009)
R.K. Jain, C.D. Bailyn, J.A. Orosz et al., Astrophys. J. Lett. **554**, L181–L184 (2001)
E. Jourdain, J.P. Roques, M. Chauvin, D.J. Clark, Astrophys. J. **761**, 27 (2012b)
E. Jourdain, J.P. Roques, J. Malzac, Astrophys. J. **744**, 64 (2012a)
E. Kalemci, J.A. Tomsick, M.M. Buxton et al., Astrophys. J. **622**, 508–519 (2005)
E. Kalemci, T. Dincer, J.A. Tomsick et al., Astrophys. J. **779**, 95 (2013). doi:10.1088/0004-637X/779/2/95
G. Kanbach, C. Straubmeier, H.C. Spruit, T. Belloni, Nature **414**, 180–182 (2001)
D. Kazanas, X.-M. Hua, L. Titarchuk, Astrophys. J. **480**, 735–740 (1997)
M. Kolehmainen, C. Done, M. Díaz Trigo, Mon. Not. R. Astron. Soc. **437**, 316–326 (2014). doi:10.1093/mnras/stt1886
E. Körding, H. Falcke, Astron. Astrophys. **414**, 795–806 (2004)
O. Kotov, E. Churazov, M. Gilfanov, Mon. Not. R. Astron. Soc. **327**, 799–807 (2001)
J.H. Krolik, Astrophys. J. Lett. **498**, L13–L16 (1998)
P. Laurent, J. Rodriguez, J. Wilms et al., Science **332**, 438 (2011)
A.P. Lightman, S.L. Shapiro, Astrophys. J. **203**, 701–703 (1976)
J.C. Ling, W.A. Wheaton, P. Wallyn et al., Astrophys. J. **484**, 375–382 (1997)
M.L. Lister, Astrophys. J. **562**, 208–232 (2001)
Y.E. Lyubarskii, Mon. Not. R. Astron. Soc. **292**, 679–685 (1997)
T.J. Maccarone, P.S. Coppi, J. Poutanen, Astrophys. J. Lett. **537**, L107–L110 (2000)
R. Mahadevan, Nature **394**, 651–653 (1998)
K. Makishima, H. Takahashi, S. Yamada et al., Publ. Astron. Soc. Jpn. **60**, 585–604 (2008)
J. Malzac, R. Belmont, Mon. Not. R. Astron. Soc. **392**, 570–589 (2009)
J. Malzac, A.M. Beloborodov, J. Poutanen, Mon. Not. R. Astron. Soc. **326**, 417–427 (2001)
J. Malzac, T. Belloni, H.C. Spruit, G. Kanbach, Astron. Astrophys. **407**, 335–345 (2003)
A.P. Marscher, Astrophys. J. **216**, 244–256 (1977)
A.P. Marscher, S.G. Jorstad, J.R. Mattox, A.E. Wehrle, Astrophys. J. **577**, 85–97 (2002)
J.E. McClintock, C.A. Haswell, M.R. Garcia et al., Astrophys. J. **555**, 477–482 (2001)

J.E. McClintock, R. Narayan, J.F. Steiner, Space Sci. Rev. (2013). doi:10.1007/s11214-013-0003-9
M. McConnell, D. Forrest, J. Ryan et al., Astrophys. J. **424**, 933–939 (1994)
M.L. McConnell, A.A. Zdziarski, K. Bennett et al., Astrophys. J. **572**, 984–995 (2002)
S. Miyamoto, S. Kitamoto, Nature **342**, 773 (1989)
C. Motch, M.J. Ricketts, C.G. Page et al., Astron. Astrophys. **119**, 171–176 (1983)
C. Motch, S.A. Ilovaisky, C. Chevalier, P. Angebault, Space Sci. Rev. **40**, 219–224 (1985)
M.P. Muno, J. Mauerhan, Astrophys. J. Lett. **648**, L135–L138 (2006)
R. Narayan, I. Yi, Astrophys. J. Lett. **428**, L13–L16 (1994)
R. Narayan, I. Yi, Astrophys. J. **452**, 710–735 (1995)
R. Narayan, R. Mahadevan, E. Quataert, in *Theory of Black Hole Accretion Disks*, ed. by M.A. Abramowicz, G. Bjornsson, J.E. Pringle (Cambridge University Press, Cambridge, 1998), pp. 148–182
S. Nayakshin, F. Melia, Astrophys. J. Suppl. Ser. **114**, 269–288 (1998)
A. Niedźwiecki, F.-G. Xie, A.A. Zdziarski, Mon. Not. R. Astron. Soc. **420**, 1195–1206 (2012)
P.L. Nolan, D.E. Gruber, J.L. Matteson et al., Astrophys. J. **246**, 494–501 (1981)
D.I. Novikov, K.S. Thorne, in *Black Holes (Les Astres Occlus)*, ed. by C. Dewitt, B.S. Dewitt (Gordon & Breach, New York, 1973), pp. 343–450
M.A. Nowak, Mon. Not. R. Astron. Soc. **318**, 361–367 (2000)
M.A. Nowak, B.A. Vaughan, J. Wilms et al., Astrophys. J. **510**, 874–891 (1999b)
M.A. Nowak, J. Wilms, J.B. Dove, Astrophys. J. **517**, 355–366 (1999a)
K. O'Brien, K. Horne, R.I. Hynes et al., Mon. Not. R. Astron. Soc. **334**, 426–434 (2002)
J.A. Orosz, J.F. Steiner, J.E. McClintock et al., Astrophys. J. **730**, 75 (2011)
F. Özel, D. Psaltis, R. Narayan, Astrophys. J. **541**, 234–249 (2000)
A.G. Pacholczyk, T.L. Swihart, Astrophys. J. **150**, 647–650 (1967)
J. Poutanen, Astrophys. J. Suppl. Ser. **92**, 607–609 (1994)
J. Poutanen, in *Theory of Black Hole Accretion Disks*, ed. by M.A. Abramowicz, G. Björnsson, J.E. Pringle (Cambridge University Press, Cambridge, 1998), pp. 100–122
J. Poutanen, Adv. Space Res. **28**, 267–280 (2001)
J. Poutanen, Mon. Not. R. Astron. Soc. **332**, 257–270 (2002)
J. Poutanen, P.S. Coppi, Phys. Scr. T **77**, 57–60 (1998)
J. Poutanen, A.C. Fabian, Mon. Not. R. Astron. Soc. **306**, L31–L37 (1999)
J. Poutanen, R. Svensson, Astrophys. J. **470**, 249–268 (1996)
J. Poutanen, I. Vurm, Astrophys. J. Lett. **690**, L97–L100 (2009)
J. Poutanen, J.H. Krolik, F. Ryde, Mon. Not. R. Astron. Soc. **292**, L21–L25 (1997)
W. Priedhorsky, G.P. Garmire, R. Rothschild et al., Astrophys. J. **233**, 350–363 (1979)
F. Rahoui, M. Coriat, S. Corbel et al., Mon. Not. R. Astron. Soc. **422**, 2202–2212 (2012)
R.A. Remillard, J.E. McClintock, Annu. Rev. Astron. Astrophys. **44**, 49–92 (2006)
M. Revnivtsev, M. Gilfanov, E. Churazov, Astron. Astrophys. **347**, 23–26 (1999)
M. Revnivtsev, M. Gilfanov, E. Churazov, Astron. Astrophys. **380**, 520–525 (2001)
M.A. Riquelme, E. Quataert, P. Sharma, A. Spitkovsky, Astrophys. J. **755**, 50 (2012)
D.M. Russell, R.P. Fender, Mon. Not. R. Astron. Soc. **387**, 713–723 (2008)
D.M. Russell, D. Maitra, R.J.H. Dunn, S. Markoff, Mon. Not. R. Astron. Soc. **405**, 1759–1769 (2010)
D.M. Russell, D. Maitra, R.J.H. Dunn, R.P. Fender, Mon. Not. R. Astron. Soc. **416**, 2311–2317 (2011)
D.M. Russell, P.A. Curran, T. Muñoz-Darias et al., Mon. Not. R. Astron. Soc. **419**, 1740–1751 (2012)
D.M. Russell, T.D. Russell, J.C.A. Miller-Jones et al., Astrophys. J. Lett. **768**, L35 (2013)
J.D. Schnittman, J. Homan, J.M. Miller, Astrophys. J. **642**, 420–426 (2006)
J. Schultz, P. Hakala, J. Huovelin, Balt. Astron. **13**, 581–595 (2004)
T. Shahbaz, R.P. Fender, C.A. Watson, K. O'Brien, Astrophys. J. **672**, 510–515 (2008)
N.I. Shakura, R.A. Sunyaev, Astron. Astrophys. **24**, 337–355 (1973)
S.L. Shapiro, A.P. Lightman, D.M. Eardley, Astrophys. J. **204**, 187–199 (1976)
M. Shidatsu, Y. Ueda, F. Tazaki et al., Publ. Astron. Soc. Jpn. **63**, 785 (2011)
M.A. Sobolewska, I.E. Papadakis, C. Done, J. Malzac, Mon. Not. R. Astron. Soc. **417**, 280–288 (2011)
T.Y. Steiman-Cameron, J.D. Scargle, J.N. Imamura, J. Middleditch, Astrophys. J. **487**, 396–401 (1997)
L. Stella, M. Vietri, Astrophys. J. Lett. **492**, L59–L62 (1998)
B.E. Stern, M.C. Begelman, M. Sikora, R. Svensson, Mon. Not. R. Astron. Soc. **272**, 291–307 (1995a)
B.E. Stern, J. Poutanen, R. Svensson et al., Astrophys. J. Lett. **449**, L13–L17 (1995b)
P. Uttley, I.M. McHardy, Mon. Not. R. Astron. Soc. **323**, L26–L30 (2001)
P. Uttley, T. Wilkinson, P. Cassatella et al., Mon. Not. R. Astron. Soc. **414**, 60–64 (2011)
A. Veledina, J. Poutanen, I. Vurm, Astrophys. J. Lett. **737**, L17 (2011a)
A. Veledina, I. Vurm, J. Poutanen, Mon. Not. R. Astron. Soc. **414**, 3330–3343 (2011b)
A. Veledina, J. Poutanen, A. Ingram, Astrophys. J. **778**, 165 (2013b)
A. Veledina, J. Poutanen, I. Vurm, Mon. Not. R. Astron. Soc. **430**, 3196–3212 (2013a)

I. Vurm, J. Poutanen, Astrophys. J. **698**, 293–316 (2009)
R.V. Wagoner, A.S. Silbergleit, M. Ortega-Rodríguez, Astrophys. J. Lett. **559**, L25–L28 (2001)
G. Wardziński, A.A. Zdziarski, Mon. Not. R. Astron. Soc. **325**, 963–971 (2001)
M.C. Weisskopf, E.H. Silver, H.L. Kestenbaum et al., Astrophys. J. Lett. **215**, L65–L68 (1977)
B.J. Wills, D. Wills, N.J. Evans II et al., Astrophys. J. **400**, 96–114 (1992)
Q. Wu, M. Gu, Astrophys. J. **682**, 212–217 (2008)
F.-G. Xie, A. Niedźwiecki, A.A. Zdziarski, F. Yuan, Mon. Not. R. Astron. Soc. **403**, 170–178 (2010)
S. Yamada, K. Makishima, C. Done et al., Publ. Astron. Soc. Jpn. **65**, 80 (2013)
Z. Yan, W. Yu, Mon. Not. R. Astron. Soc. **427**, 11–15 (2012)
F. Yuan, R. Narayan, Annu. Rev. Astron. Astrophys. (2014, submitted)
F. Yuan, A.A. Zdziarski, Mon. Not. R. Astron. Soc. **354**, 953–960 (2004)
F. Yuan, E. Quataert, R. Narayan, Astrophys. J. **598**, 301–312 (2003)
F. Yuan, W. Cui, R. Narayan, Astrophys. J. **620**, 905–914 (2005)
F. Yuan, A.A. Zdziarski, Y. Xue, X.-B. Wu, Astrophys. J. **659**, 541–548 (2007)
A.A. Zdziarski, M. Gierliński, Prog. Theor. Phys. Suppl. **155**, 99–119 (2004)
A.A. Zdziarski, W.N. Johnson, J. Poutanen et al., in *ESA SP-382: The Transparent Universe*, ed. by C. Winkler, T.J.-L. Courvoisier, P. Durouchoux (ESA, Noordwijk, 1997), pp. 373–380
A.A. Zdziarski, J. Poutanen, J. Mikolajewska et al., Mon. Not. R. Astron. Soc. **301**, 435–450 (1998)
A.A. Zdziarski, P. Lubinski, D.A. Smith, Mon. Not. R. Astron. Soc. **303**, L11–L15 (1999)
A.A. Zdziarski, J.E. Grove, J. Poutanen, A.R. Rao, S.V. Vadawale, Astrophys. J. Lett. **554**, L45–L48 (2001)
A.A. Zdziarski, P. Lubiński, M. Gilfanov, M. Revnivtsev, Mon. Not. R. Astron. Soc. **342**, 355–372 (2003)
A.A. Zdziarski, M. Gierliński, J. Mikołajewska et al., Mon. Not. R. Astron. Soc. **351**, 791–807 (2004)
A.A. Zdziarski, P. Lubiński, M. Sikora, Mon. Not. R. Astron. Soc. **423**, 663–675 (2012)

DOI 10.1007/978-1-4939-2227-7_6
Reprinted from *Space Science Reviews* Journal, DOI 10.1007/s11214-013-9986-5

Current Status of Simulations

P. Chris Fragile

Received: 21 January 2013 / Accepted: 20 April 2013 / Published online: 9 May 2013

Abstract As the title suggests, the purpose of this chapter is to review the current status of numerical simulations of black hole accretion disks. This chapter focuses exclusively on *global* simulations of the accretion process within a few tens of gravitational radii of the black hole. Most of the simulations discussed are performed using general relativistic magnetohydrodynamic (MHD) schemes, although some mention is made of Newtonian radiation MHD simulations and smoothed particle hydrodynamics. The goal is to convey some of the exciting work that has been going on in the past few years and provide some speculation on future directions.

Keywords Accretion, accretion disks · Black hole physics · Magnetohydrodynamics (MHD) · Methods: numerical

1 Introduction

Going hand-in-glove with analytic models of accretion disks, discussed in Chap. 2.1, are direct numerical simulations. Although analytic theories have been extremely successful at explaining many general observational properties of black hole accretion disks, numerical simulations have become an indispensable tool in advancing this field. They allow one to explore the full, non-linear evolution of accretion disks from a first-principles perspective. Because numerical simulations can be tuned to a variety of parameters, they serve as a sort of "laboratory" for astrophysics.

The last decade has been an exciting time for black hole accretion disk simulations, as the fidelity has become sufficient to make genuine comparisons between them and real observations. The prospects are also good that within the next decade, we will be able to include the full physics (gravity + hydrodynamics + magnetic fields + radiation) within these simulations, which will yield complete and accurate numerical representations of the accretion process. In the rest of this chapter I will review some of the most recent highlights from this field.

P.C. Fragile (✉)
Department of Physics & Astronomy, College of Charleston, Charleston, SC 29424, USA
e-mail: fragilep@cofc.edu

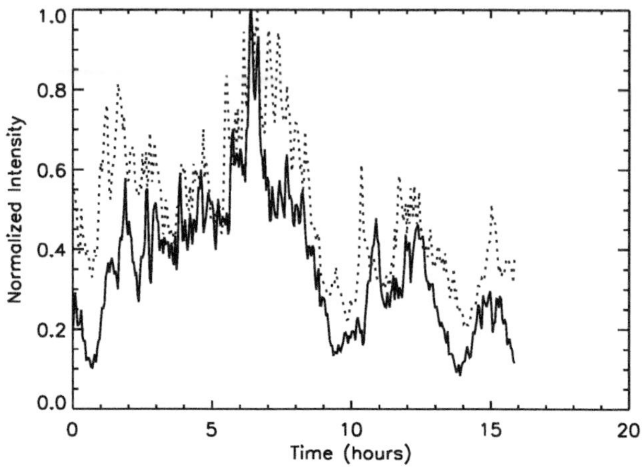

Fig. 1 Model light curve at 0.4 mm (*solid*) and accretion rate (*dotted*) at the inner boundary of a simulation from McKinney and Blandford (2009). Both quantities are scaled to their maximum value. These are compared to data from Sgr A* in Dexter et al. (2010)

2 Matching Simulations with Observations

One of the most exciting recent trends has been a concerted effort by various collaborations to make direct connections between very sophisticated simulations and observations. Of course, observers have been clamoring for this sort of comparison for years!

Perhaps the first serious attempt at this was presented in Schnittman et al. (2006). Schnittman produced a simulation similar to those in De Villiers et al. (2003) and coupled it with a ray-tracing and radiative transfer code to produce "images" of what the simulated disk would look like to a distant observer. By creating images from many time dumps in the simulation, Schnittman was able to create light curves, which were then analyzed for variability properties much the way real light curves are.

Following that same prescription, a number of groups have now presented results coupling general relativistic MHD (GRMHD) simulations with radiative modeling and ray-tracing codes (e.g. Noble et al. 2007; Mościbrodzka et al. 2009; Dexter et al. 2009, 2010). More recent models have even included polarization measurements (Shcherbakov et al. 2012). This approach is most applicable to very low-luminosity systems, such as Sgr A* and M87. A sample light curve for Sgr A* covering a 16-hour window is shown in Fig. 1. In the case of M87, modeling has focused on accounting for the prominent jet in that system (Mościbrodzka et al. 2011; Dexter et al. 2012).

Along with modeling light curves and variability, this approach can also be used to create synthetic broadband spectra from simulations (e.g. Mościbrodzka et al. 2009; Drappeau et al. 2013), which can be compared with modern multi-wavelength observing campaigns (see Chap. 3.1). This is very useful for connecting different components of the spectra to different regions of the simulation domain. For example, Fig. 2 shows that the sub-mm bump in Sgr A* is well represented by emission from relatively cool, high-density gas orbiting close to the black hole, while the X-ray emission seems to come from Comptonization by very hot electrons in the highly magnetized regions of the black hole magnetosphere or base of the jet.

3 Thermodynamics of Simulations

As important as the radiative modeling of simulations described in Sect. 2 has been, its application is very limited. This is because, in most cases, the radiative modeling has been

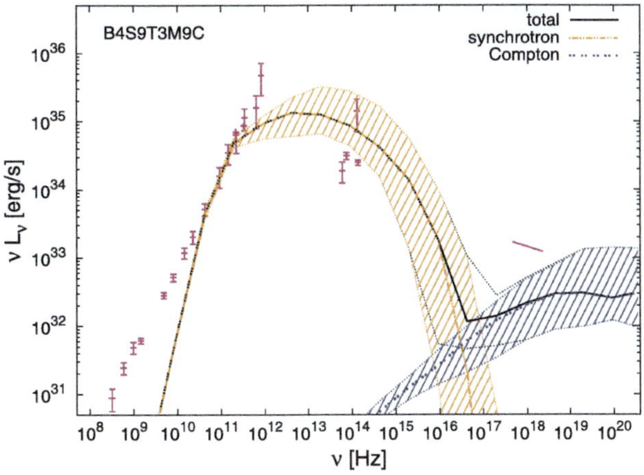

Fig. 2 Synthetic broadband spectrum created from one of the simulations presented in Dibi et al. (2012). The *pink points* represent a compilation of Sgr A* observations. Figure from Drappeau et al. (2013)

done after the fact; it was not included in the simulations themselves. Therefore, the gas in the accretion disk was not allowed to respond thermodynamically to the cooling. This calls into question how much the structure obtained from the simulation reflects the true structure of the disk. Fortunately, various groups are beginning to work on treating the thermodynamics of accretion disks within the numerical simulations with greater fidelity. Thus far, two approaches have principally been explored: (1) *ad hoc* cooling prescriptions used to artificially create *optically thick, geometrically thin* disks and (2) fully self-consistent treatments of radiative cooling for *optically thin, geometrically thick* disks. We review each of these in the next 2 sections.

3.1 Geometrically Thin Disks

For the *ad hoc* cooling prescription, cooling is assumed to equal heating (approximately) everywhere locally in the disk. Since this is the same assumption as is made in the Shakura-Sunyaev (Shakura and Sunyaev 1973) and Novikov-Thorne (Novikov and Thorne 1973) disk models, this approach has proven quite useful in testing the key assumptions inherent in these models (e.g. Shafee et al. 2008; Noble et al. 2009, 2010; Penna et al. 2010). In particular, these simulations have been useful for testing the assumption that the stress within the disk goes to zero at the innermost stable circular orbit (ISCO). A corollary to this is that the specific angular momentum of the gas must remain constant at its ISCO value inside this radius. Both of these effects have been confirmed in simulations of sufficiently thin disks (Penna et al. 2010), as shown in Fig. 3.

3.2 Self-Consistent Radiative Cooling of Optically Thin Disks

Another approach to treating the thermodynamics of accretion disks has been to include *physical* radiative cooling processes directly within the simulations. So far there has been very limited work done on this for optically thick disks, but an optically-thin treatment was introduced in Fragile and Meier (2009). Similar to the after-the-fact radiative modeling described in Sect. 2, the optically-thin requirement restricts the applicability of this approach to relatively low luminosity systems, such as the Quiescent and Low/Hard states of black hole X-ray binaries.

Recently this approach has been applied to Sgr A* (Drappeau et al. 2013), which turns out to be right on the boundary between where after-the-fact radiative modeling breaks down

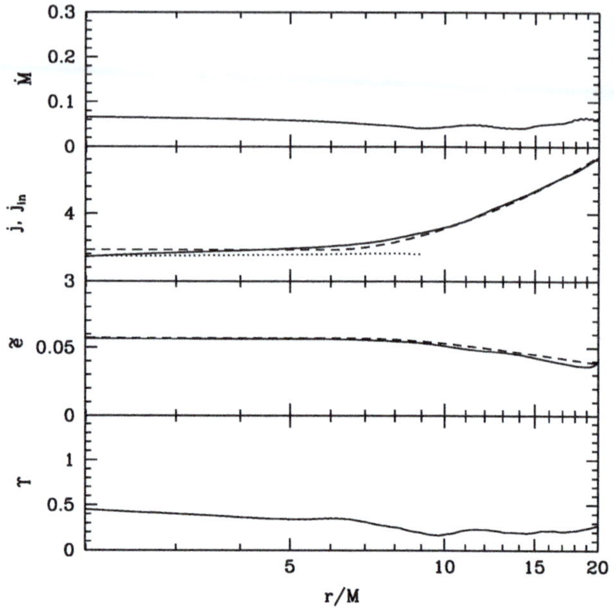

Fig. 3 Various fluxes as functions of radius for a numerical Novikov–Thorne disk simulation. *Top:* Mass accretion rate. *Second panel:* Accreted specific angular momentum. *Solid line* is simulation data; *dashed line* gives Novikov–Thorne solution; *dotted line* is ISCO value. Note that the specific angular momentum does not drop significantly inside the ISCO. *Third panel:* The "nominal" efficiency, which is the total loss of specific energy from the fluid. *Bottom panel:* Specific magnetic flux. The near constancy of this quantity inside the ISCO is an indication that magnetic stresses are not significant in this region. Figure from Penna et al. (2010)

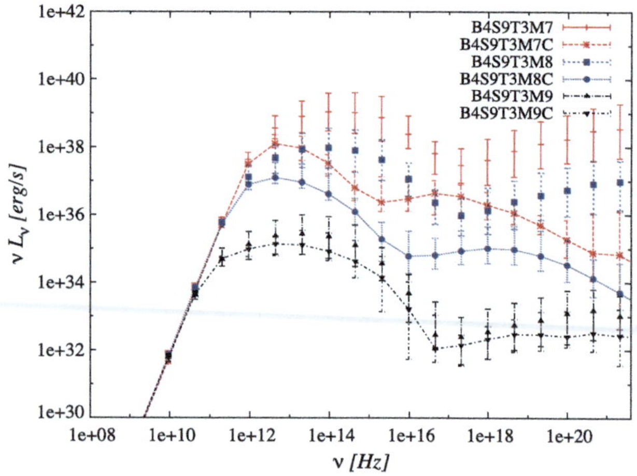

Fig. 4 Comparison of sample spectra generated for Sgr A* from numerical simulations at three different accretion rates: 10^{-9} (*black*), 10^{-8} (*blue*), and $10^{-7} M_\odot$ yr^{-1} (*red*). For each accretion rate, two simulations are shown, one that includes cooling self-consistently (model names ending in "C") and one that does not. The spectra begin to diverge noticeably at $\dot{M} \approx 10^{-8} M_\odot$ yr^{-1}. Figure from Dibi et al. (2012)

and a self-consistent treatment becomes necessary (Dibi et al. 2012). Figure 4 illustrates that this transition occurs right around an accretion rate of $\dot{M} \approx 10^{-8} M_\odot$ yr^{-1} for Sgr A*.

4 Magnetic Field Topology

Another area where a lot of interesting new results have come out is in the study of how magnetic field topology and strength affect black hole accretion.

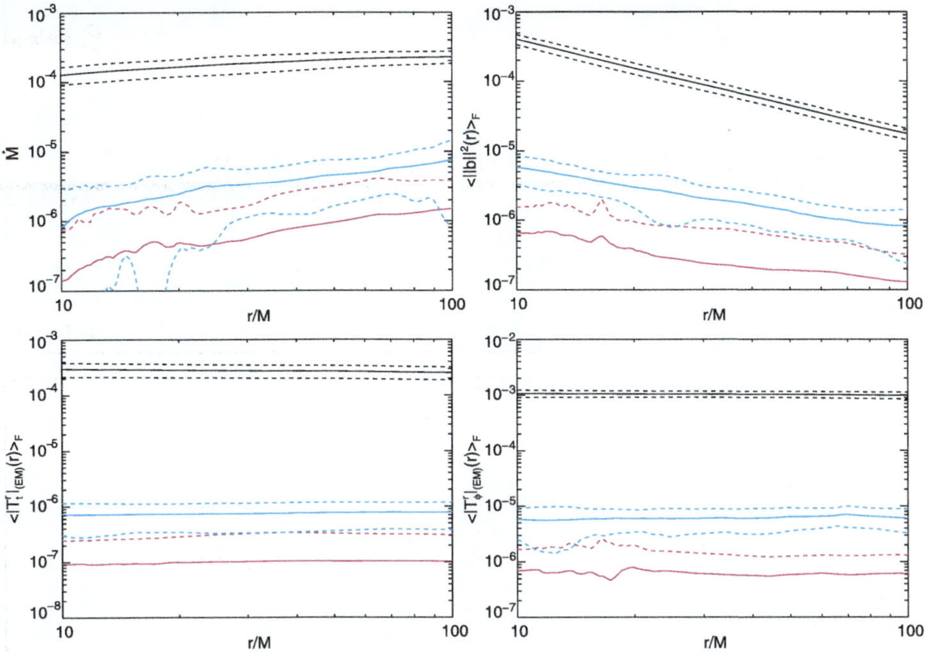

Fig. 5 Time-averaged shell integrals of data from unbound outflows as a function of radius for 3 models from Beckwith et al. (2008): a dipole magnetic field model (*black*), a quadrupole model (*cyan*), and a toroidal field model (*magenta*). Shown are mass outflow rate $\dot{M}$ (*top left*), magnetic field strength $\|b^2\|$ (*top right*), electromagnetic energy flux $|T^r_t|_{\mathrm{EM}}$ (*bottom left*), and angular momentum flux $|T^r_\phi|_{\mathrm{EM}}$ (*bottom right*). *Dashed lines* show ± 1 standard deviation from the average

4.1 Jet Power

Although there is now convincing evidence that the Blandford-Znajek mechanism (Blandford and Znajek 1977) works as predicted in powering jets (e.g. Komissarov 2001; McKinney and Gammie 2004), one lingering question is still how the accretion process supplies the required poloidal flux onto the black hole. Simulations have demonstrated that such field can, in many cases, be generated self-consistently within MRI-unstable disks (e.g. De Villiers et al. 2005; Hawley and Krolik 2006; McKinney and Gammie 2004; McKinney 2006). However, this is strongly dependent on the initial magnetic field topology, as shown in Beckwith et al. (2008). Figure 5 nicely illustrates that when there is no net poloidal magnetic flux threading the inner disk, the magnetically-driven jet can be 2 orders of magnitude less energetic than when there is. At this time it is unclear what the "natural" field topology would be, or even if there is one.

4.2 Magnetically Arrested Accretion

Although strong poloidal magnetic fields are useful for driving powerful jets, they can also create interesting feedback affects on an accretion disk. In the case where a black hole is able to accumulate field with a consistent net flux for an extended period of time, it is possible for the amassed field to eventually "arrest" the accretion process (Narayan et al. 2003). An example of an arrested state is shown in Fig. 6.

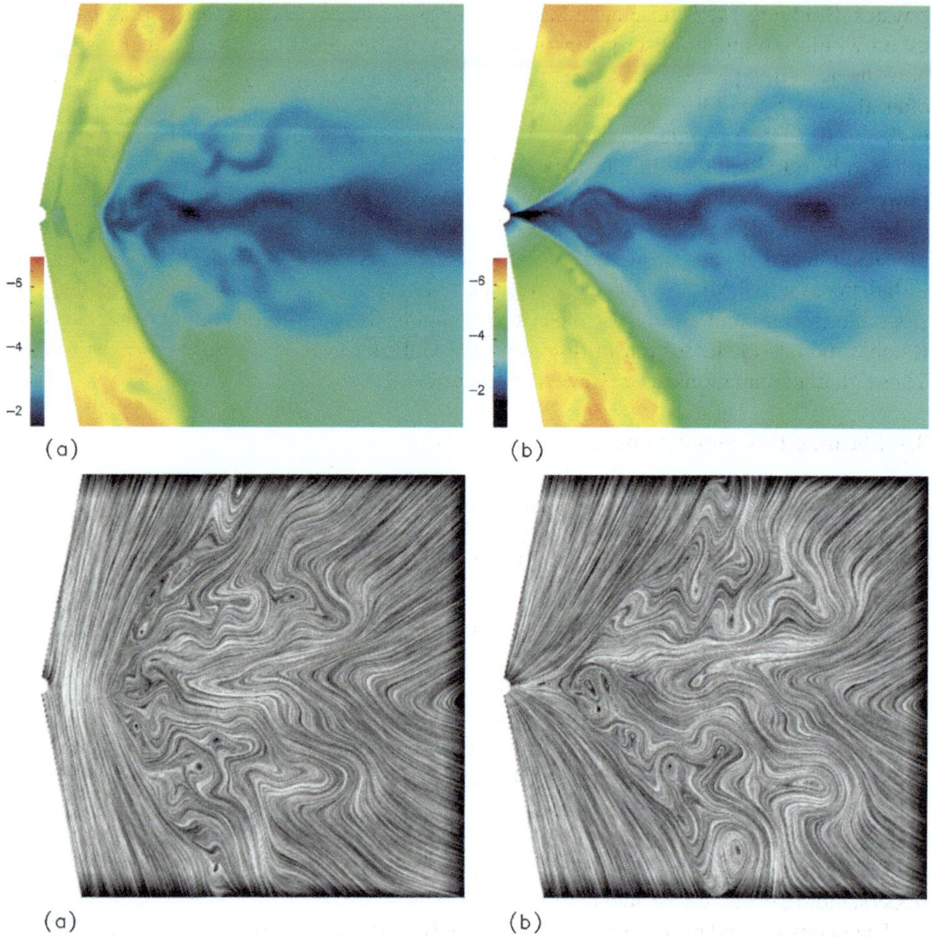

Fig. 6 Distributions of density in the meridional plane at different simulation times, showing a magnetically-arrested state (*left*) and a non-arrested state (*right*). *Bottom:* Snapshot of magnetic field lines at the same simulation times. Figure from Igumenshchev (2008)

In a two-dimensional simulation where plasma with a constant net flux is fed in from the outer boundary, a limit-cycle behavior can set in, where the mass accretion rate varies by many orders of magnitude between the arrested and non-arrested states. Figure 7 provides an example of the resulting mass accretion history. It is straightforward to show that the interval, Δt, between each non-arrested phase in this scenario grows with time according to

$$\Delta t \sim \frac{2 v_r B_z^2}{\rho} t^2, \tag{1}$$

where v_r is the radial infall velocity of the gas, B_z is the strength of the magnetic field, and ρ is the density of the gas.

In three-dimensions, the magnetic fields are no longer able to perfectly arrest the infalling gas because of a "magnetic" Rayleigh-Taylor effect. Basically, as low density, highly magnetized gas tries to support higher density gas in a gravitational potential, it becomes unstable to an interchange of the low- and high-density materials. Indeed, such a mag-

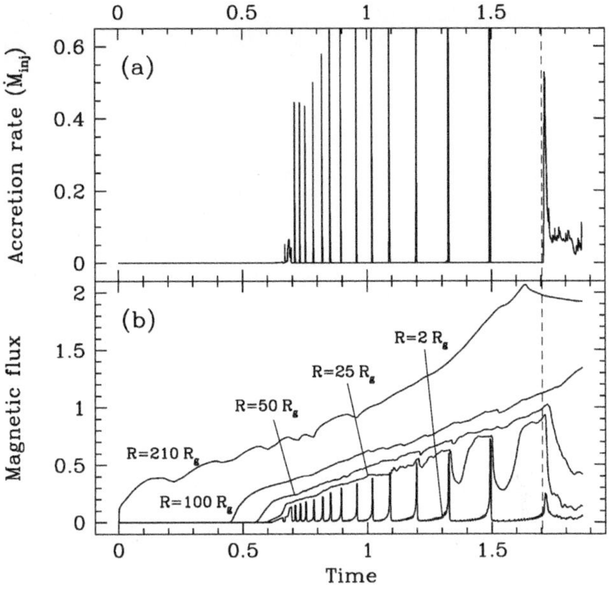

Fig. 7 Evolution of mass accretion rate and magnetic fluxes in 2D (axisymmetric) simulation of magnetically-arrested accretion. Accretion into the black hole begins at $t \approx 1.3$. Starting from $t \approx 1.4$, a pattern of cyclic accretion develops (seen as a sequence of spikes). Figure from Igumenshchev (2008)

netic Rayleigh-Taylor effect has been seen in recent simulations by Igumenshchev (2008), Tchekhovskoy et al. (2011), McKinney et al. (2012).

The results of Tchekhovskoy et al. (2011) shown in Fig. 8 are important for another reason. These were the first simulations to demonstrate a jet efficiency $\eta = (\dot{M} - \dot{E})/\langle \dot{M} \rangle$ greater than unity. Since the efficiency measures the amount of energy extracted by the jet, normalized by the amount of energy made available via accretion, a value $\eta > 1$ indicates more energy is being extracted than is being supplied by accretion. This is only possible if some other source of energy is being tapped—in this case the rotational energy of the black hole. This was the first demonstration that a Blandford-Znajek (Blandford and Znajek 1977) process *must* be at work in driving these simulated jets.

5 Tilted Disks

There is observational evidence that several black-hole X-ray binaries (BHBs), e.g. GRO J1655-40 (Orosz and Bailyn 1997), V4641 Sgr (Miller et al. 2002) and GX 339-4 (Miller et al. 2009), and active galactic nuclei (AGN), e.g. NGC 3079 (Kondratko et al. 2005), NGC 1068 (Caproni et al. 2006), and NGC 4258 (Caproni et al. 2007), may have accretion disks that are tilted with respect to the symmetry plane of their central black hole spacetimes. There are also compelling theoretical arguments that many black hole accretion disks should be tilted (Fragile et al. 2001; Maccarone 2002). This applies to both stellar mass black holes, which can become tilted through asymmetric supernovae kicks (Fragos et al. 2010) or binary captures and will remain tilted throughout their accretion histories, and to supermassive black holes in galactic centers, which will likely be tilted for some period of time after every major merger event (Kinney et al. 2000).

Close to the black hole, tilted disks may align with the symmetry plane of the black hole, either through the Bardeen-Petterson effect (Bardeen and Petterson 1975) in geometrically thin disks or through the magneto-spin alignment effect (McKinney et al. 2013) in the case

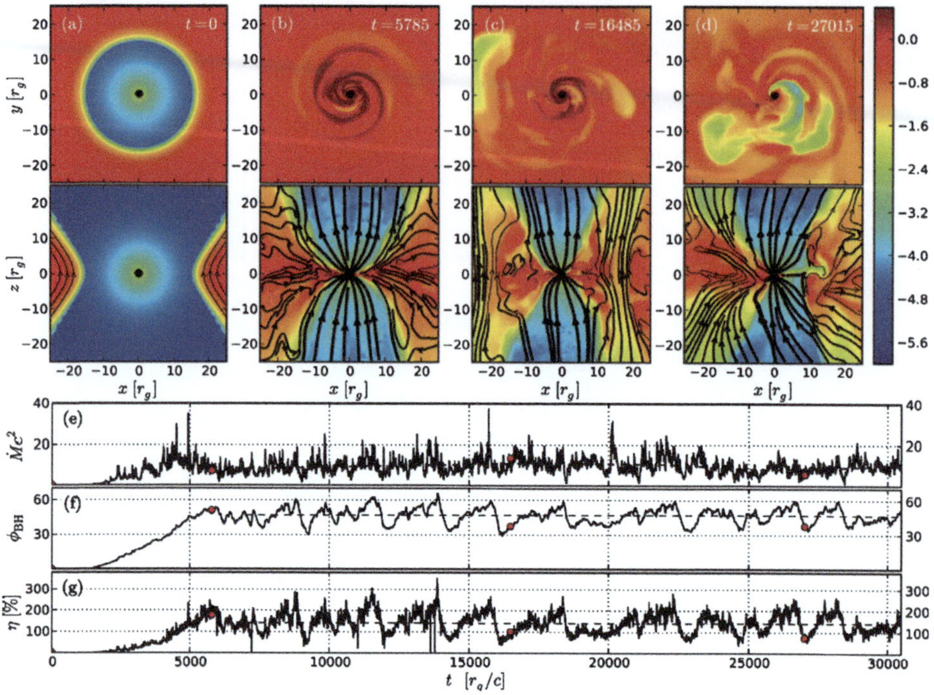

Fig. 8 (Panels **a**–**d**): the *top* and *bottom rows* show, respectively, equatorial ($z = 0$) and meridional ($y = 0$) snapshots of the gas density of a magnetically arrested flow in 3D. *Black lines* show field lines in the image plane. (Panel **e**): time evolution of the mass accretion rate. (Panel **f**): time evolution of the large-scale magnetic flux threading the BH horizon. (Panel **g**): time evolution of the energy outflow efficiency η. Figure from Tchekhovskoy et al. (2011)

of geometrically thick, magnetically-choked accretion (McKinney et al. 2012). However, for weakly magnetized, moderately thick disks ($H/r \gtrsim 0.1$), no alignment is observed (Fragile et al. 2007). In such cases, there are many observational consequences to consider.

5.1 Tilted Disks and Spin

Chapter 4.3 of this book discusses the two primary methods for estimating the spins of black holes: continuum-fitting and reflection-line modeling. Both rely on an assumed monotonic relation between the inner edge of the accretion disk (assumed to coincide with the radius of the ISCO) and black hole spin, a. This is because what both methods actually measure is the effective inner radius of the accretion disk, $r_{\rm in}$. One problem with this is that it has been shown (Fragile 2009; Dexter and Fragile 2011) that tilted disks do not follow such a monotonic behavior, at least not for disks that are not exceptionally geometrically thin. Figure 9 shows an example of the difference between how $r_{\rm in}$ depends on a for untilted and tilted simulated disks. Similar behavior has been confirmed using both dynamical (Fragile 2009) and radiative (Dexter and Fragile 2011) measures of $r_{\rm in}$. The implication is that spin can only be reliably inferred in cases where the inclination of the inner accretion disk can be independently determined and is known to be coincident with the symmetry plane of the black hole, which may be inferred from jet kinematics (Steiner and McClintock 2012).

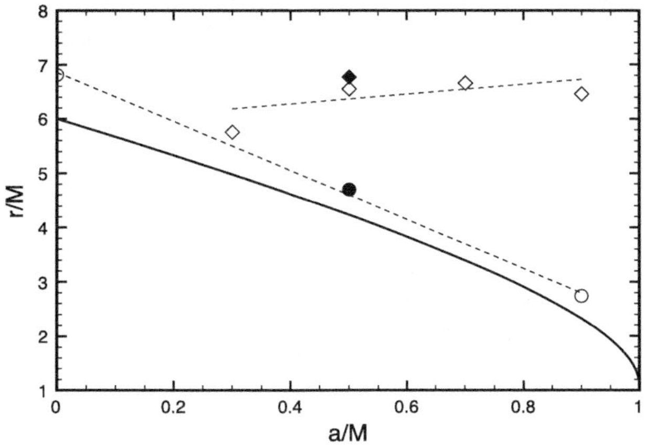

Fig. 9 Plot of the effective inner radius r_{in} of simulated untilted (*circles*) and tilted (*diamonds*) accretion disks as a function of black-hole spin using a surface density measure $\Sigma(r_{in}) = \Sigma_{max}/3e$. The *solid line* is the ISCO radius. Figure from Fragile (2009)

5.2 Tilted Disks and Sgr A* Spectral Fitting

For geometrically thin, Shakura-Sunyaev type accretion disks, the Bardeen-Petterson effect (Bardeen and Petterson 1975) may allow the inner region of the accretion disk to align with the symmetry plane of the black hole, perhaps alleviating concerns about measuring a, at least for systems in the proper state ("Soft" or "Thermally dominant") and luminosity range $L \leq 0.3L_{\mathrm{Edd}}$, where L_{Edd} is the Eddington luminosity. Extremely low luminosity systems, though, such as Sgr A*, do not experience Bardeen-Petterson alignment. Further, for a system like Sgr A* that is presumed to be fed by winds from massive stars orbiting in the galactic center, there is no reason to expect the accretion flow to be aligned with the black hole spin axis. Therefore, a tilted configuration should be expected. In light of this, Dexter and Fragile (2012) presented an initial comparison of the effect of tilt on spectral fitting of Sgr A*. Figure 10 gives one illustration of how important this effect is; it shows how the probability density distribution of four observables change if one simply accounts for the two extra degrees of freedom introduced by even a modestly tilted disk. The take away point should be clear—ignoring tilt artificially constrains these fit parameters!

5.3 Tilted Disks and Strong Shocks

One remarkable outcome of considering tilt in fitting the spectral data for Sgr A* is that tilted simulations seem able to naturally resolve a problem that had plagued earlier studies. Spectra produced from untilted simulations of Sgr A* have always yielded a deficit of flux in the near-infrared compared to what is observed. For untilted simulations, this can only be rectified by invoking additional spectral components beyond those that naturally arise from the simulations. Tilted simulations, though, produce a sufficient population of hot electrons, *without any additional assumptions*, to produce the observed near-infrared flux (see comparison in Fig. 11). They are able to do this because of another unique feature of tilted disks: the presence of standing shocks near the line-of-nodes at small radii (Fragile and Blaes 2008). These shocks are a result of epicyclic driving due to unbalanced pressure gradients in tilted disks leading to a crowding of orbits near their respective apocenters. Figure 12 shows the orientation of these shocks in relation to the rest of the inner accretion flow.

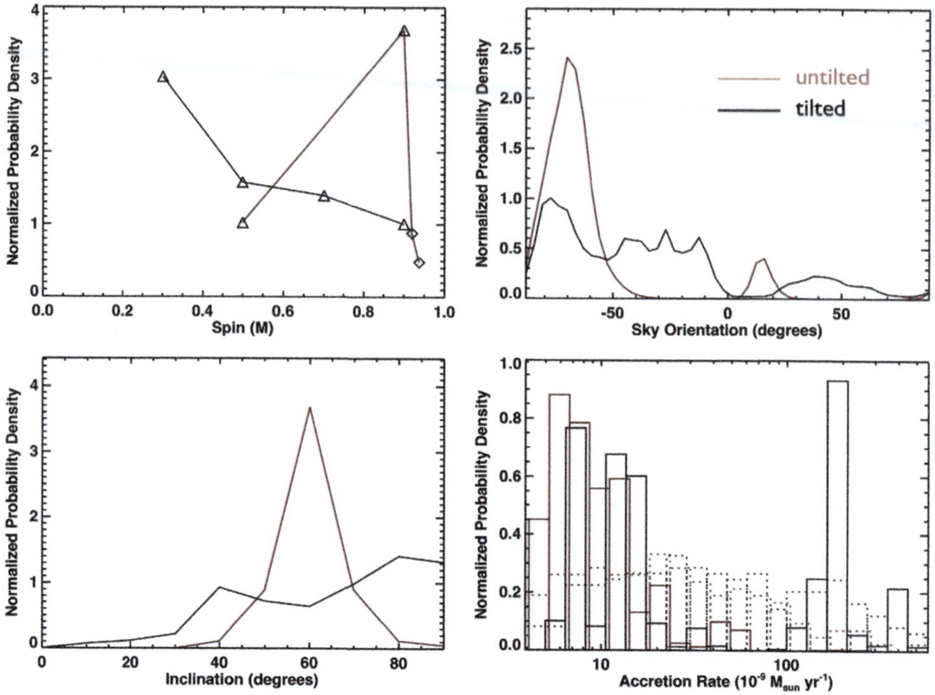

Fig. 10 Normalized probability distributions as a function of black hole spin (*top left*), sky orientation (*top right*), inclination (*bottom left*), and accretion rate (*bottom right*) for untilted (*red*) and tilted (*black*) simulations. Figure adapted from Dexter et al. (2010) and Dexter and Fragile (2012)

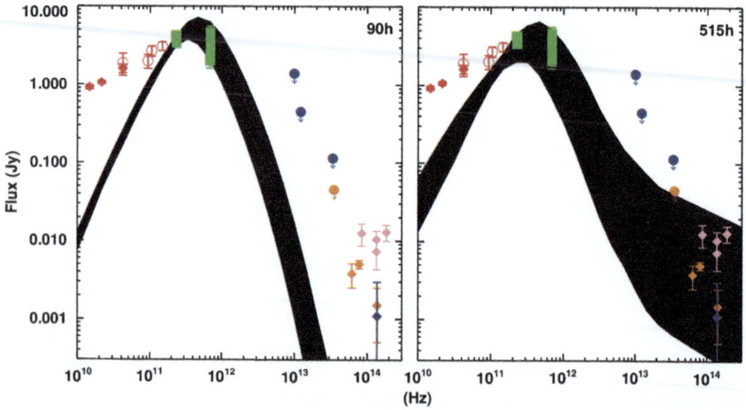

Fig. 11 Spectra from untilted (*left*) and tilted (*right*) simulations. Symbols represent Sgr A* data. In both cases the spectra are fit to the green sub-mm points. In the tilted simulations, multiple electron populations, some heated by shocks associate with the tilt, are present and can naturally produce the observed NIR emission, which is underproduced by ≈2 orders of magnitude in comparable untilted simulations. Figure from Dexter and Fragile (2012)

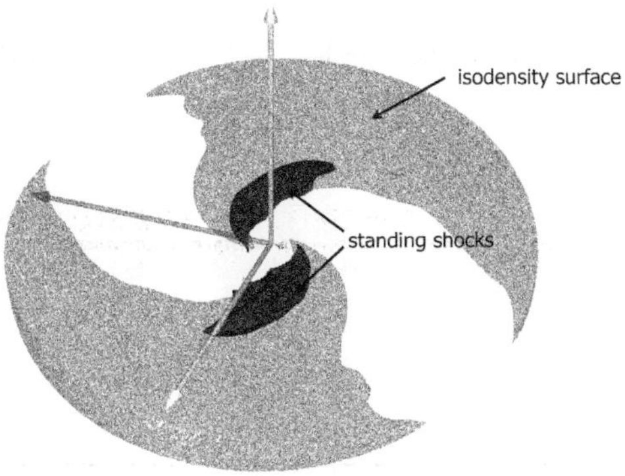

Fig. 12 Three-dimensional contours of density (*semi-transparent gray*) and Lagrangian specific entropy generation rate in arbitrary units (*solid gray*), indicating the shocks. Figure from Henisey et al. (2012)

5.4 Tilted Disks—GRMHD vs. SPH

A worthwhile future direction to pursue in this area would be a robust comparison of tilted disk simulations using both GRMHD and smoothed-particle hydrodynamics (SPH) numerical methods. The GRMHD simulations (e.g. Fragile et al. 2007, 2009; McKinney et al. 2013) enjoy the advantage of being "first principles" calculations, since they include all of the relevant physics, whereas the SPH simulations (e.g. Nelson and Papaloizou 2000; Lodato and Pringle 2007; Lodato and Price 2010; Nixon and King 2012) enjoy the advantage of being more computationally efficient, though they make certain assumptions about the form of the "viscosity" in the disk. Thus far, the GRMHD and SPH communities have proceeded separately in their studies of tilted accretion disks, and it has yet to be demonstrated that the two methods yield equivalent results. This would seem to be a relatively straightforward and important thing to check.

6 Future Direction—Radiation MHD

A few years ago, it might have been very ambitious to claim that researchers would soon be able to perform global radiation MHD simulations of black hole accretion disks, yet a lot has happened over that time, so that now it is no longer a prediction but a reality. In the realm of Newtonian simulations, a marvelous study was published by Ohsuga and Mineshige (2011), showing global (though two-dimensional) radiation MHD simulations of accretion onto a black hole in three different accretion regimes: $L \ll L_{\rm Edd}$, $L < L_{\rm Edd}$, and $L \gtrsim L_{\rm Edd}$. The remarkably different behavior of the disk in each of the simulations (illustrated in Fig. 13) is testament to how rich this field promises to be as more groups join this line of research. The specifics of this work are discussed more in Chap. 5.3.

The other big thing to happen (mostly) within the past year is that a number of groups have now tackled, for the first time, the challenge of developing codes for *relativistic* radiation MHD in black hole environments (Farris et al. 2008; Zanotti et al. 2011; Roedig et al. 2012; Fragile et al. 2012; Sądowski et al. 2013; Takahashi et al. 2013). So far none of these groups have gotten to the point of simulating accretion disks in the way Ohsuga and Mineshige (2011) did (they are still mostly at the stage of code tests and simple one- and two-dimensional problems), but with so many groups joining the chase, one can surely expect rapid progress in the near future. One result of some astrophysical interest is the study of Bondi-Hoyle (wind) accretion onto a black hole, including the effects of radiation (Zanotti et al. 2011; Roedig et al. 2012) (see Fig. 14).

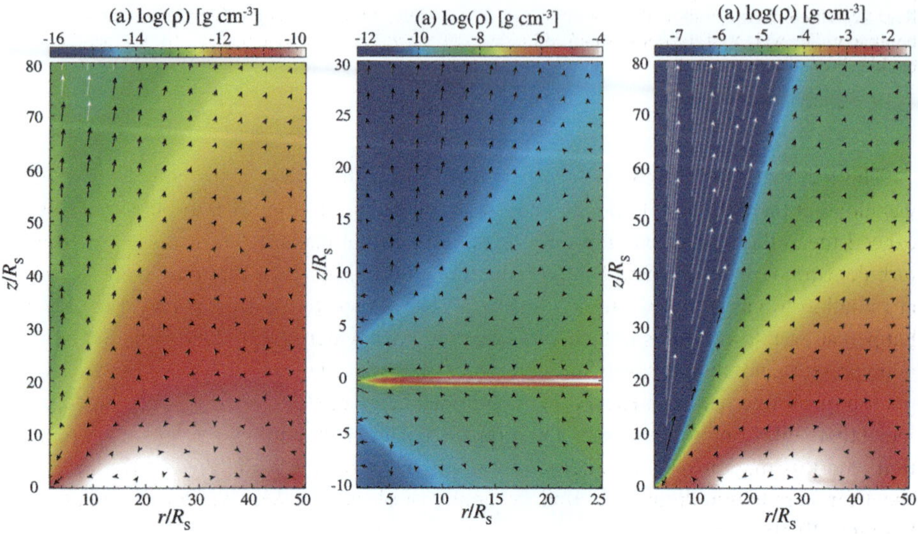

Fig. 13 Gas density overlaid with velocity vectors from three different radiation MHD simulations of a black hole accretion disk, probing luminosities of $L \sim 10^{-8}$ (*left*), 10^{-4} (*center*), and $1 L_{\mathrm{Edd}}$ (*right*). Figure adapted from Ohsuga and Mineshige (2011)

Fig. 14 Logarithm of the optical depth (*top*) and radiative flux (*bottom*) of a black hole accreting from a wind passing from left to right. Figure from Zanotti et al. (2011)

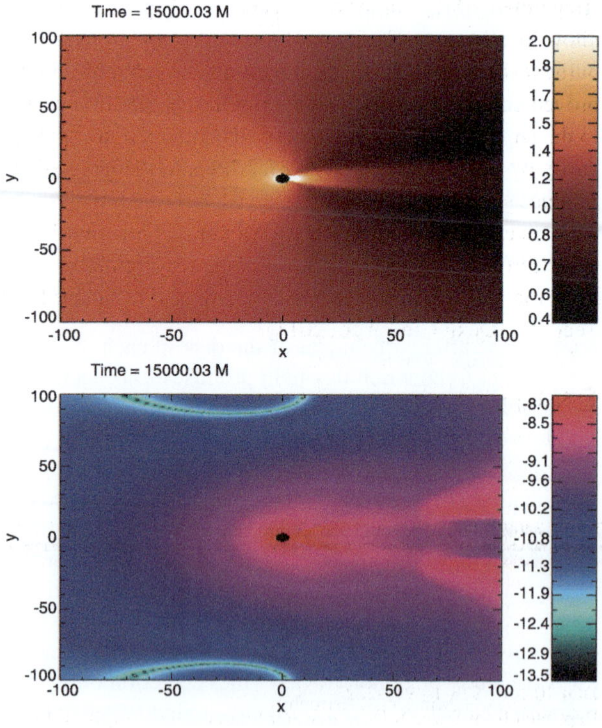

Acknowledgements Work presented in this chapter was supported in part by a High-Performance Computing grant from Oak Ridge Associated Universities/Oak Ridge National Laboratory and by the National Science Foundation under grant no. AST-0807385.

References

J.M. Bardeen, J.A. Petterson, Astrophys. J. Lett. **195**, 65 (1975)
K. Beckwith, J.F. Hawley, J.H. Krolik, Astrophys. J. **678**, 1180–1199 (2008)
R.D. Blandford, R.L. Znajek, Mon. Not. R. Astron. Soc. **179**, 433–456 (1977)
A. Caproni, Z. Abraham, H.J. Mosquera Cuesta, Astrophys. J. **638**, 120–124 (2006)
A. Caproni, Z. Abraham, M. Livio, H.J. Mosquera Cuesta, Mon. Not. R. Astron. Soc. **379**, 135–142 (2007)
J.-P. De Villiers, J.F. Hawley, J.H. Krolik, Astrophys. J. **599**, 1238–1253 (2003)
J.-P. De Villiers, J.F. Hawley, J.H. Krolik, S. Hirose, Astrophys. J. **620**, 878–888 (2005)
J. Dexter, P.C. Fragile, Astrophys. J. **730**, 36 (2011)
J. Dexter, P.C. Fragile, Mon. Not. R. Astron. Soc. (2013)
J. Dexter, E. Agol, P.C. Fragile, Astrophys. J. Lett. **703**, 142–146 (2009)
J. Dexter, J.C. McKinney, E. Agol, Mon. Not. R. Astron. Soc. **421**, 1517–1528 (2012)
J. Dexter, E. Agol, P.C. Fragile, J.C. McKinney, Astrophys. J. **717**, 1092–1104 (2010)
S. Dibi, S. Drappeau, P.C. Fragile, S. Markoff, J. Dexter, Mon. Not. R. Astron. Soc. **426**, 1928–1939 (2012)
S. Drappeau, S. Dibi, J. Dexter, S. Markoff, P.C. Fragile, Mon. Not. R. Astron. Soc. **431**, 2872–2884 (2013)
B.D. Farris, T.K. Li, Y.T. Liu, S.L. Shapiro, Phys. Rev. D **78**(2), 024023 (2008)
P.C. Fragile, Astrophys. J. Lett. **706**, 246–250 (2009)
P.C. Fragile, O.M. Blaes, Astrophys. J. **687**, 757–766 (2008)
P.C. Fragile, D.L. Meier, Astrophys. J. **693**, 771–783 (2009)
P.C. Fragile, G.J. Mathews, J.R. Wilson, Astrophys. J. **553**, 955–959 (2001)
P.C. Fragile, O.M. Blaes, P. Anninos, J.D. Salmonson, Astrophys. J. **668**, 417–429 (2007)
P.C. Fragile, C.C. Lindner, P. Anninos, J.D. Salmonson, Astrophys. J. **691**, 482–494 (2009)
P.C. Fragile, A. Gillespie, T. Monahan, M. Rodriguez, P. Anninos, Astrophys. J. Suppl. **201**, 9 (2012)
T. Fragos, M. Tremmel, E. Rantsiou, K. Belczynski, Astrophys. J. Lett. **719**, 79–83 (2010)
J.F. Hawley, J.H. Krolik, Astrophys. J. **641**, 103–116 (2006)
K.B. Henisey, O.M. Blaes, P.C. Fragile, Astrophys. J. **761**, 18 (2012)
I.V. Igumenshchev, Astrophys. J. **677**, 317–326 (2008)
N.R. Ikhsanov, L.A. Pustil'nik, N.G. Beskrovnaya, J. Phys. Conf. Ser. **372**(1), 012062 (2012)
A.L. Kinney, H.R. Schmitt, C.J. Clarke, J.E. Pringle, J.S. Ulvestad, R.R.J. Antonucci, Astrophys. J. **537**, 152–177 (2000)
S.S. Komissarov, Mon. Not. R. Astron. Soc. **326**, 41–44 (2001)
P.T. Kondratko, L.J. Greenhill, J.M. Moran, Astrophys. J. **618**, 618–634 (2005)
G. Lodato, D.J. Price, Mon. Not. R. Astron. Soc. **405**, 1212–1226 (2010)
G. Lodato, J.E. Pringle, Mon. Not. R. Astron. Soc. **381**, 1287–1300 (2007)
T.J. Maccarone, Mon. Not. R. Astron. Soc. **336**, 1371–1376 (2002)
J.C. McKinney, Mon. Not. R. Astron. Soc. **368**, 1561–1582 (2006)
J.C. McKinney, R.D. Blandford, Mon. Not. R. Astron. Soc. **394**, 126–130 (2009)
J.C. McKinney, C.F. Gammie, Astrophys. J. **611**, 977–995 (2004)
J.C. McKinney, A. Tchekhovskoy, R.D. Blandford, Science **339**, 49 (2013)
J.C. McKinney, A. Tchekhovskoy, R.D. Blandford, Mon. Not. R. Astron. Soc. **423**, 3083–3117 (2012)
J.M. Miller, A.C. Fabian, J.J.M. in't Zand, C.S. Reynolds, R. Wijnands, M.A. Nowak, W.H.G. Lewin, Astrophys. J. Lett. **577**, 15–18 (2002)
J.M. Miller, C.S. Reynolds, A.C. Fabian, G. Miniutti, L.C. Gallo, Astrophys. J. **697**, 900–912 (2009)
M. Mościbrodzka, C.F. Gammie, J.C. Dolence, H. Shiokawa, P.K. Leung, Astrophys. J. **706**, 497–507 (2009)
M. Mościbrodzka, C.F. Gammie, J.C. Dolence, H. Shiokawa, Astrophys. J. **735**, 9 (2011)
R. Narayan, I.V. Igumenshchev, M.A. Abramowicz, Publ. Astron. Soc. Jpn. **55**, 69–72 (2003)
R.P. Nelson, J.C.B. Papaloizou, Mon. Not. R. Astron. Soc. **315**, 570–586 (2000)
C.J. Nixon, A.R. King, Mon. Not. R. Astron. Soc. **421**, 1201–1208 (2012)
S.C. Noble, J.H. Krolik, J.F. Hawley, Astrophys. J. **692**, 411–421 (2009)
S.C. Noble, J.H. Krolik, J.F. Hawley, Astrophys. J. **711**, 959–973 (2010)
S.C. Noble, P.K. Leung, C.F. Gammie, L.G. Book, Class. Quantum Gravity **24**, 259 (2007)
I.D. Novikov, K.S. Thorne, in *Black Holes (Les Astres Occlus)*, ed. by C. Dewitt, B.S. Dewitt (1973), pp. 343–450
K. Ohsuga, S. Mineshige, Astrophys. J. **736**, 2 (2011)

J.A. Orosz, C.D. Bailyn, Astrophys. J. **477**, 876 (1997)
R.F. Penna, J.C. McKinney, R. Narayan, A. Tchekhovskoy, R. Shafee, J.E. McClintock, Mon. Not. R. Astron. Soc. **408**, 752–782 (2010)
C. Roedig, O. Zanotti, D. Alic, Mon. Not. R. Astron. Soc. **426**, 1613–1631 (2012)
A. Sądowski, R. Narayan, A. Tchekhovskoy, Y. Zhu, Mon. Not. R. Astron. Soc. **429**, 3533–3550 (2013)
J.D. Schnittman, J.H. Krolik, J.F. Hawley, Astrophys. J. **651**, 1031–1048 (2006)
R. Shafee, J.C. McKinney, R. Narayan, A. Tchekhovskoy, C.F. Gammie, J.E. McClintock, Astrophys. J. Lett. **687**, 25–28 (2008)
N.I. Shakura, R.A. Sunyaev, Astron. Astrophys. **24**, 337–355 (1973)
R.V. Shcherbakov, R.F. Penna, J.C. McKinney, Astrophys. J. **755**, 133 (2012)
H.C. Spruit, D.A. Uzdensky, Astrophys. J. **629**, 960–968 (2005)
J.F. Steiner, J.E. McClintock, Astrophys. J. **745**, 136 (2012)
H.R. Takahashi, K. Ohsuga, Y. Sekiguchi, T. Inoue, K. Tomida, Astrophys. J. **764**, 122 (2013)
A. Tchekhovskoy, R. Narayan, J.C. McKinney, Mon. Not. R. Astron. Soc. **418**, 79–83 (2011)
O. Zanotti, C. Roedig, L. Rezzolla, L. Del Zanna, Mon. Not. R. Astron. Soc. **417**, 2899–2915 (2011)

DOI 10.1007/978-1-4939-2227-7_7
Reprinted from *Space Science Reviews* Journal, DOI 10.1007/s11214-013-0032-4

Observational Tests of the Picture of Disk Accretion

Thomas J. Maccarone

Received: 25 September 2013 / Accepted: 3 December 2013 / Published online: 14 December 2013

Abstract In this chapter, I present a summary of observational tests of the basic picture of disk accretion. An emphasis is placed on tests relevant to black holes, but many of the fundamental results are drawn from studies of other classes of systems. Evidence is discussed for the basic structures of accretion flows. The cases of systems with and without accretion disks are discussed, as is the evidence that disks actually form. Also discussed are the hot spots where accretion streams impact the disks, and the boundary layers in the inner parts of systems where the accretors are not black holes. The nature of slow, large amplitude variability is discussed. It is shown that some of the key predictions of the classical thermal-viscous ionization instability model for producing outbursts are in excellent agreement with observational results. It is also show that there are systems whose outbursts are extremely difficult to explain without invoking variations in the rate of mass transfer from the donor star into the outer accretion disk, or tidally induced variations in the mass transfer rates. Finally, I briefly discuss recent quasar microlensing measurements which give truly independent constraints on the inner accretion geometry around black holes.

Keywords Accretion: accretion disks

1 Introduction

Observations which, in hindsight, present evidence for mass transfer have been known about since antiquity (e.g. the system Algol). Evidence for accretion power by black holes can be seen, again, in hindsight, as far back as the discovery of jets from active galactic nuclei (Curtis 1918), and the bright, broad-lined nuclei of the Seyfert galaxies (1943). Theoretical discussion of the possibility of accretion began in the 1940s (Kuiper 1941; Bondi and Hoyle 1944), with the focus on mass transfer in contact binaries, and accretion from the interstellar medium, respectively. Studies of the importance of accretion power in astrophysics first became prominent about 15 years later (Crawford and Kraft

T.J. Maccarone (✉)
Department of Physics, Texas Tech University, P.O. Box 41051, Lubbock, TX 79409-1051, USA
e-mail: thomas.maccarone@ttu.edu

1956), as it began to become clear that many of the classical, recurrent and dwarf novae were short orbital period spectroscopic binaries, often with photometric modulation at the same period (e.g. Walker 1954; Joy 1954; Joy 1956; Greenstein and Kraft 1959; Kraft 1964). The conclusion was thus drawn that mass transfer must, in some way, be responsible for the unusual phenomenology of the class of objects (see also Kopal 1959 for a synthetic discussion).

The basic elements of accretion disk theory are reviewed in the chapter in this issue by Omer Blaes, while a review of the progress in connecting the "microphysics" of magnetohydrodynamics to the "macrophysics" of classical accretion disk theory is presented in the chapter by Chris Fragile. This article will focus on showing the observational tests that have verified that simple pictures of how accretion disks should work, largely developed in the 1970s, match many of the observational constraints at some broad level. It will also discuss the observational evidence for cases where the simple picture breaks down, and more complicated models must be invokved—even though in some cases it is not clear what those more complicated models are.

A variety of means of testing accretion disk theory can be made—broadband spectroscopic measurements (see e.g. Juri Poutanen's and Jeff McClintock's articles in this issue), and measurements of rapid variability (see e.g. the article by Tomaso Belloni & Luigi Stella in this issue) are generally used to understand the processes in the inner accretion flows. In some cases, these methods also provide valuable information about the global accretion process. In this article, I will focus on discussing the observational constraints on accretion disk theory that come from other methodology—chiefly, but not exclusively variability on much longer timescales. Apart from some discussion of quasar microlensing, I will leave aside the discussion of topics such as the detailed structure of inner accretion flows around black holes, as this topic is covered elsewhere in this issue. Accretion disk theory will be discussed in this article in broadbrush strokes, to set the stage for understanding which observations will be interesting, but will be discussed in detail in the other articles in this issue.

2 A Basic Picture: Setting a Target for Observations to Test

In the standard picture of an accretion flow onto a compact object from a binary companion, we have the following structures:

1. An accretion stream which begins at the donor star (see e.g. Albright and Richards 1996 for a discussion of how evidence for streams can be found using Doppler tomography)
2. An accretion disk—a geometrically thin, optically thick flow in Keplerian rotation, with a small inward drift due to either *bona fide* viscosity, or, as has become increasingly favored in recent years, a magnetic field effect that can be well modelled on a macroscopic scale as acting like viscosity. The accretion disk is generally assumed to extend outward to the "circularization radius"—the radius where the specific angular momentum of the accreted material is the same as that for gas in a circular orbit around the accretor. If the stellar radius is larger than the circularization radius (as can happen for wind-fed systems, and for accretion by "normal" stars, rather than by compact objects), a sisk will fail to form, and the accretion stream will impact the accretor directly. A disk may also fail to form in the case of a rotating high magnetic field accretor, where the magnetic field of the central star becomes dynamically important for a circular disk outside the circularization radius (e.g. Ghosh and Lamb 1979).
3. A "hot spot" where the accretion stream from the donor star collides with the outer accretion disk, and releases its excess kinetic energy.

4. In systems without black hole accretors, a boundary layer should exist, where the excess rotational energy of the innermost part of the accretion disk is dissipated (Lynden-Bell and Pringle 1974). In systems with black hole accretors, there may be excess rotational energy at the innermost stable circular orbit, but it should be transported across the event horizon. As the focus of this issue is black holes, we will not discuss the boundary layers further, except to say that evidence for them is found in both white dwarf (e.g. Pandel et al. 2003) and neutron star (e.g. Mitsuda et al. 1984)[1] accretors, but that their spectra can be complicated, and, often, the emission does not show up as an extra blackbody component added to a disk model fit (e.g. Godon and Sion 2005; Piraino et al. 1999).

In active galactic nuclei, only the accretion disks are present. Certain types of deviations from this picture are well-studied, and represent separate topics in their own right—e.g. spectral state phenomenology (in particular states in which the accretion disk becomes geometrically thick) and the presence of jets. These topics are covered in detail in other chapters in this issue and will not be covered here in detail.

3 Classes of Accretors

A wide variety of classes of accreting objects exist in the Universe. While a large fraction of the literature on accretion involves the study of mass-transferring binary stars, some accretion takes place onto isolated objects. In particular, active galactic nuclei are generally presumed to accrete from their local interstellar medium. In many protostars, the accreting object is a single star.

Since the focus of this issue is on accretion onto black holes, the focus of this article will be on the black hole accretors themselves. Studies of stellar mass black holes are often plagued by small number statistics. Studies of active galactic nuclei suffer from difficulty in making detailed measurements, as well as long timescales of variability. It is thus useful to supplement studies of black holes with studies of other classes of accreting objects. Additionally, comparisons between black hole accretors and other classes of accretors can be excellent ways to determine which phenomenology is truly unique to black holes, rather than being generic to the process of accretion.

The classes of mass transferring binaries seen in nature include:

1. Low Mass X-ray Binaries (LMXBs). These are systems in which a neutron star or black hole accretes from a low mass main sequence star or a low mass subgiant star through Roche lobe overflow. The mass transfer rates in these systems are determined by the rate at which either the donor star expands (tdue to nuclear evolution) or the orbit shrinks (due to magnetic braking and/or gravitational radiation).
2. High Mass X-ray Binaries (HMXBs). These are systems in which a neutron star or black hole accretes from a massive star. Many of these objects have Be stars as the donor stars, and the accretion seems to take place from the equatorial wind. Others have supergiant donors, and the accretion takes place through gravitational capture of the stellar wind of the supergiant. For these fast-wind systems, the accretion disk may begin to have a circular orbit very close to the compact object, or, in some cases, a disk may not form

[1] Many other authors decompose neutron star spectra in different ways—White et al. 1986; Church and Balucińska-Church (2004), and there remains debate about the right spectral models for accreting low magnetic field neutron stars. There *is* widespread agreement that neutron star spectra usually require at least two thermal or quasi-thermal spectral components.

at all. Finally, others have much shorter orbital periods, and their donor stars are either Roche-lobe filling or nearly Roche-lobe filling. In these systems the mass transfer takes place through either a focused wind, or actual Roche lobe overflow. See Corbet (1986) for a discussion of the different classes of high mass X-ray binaries.

3. Symbiotic stars. These are systems which, based on their initial definition, show evidence for both hot and cool components in their optical spectra. Since their initial discovery, it has been realized that the cool components are red giants, and the hot components are various forms of accreting objects. Most symbiotic stars have white dwarf accretors, but a small fraction have neutron star accretors (Hynes et al. 2013 and references within). The mass transfer is generally believed to take place through capture of the red giant star wind, although some symbiotic stars are at least very close to being in Roche lobe contact. For a relatively recent review, see Sokoloski (2003).
4. Cataclysmic variables. These are systems in which mass is transferred from a low mass main sequence star or a low mass subgiant star to a white dwarf; they are the analogs of low mass X-ray binaries, for systems where the accretors is a white dwarf. Cataclysmic variables are broken into a large number of subclasses based on different observed phenomenology of variability. These sub-types are often named in terms of the prototype object, as is typical for nomenclature of variable stars. In this article, we will use nomenclature descriptive of the phenomenology, a practice which is thankfully becoming more common in the cataclysmic variable community as well.
5. Ultracompact binaries. These are binaries in which the donor star is degenerate—either a white dwarf, or a low mass degenerate helium star. In the Milky Way, these have been seen only with white dwarf or neutron star companions. When the system is a double white dwarf binary with mass transfer, it is called an AM CVn star, after the prototype object. When the accretor is a neutron star (or, potentially, a black hole) it is called an ultracompact X-ray binary. We refer the reader to Nelemans and Jonker (2010) for a review on the ultracompact X-ray binaries, and to Maccarone et al. (2007) and Zepf et al. (2008) for a discussion of the evidence for an ultracompact black hole X-ray binary in NGC 4472.
6. Various classes of close binaries with two "normal" (i.e. not compact remnant) stars show evidence for accretion. These include, for example, Algol systems (in which mass transfer takes place from an evolved star to a main sequence star) and W UMa systems (contact binaries, in which both stars overflow their Roche lobes simultaneously). We refer the reader to Thomas (1977) for a review of these systems. Accretion disks form rarely in these systems (although see Olson 1991 for evidence of a disk in one Algol, KU Cyg).

While active galactic nuclei represent one of the two major classes of accreting black holes, almost no fundamental tests of accretion theory have been put forth primarily from studies of active galactic nuclei. Such tests would be exceedingly difficult—tests based on spectroscopy would run into the problem that these systems are not fully ionized like the disks of X-ray binaries, making models of the inner accretion disks much more complicated (e.g. Done et al. 2012)—trying to understand the inner accretion disks around black holes using active galactic nuclei is considerably more difficult than trying to understand the inner disks around stellar mass black holes.

A further large part of our understanding of the outer parts of accretion disks comes from studies of variability on timescales long compared with the light crossing time at the Schwarzschild radius. In this case, the problem for using active galactic nuclei as test cases stems from the fact that their viscous timescales, even at the inner edge of the accretion disk, are expected to be much longer than the durations of any light curves assembled by astronomers. As a result, nearly all of our understanding of the outer parts of accretion disks, as well, comes from studies of stellar mass accretors. It should be noted, of course, that many fundamental advances in the studies of outflows from accretion disks have been developed with essential contributions from observations of active galactic nuclei, and in recent years, studies of lensed quasars have started to give some distinctive observational tests of the accretion geometry in active galactic nuclei.

It is thus the case that most of the fundamental constraints on accretion disk theory must come from studies of mass transferring binary systems, because they present accessible timescales, and have disks in an ionization state which is simpler than do active galactic nuclei. Our goals should be to develop fundamental theories of accretion in general, and then to determine which phenomenology is specific to black holes. As a result, it often makes sense to incorporate observational constraints from other classes of accreting objects. The systems with neutron star accretors are the most similar to those with black hole accretors, given that the radiative efficiency for accretion onto a neutron star is very similar to that for a non-rotating black hole, at least in the context of a standard Shakura-Sunyaev accretion flow.

The systems with white dwarf accretors (and particularly the cataclysmic variable stars, rather than the symbiotic binaries) are, however, the class of systems which have often provided the best constraints on accretion theory. Like black hole and neutron star accretors, the cataclysmic variables have the emission from their primary stars dominated by the accretion process,[2] rather than by core fusion, so that the radiation from the accretor can be taken as a tracer of the accretion disk's activity. That nuclear fusion contributes significantly to the emission only in low-duty cycle bursts is a fundamental difference between accreting compact objects and other kinds of accreting stars.

It may seem a bit odd for a volume on black holes to include a substantial discussion of the literature on cataclysmic variables, but in many cases, the CVs provide tighter constraints on the basic physics processes which can be expected to be generalizable to all of accretion physics. The chief advantage of cataclysmic variables relative to X-ray binaries for studying accretion is that the CVs are more numerous. About 10 times as many CVs as LMXBs are known, and the nearest CVs are about 10 times as close as the nearest low mass X-ray binaries. While in recent years, there have been a few geometric parallax measurements made of X-ray binaries in the radio (e.g. Miller-Jones et al. 2009; Reid et al. 2011), such measurements have been made for much larger samples of CVs. It also turns out that the outbursting cataclysmic variable stars tend to have shorter recurrence times than the outbursts X-ray binaries, allowing for a large class of systems which have been studied repeatedly. The other advantages of the CVs being closer is that they are brighter optically, and are observed at lower extinction. Furthermore by being more numerous, there are many more eclipsing CVs known than eclipsing neutron star, or, especially, black hole X-ray binaries.

4 Evidence for the Basic Structures of Accretion Disks

A variety of observational approaches have been used to establish the basic picture of disk accretion seen in binary systems. These include both spectroscopic observations, and variability studies. Here, we first show that accretion disks really do form in many cases, and

[2]One exception is the surface layer fusion that can take place in supersoft sources. Another exception is in classical nova explosions. Novae can actually dominate the total energy output from accreting white dwarfs, but they have very low duty cycles, and hence are negligible most of the time for most CVs.

discuss the circumstances where accretion takes place without disks. We then discuss techniques that can be used to map out the accretion geometry in binary systems.

4.1 Proof that Accretion Really Happens in Disks

Among the first things worth testing are whether accretion really does take place in disks. In fact, the paper with the first association of stellar binarity with the production of classical novae (Walker 1954) presents a key piece of evidence that the accretion process, at least in those systems must take place via a disk. It shows the properties of the eclipse of the system—relatively long ingresses and egresses and short minima for the primary eclipse, while having much weaker secondary eclipses (constrained by Walker 1956 to be less than 0.03 magnitudes, with primary eclipses of about 1 magnitude). In hindsight, these eclipse properties clearly indicate that the solid angle subtended by the accretion disk must be a small fraction of what would be subtended by a star of the same maximum radial extent. A second, more direct, but more recent piece of strong evidence that the objects believed to be accretion disks really are disks is that many of them show double-peaked emission lines (see Bailey and Ward 1981; Marsh 1988 for a discussion of cataclysmic variables; Eracleous and Halpern 2003 for a discussion of active galactic nuclei; and Soria 2002 for a discussion of X-ray binaries).

4.1.1 Systems with Accretion not Happening via Disks

At the same time, there are clear examples of systems in which there is accretion which *does not* take place through a disk. These are systems which have circularization radii smaller than the radius at which a potential accretion disk would be disrupted. The most obvious such radius would be the radius of the accreting star, and indeed for non-compact stars, accretion disks are the exception rather than the rule. For compact stars, the relevant radius will usually be the magnetospheric radius of the accretor.

The observational data on accreting objects support this theoretical scenario. There are classes of accreting white dwarfs and accreting neutron stars which lack the standard signatures of disk accretion. The accreting white dwarfs in this class are called polars. The name derives from the fact that their accretion light is often strongly polarized and that polarization stems from the fact that the accretion is channelled down the systems' magnetic poles. These systems release large fractions of their emission in the X-rays relative to other cataclysmic variables, because the emission comes mostly from an accretion column rather than accretion disk. The also frequently show periodic emission, with the modulation taking place on the rotation period of the white dwarf. An analogous class of neutron stars are the accretion-powered X-ray pulsars. In both classes of objects, cyclotron lines have been seen (e.g. Reimers and Hagen 2000 for polars; Hemphill et al. 2013 for X-ray pulsars). It is important to note that there are classes of systems which show magnetically dominated accretion and disk accretion at the same time—the intermediate polars among cataclysmic variables (Warner 1983), and both slow (La Barbera et al. 2001) and millisecond (Wijnands and van der Klis 1998) X-ray pulsars among the neutron stars.

4.2 Eclipse Mapping of Accretion Disks

At this point we have established that disks really do form in accreting objects. We also have sound theoretical reasoning, combined with empirical support, to show that the binary systems which do not have accretion disks have accretors which are fundamentally different from black holes—they either have surfaces at large radii, or they have dynamically important magnetic fields. It is thus reasonable to assume that all accretion onto black holes takes place through accretion disks, and, in this issue about black holes, to worry only about disk accretion from this point on.

Now, we can determine whether certain specific predictions of the simplest accretion disk models are in agreement with the observational data. One of the most straightforward predictions is that any disk in a steady state should have $T \propto R^{-3/4}$. This result comes from equating the differential blackbody luminosity in an annulus, $dL = 2\pi\sigma T^4 R dR$ to the differential power released by gas falling through that annulus, $\frac{GM\dot{m}}{R^2}dR$, and solving for T as a function of R. In this framework $\dot{m}$, the accretion rate, and M, the accretor mass, are constants in a steady state accretion disk, while T is the temperature of the annulus, R is the radial distance of the annulus from the compact object's center, dL is the luminosity of the annulus, and G and σ are the gravitational constant and the Stefan-Boltzmann constant, respectively.

The technique of choice for testing these models has been eclipse mapping of cataclysmic variables (e.g. Horne 1993; Baptista 2001). Eclipse mapping is one of the few means of getting geometric information about continuum emission processes of accretion disks (with quasar microlensing being the other major method). Cataclysmic variables are the system class of choice for this work because there are optically bright eclipsing cataclysmic variables. No eclipsing low mass X-ray binaries with black holes are known, and the few eclipsing neutron star X-ray binaries are not as bright as the brightest eclipsing CVs. Furthermore, bright accreting neutron stars are likely to have important effects from irradiation of the outer accretion disk by the inner accretion disk, meaning that the implications of a disagreement between the data and the standard theoretical $T \propto R^{-3/4}$ law might be expected, and could be difficult to disentangle from other effects.

It is most convenient to begin by attempting to model the "novalike" (i.e. persistently bright) cataclysmic variables, or the "dwarf nova" cataclysmic variables near the peaks of their outbursts. These are the systems that are expected to have nearly constant mass transfer rates on timescales of order the viscous propagation timescale through the accretion disks. The novalike systems have frequently failed to show $R^{-3/4}$ temperature profiles (Wood et al. 1992; Baptista et al. 1995—but see Rutten et al. 1992 for an alternative result); they often typically show much flatter temperature profiles. The quiescent dwarf novae almost universally show temperature profiles that are much flatter than $R^{-3/4}$ (e.g. Wood et al. 1986, 1992). This can be interpreted as a build-up of mass in the outer accretion disk during quiescence, a loss of mass due to winds which take mass away from the inner disk relative to what is in the outer disk, the emission of that optically thin disk wind which emits substantial light, or some combination of the different effects (e.g. Wood et al. 1986; Baptista et al. 1998).

Knigge et al. (1998) showed that the integrated spectra found simply from summing optically thick blackbodies does not describe the integrated spectrum of UX UMa, one of the prototype objects for eclipse mapping studies. Baptista et al. (1998) and Robinson et al. (1999) show that the results of eclipse mapping campaigns can be affected significantly by errors in what had previously been standard treatments—the treatment of brightness temperatures as effective temperatures (implictly assuming optically thick blackbody emission as the only source of light), and the failure to compute the effects of limb darkening properly. They find that even with careful treatment of limb darkening, the temperature profiles are flatter than $R^{-3/4}$.

Additionally, with a more careful treatment of the radiative transfer in accretion disks, it becomes possible to model the vertical structure of the disks. The standard Shakura-Sunyaev treatment yields a ratio of height H to radius R of $H/R \propto R^{1/8}$. Cataclysmic variable accretion disks typically span a range of a factor of only about 10–100 in radius between the surface of the white dwarf and the outer edge of the accretion disk, meaning that H/R should change by, at most, a factor of 1.8.

Only relatively large values of H/R can be measured using eclipse mapping techniques, so generally, attempts are made only to estimate the scale height of the outer accretion disk. The numerical values of the theoretical scale heights for the Shakura-Sunyaev model indicate that $H/R \sim 0.03$ should be typical for bright CVs. Higher values have generally been found (e.g. H/R of about 0.06 for Z Cha), indicating that some additional process is puffing up the outer disks in these systems, or that some other geometric feature or radiative transfer process is not accounted for in the existing eclipse mapping analysis (Robinson et al. 1999).

A few eclipse mapping studies of X-ray binary accretion disks have been made as well. Here, one expects irrations of the outer disk by the inner X-ray emitter to heat the outer disk, and cause it to have a larger scale height than expected in the context of the Shakura-Sunyaev disk model (e.g. Meyer and Meyer-Hofmeister 1982). At least for the source X 1822-371, the prediction of Meyer and Meyer-Hofmeister (1982) is verified (Puchnarewicz et al. 1995; Bayless et al. 2010). X-ray eclipse mapping has shown large spatial scale X-ray emission. This is sometimes interpreted in terms of the region in which the X-rays are produced being spatially very large (Church 2004), but may be due to large scale optically thin disk winds which scatter a small fraction of the X-ray emission back into the observer's line of sight. It is generally true in X-ray binaries that the disk winds seem to be more important in soft states than in hard states (Neilsen and Lee 2009); it is also true that the accretion disk corona sizes from eclipse mapping are larger in bright sources than in fainter sources (Church 2004).

To date, a single strong candidate eclipsing black hole X-ray binary is known (see Pietsch et al. 2006 for evidence of the eclipsing nature of the object and Orosz et al. 2007 for the dynamical evidence that the object is a black hole X-ray binary). This system is a high mass X-ray binary with a luminous 70 $M_{\odot}$ donor star, (Orosz et al. 2007). The combination of the brightness of the donor star relative to the accretion disk, and the fact that the donor star should have a strong wind, and hence not act as a "sharp edge" for doing eclipsing mean that eclipse mapping of this accretion disk is not particularly promising. Furthermore, the object is in M33, at a distance of about 800 kpc.

4.3 Evidence for Hot Spots

There are multiple lines of reasoning supporting the existence of hot spots where accretion streams impact the outer circular disks of accreting objects. In general, the hot spots in X-ray binaries can be quite a bit more difficult to detect than those in cataclysmic variables. This can be well understood in terms of the fraction of the total energy released as the material falls inwards. If one sets the expected luminosity of an accretion flow due to a fall through a potential from height r_{out} to height r_{in}, then $L = -GM\dot{m}(\frac{1}{r_{out}} - \frac{1}{r_{in}})$. The hot spot luminosity can be obtained by setting r_{out} to the orbital radius, and r_{in} to the circularization radius, while the total luminosity can be obtained by setting r_{in} to the radius of the compact star. For systems with orbital periods of a few hours, the few $\times 10^9$ cm radii of white dwarfs will typically be of order 10 % of the circularization radii, so a substantial fraction of the luminosity will be produced at the hot spot. For X-ray binaries, the fraction of the power produced at the hot spot will be a factor of order 1000 smaller. Hot spots can thus be detected in black hole and neutron star accretors only if either the mass transfer rate is extremely low, and hence the radiative efficiency of the inner accretion flow is extremely small, or in the more common case, the systems are transients, and are being observed during a quiescent period, in which the accretion rate into the outer accretion disk far exceeds the accretion rate onto the central compact object (McClintock et al. 1995; Froning et al. 2011). The hot spots tend to have a larger vertical height from the disk midplane than do the other parts of the outer disk, so there are also cases, for specific inclination angles, where the hot spot occults the inner disk when it is in the observers path to the compact object (White and Mason 1985). In quiescent dwarf novae, the hot spot luminosity can be a very large fraction of the total luminosity from the system, leading to strong orbital modulations as the viewing angle of the hot spot changes (e.g. Wood et al. 1989).

4.4 Spiral Structure: Evidence for Deviations from Simple Disk Models

The technique of Doppler tomography (Marsh and Horne 1988) allows a form of indirect imaging of accretion disks by looking at how line profiles change as a function of orbital phase. Steeghs (2001) shows two-armed asymmetries in a several cataclysmic variables in outburst using Doppler tomography. The evidence for such phenomena in X-ray binaries is much weaker, although it has been suggested that spiral density wave may help explain some of the large amplitude variability in GRS 1915+105 (Tagger and Pellat 1999).

5 Large Amplitude, Long Timescale Variability

One of the most important facets of the behavior of both cataclysmic variables and X-ray binaries is the presents of large amplitude, relatively smooth variations. These are often called dwarf nova outbursts in cataclysmic variables, and X-ray novae, or soft X-ray transient outbursts, in the X-ray binaries. In the X-ray binaries, these transients can lead to variations in the X-ray luminosities of the accreting systems of factors of 10^4–10^6 or more.

As it became clear that the dwarf novae and classical novae, often seen in the same objects, were fundamentally different phenomena,[3] two models emerged for explaining the dwarf novae. Both involved modulating the accretion rate onto the compact object. In one model, the mass transfer instability model, the rate at which mass *enters* the accretion disk is variable, while in the other model, the disk instability model, mass is supplied to the accretion disk at a constant rate, but there are instabilities in the way the mass flows through the disk. I will argue in this article that there is evidence supporting, if not demonstrating conclusively, the idea that both of these mechanisms apply at least some of the time.

5.1 Mechanisms for Large Variations in Luminosity

Some mechanisms for producing a mass transfer rate variations may be irradiation of the donor star by the accretor (e.g. Hameury et al. 1986; Harpaz and Rappaport 1991), magnetic activity cycles in the donor stars (e.g. Bianchini 1990), or changes in the eccentricity of the orbit of the inner mass transferring binary in a hierarchical triple system (e.g. Hut and Paczynski 1984; Maccarone 2005; Zdziarski et al. 2007). Mass transfer rates depend on the gas density at the inner Lagrange point. Since the pressure scale height in a stellar atmosphere is typically of order 10^{-4} of the orbital separation, changes in either the radius of the

[3]Classical novae are runaway nuclear fusion episodes on the surface of white dwarfs (Schatzman 1949), and hence have nothing to do with accretion disks, apart from that disks are usually the means by which the has is transported to the white dwarf.

star or the orbital separation of the binary can lead to large changes in the mass accretion rate of order 10^{-4} could lead to factors of a few changes in the mass transfer rate—thus the lack of direct evidence that these changes should not be taken as proof that the mass transfer rate is not changing via these mechanisms. Strong evidence for mass transfer variations can come in the form of finding variations in the accretion rate when one averages over timescales much longer than any reasonable timescale for mass to propagate through the accretion disk. This does not preclude variations in the mass transfer rate which happen faster than this timescale—it is just that such faster variations are likely to be extremely difficult to disentangle from disk instabilities, and as we will show in this article, considerable evidence exists that disk instabilities do explain much of the outburst phenomenology in accretion disks.

The thermal viscous instability is strongly favored as the primary source of disk instability. In this scenario, the viscosity parameter of the accretion disk is a function of the ionization state of the gas, so that when the disk is cold and neutral, the mass flow rate is smaller than when the disk is hot and ionized. Given, also, a temperature-density relationship for the disk, a limit cycle instability will develop for accretion rates within a range commonly seen in binaries with accreting compact objects, so that many of these objects are expected to show outburst cycles as predicted by the limit cycles. The leading alternative, or perhaps complement, to the disk instability model is a model in which the mass transfer rate is varied. An extensive review of the features, successes, and shortcomings of the disk instability model is given by Lasota (2001). I will summarize some of the major points presented in that work, but will focus in this section on the observational developments since that time.

5.2 The Thermal-Viscous Instability and Stability Criteria

The basic first order predictions of the disk instability model are seen to be followed pretty well by most transient accreting compact objects. By and large, the systems which have accretion rates high enough that one would expect their disks always to be fully ionized are persistent, and the systems which have accretion rates lower than that value are transient (Lasota 2001; see Coriat et al. 2012 for a confirmation that the result still holds with a much larger sample of objects). A particular system near the threshold for stability is persistently accreting, but shows large amplitude variability (Maccarone et al. 2010).

An encouraging recent result came from the measurement of a precise geometric parallax distance for SS Cyg. For quite some time it had simultaneously been held up as the prototype system for studying dwarf nova outbursts, and had been a system which appeared to have too high a mass accretion rate to allow the dwarf nova ionization instability to take place. The VLBI parallax distance found by Miller-Jones et al. (2013) indicates that the source is closer than was previously thought. The change of distance results in a mass transfer rate for SS Cyg sufficiently low that the system is below the threshold for stable accretion in the ionization instability model.

5.3 Disk Instabilities and Peak Outburst Luminosities

Warner (1987) first found the relation between peak brightness and orbital period for cataclysmic variables, and that has recently been revisited with a larger sample by Patterson (2011) who finds:

$$M_{V,peak} = 5.70 - 0.287 P_{orb,hr}. \tag{1}$$

A least-squares fit of the data from Patterson (2011) to a power law relationship finds that the peak luminosity scales with $P_{orb}^{1.2}$. These data are plotted in Fig. 1. Given the relatively

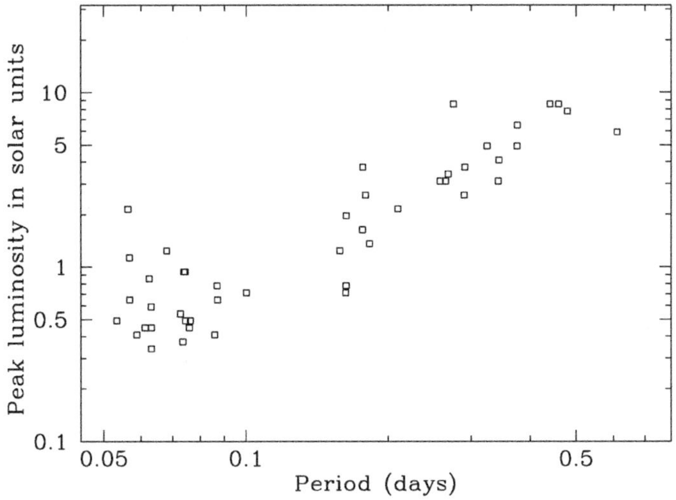

Fig. 1 The relation between peak optical luminosity and orbital period for dwarf novae. The data are taken from Patterson (2011). The relation is consistent with the $L \propto P^{4/3}$ relation expected from theory. Data points are plotted without error bars for clarity, but the typical errors are ~10–20 % on the peak luminosity, and negligible on the periods

large scatter in the data, is in fairly good agreement with the predictions for theoretical models which suggest that the whole accretion disk should be at a constant temperature in dwarf novae outbursts (e.g. Osaki 1996; Cannizzo 1998; Smak 2000), yielding a $L \propto P^{4/3}$ relationship.

Some other indications that favor disk instabilities as a baseline model come from looking at the peak luminosities seen from X-ray binaries and from cataclysmic variables. In both cases, these are well-correlated with orbital period (Shahbaz et al. 1998; Portegies Zwart et al. 2004—P04; Wu et al. 2010). Wu et al. find:

$$\frac{L_{peak}}{L_{Edd}} = -1.80 + 0.64 \log P_{orb,days} \tag{2}$$

for X-ray binaries, althought with a different treatment of the bolometric corrections, P04 found a steeper relationship for short orbital periods and a saturation at about the Eddington luminosity for long orbital periods. These data are plotted in Fig. 2. In any event, there is at least rough agreement with the finding of King and Ritter (1998) that outburst peak luminosities should scale with the radius of the accretion disk.

5.4 Outburst Durations

The durations of outbursts of many, but not all, systems, are relatively well in-line with the expected viscous timescales of the accretion disks for the black hole systems (see e.g. Chen et al. 1997; P04). The outburst durations are shorter than the viscous timescales for the CV accretion disk, where irradiation is not important, and so cooling fronts can truncate the outbursts before the entire disks are accreted (Lasota 2001). Some significant amount of the data which represent exceptions to the basic ionization instability may be explained as the result of tidal effects (see Sect. 5.5.1).

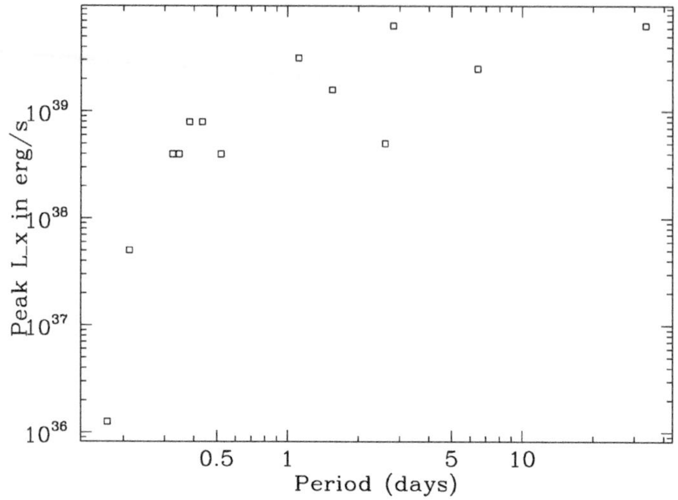

Fig. 2 A replotting of the data presented in Portegies Zwart et al. (2004), which shows that the peak X-ray luminosity for X-ray transients increases with orbital period. Unlike for the cataclysmic variables, irradiation is expected to be important, the distances are in many cases very poorly known, the bolometric corrections can be uncertain, and some systems may remain in radiatively inefficient states, so it is difficult to make quantitative comparisons between theory and data

5.5 Phenomena Which Are Hard to Explain in Terms of Disk Ionization Instabilities

While the ionization instability explains the phenomenology of X-ray binary and CV outbursts in broad brushstrokes, there are phenomena which are clearly not in agreement with that picture. In the cataclysmic variables, where the recurrence times between outbursts tend to be much shorter than in X-ray binaries, it can be clearly seen that there are variations from outburst to outburst in ways that have been fit, to date, only by adding in truncations of the inner accretion disk *and* variations of the mass transfer rate (e.g. Schreiber et al. 2003; Lasota 2012).

There are a few other cases where strong evidence for mass transfer instabilities are expected. A prime recent example among black hole candidates is the ongoing outburst of Swift J1753.5-0127, which has been in outburst since 2005, and has an orbital period of about 3.2 hours (Zurita et al. 2008). The outburst duration of 8 years (and counting) combined with the short orbital period is something that cannot be explained in terms of an ionization instability model see Fig. 3 for an illustration of this source's outburst light curve. While the observation of superhumps (see the following section for a discussion of superhumps) in this source (Zurita et al. 2008) should imply that this outburst is a "super-outburst" and hence should be longer than normal outbursts, the super-outbursts in well-studied systems are only a factor of a few longer than the normal outbursts. Several neutron star accretors have also undergone outbursts that lasted far longer than the expected viscous timescales for the systems' orbital periods (e.g. Wijnands et al. 2001 and references within).

On the flip side, the 1999 transient episode of the accreting black hole XTE J1819-254 showed strong evolution on timescales of a few hours, having reached a flux of about 12 Crab, corresponding to a super-Eddington luminosity for the source (Hjellming et al. 2000; Orosz et al. 2001). The variations were clearly too fast to be the result of some global disk instability, and too strong to have been the result of the normal variability typically seen in X-ray binaries.

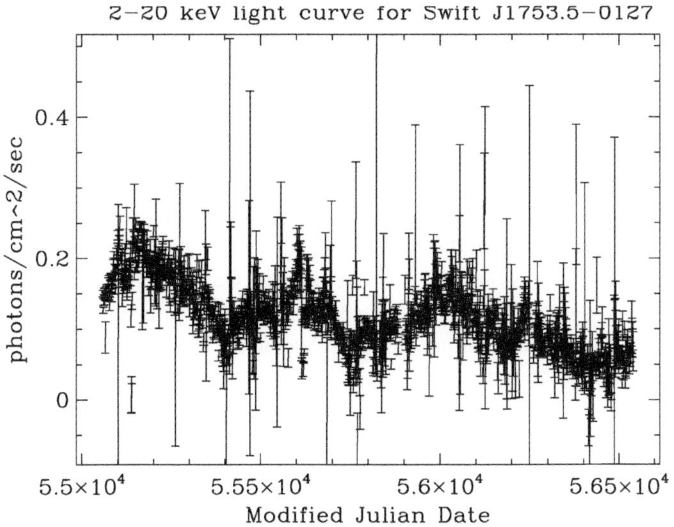

Fig. 3 The MAXI data for Swift J1753.5-0127. The system, which had not been observed prior to 2005, has clearly been a bright X-ray source for the duration of the MAXI mission. Given its orbital period of 3.2 hours, its viscous timescale should be much less than a year, and the long outburst cannot be explained in the context of the standard disk instability model

An additional line of evidence for variations in the mass transfer rates—perhaps the most direct such evidence—comes from Cantrell et al. (2010). They interpret the blue excess in the light from A 0620-00 in quiescence as coming from the accretion impact spot. In the context of that interpretation, the variability in the quiescent luminosity of the hot spot immediately implies that the rate at which matter is reaching the outer accretion disk is changing strongly as a function of time. The variation in the quiescent ultraviolet flux of the source also supports this interpretation, since the ultraviolet light can be demonstrated even more convincingly than the optical light to come from the hot spot (McClintock et al. 1995; Froning et al. 2011).

5.5.1 Tidal Effects

In an X-ray binary, the presence of a donar star means that tidal forces on the accretion disk may be substantial and variable on the orbital period. There are a few observe phenomena which have very well motivated theoretical explanations as coming from tidal effects. There are also a few phenomena which may more speculatively be associated with tidal effects.

One phenomenon which is well-associated with tidal interactions is that of superhumping. Superhumps are oscillations in the light curves of some outbursting dwarf novae and X-ray binaries. There is a critical mass ratio of 0.35 below which CVs in bright states show superhumps, and above which CVs never show superhumps (Patterson et al. 2005). The oscillations occur with a period very close to, but not exactly equal to, the orbital period of the binary. The oscillations are well-explain by a model in which an eccentric instability develops at the 3:1 resonance between the orbital period of the accretion disk and the orbital period of the binary system, and then this disk precesses due to the tidal forces (Whitehurst 1988). The limit on the mass ratio is then given by the limit that the accretion disk not be tidally truncated inside the location of the 3:1 resonance region.

The superhumping behavior is associated with "super-outbursts" of these systems. Most of the cataclysmic variables which show superhumps show a sequence of outbursts, in which some small fraction of the outbursts are significantly brighter than the rest of them. For example, in V1504 Cyg, which has been well-studied with Kepler, the superoutbursts happen about once for every 14 normal outbursts (Osaki and Kato 2013).

The prevailing view for these phenomena being so well coupled is that the thermal-tidal disk instability explains both effects (e.g. Osaki 1996). In this model, the outbursts of the disk are caused by the standard thermal instability model, discussed above. The outer edge of the accretion disk moves outwards during each normal outburst. After a series of normal outbursts, the disk is outside the 3:1 resonance radius, and the next normal outburst triggers the tidal instability, which drives in material from far out in the disk, leading to an increased peak accretion rate and outburst duration. A modification of the tidal instability model has been proposed by Truss et al. (2002) to explain the re-brightening in soft X-ray transients. In their model, they suggest that irradiation of the outer accretion disk is uneven. Tidal forces cause the location of the irradiation region to change, allowing for the originally unirradiated region to become irradiated later, allowing, at a relatively late time, a new portion of the disk to become hot and enter a high viscosity state.

Superhumps have also been seen in several short period X-ray binaries (e.g. O'Donoghue and Charles 1996; Zurita et al. 2002, 2008; Wang and Chakrabarty 2010; see also Wachter et al. 2002 for a more tentative result). The emission modulation mechanism in the CVs is thought to be tidal modulation of the viscosity, a mechanism which cannot work in low mass X-ray binaries because such a small fraction of the energy is dissipated in the outer accretion disks of X-ray binaries (e.g. Haswell et al. 2001).

Among the X-ray binaries which have shown some evidence for superhumps are short period black hole X-ray binaries (e.g. Nova Mus 1991, GRO J0422+32, XTE J1118+480 and Swift J1753.5-0127), and a few recurrent transient neutron star X-ray binaries (4U 1608-52—Wachter et al. 2002; and Aql X-1—R. Jain, private communication—and see also the discussion in Wachter et al. 2002). Interestingly, the two recurrent transient neutron stars are known to show outbursts with different amplitudes—e.g. Aql X-1 seems to show some kind of outburst roughly every 100 days, and a more major outburst every 300 or so days (Maitra and Bailyn 2008). There is also evidence for superhumping in the ultracompact X-ray binary 4U 1820-30 (Wang and Chakrabarty 2010)—this system is *persistently* bright, and at a short orbital period.

The black holes which have been seen to show prominent superhumps in recent years have not shown outbursts of varied amplitude. It may be, however, that these systems have shown *only* super-outbursts. This idea has been put forth by Maccarone and Patruno (2013) as part of the reason why some short period black hole X-ray binaries seem to be brighter than one might expect given the Wu et al. (2010) relation, and might then allow for the normal outbursts of the short period black hole X-ray binaries to manifest themselves as very faint X-ray transients (see also Knevitt et al. 2013 who discuss how short period transients may be absent from all-sky surveys). Such a scenario is strongly bolstered by finding superhumps in the bright outbursts from short-period systems. It is worth further noting that the "normal" outbursts from such systems, because of radiative inefficiency for black holes in the low hard state, should be $\sim$10 times fainter than their superoutbursts, rather than just the factor of a few difference seen for the CV superoutbursts and the candidate superoutbursts in the neutron star X-ray binaries.

5.6 Large Amplitude Variability in Active Galactic Nuclei

A much harder question to answer is whether active galactic nuclei are susceptible to the same type of accretion disk instabilities as X-ray binaries. In principle, they should be, as their accretion disks are much cooler than those of X-ray binaries, but theoretical calculations suggest that the outbursts may be quite a bit less dramatic than in X-ray binaries (Hameury et al. 2009). Determining observationally whether they do have outbursts is complicated by the fact that the typical one month timescales for outbursts of X-ray binaries, if scaled up to even the smallest, $10^6\ M_\odot$ black holes in AGN, would take place on timescales of several millenia. Evidence for a rather sharp variation in the luminosities of the Galactic Center can be seen by looking at reflection spectra from molecular clouds, which indicate that it was several orders of magnitude brighter about 100 years ago than it is now (e.g. Ponti et al. 2010)—still, there is no observational means to determine whether this variation was due to changes in the mass transfer rate into the AGN's accretion disk, or changes in the rate of flow through the disk. Körding et al. (2006) found that the spectra of active galactic nuclei are consistent with following a hysteresis loop like that followed for black hole and neutron star X-ray binaries (Maccarone and Coppi 2003), which is suggestive of the idea that the AGN pass through similar outburst curves, but this merely suggestive evidence is the strongest evidence to date that AGN actually *do* have outbursts due to disk instabilities, rather than that they merely should have such outbursts.

6 Mass Loss from Accretion Disks

The evidence for disk winds in active galactic nuclei has been well-reviewed by Ken Pounds in this issue. In X-ray binaries, similar types of evidence for disk winds—the discovery of X-ray absorption lines which appear to be dependent on inclination angle and on source spectral state (e.g. Diaz-Trigo et al. 2006; Neilsen and Lee 2009; Ponti et al. 2012). X-ray binaries provide an additional means of searching for evidence for disk winds. The mass transfer rates can be estimated from both the luminosities of the hot spots in quiescence, and from binary evolution modelling. These can then be compared with the long-term mean luminosities of the system, which provide an independent estimate of the mass accretion rate. If the mean mass transfer rate from the donor star is substantially larger than the mean mass accretion rate by the accretor, then there is additional evidence in support of winds being important. This methodology, too, shows that substantial mass loss is likely to be taking place from accretion disks around black holes (e.g. Froning et al. 2011).

7 Gravitationally Lensed Quasars

Ideally, one would like to learn about the structures of accretion disks by imaging them, rather than by testing models of their spectral or variability properties. Good prospects exist for making millimeter VLBI images of a very small number of very nearby galactic nuclei (see Heino Falcke's article in this issue). Relatively little hope exists in the near term for making direct images of a large sample of black hole accretion disks spanning a range of flux levels, and relatively little hope exists for doing small scale direct imaging at frequencies other than in the millimeter through sub-millimeter band.[4]

[4] In the longer wavelength radio bands, an alternative indirect imaging technique has recently been applied—the examination of the details of interstellar scintillation properties of a source (Macquart et al. 2013). This

A technique has been exploited for making indirect imaging measurements of the accretion flows around active galactic nuclei—namely using gravitationally lensed quasars (see e.g. Chen et al. 2012 for an extensive discussion of recent results; Chang and Refsdahl 1979 for the first discussion of the possibility). When a quasar is behind a galaxy or group or cluster of galaxies, two types of lensing take place. The first is that the effect of the "smooth mass" of the lens (i.e. the lens's dark matter halo plus the sum of the stars) leads to the production of multiple images of the background object. The second is that individual stars in the lensing galaxies may microlens the background object. For cases where the lensed background source is much larger than the stars, the effects of the microlensing process are small, since then only a small part of the object is microlensed at a time. As a result, the magnitude of variability due to microlensing can be used to probe the size scale of the background object relative to the sizes of the stars doing the lensing.

By observing the amplification factors due to microlensing at different wavelengths, one can map out the size scale of the accretion disks versus wavelength. The technique of choice for such work is time series analysis of microlensing. In the ideal case, the system will be well enough studied that the time delays due to the different path lengths light travels to form each of the observed multiple images are known. Then one can correct for these time delays, and remove the variability instrinsic to the quasar, so that the variability due to microlensing can be isolated (e.g. Morgan et al. 2008, 2010).

It can be difficult to arrange large numbers of epochs of monitoring data in the X-rays, particularly when arcsecond angular resolution is needed, and only Chandra can provide the necessary data. The optical monitoring data are more readily obtained. As a result, methods which invoke less intensive X-ray coverage are desirable. Pooley et al. (2006) show that optical data can be used to determine the magnitude of the optical microlensing anomalies, and then a single X-ray epoch can be used to estimate the variance in the X-ray magnifications due to microlensing in a statistical sense, from the variance in the X-ray images' brightnesses. This can still often yield important information about the size scale of the X-ray emitting region, while using considerably less time on the most oversubscribed telescopes.

A few key results come from Morgan et al. (2008; 2010)'s studies: that the optical continuum comes from regions with spatial scales of $\sim 100 r_g$, that the X-ray continuum comes from regions with spatial scales of $\sim 10 r_g$, and that the Fe Kα emission typically comes from regions even smaller than the X-ray continuum. Microlensing of the broad line regions of quasars—by the sheer fact that any microlensing is detected at all—indicates that the broad line regions are not spherically symmetric (Sluse et al. 2012).

Chen et al. (2013) have shown that the gravitational lensing by the quasar's own black hole can lead to factor of ~ 2 systematic errors on the estimates of the spatial scales. This result applies primarily to the small X-ray emission regions, since for much spatial scales of more than tens of Schwarzschild radii, the effects of light bending by the black hole are very small. Usually, the size of the emission region will be under-estimated, but the direction and magnitude of the effect depend on the inclination angle of the accretion disk, the spin of the black hole, and the emissivity profile of the accretion disk. These errors are of the same order as errors in black hole masses from most techniques used for active galactic nuclei. Chen et al. (2013) also find that subtle differences in the time delays for different images should be detectable with excellent microlensing campaigns. With good enough data, the inclination

technique is useful only for very compact radio bright objects—i.e. core dominated active galactic nucleus jets—and since this article is concerned with disks, we do not discuss the technique except to point out that it exists.

angles and spins of black holes in quasars might be measureable using microlensing—giving measurements independent of those which come e.g. from iron line measurements.

It is important to note that low luminosity active galactic nuclei have not yet been well-studied using these techniques, and may have different accretion geometries—the systems which have been analyzed are all bright quasars. The finding of very small X-ray emission regions thus does *not* have any clear implications for the controversy about whether the thin accretion disks around black holes in X-ray binaries are truncated, with an inner advection dominated region emitting most of the hard X-rays (see e.g. Rykoff et al. 2007; Kolehmainen et al. 2013). These data thus *do not* help to resolve the controversies about whether the inner accretion disks in low/hard states are truncated, or extend in to the innermost stable circular orbits—although the technique is, in principle, useful for resolving such controversies in low luminosity AGN, if lensed LLAGN can be discovered which are bright enough to perform such studies. While there has been a recent discovery of a candidate low luminosity AGN with short time delays between the different images and flux ratio anomalies (Anguita et al. 2009), this particular object is extremely faint (the brightest image is seen at magnitude 24.6 in the 606 W and 814 W filters with the Hubble Space Telescope), and it is unlikely that it will be useful for understanding the X-ray geometry of low luminosity AGN.

8 Summary

Several key pieces of accretion disk theory show good agreement between models and observations—the basic structures of the accretion disks as geometrically thin, and optically thick in their outer parts, and the existence of hot spots and spiral arms are all well established. At the same time, there is much about accretion disks which is of great importance which is not fully understood—particularly the development of a theory of why and how accretion disks vary. A great deal of the observed phenomenology agrees with the basic picture of the hydrogen ionization instability model. Particularly, the accretion rate above which systems becomes stable is in good agreement with the predictions of that model, as is the relationship between outburst peak luminosities and system orbital periods. At the same time, there are observational results, such as the duration of the outbursts of several short period X-ray binaries, that probably require mass transfer variations as well, and the detailed shapes of the outbursts are not always well matched by models.

Acknowledgements The author thanks the conference organizers for having promoted a series of stimulating discussions. He also thanks the Avett Brothers,, whose sublime *Magpie and the Dandelion* made the process of finalizing this manuscript far more enjoyable than it would have been otherwise.

References

G.E. Albright, M.T. Richards, Astrophys. J. **459**, L99 (1996)

T. Anguita, C. Faure, J.-P. Kneib, J. Wambsganss, C. Knobel, A.M. Koekemoer, M. Limousin, Astron. Astrophys. **507**, 35 (2009)

J. Bailey, M. Ward, Mon. Not. R. Astron. Soc. **194P**, 17 (1981)

R. Baptista, Lect. Notes Phys. **572**, 307 (2001)

R. Baptista, K. Horne, R. Hilditch, K.O. Mason, J.E. Drew, Astrophys. J. **448**, 395 (1995)

R. Baptista, K. Horne, R.A. Wade, I. Hubeny, K. Long, R.G.M. Rutten, Mon. Not. R. Astron. Soc. **298**, 1079 (1998)

A.J. Bayless, E.L. Robinson, R.I. Hynes, T.A. Ashcraft, M.E. Cornell, The Structure of the Accretion Disk in the Accretion Disk Corona X-ray Binary 4U 1822-371 at Optical and Ultraviolet Wavelengths. Astrophys. J. **709**, 251–262 (2010). doi:10.1088/0004-637X/709/1/251. http://adsabs.harvard.edu/abs/2010ApJ...709..251B

A. Bianchini, Astron. J. **99**, 1941 (1990)
H. Bondi, F. Hoyle, Mon. Not. R. Astron. Soc. **104**, 273 (1944)
J.K. Cannizzo, Astrophys. J. **493**, 426 (1998)
A.G. Cantrell et al., Astrophys. J. **710**, 1127 (2010)
K. Chang, S. Refsdahl, Nature **282**, 561 (1979)
W. Chen, C.R. Shrader, M. Livio, Astrophys. J. **491**, 312 (1997)
B. Chen, X. Dai, C.S. Kochanek, G. Chartas, J.A. Blackburne, C.W. Morgan, Astrophys. J. **755**, 24 (2012)
B. Chen, X. Dai, E. Baron, R. Kantowski, Astrophys. J. **769**, 131 (2013)
M.J. Church, Rev. Mex. Astron. Astrofís. **20**, 143 (2004)
M.J. Church, M. Balucińska-Church, Mon. Not. R. Astron. Soc. **348**, 955 (2004)
R.H.D. Corbet, Mon. Not. R. Astron. Soc. **220**, 1047 (1986)
M. Coriat, R.P. Fender, G. Dubus, Mon. Not. R. Astron. Soc. **424**, 1991 (2012)
J.A. Crawford, R.P. Kraft, Astrophys. J. **123**, 44 (1956)
H. Curtis, Publ. Lick. Obs. **13**, 55 (1918)
M. Diaz-Trigo, A.N. Parmar, L. Boirin, M. Mendez, J.S. Kaastra, Astron. Astrophys. **445**, 179 (2006)
C. Done, S.W. Davis, C. Jin, O. Blaes, M. Ward, Mon. Not. R. Astron. Soc. **420**, 1848 (2012)
M. Eracleous, J.P. Halpern, Astrophys. J. **599**, 886 (2003)
C.S. Froning et al., Astrophys. J. **743**, 26 (2011)
P. Ghosh, F.K. Lamb, Astrophys. J. **234**, 296 (1979)
P. Godon, E.M. Sion, Mon. Not. R. Astron. Soc. **361**, 809 (2005)
J.L. Greenstein, R.P. Kraft, Astrophys. J. **130**, 99 (1959)
J.-M. Hameury, A.R. King, J.-P. Lasota, Astron. Astrophys. **162**, 71 (1986)
J.-M. Hameury, M. Viallet, J.-P. Lasota, Astron. Astrophys. **496**, 413 (2009)
A. Harpaz, S. Rappaport, Astrophys. J. **383**, 739 (1991)
C.A. Haswell, A.R. King, J.R. Murray, P.A. Charles, Mon. Not. R. Astron. Soc. **321**, 475 (2001)
P.B. Hemphill, R.E. Rothschild, I. Caballero, K. Pottschmidt, M. Kühnel, F. Fürst, J. Wilms, Astrophys. J. **777**, 61 (2013)
R. Hjellming et al., Astrophys. J. **544**, 977 (2000)
K. Horne, in *Accretion Disks in Compact Stellar Systems*, ed. by J.C. Wheeler (World Scientific, Singapore, 1993), p. 117
P. Hut, B. Paczynski, Astrophys. J. **284**, 675 (1984)
R.I. Hynes et al., Astrophys. J. (2013, in press)
A.H. Joy, Astrophys. J. **120**, 377 (1954)
A.H. Joy, Astrophys. J. **124**, 317 (1956)
A.R. King, H. Ritter, Mon. Not. R. Astron. Soc. **293**, 42 (1998)
G. Knevitt, G. Wynn, S. Vaughan, M. Watson, Mon. Not. R. Astron. Soc. (2013, in press)
C. Knigge, K.S. Long, R.A. Wade, R. Baptista, K. Horne, I. Hubeny, R.G.M. Rutten, Astrophys. J. **499**, 414 (1998)
M. Kolehmainen, C. Done, M. Diaz Trigo, Mon. Not. R. Astron. Soc. (2013, in press)
Z. Kopal, in *Close binary systems*. The International Astrophysics Series (Chapman & Hall, London, 1959)
E.G. Körding, S. Jester, R. Fender, Mon. Not. R. Astron. Soc. **372**, 1366 (2006)
R.P. Kraft, Astrophys. J. **139**, 4578 (1964)
G.P. Kuiper, Astrophys. J. **93**, 133 (1941)
A. La Barbera, L. Burderi, T. Di Salvo, R. Iaria, N. Robba, Astrophys. J. **553**, 375 (2001)
J.-P. Lasota, New Astron. Rev. **45**, 449 (2001)
J.-P. Lasota, Disc outbursts in various types of binary systems. Mem. Soc. Astron. Ital. **83**, 469 (2012). http://adsabs.harvard.edu/abs/2012MmSAI..83..469L
D. Lynden-Bell, J.E. Pringle, The evolution of viscous discs and the origin of the nebular variables. Mon. Not. R. Astron. Soc. **168**, 603–637 (1974). http://adsabs.harvard.edu/abs/1974MNRAS.168..603L
T.J. Maccarone, An explanation for long flares from extragalactic globular cluster X-ray sources. Mon. Not. R. Astron. Soc., **364**, 971–976 (2005). doi:10.1111/j.1365-2966.2005.09623.x. http://adsabs.harvard.edu/abs/2005MNRAS.364..971M
T.J. Maccarone, P.S. Coppi, Mon. Not. R. Astron. Soc. **338**, 189 (2003)
T.J. Maccarone, A. Patruno, Mon. Not. R. Astron. Soc. **428**, 1335 (2013)
T.J. Maccarone, A. Kundu, S.E. Zepf, K.L. Rhode, Nature **445**, 183 (2007)
T.J. Maccarone, K.S. Long, C. Knigge, A. Dieball, D.R. Zurek, Mon. Not. R. Astron. Soc. **406**, 2087 (2010)
J.-P. Macquart, L.E.H. Godfrey, H.E. Bignall, J.A. Hodgson, Astrophys. J. **765**, 142 (2013)
D. Maitra, C.D. Bailyn, Astrophys. J. **688**, 537 (2008)
T.R. Marsh, Mon. Not. R. Astron. Soc. **231**, 1117 (1988)
T.R. Marsh, K. Horne, Images of accretion discs. II—Doppler tomography. Mon. Not. R. Astron. Soc. **235**, 269–286 (1988). http://adsabs.harvard.edu/abs/1988MNRAS.235..269M

J.E. McClintock, K. Horne, R.A. Remillard, Astrophys. J. **442**, 358 (1995)
F. Meyer, E. Meyer-Hofmeister, Vertical structure of accretion disks. Astron. Astrophys. **106**, 34–42 (1982). http://adsabs.harvard.edu/abs/1982A%26A...106...34M
J.C.A. Miller-Jones, P.G. Jonker, V. Dhawan, W. Brisken, M.P. Rupen, G. Nelemans, E. Gallo, Astrophys. J. **706L**, 230 (2009)
J.C.A. Miller-Jones, G.R. Sivakoff, C. Knigge, E. Körding, M. Templeton, E.O. Waagen, Science **340**, 950 (2013)
K. Mitsuda et al., Publ. Astron. Soc. Jpn. **36**, 741 (1984)
C.W. Morgan, C.S. Kochanek, X. Dai, N.D. Morgan, E.E. Falco, Astrophys. J. **689**, 755 (2008)
C.W. Morgan, C.S. Kochanek, N.D. Morgan, E.E. Falco, Astrophys. J. **712**, 1129 (2010)
J. Neilsen, J.C. Lee, Nature **458**, 481 (2009)
G. Nelemans, P.G. Jonker, Ultra-compact (X-ray) binaries. New Astron. Rev. **54**, 87–92 (2010). doi:10.1016/j.newar.2010.09.021. http://adsabs.harvard.edu/abs/2010NewAR..54...87N
D. O'Donoghue, P.A. Charles, Mon. Not. R. Astron. Soc. **282**, 191 (1996)
E.C. Olson, Astron. J. **102**, 1423 (1991)
J.A. Orosz et al., Astrophys. J. **555**, 489 (2001)
J.A. Orosz et al., Nature **449**, 872 (2007)
Y. Osaki, Publ. Astron. Soc. Pac. **108**, 39 (1996)
Y. Osaki, T. Kato, Publ. Astron. Soc. Jpn. (2013, in press)
D. Pandel, F.A. Cordova, S.B. Howell, Mon. Not. R. Astron. Soc. **346**, 1231 (2003)
J. Patterson, Mon. Not. R. Astron. Soc. **411**, 2695 (2011)
J. Patterson, J. Kemp, D.A. Harvey, R.E. Fried, R. Rea, B. Monard, L.M. Cook, D.R. Skillman, T. Vanmunster, G. Bolt, E. Armstrong, J. McCormick, T. Krajci, L. Jensen, J. Gunn, N. Butterworth, J. Foote, M. Bos, G. Masi, P. Warhurst, Superhumps in Cataclysmic Binaries. XXV. q_{crit}, ϵ(q), and Mass-Radius. Publ. Astron. Soc. Pac. **117**, 1204–1222 (2005) doi:10.1086/447771. http://adsabs.harvard.edu/abs/2005PASP..117.1204P
W. Pietsch et al., Astrophys. J. **646**, 420 (2006)
S. Piraino, A. Santangelo, E.C. Ford, P. Kaaret, Astron. Astrophys. **349L**, 77 (1999)
G. Ponti, R. Terrier, A. Goldwurm, G. Belanger, G. Trap, Astrophys. J. **714**, 732 (2010)
G. Ponti, R.P. Fender, M.C. Begelman, R.J.H. Dunn, J. Neilsen, M. Coriat, Mon. Not. R. Astron. Soc. **422**, L11 (2012)
D. Pooley, J.A. Blackburne, S. Rappaport, P.L. Schecheter, W-f. Fong, Astrophys. J. **648**, 67 (2006)
S.F. Portegies Zwart, J. Dewi, T. Maccarone, Mon. Not. R. Astron. Soc. **355**, 413 (2004)
E.M. Puchnarewicz, K.O. Mason, F.A. Cordova, HST UV observations of the accretion disk corona X-ray binary X1822-371. Adv. Space Res. **16**, 65–68 (1995). doi:10.1016/0273-1177(95)00045-G. http://adsabs.harvard.edu/abs/1995AdSpR..16...65P
M.J. Reid, J.E. McClintock, R. Narayan, L. Gou, R.A. Remillard, J.A. Orosz, Astrophys. J. **742**, 83 (2011)
D. Reimers, H.-J. Hagen, Astron. Astrophys. **358L**, 45 (2000)
E.L. Robinson, J.H. Wood, R.A. Wade, Astrophys. J. **514**, 952 (1999)
R.G.M.R. Rutten, J. van Paradijs, J. Tinbergen, Astron. Astrophys. **260**, 213 (1992)
E.S. Rykoff, J.M. Miller, D. Steeghs, M.A.P. Torres, Astrophys. J. **666**, 1129 (2007)
E. Schatzman, Anal. Appl. **12**, 281 (1949)
M.R. Schreiber, J.-M. Hemury, J.-p. Lasota, Astron. Astrophys. **410**, 239 (2003)
C.K. Seyfert, Astrophys. J. **97**, 28 (1943)
T. Shahbaz, P.A. Charles, A.R. King, Mon. Not. R. Astron. Soc. **301**, 382 (1998)
D. Sluse, D. Hutsemekers, F. Courbin, G. Meylan, J. Wambsganss, Astron. Astrophys. **544A**, 62 (2012)
J.I. Smak, Acta Astron. **50**, 399 (2000)
J. Sokoloski, J. Am. Assoc. Var. Star Obs. **31**, 89 (2003)
R. Soria, in *Proceedings of the MGIX MM Meeting*, ed. by R.T. Jantzen, R. Ruffini (World Scientific, Singapore, 2002)
D. Steeghs, Lect. Notes Phys. **573**, 45 (2001)
M. Tagger, R. Pellat, An accretion-ejection instability in magnetized disks. Astron. Astrophys. **349**, 1003–1016 (1999). http://adsabs.harvard.edu/abs/1999A%26A...349.1003T
H.-C. Thomas, Annu. Rev. Astron. Astrophys. **15**, 127 (1977)
M.R. Truss, G.A. Wynn, J.A. Murray, A.R. King, Mon. Not. R. Astron. Soc. **337**, 1329 (2002)
S. Wachter, D.W. Hoard, C.D. Bailyn, S. Corbel, P. Kaaret, Astrophys. J. **568**, 901 (2002)
M.F. Walker, Publ. Astron. Soc. Pac. **66**, 230 (1954)
M.F. Walker, Astrophys. J. **123**, 68 (1956)
Z. Wang, D. Chakrabarty, Astrophys. J. **712**, 653 (2010)
B. Warner, Astrophys. Space Sci. Libr. **101**, 155 (1983)
B. Warner, Mon. Not. R. Astron. Soc. **227**, 23 (1987)

N.E. White, K.O. Mason, Space Sci. Rev. **40**, 167 (1985)
N.E. White, A. Peacock, G. Hasinger, K.O. Mason, G. Manzo, B.G. Taylor, G. Branduardi-Raymont, Mon. Not. R. Astron. Soc. **218**, 129 (1986)
R. Whitehurst, Mon. Not. R. Astron. Soc. **232**, 35 (1988)
R. Wijnands, M. van der Klis, Nature **394**, 344 (1998)
R. Wijnands, J.M. Miller, C. Markwardt, W.H.G. Lewin, M. van der Klis, Astrophys. J. **560L**, 159 (2001)
J. Wood, K. Horne, G. Berriman, R. Wade, D. O'Donoghue, B. Warner, **219**, 629 (1986)
J.H. Wood, K. Horne, G. Berriman, R.A. Wade, **341**, 974 (1989)
J.H. Wood, T.M.C. Abbott, A.W. Shafter, Astrophys. J. **393**, 729 (1992)
Y.X. Wu, W. Yu, T.P. Li, T.J. Maccarone, X.D. Li, Astrophys. J. **718**, 620 (2010)
A.A. Zdziarski, L. Wen, M. Gierlinski, Mon. Not. R. Astron. Soc. **377**, 1006 (2007)
S.E. Zepf, D. Stern, T.J. Maccarone, A. Kundu, M. Kamionkowski, K.L. Rhode, J.J. Salzer, R. Ciardullo, C. Gronwall, Astrophys. J. Lett. **683**, 139 (2008)
C. Zurita et al., Mon. Not. R. Astron. Soc. **333**, 791 (2002)
C. Zurita, M. Durant, M.A.P. Torres, T. Shahbaz, J. Casares, D. Steeghs, Astrophys. J. **681**, 1458 (2008)

DOI 10.1007/978-1-4939-2227-7_8
Reprinted from *Space Science Reviews* Journal, DOI 10.1007/s11214-014-0071-5

Observational Appearance of Black Holes in X-Ray Binaries and AGN

Marat Gilfanov · Andrea Merloni

Received: 29 April 2014 / Accepted: 9 July 2014 / Published online: 30 July 2014

Abstract Accretion onto black holes powers most luminous compact sources in the Universe. Black holes are found with masses extending over an extraordinary broad dynamic range, from several to a few billion times the mass of the Sun. Depending on their position on the mass scale, they may manifest themselves as X-ray binaries or active galactic nuclei. X-ray binaries harbor stellar mass black holes—endpoints of the evolution of massive stars. They have been studied by X-ray astronomy since its inception in the early 60-ies, however, the enigma of the most luminous of them—ultra-luminous X-ray sources, still remains unsolved. Supermassive black holes, lurking at the centers of galaxies, are up to hundreds of millions times more massive and give rise to the wide variety of different phenomena collectively termed "Active Galactic Nuclei". The most luminous of them reach the Eddington luminosity limit for a few billions solar masses object and are found at redshifts as high as $z \geq 5$–7. Accretion onto supermassive black holes in AGN and stellar- and (possibly) intermediate mass black holes in X-ray binaries and ultra-luminous X-ray sources in star-forming galaxies can explain most, if not all, of the observed brightness of the cosmic X-ray background radiation. Despite the vast difference in the mass scale, accretion in X-ray binaries and AGN is governed by the same physical laws, so a degree of quantitative analogy among them is expected. Indeed, all luminous black holes are successfully described by the standard Shakura-Sunyaev theory of accretion disks, while the output of low-luminosity accreting black holes in the form of mechanical and radiative power of the associated jets obeys to a unified scaling relation, termed as the "fundamental plane of black holes". From that standpoint, in this review we discuss formation of radiation in X-ray binaries and AGN,

M. Gilfanov (✉)
Max-Planck-Institute for Astrophysics, Garching, Germany
e-mail: gilfanov@mpa-garching.mpg.de

M. Gilfanov
Space Research Institute, Moscow, Russia

A. Merloni
Max-Planck-Institute for Extraterrestrial Physics, Garching, Germany
e-mail: am@mpe.mpg.de

emphasizing their main similarities and differences, and examine our current knowledge of the demographics of stellar mass and supermassive black holes.

Keywords Black holes · Accretion · X-ray binaries · Active galactic nuclei

1 Introduction

Accretion onto black holes (BH), due to their compactness, releases a large fraction of the rest mass energy of the accreting matter, $L = \eta \dot{M} c^2$ with $\eta \sim 0.1$–0.2. A significant fraction of released energy is converted into electro-magnetic emission close to the compact object, and studies of the radiation produced in the innermost region of accretion flows are unique tools to probe General Relativity in the strong field regime.

Stellar mass black holes, with masses $\sim 10 M_\odot$, are endpoints of the evolution of massive stars (e.g. Verbunt and van den Heuvel 1995; Postnov and Yungelson 2006). If a compact object—black hole or a neutron star, is a member of a close binary system in which the companion star either fills its Roche lobe or intensely looses mass via stellar wind, accretion of its material leads to appearance of a bright X-ray source—an X-ray binary (XRB). Their luminosities can reach up to $\log(L_X) \sim 39$–40, of the order of the Eddington luminosity[1] ($L_{\rm Edd}$) of a few tens of solar masses object. XRBs are the brightest stellar compact sources in galaxies, with a typical galaxy containing from a few dozen to a few hundred of them (e.g. Gilfanov 2004). On the other hand, currently there are no confirmed cases of isolated black holes accreting material from surrounding interstellar medium at the rate sufficient to provide detectable emission at any wavelength (Fender et al. 2013).

Accretion onto supermassive black holes lurking at the center of galaxies (SMBH, with masses approximately between fractions of a million to billions times that of the sun) manifests itself in a wide variety of different phenomena, collectively termed "Active Galactic Nuclei" (AGN). Their luminosity can reach values orders of magnitude larger than the collective radiative output of all stars in a galaxy, as in the case of powerful Quasars (QSO), reaching the Eddington limit for black holes of a few billion solar masses. On the other hand, massive black holes in galactic nuclei can be exceedingly faint, like in the case of Sgr A* in the nucleus of the Milky Way, which radiates at less than a billionth of the Eddington luminosity of the $4.3 \times 10^6 M_\odot$ BH harboured there (Ghez et al. 2008; Gillessen et al. 2009). Such a wide range in both black hole masses and accretion rates of SMBH results in a much wider, and observationally more complex, observational phenomenology.

Some fundamental similarities in the appearance of XRB and AGN do however remain, strongly suggesting a common physical framework for the description of accretion process in black holes of all masses. Optically thick and geometrically thin accretion disc solutions (Shakura and Sunyaev 1973) appear to describe well the state of all accreting black holes near the Eddington luminosity; relativistic jets are also ubiquitous in accreting compact objects, and appear to obey scale-invariant solutions under a change of black hole mass and/or accretion rate[2] (Sams et al. 1996; Heinz and Sunyaev 2003); finally, all accreting

[1] The Eddington luminosity is defined as $L_{\rm Edd} = 4\pi G M_{\rm BH} m_{\rm p} c/\sigma_{\rm T} \simeq 1.3 \times 10^{38} (M_{\rm BH}/M_\odot)$ ergs s^{-1}, where G is the Newton constant, $m_{\rm p}$ is the proton mass, c the speed of light and $\sigma_{\rm T}$ the Thomson scattering cross section. Throughout this review we will also use the Eddington ratio defined as $\lambda \equiv L/L_{\rm Edd}$.

[2] A thorough discussion of the role played by jets and outflows is presented by S. Heinz elsewhere in this issue.

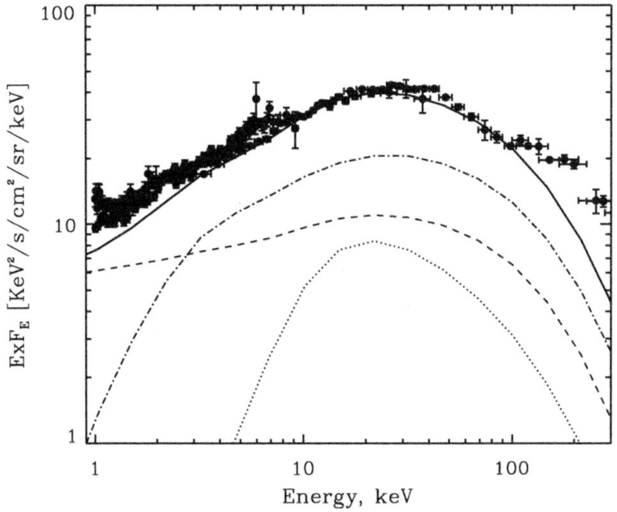

Fig. 1 Contribution of supermassive black holes in AGN to the spectral intensity of the Cosmic X-ray background. *Points* with *error bars* show the observed CXRB brightness (as compiled from Gilli et al. 2007). The *solid line* is the overall contribution of AGN, which is the sum of the contributions from: i) un-obscured AGN (i.e. those with absorbing column density $N_H < 10^{22}$ cm^2, *dashed line*); ii) obscured, Compton-Thin AGN (with $10^{22} < N_H < 10^{24}$ cm^2, *dot-dashed line*) and iii) heavily obscured, Compton-Thick sources ($N_H > 10^{24}$ cm^2, *dotted line*). The computation is based on the Gilli et al. (2007) synthesis model

black holes release a substantial fraction of their power in the X-ray band, with spectral shapes suggestive of common physical processes being at play in BH of all masses. As the fraction of X-ray to bolometric emission is negligible in the case of main sequence stars, X-ray observations are among the most effective tools to reveal black holes in the Universe.

In particular, the cosmic X-ray background (CXRB) radiation, first discovered by Giacconi with collaborators in 1962 can be considered as the ultimate inventory of the energy released by the process of accretion onto black holes throughout the history of the Universe. Historically, detailed modelling of the CXRB has proceeded hand in hand with our deeper understanding of the physical properties of accreting black holes, and of their cosmological evolution. Indeed, deep extragalactic surveys with X-ray satellites, including Chandra and XMM-Newton, resolved about ~ 80–$90\,\%$ of the CXRB. These observations have shown that a similar fraction of the CXRB is provided by the emission of supermassive black holes; moreover, the hard slope of the CXRB spectrum (well described by a power-law with photon index $\Gamma_{\mathrm{CXRB}} \simeq 1.4$ at $E < 10$ keV, see Fig. 1), and the prominent peak observed at about 30 keV are best accounted for by assuming that a significant fraction of active galactic nuclei are in fact obscured (Setti and Woltjer 1989; Comastri et al. 1995, and Fig. 1). Interestingly, the average redshift of the objects contributing mostly to the observed background is $z \sim 1$.

Although there is now direct observational confirmation, calculations show that the major fraction of the remaining ~ 10–$15\,\%$ can be comfortably provided by the emission of X-ray binaries (Dijkstra et al. 2012) (Fig. 2). Of these the absolute majority is supplied by stellar mass black holes accreting material from a massive companion in binary systems (so-called High-Mass X-ray Binaries, HMXB) in star-forming galaxies. Although at the sensitivity level achieved in the current deepest extragalactic surveys, $F_X \sim 10^{-17}$ erg/s/cm^2, star-forming galaxies constitute about $\sim 1/4$ of all sources, extrapolation of source counts

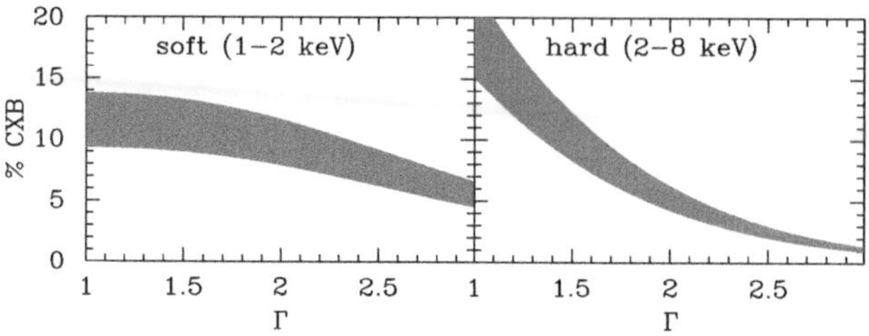

Fig. 2 Fraction of the total Cosmic X-ray background that can be produced by star-forming galaxies as a function of the average photon index Γ of their spectral energy distribution, assuming a power law SED. About $\sim 2/3$ of the 0.5–8 keV emission from star-forming galaxies is produced by accreting compact objects in high-mass X-ray binaries (Mineo et al. 2014). Depending on the actual value of Γ, star-forming galaxies can account for ~ 5–$15\,\%$ of the total observed CXB brightness in the soft X-ray band and up to $\sim 20\,\%$ in the hard band. Adopted from Dijkstra et al. (2012). Recent NuStar observations of ultra-luminous X-ray sources (e.g. Bachetti et al. 2013) suggest that appropriate are values of $\Gamma \sim 2$ for the soft band and $\Gamma \geq 2$–3 for the hard band

to deeper fluxes shows that at the ~ 2–5 times deeper fluxes they will become the dominant (in numbers) class of objects.

The rest of the article is organised as follows: We start with discussing emission mechanisms, geometry of the accretion flow and spectral distribution of emitted energy in black holes of all masses in Sect. 2. We then proceed with the demographics of stellar mass and supermassive black holes in Sects. 3 and 4 correspondingly. We will use CGS units unless otherwise mentioned.

2 Emission Mechanisms and SED

The gravitational energy of matter dissipated in the accretion flow around a black hole is primarily converted to photons of UV and X-ray wavelengths. The lower limit on the characteristic temperature of the spectral energy distribution of the emerging radiation can be estimated assuming the most radiatively efficient configuration—optically thick accretion flow. Taking into account that the size of the emitting region is $r \sim 20 r_g$ (r_g is the gravitational radius) and assuming a black body emission spectrum one obtains:

$$kT_{bb} = \left(\frac{L_{bol}}{\sigma_{SB}\pi r^2}\right)^{1/4} \approx 1.4\left(\frac{L_{bol}}{10^{38}}\right)^{1/4}\left(\frac{M_{BH}}{10}\right)^{-1/2} \text{keV}$$
$$\approx 14\left(\frac{L_{bol}}{10^{44}}\right)^{1/4}\left(\frac{M_{BH}}{10^8}\right)^{-1/2} \text{eV} \tag{1}$$

It is interesting (and well known) that T_{bb} scales as $\propto M_{BH}^{-1/2}$. This is confirmed very well by the measurements of the disk emission temperature in stellar mass systems and around supermassive black holes in AGN. It is also illustrated, albeit less dramatically, by the comparison of the soft state spectra of black holes and neutron stars.

The upper end of the relevant temperature range is achieved in the limit of optically thin emission. It is not unreasonable to link it to the virial temperature of particles near the black hole, $kT_{vir} = GM_{BH}m/r \propto mc^2/(r/r_g)$. Unlike the black body temperature this quantity does not depend on the mass of the compact object, but does depend on the mass of the particle m. For electrons $T_{vir} \sim 51(r/10r_g)^{-1}$ keV and it is correspondingly $m_p/m_e = 1836$ times higher for protons. Protons and ions are the main energy reservoir in the accretion

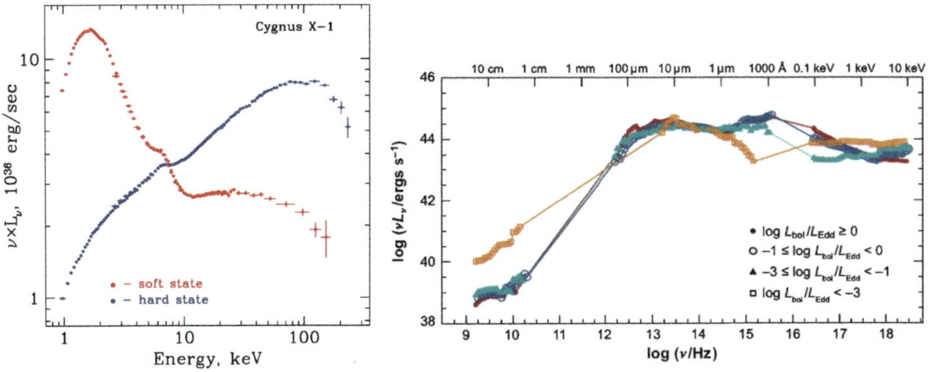

Fig. 3 Energy spectra of a black hole Cyg X-1 (*left panel*, adapted from Gilfanov et al. 2000) and from a compilation of AGN at different Eddington-scaled luminosities (*right panel*, from Ho 2008)

flow, but for all plausible mechanisms of spectral formation it is the temperature of electrons that determines the spectral energy distribution of the emerging radiation. The latter depends on the poorly constrained efficiency of the energy exchange between electrons and protons in the plasma near the compact object. The values of the electron temperature typically derived from the spectral fits to the hard spectral component in accreting black holes, $kT_e \sim 50$–150 keV, are comfortably within the range defined by the two virial temperatures. However, differently from the case of the optically thick soft component temperature, the precise value of kT_e and its universality in a broad range of black hole masses and luminosities still remains unexplained from first principles. In addition, non-thermal processes in optically thin media (e.g. Comptonization on the non-thermal tail of the electron distribution) may also contribute to the X-ray emission from black holes in some spectral states, further complicating our basic picture.

The spectrum formed by the unsaturated Comptonization of low frequency seed photons with characteristic temperature T_{bb} on hot electrons with temperature T_e has a nearly power law shape in the energy range from $\sim 3kT_{bb}$ to $\sim kT_e$ (Sunyaev and Titarchuk 1980). For the parameters typical for black holes in AGN and X-ray binaries in the hard spectral state, this corresponds to the energy range from $\sim$ a few tens of eV–1 keV to ~ 50–100 keV (see e.g. the top panel of Fig. 4). The photon index Γ of the Comptonized spectrum depends in a rather complicated way on the parameters of the Comptonizing media, primarily on the electron temperature and the Thompson optical depth (Sunyaev and Titarchuk 1980). It is more meaningful to relate Γ to the Comptonization parameter y or, nearly equivalently, to the Compton amplification factor A. The latter describes the energy balance in the optically thin medium and is defined as the ratio of the energy deposition rate into hot electrons and the energy flux brought into the Comptonization region by soft seed photons. The exact shape of the $\Gamma(A)$ relation depends on the ratio T_{bb}/T_e of the temperatures of the seed photons and the electrons, the Thomson optical depth and the geometry, but broadly speaking, the higher the Compton amplification factor, the harder is the Comptonized spectrum (Sunyaev and Titarchuk 1989; Dermer et al. 1991; Haardt and Maraschi 1993).

Broadly speaking, significant part of, if not the entire diversity of the spectral behavior observed in accreting black holes of stellar mass can be explained by the changes in the proportions in which the gravitational energy of the accreting matter is dissipated in the optically thick and optically thin parts of the accretion flow. This is less so for supermassive

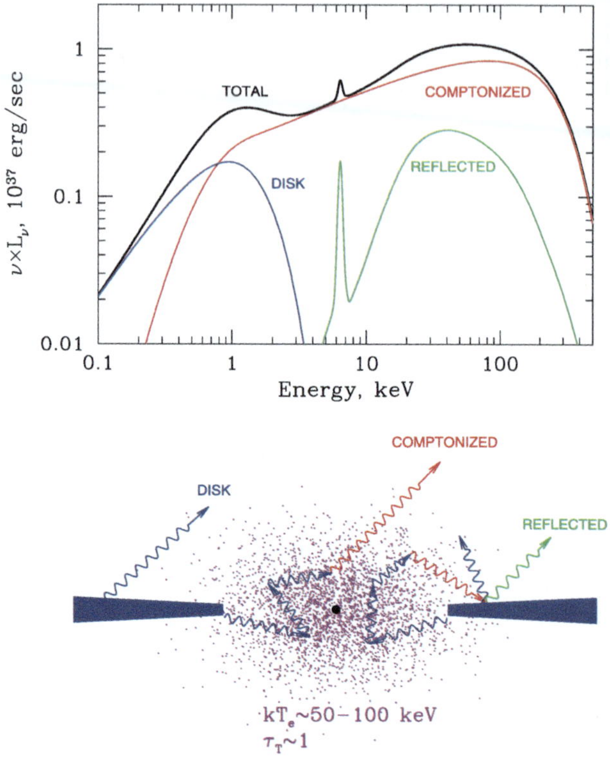

Fig. 4 The three main components of the X-ray emission from an accreting black hole (*top*) and a plausible geometry of the accretion flow in the hard spectral state (*bottom*)

black holes in AGN, where emission sites other than the accretion disk and hot corona may play significant role (e.g. broad and narrow emission line regions, see later in this chapter). The particular mechanism driving these changes is however unknown—despite significant progress in MHD simulations of the accretion disk achieved in recent years (Ohsuga and Mineshige 2011; Schnittman et al. 2013; Jiang et al. 2014) there is no accepted global model of accretion onto a compact object able to fully explain all the different spectral energy distributions observed, nor the transitions among them.

2.1 X-Ray Binaries: Geometry and Spectral Components

The contributions of optically thick and optically thin emission mechanisms can be easily identified in the observed spectra of X-ray binaries as soft and hard spectral components (Fig. 3). Depending on the spectral state of the source one of these components may dominate the spectrum or they can coexist giving comparable contribution to the total emission.

2.1.1 XRB Accretion Discs

The soft component is believed to originate in the geometrically thin and optically thick accretion disk of the Shakura-Sunyaev type (Shakura and Sunyaev 1973). If the Eddington ratio is high enough, the formation of such a disk seems unavoidable, and is indeed confirmed by the observed $L_{\rm disk} \propto T_{\rm bb}^4$ relation between disk luminosity and temperature in luminous XRB (Davis et al. 2006; Dunn et al. 2011). The expected spectrum is, to a first approximation, a superposition of blackbody spectra of different temperatures. For the zero-torque inner boundary condition, the local temperature at the radius r (for a non-spinning black hole) is given by:

$$T(r) = \left[\frac{3GM_{\rm BH}\dot{M}}{8\pi\sigma_{\rm SB}r^3}\left(1-\sqrt{\frac{6r_{\rm g}}{r}}\right)\right]^{1/4} \text{K} \tag{2}$$

A popular variation of the standard disk model used to interpret observations of X-ray binaries as well as AGN is the so-called "multicolor disk blackbody model" introduced by Mitsuda et al. (1984). Neglecting the second term in the parenthesis, and therefore valid for one particular inner torque boundary condition allowing easy integration of the total flux, this model has been widely and efficiently used in the era of more limited computer resources due to its simplicity, speed and early integration in the `XSPEC` spectral fitting package. To achieve a higher degree of accuracy one would need to consider full Eq. (2) or its analogs for different inner torque boundary conditions, deviations of the gravitational potential from the Newtonian and to account for such effects as distortion of the black body spectrum due to Thomson scatterings in the upper layers and in the atmosphere of the disk, Doppler effect due to rotation of the matter in the disk, etc. A number of models has been proposed to include these effects (Ebisawa et al. 1991; Shimura and Takahara 1995; Ross and Fabian 1996; Davis et al. 2006), many of them currently implemented in commonly used spectral fitting packages.

Conversely, having a well-sampled spectral energy distribution for the emission produced by a standard, Shakura and Sunyaev (1973) accretion disc around a black hole of known mass (and inclination), could in principle lead to useful constraints on the overall radiative efficiency of the accretion process, and therefore on the nature of the inner boundary condition of the accretion disc and on the black hole spin itself (see McClintock et al. 2014 contribution in this issue).

2.1.2 XRB Coronae

The Comptonization site—cloud of hot (thermal or non-thermal) electrons is often referred to as a "corona". Although it is generally accepted that the Comptonizing corona has to be located in the close vicinity of the compact object, there is currently no broad consensus on the detailed geometry of the region.

Nevertheless, it is clear that, due to heating of the disk by the Comptonized radiation from the corona and soft photon feedback, a *uniform* thermal corona above an optically thick accretion disk ('sandwich-like configuration') cannot explain observed hard spectra, with typical photon index of $\Gamma \sim 1.5$–2.0. Indeed, it has been first shown by Sunyaev and Titarchuk (1989) that the presence of a cool optically thick medium in the vicinity of the Comptonization region, e.g. an accretion disk or the surface of a neutron star, will affect the parameters of the latter and, consequently, the shape of the outgoing radiation. Some fraction of the Comptonized radiation will be returned to the accretion disk, increasing its temperature and, consequently, the soft photon flux through the Comptonization region. This in turn will increase the cooling rate, and will decrease the electron temperature in the corona, leading to softer and steeper spectra. We will further illustrate this point with the following simple quantitative consideration proposed in early 90-ies by E. Churazov and M. Gilfanov (Churazov et al. 1993; Gilfanov 2010). In a sandwich-like geometry, assuming moderate Thompson optical depth of the corona, $\tau_{\rm T} \sim 1$, a fraction $f \sim 1/2$ of the Comptonized emission will be returned to the accretion disk. Of this, a fraction of $1 - a$ (with $a \sim 0.2$ being the disc albedo) will be absorbed and will contribute to the heating of the accretion disk, adding to its heating due to the gravitational energy release. Ignoring the latter, the luminosity enhancement factor in the Comptonization region, defined as a ratio of its total luminosity to the luminosity of the seed photons will be $A \approx (1-a)^{-1} f^{-1} < 2.5$. As it is well known (Dermer et al. 1991), the luminosity enhancement factor is intimately related to the Comptonization parameter y and the photon index Γ of the Comptonized radiation. The above constrain on A implies $\Gamma > 2.3$, steeper than the hard state spectra typically observed in black holes.[3] This conclusion is confirmed by the full treatment of the Comptonization problem in "sandwich" geometry (Haardt and Maraschi 1993). In order to produce a harder Comptonized spectrum, the value of the feedback coefficient f needs to be reduced. This is achieved, generically, in "photon starved" geometries, i.e. in configurations where the Comptonization region only intercepts a fraction of the reprocessed disc flux. Examples of such kind of geometry are the "sombrero" configuration (Poutanen et al. 1997) (also known as "truncated disc"), or a non-uniform, patchy and/or non-stationary corona. Another example suggested by Beloborodov (1999) involves bulk motion of the corona with mildly relativistic velocity away from the disk reducing the feedback coefficient due to the aberration effect.[4]

One of the variants of the "sombrero" configuration is depicted in the bottom panel of Fig. 4. It is assumed that outside some truncation radius the accretion takes place predominantly via an optically thick and geometrically thin accretion disk, whereas closer to the compact object the accretion disk is transformed into a hot optically thin and geometrically thick flow with the aspect ratio of $H/R \sim 0.5$–1. The soft (optically thick) and hard (optically thin) spectral components are formed in the accretion disk and the hot inner flow correspondingly. The value of the truncation radius can be inferred from observations, both in the energy and time domains (e.g. Gilfanov et al. 2000). Although their interpretation is not unique and unambiguous, the plausible range of values is between ~ 3 and a few hundred gravitational radii.

There is no commonly accepted mechanism of truncation of the disk and formation of the corona, with a number of plausible scenarios having been investigated in the literature. Among the more promising ones is the evaporation of the accretion disk under the effect of the heat conduction. It was initially suggested to explain quiescent X-ray emission from cataclysmic variables by Mayer and Meyer-Hofmeister (1994) and was later applied to the case of accretion onto black holes and neutron stars (Meyer et al. 2000). It not only provides a physically motivated picture describing formation of the corona and destruction of the optically thick disk but also correctly predicts the ordering of spectral states vs. the mass accretion rate. Namely, it explains the fact that hard spectra indicating prevalence of the hot optically thin flow are associated with lower $\dot{M}$ values, whereas the optically thick disk appears to dominate the photon production in the accretion flow at higher $\dot{M}$. This is contrary to the expectations of the early models which ascribed the presence of the hot optically thin plasma in the vicinity of the compact object to disk instabilities occurring at *high* accretion rates (Lightman and Eardley 1974).

Another geometrical configuration considered in the context of hard X-ray emission from black holes is a non-stationary and non-uniform (patchy) corona above the optically thick accretion disk. This scenario has been largely inspired by the suggestion of Galeev et al.

[3] Considerations of a similar kind involving the neutron star surface can explain the fact that the neutron star spectra are typically softer than those of black holes (e.g. Sunyaev and Titarchuk 1989).

[4] A uniform stationary corona above the accretion disk can still, in principle, be responsible for the steep power law component often detected in the soft state, although a non-thermal electron distribution may be a more plausible explanation.

(1979), that a magnetic field amplified in the hot inner disk by turbulence and differential rotation may reach the equipartition value and emerge from the disk in the form of buoyant loop-like structures of solar type above its surface. These structures may lead to the formation of a hot magnetically confined structured corona similar to the solar corona, which may produce hard emission via inverse Compton and bremsstrahlung mechanisms. The model could also explain the faintness of the hard emission in the soft state as a result of efficient cooling of plasma in the magnetic loops via inverse Compton effect due to increased flux of soft photons at higher $\dot{M}$.

Although the original Galeev, Rosner and Vaiana paper was focused on the thermal emission from hot plasma confined in buoyant magnetic loops, the latter are a plausible site of particle acceleration responsible for the non-thermal component in the electron distribution. This is expected on theoretical grounds, and is illustrated very well observationally by the presence of the non-thermal emission component in the spectra of solar flares (see e.g. Coppi 1999, and references therein). On the other hand, a power law-shaped hard X-ray component is commonly observed in the spectra of black hole candidates. It reveals itself most graphically in the soft spectral states, but may be as well present in the hard state, along with the thermal Comptonized spectrum (e.g. McConnell et al. 2002). This power law component has a photon index of $\Gamma \sim 2$–3 and, although it may extend into the several hundred keV–MeV range (e.g. Sunyaev et al. 1992), it is relatively unimportant energetically, contributing a small fraction to the total radiation output of the black hole. This is in contrast with the hard spectral component produced by thermal Comptonization in the hard spectral state, which accounts for dominant fraction of the source luminosity.

2.2 X-Ray Binaries: Spectral States

The existence of different spectral states is a distinct feature of X-ray binaries, independently of the nature of the compact object (Fig. 3). Although their phenomenology is far richer, including clear signs of hysteresis behaviour (Fender et al. 2004; Dunn et al. 2010), for the purpose of this discussion we will restrict ourselves to the simple dichotomy between soft (high) and hard (low) spectral states and refer to other chapters of this book for a more detailed discussion. As no global self-consistent theory/model of accretion state transitions exists, all theories explaining spectral states have to retreat to qualitative considerations. These considerations, although phenomenological in nature, usually are based on numerous observations of black hole systems, simple theoretical arguments and some simplified solutions and simulations of the accretion problem. Described below is a plausible, although neither unique nor unanimously accepted scenario of this kind based on the "sombrero" geometry of the accretion flow. There is a number of cartoons and geometry sketches, illustrating this and other scenarios which we will not repeat here and will refer the interested reader to original works (e.g. Zdziarski and Gierlinski 2004; Done et al. 2007).

As obvious from the left panel of Fig. 3, the spectral states phenomenon is related to the redistribution of the energy released in the optically thick and optically thin components of the accretion flow. In the sombrero configuration one may associate spectral state transitions with change of the disk truncation radius—the boundary between the outer optically thick accretion disk and the inner optically thin hot flow.

In the soft (aka disk-dominated) spectral state, the optically thick accretion disk extends close to the compact object, possibly to the last marginally stable Keplerian orbit ($r = 6r_g$ for a Schwarzschild black hole), leaving no "room" for the hot optically thin flow. Therefore the major fraction of the accretion energy is emitted in the optically thick accretion disk giving rise to a soft spectrum of the multicolor blackbody type. The magnetic activity at the disk surface may (or may not) produce a hard power-law like tail due to non-thermal Comptonization in the corona. As discussed in the previous section, this power law component has a steep slope $\Gamma \sim 2$–3 and is relatively insignificant energetically.

In the hard (aka corona-dominated) spectral state, the accretion disk truncates at larger distances from the black hole, at ~ 20–$100 r_g$ or further out, depending on the luminosity (but see Reynolds and Miller 2013, for a contrasting view). The major fraction of the gravitational energy is released in the hot inner flow. Comptonization of soft photons emitted by the accretion disk on the hot thermal electrons of the inner flow leads to the formation of the hard spectrum of the shape characteristic for unsaturated thermal Comptonization. The typical parameters in the Comptonization region—hot inner flow, inferred from observations are: electron temperature of $T_e \sim 100$ keV and Thompson optical depth of $\tau_T \sim 1$. The significance of the soft blackbody-like emission from the optically thick disk as well as of the non-thermal emission due to magnetic flares at its surface varies depending on the disk truncation radius, increasing as the disk moves inwards. There is evidence that both thermal and non-thermal hard components may co-exist in the hard state in the certain range of the disk truncation radii, as suggested in Ibragimov et al. (2005).

As a side note, for the purpose of this chapter, we are limiting ourselves to rather simple and general considerations aimed to convey the "large-scale picture". However, observational data amassed to date permits to go much further and to increase complexity and level of sophistication of theoretical models, allowing comprehensive tests of models against observations in the multi-wavelength spectral and timing domains. Such an attempt to go beyond simplified qualitative description and to develop a detailed model of spectral formation in the hard state of X-ray binaries extensively testing it against spectral and timing data is presented in one of the chapters in this book by Poutanen and Veledina (2014). Furthermore, in this chapter we do not discuss the possible contribution of jets and disk winds. Although they may play an important role in some configurations of the accretion flow both in stellar mass and supermassive black holes, their contribution does not change the overall picture in the objects, spectral states and wavelength ranges considered here. These phenomena are discussed in detail in the corresponding section of this book.

2.3 AGN Spectral Energy Distribution: From Phenomenology to Physics

In the case of AGN, the observational characterization of accretion is hampered by the uniquely complex multi-scale nature of the problem. Such a complexity greatly affects our ability to extract reliable information on the nature of the accretion processes in AGN and does often introduce severe observational biases, that need to be accounted for when trying to recover the underlying physics from observations at various wavelengths, either of individual objects or of large samples (see Sect. 4 below).

Like any accreting black hole, an AGN releases most of its energy (radiative or kinetic) on the scale of a few Schwarzschild radii ($\sim 10^{-5}$ pc for a $10^8 M_\odot$ BH). However, the mass inflow rate (accretion rate) is not determined by stellar (and binary) evolution physics, but rather determined by the galaxy ISM properties at the Bondi radius (i.e. at the location where the gravitational influence of the central black hole starts dominating the dynamics of the intergalactic gas), some 10^5 times further out. Broad permitted atomic emission lines, used to estimate SMBH masses in QSOs (Peterson et al. 2004), are produced at ~ 0.1–1 pc (Broad Line Region, BLR), while, on the parsec scale, and outside the sublimation radius, a dusty, large-scaleheight, possibly clumpy, medium obscures the view of the inner engine (Elitzur 2008) crucially determining the observational properties of the AGN (Netzer 2008); on the

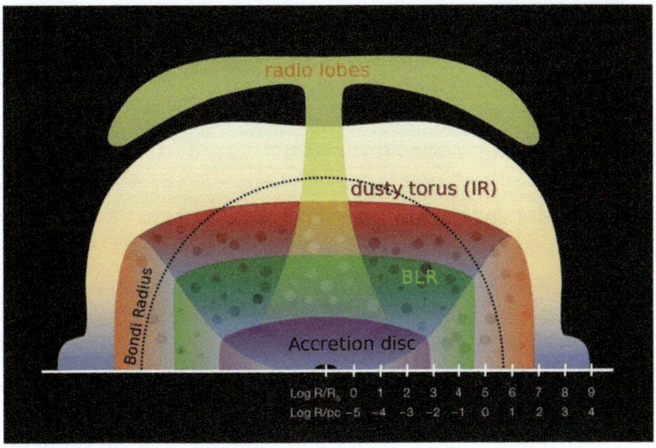

Fig. 5 A schematic, logarithmic view of an AGN-galaxy system. *Scales* on the *bottom axes* are in units of Schwarzschild radii and parsecs, and are inferred for a $\sim 10^{11} M_{\odot}$ galaxy containing a $\sim 10^{8} M_{\odot}$ black hole. The approximate location of the Bondi radius is marked by a *thick dashed line*. See text for details. Courtesy A. Merloni, ESO Graphics

same scale, powerful star formation might be triggered by the self-gravitational instability of the inflowing gas (Goodman 2003). Finally, AGN-generated outflows (either in the form of winds or relativistic jets) are observed on galactic scales and well above (from a few to a few hundreds kpc, some $\sim 10^{8}$–10^{10} times r_g!), often carrying substantial amounts of energy that could dramatically alter the (thermo-)dynamical state of the inter-stellar and inter-galactic medium.

A schematic logarithmic map of a AGN-galaxy system is shown in Fig. 5, where we have highlighted the various physical regions of interest. Bringing all of the above into a coherent framework is indeed a formidable challenge: each physical process active on each different physical scale has a signature in a different part of the electromagnetic spectrum (so that different instruments are needed to unveil it).

2.3.1 AGN SED Phenomenology

If AGN power comes from release of gravitational energy of the accreting matter, we need to identify in the SED phenomenology the various ingredients suggested by accretion theory: in particular, we focus here on the ubiquitous optically thick and geometrically thin accretion disc, and on the hot, X-ray emitting corona.

Up until recent years accurate SED of accreting SMBH were constructed mainly from bright un-obscured (type-1) QSO samples. Setting the standard for almost 20 years, the work of Elvis et al. (1994), based on a relatively small number (47) of UV/X-ray selected quasars, has been used extensively as a template for the search and characterization of nearby and distant AGN. The Elvis et al. SED is dominated by AGN accreting at the highest Eddington ratio. As shown by the blue and red curves in the right panel Fig. 3, this spectral energy distribution is characterized by a relative flatness across many decades in frequency, with superimposed two prominent broad peaks: one in the UV part of the spectrum (the so-called Big Blue Bump; BBB), one in the Near-IR, separated from an inflection point at about 1 μm. The former is believed to be associated to the emission from the optically thick (and geometrically thin) accretion disc, while the latter from the reprocessed emission from the surrounding dusty, molecular material responsible for the nuclear extinction and absorption in obscured (type-2) AGN.

Subsequent investigations based on large, optically selected QSO samples (most importantly the SDSS one, Richards 2006) have substantially confirmed the picture emerged from the Elvis et al. (1994) study. Apart from a difference in the mean X-ray-to-optical ratio (optically selected samples tend to be more optically bright than X-ray selected ones, as expected), the SDSS quasars have indeed a median SED similar to those shown in Fig. 3 for AGN accreting at $L/L_{\rm Edd} > 0.1$, despite the difference in redshift and sample size.

2.3.2 AGN Accretion Discs

It is a major success of the Shakura and Sunyaev (1973) theory that, for typical AGN masses and luminosities (and thus accretion rates), the expected spectrum of the accretion disc should peak in the optical-UV bands (see Eq. (2) above), as observed. Indeed, a primary goal of AGN astrophysics in the last decades has been to model accurately the observed shape of the BBB in terms of standard accretion disc models, and variations thereof.

The task is complicated by at least three main factors. First of all, standard accretion disc theory, as formulated by Shakura and Sunyaev (1973), needs to be supplemented by a description of the disc vertical structure and, in particular, of its atmosphere, in order to accurately predict spectra. This, in turn, depends on the exact nature of viscosity and on the micro-physics of turbulence dissipation within the disc. As in the case of XRB, models for geometrically thin and optically thick AGN accretion discs has been calculated to increasing levels of details, from the simple local blackbody approximation to stellar atmosphere-like models where the vertical structure and the local spectrum are calculated accounting for the major radiative transfer processes (e.g. the TLUSTY code of Hubeny et al. 2000)

A second complicating effect, a purely observational one, is the fact that the intrinsic disc continuum emission is often buried underneath a plethora of emission lines, many of which broadened significantly by gas motions in the vicinity of the central black hole, that can reach non-negligible fraction of the speed of light. In particularly favorable geometrical conditions, however, optical spectra in polarised light, removing "contaminating" line emission, show a broad dip possibly corresponding to the Balmer edge absorption expected from an accretion disc atmosphere (Kishimoto et al. 2003). Extending the polarised continuum into the near-IR reveals the classic long wavelength $\nu^{1/3}$ spectrum expected from simple accretion disc models (Kishimoto et al. 2008).

Finally, the real physical condition in the inner few hundreds of Schwarzschild radii of an AGN might be more complicated than postulated in the standard accretion disc model: for example, density inhomogeneities resulting in cold, thick clouds which reprocess the intrinsic continuum have been considered at various stages as responsible for a number of observed mismatches between the simplest theory and the observations (see e.g. Guilbert and Rees 1988; Merloni et al. 2006; Lawrence 2012, and references therein).

The above mentioned problems are particularly severe in the UV part of the spectrum, where observations are most challenging. Shang et al. (2005) compared broad-band UV-optical accretion disc spectra from observed quasars with accretion disc models. They compiled quasi-simultaneous QSO spectra in the rest-frame energy range 900–9000 Å and fitted their continuum emission with broken power-law models, and then compared the behavior of the sample to those of non-LTE thin-disk models covering a range in black hole mass, Eddington ratio, disk inclination, and other parameters. The results are far from conclusive: on the one hand, the observed slopes are in general consistent with the expectations of sophisticated accretion disc models. On the other hand, the spectral UV break appears to always be around 1100 Å, and does not scale with the black hole mass in the way expected.

Jin et al. (2012), on the other hand, have looked at detailed continuum fits to the joint optical-UV-X-ray SED of 51 nearby AGN with known black hole masses, and studied the variation of the mean SED as a function of various parameters, such as X-ray spectral index Γ, black hole mass, bolometric luminosity and Eddington ratio. They found global trends in the basic disc properties (such as its peak temperature), in correlation with the main parameter of the accretion flow. From a Principal Component analysis, Jin et al. (2012) found that the first two eigenvectors contain $\sim 80\,\%$ of all correlations in the matrix, with the first one strongly correlating with black hole mass, and the second one with the bolometric luminosity, while both correlate with the Eddington ratio. Interestingly, this turns out to be consistent with the results of a principal component analysis of the emission line-dominated QSO spectra (Boroson 2002).

As in the case of XRB, a good modelling of the accretion disc could be employed in order to constrain BH spin from the disc continuum measurements. However, in a typical AGN, the above-mentioned observational intricacies need to be dealt with, together with the fact that the BBB is much worse sampled than in a stellar mass black hole. On the other hand, the uncertainty in the distance to the object, that plagues the studies of galactic black holes is not an issue for QSOs with measured spectroscopic redshift.

Davis and Laor (2011) have made a first systematic attempt to estimate the radiative efficiencies in a sample of QSOs. In individual AGN, thin accretion disk model spectral fits can be used to deduce the absolute accretion rate $\dot{M}$, if the black hole mass $M_{\rm BH}$ is known. In fact, by measuring the continuum disk luminosity in the optical band (i.e. in the Rayleigh-Taylor part of the optically thick multi-color disc spectrum), the accretion rate estimates are relatively insensitive to the actual model of the disc atmosphere.

The radiative efficiency $\epsilon_{\rm rad}$ is then set by the ratio of the bolometric luminosity $L_{\rm bol}$ to $\dot{M}c^2$, and the main task is the non-trivial one of measuring accurately the intrinsic bolometric luminosity of the accretion disc. The results of Davis and Laor (2011) indicate an average radiative efficiency of about $10\,\%$ for SMBH of about $10^8 M_\odot$. Because of the large systematics related to the simultaneous measures of the black hole mass and the bolometric luminosity, it is still premature to speculate on possible trends in the accretion efficiency as a function of either black hole mass or luminosity. Larger samples of AGN with good coverage of the UV-optical SED across a wide range of redshift and luminosity will be needed to make further progress.

2.3.3 AGN Coronae

As in the case of XRBs, the exact mechanism responsible for AGN X-ray emission and its physical location are not fully understood yet. Generically, the X-ray spectra are dominated by a power-law in the 2–10 keV energy range, with a relative narrow distribution of slopes: $\langle \Gamma \rangle = 1.8 \pm 0.2$ (Nandra and Pounds 1994; Steffen et al. 2006; Young et al. 2009), consistent with the expectations of Comptonization models discussed above. Superimposed on it, a narrow iron Kα emission line is the most prominent feature in AGN X-ray spectra. Such a line, produced by cold, distant material is clearly dependent on luminosity, with more luminous sources having smaller equivalent widths (the so-called Iwasawa-Taniguchi effect, Iwasawa and Taniguchi 1993). Also, a clear reflection component (George and Fabian 1991; Reynolds 1998, and references therein) is observed in a number of nearby AGN (Fig. 4, top panel), and further required by CXRB synthesis models (Gilli et al. 2007). We will discuss further the relationship between reflection component and X-ray spectra in Sect. 2.4.

Another key question we would like to address with AGN SED studies is to what extent the overall properties of the accretion disc-corona systems around SMBH can be traced to those of the known spectral states of XRB. Because of the complexities of galactic nuclei discussed above, and because the discs and coronae of AGN emit in distinct parts of the electromagnetic spectrum, it is much more difficult to clearly distinguish between different

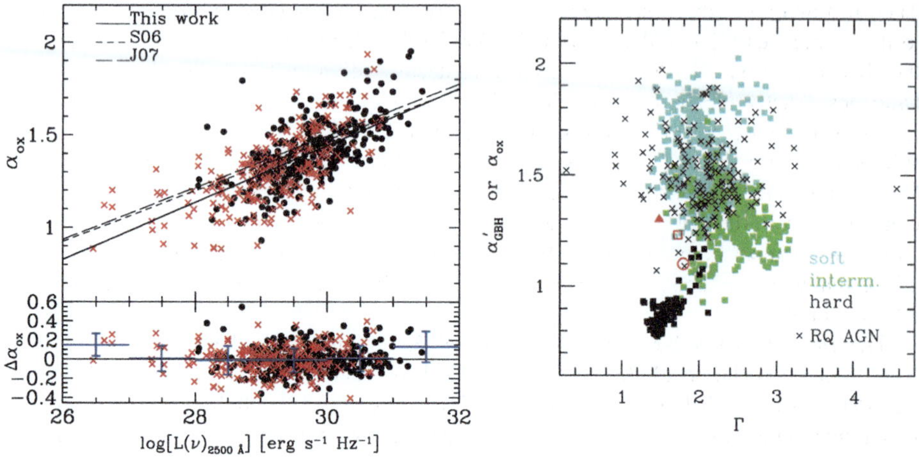

Fig. 6 *Left*: The optical-to-X-ray spectral slope $\alpha_{\rm ox}$ as a function of luminosity density at 2500 Å. From Lusso et al. (2010). *Right*: Relation between the photon index, Γ, and a "disc-to-Comptonization" index: $\alpha_{\rm ox}$ for radio-quiet quasars (*black crosses*) and the analogous (rescaled) quantity for Galactic black holes in a hard (*black*), intermediate (*dark grey/green*) and soft (*light grey/cyan*) spectral state. From Sobolewska et al. (2009)

spectral states in terms of a simple power ratios between the two main spectral components. Nevertheless, as a very general diagnostic, the "X-ray loudness", usually characterized by the $\alpha_{\rm ox}$ parameter, i.e. the slope of the spectrum between 2500 Å = 5 eV and 2 keV: $\alpha_{\rm ox} = 0.3838 \log(F_{\rm 2keV}/F_{2500})$ can be used to characterize the fraction of bolometric light carried away by high-energy X-ray photons. Recent studies of large samples of both X-ray and optical selected AGN have clearly demonstrated that $\alpha_{\rm ox}$ is itself a function of UV luminosity, with less luminous objects being more X-ray bright (see e.g. Steffen et al. 2006; Lusso et al. 2010; Jin et al. 2012 and Fig. 6). In very general terms, this might point towards a connection between accretion disc physics and the mechanism(s) of coronae generation in AGN (see e.g. Merloni 2003; Wang et al. 2004, for a possible scenario).

In the right panel of Fig. 6 the $\alpha_{\rm ox}$ parameter for a sample of radio-quiet type-1 (un-obscured) QSOs is compared with the (appropriately rescaled to account for the black hole mass difference) analogous quantity for XRBs in different states. Clearly, bright un-obscured QSOs are consistent with being (scaled) analogous of stellar mass systems in soft or intermediate spectral states (Sobolewska et al. 2009).

At even lower accretion rates, the precise determination of AGN SED is severely hampered by the contamination from stellar light, and high resolution imaging is needed to robustly separate the accretion-related emission. As a consequence, only for a handful of objects it has been possible to derive accurate estimates of the spectral shape of Low-Luminosity AGN (LLAGN) across the electromagnetic spectrum (Ho 2008, and references therein). Nevertheless, understanding the accretion state of these weak AGN is important. The ubiquity of SMBH in the nuclei of nearby galaxies implies that, in the local Universe, AGN of low and very low luminosity vastly outnumber their bright and active counterparts. An important step towards the classification of AGN in terms of their specific modes of accretion was taken by Merloni et al. (2003) and Falcke et al. (2004), whereby a "fundamental plane" relation between mass, X-ray and radio core luminosity of active black holes was discovered and characterized in terms of accretion flows (see review of E. Koerding in this issue). In particular, the observed scaling between radio luminosity, X-ray luminosity and

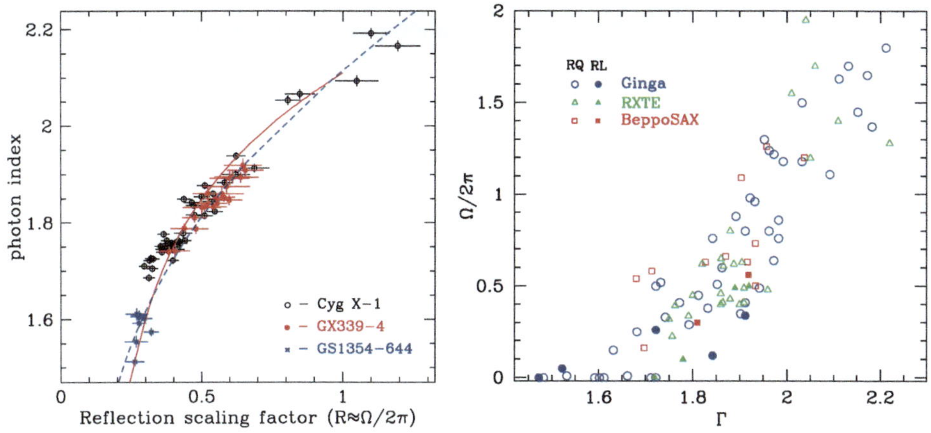

Fig. 7 The R–Γ correlations between the photon index Γ of the Comptonized radiation and the relative amplitude of the reflected component R in X-ray binaries and AGN. The *left panel* shows results for three "well-behaved" black hole systems Cyg X-1, GX339–4 and GS1354–644. The *solid* and *dashed lines* show the dependence $\Gamma(R)$ expected in the disk-spheroid and in the plasma ejection models. Adopted from Gilfanov (2010). The *right panel* shows R–Γ relation for supermassive black holes, from Zdziarski et al. (2003). Note that *horizontal* and *vertical axis* are transposed in the *left* and *right panels*

BH mass implies that the output of low-luminosity AGN is dominated by kinetic energy rather than by radiation (Sams et al. 1996; Merloni and Heinz 2007; Hardcastle et al. 2007; Best and Heckman 2012). This is in agreement with the average SED of LLAGN (Ho 2008; Nemmen et al. 2014) displaying a clear lack of thermal (BBB) emission associated to an optically thick accretion disc, strongly suggestive of a "truncated disc" scenario and of a radiative inefficient inner accretion flow. If that is the case, the total kinetic energy output from LLAGN could provide a sizable, if not dominant, fraction of the total AGN feedback in the local Universe (Merloni and Heinz 2008; Cattaneo et al. 2009).

2.4 R–Γ Correlations in X-Ray Binaries and AGN

Observations show that spectral and timing parameters of accreting black holes often change in a correlated way (e.g. Gilfanov et al. 1999; Revnivtsev et al. 2001; Pottschmidt et al. 2003). One of the most significant correlations is the one between the photon index of the Comptonized spectrum, the amplitude of the reflected component and the characteristic frequencies of aperiodic variability (Fig. 7). The correlation between spectral slope and reflection amplitude is also known as R–Γ correlation (Gilfanov et al. 1999; Zdziarsk et al. 1999). Its importance is further amplified by the fact that it is also valid for supermassive black holes (Fig. 7) (Zdziarski et al. 2003).

The R–Γ correlation allows a simple and physically motivated interpretation, as described below. The strength of the reflected component in the spectrum depends on the fraction of the Comptonized radiation intercepted by the accretion disk (e.g. Gilfanov 2010). The latter is defined by the geometry of the accretion flow, namely, by the solid angle Ω_{disk} subtended by the accretion disk as seen from the corona. In addition, the spectrum of the reflected emission depends on the ionization state of the disk, in particular its low energy part which is formed by the interplay between Thomson scattering and photoabsorption and fluorescence by metals. The problem is further complicated by the fact that the ionization state of the disk can be modified by the Comptonized radiation.

Observations show that there is a clear correlation between the photon index of the Comptonized radiation Γ (i.e. the Comptonization parameter) and the relative amplitude of the reflected component R (Fig. 7). Softer spectra (lower value of the Comptonization parameter y and of the Compton amplification factor A) have stronger reflected component, revealing itself, for example, via a larger equivalent width of the iron fluorescent line. This suggests that there is a positive correlation between the fraction of the Comptonized radiation intercepted by the accretion disk and the energy flux of the soft seed photons to the Comptonization region (Zdziarsk et al. 1999). This is a strong argument in favor of the accretion disk being the primary source of soft seed photons to the Comptonization region (however, see Veledina et al. 2011 and the contribution of Poutanen and Veledina 2014 in this issue for an alternative point of view). Indeed, in the absence of strong beaming effects a correlation between $\Omega_{\rm disk}$ and the seed photons flux should be expected since an increase of the solid angle of the disk seen by the hot electrons ($=\Omega_{\rm disk}$) should generally lead to the increase of the fraction of the disk emission reaching the Comptonization region.

3 Populations of X-Ray Binaries and Stellar Mass Black Holes in External Galaxies

X-ray binaries are the most luminous type of compact X-ray sources in the Milky Way and other normal (i.e. without significant nuclear activity) galaxies, providing the dominant contribution to their total X-ray output (see Fabbiano 2006, for a review). In this section we discuss their demographics. In particular, we consider how their populations are related to the fundamental characteristics of their host galaxies, such as stellar mass and star-formation history and how these scaling relations can be used to shed light on formation and evolution of their different types, including ultra-luminous X-ray sources.

3.1 Scaling Relations for X-Ray Binaries

3.1.1 General Considerations

Depending on the mass of the optical companion, X-ray binaries are subdivided in to two classes—high- and low-mass X-ray binaries, separated by a thinly populated region between $\sim 1M_\odot$ and $\sim 5M_\odot$, where there are practically no bright persistent sources. The difference in the mass of the donor star determines the difference in the characteristic evolutionary time scales of these two types of X-ray binaries. In the case of a massive donor, the longest time scale is determined by the nuclear evolution time scale of the donor star and does not exceed a few tens of Myrs (Verbunt and van den Heuvel 1995). This time scale is comparable to the characteristic time scale of the star-formation episode, therefore one may expect that the number of such systems in a galaxy is proportional to its star-formation rate (Sunyaev et al. 1978; Grimm et al. 2003; Mineo et al. 2012):

$$N_{\rm HMXB}, L_{\rm X,HMXB} \propto {\rm SFR} \tag{3}$$

Evolution of Low-Mass X-ray Binaries (LMXB), on the contrary, is determined by the rate of loss of the orbital angular momentum of the binary system or by the nuclear evolution of the low-mass star, both of which are typically in the ~ 1–10 Gyrs range (Verbunt and van den Heuvel 1995). Correspondingly, one may expect that their population scales with the total mass of stars in the host galaxy (Gilfanov 2004):

$$N_{\rm LMXB}, L_{\rm X,LMXB} \propto {\rm M}_* \tag{4}$$

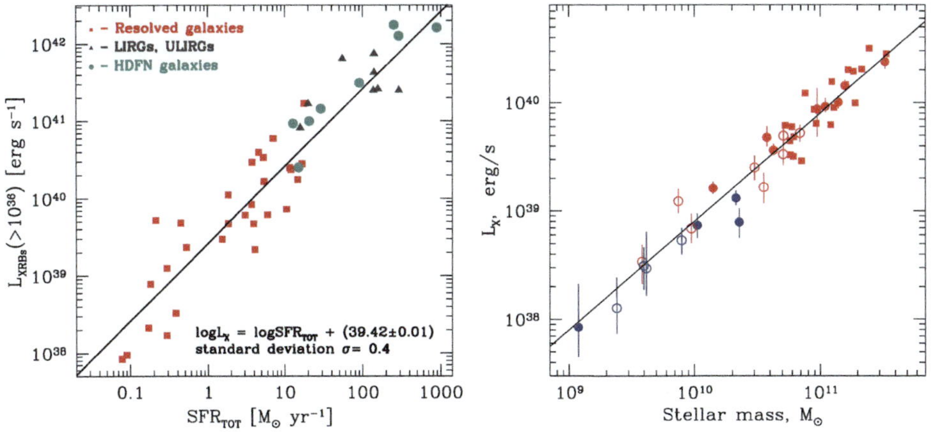

Fig. 8 Dependence of the total X-ray luminosity of X-ray binaries on the star-formation rate (*left panel*) and stellar mass (*right panel*) of the host galaxy. *Left panel* shows star-forming galaxies, young stellar population of which is dominated by massive X-ray binaries; their population is roughly proportional to the star-formation rate of the host galaxy. *Right panel* shows data for elliptical galaxies where star-formation mostly stopped at least several Gyrs ago and only low-mass X-ray binaries are left. Their population is determined by the total stellar mass of the host galaxy. *Solid lines* show approximation of the data by the linear laws. In the *left panel* we also show the data for ULIRGs (*triangles*) and star-forming galaxies from the Chandra Deep Fields. These galaxies are not resolved by Chandra, therefore the total luminosity is shown, including contribution of faint unresolved compact sources and diffuse emission. Adopted from Gilfanov (2004); Mineo et al. (2012); Zhang et al. (2012)

3.1.2 Chandra Results

Chandra observatory, thanks to its sub-arcsec angular resolution, opened a new era in studying X-ray binary populations in nearby galaxies. For the first time an opportunity was presented to observe compact sources in nearby galaxies (out to ~ 30–100 Mpc) in a nearly confusion free regime and to measure luminosity functions and to determine total luminosities of different populations of compact sources. Chandra observation of large number (~ 100) of nearby galaxies have demonstrated that indeed, populations of LMXBs and HMXBs in a galaxy scale proportionally to its stellar mass and star formation rate respectively (Fig. 8):

$$L_{X,\mathrm{HMXB}} \approx 2.5 \cdot 10^{39} \times \mathrm{SFR}, \qquad N_{\mathrm{HMXB}} \approx 13 \times \mathrm{SFR} \tag{5}$$

$$L_{X,\mathrm{LMXB}} \approx 1.0 \cdot 10^{39} \times \frac{\mathrm{M}_*}{10^{10} M_\odot}, \qquad N_{\mathrm{LMXB}} \approx 14 \times \frac{\mathrm{M}_*}{10^{10} M_\odot} \tag{6}$$

where L_X is the total X-ray luminosity of X-ray binaries of the given type in the 0.5–8 keV energy band, N_X the number of X-ray binaries with luminosity exceeding $L_X \geq 10^{37}$ erg/s, SFR is the star-formation rate in $M_\odot$/yr, and M_* is the stellar mass of the galaxy (Grimm et al. 2003; Mineo et al. 2012; Gilfanov 2004). Consistent scaling relations were obtained in several other independent studies (e.g. Ranalli et al. 2003; Colbert et al. 2004; Kim and Fabbiano 2004; Lehmer et al. 2010).

Depending on the sample selection and the significance of the observer bias, in the low SFR and stellar mass regime the L_X–SFR and L_X–M_* relations for a small sample of randomly selected galaxies may be modified by the effects of small number statistics (Gilfanov et al. 2004). The most pronounced these effects are for HMXBs (Grimm et al. 2003). Note that these effects are not seen in Fig. 8 because of the observer bias introduced by selection

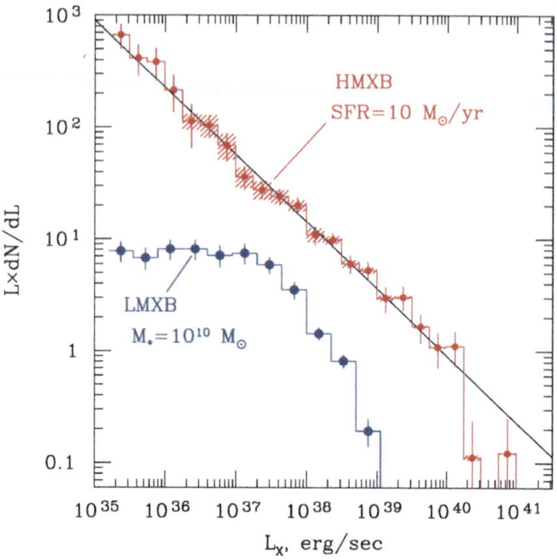

Fig. 9 Average X-ray luminosity functions of compact X-ray sources in star-forming (marked "HMXB") and elliptical (marked "LMXB") galaxies. In star-forming galaxies high-mass X-ray binaries dominate, whereas in old elliptical galaxies the dominant component of X-ray populations are low-mass X-ray binaries. Luminosity functions are normalized to star-formation rate of SFR $= 10 M_{\odot}$/yr and stellar mass of $M_* = 10^{10} M_{\odot}$ respectively. Based on results of Gilfanov (2004) and Mineo et al. (2012)

of galaxies—targets for Chandra observations (for example, the proposers may have predominantly selected galaxies that were known to have an enhanced level of X-ray emission, e.g. based on the ROSAT All-Sky Survey) (Mineo et al. 2012).

3.2 X-Ray Luminosity Functions of X-Ray Binaries

Chandra observations of a large number of nearby galaxies showed that X-ray luminosity functions of compact X-ray sources in different galaxies have similar shape, differing only in normalization. These shapes are different in star-forming and elliptical galaxies, i.e. for high- and low-mass X-ray binaries. As it should be obvious from the existence of scaling relations (Sect. 3.1), normalization of the HMXB and LMXB luminosity functions scale proportionally to the star-formation rate and stellar mass of the host galaxy. Thus, to the first approximations, luminosity distributions of X-ray binaries can be described by universal luminosity functions. These are plotted in Fig. 9.

The shapes of the universal XLFs of high- and low-mass X-ray binaries are qualitatively different. The difference is mainly caused by the difference in the mass transfer regime in high and low-mass X-ray binaries. Indeed, in the majority of the former the compact object accretes material from the wind of the massive donor star. Therefore the luminosity function is determined primarily by the mass distribution of the donors in high-mass X-ray binaries (Postnov 2003) which leads to the formation of the observed power law luminosity distribution (Grimm et al. 2003; Mineo et al. 2012):

$$\frac{dN_{\mathrm{HMXB}}}{dL_{\mathrm{X}}} \propto \mathrm{SFR} \times L^{-1.6} \tag{7}$$

In the case of low-mass X-ray binaries, on the contrary, the mass transfer occurs via donor star Roche lobe overflow through the inner Lagrangian point of the binary system and the X-ray luminosity function of these systems is determined by the orbital parameter distribution of semi-detached binary systems in the galaxy. This leads to formation of the luminosity distribution of a complex shape, with two breaks at $\log L_X \sim 38.5$ and $\log L_X \sim 37$–37.5 (Fig. 9) (Gilfanov 2004; Kim and Fabbiano 2010). The first break is located near Eddington luminosity of the neutron star and is likely related with the existence of the luminosity limit for an accreting neutron star—compact objects in more luminous systems are black holes whose occurrence rates in the population are smaller. The nature of the second break is still not clear.

3.3 Ultraluminous X-Ray Sources

An unusual class of compact sources—ultraluminous X-ray sources, has been discovered in nearby galaxies about two decades ago (Colbert and Mushotzky 1999; Fabbiano 2006). Although bright, $L_X > 10^{39}$ erg/s, point-like sources are found both in young star-forming galaxies and in old stellar population of elliptical and S0 galaxies (Zhang et al. 2012), the most luminous and exotic objects are associated with actively star-forming galaxies. Their nature and relation to more ordinary X-ray binaries is still a matter of a significant debate. Based on a simple Eddington luminosity argument, they appear to be powered by accretion onto an intermediate mass object—a black hole with the mass in the hundreds-thousands solar masses range (Fabbiano 2006; Miller et al. 2003). However, a number of alternative models have been considered as well—from collimated radiation (Koerding et al. 2002) to ~stellar mass black holes, representing the high mass tail of the standard stellar evolution sequence and accreting in the near- or slightly super-Eddington regime (King et al. 2001; Grimm et al. 2003).

As it is well known, the maximum mass of a black hole produced in the course of standard stellar evolution at solar abundance of elements is limited to $\approx 10\text{–}20M_\odot$, whereas formation of more massive black holes with mass exceeding ~ 100 is only possible at virtually zero abundance of metals (Zhang et al. 2008). It is possible in principle that the most luminous sources are accreting intermediate mass black holes—descendants of Pop III stars, which acquired a massive companion in star-forming regions. Obviously, the abundance of such systems should be significantly smaller than the abundance of normal high-mass X-ray binaries formed in the course of standard stellar evolution. Therefore there must be a break in the luminosity function at the transition between "normal" X-ray binaries and these objects. However, observations show that the luminosity distribution of compact X-ray sources in star forming galaxies smoothly extends up to the luminosities of $\log L_X \sim 40\text{–}40.5$, without any significant features or slope changes. In particular, unlike the LMXB XLF, it does not have any significant features at the luminosities corresponding to the Eddington limit of a neutron star ($\log L_X \sim 38.3$) or of a black hole ($\log L_X \sim 39\text{–}39.5$). On the other hand, it breaks at the luminosity $\log L_X \approx 40.0\text{–}40.5$ (Fig. 10), corresponding to the Eddington luminosity of a $\sim 100M_\odot$ object. Because of such a smooth shape of their XLF, it appears most likely that systems with luminosity $\log L_X \lesssim 40\text{–}40.5$ are "normal" X-ray binaries formed in the course of standard stellar evolution and represent the tail of the distribution of black hole masses and mass accretion rates. We note here that luminosities exceeding the Eddington limit by several times are possible in the standard accretion model (Shakura and Sunyaev 1973; Grimm et al. 2003). The break in the HMXB XLF observed at $\log L_X \sim 40.5$ (Figs. 9, 10) may indicate the transition to a different population of X-ray sources. The few known sources with luminosities exceeding this value may indeed be intermediate mass black holes—result of the evolution of Pop III stars.

It is interesting to estimate the fraction of black holes experiencing the ultra-luminous X-ray source phase in the course of their evolution (Mineo et al. 2012). We will use here a standard definition of the ULX as a source with $\log(L_X) > 39$. Number of ULXs in a galaxy with star-formation rate *SFR* can be estimated from the universal XLF of star-forming galaxies:

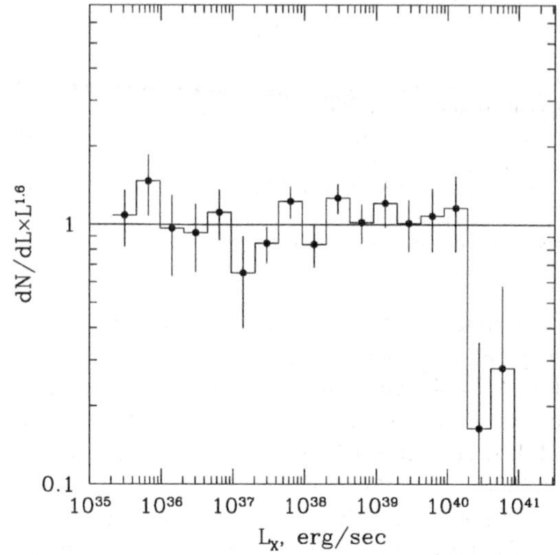

Fig. 10 Detailed shape of the X-ray luminosity function of compact X-ray sources in star-forming galaxies. The figure shows the ratio of the X-ray luminosity to a power law with slope of 1.6. Based on results of Mineo et al. (2012)

$$N_{\rm ULX}(>10^{39}\ {\rm erg/s}) \approx 0.48 \times {\rm SFR} \tag{8}$$

On the other hand, their number can be expressed through the formation rate of black holes $\dot{N}_{\rm BH}$ and average ULX lifetime:

$$N_{\rm ULX} \sim \dot{N}_{\rm BH} \sum_k f_{{\rm X},k}\, \tau_{{\rm X},k} \sim \dot{N}_{\rm BH}\, f_{\rm X}\, \bar{\tau}_{\rm X} \tag{9}$$

The formation rate of black holes approximately equals to the formation rate of massive stars $\dot{N}_{\rm BH} \approx \dot{N}_{\star}(M > 25\, M_{\odot})$. Assuming Salpeter IMF[5] we obtain:

$$\dot{N}_{*}(M > 25\, M_{\odot}) \approx 7.4 \cdot 10^{-3} \times {\rm SFR} \tag{10}$$

Summation in Eq. (9) is done over various types of binary systems. The quantity $f_{{\rm X},k}$ is the fraction of black holes formed in the galaxy which became ULXs due to accretion in a system of the given type, $\tau_{\rm X,k}$ is the mean life time (as ULX) of such systems. The quantity $f_{\rm X} = \sum_k f_{\rm X,k}$ is the total fraction of black holes which passed through the ULX phase and $\bar{\tau}_{\rm X}$ is average duration of their ULX phase. Taking into account that ULXs are primarily fed by the supergiant donors, i.e. $\bar{\tau}_{\rm ULX} \sim 10^4$ yrs, we obtain

$$f_{\rm ULX} \sim 3.5 \cdot 10^{-2} \times \left(\frac{\bar{\tau}_{\rm ULX}}{10^4\ {\rm yr}}\right)^{-1} \tag{11}$$

Thus, a few per cent of all black holes formed in a galaxy become luminous X-ray sources with luminosity $\geq 10^{39}$ erg/s, thus maintaining the observed population of ULXs.

4 Finding Supermassive Black Holes: Surveys, Biases, Demographics

We conclude this review with a discussion about the demographics of SMBH and AGN. The story of the study of the evolution of AGN is as old as the story of relativistic astrophysics

[5]We use Salpeter IMF to be consistent with the definition of star-formation rate, see discussion in Mineo et al. (2012) for details.

itself. As for the case of XRB in nearby galaxies, the major progresses in the census of distant supermassive black holes, however, had to wait until the current generation of powerful focusing X-ray telescopes. We are now in the position of reconstructing the history of black holes growth for more than 3/4 of the objects over more than 3/4 of cosmic history. As we will show below, this allows non-trivial independent verification of the relativistic nature of the accretion process.

4.1 Accreting Black Holes in Galactic Nuclei

Distinguishing nuclear AGN emission from the surrounding galactic one requires high spatial resolution observations, but these are challenging, particularly because the resolving power of telescopes varies widely across the electromagnetic spectrum. Indeed, for more distant AGN, combining observations at the highest possible resolution at different wavelength on large, statistically significant samples, is nearly impossible, and one often resorts to less direct means of separating nuclear from galactic light. There is, however, no simple prescription for efficiently performing such a disentanglement: the very existence of scaling relations between black holes and their host galaxies (e.g. Gültekin et al. 2009, and references therein) and the fact that, depending on the specific physical condition of the nuclear region of a galaxy at different stages of its evolution, the amount of matter captured within the Bondi radius can vary enormously, imply that growing black holes will always display a large range of "contrast" with the host galaxy light.

More specifically, let us consider an AGN with optical B-band luminosity given by $L_{\mathrm{AGN,B}} = \lambda L_{\mathrm{Edd}} f_{\mathrm{B}}$, where we have introduced the Eddington ratio ($\lambda \equiv L_{\mathrm{bol}}/L_{\mathrm{Edd}}$), and a bolometric correction $f_{\mathrm{B}} \equiv L_{\mathrm{AGN,B}}/L_{\mathrm{bol}} \approx 0.1$ (Richards 2006). Assuming a bulge-to-black hole mass ratio of 0.001 and a bulge-to-total galactic stellar mass ratio of (B/T), the contrast between nuclear AGN continuum and host galaxy blue light is given by:

$$\frac{L_{\mathrm{AGN,B}}}{L_{\mathrm{host,B}}} = \frac{\lambda}{0.1}\frac{(M_*/L_{\mathrm{B}})_{\mathrm{host}}}{3(M_\odot/L_\odot)}\left(\frac{B}{T}\right) \tag{12}$$

Thus, for typical mass-to-light ratios, the AGN will become increasingly diluted by the host stellar light in the UV-optical-IR bands at Eddington ratios λ smaller than a few per cent.

The most luminous QSOs (i.e. AGN shining at bolometric luminosity larger than a few times 10^{45} erg/s, and thus with the highest Eddington ratios), represent just the simplest case, as their light out-shines the emission from the host galaxy, resulting in point-like emission with peculiar colors. Less luminous, Seyfert-like, AGN will have a global SED with a non-negligible contribution from the stellar light of host. As a result, unbiased AGN samples extending to lower-luminosities, will inevitably have optical-NIR colors spanning a large range of intermediate possibilities between purely accretion-dominated and purely galaxy-dominated.

The fact that optical (and, in fact, NIR) surveys easily pick up AGN at high Eddington ratio, and thus, potentially, all members of a relatively homogeneous class of accretors, is not surprising. At the end of the day, Eq. (12) clearly shows that the "galaxy dilution bias", i.e. the inefficiency of selecting accreting black holes because of the "contamination" from (stellar) galaxy light in distant systems is linearly correlated to the Eddington ratio. Only deep X-ray and radio surveys can circumvent such biases.

Indeed, large multi-wavelength galaxy survey and extensive follow-up campaigns of medium-wide and deep X-ray surveys (such as the Chandra Deep Field, Giacconi et al. 2002; the COSMOS field, Hasinger et al. 2007; or the X-Bootes field, Hickox et al. 2009) have allowed to extend AGN SED systematic studies to a wide variety of Eddington ratios and AGN-galaxy relative contributions. Lusso (2011) (see also Lusso et al. 2010 and Elvis et al. 2012) analysed an X-ray selected sample of AGN in the COSMOS field, the largest fully identified and redshift complete AGN sample to date. When restricted to a "pure" QSO sample (i.e. one where objects are pre-selected on the basis of a minimal estimated galaxy contamination of $< 10\,\%$), the SED of the COSMOS X-ray selected AGN is reminiscent of the Elvis et al. (1994) and Richards (2006) ones, albeit with a less pronounced inflection point at 1 μm. The mean (and median) SED for the whole sample, however, apart from having a lower average luminosity, is also characterized by much less pronounced UV and NIR peaks. This is indeed expected whenever stellar light from the host galaxy is mixed in with the nuclear AGN emission.

Figure 11 (Bongiorno et al. 2012; Hao et al. 2013) further illustrates this point. It displays the slope of the rest-frame SED in the optical ($\alpha_{\rm OPT}$, between 0.3 and 1 μm) and NIR ($\alpha_{\rm NIR}$ between 1 and 3 μm) bands, i.e. long- and short-wards of the $\sim$ 1 μm inflection point. Pure QSOs, i.e., objects in which the overall SED is dominated by the nuclear (AGN) emission would lie close to the empty blue star in the lower right corner (positive optical slope and negative NIR slope). The location of the X-ray selected AGN in Fig. 11 clearly shows instead that, in order to describe the bulk of the population, one needs to consider both the effects of obscuration (moving each pure QSO in the direction of the orange arrow) and an increasing contribution from galactic stellar light (moving the objects towards the black stars in the upper part of the diagram).

4.2 Bolometric AGN Luminosity Functions and the History of Accretion

The deepest surveys so far carried out in the soft X-ray energy range (0.5–2 keV), supplemented by the painstaking work of optical identification and redshift determination of the detected sources have provided the most accurate description of the overall evolution of the AGN luminosity function. Neither pure luminosity nor pure density evolution provide a satisfactory description of the X-ray LF evolution, with a good fit to the data achieved with a "Luminosity Dependent Density Evolution" (LDDE) model, or variations thereof. In their influential work, Hasinger et al. (2005) unambiguously demonstrated that in the observed soft X-ray energy band more luminous AGN peaked at higher redshift than lower luminosity ones.

Thus, a qualitatively consistent picture of the main features of AGN evolution is emerging from the largest surveys of the sky in various energy bands. Strong (positive) redshift evolution of the overall number density, as well as marked differential evolution (with more luminous sources being more dominant at higher redshift) characterize the evolution of AGN.

A thorough and detailed understanding of the AGN SED as a function of luminosity could in principle allow us to compare and cross-correlate the information on the AGN evolution gathered in different bands. A luminosity dependent bolometric correction is required in order to match type I (unabsorbed) AGN luminosity functions obtained by selecting objects in different bands. This is, in a nutshell, a direct consequence of the observed trend of the relative contribution of optical and X-ray emission to the overall SED (the $\alpha_{\rm ox}$ parameter) as a function of luminosity (see the left panel of Fig. 6).

Adopting a general form of luminosity-dependent bolometric correction, and with a relatively simple parametrization of the effect of the obscuration bias on the observed LF, Hopkins et al. (2007) were able to project the different observed luminosity functions in various bands into a single bolometric one, $\phi(L_{\rm bol})$. As a corollary from such an exercise, we can then provide a simple figure of merit for AGN selection in various bands by measuring the bolometric energy density associated with AGN selected in that particular band as a

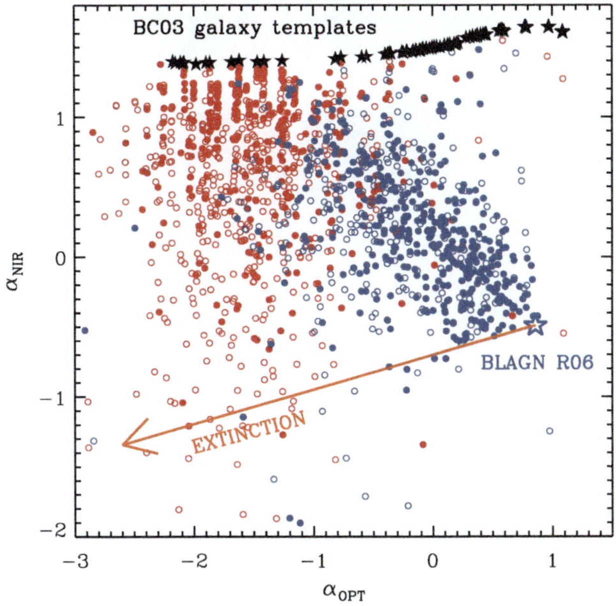

Fig. 11 Observed rest-frame SED slopes in the optical (α_{OPT}, between 0.3 and 1 μm) and NIR (α_{NIR} between 1 and 3 μm) for all (∼ 1650) X-ray selected AGN in the COSMOS survey. *Blue filled circles* denote spectroscopically confirmed type 1 (broad lined) AGN, *blue empty circles* denote candidate type 1 AGN from the photo-z sample. *Red filled circles* are spectroscopically confirmed type 2 (narrow lined) AGN, *empty red circles* are candidate type 2 AGN from the photo-z sample. The *empty blue star* marks the colors of a pure intrinsic type 1 quasar SED (from Richards 2006), while *black stars* are the loci of synthetic spectral templates of galaxies, with increasing levels of star formation from the *left* to the *right*. Nuclear obscuration moves every pure type 1 AGN along the direction of the *orange arrow*. From Bongiorno et al. (2012)

function of redshift. We show this in the left panel of Fig. 12 for four specific bands (hard X-rays, soft X-rays, UV, and mid-IR). From this, it is obvious that the reduced incidence of absorption in the 2–10 keV band makes the hard X-ray surveys recover a higher fraction of the accretion power generated in the universe than any other method.

While optical QSO surveys miss more than three quarters of all AGN of any given $L_{\rm bol}$, hard X-ray selection only fails to account for about one third (up to 50 %) of all AGN, the most heavily obscured (Compton Thick) ones, as shown in the right panel of Fig. 12. It is important to note that the high missed fraction for mid-IR selected AGN is a direct consequence of the need for (usually optical) AGN identification of the IR sources, so that optically obscured active nuclei are by and large missing in the IR AGN luminosity functions considered here.

Figure 13 shows the evolution of the parameters of the analytic fit to the bolometric LF data. They encompass our global knowledge of the evolution of accretion power onto nuclear black holes throughout the history of the universe. The three bottom panels reveal the overall increase in AGN activity with redshift, up to $z \approx 2$, and the mirroring high-redshift decline. At the center, the total integrated luminosity density evolution mark the epochs of rapid build-up of the SMBH mass density. On the lower left, the evolution in the break luminosity $L_{\rm bol,*}$ indicates that the "typical" accreting black hole was significantly more luminous at $z \approx 2$ than now, a different way of looking at AGN "downsizing". This is accompanied by a progressive steepening of the faint end slope of the LF (upper left panel):

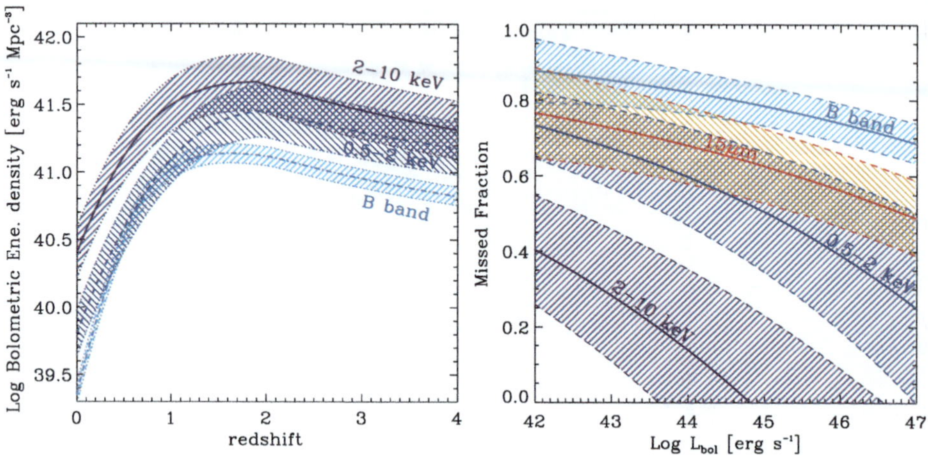

Fig. 12 *Left*: The redshift evolution of the bolometric energy density for AGN selected in different bands. Bolometric corrections from Hopkins et al. (2007) have been used, and the *shaded areas* represent the uncertainty coming from the bolometric corrections only. *Right*: The fraction of AGN missed by observations in any specific band as a function of the intrinsic bolometric luminosity of the AGN. *Red*, *light blue*, *dark blue* and *purple shaded areas* correspond to rest-frame mid-IR (15 μm), UV (B-band), soft X-rays (0.5–2 keV) and hard X-rays (2–10 keV), respectively. The uncertainty on the missed fractions depend on the uncertainties of the bolometric corrections and on the shape of the observed luminosity functions only

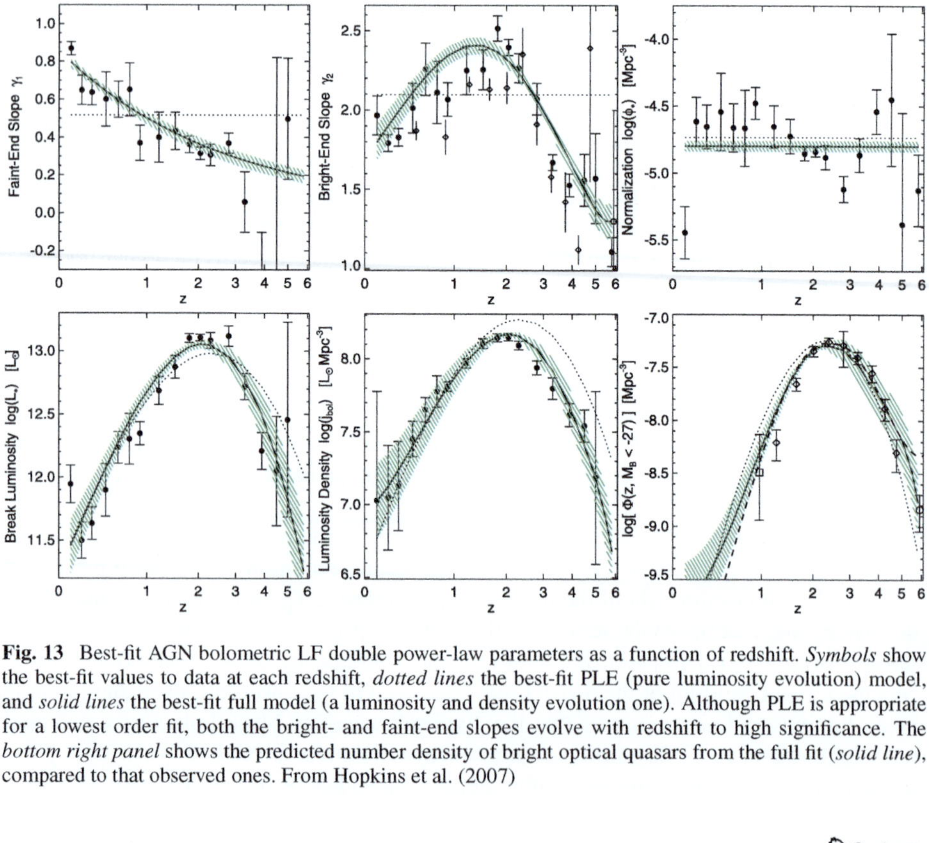

Fig. 13 Best-fit AGN bolometric LF double power-law parameters as a function of redshift. *Symbols* show the best-fit values to data at each redshift, *dotted lines* the best-fit PLE (pure luminosity evolution) model, and *solid lines* the best-fit full model (a luminosity and density evolution one). Although PLE is appropriate for a lowest order fit, both the bright- and faint-end slopes evolve with redshift to high significance. The *bottom right panel* shows the predicted number density of bright optical quasars from the full fit (*solid line*), compared to that observed ones. From Hopkins et al. (2007)

low-luminosity AGN become more and more dominant in the overall number density of AGN as time progresses.

4.3 The Soltan Argument

A reliable census of the bolometric energy output of growing supermassive black holes (see, e.g., the central bottom panel of Fig. 13) allows a more direct estimate of the global rate of mass assembly in AGN, and an interesting comparison with that of stars in galaxies. Together with the tighter constraints on the "relic" SMBH mass density in the local universe, $\rho_{BH,0}$, provided by careful application of the scaling relations between black hole masses and host spheroids, this enables meaningful tests of the classical 'Soltan argument' (Soltan 1982), according to which the local mass budget of black holes in galactic nuclei should be accounted for by integrating the overall energy density released by AGN, with an appropriate mass-to-energy conversion efficiency.

Many authors have carried out such a calculation, either using the CXRB as a "bolometer" to derive the total energy density released by the accretion process (Fabian and Iwasawa 1999), or by considering evolving AGN luminosity functions (Yu and Tremaine 2002; Marconi et al. 2004; Merloni and Heinz 2008). Despite some tension among the published results that can be traced back to the particular choice of AGN LF and/or scaling relation assumed to derive the local mass density, it is fair to say that this approach represents a major success of the standard paradigm of accreting black holes as AGN power-sources, as the radiative efficiencies needed to explain the relic population are within the range $\approx 0.06 \div 0.40$, predicted by standard accretion disc theory (Shakura and Sunyaev 1973).

In general, we can summarize our current estimate of the (mass-weighted) average radiative efficiency, $\langle\epsilon_{\rm rad}\rangle$, together with all the systematics uncertainties, within one formula (see Merloni and Heinz 2008), relating $\langle\epsilon_{\rm rad}\rangle$ to various sources of systematic errors in the determination of supermassive black hole mass density. From the integrated bolometric luminosity function, we get:

$$\frac{\langle\epsilon_{\rm rad}\rangle}{1-\langle\epsilon_{\rm rad}\rangle} \approx 0.075\big[\xi_0(1-\xi_i-\xi_{\rm CT}+\xi_{\rm lost})\big]^{-1} \tag{13}$$

where $\xi_0 = \rho_{{\rm BH},z=0}/4.2\times10^5 M_\odot\,{\rm Mpc}^{-3}$ is the local ($z=0$) SMBH mass density in units of $4.2\times10^5 M_\odot\,{\rm Mpc}^{-3}$ (Marconi et al. 2004); ξ_i is the mass density of black holes at the highest redshift probed by the bolometric luminosity function, $z\approx 6$, in units of the local one, and encapsulate our uncertainty on the process of BH formation and seeding in proto-galactic nuclei (see e.g. Volonteri 2010); $\xi_{\rm CT}$ is the fraction of SMBH mass density (relative to the local one) grown in unseen, heavily obscured, Compton Thick AGN, still missing from our census; finally, $\xi_{\rm lost}$ is the fraction black hole mass contained in "wandering" objects, that have been ejected from a galaxy nucleus following, for example, a merging event and the subsequent production of gravitational wave, the net momentum of which could induce a kick capable of ejecting the black hole from the host galaxy. In the real evolution of SMBH, all these terms must be linked at some level, as the radiative efficiency of accretion depends on the location of the innermost stable circular orbit, thus on the black hole spin, which itself evolves under the effects of both accretion and BH-BH mergers. Accurate models which keep track of both mass and spin evolution of SMBH could in principle be used, together with observational constraints from the AGN luminosity functions, to put constraints on those unknowns, providing a direct link between the relativistic theory of accretion and structure formation in the Universe (see e.g. Volonteri et al. 2013; Sesana et al. 2014).

Acknowledgements We thank M. Falanga and the organizers of the ISSI Workshop "The physics of accretion onto black holes", where this work was initiated. We also thank R. Gilli, R. Hickox, B. Lehmer, E. Lusso, S. Mineo, M. Volonteri for useful discussions. A.M. research was partly supported by the DFG cluster of excellence "Origin and Structure of the Universe" (www.universe-cluster.de).

References

M. Bachetti et al., Astrophys. J. **778**, 163 (2013)
A.M. Beloborodov, Astrophys. J. **510**, L123 (1999)
P.N. Best, T.M. Heckman, Mon. Not. R. Astron. Soc. **421**, 1569 (2012)
A. Bongiorno et al., Mon. Not. R. Astron. Soc. **427**, 3103 (2012)
T.A. Boroson, Astrophys. J. **565**, 78 (2002)
A. Cattaneo et al., Nature **460**, 213 (2009)
E. Churazov et al., Astrophys. J. **407**, 752 (1993)
E.J.M. Colbert, R.F. Mushotzky, Astrophys. J. **519**, 89 (1999)
E.J.M. Colbert et al., Astrophys. J. **602**, 231 (2004)
A. Comastri et al., Astron. Astrophys. **296**, 1 (1995)
P.S. Coppi, in *High Energy Processes in Accreting Black Holes*. ed. by J. Poutanen, R. Svensson. ASP Conference Series, vol. 161 (1999), p. 375
S.W. Davis, A. Laor, Astrophys. J. **728**, 98 (2011)
S.W. Davis, C. Done, O.M. Blaes, Astrophys. J. **647**, 525 (2006)
C.D. Dermer, E.P. Liang, E. Canfield, Astrophys. J. **369**, 410 (1991)
M. Dijkstra, M. Gilfanov, A. Loeb, R. Sunyaev, Mon. Not. R. Astron. Soc. **421**, 213 (2012)
C. Done, M. Gierlinski, A. Kubota, Annu. Rev. Astron. Astrophys. **15**, 1 (2007)
R.J.H. Dunn et al., Mon. Not. R. Astron. Soc. **403**, 61 (2010)
R.J.H. Dunn et al., Mon. Not. R. Astron. Soc. **411**, 337 (2011)
K. Ebisawa, K. Mitsuda, T. Hanawa, Astrophys. J. **367**, 213 (1991)
M. Elitzur, New Astron. Rev. **52**, 274 (2008)
M. Elvis et al., Astrophys. J. Suppl. Ser. **95**, 1 (1994)
M. Elvis et al., Astrophys. J. **759**, 6 (2012)
G. Fabbiano, Annu. Rev. Astron. Astrophys. **44**, 323 (2006)
A.C. Fabian, K. Iwasawa, Mon. Not. R. Astron. Soc. **303**, L34 (1999)
H. Falcke, E. Koerding, S. Markoff, Astron. Astrophys. **414**, 895 (2004)
R.P. Fender, T. Belloni, E. Gallo, Mon. Not. R. Astron. Soc. **355**, 1105 (2004)
R.P. Fender, T.J. Maccarone, A. Haywood, Mon. Not. R. Astron. Soc. **430**, 1538 (2013)
A.A. Galeev, R. Rosner, G.S. Vaiana, Astrophys. J. **229**, 318 (1979)
I.M. George, A.C. Fabian, Mon. Not. R. Astron. Soc. **249**, 35 (1991)
A.M. Ghez et al., Astrophys. J. **689**, 1044 (2008)
R. Giacconi et al., Astrophys. J. Suppl. Ser. **139**, 369 (2002)
M. Gilfanov, Mon. Not. R. Astron. Soc. **349**, 146 (2004)
M. Gilfanov, *The Jet Paradigm*. Lecture Notes in Physics, vol. 794 (Springer, Berlin, 2010), p. 17
M. Gilfanov, E. Churazov, M. Revnivtsev, Astron. Astrophys. **352**, 182 (1999)
M. Gilfanov, E. Churazov, M. Revnivtsev, Mon. Not. R. Astron. Soc. **316**, 923 (2000)
M. Gilfanov, H.-J. Grimm, R. Sunyaev, Mon. Not. R. Astron. Soc. **347**, L57 (2004)
S. Gillessen et al., Astrophys. J. **692**, 1075 (2009)
R. Gilli, A. Comastri, G. Hasinger, Astron. Astrophys. **463**, 79 (2007)
J. Goodman, Mon. Not. R. Astron. Soc. **339**, 937 (2003)
H.J. Grimm, M. Gilfanov, R. Sunyaev, Mon. Not. R. Astron. Soc. **339**, 793 (2003)
P.W. Guilbert, M.J. Rees, Mon. Not. R. Astron. Soc. **233**, 475 (1988)
K. Gültekin et al., Astrophys. J. **698**, 198 (2009)
F. Haardt, L. Maraschi, Astrophys. J. **413**, 507 (1993)
H. Hao et al., Mon. Not. R. Astron. Soc. **434**, 3104 (2013)
M.J. Hardcastle, D.A. Evans, J.H. Croston, Mon. Not. R. Astron. Soc. **376**, 1849 (2007)
G. Hasinger, T. Miyaji, M. Schmidt, Astron. Astrophys. **441**, 417 (2005)
G. Hasinger et al., Astrophys. J. Suppl. Ser. **172**, 29 (2007)
S. Heinz, R. Sunyaev, Mon. Not. R. Astron. Soc. **343**, L59 (2003)
R. Hickox et al., Astrophys. J. **696**, 891 (2009)
L.C. Ho, Annu. Rev. Astron. Astrophys. **46**, 475 (2008)
P.F. Hopkins, G.T. Richards, L. Hernquist, Astrophys. J. **654**, 731 (2007)

I. Hubeny et al., Astrophys. J. **533**, 710 (2000)
A. Ibragimov et al., Mon. Not. R. Astron. Soc. **362**, 1435 (2005)
K. Iwasawa, Y. Taniguchi, Astrophys. J. Lett. **413**, L15 (1993)
Y.F. Jiang, J.M. Stone, S.W. Davis, Astrophys. J. **784**, 169 (2014)
C. Jin et al., Mon. Not. R. Astron. Soc. **425**, 907 (2012)
D.-W. Kim, G. Fabbiano, Astrophys. J. **611**, 846 (2004)
D.-W. Kim, G. Fabbiano, Astrophys. J. **721**, 1523 (2010)
A.R. King et al., Astrophys. J. **552**, 109 (2001)
M. Kishimoto, R. Antonucci, O.J. Blaes, Mon. Not. R. Astron. Soc. **345**, 253 (2003)
M. Kishimoto et al., Nature **454**, 492 (2008)
E. Koerding, H. Falcke, S. Markoff, Astron. Astrophys. **382**, L13 (2002)
A. Lawrence, Mon. Not. R. Astron. Soc. **423**, 451 (2012)
B. Lehmer et al., Astrophys. J. **724**, 559 (2010)
A.P. Lightman, D.M. Eardley, Astrophys. J. **178**, L1 (1974)
E. Lusso, PhD Thesis, University of Bologna (2011)
E. Lusso et al., Astron. Astrophys. **512**, A34 (2010)
A. Marconi et al., Mon. Not. R. Astron. Soc. **351**, 169 (2004)
F. Mayer, E. Meyer-Hofmeister, Astron. Astrophys. **288**, 175 (1994)
J.E. McClintock, R. Narayan, J.F. Steiner, arXiv:1303.1583 (2014 this issue). doi:10.1007/s11214-013-0003-9
M.L. McConnell et al., Astrophys. J. **572**, 984 (2002)
A. Merloni, Mon. Not. R. Astron. Soc. **341**, 1051 (2003)
A. Merloni, S. Heinz, Mon. Not. R. Astron. Soc. **381**, 589 (2007)
A. Merloni, S. Heinz, Mon. Not. R. Astron. Soc. **388**, 1011 (2008)
A. Merloni, S. Heinz, T. Di Matteo, Mon. Not. R. Astron. Soc. **345**, 1057 (2003)
A. Merloni et al., Mon. Not. R. Astron. Soc. **370**, 1699 (2006)
F. Meyer, B.F. Liu, E. Meyer-Hofmeister, Astron. Astrophys. **354**, 67 (2000)
J.M. Miller et al., Astrophys. J. **585**, 37 (2003)
S. Mineo, M. Gilfanov, R. Sunyaev, Mon. Not. R. Astron. Soc. **419**, 2095 (2012)
S. Mineo, M. Gilfanov, B.D. Lehmer, G.E. Morrison, R. Sunyaev, Mon. Not. R. Astron. Soc. **437**, 1698 (2014)
K. Mitsuda et al., Publ. Astron. Soc. Jpn. **36**, 741 (1984)
K. Nandra, K.A. Pounds, Mon. Not. R. Astron. Soc. **268**, 405 (1994)
R.S. Nemmen, T. Storchi-Bernardi, M. Eracleous, Mon. Not. R. Astron. Soc. **438**, 2804 (2014)
H. Netzer, New Astron. Rev. **52**, 257 (2008)
K. Ohsuga, S. Mineshige, Astrophys. J. **736**, 2 (2011)
B.M. Peterson et al., Astrophys. J. **613**, 682 (2004)
K. Postnov, Astron. Lett. **29**, 372 (2003)
K. Postnov, L. Yungelson, Living Rev. Relativ. **9**, 6 (2006)
K. Pottschmidt et al., Astron. Astrophys. **407**, 1039 (2003)
J. Poutanen, A. Veledina, arXiv:1312.2761 (2014 this issue). doi:10.1007/s11214-013-0033-3
J. Poutanen, J.H. Krolik, F. Ryde, Mon. Not. R. Astron. Soc. **292**, L21 (1997)
P. Ranalli, A. Comastri, G. Setti, Astron. Astrophys. **399**, 39 (2003)
M. Revnivtsev, M. Gilfanov, E. Churazov, Astron. Astrophys. **380**, 520 (2001)
C.S. Reynolds, Invited review at the workshop on "High Energy Processes in Accreting Black Holes", ed. by J. Poutanen, R. Svensson (1998). arXiv:astro-ph/9810018
M.T. Reynolds, J.M. Miller, Astrophys. J. **769**, 16 (2013)
G.T. Richards, Astrophys. J. Suppl. Ser. **166**, 470 (2006)
R. Ross, A.C. Fabian, Mon. Not. R. Astron. Soc. **281**, 637 (1996)
B. Sams, A. Eckart, R. Sunyaev, Nature **382**, 47 (1996)
J.D. Schnittman, J.H. Krolik, S.D. Noble, Astrophys. J. **769**, 156 (2013)
A. Sesana et al., Astrophys. J. (2014 submitted). arXiv:1402.7088
G. Setti, L. Woltjer, Astron. Astrophys. **224**, L21 (1989)
N. Shakura, R. Sunyaev, Astron. Astrophys. **24**, 337 (1973)
Z. Shang et al., Astrophys. J. **619**, 41 (2005)
T. Shimura, F. Takahara, Astrophys. J. **331**, 780 (1995)
M.A. Sobolewska et al., Mon. Not. R. Astron. Soc. **394**, 1640 (2009)
A. Soltan, Mon. Not. R. Astron. Soc. **200**, 115 (1982)
A.T. Steffen et al., Astron. J. **131**, 2826 (2006)
R. Sunyaev, L. Titarchuk, Astron. Astrophys. **86**, 121 (1980)

R. Sunyaev, L. Titarchuk, in *The 23rd ESLAB Symposium on Two Topics in X Ray Astronomy*, vol. 1 (1989), p. 627
R. Sunyaev, B. Tinslew, D. Meier, Comments Astrophys. **7**, 183 (1978)
R. Sunyaev et al., Astrophys. J. **389**, 75 (1992)
A. Veledina, I. Vurm, J. Poutanen, Mon. Not. R. Astron. Soc. **414**, 3330 (2011)
F. Verbunt, E. van den Heuvel, in *X-Ray Binaries* (Cambridge University Press, Cambridge, 1995), p. 457
M. Volonteri, Annu. Rev. Astron. Astrophys. **18**, 279 (2010)
M. Volonteri et al., Astrophys. J. **775**, 94 (2013)
J.M. Wang, K.-Y. Watarai, S. Mineshige, Astrophys. J. Lett. **607**, L107 (2004)
M. Young, M. Elvis, G. Risaliti, Astrophys. J. Suppl. Ser. **183**, 17 (2009)
Q. Yu, S. Tremaine, Mon. Not. R. Astron. Soc. **335**, 965 (2002)
A.A. Zdziarsk, P. Lubinski, D.A. Smith, Mon. Not. R. Astron. Soc. **303**, L11 (1999)
A.A. Zdziarski, M. Gierlinski, Prog. Theor. Phys. Suppl. **155**, 99 (2004)
A.A. Zdziarski et al., Mon. Not. R. Astron. Soc. **342**, 355 (2003)
W. Zhang, S. Woosley, A. Heger, Astrophys. J. **679**, 639 (2008)
Z. Zhang, M. Gilfanov, A. Bogdan, Astron. Astrophys. **546**, 36 (2012)

DOI 10.1007/978-1-4939-2227-7_9
Reprinted from *Space Science Reviews* Journal, DOI 10.1007/s11214-014-0056-4

Scaling Relations from Stellar to Supermassive Black Holes

Elmar Körding

Received: 10 February 2014 / Accepted: 5 June 2014 / Published online: 28 June 2014

Abstract Accretion is a ubiquitous phenomenon—it is seen in sources ranging from young stars to accreting supermassive black holes in the centres of galaxies. Here, we present the known empirical connections between stellar mass X-ray binaries and active galactic nuclei. We argue that this implies that both the accretion disc and the jet are scale invariant with respect to the black hole mass. Finally, we show that also accretion discs and jets in sources with a different accretor can be connected empirically to accreting black holes, hinting towards a common mechanism of accretion in all sources.

Keywords Black holes · Accretion · Compact objects

1 Introduction

Accretion onto compact objects powers a large variety of different astrophysical phenomena. It is thought to provide the energy of Quasars, accreting supermassive black holes, X-ray binaries (XRBs) and Gamma-ray bursts. It can be observed with the naked eye in accreting white dwarf systems. Accretion is thus observed in objects of vastly different size and accretion rate. Here, we will discuss the effects of these different scales on the signatures of the accreting sources.

The most prominent case of accreting compact objects are black holes. Black holes are thought to exist over a large range of masses. Stellar black holes are thought to be remnants of massive stars (e.g., Woosley and Weaver 1995; Heger et al. 2003). Investigating the masses of stellar black holes, Özel et al. (2010) find that the known masses of black hole XRBs are very narrowly distributed around 7.8 solar masses, with a width of 1.2 solar masses. The narrow distribution might be due to a particular evolutionary channel followed by low-mass XRBs. Black holes of a much higher mass are supermassive black holes, which

E. Körding (✉)
Department of Astrophysics/IMAPP, Radboud University Nijmegen, P.O. Box 9010, 6500 GL, Nijmegen, The Netherlands
e-mail: e.koerding@astro.ru.nl

are found in the centers of galaxies. These sources have masses ranging from 10^5 to 10^{10} solar masses (e.g., Lynden-Bell 1969). Besides these stellar and supermassive black holes, "intermediate" mass black holes have been postulated, but besides one possible source (HLX-1, e.g., Farrell et al. 2009) the evidence seems still to be fairly inconclusive (see Casares & Jonker in this issue).

The idea that accretion onto stellar and supermassive black holes is similar and that some scaling relations between the parameters describing the accretion phenomena in both types of objects exist, is old and has been discussed for a long time (see e.g., White et al. 1984). Both active galactic nuclei (AGN) as well as black hole XRBs contain the same three basic ingredients: a black hole, an accretion disc, and relativistic collimated jets (during some stages of their lifetime). Using similarities in the radio morphology and the temporal behaviour of the radio emission, Mirabel and Rodriguez (1994) used the term 'Microquasar' to highlight the connection between stellar and supermassive black holes. Simple scaling laws govern the physics of accretion discs around black holes: The basic scale of the system is the radius of the black hole, and thus its mass (see e.g., Rees 1998). If everything is described in units of this scale, one can expect to describe the system in the vicinity of the black hole in a scale invariant form. On a larger scale environmental effects will start to play a role. Nevertheless, some scaling is typically seen: While one can find AGN with jets extending millions of parsecs, those of XRBs will only have parsec scales.

A black hole only has 3 basic parameters: its mass, spin and electric charge. The charge of an astrophysical black hole will be immediately neutralized. Thus, only the mass and the spin of the black hole remain as parameters. In addition to these two parameters, an accreting black hole also has the accretion rate as a defining parameter. As discussed above, the mass can change by more than 8 orders of magnitude. The accretion rate can also change over similar orders of magnitude. While the spin may have an important effect on the efficiency of accretion and of the efficiency in launching relativistic jets, its overall effect on the total energetics of an accreting black hole are comparatively small with respect to the possible changes in mass and accretion rate. This is illustrated by the fact that the difference between the innermost stable orbit of a maximally rotating Kerr black hole and that of a Schwarzschild black hole is only a factor 6. Thus, we can expect that in first order approximation, the black hole mass and its accretion rate will be the main parameters of an accreting black hole system.

One aspect that complicates any scaling law between black holes of different sizes is that we know from stellar X-ray binaries that the same system can exhibit very different properties at similar luminosities, i.e. for a given accretion rate the source does not have a unique appearance (see e.g., Remillard and McClintock 2006; Homan and Belloni 2005, Belloni & Stella in these proceedings). These differences are described by a set of accretion states. The two main states are the hard state, characterized by a hard power law spectrum in the X-rays, and the soft state, where the X-ray spectrum is dominated by soft black body radiation. Besides these two states, there are several intermediate states, for a discussion see either Homan and Belloni (2005) or Remillard and McClintock (2006). If any scaling relations exist between stellar and supermassive black holes, one has to expect that also AGN show something similar to those accretion states. The analogue of hard state XRBs amid AGN are often identified with low ionization sources like Low Ionization Nuclear Emission Region sources (LINERs), see for example Ho (2005). The analogue of soft state XRBs amid AGN are most likely bright radio-quiet quasars, as they have very strong emission from an accretion disc. Given the existence of these different accretion states, one has to ensure that any attempt to study scaling laws between XRBs and AGN takes the accretion states into account.

In this paper, we first present connections between stellar and supermassive black holes based on variability properties of the sources. Then, we look at the scaling behaviour of relativistic jets. Finally, we explore how these scaling relations can be extended to other source classes like neutron star X-ray binaries or white dwarf systems.

2 Connecting the Variability Properties of Accreting Objects

In order to test the idea that accreting black holes are scale invariant with respect to the black hole mass, one has to look for properties that can be measured well in stellar and supermassive black holes. One such property can be obtained from the power spectral density (PSD) of their light-curves. Both XRBs and AGN are highly variable in the X-ray band. If one looks at the PSD of such sources, the PSD of an AGN looks remarkably similar to that of an XRB (McHardy 1988; Uttley et al. 2002; Markowitz et al. 2003), albeit with longer variability timescales.

The X-ray emission of both AGN and XRBs originates from the inner parts of the accretion flow/jet system (Elvis et al. 1978). In both cases, one often assumes that the X-ray emission is produced either by Comptonization in a hot corona or by the relativistic jet (e.g., Elvis et al. 1986; Nowak et al. 1999; Fossati et al. 1998). The relativistic jet dominates the overall spectrum of BL Lac objects and other Blazars and may also contribute to the X-ray emission from low luminosity AGN (Fossati et al. 1998). In most other AGN, especially radio quiet quasars and Seyfert objects, one typically believes that the X-ray emission is dominated by the Comptonization in a hot phase of the accretion flow (the corona). In XRBs, the picture is less clear. In soft state XRBs, one can detect the multi color black body emission from the standard accretion disc, but there is also a non-thermal component in the spectrum. In hard state XRBs, one often explains the hard state spectrum with Comptonization (see e.g., Nowak et al. 1999), but especially for low luminosity hard state XRBs the jet may contribute to the total emission (Markoff et al. 2005). Thus, it might well be that the emission process is different for the different sources. The similarity of the PSD hints towards a common driver that drives the variability of the source, e.g., a fluctuating accretion disc (Lyubarskii 1997). At the inner edge of the accretion flow, either the jet or the corona picks up the fluctuations and translates them into variable X-ray emission.

The PSD of an XRB is strongly dependent on the accretion state of the source (see other chapters in this book). In its hard state, the PSD can be well fitted by a superposition of Lorentzians (Belloni et al. 2002). Their centroid frequencies also depend on the accretion rate or the spectral hardness (e.g., Pottschmidt et al. 2003). At high frequencies, the PSD is well described by a power law with spectral index around -2 that flattens towards the lower frequencies near the peak frequency of the highest frequency Lorentzian. In the soft state, the source is generally less variable and sometimes well described by a broken power law (Cui et al. 1997).

In AGN one finds a very similar shape (Green et al. 1993). However, due to lower luminosities and longer timescales involved, PSDs of AGN are typically less well sampled compared to bright XRBs. For AGN, one can often only approximate the PSD with a broken power law, where again the high frequency slope has an index around -2. In some cases one can constrain Lorentzian components also for AGN (e.g., McHardy et al. 2007). The break in the PSD, where the high-frequency slope flattens towards lower frequencies, is visible in both XRBs and AGN. This break will be referred to as the characteristic frequency or timescale.

Most of the well studied AGN are radio quiet nearby Seyfert objects and quasars, and thus have a bright accretion disc ('big blue bump'). These AGN would need to be classified

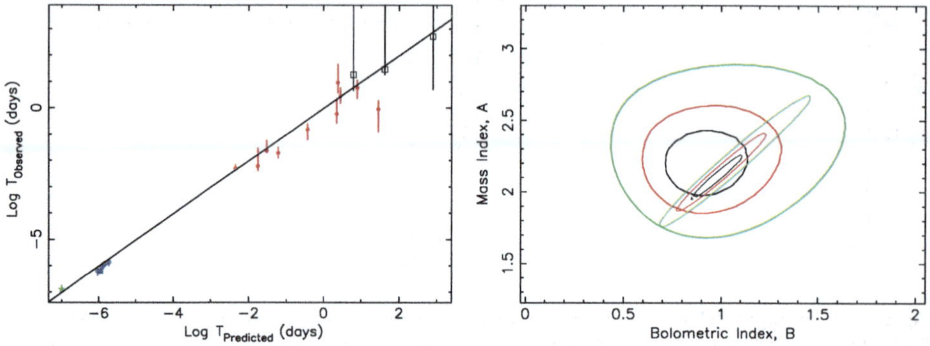

Fig. 1 *Left*: Projection of the variability correlation. The *vertical axis* denotes the measured break timescale, while the *horizontal axis* denotes the expectation of scale invariant black hole accretio. The *solid line* is a fit to the XRB and AGN data. *Right side*: χ^2 map of the fit parameters. The *vertical axis* denotes the black hole mass coefficient and the *horizontal axis* indicates the accretion rate coefficient for the multivariate correlation. The large contours are the 1, 2 and 3σ contours for the fit to AGN alone. When XRBs are included, in the fit the contours contract to the small contours. Plots reproduced with permission from McHardy et al. (2006)

to belong to the analogue of soft state XRBs. As the characteristic timescale depends on the accretion state, one needs to compare Seyferts and Quasars to soft state XRBs.

In its most simple incarnation, the idea of scale invariant black hole accretion suggests that the characteristic timescale should change linearly with the black hole mass (e.g., McHardy 1988; Uttley et al. 2002; Uttley and McHardy 2005). While there is a correlation of the characteristic timescale with the black hole mass, this correlation shows significant scatter (see e.g., Uttley and McHardy 2005). Uttley and McHardy (2005) found that the most prominent outliers of the correlation are those AGN with the most extreme Eddington ratios. Thus, it seems that besides the black hole mass also the accretion rate plays a role in determining the characteristic timescales.

The idea that the accretion rate might influence the characteristic timescales is also supported by observations of neutron star and black hole XRBs. In neutron star XRBs one observes parallel tracks in the quasi periodic oscillation (QPO) frequency luminosity plane (see e.g., Kuulkers and van der Klis 1996; Wijnands et al. 1998) and it has been suggested that the frequencies might be better tracers of the accretion rate than the luminosity itself (van der Klis 2001). Thus, one should include a proxy of the accretion rate, besides the mass of the accretor, in a possible relation describing the characteristic timescales.

One possible tracer of the accretion rate in AGN is the bolometric luminosity. For an efficiently accreting object (as e.g. Seyfert objects are thought to be) one can roughly obtain accretion rates from bolometric luminosities as $\dot{M} = L_{\mathrm{bol}}/(0.1c^2)$, where we assumed a radiative efficiency of 0.1. If one looks for a correlation for AGN in the three dimensional space given by the black hole mass, the bolometric luminosity and the characteristic frequency, one finds a good correlation (see Fig. 1). Already with AGN data alone, one can rule out the hypothesis that the characteristic timescale only depends on the black hole mass, it also depends on the bolometric luminosity. If, in a second step, one also includes data from soft-state XRBs, the χ^2 contours get smaller (see Fig. 1) and stay in agreement with the relation found for AGN alone.

The found relationship is in agreement with the most simple scale invariant description of a characteristic timescale:

$$\nu \propto \frac{\dot{M}}{M^2} \quad \text{or} \quad M\nu \propto \frac{\dot{M}}{M}, \tag{1}$$

where $\dot{M}$ denotes the accretion rate, M the black hole mass and ν the characteristic frequency. The fitted indices on the mass and the accretion rate dependence stay very close to the integer values expected from simple scale invariance. This indicates that the accretion discs of black holes obey in first order approximation the simple prescription of scale invariance.

We note that this relation can be found using supermassive black holes alone, i.e. the relation is not artificially created by combining two classes of unrelated objects. In addition, the distance of the objects has a fairly small effect on this relation, as it does not enter the measurement of the black hole mass and the characteristic timescale. Thus, this scaling relation between XRBs and AGN cannot be explained by the huge difference in distance, and we can conclude that our naive idea of a scale invariant accretion flow around black holes seems to work.

To measure the characteristic timescale in AGN one needs long and well sampled light-curves. If one does not have all that data available one can look at other methods to describe the variability properties. One method is to look at the excess variance (e.g., Nandra et al. 1997; Ponti et al. 2012), i.e. the variance in excess of those due to measurement uncertainties. Here, Ponti et al find a highly significant and tight (0.7 dex) correlation between the excess variance and the mass of the black holes. The accretion rate excess variance correlation is however less than expected from the above mentioned PSD break scaling. The authors suggest that one can reconcile both findings if both the PSD high frequency break and the normalisation depend on accretion rate in such a way that they almost completely counterbalance each other.

Kelly et al. (2013) have recently developed a new statistical method to estimate the parameters of a PSD from a light-curve of photon counts with arbitrary sampling, eliminating the need to bin a lightcurve to achieve Gaussian statistics, and use this technique to estimate the X-ray variability parameters for a faint AGN sample. They find that the normalization of the high-frequency X-ray PSD is inversely proportional to black hole mass.

2.1 Hard State XRBs

While the bolometric luminosity is a good measure of the accretion rate for soft state objects, this is not the case for the hard state. For hard state objects, one typically assumes that the accretion flow is inefficient, and thus one cannot simply use the bolometric luminosity as a tracer of the accretion rate. Theory predicts that the luminosity depends quadratically on the accretion rate in this case (see e.g., Narayan and Yi 1994).

For hard state objects, it has been proposed that it is possible to deduce the accretion rate indirectly from emission of the jet. The underlying idea is that the jet power is linearly coupled to the accretion rate, and one can estimate the jet power from the core radio emission. For details see Körding et al. (2007a). Using this measure of the accretion rate one can add hard state XRBs to the correlation between the characteristic timescale, mass and accretion rate. The same linear dependence on the accretion rate as for soft state objects is found for hard state objects, albeit with a different normalization. The resulting correlation is shown in Fig. 2.

The different normalization of hard state objects and soft state objects is not surprising. We have mentioned that the centroid frequencies of the Lorentzians, and thus the characteristic timescale discussed here, depend strongly on spectral hardness and thus on the spectral state of the XRB. One object (XTE J1550-564) has been observed near a state transition, and in Fig. 2 one can see it moving from the hard state normalization to the soft state one.

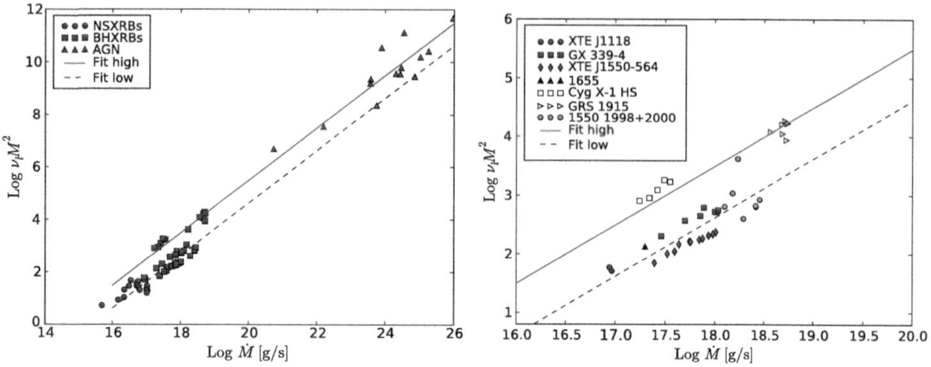

Fig. 2 Variability correlation including hard state XRBs as well as soft state XRBs and AGN. *Left side*: full projection of all sources. *Right side*: Zoom on the stellar sources. Hard state objects have a different normalisation from soft state sources. Plots reproduced with permission from Körding et al. (2007b)

2.2 Quasi-Periodic Oscillations

Quasi-periodic oscillations (QPOs) are a common phenomenon of neutron star and black hole XRBs (van der Klis 1989). As mentioned above, QPOs have been used to argue that the characteristic frequencies of XRBs are correlated with the luminosity. In addition, they correlate well with spectral properties of the source (hardness, see e.g., Pottschmidt et al. 2003). For XRBs, one has classified QPOs into different types (e.g., Casella et al. 2005) depending on the width of the QPO feature as well as their relative strength. The different types (called type A, B and C) arise during different times during an outburst. Especially in intermediate states, QPOs can be extremely prominent in the PSD.

As these features are so common in XRBs, they have been long searched in AGN. However, while there were some claims of AGN QPOs there had been—until recently—no significant detection of a QPO (Vaughan and Uttley 2006). This has been finally settled by the work of Gierliński et al. (2008), who report the first firm detection of a QPO (see Fig. 3). The QPO frequency is in agreement with the value expected from scaling linearly with black hole mass from XRBs. Thus, not only can one scale the broad band PSD from XRBs of different mass, one can also find similar quasi periodicities in both classes of black holes. However, it is not yet clear to which QPO type this QPO belongs.

3 Scale Invariance of the Relativistic Jet

Having seen that AGN and XRBs seem to obey a scaling relation for the variability properties, if one accounts for the mass and the accretion rate, one can search for different correlations connecting AGN and XRBs. We have argued that the variability properties of accretion flows seem to indicate that the accretion flow is scale invariant. As outlined in the introduction, AGN and XRBs show jets that are coupled to the accretion flow. To test whether the jet is scale invariant we need a property of the jet that can be observed in both stellar and supermassive accreting black holes. One key feature of most jets is their radio emission.

Observations of the radio emission of AGN and XRBs reveal a flat radio spectrum (spectral index around 0 to -0.3) up to a turnover frequency often referred to as the jet break (e.g., Blandford and Kőnigl 1979; Falcke and Biermann 1995). After this turnover frequency, one

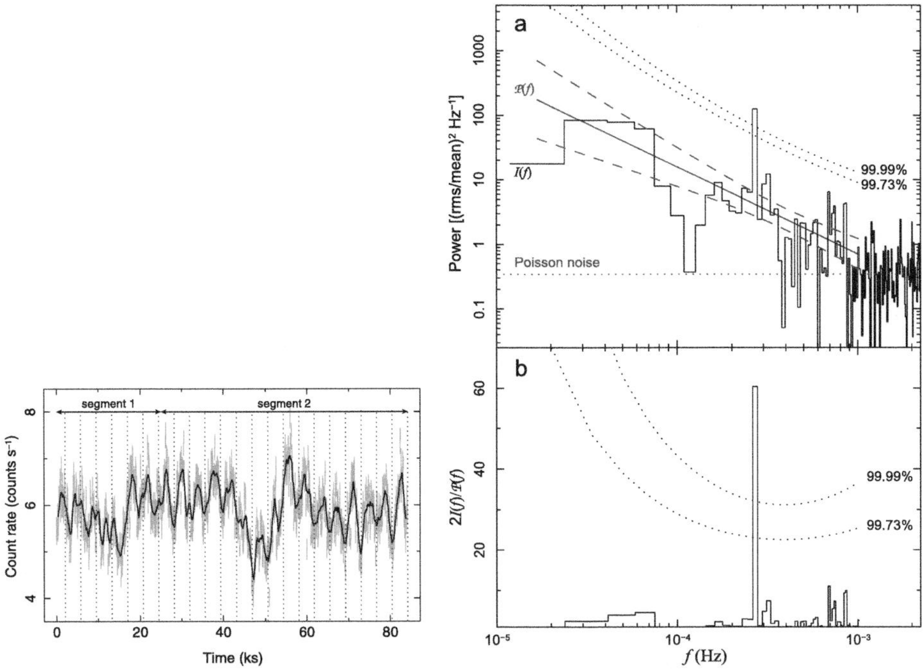

Fig. 3 Quasi-periodic oscillations in an AGN: The *left panel* shows the light-curve together with an indication of the periodicity. The *right panel* presents the power spectral density and the significance of any peaks. Plots reproduced with permission from Gierliński et al. (2008)

observes a steep spectrum with a power law index around 0.6, which is consistent with optically thin synchrotron emission. This flat spectrum (below the turnover frequency) is typically explained with a conical jet model that can reproduce the flat spectrum assuming continuous re-acceleration (Blandford and Kőnigl 1979; Falcke and Biermann 1995; Heinz and Sunyaev 2003). Newer studies indicate that the flat spectrum is hard to produce if one includes adiabatic cooling (Kaiser 2006), but it has been found that one can produce such a spectrum with internal shock models (e.g., Jamil et al. 2010).

Using these conical jet models, one can calculate the radio luminosity for a given jet power. In the flat part of the spectrum one can approximate the radio luminosity as

$$L_{\mathrm{Rad}} \propto P_{\mathrm{jet}}^{17/12} \tag{2}$$

See for example (Blandford and Kőnigl 1979; Falcke and Biermann 1995). The radio luminosity of the flat spectrum radio core thus only depends on the jet power and not on the black hole mass or other properties of the accreting system. It is therefore a good candidate to look for connections between stellar XRBs and AGN.

The theoretically predicted dependency of the radio luminosity on jet power can be observed in XRBs. One observes a tight radio/X-ray correlation that seems to be valid for a number of hard state XRBs (Corbel et al. 2000; Gallo et al. 2003). This correlation can be well explained using a conical jet model together with the assumption that the X-ray emission originates either in the jet or in an inefficient accretion flow, see e.g. Markoff et al. (2003), Merloni et al. (2003). Only recently, significant outliers have been found; there seems to be a second class of hard state XRBs that show a much steeper correlation (Coriat

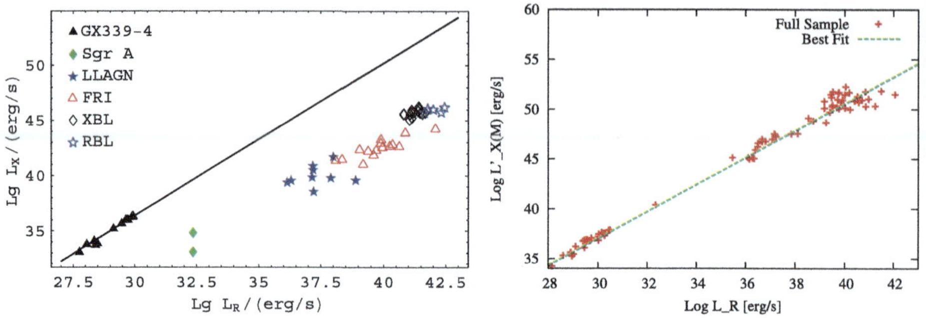

Fig. 4 Fundamental plane of accreting black holes. *Left side*: Radio vs. X-ray for XRBs and AGN. AGN are more radio loud than XRBs. *Right side*: Projection of the fundamental plane. Both XRBs and AGN follow a single correlation between radio/X-ray luminosity and black hole mass. Plots reproduced with permission from Falcke et al. (2004), Körding et al. (2006a)

et al. 2011; Gallo et al. 2012). For a more detailed discussion of this topic, see Gallo et al. in this issue.

If one plots the observed core radio luminosity of stellar and supermassive black holes against their X-ray luminosity, the radio luminosity of AGN is significantly higher when compared to the radio/X-ray correlation of XRBs (Merloni et al. 2003; Falcke et al. 2004). However, as for the variability correlation, the black hole mass will likely influnce the expected radio and X-ray luminosity. One thus expects a correlation in the space given by the radio luminosity, X-ray luminosity and black hole mass. Using the sample of Körding et al. (2006a), one finds:

$$\log L_X \approx 1.4 \log L_R - 0.8M \quad (3)$$

Thus, there is a non-linear relation between the two luminosities and the black hole mass, see Fig. 4. This relation is often referred to as the fundamental plane of black hole activity (Merloni et al. 2003; Falcke et al. 2004). The exact fit parameters depend on the choice of the sample as well as the fitting method.

While for the above mentioned relation between the characteristic timescale, the accretion rate and the black hole mass only the accretion rate depends on the distance of the source, for the current relation both luminosities depend strongly on the distance. This could lead to a false correlation purely driven by the vastly different distances of XRBs (10s of kpc) and AGN (Mpcs). The fact that the found relation is non-linear hints towards a real correlation (a correlation driven by distance differences tends to bias towards a linear correlation between the two luminosities). A more detailed statistical analysis has confirmed this hint and has shown that the found correlation is indeed significant (Merloni et al. 2006).

Subsequently, the parameters for the fundamental plane have been refined using standard linear regression taking into account measured uncertainties in all variables (Körding et al. 2006a). The authors showed that when one selects only sources belonging to an analogue of the hard state (hard state XRBs + LINERS), the overall scatter around the fundamental plane is small and in agreement with just the estimated measurement errors. Higher accreting sources (Seyferts, Quasars) roughly follow the correlation as well, but the scatter increases significantly, indicating that these highly accreting sources do not belong to the hard state. This further enhances the link between the hard state and low luminosity AGN.

Using Bayesian methods, Plotkin et al. (2012) found that relativistically beamed BL Lac objects fit well onto the fundamental plane if one accounts for the relativistic beaming. The parameters of the fundamental plane fitted with Bayesian methods are in agreement with those of a more conventional fitting method. The authors found that the fitting method can bias the obtained parameters, perhaps leading to incorrect inferences of the X-ray radiation mechanism. The original fundamental plane has also been improved by Gültekin et al. (2009), who study a volume limited sample of AGN with dynamically measured black hole masses. This approach eliminates the uncertainties due to black hole mass measurements, which are a major part of the total error budget. The authors find fit parameters that are broadly in agreement with the previous studies.

The connection between black hole mass, radio luminosity and X-ray luminosity has also been studied at 1.4 GHz (Bonchi et al. 2013). A comparison between this work and the papers of the fundamental plane mentioned above is not directly possible. While Bonchi et al. (2013) uses many more sources, these are not selected according to their accretion state. In fact, the majority are most likely highly accreting sources such as Seyferts. More importantly, at 1.4 GHz it is not possible to identify a core radio flux that originated in the flat spectrum radio core with VLA data, which is needed for the fundamental plane. At 1.4 GHz, the lobes start to contribute to the measured 'nuclear' emission at VLA scales, and the radio fluxes will differ from those which are usable in the fundamental plane (which are determined either with very long baseline interferrometry or at 15 GHz). It is therefore not surprising that the authors find different correlation parameters.

The fundamental plane has been established using a sample of sources, unlike the radio/X-ray correlation seen in XRBs. The latter correlation was first established for an individual source, GX339-4 (Corbel et al. 2000). For AGN, several authors have also tried to find a radio/X-ray correlation for individual sources. Bell et al. (2011) found that NGC 7213 is in agreement with the prediction of the fundamental plane. For M81, Miller et al. (2010) found that on short timescales an individual source might not be rigidly governed by the fundamental plane, and only follow the fundamental plane relation in a time-averaged sense. As the emission regions producing the radio and the X-ray emission are most likely different, there may be time differences between the emission in both regions.

3.1 Interpretation and Extension to Neutron Stars

The fundamental plane can be explained both with a pure jet model and with a disc/jet model, i.e. it is hard to infer the emission mechanism of the observed radiation with just the fundamental plane (see e.g., Merloni et al. 2003; Falcke et al. 2004; Yuan and Cui 2005). Both suggested emission mechanisms, synchrotron emission as well as Comptonization, depend on the energy density of either the magnetic field or the photon field in the emission region (see e.g., Rybicki and Lightman 1979). If both energy densities scale similarly, one cannot determine the emission mechanism from the scaling laws of the fundamental plane.

Neutron star XRBs show radio emission similar to that of black hole XRBs. Fender and Kuulkers (2001), Migliari et al. (2005) have shown that the radio emission of neutron star XRBs is typically significantly lower than that of a black hole XRB for a given X-ray luminosity. Neutron star XRBs show imageable jets (e.g., Fender et al. 2004). Due to the radio spectrum and the fact that the jet can sometimes be imaged, one usually assumes that the radio emission, as in the black hole XRB case, is created by a relativistic jet. The lower radio emission was therefore often interpreted as an indication that neutron star XRBs are less efficient in producing radio jets than black hole XRB systems.

The fundamental plane of accreting black holes suggests that the X-ray emission of hard state XRBs most likely originates in an inefficient accretion flow (or a jet, in which case the

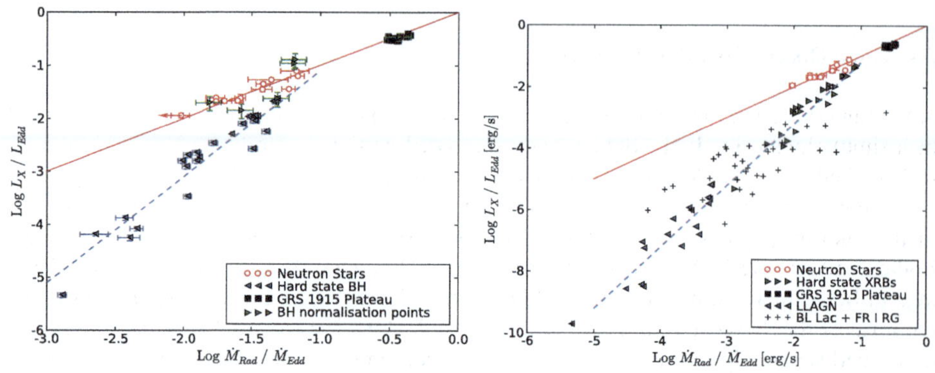

Fig. 5 Dependence of the bolometric luminosity on the accretion rate as measured from the radio emission. *Left*: While the neutron star bolometric luminosity depends linearly on the accretion rate, hard state black holes have a quadratic scaling. *Right*: Low luminosity AGN follow the quadratic dependence on the accretion rate. Plots reproduced with permission from Körding et al. (2006b)

accretion flow would be even more inefficient). When comparing the jet power of neutron stars with those of black holes one has to take the inefficiency accretion onto the black hole into account. Most models of inefficient accretion require a black hole to work. For example, all advective solutions will simply deposit the energy onto the stellar surface/boundary layer of a neutron star (Narayan and Yi 1994), and this energy has to be radiated or carried away (in jets/winds). Thus, neutron stars will most likely be efficient accretors.

To obtain jet powers from the radio luminosity of compact jets, one can use the cavities inflated by the jets as a calorimeter. This has been done for Cyg X-1 (Gallo et al. 2005) as well as for larger samples of AGN (e.g., Allen et al. 2006). Independently, Körding et al. (2006b), Heinz et al. (2007) obtained jet power estimates from radio core emission (i.e. one needs to obtain the normalization of Eq. (2)). Having estimated jet powers from the radio luminosity allows to compare the jet power of neutron star XRBs with that of black hole XRBs.

Figure 5 shows the bolometric luminosity of neutron star XRBs and black hole XRBs as a function of the accretion rate, as estimated from the radio jet luminosity. One finds the naively expected picture. The black hole systems follow an inefficient accretion track, while the neutron star XRBs show efficient accretion flows. The X-axis of this plot denotes the accretion rate as estimated from the jet power of a source. Thus, one can turn this presentation around and find that if neutron stars are efficient accretors while hard state XRBs show inefficient accretion, then neutron star XRBs are as efficient in creating jets as black hole XRBs.

Using the relation between the radio luminosity and the jet power, we can reformulate the fundamental plane of stellar and supermassive black holes as

$$L_X \propto \left(\frac{\dot{M}}{M}\right)^2 M M^{0.14} \tag{4}$$

Up to a small correction factor, which may be due to the bolometric correction terms or other second order effects, this relation is again scale invariant. Thus, the low luminosity AGN that follow the fundamental plane follow the inefficient branch, further supporting the idea that these sources are the analogue of the hard state XRBs.

3.2 Conclusions

We have presented two empirical correlations between stellar and supermassive black holes. The correlation between the characteristic timescale and the black hole mass and accretion rate indicates that the inner part of the accretion disc is scale invariant with respect to the black hole mass. The fundamental plane of accreting black holes, which links the radio luminosity, X-ray luminosity and black hole mass, can be used to argue for the scale invariance of the jet/accretion flow system. It is possible to add neutron stars onto these correlations in a way that makes sense intuitively (taking into account that neutron stars should be efficiently accreting as they have a stellar surface). This suggests that the accretion disc/jet system does not depend on the exact type of accretor and can be scaled between different source classes.

References

S.W. Allen, R.J.H. Dunn, A.C. Fabian, G.B. Taylor, C.S. Reynolds, The relation between accretion rate and jet power in x-Ray luminous elliptical galaxies. Mon. Not. R. Astron. Soc. **372**, 21–30 (2006). doi:10.1111/j.1365-2966.2006.10778.x

M.E. Bell, T. Tzioumis, P. Uttley, R.P. Fender, P. Arévalo, E. Breedt, I. McHardy, D.E. Calvelo, O. Jamil, E. Körding, X-ray and radio variability in the low-luminosity active galactic nucleus NGC 7213. Mon. Not. R. Astron. Soc. **411**, 402–410 (2011). doi:10.1111/j.1365-2966.2010.17692.x

T. Belloni, D. Psaltis, M. van der Klis, A unified description of the timing features of accreting X-ray binaries. Astrophys. J. **572**, 392–406 (2002)

R.D. Blandford, A. Kőnigl, Relativistic jets as compact radio sources. Astrophys. J. **232**, 34–48 (1979). http://adsabs.harvard.edu/cgi-bin/nph-bib_query?bibcode=1979ApJ...232...34B&db_key=AST

A. Bonchi, F. La Franca, G. Melini, A. Bongiorno, F. Fiore, On the radio luminosity distribution of active galactic nuclei and the black hole fundamental plane. Mon. Not. R. Astron. Soc. **429**, 1970–1980 (2013). doi:10.1093/mnras/sts456

P. Casella, T. Belloni, L. Stella, The ABC of low-frequency quasi-periodic oscillations in black hole candidates: analogies with Z sources. Astrophys. J. **629**, 403–407 (2005). doi:10.1086/431174

S. Corbel, R.P. Fender, A.K. Tzioumis, M. Nowak, V. McIntyre, P. Durouchoux, R. Sood, Coupling of the X-ray and radio emission in the black hole candidate and compact jet source GX 339-4. Astron. Astrophys. **359**, 251–268 (2000)

M. Coriat, S. Corbel, L. Prat, J.C.A. Miller-Jones, D. Cseh, A.K. Tzioumis, C. Brocksopp, J. Rodriguez, R.P. Fender, G.R. Sivakoff, Radiatively efficient accreting black holes in the hard state: the case study of H1743-322. Mon. Not. R. Astron. Soc. **414**, 677–690 (2011). doi:10.1111/j.1365-2966.2011.18433.x

W. Cui, W.A. Heindl, R.E. Rothschild, S.N. Zhang, K. Jahoda, W. Focke, Rossi X-ray Timing Explorer observation of Cygnus X-1 in its high state. Astrophys. J. Lett. **474**, 57 (1997). doi:10.1086/310419

M. Elvis, T. Maccacaro, A.S. Wilson, M.J. Ward, M.V. Penston, R.A.E. Fosbury, G.C. Perola, Seyfert galaxies as X-ray sources. Mon. Not. R. Astron. Soc. **183**, 129–157 (1978)

M. Elvis, R.F. Green, J. Bechtold, M. Schmidt, G. Neugebauer, B.T. Soifer, K. Matthews, G. Fabbiano, X-ray spectra of PG quasars. I—The continuum from X-rays to infrared. Astrophys. J. **310**, 291–316 (1986). doi:10.1086/164683

H. Falcke, P.L. Biermann, The jet-disk symbiosis. I. Radio to X-ray emission models for quasars. Astron. Astrophys. **293**, 665–682 (1995). http://esoads.eso.org/cgi-bin/nph-bib_query?bibcode=1995A%26A...293..665F&db_key=AST

H. Falcke, E. Kőrding, S. Markoff, A scheme to unify low-power accreting black holes. Jet-dominated accretion flows and the radio/X-ray correlation. Astron. Astrophys. **414**, 895–903 (2004)

S.A. Farrell, N.A. Webb, D. Barret, O. Godet, J.M. Rodrigues, An intermediate-mass black hole of over 500 solar masses in the galaxy ESO243-49. Nature **460**, 73–75 (2009). doi:10.1038/nature08083

R.P. Fender, E. Kuulkers, On the peak radio and X-ray emission from neutron star and black hole candidate X-ray transients. Mon. Not. R. Astron. Soc. **324**, 923–930 (2001). doi:10.1046/j.1365-8711.2001.04345.x

R. Fender, K. Wu, H. Johnston, T. Tzioumis, P. Jonker, R. Spencer, M. van der Klis, An ultra-relativistic outflow from a neutron star accreting gas from a companion. Nature **427**, 222–224 (2004)

G. Fossati, L. Maraschi, A. Celotti, A. Comastri, G. Ghisellini, A unifying view of the spectral energy distributions of blazars. Mon. Not. R. Astron. Soc. **299**, 433–448 (1998). doi:10.1046/j.1365-8711.1998.01828.x

E. Gallo, R.P. Fender, G.G. Pooley, A universal radio-X-ray correlation in low/hard state black hole binaries. Mon. Not. R. Astron. Soc. **344**, 60–72 (2003)

E. Gallo, R. Fender, C. Kaiser, D. Russell, R. Morganti, T. Oosterloo, S. Heinz, A dark jet dominates the power output of the stellar black hole Cygnus X-1. Nature **436**, 819–821 (2005). doi:10.1038/nature03879

E. Gallo, B.P. Miller, R. Fender, Assessing luminosity correlations via cluster analysis: evidence for dual tracks in the radio/X-ray domain of black hole X-ray binaries. Mon. Not. R. Astron. Soc. **423**, 590–599 (2012). doi:10.1111/j.1365-2966.2012.20899.x

M. Gierliński, M. Middleton, M. Ward, C. Done, A periodicity of $\sim$1 hour in X-ray emission from the active galaxy RE J1034+396. Nature **455**, 369–371 (2008). doi:10.1038/nature07277

A.R. Green, I.M. McHardy, H.J. Lehto, On the nature of rapid X-ray variability in active galactic nuclei. Mon. Not. R. Astron. Soc. **265**, 664 (1993)

K. Gültekin, E.M. Cackett, J.M. Miller, T. Di Matteo, S. Markoff, D.O. Richstone, The fundamental plane of accretion onto black holes with dynamical masses. Astrophys. J. **706**, 404–416 (2009). doi:10.1088/0004-637X/706/1/404

A. Heger, C.L. Fryer, S.E. Woosley, N. Langer, D.H. Hartmann, How massive single stars end their life. Astrophys. J. **591**, 288–300 (2003). doi:10.1086/375341

S. Heinz, R.A. Sunyaev, The non-linear dependence of flux on black hole mass and accretion rate in core-dominated jets. Mon. Not. R. Astron. Soc. **343**, 59–64 (2003). doi:10.1046/j.1365-8711.2003.06918.x

S. Heinz, A. Merloni, J. Schwab, The kinetic luminosity function and the jet production efficiency of growing black holes. Astrophys. J. Lett. **658**, 9–12 (2007). doi:10.1086/513507

L.C. Ho, "Low-state" black hole accretion in nearby galaxies. Astrophys. Space Sci. **300**, 219–225 (2005). doi:10.1007/s10509-005-1198-3

J. Homan, T. Belloni, The evolution of black hole states. Astrophys. Space Sci. **300**, 107–117 (2005). doi:10.1007/s10509-005-1197-4

O. Jamil, R.P. Fender, C.R. Kaiser, iShocks: X-ray binary jets with an internal shocks model. Mon. Not. R. Astron. Soc. **401**, 394–404 (2010). doi:10.1111/j.1365-2966.2009.15652.x

C.R. Kaiser, The flat synchrotron spectra of partially self-absorbed jets revisited. Mon. Not. R. Astron. Soc. **367**, 1083–1094 (2006). doi:10.1111/j.1365-2966.2006.10030.x

B.C. Kelly, T. Treu, M. Malkan, A. Pancoast, J.-H. Woo, Active galactic nucleus black hole mass estimates in the era of time domain astronomy. Astrophys. J. **779**, 187 (2013). doi:10.1088/0004-637X/779/2/187

E. Körding, H. Falcke, S. Corbel, Refining the fundamental plane of accreting black holes. Astron. Astrophys. **456**, 439–450 (2006a). doi:10.1051/0004-6361:20054144

E.G. Körding, R.P. Fender, S. Migliari, Jet-dominated advective systems: radio and X-ray luminosity dependence on the accretion rate. Mon. Not. R. Astron. Soc. **369**, 1451–1458 (2006b). doi:10.1111/j.1365-2966.2006.10383.x

E.G. Körding, S. Jester, R. Fender, Measuring the accretion rate and kinetic luminosity functions of supermassive black holes. Mon. Not. R. Astron. Soc. **1100** (2007a). doi:10.1111/j.1365-2966.2007.12529.x

E.G. Körding, S. Migliari, R. Fender, T. Belloni, C. Knigge, I. McHardy, The variability plane of accreting compact objects. Mon. Not. R. Astron. Soc. **380**, 301–310 (2007b). doi:10.1111/j.1365-2966.2007.12067.x

E. Kuulkers, M. van der Klis, GX340+0 with EXOSAT: its correlated X-ray spectral and timing behaviour. Astron. Astrophys. **314**, 567–575 (1996)

D. Lynden-Bell, Galactic nuclei as collapsed old quasars. Nature **223**, 690–694 (1969). doi:10.1038/223690a0

Y.E. Lyubarskii, Flicker noise in accretion discs. Mon. Not. R. Astron. Soc. **292**, 679 (1997)

S. Markoff, M. Nowak, S. Corbel, R. Fender, H. Falcke, Exploring the role of jets in the radio/X-ray correlations of GX 339-4. Astron. Astrophys. **397**, 645–658 (2003)

S. Markoff, M.A. Nowak, J. Wilms, Going with the flow: can the base of jets subsume the role of compact accretion disk coronae? Astrophys. J. **635**, 1203–1216 (2005). doi:10.1086/497628

A. Markowitz, R. Edelson, S. Vaughan, P. Uttley, I.M. George, R.E. Griffiths, S. Kaspi, A. Lawrence, I. McHardy, K. Nandra, K. Pounds, J. Reeves, N. Schurch, R. Warwick, X-ray fluctuation power spectral densities of Seyfert 1 galaxies. Astrophys. J. **593**, 96–114 (2003)

I. McHardy, EXOSAT observations of variability in active galactic nuclei. Mem. Soc. Astron. Ital. **59**, 239–259 (1988)

I.M. McHardy, E. Körding, C. Knigge, P. Uttley, R.P. Fender, Active galactic nuclei as scaled-up Galactic black holes. Nature **444**, 730–732 (2006). doi:10.1038/nature05389

I.M. McHardy, P. Arévalo, P. Uttley, I.E. Papadakis, D.P. Summons, W. Brinkmann, M.J. Page, Discovery of multiple Lorentzian components in the X-ray timing properties of the Narrow Line Seyfert 1 Ark 564. Mon. Not. R. Astron. Soc. **382**, 985–994 (2007). doi:10.1111/j.1365-2966.2007.12411.x

A. Merloni, S. Heinz, T. Di Matteo, A fundamental plane of black hole activity. Mon. Not. R. Astron. Soc. **345**, 1057–1076 (2003)

A. Merloni, E. Körding, S. Heinz, S. Markoff, T. Di Matteo, H. Falcke, Why the fundamental plane of black hole activity is not simply a distance driven artifact. New Astron. **11**, 567–576 (2006). doi:10.1016/j.newast.2006.03.002

S. Migliari, R.P. Fender, M. van der Klis, Correlation between radio luminosity and X-ray timing frequencies in neutron star and black hole X-ray binaries. Mon. Not. R. Astron. Soc. **799** (2005). doi:10.1111/j.1365-2966.2005.09412.x

J.M. Miller, M. Nowak, S. Markoff, M.P. Rupen, D. Maitra, Exploring accretion and disk-jet connections in the LLAGN M81*. Astrophys. J. **720**, 1033–1037 (2010). doi:10.1088/0004-637X/720/2/1033

I.F. Mirabel, L.F. Rodriguez, A superluminal source in the galaxy. Nature **371**, 46 (1994). http://adsabs.harvard.edu/cgi-bin/nph-bib_query?bibcode=1994Natur.371...46M&db_key=AST

K. Nandra, I.M. George, R.F. Mushotzky, T.J. Turner, T. Yaqoob, ASCA observations of Seyfert 1 galaxies. I. Data analysis, imaging, and timing. Astrophys. J. **476**, 70–82 (1997)

R. Narayan, I. Yi, Advection-dominated accretion: a self-similar solution. Astrophys. J. Lett. **428**, 13–16 (1994)

M.A. Nowak, J. Wilms, B.A. Vaughan, J.B. Dove, M.C. Begelman, Rossi X-ray Timing Explorer observation of Cygnus X-1. III. Implications for Compton corona and advection-dominated accretion flow models. Astrophys. J. **515**, 726–737 (1999)

F. Özel, D. Psaltis, R. Narayan, J.E. McClintock, The black hole mass distribution in the Galaxy. Astrophys. J. **725**, 1918–1927 (2010). doi:10.1088/0004-637X/725/2/1918

R.M. Plotkin, S. Markoff, B.C. Kelly, E. Körding, S.F. Anderson, Using the fundamental plane of black hole activity to distinguish X-ray processes from weakly accreting black holes. Mon. Not. R. Astron. Soc. **419**, 267–286 (2012). doi:10.1111/j.1365-2966.2011.19689.x

G. Ponti, I. Papadakis, S. Bianchi, M. Guainazzi, G. Matt, P. Uttley, N.F. Bonilla, CAIXA: a catalogue of AGN in the XMM-Newton archive. III. Excess variance analysis. Astron. Astrophys. **542**, 83 (2012). doi:10.1051/0004-6361/201118326

K. Pottschmidt, J. Wilms, M.A. Nowak, G.G. Pooley, T. Gleissner, W.A. Heindl, D.M. Smith, R. Remillard, R. Staubert, Long term variability of Cygnus X-1. I. X-ray spectral-temporal correlations in the hard state. Astron. Astrophys. **407**, 1039–1058 (2003)

M.J. Rees, Astrophysical evidence for black holes, in *Black Holes and Relativistic Stars*, ed. by R.M. Wald, 1998, p. 79

R.A. Remillard, J.E. McClintock, X-ray properties of black-hole binaries. Annu. Rev. Astron. Astrophys. **44**, 49–92 (2006). doi:10.1146/annurev.astro.44.051905.092532

G.B. Rybicki, A.P. Lightman, *Radiative Processes in Astrophysics* (Wiley-Interscience, New York, 1979), 393 pp. http://esoads.eso.org/cgi-bin/nph-bib_query?bibcode=1979rpa..book.....R&db_key=AST

P. Uttley, I.M. McHardy, X-ray variability of NGC 3227 and 5506 and the nature of active galactic nucleus 'states'. Mon. Not. R. Astron. Soc., 818 (2005). doi:10.1111/j.1365-2966.2005.09475.x

P. Uttley, I.M. McHardy, I.E. Papadakis, Measuring the broad-band power spectra of active galactic nuclei with RXTE. Mon. Not. R. Astron. Soc. **332**, 231–250 (2002)

M. van der Klis, Quasi-periodic oscillations and noise in low-mass X-ray binaries. Annu. Rev. Astron. Astrophys. **27**, 517–553 (1989)

M. van der Klis, A possible explanation for the "Parallel tracks" phenomenon in low-mass X-ray binaries. Astrophys. J. **561**, 943–949 (2001). doi:10.1086/323378

S. Vaughan, P. Uttley, Detecting X-ray QPOs in active galaxies. Adv. Space Res. **38**, 1405–1408 (2006). doi:10.1016/j.asr.2005.02.064

N.E. White, A.C. Fabian, R.F. Mushotzky, X-ray spectral signatures of accreting black holes. Astron. Astrophys. **133**, 9–11 (1984)

R. Wijnands, M. Mendez, M. van der Klis, D. Psaltis, E. Kuulkers, F.K. Lamb, Discovery of kilohertz quasi-periodic oscillations in the Z source GX 5-1. Astrophys. J. Lett. **504**, 35 (1998). doi:10.1086/311564

S.E. Woosley, T.A. Weaver, The evolution and explosion of massive stars. II. Explosive hydrodynamics and nucleosynthesis. Astrophys. J. Suppl. Ser. **101**, 181 (1995). doi:10.1086/192237

F. Yuan, W. Cui, Radio-X-ray correlation and the "quiescent state" of black hole sources. Astrophys. J. **629**, 408–413 (2005). doi:10.1086/431453

DOI 10.1007/978-1-4939-2227-7_10
Reprinted from *Space Science Reviews* Journal, DOI 10.1007/s11214-013-0015-5

Menus for Feeding Black Holes

Bence Kocsis · Abraham Loeb

Received: 2 April 2013 / Accepted: 12 August 2013 / Published online: 26 September 2013

Abstract Black holes are the ultimate prisons of the Universe, regions of spacetime where the enormous gravity prohibits matter or even light to escape to infinity. Yet, matter falling toward the black holes may shine spectacularly, generating the strongest source of radiation. These sources provide us with astrophysical laboratories of extreme physical conditions that cannot be realized on Earth. This chapter offers a review of the basic menus for feeding matter onto black holes and discusses their observational implications.

Keywords Black holes · Black hole binaries · Accretion and accretion disks · Plasmas—astrophysical

1 Introduction

There are several avenues for feeding matter onto black holes. Black holes can accrete ambient gas from the interstellar medium or from an attached gaseous disk, they may be fed by winds of a binary companion star or directly accrete from a binary companion through Roche lobe overflow, they can directly swallow or tidally disrupt stellar objects that approach them, or they may grow by merging with other black holes. Here we review these possibilities in detail and describe the corresponding astrophysical models of accretion.

We start with an overview of the standard formation and evolutionary scenarios of the three main classes of black holes: stellar mass, intermediate mass, and supermassive black holes. In subsequent sections we discuss the main classes of accretion including spherical accretion, disk accretion, and accretion due to tidal disruptions of stars.

B. Kocsis (✉) · A. Loeb
Institute for Advanced Study, Princeton, NJ, USA
e-mail: bkocsis@ias.edu

A. Loeb
e-mail: aloeb@cfa.harvard.edu

2 Black Hole Formation and Evolution

2.1 Collapse of Massive Stars

Compact objects may form when the core of a massive star consumes its nuclear fuel and loses pressure support, resulting in a catastrophic collapse. Whether the final state is a neutron star or a black hole depends on the mass of the star, its composition (i.e. metallicity) as depicted in Fig. 1, and rotation speed. Stars with masses between ~8–25 often explode as a Type II supernova and leave a neutron star/pulsar remnant. For more massive stars, in the range 25–35 $M_\odot$, the fall back of the carbon-oxygen core may be significant to crush the remnant into a black hole (Fryer 1999). For even more massive stars, the complete hydrogen envelope is lost to strong winds, leaving an isolated Wolf-Rayet star. This star then explodes as a Type-Ib/c supernova. Depending on the less-understood Wolf-Rayet mass loss rates, the final remnant is either a neutron star or a black hole. Recent work (e.g., Ugliano et al. 2012) shows that black holes can form across a somewhat wider range of initial masses ($M \gtrsim 15\ M_\odot$), and neutron stars can form from much higher mass progenitors ($M \gg 25\ M_\odot$). This is also hinted by the young progenitors inferred for observed magnetars ($M > 40\ M_\odot$; Muno et al. 2006).

The mass of the remnant black hole depends on the amount of wind loss, fall back, and metallicity. It is typically estimated to be between 5–15 $M_\odot$ for solar metallicities and between 5–80 $M_\odot$ in globular clusters or metal-poor galaxies with 0.01 $Z_\odot$ (Belczynski et al. 2010). Higher mass black holes may form during the collapse of the first stars, the so-called Population III stars, discussed in Sect. 2.4.

2.2 Black Holes in Stellar Binaries

Many stellar-mass black holes and neutron stars are found in close binary systems. In the local Universe, black-hole X-ray binaries come in two flavors, depending on the mass of

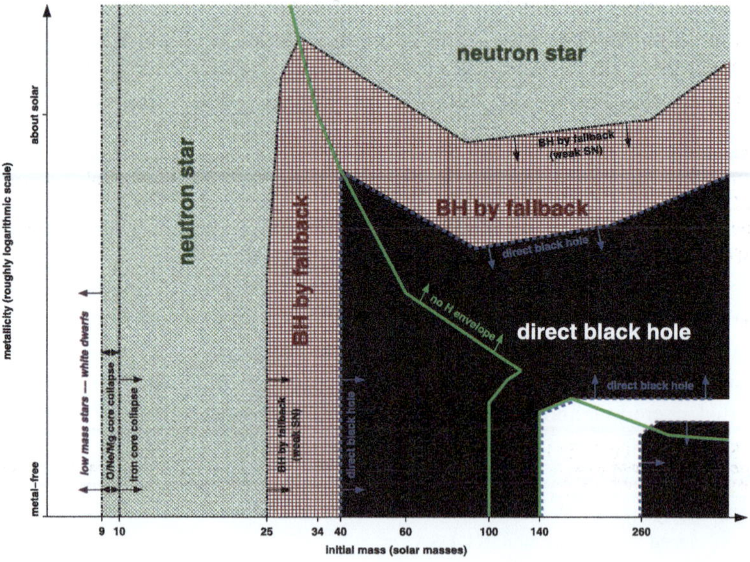

Fig. 1 Types of stellar remnants of massive stars for different initial stellar masses and metallicity. Figure credit: Heger et al. (2003)

the companion star: *low-mass X-ray binaries* where a low-mass companion transfers mass owing to the tidal force exerted by the black hole, and *high-mass X-ray binaries (BH-HMXB)* where the companion is a massive star which could also transfer mass to the black hole through a wind. At redshifts $z \gtrsim 6$ when the age of the Universe was short, BH-HMXB were probably most important since they are known to produce their X-rays over a short lifetime ($\sim 10^8$ yr). The cumulative X-ray emission from BH-HMXBs is expected to be proportional to the star formation rate (Grimm et al. 2003; Ranalli et al. 2003; Mineo et al. 2012). If indeed the early population of stars was tilted towards high masses and binaries were common, BH-HMXB may have been more abundant per star formation rate in high redshift galaxies. The X-rays produced by BH-HMXBs may have had important observable effects as they catalyzed H_2 formation, heated the intergalactic medium, and modified the 21-cm signal from neutral hydrogen. Their overall influence was, however, limited: hydrogen could not have been reionized by X-ray sources based on current limits on the unresolved component of the X-ray background.

The formation of BH binaries may be linked to the evolution of binary stars, transferring mass during their lives. X-ray binaries in particular, are believed to be affected by mass transfer in the giant phase. The standard evolutionary scenario leading to a HMXB is depicted in Fig. 2. Here at least one of the stars in the binary must be a massive star. After a few million years, it exhausts its hydrogen, expands into a red giant, and transfers its entire envelope to the secondary. At this stage, the primary and secondary become a Wolf-Rayet star and a massive O/B star. The primary then explodes as a type Ib supernova, and becomes a neutron star or a black hole depending on its initial mass (see Fig. 1). The powerful winds of the massive secondary star sources accretion for the companion compact object. Later, as the secondary runs out of hydrogen and expands to within the *Roche lobe* of the compact object where the gravity of the compact object dominates,

$$r_R = a(1-e)\left(\frac{m}{2M}\right)^{1/3} \tag{1}$$

the compact object will accrete continuously from the secondary. Here M, m, a, e are the primary and secondary masses, semimajor axis, and eccentricity. As the secondary becomes a red giant, it may completely engulf the orbit. In this common envelope stage, the angular momentum of the binary can quickly decrease as the envelope is heated and ejected. If the core of the giant merges with the neutron star, it may form a Thorne-Zytkow object (Thorne and Zytkow 1977), or if it is ejected completely, it leaves behind a black hole or neutron star primary orbiting closely around a Wolf-Rayet star secondary. Next, the secondary explodes as a Type-Ib supernova. If the kick from the supernova explosion is larger than the gravitational binding energy, the two compact objects may fly apart, as single black holes or neutron stars. Alternatively, if the kick is weaker, it leaves behind a relativistic eccentric double black hole, double neutron star, or black hole-neutron star binary.

These double compact object binaries then gradually loose energy and angular momentum due to gravitational wave emission, which leads to the shrinking of the orbital separation and eccentricity. The existence of gravitational waves is a generic prediction of Einstein's theory of gravity. They represent ripples in space-time generated by the motion of the two black holes as they move around their common center of mass in a tight binary. The energy carried by the waves is taken away from the kinetic energy of the binary, which therefore tightens with time. Ultimately, the gravitational waves emitted close to merger will be de-

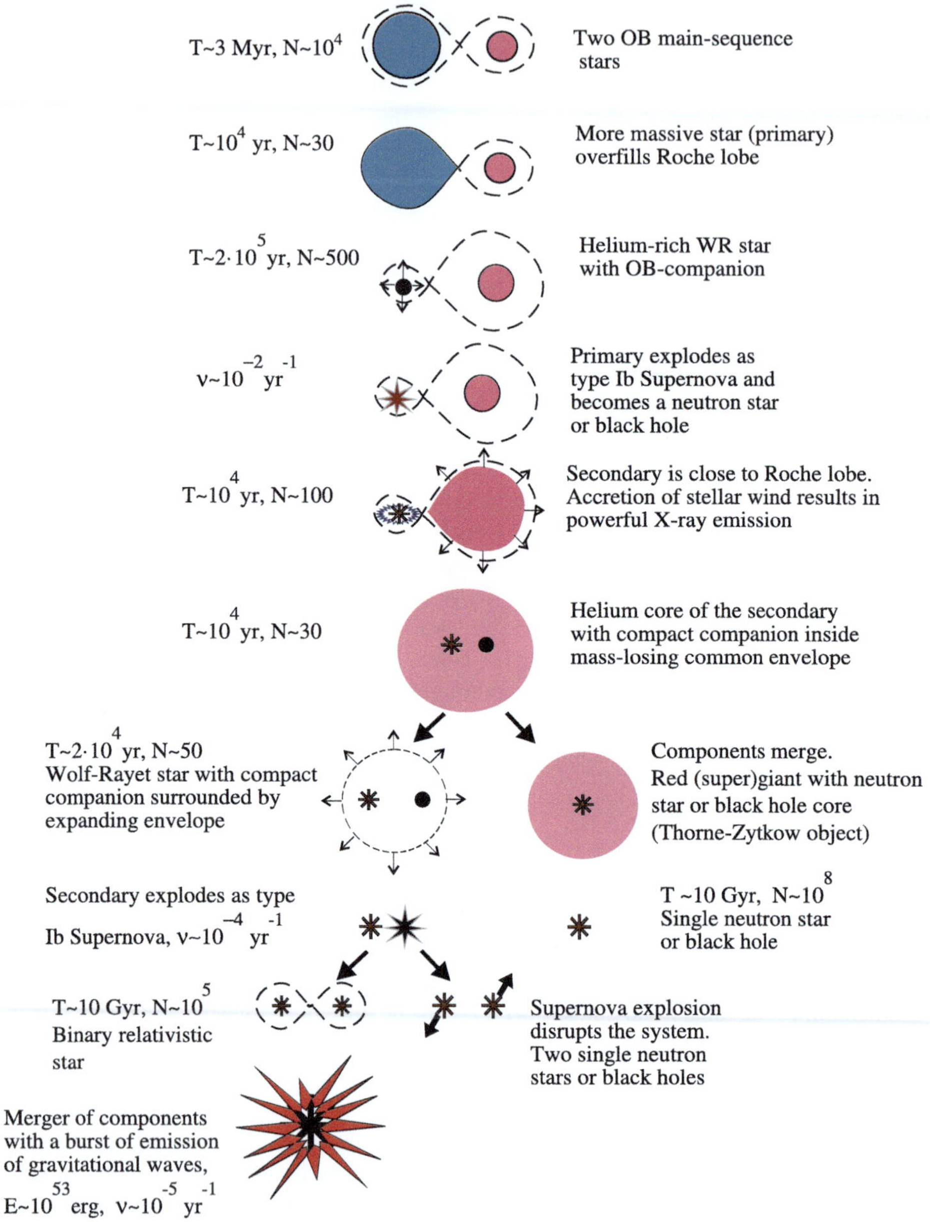

Fig. 2 Evolutionary stages of high mass binary stars from top to bottom. The initially more massive star (primary) is shown *on the left*. A black hole or neutron star forms in the 4th stage during a type Ib supernova possibly accompanied by a long GRB. The 5th stage represents a typical HMXB. Once the secondary fills its Roche lobe, it may supply super-Eddington accretion and lead to a super-soft X-ray source. Later the binary inspirals in a common envelope, ejecting the envelope. This may lead to a second supernova, leaving a relativistic double BH, NS, or BH/NS binary. These objects merge after gravitational wave emission, possibly generating a short GRB. The approximate timescale and number of such binaries in the Galaxy are labeled in each evolutionary stage. Figure credit: Postnov and Yungelson (2006)

tectable by Advanced LIGO[1] and VIRGO.[2] Additionally, if at least one of the components is a neutron star, the merger may produce a bright gamma ray burst (GRB).

The above scenario involves massive stars. The analogous evolutionary scenario for less massive binary stars leads to a white dwarf, neutron star, or black hole with a moderate to low-mass main sequence star, and ultimately a double compact object binary (e.g. double degenerate white dwarf binary). In this case, the separation is much larger than commonly observed in low mass X-ray binaries (LMXBs). The formation of LMXBs is less understood. Many-body interactions, especially in globular clusters may lead to LMXBs. Alternatively, a massive binary system may form LMXBs if there is a rapid super-Eddington wind phase and angular momentum evolution in between the 5th and 6th stage in Fig. 2. Black holes may form in low mass binaries during an accretion-induced collapse if a neutron star is feeding through Roche lobe overflow from a companion (Dermer and Atoyan 2006). Another possibility is the merger of a WD with a NS (Paschalidis et al. 2009), although in this case it is unclear whether sufficient mass will accrete onto the NS if the WD material undergoes explosive nuclear burning (Fernández and Metzger 2013).

Neutron star or black hole binaries that accrete from a strong wind of the companion star, may be modeled using Bondi-Hoyle-Lyttleton accretion. Compact objects that accrete from the companion star through a Roche-lobe overflow, provide examples of disk accretion. We discuss Bondi and disk accretion in Sects. 3 and 4, respectively.

A neutron star in a LMXB that undergoes mass transfer from a secondary star is expected to be spun up during the accretion to millisecond periods, leaving a millisecond pulsar. Since black holes are more massive with a larger angular momentum, they cannot be spun up significantly by a lower mass companion.

2.3 Supermassive Black Holes

Supermassive black holes of masses between 10^6–10^{10} $M_\odot$ are observed in the centers of most if not all nearby galaxies. As they accrete gas, the gas shines as a bright point-like quasi-stellar object, called a quasar. These objects are observed from cosmological distances up to high redshifts, where the Universe was less than a billion years old. How could such massive black holes form in such a short amount of time?

Growing a supermassive black hole very quickly is difficult. Accretion of collisionless dark matter particles is negligible and can be ignored (Shapiro and Teukolsky 1983a). For gas accretion—there is a maximum luminosity at which the environment of a black hole of mass M_{BH} may shine and still accrete gas, called the Eddington limit, L_{Edd}. This limit is obtained by setting the outward continuum radiation pressure equal to the inward gravitational force. Denoting the gravitational potential with Φ, pressure with p, density with ρ,

$$\nabla \Phi = -\frac{\nabla p}{\rho} = \frac{\kappa}{c} \boldsymbol{F}_{\text{rad}}, \tag{2}$$

where in the last equality we assumed that the pressure is dominated by radiation pressure which is associated to a radiation flux $\boldsymbol{F}_{\text{rad}}$. Here κ, is the opacity. There are two primary sources of opacity for the typical densities and temperatures here: Thomson electron scattering and bremsstrahlung (i.e labeled free-free absorption) with

$$\kappa_{\text{es}} = \frac{\sigma_{\text{T}}}{\text{m}_{\text{p}}} = 0.4\ \text{cm}^2\,\text{g}^{-1} \quad \text{and} \quad \kappa_{\text{ff}} \approx 8 \times 10^{22}\ \text{cm}^2\,\text{g}^{-1} \left(\frac{\rho}{\text{g}\,\text{cm}^{-3}}\right)\left(\frac{T}{\text{K}}\right)^{-7/2}, \tag{3}$$

[1] http://www.ligo.caltech.edu/.

[2] http://www.ego-gw.it/.

where we assumed a pure hydrogen plasma for simplicity, where m_p denotes proton mass and σ_T denotes the Thomson cross section. Substituting Eq. (2) into the definition of the luminosity of the source bounded by a surface S, we get

$$L = \int_S \boldsymbol{F}_{\rm rad} \cdot {\rm d}\boldsymbol{S} = \frac{c}{\kappa} \int_S \nabla \Phi \cdot {\rm d}\boldsymbol{S}. \tag{4}$$

Using Gauss's theorem, Poisson's equation $\nabla^2 \Phi = 4\pi G \rho$, and the definition of the mass,

$$L_{\rm Edd} = \frac{c}{\kappa} \int_V \nabla^2 \Phi {\rm d}V = \frac{4\pi G c}{\kappa} \int_V \rho {\rm d}V = \frac{4\pi G M c}{\kappa} = 1.3 \times 10^{44} \left(\frac{M_{\rm BH}}{10^6\ {\rm M}_\odot} \right) {\rm erg\, s^{-1}}. \tag{5}$$

This sets the maximum luminosity of a source in hydrostatic equilibrium, $L_{\rm Edd}$ denotes Eddington luminosity. Note that we did not assume spherical geometry. While the effects of rotation can somewhat change this theoretical limit in an accretion disk, observed quasars for which black hole masses can be measured by independent methods appear to respect this limit.

The total luminosity from gas accreting onto a black hole, L, can be written as some radiative efficiency ϵ times the mass accretion rate $\dot{M}$,

$$L = \epsilon \dot{M} c^2, \tag{6}$$

with the black hole accreting the non-radiated component, $\dot{M}_{\rm BH} = (1 - \epsilon)\dot{M}$. The equation that governs the growth of the black hole mass is then

$$\dot{M}_{\rm BH} = \frac{M_{\rm BH}}{t_{\rm Edd}}, \tag{7}$$

where (after substituting all fundamental constants),

$$t_{\rm Edd} = 4 \times 10^7 \left(\frac{\epsilon/(1-\epsilon)}{0.1} \right) \left(\frac{L}{L_{\rm Edd}} \right)^{-1} {\rm yr}. \tag{8}$$

We therefore find that as long as fuel is amply supplied, the black hole mass grows exponentially in time, $M_{\rm BH} \propto \exp(t/t_{\rm Edd})$, with an e-folding time $t_{\rm Edd}$. Since the growth time in Eq. (8) is significantly shorter than the $\sim 10^9$ years corresponding to the age of the Universe at a redshift $z \sim 6$—where black holes with a mass $\sim 10^9\ {\rm M}_\odot$ are found—one might naively conclude that there is plenty of time to grow the observed black hole masses from small seeds. For example, a seed black hole from a Population III star of $100\ {\rm M}_\odot$ can grow in less than a billion years up to $\sim 10^9\ {\rm M}_\odot$ for $\epsilon \sim 0.1$ and $L \sim L_{\rm Edd}$. However, the intervention of various processes makes it unlikely that a stellar mass seed will be able to accrete continuously at its Eddington limit without interruption.

Mergers are very common in the early Universe. Every time two gas-rich galaxies come together, their black holes are likely to coalesce. The coalescence is initially triggered by "dynamical friction" from the surrounding gas and stars (see Fig. 3), and is completed—when the binary gets tight—as a result of the emission of gravitational radiation (Begelman et al. 1980). Computer simulations reveal that when two black holes with unequal masses merge to make a single black hole, the remnant gets a kick due to the non-isotropic emission of gravitational radiation at the final plunge. This kick was calculated recently using advanced computer codes that solve Einstein's equations, a task that was plagued for decades with numerical instabilities (Pretorius 2005; Campanelli et al. 2006;

Baker et al. 2006). The typical kick velocity is hundreds of kilometer per second (and up to ten times more for special spin orientations), much larger than the escape speed from the first dwarf galaxies. This implies that continuous accretion was likely punctuated by black hole ejection events (Blecha and Loeb 2008; Tanaka and Haiman 2009), forcing the merged dwarf galaxy to grow a new black hole seed from scratch. These black hole recoils might have left observable signatures in the local Universe. For example, the halo of the Milky Way galaxy may include hundreds of freely-floating ejected black holes with compact star clusters around them, representing relics of the early mergers that assembled the Milky Way out of its original building blocks of dwarf galaxies (O'Leary and Loeb 2009).

2.4 Seeds of Supermassive Black Holes

Supermassive black holes with masses exceeding 10^9 $M_\odot$ are observed at cosmological distances where the Universe was less than a billion years old. Assuming that their luminosity does not exceed the Eddington limit and the radiative efficiency is $\sim$10 %, one is driven to the conclusion that the black hole seeds must have started more massive than the stellar regime of $\lesssim$100 $M_\odot$. The needed seeds may originate from *supermassive stars*, defined as hydrostatic configurations with masses 10^3–10^8 $M_\odot$ (Loeb and Rasio 1994; Bromm and Larson 2004). Lacking carbon, nitrogen, oxygen, and iron, these first stars do not drive powerful stellar winds as present-day massive stars. Such systems have not been observed as of yet. Theoretically, they are expected to be supported almost entirely by radiation pressure and hence their luminosity equals the Eddington limit (Eq. (5)). Supermassive stars steadily contract and convert their gravitational binding energy to radiation with a total lifetime $\lesssim 10^6$ yr before they collapse to a black hole.

Is it possible to make such stars in early galaxies? Numerical simulations indicate that stars of mass $\sim 10^6$ $M_\odot$ could have formed at the centers of early dwarf galaxies that were barely able to cool their gas through transitions of atomic hydrogen, having $T_{\rm vir} \sim 10^4$ K and no H_2 molecules (which were suppressed by a Lyman-Werner background, Bromm and Loeb 2003; Dijkstra et al. 2008; Regan and Haehnelt 2009). Such systems have a total mass that is several orders of magnitude higher than the earliest Jeans-mass condensations. In both cases, the gas lacks the ability to cool well below $T_{\rm vir}$, and so it fragments into one or two major clumps. The simulation results in clumps of several million solar masses, which inevitably form massive black holes. The existence of such massive seeds would have given a jump start to the black hole growth process.

The nearly uniform entropy established by convection makes the structure of supermassive stars simple (equivalent to a so-called polytrope with an index $n = 3$) with a unique relation between their central temperature T_c and central density ρ_c,

$$T_c = 2 \times 10^6\ \mathrm{K} \left(\frac{\rho_c}{1\ \mathrm{g\,cm^{-3}}} \right)^{1/3} \left(\frac{M_\star}{10^6\ \mathrm{M_\odot}} \right)^{1/6}. \quad (9)$$

Because of this modest temperature, nuclear reactions are insignificant in metal-poor stars with masses $M_\star > 10^5$ $M_\odot$. General relativistic corrections make the star unstable to direct collapse to a black hole as soon as its radius contracts to a value

$$R_\star < R_{\rm crit} = 1.59 \times 10^3 \left(\frac{M_\star}{10^6\ \mathrm{M_\odot}} \right)^{1/2} \left(\frac{G M_\star}{c^2} \right). \quad (10)$$

Rotation can stabilize supermassive stars to smaller radii, but even rotating stars are expected to eventually collapse to a black hole after shedding their angular momentum through a wind

(Saijo et al. 2002). If the supermassive star is made of pre-enriched gas, then powerful winds will inevitably be driven at its surface where the opacity due to lines from heavy elements far exceeds the Thomson value, making the outward radiation force stronger than gravity.

We note that the infall of a sufficiently dense, optically-thick spherical envelope of gas cannot be prevented by radiation pressure even if the radiation production rate exceeds the Eddington limit near the center. So long as the mass infall rate is sufficiently high, the Eddington limit will not apply because the photons will be trapped in the flow. Super-Eddington accretion can grow a seed black hole rapidly (as in the case of stellar collapse), as long as the blanket of infalling gas advects the radiation inwards as it accretes onto the black hole. This "obscured" mode of black hole accretion (which is hidden from view) could be particularly important at high redshifts when the gas density and infall rate onto galaxies obtain their highest values (Wyithe and Loeb 2012). We discuss the physics of this possibility in Sect. 3 below.

2.5 Intermediate Mass Black Holes

Up to this point we have discussed the two well known classes of black holes: the stellar-mass and supermassive black holes. An intriguing question is whether there is an intermediate population of black holes in the mass range $M_{\rm imbh} \sim 100$–10^5 $M_\odot$ (Miller and Colbert 2004). These objects, referred to as intermediate mass black holes (IMBHs), are interesting for several reasons: they may represent seeds for forming SMBHs through accretion, they could stabilize globular clusters against core collapse (Baumgardt et al. 2004), they can lead to dark matter overdensities which will cause excessive dark matter annihilation signals (Bertone et al. 2005, 2009), they may have participated in cosmic reionization (Madau et al. 2004), and may provide sources of gravitational waves for direct detection (Wyithe and Loeb 2003, 2004; Mandel et al. 2008; O'Leary et al. 2009; Kocsis et al. 2012c).

Several theoretical arguments have been put forth for forming IMBHs. The collapse of the first supermassive Pop-III stars may form IMBHs. This process can add $\sim 50(M_{\rm imbh}/150\ M_\odot)$ IMBHs to galactic centers (Madau and Rees 2001). Secondly, runaway collisions of stars (Portegies Zwart and McMillan 2002; Freitag et al. 2006) or collisions with stellar black holes (O'Leary et al. 2006) in dense star clusters can also produce IMBHs. The clusters sink to the galactic nucleus due to dynamical friction (Sect. 3.3) and deposit their IMBHs. This channel may supply an additional 50 IMBHs of mass 10^3 $M_\odot$ within the inner 10 pc (Portegies Zwart et al. 2006). IMBHs can also form and migrate inwards in the accretion disks of SMBHs (Goodman and Tan 2004; Levin 2007; Kocsis et al. 2011; McKernan et al. 2012).

Ultraluminous X-ray sources (ULXs) provide the best observational candidates for IMBHs (see Sect. 4 below). In particular, the estimated IMBH mass in ESO 243-49 HLX-1 based on its X-ray thermal accretion disk spectrum and radio signal is in the range 9×10^3–9×10^4 $M_\odot$ (Davis et al. 2011; Webb et al. 2012). IMBHs may also represent low luminosity X-ray sources in the Galactic bulge accreting gas from the interstellar medium or infalling gas clouds (Bartos et al. 2013).

3 Bondi-Hoyle-Lyttleton Accretion

Bondi-Hoyle-Lyttleton accretion denotes a general class of black hole accretion, where the black hole is completely embedded in a gaseous medium and the inflow is spherical without a significant rotation. This occurs if black holes accrete from the ambient interstellar medium, as they pass through a dense gas cloud, become embedded in the envelope of a giant star, or if they accrete from a powerful wind of a binary companion star. Stellar mass black holes embedded in the accretion disk of a supermassive black hole may also accrete from the gaseous medium through this mode of accretion. Here we review the physical principles of Bondi-Hoyle-Lyttleton accretion and its observational implications.

3.1 Simple Model for Spherical Accretion

Consider a black hole of mass M is moving with relative velocity V in an ambient medium of density ρ_0 and temperature T_0. The root-mean-square velocity of thermal protons in the gas relative to the black hole is roughly $v_{\rm rms} = \sqrt{c_s^2 + V^2}$, where $c_s \approx \sqrt{\mathrm{k_B} T/\mathrm{m_p}}$ is the sound speed. The gas particles interior to the Bondi radius,

$$r_{\rm B} = \frac{2GM}{v_{\rm rms}^2} \tag{11}$$

are gravitationally bound to the black hole and are accreted. The steady mass flux of particles entering this radius is $\rho_0 v_{\rm rms}$. Multiplying this flux by the surface area associated with the Bondi radius, $\pi r_{\rm B}^2$, gives the supply rate of gas,

$$\dot{M}_{\rm B} = 4\pi\rho_0 \frac{G^2M^2}{v_{\rm rms}^3} = 78\left(\frac{M}{10^8\ \mathrm{M_\odot}}\right)^2\left(\frac{n_H}{1\ \mathrm{cm^{-3}}}\right)\left(\frac{T_0}{10^4\ \mathrm{K}}\right)^{-3/2}\ \mathrm{M_\odot\ yr^{-1}}, \tag{12}$$

where the second equality assumes a static medium ($v_{\rm rms} = c_s$) and $n_H = \rho_0/\mathrm{m_p}$. In a steady state this supply rate equals the mass accretion rate into the black hole.

This simple estimate for quasi-spherical accretion onto black holes is consistent with a range of simplified models. Hoyle and Lyttleton (1939) derived the accretion onto a point mass in the limit that the gas pressure is negligible ($c_s = 0$) and particles move on ballistic orbits. Bondi and Hoyle (1944) extended the analysis to include accretion from an axisymmetric wake behind the black hole, which led to a similar result, modified by a factor $\sim 1/2$. Bondi (1952) solved the Euler and continuity equations for a spherically symmetric steady-state adiabatic flow of gas assuming that the accreting object does not move relative to the medium ($V = 0$). Well inside the sonic radius (i.e., the point at which the infall and sound speeds are equal), the velocity is close to free-fall $v \sim (2GM/r)^{1/2}$ and the gas density is $\rho(r) \sim \rho_0(r/r_{\rm B})^{-3/2}$. If the black hole moves with a supersonic velocity $V > c_s$, then a shock wave forms behind the hole, and the accretion occurs primarily from the shocked gas in a column trailing the black hole. The Bondi solution is then appropriate interior to the shock front and the Bondi radius $r_{\rm B}$. The correction factor of this detailed calculation for the accretion rate relative to Eq. (12) is of order unity (see Ruffert 1994 for an updated formula).

3.2 Luminosity

The luminosity of an accretion flow is generally expressed in terms of the radiative efficiency ϵ for converting rest mass to radiation as, $L = \epsilon \dot{M} c^2$. The accretion rate for Bondi-Hoyle-Lyttleton accretion, $\dot{M}_{\rm B}$, is given by Eq. (12). The radiative efficiency, ϵ is very small during spherical accretion since most of the energy, in the form of kinetic energy, heat, and radiation, is advected into the black hole rather than escape to infinity. This is due to the fact that the cooling time is longer than its accretion (free-fall) time or the diffusion time of the radiation outwards is much longer than the free-fall time and the radiation is trapped in the flow (Rees 1978; Begelman 1979; Blondin 1986). The infall time of the fluid element from the Bondi radius to the BH is approximately, $t_{\rm ff}(r) = r_{\rm B}/v_{\rm rms}$ and the photon diffusion time in the optically thick limit (i.e. where the optical depth satisfies $\tau = \kappa\rho r_{\rm B} \gg 1$) is

$$t_{\rm diff}(r) = \frac{1}{2c}\int_0^{r_{\rm B}} \kappa\rho(\xi)\xi\,{\rm d}\xi = \frac{\kappa}{\pi c}\frac{\dot{M}_{\rm B}}{v_{\rm rms}} \tag{13}$$

where we have used Eqs. (11) and (12), $\rho(r) \propto r^{-3/2}$ for Bondi accretion, and κ denotes the opacity. Thus, the diffusion time is larger than the infall time if

$$\dot{M}_{\rm B} \gtrsim \frac{\pi r_{\rm B} c}{\kappa} \simeq \frac{2\pi GMc}{\kappa v_{\rm rms}^2} = \frac{\dot{M}_{\rm Edd}}{2}\left(\frac{v_{\rm rms}}{c}\right)^{-2}, \tag{14}$$

where in the last equality we used the definition of the Eddington luminosity, $L_{\rm Edd}$ and the Eddington accretion rate $\dot{M}_{\rm Edd} = L_{\rm Edd}/c^2$. The radiation is trapped in the flow for hyper-Eddington accretion rates that satisfy Eq. (14). For smaller $\dot{M}_{\rm B}$, the inflowing gas can cool radiatively. Radiation-hydrodynamical simulations show that the flow remains radiatively inefficient, and the luminosity significantly sub-Eddington even if the accretion is super-Eddington. For $300 \geq \dot{m}_{\rm B} \equiv \dot{M}_{\rm B}/\dot{M}_{\rm Edd} \gg 1$, the radiative efficiency is estimated to be in the range between $10^{-6} \lesssim \epsilon \lesssim 10^{-3}$ (Fragile et al. 2012) and $\epsilon \sim 10^{-2}$ (Roedig et al. 2012). If the accretion rate is very sub-Eddington, $\dot{m}_{\rm B} \ll 0.1$, the radiative efficiency is expected to follow $\epsilon = 10\dot{m}_{\rm B}$ (see Sect. 4.2 below).

3.3 Dynamical Friction

In addition to mass accretion, a gravitating object experiences a drag when moving through a medium. The drag arises due to the gravitational focusing of material forming a wake behind the accretion. This process in collisionless astronomical systems is called dynamical friction (Chandrasekhar 1943; Binney and Tremaine 2008). It is responsible for delivering massive black holes to the centers of galaxies after galaxy mergers (Begelman et al. 1980), sinking satellites in dark matter galaxy halos and thereby transporting stars and possibly intermediate mass black holes to the galactic nucleus (Portegies Zwart et al. 2006). A similar effect arises in a gaseous medium (Rephaeli and Salpeter 1980; Ostriker 1999). The steady-state drag force is

$$F_{\rm DF} = -4\pi\rho_0\frac{GM}{V^2}I \tag{15}$$

where

$$I_{\rm supersonic} = \ln\left(\frac{r_{\max}}{r_{\min}}\right) + \frac{1}{2}\ln\left(1 - \frac{1}{\mathcal{M}^2}\right) \quad \text{if } \mathcal{M} > 1, \tag{16}$$

$$I_{\rm subsonic} = \frac{1}{2}\ln\left(\frac{1+\mathcal{M}}{1-\mathcal{M}}\right) - \mathcal{M} \quad \text{if } \mathcal{M} < 1, \tag{17}$$

and $\mathcal{M} = V/c_s$ is the Mach number, $r_{\max}$ denotes the maximum size of the medium and $r_{\min}$ denotes the size of the perturber. For black holes $r_{\min}$ is of order the gravitational radius, GM/c^2. In the limit $\mathcal{M} \gg 1$, $F_{\rm DF} \approx -\dot{M}_{\rm B} V \ln(r_{\max}/r_{\min})$, and for $\mathcal{M} \ll 1$, $F_{\rm DF} \approx -\frac{1}{3}\dot{M}_{\rm B} V$.

3.4 Vorticity, Turbulence, and Radiation Effects

The simple model of Bondi-Hoyle-Lyttleton accretion neglects rotation, gas cooling, radiative effects, turbulence, and relativistic effects, which may modify the results significantly.

These effects may be studied using state-of-the-art numerical simulations, which is becoming possible only recently.

Krumholz et al. (2005) examined Bondi accretion if the gas has a non-zero angular momentum beyond the Bondi radius. Neglecting radiation and thermodynamic effects, their simulation results are consistent with the fitting formula

$$\dot{M}_{\rm vort} = \dot{M}_{\rm B} \times \begin{cases} 0.4 & \text{if } \omega_* \ll 1 \\ 0.08\,\omega_*^{-1}\ln(16\omega_*) & \text{if } \omega_* \gg 1 \end{cases} \tag{18}$$

where $\omega_* = \omega r_{\rm B}/(2c_s)$ is the vorticity parameter, where ω is the angular velocity of the inflowing gas at the outer boundary with respect to the black hole. This shows that the accretion rate is significantly suppressed by vorticity.

Further, Krumholz et al. (2006) have shown that the accretion rate is significantly modified in a turbulent medium over the estimate one obtains by using the turbulent velocity dispersion in Eq. (12). In this case the accretion follows a lognormal distribution with a median that is roughly given by

$$\dot{M}_{\rm turb} = \left[\dot{M}_{\rm B}^{-2} + \dot{M}_{\rm vort}^{-2}\right]^{-1/2}. \tag{19}$$

These estimates assumed that the velocity of the object is much smaller than the turbulent velocity dispersion.

Incorporating radiation and realistic heating and cooling processes between multiple components (electrons, ions, and photons) poses a challenge to realistic numerical simulations of Bondi-Hoyle-Lyttleton accretion. Recently, Fragile et al. (2012) have examined optically thick Bondi accretion using a general relativistic radiation magneto-hydrodynamic (GR-R-MHD) simulation with $10 \le \dot{M}/\dot{M}_{\rm Edd} \le 300$. They assumed two components: radiation and gas, with two gas-cooling mechanisms, Thomson scattering and thermal bremsstrahlung. The ambient gas temperature was between 10^5 and 10^7 K. They found that the radiation pressure remained sub-dominant throughout the flow, and the luminosity remained very sub-Eddington despite the super-Eddington accretion rates such that the radiative efficiency was $\epsilon = L/\dot{M}c^2 \lesssim 10^{-4}$. These simulations suggest that radiation does not have a major effect on the results. The opposite conclusion was reached based an independent code by Zanotti et al. (2011) and Roedig et al. (2012). These simulations also assumed two components (gas + radiation) with Thomson scattering and bremsstrahlung. In these simulations, radiation pressure dominated over gas pressure, and suppressed accretion by two orders of magnitude. The accretion rate was still super-Eddington in the simulation, while the luminosity was around the Eddington limit with $\epsilon \sim 10^{-2}$. These simulations showed that radiation pressure may serve to further stabilize the flow against a flip-flop instability (see Sect. 3.5). While this progress is exciting, further improvements are necessary for a more secure estimate of the radiative efficiency of Bondi-Hoyle-Lyttleton accretion: primarily by incorporating the cooling effects of electrons (synchrotron emission and Comptonization) and extending calculations to smaller temperatures and higher resolution.

3.5 Instabilities

The Bondi-Hoyle-Lyttleton flow in not stable. Two dimensional numerical simulations have shown that the wake oscillates back and forth, exhibiting a "flip-flop instability" (Anzer et al. 1987; Fryxell and Taam 1988; Blondin and Pope 2009; Lora-Clavijo and Guzmán 2013), however, the structure of the wake is more regular in 3D simulations (Ruffert 1999).

At very high Mach numbers, the accretion line behind the black hole is subject to a longitudinal (Cowie 1977) and transverse instabilities (Soker 1990, 1991). Furthermore, Nakayama (1994) has shown that an axisymmetric shock is susceptible to a radial instability if the flow is accelerated after the shock. Ruffert (1999) investigated the Kelvin-Helmholtz and Rayleigh-Taylor instabilities, and concluded that these are not sufficient to explain the simulations without a feedback mechanism.

Bondi-Hoyle-Lyttleton accretion in a supersonic medium is subject to the standing accretion shock instability (SASI). The physical reason for this instability is the advective-acoustic cycle (Foglizzo 2002; Foglizzo et al. 2005; Guilet and Foglizzo 2012): entropy/vorticity perturbations are generated at the shock and advected to the sonic point, where an acoustic wave is excited and propagated back to the shock, leading to the growth of the entropy/vorticity perturbation. SASI helps drive supernova explosions (Burrows et al. 2006) and causes the emission of gravitational waves (Kotake et al. 2007).

Recent three dimensional radiation-hydrodynamical simulations revealed strong transient oscillations as the radiation pressure increases and becomes comparable to the gas pressure, eventually reversing the shock from the downstream to the upstream domain (Roedig et al. 2012).

4 Disk Accretion

4.1 Thin Disk Accretion

If the inflow is endowed with rotation with respect to the black hole, it reaches a centrifugal barrier from where it cannot accrete farther inwards unless its angular momentum is transported away. Near the centrifugal barrier, where the gas is held against gravity by rotation, an accretion disk forms around the black hole, centered on the plane perpendicular to the rotation axis. The accretion time is dictated by the rate at which angular momentum is transported through viscous stress which could be significantly longer than the free-fall time for a non-rotating flow (as in the Bondi solution described in Sect. 3). In the absence of radiative processes, the dissipation of the gas's kinetic energy into heat would make the disk thick and hot, with a kinetic energy of the gas close to half the gravitational potential energy (virial equilibrium). For protons at distance r, this corresponds to a temperature $\sim 10^{13}\ \mathrm{K}(r/r_{\mathrm{g}})^{-1}$ where $r_{\mathrm{g}} = GM/c^2$ is the gravitational radius. However, if the cooling time of the gas is shorter than the viscous time, then a thin disk would form. This latter regime is realized for the high gas inflow rate during quasar activity (Sect. 2.3), as well as for stellar mass black holes that emit in the soft thermal state in LMXBs (Sect. 2.2). To better understand such objects, we start by exploring the structure of thin disks that characterize the high accretion rate of quasars (Shakura and Sunyaev 1973; Novikov and Thorne 1973).

We imagine a planar thin disk of cold gas orbiting a central black hole. Each gas element orbits at the local Keplerian velocity $v_\phi = r\Omega = (GM/r)^{1/2}$ and spirals slowly inwards with radial velocity $v_r \ll v_\phi$ as viscous torques transport its angular momentum to the outer part of the disk. The associated viscous stress generates heat, which is radiated away locally from the disk surface. We assume that the disk is fed steadily and so it manifests a constant mass accretion rate at all radii. Mass conservation implies that

$$\dot{M} = 2\pi r \Sigma v_r = \text{constant} \tag{20}$$

independent of radius and time, where $\Sigma(r)$ is the surface mass density of the disk. The accretion rate $\dot{M}$ is a free parameter of the model, which is often expressed with the accretion rate corresponding to the Eddington luminosity as $\dot{m} = \dot{M}/\dot{M}_{\rm E}$. The corresponding angular momentum flux is $\dot{M}r^2\Omega + \text{constant}$.

In the limit of a geometrically thin disk with a scale height $H \ll r$, the hydrodynamic equations decouple in the radial and vertical directions. In the radial direction, the Keplerian velocity profile introduces shear which dissipates heat as neighboring fluid elements rub against each other. This power provides a local radiative flux that leaves the system vertically from the top and bottom surfaces of the disk,

$$F_{\rm rad} = \frac{3}{8\pi}\dot{M}\Omega^2\left(1 - \frac{r^2\Omega}{[r^2\Omega]_{\rm ISCO}}\right) = \frac{3}{8\pi}\frac{GM\dot{M}}{r^3}\left[1 - \left(\frac{r_{\rm ISCO}}{r}\right)^{1/2}\right], \tag{21}$$

where we assumed that the torque vanishes at the innermost stable circular orbit (ISCO), from where the gas plunges into the black hole over a free fall time. This sets the inner boundary of the disk. Here $r_{\rm ISCO} = 6r_{\rm g}$ for a non-spinning black hole, and $r_{\rm g} \le r_{\rm ISCO} \le 9r_{\rm g}$ for a spinning black hole, where $r_{\rm g} = GM/c^2$. The total luminosity emitted by both faces of the disk is given by

$$L = \int_{r_{\rm ISCO}}^{\infty} 2F_{\rm rad} \times 2\pi r \mathrm{d}r = \frac{1}{2}\frac{GM\dot{M}}{r_{\rm ISCO}} = \frac{r_{\rm g}}{2r_{\rm ISCO}}\dot{M}c^2, \tag{22}$$

where we have ignored the radiation emitted inside the ISCO. This shows that the radiative efficiency of the disk is $\epsilon = r_{\rm g}/(2r_{\rm ISCO}) = 1/12 = 8.3\ \%$ for a nonspinning black hole and $50\ \%$ for a maximally spinning black hole in the prograde direction. Note that this simple estimate neglected general-relativistic corrections. A more detailed calculation gives similar results: $\epsilon = 6\ \%$ for nonspinning and $42\ \%$ for maximally spinning black holes (Shapiro and Teukolsky 1983b).

In local thermodynamic equilibrium, the emergent flux from the surface of the disk (Eq. (21)) can be written in terms of the disk surface temperature as $F_{\rm rad} = \sigma_{\rm SB}T_{\rm d}^4$, where $\sigma_{\rm SB}$ is the Stephan-Boltzmann constant. This yields the following radial profile for the surface temperature of the disk,

$$T_{\rm d} = \left(\frac{F_{\rm rad}}{\sigma_{\rm SB}}\right)^{1/4} = 10^5\ {\rm K}\ M_8^{-1/4}\dot{m}_{-1}^{1/4}r_1^{-3/4}\left[1 - \left(\frac{r}{r_{\rm ISCO}}\right)^{1/2}\right], \tag{23}$$

where $\dot{m}_{-1} = \dot{m}/0.1$, $r_1 = r/(10r_{\rm g})$, and $\dot{m} = \dot{M}/\dot{M}_{\rm Edd}$. The corresponding thermal black-body spectrum peaks in the UV bands for quasars, and in the X-ray band for stellar-mass black holes. (Non-thermal X-ray emission from a hot corona or a jet can supplement this disk emission.) Stellar-mass black holes can therefore be important X-ray sources at high redshifts, especially if they are incorporated in a binary system where they accrete gas from a companion star (see Sect. 2.2).

Up to this point we did not need to adopt a model for the angular momentum transport (or an effective viscosity) in the disk. Indeed, the main observables, total luminosity and thermal spectrum, are very robustly set for a radiatively efficient thin disk by three parameters: M, $\dot{M}$, and the black hole spin which determines $r_{\rm ISCO}$. Other details, involving the disk thickness, the radial surface density profile, and opacity however do depend on the model of viscosity and the relative contribution of radiation to gas pressure. Assuming that (i) the angular momentum flux is proportional to the pressure in the disk with a dimensionless constant of proportionality α, (ii) the disk is very thin so that it can be well approximated by vertically averaged quantities (one-zone model), (iii) the disk is optically thick in the vertical direction so that the gas and radiation are in thermal equilibrium, and (iv) the self-gravity of the gas is negligible, one can algebraically express all physical properties of the disk analytically (Goodman and Tan 2004). The surface density and mid-plane temperature are

$$\Sigma = \frac{8\,(\mu \mathrm{m_p}/\mathrm{k_B})^{4/5}\sigma_{\mathrm{SB}}^{1/5}}{3^{9/5}\alpha^{4/5}\kappa^{1/5}} \frac{\beta^{(1-b)4/5}}{\Omega^{4/5}} F_{\mathrm{rad}}^{3/5}, \tag{24}$$

$$T_{\mathrm{c}} = \frac{(\mu \mathrm{m_p}/\mathrm{k_B})^{1/5}\kappa^{1/5}}{(3\alpha\sigma_{\mathrm{SB}})^{1/5}} \frac{\beta^{(1-b)/5}}{\Omega^{1/5}} F_{\mathrm{rad}}^{2/5}, \tag{25}$$

where $\beta = p_{\mathrm{gas}}/(p_{\mathrm{gas}} + p_{\mathrm{rad}})$ is the gas to total pressure given by

$$\frac{\beta^{(1/2)+(b-1)/10}}{1-\beta} = \frac{c\,[k/(\mu \mathrm{m_p})]^{2/5}}{(3\alpha\sigma_{\mathrm{SB}})^{1/10}\kappa^{9/10}} \frac{\Omega^{9/10}}{F_{\mathrm{rad}}^{4/5}}. \tag{26}$$

Here $b = 0$ and 1 denote models where viscosity is proportional to the total pressure or only the gas pressure, respectively. Equations (24)–(26) along with equation (21) define the disk model as a function of radius. Equation (26) shows that in the inner regions $\beta \ll 1$, implying that radiation pressure dominates over gas pressure. In practice, radiation pressure dominates within $600\alpha_1^{2/21} M_5^{2/21} r_{\mathrm{g}}$ if $\dot{M} = 0.1\dot{M}_{\mathrm{Edd}}$ with $\epsilon = 10\ \%$ radiative efficiency. The opacity is dominated by electron scattering κ_{es} in the inner region and free-free absorption in the outer regions κ_{ff} (Eq. (3)).

Vertical hydrostatic equilibrium implies $H = c_s/\Omega$, where $c_s = \sqrt{p/\rho}$ is the sound speed, where $\rho = \Sigma/(2H)$ is the density, $p = p_{\mathrm{gas}} + p_{\mathrm{rad}}$ is the pressure, $p_{\mathrm{gas}} = \rho \mathrm{k_B} T_{\mathrm{c}}/(\mu \mathrm{m_p})$, and $p_{\mathrm{rad}} = \frac{1}{3}a_{\mathrm{rad}}T_c^4$. Here μ is the mean molecular weight, which is unity for a hydrogen plasma, and $\mu = 0.615$ for 25 % helium and 75 % hydrogen by mass. Combining these equations, leads to $H = \kappa F_{\mathrm{rad}}/[c\Omega^2(1-\beta)] \sim (3\kappa/8\pi c)\dot{M}(1-\beta)^{-1}$. When the accretion rate approaches the Eddington limit, $\beta \approx 0$, and the disk bloats and H approaches r in the inner regions, violating the thin-disk assumption. The model also breaks down if the accretion rate is several orders of magnitude below Eddington because the disk becomes optically thin $\tau = \frac{1}{2}\kappa\Sigma < 1$.

The disk model described above ignores the self-gravity of the disk. This assumption is inevitably violated at large radii, where the disk becomes unstable to fragmentation due to its self-gravity, where the Toomre parameter, $Q = (c_s\Omega/\pi G\Sigma)$, drops below unity. Outside this radius, typically $\sim 2 \times 10^4 r_{\mathrm{g}}(\alpha/0.1)^{0.6}(M/10^8\ \mathrm{M}_\odot)^{1.2}$, the gas in quasar accretion disks would fragment into stars, and the stars may migrate inwards as the gas accretes onto the black hole. The energy output from stellar winds and supernovae would supplement the viscous heating of the disk and might regulate the disk to have $Q \sim 1$ outside the above boundary. We therefore conclude that star formation will inevitably occur on larger scales, before the gas is driven into the accretion disk that feeds the central black hole. Indeed, the broad emission lines of quasars display very high abundances of heavy elements in the spectra out to arbitrarily high redshifts. Since the total amount of mass in the disk interior to this radius makes only a small fraction of the mass of the supermassive black hole, quasars must be fed by gas that crosses this boundary after being vulnerable to fragmentation (Milosavljević and Loeb 2004).

4.2 Advection Dominated Accretion Flow

When the accretion rate is considerably lower than the Eddington limit ($\dot{m} \lesssim 10^{-2}$), the gas inflow switches to a different mode, called a *Radiatively Inefficient Accretion Flow* (RIAF) or an *Advection Dominated Accretion Flow* (ADAF), in which either the cooling time or the photon diffusion time is much longer than the accretion time of the gas and heat is mostly advected with the gas into the black hole (Narayan and Yi 1994). At the low gas densities and high temperatures characterizing this accretion mode, the Coulomb coupling is weak and the electrons do not heat up to the proton temperature even with the aid of plasma instabilities. Viscosity heats primarily the protons since they carry most of the momentum. The other major heat source, compression of the gas, also heats the protons more effectively than the electrons. As the gas falls inward and its density ρ rises, the temperature of each species T increases adiabatically as $T \propto \rho^{\gamma-1}$, where γ is the corresponding adiabatic index. At radii $r \lesssim 10^2 r_{\rm Sch}$, the electrons are relativistic with $\gamma = 4/3$ and so their temperature rises inwards with increasing density as $T_e \propto \rho^{1/3}$ while the protons are non-relativistic with $\gamma = 5/3$ and so $T_{\rm p} \propto \rho^{2/3}$, yielding a two-temperature plasma with the protons being much hotter than the electrons. Typical analytic models (Narayan and Yi 1994; Narayan and McClintock 2008, and references therein) yield $T_{\rm p} \sim 10^{12}\ {\rm K}(r/r_{\rm Sch})^{-1}$, $T_e \sim \min(T_{\rm p}, 10^{9-11}\ {\rm K})$. Since the typical sound speed is comparable to the Keplerian speed at each radius, the geometry of the flow is thick—making RIAFs the viscous analogs of Bondi accretions.

Analytic models imply a radial velocity that is a factor of $\sim\alpha$ smaller than the free-fall speed and an accretion time that is a factor of $\sim\alpha$ longer than the free-fall time. However, since the sum of the kinetic and thermal energies of a proton is comparable to its gravitational binding energy, RIAFs are expected to be associated with strong outflows.

The radiative efficiency of RIAFs is smaller than the thin-disk value (ϵ). While the thin-disk value applies to high accretion rates above some critical value, $\dot{m} > \dot{m}_{\rm crit}$, where $\dot{m}$ is the accretion rate in Eddington units, analytic RIAF models typically admit a radiative efficiency of

$$\epsilon \approx 10\dot{m} \quad \text{if}\ \dot{m} \lesssim \dot{m}_{\rm crit}, \tag{27}$$

with $\dot{m}_{\rm crit} \sim 0.01$.

The nature of the flow can change when the disk becomes radiatively inefficient, since thermal convection may compete with MHD turbulence in transporting energy and angular momentum. In particular, the standard notion that viscosity is transported outwards in convective disks has been challenged in RIAFs (Narayan et al. 2000). In such convection-dominated disks the density profile is altered considerably from the standard ADAF models.

Since at low redshifts mergers are rare and much of the gas in galaxies has already been consumed in making stars, most local supermassive black holes are characterized by a very low accretion rate. The resulting low luminosity of these dormant black holes, such as the $4 \times 10^6\ {\rm M}_\odot$ black hole lurking at the center of the Milky Way galaxy, is often described using RIAF/ADAF models. Although this mode of accretion is characterized by a low mass infall rate, it could persist over a period of time that is orders of magnitude longer than the quasar mode discussed earlier, so its contribution to the growth of black holes in galactic nuclei may not be negligible.

4.3 Circumbinary SMBH Disks

Galaxy mergers naturally lead to SMBH binaries, which can also produce bright emission. Direct X-ray imaging of active nuclei have revealed several SMBH binaries at separations of $\sim$3 kpc (Komossa et al. 2003; Ballo et al. 2004; Bianchi et al. 2008; Civano et al. 2012), 400 pc (Paggi et al. 2013), and 150 pc (Fabbiano et al. 2011). Additional evidence comes in the form of a radio galaxy with a double core with a projected separation of 10 pc (Rodriguez et al. 2006), and several other observations of radio galaxies, such as the wiggled shape of jets indicating precession (Roos et al. 1993), the X-shaped morphologies of radio lobes (Merritt and Ekers 2002; Liu 2004), the interruption and recurrence of activity in double-double radio galaxies (Schoenmakers et al. 2000; Liu et al. 2003), and the elliptical motion of the unresolved core of 3C66B (Sudou et al. 2003), which have all been interpreted as indirect evidence for SMBH binaries down to subparsec scales. Velocity offsets in broad line spectra have also been observed in a handful of AGN which may be attributed to the orbital motion of the SMBH binary around the center of mass (Eracleous et al. 2012; Ju et al. 2013).

The structure of the accretion flow around close SMBH binaries is less understood. Generally speaking, the dense nuclear gas around the binary is expected to cool rapidly and settle in a rotationally supported circumbinary disk (Escala et al. 2005). If the binary plane is initially misaligned with the disk, the gravitational torques cause the disk to warp and twist. Viscosity then causes the disk to align with the binary on a timescale $t_{\rm al} \sim \max[\alpha(H/r)^{-2} t_{\rm orb}, (H/r)^{-1} t_{\rm orb}]$ where $t_{\rm orb}$ is the binary orbital period, H is the scale-height, and α is the dimensionless viscosity parameter of a thin disk (Ivanov et al. 1999, and see Sect. 4). For nearly equal masses, the accretion flow may then develop a triple-disk structure, forming individual accretion disks around the two binary components, and a larger circumbinary disk feeding these smaller disks (Hayasaki et al. 2008). For unequal mass binaries $m/M \ll 0.1$, the accretion disk geometry may be similar to a protoplanetary disk with a planet orbiting in the disk. The secondary excites spiral density waves in the disk which carry away energy and angular momentum from the secondary. This acts to reduce the density in a radial annulus of comparable to the Roche lobe R_R (Eq. (1)), regulates the inflow rate across the secondary orbit, and leads to the inward migration of the secondary. If the secondary is capable of quenching the inflow efficiently, a hollow cavity may form interior to its orbit with negligible accretion onto the primary. The gradual accretion from the outer regions will make the gas pile up outside the secondary's orbit similar to a river dam. The gas density and pressure increases significantly, until the gas overflows in non-axisymmetric streams, filling the cavity.

There are several interesting limiting cases for the circumbinary disk structure. For extreme binary mass ratios, $q \lesssim 10^{-4}$, the secondary cannot significantly modify the disk, which is then described by a standard thin disk assuming that the accretion rate is in the range $0.01 < \dot{m} < 0.3$ (see Sect. 4). The spiral waves generated in the disk by the secondary carry away angular momentum very efficiently, shrinking the secondary's orbit quickly on the so-called Type-1 migration timescale ($\sim 10^4$–10^5 yr, Goldreich and Tremaine 1980). For comparable binary masses, $q \gtrsim 0.1$, a complete cavity forms, and the secondary is driven inwards on the viscous timescale ($\sim 10^7$ yr) as the gas is flowing in from the outer regions (Type-2 migration). However, typically the local disk mass is much smaller than the secondary SMBH mass, implying that the gas piles up before it can drive the binary in. The accretion disk acquires a self-similar structure well outside the secondary as long as there is no inflow across the gap (Ivanov et al. 1999; Rafikov 2012). If the inflow into the cavity is significant, the disk settles in a steady state profile and the migration rate may be much slower (Type-1.5 migration, Kocsis et al. 2012b).

The accretion disk luminosity depends on whether the inner disk is truncated by tidal effects of the secondary and depends on the amount of gas pile up outside the secondary's orbit. The radiative efficiency may be larger than Eq. (22) due to the energy release corresponding to the inward migration of the secondary. This excess power comes from the part of the accretion disk outside the secondary's orbit where the gas pile up is significant. Figure 3 shows that this can make the disk significantly brighter in the optical bands for $q < 0.01$ (Kocsis et al. 2012a). Furthermore, hydrodynamical simulations show that the accretion flow into the inner regions is expected to display periodic variability on roughly the

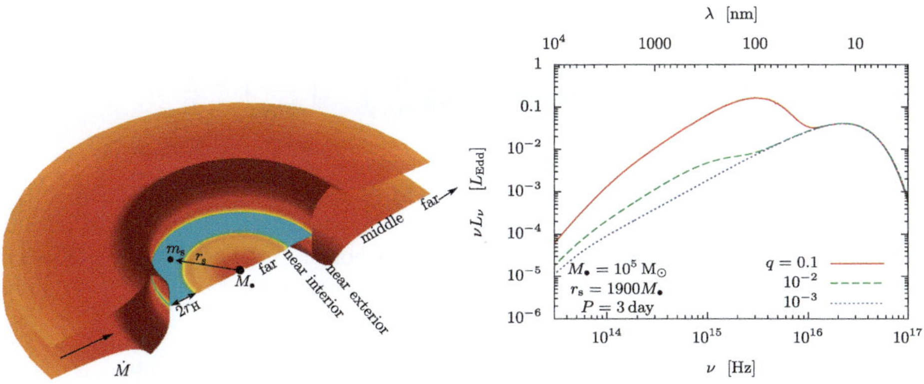

Fig. 3 Gas pile up due to a secondary in the accretion disk (*left*) and the corresponding thermal continuum spectrum for mass ratios $q = 0.1$, 0.01, and 0.001 (*right*). Figure credit: Kocsis et al. (2012a)

orbital timescale. The variability power spectrum has discrete peaks, where the frequencies and harmonic weights depend on the binary mass ratio, accretion rate, and the α parameter (MacFadyen and Milosavljević 2008; Cuadra et al. 2009; Shi et al. 2012; Noble et al. 2012; D'Orazio et al. 2012). However, further improvements are needed to include radiation pressure and extended to cover several viscous timescales to make more quantitative predictions on circumbinary accretion.

Supermassive black hole binaries may be identified in future surveys based on the velocity offsets in their broad line spectra (Shen and Loeb 2010; Eracleous et al. 2012; Ju et al. 2013), the expected periodic variability of the accretion luminosity on weeks-to-months timescales (Haiman et al. 2009), the excess brightness in the optical bands of the thermal spectrum corresponding to the pile up (Kocsis et al. 2012a), the variability of relativistic iron lines due to the orbital motion of the secondary and the effects of a gap in the disk (McKernan et al. 2012), or through the gravitational waves emitted by the binary (Kocsis et al. 2008; Lang et al. 2011).

5 Feeding Black Holes with Stars

5.1 Tidal Disruption Events

Most supermassive black holes in the local Universe are not accreting gas. They are typically lurking at the centers of galaxies, surrounded by a dense population of stars. A star that comes close to the black hole is torn apart by tidal gravity. The spaghetti-like stellar debris falling back onto the black hole forms an accretion disk and produces a luminous electromagnetic transient lasting weeks to years, observable from cosmological distances (Frank and Rees 1976). Around 25 tidal disruption events have been observed to date in X-ray, UV, optical, and radio bands (Komossa 2012; Gezari 2012, and references therein). These observations imply a stellar disruption rate of $3 \times 10^{-5}\ \mathrm{yr}^{-1}\ \mathrm{galaxy}^{-1}$ (van Velzen and Farrar 2012), consistent with theoretical rate estimates (Rees 1988; Wang and Merritt 2004).

The characteristic light-curve produced by tidal disruption events is derived as follows. The maximum distance at which the tidal field of the black hole is capable of unbinding a star is the tidal radius,

$$R_t = \left(\frac{M_{\rm BH}}{M_*}\right)^{1/3} R_* = 46 r_{\rm g} \left(\frac{M_{\rm BH}}{10^6\,{\rm M}_\odot}\right)^{-2/3} \left(\frac{M_*}{{\rm M}_\odot}\right)^{-1/3} \frac{R_*}{{\rm R}_\odot}. \tag{28}$$

The disruption occurs outside of the black hole horizon for $M_{\rm BH} < 10^8\,{\rm M}_\odot$ for main sequence stars, $\lesssim 10^6\,{\rm M}_\odot$ for white dwarfs and $\lesssim 5\,{\rm M}_\odot$ for neutron stars (see Kesden 2012b, for highly spinning black holes where these bounds may be a factor 10 larger). After tidal disruption, approximately half of the stellar mass is bound to the black hole and half is unbound. The most bound material returns to pericenter after a time

$$t_{\rm fb} = \frac{2\pi}{6^{3/2}} \left(\frac{r_p}{R_*}\right)^{3/2} t_p = 20 \left(\frac{M_{\rm BH}}{10^6\,{\rm M}_\odot}\right)^{3/2} \left(\frac{r_p}{6 r_{\rm g}}\right)^{3} \left(\frac{R_*}{{\rm R}_\odot}\right)^{-3/2} {\rm min} \tag{29}$$

where R_p is the pericenter distance, $t_p = (G M_{\rm BH}/R_p^3)^{-1/2}$ is the pericenter timescale. The fallback rate of bound material is then (Rees 1988)

$$\dot{M}_{\rm fb} = \frac{{\rm d}M}{{\rm d}E}\frac{{\rm d}E}{{\rm d}t} = \frac{1}{3}\frac{M_*}{t_{\rm fb}}\left(\frac{t}{t_{\rm fb}}\right)^{-5/3}, \tag{30}$$

where $E \sim G M_{\rm BH}/a$ is the orbital energy, and $a \propto t^{-3/2}$ from Kepler's law, and we assumed that ${\rm d}M/{\rm d}E$ is constant. The timescale for the light-curve to reach the $t^{-5/3}$ limit depends on the stellar structure and spin through ${\rm d}M/{\rm d}E$ (Lodato et al. 2009; MacLeod et al. 2012; Guillochon and Ramirez-Ruiz 2013; Stone et al. 2012). This Newtonian estimate is modified for smaller stellar orbital eccentricities and by GR corrections, which depend on the spin of the black hole (Kesden 2012a; Hayasaki et al. 2012; Dai et al. 2013).

Once the infalling gas returns to pericenter, it is expected to shock and form an accretion disk around the black hole. Initially, weeks to months after the disruption, the fallback rate is very super-Eddington. This phase is probably described by an advective slim disk with powerful outflows (Strubbe and Quataert 2009; De Colle et al. 2012; Tchekhovskoy et al. 2013). Months to a year later, the fallback rate becomes sub-Eddington, and the disk may radiatively cool efficiently to form a thin disk (Sect. 4). Finally, several years after the disruption, the fallback rate becomes very sub-Eddington $\dot{M}_{\rm fb} \lesssim 10^{-2}\dot{M}_{\rm Edd}$, and the disk becomes radiatively inefficient and hot (Sect. 4.2) which may be accompanied by a jet. Two possible tidal disruption event sources have been observed to display jet activity (Swift J1644+57 and Swift J2058.4+0516, Burrows et al. 2011; Cenko et al. 2012). The lack of evidence for Lense-Thirring precession of the jet away from the observer suggests that the jet is aligned with the SMBH spin and not the accretion disk in these events (Stone and Loeb 2012a).

The bolometric luminosity of the system is $L = \epsilon \dot{M}_{\rm fb} c^2 \propto t^{-5/3}$. In the thin disk phase, the multicolor blackbody spectrum of the accretion disk peaks in the soft X-ray bands. The light-curve in the X-ray bands follows $t^{-5/3}$ for several years, indeed consistent with many observed sources (Komossa 2012). Optical and UV wavelengths initially fall on the Rayleigh-Jeans tail of the disk spectrum, and the luminosity is initially proportional to the temperature $L_\nu \propto T_\nu \propto \dot{M}_{\rm fb}^{1/4} \propto t^{-5/12}$, and eventually converges to the $t^{-5/3}$ profile as the disk cools (Lodato and Rossi 2011). Such a shallower decline has been observed in the optical and UV bands for Swift J2058 (Cenko et al. 2012).

Upcoming surveys will detect more hundreds to thousands of tidal disruption events (TDEs). Multi-wavelength observations will enable us to probe the supermassive and intermediate-mass black hole populations, allowing us to measure their masses and spins (Kesden 2012a). Supermassive black hole binaries may imply a transient very high TDE rate of 1 yr^{-1} and account overall for 10 % of the cosmic TDE rate (Chen et al. 2011). Detecting multiple TDEs in the same galaxy within a few years time could be a signpost for a SMBH binary. Merging SMBHs receive a gravitational wave recoil kick due to anisotropic gravitational radiation. The recoiling black hole could produce tidal disruption flares spatially offset from the galactic center, which might contribute 1 % of the TDE rates (Stone and Loeb 2012b). Finally, white dwarfs passing close to IMBHs may be detonated by the tidal gravity which would appear as underluminous supernovae (Carter and Luminet 1982; Rosswog et al. 2009; Haas et al. 2012).

5.2 Fueling Active Galactic Nuclei Accretion Disks with Stars

In the standard theory of AGN accretion, the effective viscosity responsible for delivering material onto the black hole is believed to be generated by the magneto-rotational instability (Sect. 4). An interesting alternative idea was examined by Goodman and Rafikov (2001) in which massive objects embedded in the accretion disk serve as an effective source of viscosity. These objects drive density waves in the disk which transport energy and angular momentum like a kinematic viscosity. In particular, stars in the stellar cluster around the supermassive black hole that cross the disk become captured in the disk due to accretion and hydrodynamical drag. Miralda-Escudé and Kollmeier (2005) have argued that the matter of these stars is sufficient to replenish the disk and fuel AGN. This process can also explain the observed correlation between the black hole mass and velocity dispersion of the stellar cluster (M-σ relation), although other explanations based on AGN feedback are also possible (see Chap. 5.6 in this book). Additionally, stars may also form in the outer regions of accretion disks due to gravitational fragmentation, and migrate inwards as they interact with the disk to fuel the AGN (Levin 2007).

The fate of the stars embedded in accretion disks is not well understood. Once they approach the black hole within the tidal radius (Eq. (28)), they are tidally disrupted. However, their structure might change considerably before reaching this radius if they migrate across a gaseous disk. Accretion from the disk and mergers with other stars can lead to the formation of a supermassive star or an intermediate mass black hole already at large distances from the black hole (Goodman and Tan 2004; McKernan et al. 2012). If so, these objects orbiting around supermassive black holes will generate gravitational waves, and provide a sensitive new probe of the accretion disk (Kocsis et al. 2011).

References

U. Anzer, G. Boerner, J.J. Monaghan, Numerical studies of wind accretion. Astron. Astrophys. **176**, 235–244 (1987)

J.G. Baker, J. Centrella, D.-I. Choi, M. Koppitz, J. van Meter, Gravitational-wave extraction from an inspiraling configuration of merging black holes. Phys. Rev. Lett. **96**(11), 111102 (2006). doi:10.1103/PhysRevLett.96.111102

L. Ballo, V. Braito, R. Della Ceca, L. Maraschi, F. Tavecchio, M. Dadina, Arp 299: a second merging system with two active nuclei? Astrophys. J. **600**, 634–639 (2004). doi:10.1086/379887

I. Bartos, Z. Haiman, B. Kocsis, S. Marka, G2 can illuminate the black hole population near the galactic center. ArXiv e-prints (2013)

H. Baumgardt, J. Makino, T. Ebisuzaki, Massive black holes in star clusters. II. Realistic cluster models. Astrophys. J. **613**, 1143–1156 (2004). doi:10.1086/423299

M.C. Begelman, Can a spherically accreting black hole radiate very near the Eddington limit. Mon. Not. R. Astron. Soc. **187**, 237–251 (1979)

M.C. Begelman, R.D. Blandford, M.J. Rees, Massive black hole binaries in active galactic nuclei. Nature **287**, 307–309 (1980). doi:10.1038/287307a0

K. Belczynski, T. Bulik, C.L. Fryer, A. Ruiter, F. Valsecchi, J.S. Vink, J.R. Hurley, On the maximum mass of stellar black holes. Astrophys. J. **714**, 1217–1226 (2010). doi:10.1088/0004-637X/714/2/1217

G. Bertone, A.R. Zentner, J. Silk, New signature of dark matter annihilations: gamma rays from intermediate-mass black holes. Phys. Rev. D **72**(10), 103517 (2005). doi:10.1103/PhysRevD.72.103517

G. Bertone, M. Fornasa, M. Taoso, A.R. Zentner, Dark matter annihilation around intermediate mass black holes: an update. New J. Phys. **11**(10), 105016 (2009). doi:10.1088/1367-2630/11/10/105016

S. Bianchi, M. Chiaberge, E. Piconcelli, M. Guainazzi, G. Matt, Chandra unveils a binary active galactic nucleus in Mrk 463. Mon. Not. R. Astron. Soc. **386**, 105–110 (2008). doi:10.1111/j.1365-2966.2008.13078.x

J. Binney, S. Tremaine, *Galactic Dynamics*, 2nd edn. (Princeton University Press, Princeton, 2008)

L. Blecha, A. Loeb, Effects of gravitational-wave recoil on the dynamics and growth of supermassive black holes. Mon. Not. R. Astron. Soc. **390**, 1311–1325 (2008). doi:10.1111/j.1365-2966.2008.13790.x

J.M. Blondin, Hypercritical spherical accretion onto compact objects. Astrophys. J. **308**, 755–764 (1986). doi:10.1086/164548

J.M. Blondin, T.C. Pope, Revisiting the "flip-flop" instability of Hoyle-Lyttleton accretion. Astrophys. J. **700**, 95–102 (2009). doi:10.1088/0004-637X/700/1/95

H. Bondi, On spherically symmetrical accretion. Mon. Not. R. Astron. Soc. **112**, 195 (1952)

H. Bondi, F. Hoyle, On the mechanism of accretion by stars. Mon. Not. R. Astron. Soc. **104**, 273 (1944)

V. Bromm, R.B. Larson, The first stars. Annu. Rev. Astron. Astrophys. **42**, 79–118 (2004). doi:10.1146/annurev.astro.42.053102.134034

V. Bromm, A. Loeb, Formation of the first supermassive black holes. Astrophys. J. **596**, 34–46 (2003). doi:10.1086/377529

A. Burrows, E. Livne, L. Dessart, C.D. Ott, J. Murphy, A new mechanism for core-collapse supernova explosions. Astrophys. J. **640**, 878–890 (2006). doi:10.1086/500174

D.N. Burrows, J.A. Kennea, G. Ghisellini, V. Mangano, B. Zhang, K.L. Page, M. Eracleous, P. Romano, T. Sakamoto, A.D. Falcone, J.P. Osborne, S. Campana, A.P. Beardmore, A.A. Breeveld, M.M. Chester, R. Corbet, S. Covino, J.R. Cummings, P. D'Avanzo, V. D'Elia, P. Esposito, P.A. Evans, D. Fugazza, J.M. Gelbord, K. Hiroi, S.T. Holland, K.Y. Huang, M. Im, G. Israel, Y. Jeon, Y.-B. Jeon, H.D. Jun, N. Kawai, J.H. Kim, H.A. Krimm, F.E. Marshall, P. Mészáros, H. Negoro, N. Omodei, W.-K. Park, J.S. Perkins, M. Sugizaki, H.-I. Sung, G. Tagliaferri, E. Troja, Y. Ueda, Y. Urata, R. Usui, L.A. Antonelli, S.D. Barthelmy, G. Cusumano, P. Giommi, A. Melandri, M. Perri, J.L. Racusin, B. Sbarufatti, M.H. Siegel, N. Gehrels, Relativistic jet activity from the tidal disruption of a star by a massive black hole. Nature **476**, 421–424 (2011). doi:10.1038/nature10374

M. Campanelli, C.O. Lousto, P. Marronetti, Y. Zlochower, Accurate evolutions of orbiting black-hole binaries without excision. Phys. Rev. Lett. **96**(11), 111101 (2006). doi:10.1103/PhysRevLett.96.111101

B. Carter, J.P. Luminet, Pancake detonation of stars by black holes in galactic nuclei. Nature **296**, 211–214 (1982). doi:10.1038/296211a0

S.B. Cenko, H.A. Krimm, A. Horesh, A. Rau, D.A. Frail, J.A. Kennea, A.J. Levan, S.T. Holland, N.R. Butler, R.M. Quimby, J.S. Bloom, A.V. Filippenko, A. Gal-Yam, J. Greiner, S.R. Kulkarni, E.O. Ofek, F.E. Olivares, P. Schady, J.M. Silverman, N.R. Tanvir, D. Xu, Swift J2058.4+0516: discovery of a possible second relativistic tidal disruption flare? Astrophys. J. **753**, 77 (2012). doi:10.1088/0004-637X/753/1/77

S. Chandrasekhar, Dynamical friction. I. General considerations: the coefficient of dynamical friction. Astrophys. J. **97**, 255 (1943). doi:10.1086/144517

X. Chen, A. Sesana, P. Madau, F.K. Liu, Tidal stellar disruptions by massive black hole pairs. II. Decaying binaries. Astrophys. J. **729**, 13 (2011). doi:10.1088/0004-637X/729/1/13

F. Civano, M. Elvis, G. Lanzuisi, T. Aldcroft, M. Trichas, A. Bongiorno, M. Brusa, L. Blecha, A. Comastri, A. Loeb, M. Salvato, A. Fruscione, A. Koekemoer, S. Komossa, R. Gilli, V. Mainieri, E. Piconcelli, C. Vignali, Chandra high-resolution observations of CID-42, a candidate recoiling supermassive black hole. Astrophys. J. **752**, 49 (2012). doi:10.1088/0004-637X/752/1/49

L.L. Cowie, An investigation of the stability of the Bondi-Hoyle model of accretion flow. Mon. Not. R. Astron. Soc. **180**, 491–494 (1977)

J. Cuadra, P.J. Armitage, R.D. Alexander, M.C. Begelman, Massive black hole binary mergers within subparsec scale gas discs. Mon. Not. R. Astron. Soc. **393**, 1423–1432 (2009). doi:10.1111/j.1365-2966.2008.14147.x

L. Dai, A. Escala, P. Coppi, The impact of bound stellar orbits and general relativity on the temporal behavior of tidal disruption flares. ArXiv e-prints (2013)

S.W. Davis, R. Narayan, Y. Zhu, D. Barret, S.A. Farrell, O. Godet, M. Servillat, N.A. Webb, The cool accretion disk in ESO 243-49 HLX-1: further evidence of an intermediate-mass black hole. Astrophys. J. **734**, 111 (2011). doi:10.1088/0004-637X/734/2/111

F. De Colle, J. Guillochon, J. Naiman, E. Ramirez-Ruiz, The dynamics, appearance, and demographics of relativistic jets triggered by tidal disruption of stars in quiescent supermassive black holes. Astrophys. J. **760**, 103 (2012). doi:10.1088/0004-637X/760/2/103

C.D. Dermer, A. Atoyan, Collapse of neutron stars to black holes in binary systems: a model for short gamma-ray bursts. Astrophys. J. Lett. **643**, 13–16 (2006). doi:10.1086/504895

M. Dijkstra, Z. Haiman, A. Mesinger, J.S.B. Wyithe, Fluctuations in the high-redshift Lyman-Werner background: close halo pairs as the origin of supermassive black holes. Mon. Not. R. Astron. Soc. **391**, 1961–1972 (2008). doi:10.1111/j.1365-2966.2008.14031.x

D.J. D'Orazio, Z. Haiman, A. MacFadyen, Accretion into the central cavity of a circumbinary disk. ArXiv e-prints (2012)

M. Eracleous, T.A. Boroson, J.P. Halpern, J. Liu, A large systematic search for close supermassive binary and rapidly recoiling black holes. Astrophys. J. Suppl. Ser. **201**, 23 (2012). doi:10.1088/0067-0049/201/2/23

A. Escala, R.B. Larson, P.S. Coppi, D. Mardones, The role of gas in the merging of massive black holes in galactic nuclei. II. Black hole merging in a nuclear gas disk. Astrophys. J. **630**, 152–166 (2005). doi:10.1086/431747

G. Fabbiano, J. Wang, M. Elvis, G. Risaliti, A close nuclear black-hole pair in the spiral galaxy NGC3393. Nature **477**, 431–434 (2011). doi:10.1038/nature10364

R. Fernández, B.D. Metzger, Nuclear dominated accretion flows in two dimensions. I. Torus evolution with parametric microphysics. Astrophys. J. **763**, 108 (2013). doi:10.1088/0004-637X/763/2/108

T. Foglizzo, Non-radial instabilities of isothermal Bondi accretion with a shock: vortical-acoustic cycle vs. post-shock acceleration. Astron. Astrophys. **392**, 353–368 (2002). doi:10.1051/0004-6361:20020912

T. Foglizzo, P. Galletti, M. Ruffert, A fresh look at the unstable simulations of Bondi-Hoyle-Lyttleton accretion. Astron. Astrophys. **435**, 397–411 (2005). doi:10.1051/0004-6361:20042201

P.C. Fragile, A. Gillespie, T. Monahan, M. Rodriguez, P. Anninos, Numerical simulations of optically thick accretion onto a black hole. I. Spherical case. Astrophys. J. Suppl. Ser. **201**, 9 (2012). doi:10.1088/0067-0049/201/2/9

J. Frank, M.J. Rees, Effects of massive central black holes on dense stellar systems. Mon. Not. R. Astron. Soc. **176**, 633–647 (1976)

M. Freitag, M.A. Gürkan, F.A. Rasio, Runaway collisions in young star clusters—II. Numerical results. Mon. Not. R. Astron. Soc. **368**, 141–161 (2006). doi:10.1111/j.1365-2966.2006.10096.x

C.L. Fryer, Mass limits for black hole formation. Astrophys. J. **522**, 413–418 (1999). doi:10.1086/307647

B.A. Fryxell, R.E. Taam, Numerical simulation of nonaxisymmetric adiabatic accretion flow. Astrophys. J. **335**, 862–880 (1988). doi:10.1086/166973

S. Gezari, Ultraviolet and optical observations of tidal disruption events, in *European Physical Journal Web of Conferences*, vol. 39, 2012, p. 3001. doi:10.1051/epjconf/20123903001

P. Goldreich, S. Tremaine, Disk-satellite interactions. Astrophys. J. **241**, 425–441 (1980). doi:10.1086/158356

J. Goodman, R.R. Rafikov, Planetary torques as the viscosity of protoplanetary disks. Astrophys. J. **552**, 793–802 (2001). doi:10.1086/320572

J. Goodman, J.C. Tan, Supermassive stars in quasar disks. Astrophys. J. **608**, 108–118 (2004). doi:10.1086/386360

H.-J. Grimm, M. Gilfanov, R. Sunyaev, High-mass X-ray binaries as a star formation rate indicator in distant galaxies. Mon. Not. R. Astron. Soc. **339**, 793–809 (2003). doi:10.1046/j.1365-8711.2003.06224.x

J. Guilet, T. Foglizzo, On the linear growth mechanism driving the standing accretion shock instability. Mon. Not. R. Astron. Soc. **421**, 546–560 (2012). doi:10.1111/j.1365-2966.2012.20333.x

J. Guillochon, E. Ramirez-Ruiz, Hydrodynamical simulations to determine the feeding rate of black holes by the tidal disruption of stars: the importance of the impact parameter and stellar structure. Astrophys. J. **767**, 25 (2013). doi:10.1088/0004-637X/767/1/25

R. Haas, R.V. Shcherbakov, T. Bode, P. Laguna, Tidal disruptions of white dwarfs from ultra-close encounters with intermediate-mass spinning black holes. Astrophys. J. **749**, 117 (2012). doi:10.1088/0004-637X/749/2/117

Z. Haiman, B. Kocsis, K. Menou, The population of viscosity- and gravitational wave-driven supermassive black hole binaries among luminous active galactic nuclei. Astrophys. J. **700**, 1952–1969 (2009). doi:10.1088/0004-637X/700/2/1952

K. Hayasaki, S. Mineshige, L.C. Ho, A supermassive binary black hole with triple disks. Astrophys. J. **682**, 1134–1140 (2008). doi:10.1086/588837

K. Hayasaki, N. Stone, A. Loeb, Finite, intense accretion bursts from tidal disruption of stars on bound orbits. ArXiv e-prints (2012)

A. Heger, C.L. Fryer, S.E. Woosley, N. Langer, D.H. Hartmann, How massive single stars end their life. Astrophys. J. **591**, 288–300 (2003). doi:10.1086/375341

F. Hoyle, R.A. Lyttleton, The effect of interstellar matter on climatic variation. Proc. Camb. Philos. Soc. **35**, 405 (1939). doi:10.1017/S0305004100021150

P.B. Ivanov, J.C.B. Papaloizou, A.G. Polnarev, The evolution of a supermassive binary caused by an accretion disc. Mon. Not. R. Astron. Soc. **307**, 79–90 (1999). doi:10.1046/j.1365-8711.1999.02623.x

W. Ju, J.E. Greene, R.R. Rafikov, S.J. Bickerton, C. Badenes, Search for supermassive black hole binaries in the Sloan digital sky survey spectroscopic sample (2013). arXiv:1306.4987

M. Kesden, Black-hole spin dependence in the light curves of tidal disruption events. Phys. Rev. D **86**(6), 064026 (2012a). doi:10.1103/PhysRevD.86.064026

M. Kesden, Tidal-disruption rate of stars by spinning supermassive black holes. Phys. Rev. D **85**(2), 024037 (2012b). doi:10.1103/PhysRevD.85.024037

B. Kocsis, Z. Haiman, K. Menou, Premerger localization of gravitational wave standard sirens with LISA: triggered search for an electromagnetic counterpart. Astrophys. J. **684**, 870–887 (2008). doi:10.1086/590230

B. Kocsis, N. Yunes, A. Loeb, Observable signatures of extreme mass-ratio inspiral black hole binaries embedded in thin accretion disks. Phys. Rev. D **84**(2), 024032 (2011). doi:10.1103/PhysRevD.84.024032

B. Kocsis, Z. Haiman, A. Loeb, Gas pile-up, gap overflow and type 1.5 migration in circumbinary discs: application to supermassive black hole binaries. Mon. Not. R. Astron. Soc. **427**, 2680–2700 (2012a). doi:10.1111/j.1365-2966.2012.22118.x

B. Kocsis, Z. Haiman, A. Loeb, Gas pile-up, gap overflow and type 1.5 migration in circumbinary discs: general theory. Mon. Not. R. Astron. Soc. **427**, 2660–2679 (2012b). doi:10.1111/j.1365-2966.2012.22129.x

B. Kocsis, A. Ray, S. Portegies Zwart, Mapping the galactic center with gravitational wave measurements using pulsar timing. Astrophys. J. **752**, 67 (2012c). doi:10.1088/0004-637X/752/1/67

S. Komossa, Tidal disruption of stars by supermassive black holes: the X-ray view, in *European Physical Journal Web of Conferences*, vol. 39, 2012, p. 2001. doi:10.1051/epjconf/20123902001

S. Komossa, V. Burwitz, G. Hasinger, P. Predehl, J.S. Kaastra, Y. Ikebe, Discovery of a binary active galactic nucleus in the ultraluminous infrared galaxy NGC 6240 using Chandra. Astrophys. J. Lett. **582**, 15–19 (2003). doi:10.1086/346145

K. Kotake, N. Ohnishi, S. Yamada, Gravitational radiation from standing accretion shock instability in core-collapse supernovae. Astrophys. J. **655**, 406–415 (2007). doi:10.1086/509320

M.R. Krumholz, C.F. McKee, R.I. Klein, Bondi accretion in the presence of vorticity. Astrophys. J. **618**, 757–768 (2005). doi:10.1086/426051

M.R. Krumholz, C.F. McKee, R.I. Klein, Bondi-Hoyle accretion in a turbulent medium. Astrophys. J. **638**, 369–381 (2006). doi:10.1086/498844

R.N. Lang, S.A. Hughes, N.J. Cornish, Measuring parameters of massive black hole binaries with partially aligned spins. Phys. Rev. D **84**(2), 022002 (2011). doi:10.1103/PhysRevD.84.022002

Y. Levin, Starbursts near supermassive black holes: young stars in the galactic centre, and gravitational waves in LISA band. Mon. Not. R. Astron. Soc. **374**, 515–524 (2007). doi:10.1111/j.1365-2966.2006.11155.x

F.K. Liu, X-shaped radio galaxies as observational evidence for the interaction of supermassive binary black holes and accretion disc at parsec scale. Mon. Not. R. Astron. Soc. **347**, 1357–1369 (2004). doi:10.1111/j.1365-2966.2004.073

F.K. Liu, X.-B. Wu, S.L. Cao, Double-double radio galaxies: remnants of merged supermassive binary black holes. Mon. Not. R. Astron. Soc. **340**, 411–416 (2003). doi:10.1046/j.1365-8711.2003.06235.x

G. Lodato, E.M. Rossi, Multiband light curves of tidal disruption events. Mon. Not. R. Astron. Soc. **410**, 359–367 (2011). doi:10.1111/j.1365-2966.2010.17448.x

G. Lodato, A.R. King, J.E. Pringle, Stellar disruption by a supermassive black hole: is the light curve really proportional to $t^{-5/3}$? Mon. Not. R. Astron. Soc. **392**, 332–340 (2009). doi:10.1111/j.1365-2966.2008.14049.x

A. Loeb, F.A. Rasio, Collapse of primordial gas clouds and the formation of quasar black holes. Astrophys. J. **432**, 52–61 (1994). doi:10.1086/174548

F.D. Lora-Clavijo, F.S. Guzmán, Axisymmetric Bondi-Hoyle accretion on to a Schwarzschild black hole: shock cone vibrations. Mon. Not. R. Astron. Soc. **429**, 3144–3154 (2013). doi:10.1093/mnras/sts573

A.I. MacFadyen, M. Milosavljević, An eccentric circumbinary accretion disk and the detection of binary massive black holes. Astrophys. J. **672**, 83–93 (2008). doi:10.1086/523869

M. MacLeod, J. Guillochon, E. Ramirez-Ruiz, The tidal disruption of giant stars and their contribution to the flaring supermassive black hole population. Astrophys. J. **757**, 134 (2012). doi:10.1088/0004-637X/757/2/134

P. Madau, M.J. Rees, Massive black holes as population III remnants. Astrophys. J. Lett. **551**, 27–30 (2001). doi:10.1086/319848

P. Madau, M.J. Rees, M. Volonteri, F. Haardt, S.P. Oh, Early reionization by miniquasars. Astrophys. J. **604**, 484–494 (2004). doi:10.1086/381935

I. Mandel, D.A. Brown, J.R. Gair, M.C. Miller, Rates and characteristics of intermediate mass ratio inspirals detectable by advanced LIGO. Astrophys. J. **681**, 1431–1447 (2008). doi:10.1086/588246
B. McKernan, K.E.S. Ford, W. Lyra, H.B. Perets, Intermediate mass black holes in AGN discs—I. Production and growth. Mon. Not. R. Astron. Soc. **425**, 460–469 (2012). doi:10.1111/j.1365-2966.2012.21486.x
D. Merritt, R.D. Ekers, Tracing black hole mergers through radio lobe morphology. Science **297**, 1310–1313 (2002). doi:10.1126/science.1074688
M.C. Miller, E.J.M. Colbert, Intermediate-mass black holes. Int. J. Mod. Phys. D **13**, 1–64 (2004). doi:10.1142/S0218271804004426
M. Milosavljević, A. Loeb, The link between warm molecular disks in maser nuclei and star formation near the black hole at the galactic center. Astrophys. J. Lett. **604**, 45–48 (2004). doi:10.1086/383467
S. Mineo, M. Gilfanov, R. Sunyaev, X-ray emission from star-forming galaxies—I. High-mass X-ray binaries. Mon. Not. R. Astron. Soc. **419**, 2095–2115 (2012). doi:10.1111/j.1365-2966.2011.19862.x
J. Miralda-Escudé, J.A. Kollmeier, Star captures by quasar accretion disks: a possible explanation of the M-σ relation. Astrophys. J. **619**, 30–40 (2005). doi:10.1086/426467
M.P. Muno, J.S. Clark, P.A. Crowther, S.M. Dougherty, R. de Grijs, C. Law, S.L.W. McMillan, M.R. Morris, I. Negueruela, D. Pooley, S. Portegies Zwart, F. Yusef-Zadeh, A neutron star with a massive progenitor in Westerlund 1. Astrophys. J. Lett. **636**, 41–44 (2006). doi:10.1086/499776
K. Nakayama, Dynamical instability of standing shock waves in adiabatic accretion flows and wind flows. Mon. Not. R. Astron. Soc. **270**, 871 (1994)
R. Narayan, J.E. McClintock, Advection-dominated accretion and the black hole event horizon. New Astron. Rev. **51**, 733–751 (2008). doi:10.1016/j.newar.2008.03.002
R. Narayan, I. Yi, Advection-dominated accretion: a self-similar solution. Astrophys. J. Lett. **428**, 13–16 (1994). doi:10.1086/187381
R. Narayan, I.V. Igumenshchev, M.A. Abramowicz, Self-similar accretion flows with convection. Astrophys. J. **539**, 798–808 (2000). doi:10.1086/309268
S.C. Noble, B.C. Mundim, H. Nakano, J.H. Krolik, M. Campanelli, Y. Zlochower, N. Yunes, Circumbinary magnetohydrodynamic accretion into inspiraling binary black holes. Astrophys. J. **755**, 51 (2012). doi:10.1088/0004-637X/755/1/51
I.D. Novikov, K.S. Thorne, Astrophysics of black holes, in *Black Holes (Les Astres Occlus)*, ed. by A. Giannaras, 1973, pp. 343–450
R.M. O'Leary, A. Loeb, Star clusters around recoiled black holes in the Milky Way halo. Mon. Not. R. Astron. Soc. **395**, 781–786 (2009). doi:10.1111/j.1365-2966.2009.14611.x
R.M. O'Leary, F.A. Rasio, J.M. Fregeau, N. Ivanova, R. O'Shaughnessy, Binary mergers and growth of black holes in dense star clusters. Astrophys. J. **637**, 937–951 (2006). doi:10.1086/498446
R.M. O'Leary, B. Kocsis, A. Loeb, Gravitational waves from scattering of stellar-mass black holes in galactic nuclei. Mon. Not. R. Astron. Soc. **395**, 2127–2146 (2009). doi:10.1111/j.1365-2966.2009.14653.x
E.C. Ostriker, Dynamical friction in a gaseous medium. Astrophys. J. **513**, 252–258 (1999). doi:10.1086/306858
A. Paggi, G. Fabbiano, G. Risaliti, J. Wang, M. Elvis, Two Compton-thick active nuclei in Arp 220? ArXiv e-prints (2013)
V. Paschalidis, M. MacLeod, T.W. Baumgarte, S.L. Shapiro, Merger of white dwarf-neutron star binaries: prelude to hydrodynamic simulations in general relativity. Phys. Rev. D **80**(2), 024006 (2009). doi:10.1103/PhysRevD.80.024006
S.F. Portegies Zwart, S.L.W. McMillan, The runaway growth of intermediate-mass black holes in dense star clusters. Astrophys. J. **576**, 899–907 (2002). doi:10.1086/341798
S.F. Portegies Zwart, H. Baumgardt, S.L.W. McMillan, J. Makino, P. Hut, T. Ebisuzaki, The ecology of star clusters and intermediate-mass black holes in the galactic bulge. Astrophys. J. **641**, 319–326 (2006). doi:10.1086/500361
K.A. Postnov, L.R. Yungelson, The evolution of compact binary star systems. Living Rev. Relativ. **9**, 6 (2006)
F. Pretorius, Evolution of binary black-hole spacetimes. Phys. Rev. Lett. **95**(12), 121101 (2005). doi:10.1103/PhysRevLett.95.121101
R.R. Rafikov, Structure and evolution of circumbinary disks around supermassive black hole (SMBH) binaries. ArXiv e-prints (2012)
P. Ranalli, A. Comastri, G. Setti, The 2-10 keV luminosity as a star formation rate indicator. Astron. Astrophys. **399**, 39–50 (2003). doi:10.1051/0004-6361:20021600
M.J. Rees, Accretion and the quasar phenomenon. Phys. Scr. **17**, 193–200 (1978). doi:10.1088/0031-8949/17/3/010
M.J. Rees, Tidal disruption of stars by black holes of 10 to the 6th-10 to the 8th solar masses in nearby galaxies. Nature **333**, 523–528 (1988). doi:10.1038/333523a0
J.A. Regan, M.G. Haehnelt, Pathways to massive black holes and compact star clusters in pre-galactic dark matter haloes with virial temperatures $\geq$ 10,000 K. Mon. Not. R. Astron. Soc. **396**, 343–353 (2009). doi:10.1111/j.1365-2966.2009.14579.x

Y. Rephaeli, E.E. Salpeter, Flow past a massive object and the gravitational drag. Astrophys. J. **240**, 20–24 (1980). doi:10.1086/158202

C. Rodriguez, G.B. Taylor, R.T. Zavala, A.B. Peck, L.K. Pollack, R.W. Romani, A compact supermassive binary black hole system. Astrophys. J. **646**, 49–60 (2006). doi:10.1086/504825

C. Roedig, O. Zanotti, D. Alic, General relativistic radiation hydrodynamics of accretion flows—II. Treating stiff source terms and exploring physical limitations. Mon. Not. R. Astron. Soc. **426**, 1613–1631 (2012). doi:10.1111/j.1365-2966.2012.21821.x

N. Roos, J.S. Kaastra, C.A. Hummel, A massive binary black hole in 1928 + 738? Astrophys. J. **409**, 130–133 (1993). doi:10.1086/172647

S. Rosswog, E. Ramirez-Ruiz, W.R. Hix, Tidal disruption and ignition of white dwarfs by moderately massive black holes. Astrophys. J. **695**, 404–419 (2009). doi:10.1088/0004-637X/695/1/404

M. Ruffert, Three-dimensional hydrodynamic Bondi-Hoyle accretion. 1: code validation and stationary accretors. Astrophys. J. **427**, 342–350 (1994). doi:10.1086/174144

M. Ruffert, Non-axisymmetric wind-accretion simulations. II. Density gradients. Astron. Astrophys. **346**, 861–877 (1999)

M. Saijo, T.W. Baumgarte, S.L. Shapiro, M. Shibata, Collapse of a rotating supermassive star to a supermassive black hole: post-Newtonian simulations. Astrophys. J. **569**, 349–361 (2002). doi:10.1086/339268

A.P. Schoenmakers, A.G. de Bruyn, H.J.A. Röttgering, H. van der Laan, C.R. Kaiser, Radio galaxies with a 'double-double morphology'—I. Analysis of the radio properties and evidence for interrupted activity in active galactic nuclei. Mon. Not. R. Astron. Soc. **315**, 371–380 (2000). doi:10.1046/j.1365-8711.2000.03430.x

N.I. Shakura, R.A. Sunyaev, Black holes in binary systems. Observational appearance. Astron. Astrophys. **24**, 337–355 (1973)

S.L. Shapiro, S.A. Teukolsky, Black holes, white dwarfs, and neutron stars: the physics of compact objects, in *14.2. Collisionless Spherical Accretion* (1983a)

S.L. Shapiro, S.A. Teukolsky, Black holes, white dwarfs, and neutron stars: the physics of compact objects, in *12. Black Holes (eq. 12.4.30 and 12.7.25)* (1983b)

Y. Shen, A. Loeb, Identifying supermassive black hole binaries with broad emission line diagnosis. Astrophys. J. **725**, 249–260 (2010). doi:10.1088/0004-637X/725/1/249

J.-M. Shi, J.H. Krolik, S.H. Lubow, J.F. Hawley, Three-dimensional magnetohydrodynamic simulations of circumbinary accretion disks: disk structures and angular momentum transport. Astrophys. J. **749**, 118 (2012). doi:10.1088/0004-637X/749/2/118

N. Soker, Stability analysis of the accretion line. Astrophys. J. **358**, 545–550 (1990). doi:10.1086/169007

N. Soker, Nonlinear instability of the accretion line. Astrophys. J. **376**, 750–756 (1991). doi:10.1086/170322

N. Stone, A. Loeb, Observing Lense-Thirring precession in tidal disruption flares. Phys. Rev. Lett. **108**(6), 061302 (2012a). doi:10.1103/PhysRevLett.108.061302

N. Stone, A. Loeb, Tidal disruption flares of stars from moderately recoiled black holes. Mon. Not. R. Astron. Soc. **422**, 1933–1947 (2012b). doi:10.1111/j.1365-2966.2012.20577.x

N. Stone, R. Sari, A. Loeb, Consequences of strong compression in tidal disruption events. ArXiv e-prints (2012)

L.E. Strubbe, E. Quataert, Optical flares from the tidal disruption of stars by massive black holes. Mon. Not. R. Astron. Soc. **400**, 2070–2084 (2009). doi:10.1111/j.1365-2966.2009.15599.x

H. Sudou, S. Iguchi, Y. Murata, Y. Taniguchi, Orbital motion in the radio galaxy 3C 66B: evidence for a supermassive black hole binary. Science **300**, 1263–1265 (2003). doi:10.1126/science.1082817

T. Tanaka, Z. Haiman, The assembly of supermassive black holes at high redshifts. Astrophys. J. **696**, 1798–1822 (2009). doi:10.1088/0004-637X/696/2/1798

A. Tchekhovskoy, B.D. Metzger, D. Giannios, L.Z. Kelley, Swift J1644+57 gone MAD: the case for dynamically-important magnetic flux threading the black hole in a jetted tidal disruption event. ArXiv e-prints (2013)

K.S. Thorne, A.N. Zytkow, Stars with degenerate neutron cores. I—structure of equilibrium models. Astrophys. J. **212**, 832–858 (1977). doi:10.1086/155109

M. Ugliano, H.-T. Janka, A. Marek, A. Arcones, Progenitor-explosion connection and remnant birth masses for neutrino-driven supernovae of iron-core progenitors. Astrophys. J. **757**, 69 (2012). doi:10.1088/0004-637X/757/1/69

S. van Velzen, G.R. Farrar, The rate of stellar tidal disruption flares from sdss data, in *European Physical Journal Web of Conferences*, vol. 39, 2012, p. 8002. doi:10.1051/epjconf/20123908002

J. Wang, D. Merritt, Revised rates of stellar disruption in galactic nuclei. Astrophys. J. **600**, 149–161 (2004). doi:10.1086/379767

N. Webb, D. Cseh, E. Lenc, O. Godet, D. Barret, S. Corbel, S. Farrell, R. Fender, N. Gehrels, I. Heywood, Radio detections during two state transitions of the intermediate-mass black hole HLX-1. Science **337**, 554 (2012). doi:10.1126/science.1222779

J.S.B. Wyithe, A. Loeb, Low-frequency gravitational waves from massive black hole binaries: predictions for LISA and pulsar timing arrays. Astrophys. J. **590**, 691–706 (2003). doi:10.1086/375187

J.S.B. Wyithe, A. Loeb, Detection of gravitational waves from the coalescence of population III remnants with advanced LIGO. Astrophys. J. **612**, 597–601 (2004). doi:10.1086/422183

J.S.B. Wyithe, A. Loeb, Photon trapping enables super-Eddington growth of black hole seeds in galaxies at high redshift. Mon. Not. R. Astron. Soc. **425**, 2892–2902 (2012). doi:10.1111/j.1365-2966.2012.21127.x

O. Zanotti, C. Roedig, L. Rezzolla, L. Del Zanna, General relativistic radiation hydrodynamics of accretion flows—I. Bondi-Hoyle accretion. Mon. Not. R. Astron. Soc. **417**, 2899–2915 (2011). doi:10.1111/j.1365-2966.2011.19451.x

DOI 10.1007/978-1-4939-2227-7_11
Reprinted from *Space Science Reviews* Journal, DOI 10.1007/s11214-014-0067-1

Massive Binary Black Holes in Galactic Nuclei and Their Path to Coalescence

Monica Colpi

Received: 20 February 2014 / Accepted: 25 June 2014 / Published online: 15 July 2014

Abstract Massive binary black holes (10^5 M$_\odot$–10^9 M$_\odot$) form at the centre of galaxies that experience a merger episode. They are expected to coalesce into a larger black hole, following the emission of gravitational waves. Coalescing massive binary black holes are among the loudest sources of gravitational waves in the Universe, and the detection of these events is at the frontier of contemporary astrophysics. Understanding the black hole binary formation path and dynamics in galaxy's mergers is therefore mandatory. A key question poses: during a merger, will the black holes descend over time on closer orbits, form a Keplerian binary and coalesce shortly after? Here we review progress discussing the fate of black holes in different environments: from major mergers of collisionless galaxies to major and minor mergers of gas-rich disc galaxies, from smooth and clumpy circum-nuclear discs to circum-binary discs present on the smallest scales inside galactic nuclei.

Keywords Black hole physics · Dynamics · Galaxy mergers · Black hole binaries

1 Massive Binary Black Holes as Tracers of Black Hole Seed Formation and Galaxy Assembly, Along Cosmic History

In the universe, *black holes* come in *two flavours*: the "stellar black holes" relic of the most massive stars, weighing $\sim$ 5–30 M$_\odot$ (Özel et al. 2010), and the "super-massive black holes" residing in the nuclear regions of galaxies which carry large masses, typically in excess of 10^8 M$_\odot$ (Vestergaard et al. 2008). The black holes of stellar origin are observed in X-ray binaries as accreting objects, while the super-massive ones power the bright QSOs and the

M. Colpi (✉)
Department of Physics G. Occhialini, University of Milano Bicocca, Piazza della Scienza 3, 20123, Milano, Italy
e-mail: monica.colpi@unimib.it

M. Colpi
Istituto Nazionale di Fisica Nucleare (INFN)—Milano Bicocca, Piazza della Scienza 3, 20123, Milano, Italy

less luminous active galactic nuclei (AGN) (Merloni and Heinz 2013). Super-massive black holes are also observed in their quiescent state as massive dark objects in nearby galaxy spheroids (Gültekin et al. 2009; Kormendy and Ho 2013), and a compelling case is that of the Milky Way housing an (almost) inactive black hole of 4×10^6 $M_\odot$ (Ghez et al. 2008; Gillessen et al. 2009).

A black hole *desert* exists between ~ 30 $M_\odot$ and $\sim 10^6$ $M_\odot$. These black holes are often called *intermediate mass* or *middleweight* black holes with boundaries of the desert zone that are not physically constrained, due to the lack of observations. The *maximum* mass of a black hole of stellar origin can be as large as a few $\times 10^2$ $M_\odot$, according to theoretical studies, depending on the metallicity of the collapsing progenitor stars and on the role of radiative feed-back in limiting the final mass (Omukai and Palla 2001; Heger et al. 2003). The *minimum* mass of super-massive black holes is not constrained as unknown is the process of formation of these black holes (Volonteri 2010; Schleicher et al. 2013).

Limits on the density of the X-ray background light (resulting from accretion of an unresolved population of massive black holes) and on the local black hole mass density (Marconi et al. 2004; Merloni and Heinz 2013), suggest that super-massive black holes have grown their mass across cosmic ages through repeated episodes of accretion and via coalescences driven by galaxy mergers. This has led to the concept of black hole *seed* and black hole growth from seeds in concordance with the rise of cosmic structure. The characteristic mass or mass interval of the seed population is unknown and weakly constrained theoretically. Thus, aim of contemporary astrophysics is to disclose the mechanism of black hole seed formation through the detection of middleweight black holes in galaxies (Reines et al. 2013).

The discovery of tight correlations between the black hole mass $M_\bullet$ and stellar velocity dispersion σ_*, and between $M_\bullet$ and the stellar mass of the spheroid M_* ($M_\bullet/M_* \sim 10^{-3}$) highlighted the existence of a process of symbiotic evolution between black holes and galaxies (Marconi and Hunt 2003; Häring and Rix 2004; Ferrarese and Ford 2005; Gültekin et al. 2009; Kormendy and Ho 2013). The current interpretation is that the huge power emitted by the central black hole when active may have affected the rate of star formation in the host, turing the galaxy into a red and dead elliptical (Mihos and Hernquist 1996; Di Matteo et al. 2005; Hopkins et al. 2006). At present, there is a live debate on whether the correlation extends to bulge-less disc galaxies or in general to lower mass galaxies (Kormendy and Ho 2013). Lower mass galaxies are expected to house lighter super-massive black holes, according to the above relations. Thus low-mass galaxies are the preferred sites for the search of the middleweight black holes in the desert zone. Many galaxies (up to 75 %) host at their centre a Nuclear Star Cluster, a compact sub-system of stars with mass $M_{\rm NSC}$ typically $\lesssim 10^7$ $M_\odot$, and half-mass radius of $\lesssim 10$ pc. In a number of Nuclear Star Clusters a central middleweight black hole has been discovered which co-habit the cluster (Ferrarese et al. 2006). Less tight correlations have been found between $M_{\rm NSC}$ and the mass M_* of the host galaxy (Scott and Graham 2013), indicating that while in bright spheroids the presence of a central black hole appears to be compulsory,[1] in less bright (disc) galaxies a Nuclear Star Cluster, with or without a central black hole, is preferred.

The search for middleweight black holes in less massive disc galaxies will be central for understanding the process of formation and co-evolution of black holes and galaxies (Reines et al. 2013; Kormendy and Ho 2013). But, this is not the only strategy. A powerful and promising new route exists to unveil infant black holes, forming at high redshift $z \lesssim 15$: this is the route proposed by *The Gravitational Universe*, the science theme selected by ESA

[1] See Gerosa and Sesana (2014) for missing black holes in the brightest cluster galaxies following black hole coalescence and ejection by gravitational recoil.

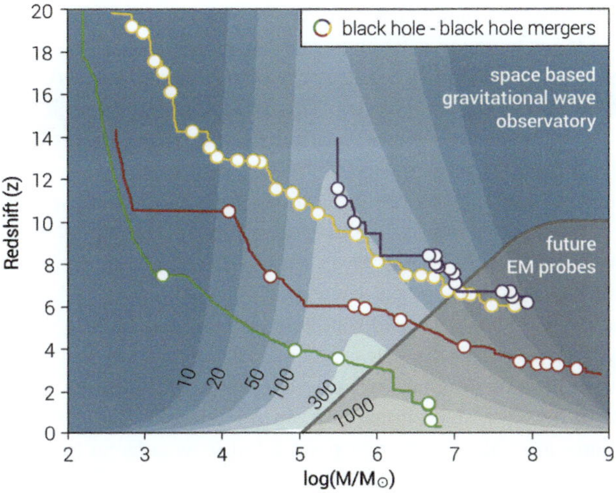

Fig. 1 Paths of black holes forming at high redshift from light (10^{2-3} $M_\odot$) and heavy (10^{5-6} $M_\odot$) seeds. The black holes evolve along tracks, in the mass versus redshift diagram, as they experience accretion episodes and coalescences with other black holes. *Circles* mark the loci of black hole coalescences. Four paths are selected: two ending with a black hole powering a $z \sim 6$ QSO (starting from a massive seed, *blue curve*, and from a seed resulting from the collapse of a massive metal-free star, *yellow curve*); a third ending with a typical 10^9 $M_\odot$ black hole in a giant elliptical galaxy (*red curve*); and finally the forth ending with the formation of a Milky Way-like black hole (*green curve*). The tracks are obtained using state-of-the-art semi-analytical merger tree models. The *grey* transparent area in the *bottom right corner* roughly identifies the parameter space accessible by future electromagnetic probes which will observe black holes powered by accretion. Over-lied are contour levels of constant sky and polarization angle-averaged Signal-to-Noise-Ratios (SNRs) for eLISA, for equal mass non-spinning binaries as a function of their total rest frame mass (Consortium 2013). It is remarkable that black hole mergers can be detected by eLISA with a very high SNR across all cosmic ages. Courtesy of Consortium (2013)

for the next large mission L3, *for the search and detection of low frequency gravitational waves from coalescing binary black holes in merging galactic halos* (Consortium 2013).

According to the current paradigm of the Λ-CDM cosmology, galaxies form following the baryonic infall of gas into collapsing dark matter halos and assemble hierarchically through mergers of (sub-)galactic units (White and Rees 1978). Black hole seeds growing in these pristine merging halos inescapably undergo coalescences (Volonteri et al. 2003). This is illustrated in Fig. 1, where we show characteristic tracks of black holes along cosmic history computed from semi-analytical models of galaxy formation (Volonteri and Natarajan 2009). The tracks are reported in the mass versus redshift plane, from 10^2 $M_\odot$ up to 10^9 $M_\odot$ and for $0 \lesssim z \lesssim 20$: black holes form from seeds of different masses (according to different seed formation models) and grow via accretion and mergers, the last denoted as circles in the diagram. The gravitational wave signal from these mergers will be detected with a high signal-to-noise ratio from the forthcoming science mission of *The Gravitational Universe*, eLISA (Consortium 2013). This will allow to explore black hole seed formation and evolution as early as $z \gtrsim 10$, just at the end of the dark ages but well before the epoch of cosmic re-ionisation of the intergalactic hydrogen.

Black holes come in binaries when two galaxies merge, and the gravitational wave signal emitted at coalescence offer a unique environment to measure, with exquisite precision, the black hole masses and spins, every time there is a merger (Amaro-Seoane et al. 2013). Thus, it has become mandatory to study how and when binary black holes form and evolve inside galactic halos, during the formation of cosmic structures. This is a multi-face problem crossing the boundaries between astrophysics and cosmology.

This Chapter aims at describing one aspect of the astrophysics of binary black holes: that of their dynamics in merging galaxies. This is a central problem if we want to consider binary black holes as powerful sources of gravitational waves and as unique tracers of the cosmic assembly of galaxies, as proposed in *The Gravitational Universe* (Consortium 2013). This carries some resemblance to the problem of formation of close binary neutron stars fated to coalesce: rapid shrinking of the star's orbits occurs through phases of unstable mass transfer and common envelope evolution to avoid the risk of supernova disruption, before the two compact stars reach the phase of inspiral by gravitational waves. Likewise, binary black holes experience a phase of rapid sinking by dynamical friction when the two galaxies merge, but only after experiencing close encounters with stars and strong coupling with gas, they enter the phase of gravitational wave driven inspiral. In the next sections, we will describe the fate of massive black holes in merging galaxies, indicating whether they form close binaries ready to coalesce or wide pairs fated to wander in the host galaxy, and the conditions for this to happen.

Section 2 starts with a brief historical recollection of the problem of black hole dynamics in stellar environments, and expands these early findings to account for recent advances in this field. Section 3 addresses the problem of black hole dynamics in gas-rich galaxy major and minor mergers, while Sect. 4 describes the fate of black holes in gaseous circum-nuclear and circum-binary gas discs. Section 5 presents a short summary of the timescales along the path to coalescence, and Sect. 6 contains the main conclusion and the directions in which the field may evolve into.

2 Binary Black Holes in Stellar Environments

2.1 Super-Massive Black Hole Binaries in Galactic Nuclei

In a pioneering Letter to *Nature*, Begelman et al. (1980) write: "There are straightforward reasons for surmising that super-massive black hole binaries exist: mergers between galaxies appear to be frequent; cD galaxies in clusters or groups quite probably formed in this manner; and there is direct evidence that the near-by active galaxy Centaurus A is a merger product." In particular, they correlate the bending and apparent precession of radio jets, observed in a number of active nuclei, with the presence of two black holes in a binary, exploring the dynamics of its formation inside the violently relaxed *stellar core* of the newly formed galaxy.

Begelman et al. depicted the existence of three main phases along the path to coalescence: an early phase of *pairing* (phase I) under dynamical friction in the stellar bulge of the post-merger galaxy, ending with the formation of a close *Keplerian binary*; a phase of *hardening* (phase II) during which the binary separation decreases due to energy loss by close encounters with single stars plunging on nearly radial orbits on the binary; a phase of *gravitational wave inspiral* (phase III), ending with the coalescence of the two black holes due to the emission of gravitational waves. In all the phases, gravitational torques act to decrease the orbital angular momentum and energy of the black holes, to promote their pairing and sinking toward more bound states.

The black hole that forms has a new mass, new spin according to mass-energy conservation (Rezzolla et al. 2008) and because gravitational waves carry away linear momentum, the new black hole receives a gravitational recoil that can be as large as $\lesssim 5000\ \mathrm{km\,s^{-1}}$

depending on the orientation and magnitude of the black hole spins and orbital angular momentum at the time of binary coalescence (Lousto and Zlochower 2013). Thus, an additional phase—phase IV—subsequent to merging should be considered corresponding to a *recoiling* black hole moving-inside or escaping-from it host galaxy (Gualandris and Merritt 2008; Merritt et al. 2009; Devecchi et al. 2009). This phase and the relation between spin and recoil are not considered here (Bogdanović et al. 2007; Dotti et al. 2010) as well as the observability of dual black holes along the path, and we refer to a number of papers and reviews for completeness (Komossa 2012; Colpi and Dotti 2011; Schnittman 2011; Centrella et al. 2010; Bode et al. 2010; Komossa 2006; Eracleous et al. 2011; Decarli et al. 2013; Lusso et al. 2014).

Returning to phase I, it is known that dynamical friction against the stars acts on each black hole individually to cause their progressive sinking (Chandrasekhar 1943; Begelman et al. 1980; Colpi et al. 1999; Yu 2002), until they come close enough to form a *Keplerian binary*. As dynamical friction is proportional to the background density of stars and to the square of the black hole mass, more massive black holes in denser environments sink more rapidly. In a stellar background of N stars described by a singular isothermal sphere, with density profile $\rho_* = \sigma_*^2/(2\pi G r^2)$ and one-dimensional (1D) velocity dispersion σ_*, a black hole of mass $m_\bullet$ at distance r sinks by dynamical friction on a timescale

$$\tau_{\rm df} \sim 2 \times 10^8 \ln^{-1} N \left(\frac{10^6\,{\rm M}_\odot}{m_\bullet}\right)\left(\frac{r}{100\,{\rm pc}}\right)^2\left(\frac{\sigma_*}{100\,{\rm km\,s^{-1}}}\right)\,{\rm yr}. \tag{1}$$

This timescale decreases with decreasing distance from the galaxy's nucleus, so that dynamical friction becomes more and more rapid with orbital decay. Eventually, the black holes end forming a Keplerian system.

Binary formation occurs approximately when the mass in stars enclosed in their orbit drops below twice the total mass of the binary $m_{\bullet,{\rm t}} = m_{\bullet,1} + m_{\bullet,2}$; hereon $m_{\bullet,1}$ ($m_{\bullet,2}$) is the mass of the primary (secondary) black hole, and $q = m_{\bullet,2}/m_{\bullet,1} \le 1$ the mass ratio. In a singular isothermal sphere, a Keplerian binary forms when $a^*_{\rm binary} \simeq G m_{\bullet,{\rm t}}/\sigma_*^2$, i.e. at a separation comparable to the gravitational sphere of influence of the black holes viewed as a single point mass $m_{\bullet,{\rm t}}$. Dynamical friction guides the inspiral, with no significant amplification of the eccentricity (Colpi et al. 1999), approximately down to $a^*_{\rm hard}$, defined as the binary separation at which the kinetic energy per unit mass of the binary equals the kinetic energy per unit mass of the stars in the galactic potential. The weakening of dynamical friction is due to the high velocity that the black holes acquire during inspiral, being the drag inversely proportional to the square of the orbital velocity. Phase I ends when the binary separation a has decayed below

$$a^*_{\rm hard} = a^*_{\rm binary}\frac{\mu}{3m_{\bullet,{\rm t}}} \sim \frac{G\mu}{3\sigma_*^2} \sim 0.1\frac{q}{(1+q)^2}\left(\frac{m_{\bullet,{\rm t}}}{10^6\,{\rm M}_\odot}\right)\left(\frac{100\,{\rm km\,s^{-1}}}{\sigma_*}\right)^2\,{\rm pc}, \tag{2}$$

where $\mu = m_{\bullet,{\rm t}} q/(1+q)^2$ is the reduced mass of the binary (Quinlan 1996; Yu 2002; Merritt and Milosavljević 2005).

During the hardening phase II, the black hole orbital energy and angular momentum are extracted via scattering of single stars off the binary, in close three-body encounters. As a single star impinging on the binary causes a fractional energy change of the order of $\sim \xi m_*/m_{\bullet,{\rm t}}$ (where $\xi \sim 0.2$–1 is a coefficient calculated after averaging over many star-binary scattering experiments), a large number of stars, or the order of $\sim m_{\bullet,{\rm t}}/m_*$, is necessary for a sizeable change of the binary binding energy $E_\bullet = G m_{\bullet,1} m_{\bullet,2}/2a$. The binary offers a cross section $A \sim \pi a G m_{\bullet,{\rm t}}/\sigma_*^2$ to the incoming flow of stars and this leads to a hardening rate $s \equiv d(1/a)/dt \sim \xi \pi G \rho_* / \sigma_*$ for the semi-major axis a, and a corresponding hardening time (independent on the number N of stars in the galaxy)

$$\tau^*_{\rm hard} \sim \frac{\sigma_*}{\pi G \rho_* a} \sim 70 \left(\frac{\sigma_*}{100\ {\rm km\,s^{-1}}} \right) \left(\frac{10^4\ {\rm M}_\odot\ {\rm pc}^{-3}}{\rho_*} \right) \left(\frac{10^{-3}\ {\rm pc}}{a} \right) {\rm Myr}. \quad (3)$$

Opposite to $\tau_{\rm df}$, the hardening time $\tau^*_{\rm hard}$ increases with decreasing a, indicating that it is more difficult to harden the binary as a decays with time (Yu 2002; Merritt and Milosavljević 2005).

Phase III starts when the coalescence time driven by gravitational wave emission

$$\tau_{\rm gw} \sim 5.4 \times 10^8 f(e)^{-1} \frac{(1+q)^2}{q} \frac{a^4}{m^3_{\bullet,{\rm t}}} \left(\frac{1}{0.001\ {\rm pc}} \right)^4 \left(\frac{10^6\ {\rm M}_\odot}{m_{\bullet,{\rm t}}} \right)^3 {\rm yr} \quad (4)$$

drops below $\tau^*_{\rm hard}$, where $f(e) = [1 + (73/24)e^2 + (37/96)e^4](1 - e^2)^{-7/2}$. The crossing condition, $\tau^*_{\rm hard} = \tau_{\rm gw}$ thus provides the binary separation at which the black holes transits from phase II into III:

$$a^*_{\rm II \to III} = \left(\frac{G^2}{c^5} \frac{256}{5\pi} \right)^{1/5} \left(\frac{\sigma_*}{\rho_*} \right)^{1/5} f^{1/5}(e) \left(\frac{q}{(1+q)^2} \right)^{1/5} m^{3/5}_{\bullet,{\rm t}}. \quad (5)$$

If $\tau_{\rm gw}$ evaluated at $a^*_{\rm II \to III}$ exceeds the age of the universe, than the binary *stalls* and does not reach coalescence. From Eq. (4), we can define as $a_{\rm gw}$ the distance at which the coalescence time $\tau_{\rm gw}$ equals the Hubble time $\tau_{\rm Hubble}$:

$$a_{\rm gw} = 2 \times 10^{-3} f(e)^{1/4} \frac{q^{1/4}}{(1+q)^{1/2}} \left(\frac{m_{\bullet,{\rm t}}}{10^6\ {\rm M}_\odot} \right)^{3/4} \left(\frac{\tau_{\rm Hubble}}{13.6\ {\rm Gyr}} \right)^{1/4} {\rm pc}. \quad (6)$$

Expressed in units of the Schwartzschild radius $r_{\rm S} = 2Gm_{\bullet,{\rm t}}/c^2$ associated to $m_{\bullet,{\rm t}}$, $a_{\rm gw} = 1.4 \times 10^4 (m_{\bullet,{\rm t}}/10^6\ {\rm M}_\odot)^{-1/4} r_{\rm S}$ for the case of an equal mass circular binary. Coalescence occurs as long as $a^*_{\rm II \to III} < a_{\rm gw}$.

According to Eq. (3), for a wide interval of stellar densities and velocity dispersions, the coalescence time $\tau_{\rm gw}$, evaluated at $a^*_{\rm II \to III}$, is less than the Hubble time, so that the binary is excepted to enter the gravitational wave driven regime shortly after it has become hard. However, the estimate of $\tau^*_{\rm hard}$, in Eq. (3), severely *underestimates the true hardening time* since a large number of stars in "loss cone" orbits is necessary to drive the binary down to phase III. The loss cone in the black hole binary system is identified as the domain, in phase-space, populated by stars with sufficiently low angular momentum, $J^2 \lesssim J^2_{\rm lc} \lesssim 2Gm_{\bullet,{\rm t}} a$, to interact with the binary. If hardening occurs at a constant rate s, the number of stars necessary to complete the hardening phase is as large as $N^{\rm lc} \sim (\mu/m_*) \ln(a^*_{\rm hard}/a_{\rm gw})$, comparable to the mass of the binary. In the case of massive black holes ($m_{\bullet,{\rm t}} > 10^8\ {\rm M}_\odot$) in elliptical galaxies and spheroids, such a large reservoir of stars may *not* be available (Merritt 2013b).

At the end of phase I, when stellar encounters begin to control the contraction of the newly formed binary, the black holes start ejecting stars from the loss cone at a high clearing rate. Refilling of stars in the phase-space requires a lapse time comparable to the two-body relaxation timescale $\tau_{\rm rel} \propto N$ which in galactic nuclei, viewed as spherical systems, is often longer than the Hubble time (Yu 2002). Thus, the lack of stars in phase-space causes the binary to *stall*, at a separation $a^*_{\rm stall}$ typically of ~ 0.1–1 pc, much larger than $a_{\rm gw}$ (Eq. (6)). Thus the binary can not reach coalescence in a Hubble time, and this is referred to as *the last parsec problem.* This represents an obstacle to the path to coalescence during the transit across phases II and III, for a large range of black hole masses, and mass ratios q (Yu 2002).

We are therefore left with a major uncertainty on the estimate of the *true hardening time* τ_H which is expected to be closer to τ_{rel}, in the case of empty loss cone, and to τ^*_{hard} as given by Eq. (3), in the case of full loss cone: thus, $\tau^*_{hard} < \tau_H < \tau_{rel}$.

The binary is a source of kinetic energy as it deposits in the stellar bath an energy

$$\Delta E_\bullet \sim E_\bullet(a_{gw}) \sim 2 \times 10^{55} f(e)^{-1/4} \frac{q^{3/4}}{(1+q)^{3/2}} \left(\frac{m_{\bullet,t}}{10^6 \mathrm{M}_\odot}\right)^{5/4} \left(\frac{\tau_{Hubble}}{13.6 \mathrm{Gyr}}\right)^{-1/4} \mathrm{erg}, \quad (7)$$

in order to enter phase III. Compared to the binding energy of a stellar bulge of mass M_*, $(3/2)M_*\sigma_*^2$, energy deposition accounts ~ 10 % of the total energy of the system, if one assumes $m_{\bullet,t} \sim 10^{-3} M_*$, a value of $\sigma_* \sim 100 \mathrm{~km~s^{-1}}$, and an equal mass binary of 10^6 M$_\odot$. Binary energy deposition via encounters with single stars can create a stellar core in an otherwise steep density profile, due to star's ejection. Stellar scouring has been observed in a number of core, missing-light elliptical galaxies that are at present indirect candidates for black hole mergers (Milosavljević and Merritt 2001; Kormendy and Ho 2013; Merritt 2013a). The binary carries a larger angular momentum at a^*_{hard} compared to the angular momentum at the onset of gravitational wave inspiral,

$$\frac{J_\bullet(a^*_{hard})}{J_\bullet(a_{gw})} \sim 10\left(1-e^2\right)^{7/16} \frac{q^{3/8}}{(1+q)^{3/4}} \left(\frac{m_{\bullet,t}}{10^6 M_\odot}\right)^{1/8} \frac{100 \mathrm{~km~s^{-1}}}{\sigma_*} \left(\frac{\tau_{Hubble}}{13.6 \mathrm{Gyr}}\right)^{-1/8}. \quad (8)$$

From the equation it is clear that the binary can reduce $J_\bullet$ when transiting from phase II to III, increasing the eccentricity during orbital decay.

After Begelman et al., the last parsec problem has been considered a major bottleneck to the path of binary coalescence, and has motivated many studies (Milosavljević and Merritt 2001; Yu 2002; Merritt and Milosavljević 2005). Direct N-Body simulations of binary inspiral in isotropic, spherical galaxy models confirmed, on solid grounds, the stalling of the binary: the binary hardening rate s was found to be proportional to the rate of repopulation of loss cone orbits which in turn depends on N. Simulations with a lower number of particles N (corresponding to shorter two-body relaxation times τ_{rel}) show rapid binary decay. By contrast, more realistic simulations with larger N (longer τ_{rel}) display a much lower hardening rate s for the binary (Preto et al. 2011). The extrapolation of the result to the limit of very large N, as in elliptical galaxies or bulges of spirals, leads to stalling of the massive binary over a Hubble time.

Yu noticed that if one drops the assumption of sphericity, the hardening time τ_H is lower and can be less than the Hubble time in the case of less massive (power-law) galaxies which have a shorter τ_{rel} (Yu 2002). Spherical galaxies have all stars on centrophobic orbits, whereas galaxies with a high degree of axisymmetry and triaxiality host a significant fraction of stars on centrophilic orbits, such box orbits, which pass arbitrarily close to the binary and have low angular momentum. Furthermore, chaotic orbits in steep triaxial potentials can enhances the mass flux into the loss cone region (Merritt and Poon 2004). Some non axisymmetric potential can also excite bar instabilities causing a flow of stars toward the binary (Berczik et al. 2006).

Binary stalling has been recently challenged in models of galaxy's mergers. A number of direct N-Body simulations indicate that the end-product of a merger is not a spherical galaxy (Berczik et al. 2006; Khan et al. 2011; Preto et al. 2011; Khan et al. 2013; Wang et al. 2014). The new galaxy retains substantial amount of rotation or/and a large degree of asphericity or triaxiality such that the binary is seen to harden at a rate independent of N, as if the loss cone were fully refilled, or as if an N-independent mechanism (collisionless relaxation) provides a supply of stars in loss cone orbits. In light of these findings the last parsec problem appears today as an artefact of the oversimplifying assumption of sphericity of the relic galaxy, and that more realistic models, simulated starting from *ab initio* conditions, point in the direction of hardening times ranging between 0.1 and a few Gyrs, for the models explored.[2] An interesting corollary of these investigations is that in non-spherical models the binary eccentricity e is seen to increase to values very close to 1 (Preto et al. 2011; Khan et al. 2011), indicating rapid transfer of angular momentum to stars from the cumulative action of many scatterings (Sesana 2010; Dotti et al. 2012).

As final remark, alternative mechanisms exist that can cause the contraction of the binary orbit. Recycling of stars ejected by the binary on returning eccentric orbits (Milosavljević and Merritt 2003), massive perturbers scattering stars into loss cone orbits (Perets et al. 2007), and a third black hole in a triple encounter. In the latter case, a third closely interacting black hole can hard the binary due to eccentricity oscillations, or repopulation of the loss cone induced by the perturbation of third black hole in the galactic potential, or hardening during the three-body encounter (Blaes et al. 2002; Hoffman and Loeb 2007; Kulkarni and Loeb 2012).

So far, we considered the hardening of massive black hole binaries in massive galaxies. But, a question to investigate is related to the evolution of middleweight black holes of $\sim 10^{3-4}\ M_{\odot}$ which tend to inhabit smaller mass halos with shorter relaxation timescales, and that form at high redshift when the universe was younger. This narrows down the interval of time accessible for hardening. Is there a last parsec problem? Extrapolating the results to middleweight black hole masses may not be straightforward, and in the next subsection we shortly explore the hardening in this regime.

2.2 Middleweight Binary Black Holes in Stellar Environments: Hardening or Stalling in High Redshift Nuclei?

Studying the hardening and coalescence of middleweight black hole binaries with $m_{\bullet,\mathrm{t}} \gtrsim 10^4\ M_{\odot}$ is of importance as black holes of this mass are primary sources for eLISA, as illustrated in Fig. 1. In the figure, black hole coalescences occur at a rate equal to the rate of merging of their parent dark matter halos controlled by dynamical friction only. The underlying assumption is there is no or negligible delay between the merger of the halo and that of the nested black holes, caused by the potential stalling of the binary. At present, whether black hole binaries at very high redshift are able to reach coalescence in the short cosmological time lapse between black hole seed formation and halo mergers is unclear (paper in preparation).

As an exercise and for illustrative purposes, one can compute limits upon the density ρ_* and velocity dispersion σ_* that a massive stellar cluster should have to allow rapid hardening of the binary during phase II. In star clusters the density and velocity dispersion are functions of distance. Thus, one should consider ρ^* and σ_* are characteristic values of the central region of a massive star cluster in an hypothetical galactic nucleus.

[2] This view, however, has been criticised by Vasiliev et al. (2013) who compared the evolution of binary black holes in spherical, axisymmetric and triaxial equilibrium galaxy models. They find that the rate of binary hardening exhibits a significant N-dependence in all the models, in the investigated range of $10^5 \le N \le 10^6$. Their hardening rates are substantially lower than those expected if the binary loss cone remained full, with rates between the spherical and non-spherical models differing in less than a factor of two. This finding seems to cast doubt on claims that triaxiality or axisymmetry alone are capable of solving the final-parsec problem. Vasiliev and co-authors invite caution in extrapolating results to galaxies with high values of N until all discrepancies or intrinsic differences between equilibrium models and merged galaxy models are not understood deeply.

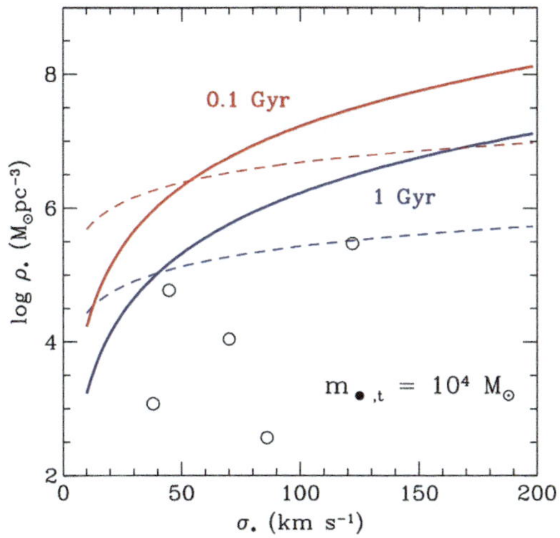

Fig. 2 Stellar mass density (in units of $M_\odot\,pc^{-3}$) versus velocity dispersion (in units of $km\,s^{-1}$) of hypothetical nuclear star clusters hosting middleweight black hole binaries on their path to coalescence which harden via single-binary encounters with solar mass stars: *solid lines* refer to the loci in the (σ_*, ρ_*) plane given by Eq. (9) where the central relaxation time τ_{rel} equals $\Delta\tau_{lapse}$. The *upper-red* (*lower blue*) *solid line* refers to $\tau_{rel} = \Delta\tau_{lapse} = 0.1$ Gyr (1 Gyr). *Dashed lines* refer to an equal-mass black hole binary of $10^4\ M_\odot$ and refer to the loci where the hardening time τ^*_{hard}, given by Eq. (3), equals $\Delta\tau_{lapse}$. The hardening time is computed at a black hole binary separation a_{gw} given by Eq. (6). The upper (lower) dashed line refers to $\tau^*_{hard} = \Delta\tau_{lapse} = 0.1$ Gyr (1 Gyr). *Empty circles* refer to mean stellar densities and velocity dispersion (calculated using the virial theorem) for five Nuclear Star Clusters of known mass and half-mass radii (Seth et al. 2008; Merritt 2013b)

Focus on the case of a black hole forming at z_{form} (e.g. ~ 20) and coalescing with another black hole, following a halo-halo merger at z_{coal} (~ 15), or the case of two adjacent mergers between redshift z_1 and z_2 over a short interval of cosmic time. The time lapse $\Delta\tau_{lapse}$ can be as short as $\lesssim 0.1$ Gyr, or more conservatively as short as 1 Gyr. In Fig. 2 we plot, in the $\sigma_* - \rho_*$ plane, the lines of constant $\tau_{rel} = 0.34\sigma_*^3/(G^2 m_* \rho_* \ln \Lambda)$ (with $m_* = 1\ M_\odot$ and $\ln \Lambda \sim 10$) corresponding to a cosmic time lapse $\Delta\tau_{lapse}$ equal to 0.1 Gyr and 1 Gyr, respectively. The solid lines in Fig. 2 refer to the loci where $\tau_{rel} = \Delta\tau_{lapse}$, so that characteristic densities higher than

$$\left(\frac{\rho_*}{1.6 \times 10^7\ M_\odot\,pc^{-3}}\right)_{rel} \gtrsim \left(\frac{\sigma_*}{100\ km\,s^{-1}}\right)^3 \left(\frac{0.1\ Gyr}{\Delta\tau_{lapse}}\right) \tag{9}$$

are requested, at a fixed σ_*, to allow for binary hardening on the relaxation timescale (corresponding to the empty loss cone regime). The dashed lines in Fig. 2 refer instead to the loci where $\tau^*_{hard}(a_{gw}) = \Delta\tau_{lapse}$, as given by Eq. (3) (corresponding to the full loss cone regime), for a black hole binary of $m_{\bullet,t} = 10^4\ M_\odot$ (upper dashed curve for $\Delta\tau_{lapse}$ equal to 0.1 Gyr, lower dashed curve for 1 Gyr). This condition implies

$$\left(\frac{\rho_*}{6 \times 10^6\ M_\odot\,pc^{-3}}\right)_{hard} \gtrsim \left(\frac{\sigma_*}{100\ km\,s^{-1}}\right) \left(\frac{10^4\ M_\odot}{m_{\bullet,t}}\right)^{3/4} \left(\frac{0.1\ Gyr}{\Delta\tau_{lapse}}\right)^{5/4}. \tag{10}$$

Fig. 2 shows that true hardening, at early cosmic epochs, requires densities in excess of $\gtrsim 10^{6-8}\ M_\odot\,pc^{-3}$ and comparatively low dispersion velocities $\lesssim 70\ km\,s^{-1}$ to meet the conditions for coalescence in the short time lapse $\Delta\tau_{lapse}$ of 0.1 or 1 Gyr. We do not know if such dense stellar environment were present at the centre of unstable pre-galactic discs. Today Nuclear Star Clusters, plausible candidates to harbour a central middleweight black hole, have all densities and velocity dispersions that do not meet this condition. Non-equilibrium conditions and/or the presence of gas, abundant in pre-galactic discs, may be instrumental in taxing the black holes to small separations, in this interval of masses, and at earlier cosmic epochs. Thus, a key question to pose is whether *gas* can *fasten* the transition along the three phases of pairing, hardening and gravitational wave driven inspiral and this question will be addressed in the incoming sections.

3 Black Holes Dynamics in Gas-Rich Mergers

Merging galaxies which are the sites of formation of binary black holes are expected to contain large concentrations of cold gas (unless one considers mergers between today elliptical galaxies only). This inevitable abundance of gas, in particular in high redshift disc galaxies, motivated us to inquiry into the role of gas dynamics as an alternative in the process of black hole hardening and coalescence.

In this section, we review the pairing of black holes in gas-rich merging galaxies following their dynamics *ab initio* to highlight the key role played by gas in affecting the black hole inspiral and the remarkable difference between *major* and *minor* mergers. Major mergers refer to interactions between galaxies of comparable mass, while minor mergers refer to interactions between a primary massive galaxy and a less massive galaxy, typically with mass ratio 1:10, and below. Boundaries among major or minor mergers are not sharp, as in many cases, the various outcomes depend also on the internal structure and gas content of the interacting galaxies.

3.1 Major Mergers and the Formation of a Keplerian Binary

The study of black hole dynamics in *gas-rich mergers* dates back to Mayer et al. (2007), yet it is still in its infancy. The rich physics involved and the high computational demand require state-of-the-art simulations, and the body of data is still inhomogeneous, fragmented and incomplete. While black hole dynamics in collisionless mergers of spherical galaxies has been explored (with direct N-Body codes) starting from galaxies on close elliptical bound orbits and followed mainly during the hardening phase (Khan et al. 2011), black hole dynamics in mergers between gas-rich disc galaxies has been studied starting from cosmologically motivated (parabolic) orbits, during the pairing phase over separations $\gtrsim$ 10 kpc, down to the scale ($\lesssim$ 10 pc) when the black holes form a Keplerian binary (Mayer et al. 2007; Colpi et al. 2009; Colpi and Dotti 2011; Chapon et al. 2013; Mayer 2013). Further hardening has been later explored in dedicated simulations (Escala et al. 2005; Dotti et al. 2006, 2007, 2009; Fiacconi et al. 2013).

Disc galaxies, as observed at low redshifts, are multi-component systems comprising a collisionless dark matter halo, a stellar disc which coexists with a multi-phase gaseous disc, and a central bulge housing (when present) a massive black hole. Simulating a collision between two disc galaxies with central black holes thus requires simulating the dynamics of the collisionless components (dark matter and stars) jointly with that of gas which is dissipative, and thus subject to cooling, star formation, shock heating and stellar feed-back.

There are many simulations of disc galaxy mergers in the literature (e.g. Hopkins et al. 2013), but there exists only a limited number in which the black hole dynamics is followed self-consistently from the $\gtrsim$ kpc scale down to scales $\lesssim$ 10 pc. When two galaxies merge, the two black holes are customarily assumed to merge promptly and form a single black hole. A recent set of N-Body/Smooth Particle Hydrodynamic simulations exists which follow the dynamics of black holes from the 100 kpc scale, typical of a merger, down to a scale $\lesssim 10$ pc, and which include star formation and feed-back (Van Wassenhove et al. 2012, 2014). There exists a further class of SPH or/and Adaptive Mesh Refinment (AMR) simulations which have enough resolution to witness the formation of a Keplerian binary on the ~ 1 pc scale, but which treat the gas thermodynamics via a phenomenological energy equation, in the form of a polytrope (Mayer et al. 2007; Chapon et al. 2013).

Equal mass mergers are disruptive for both progenitor galaxies. The galaxies first experience a few close fly-by during which tidal forces start to tear the galactic discs apart, generating tidal tails and plumes. The discs sink by dynamical friction against the dark matter background, and the massive black holes follow passively the dynamics of the bulge and disc they inhabit. Prior to merging, during the second pericentre passage, strong spiral patterns appear in both the stellar and gaseous discs: non axisymmetric torques redistribute angular momentum so that as much as 60 % of the gas originally present in each disc of the parent galaxy is funnelled inside the inner few hundred parsecs of the individual galaxy centres. The black holes, still in the pairing phase, are found to be surrounded by a rotating stellar and gaseous disc.

Later, the gasoues discs eventually merge in *a single massive rotationally supported nuclear disc* of $\lesssim 100$ pc in size, now weighing $\sim 10^9\ M_\odot$. The disc develops gravo-turbulence (with velocities ~ 60–$100\ \mathrm{km\,s^{-1}}$) that guarantees a Toomre parameter $Q \gtrsim 2$. This prevents fragmentation of gas into stars on the timescale, a few Myr, necessary for the black holes to form a Keplerian binary. This short sinking timescale comes from the combination of two facts: that gas densities are higher than stellar densities due to the dissipative nature of the interaction, and that the black holes move relative to the background with mild supersonic velocities. Under these conditions, the hydro-dynamical drag is the highest (Ostriker 1999). The subsequent evolution is described in Sect. 4.

During final revision of this Chapter, a new dedicated N-Body/SPH simulation by Roškar et al. (2014) of two Milky-Way-like galaxy discs with moderate gas fractions, has been carried out at parsec-scale resolution, including a new model for radiative cooling and heating in a multi-phase medium, star formation and feedback from supernovae. The massive black holes weighing $\sim 10^6\ M_\odot$ are form a pair at a separation of ~ 100 pc which gradually spirals inward. However, due to the strong starburst triggered by the merger, the gas in the centre most region is evacuated, requiring ~ 10 Myr for the nuclear disc to rebuild. The clumpy nature of the interstellar medium has a major impact on the dynamical evolution of the pair now subjected to stochastic torquing by both clouds and spiral modes in the disc. These effects combine to delay the orbital decay of the two black holes, just in phase I of gas-dominated dynamical friction. An inspiral timescale of ~ 100 Myr is found in this simulation which is smaller compared to that estimated in collisionless mergers, but longer of a factor at least 10 compared to the case of mergers with a single-phase gas. The result is in line with what found in Fiacconi et al. (2013) (see Sect. 4.1.2) who describes black hole dynamics in clumpy nuclear discs. We notice however that a single run may not suffice to pin down the characteristic gas-dynamical friction timescale in dissipative mergers, and that the perturbations induced by a population of massive clumps in the stellar component may alter the star's dynamics, prompting rapid refilling of the loss cone region around the two black holes, an effect that these simulations can not capture.

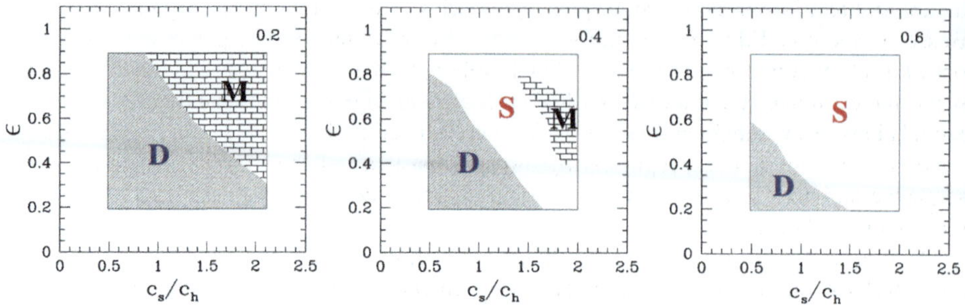

Fig. 3 Outcome of simulations of unequal mass mergers between spherical (*dark matter*) halos described with a NFW profile (Navarro et al. 1996), and described in the plane circularity ε versus relative concentration $c_{\mathrm{s}}/c_{\mathrm{h}}$ (measuring the ratio between the scale radii of the two halos, defined as in Navarro et al. 1996; Taffoni et al. 2003). The figure depicts the life diagram of a satellite halo with mass $M_{\mathrm{s}}/M_{\mathrm{h}} = 0.01$. Each plot is labelled by the value $x(E)$, the radius of the circular orbit (in units of the half mass radius of the main halo) at the onset of dynamical evolution. We identify the regions corresponding to merger (M) of the satellite into the main host halo, disruption against the background (D), and survival (S) of part of the satellite in the periphery of the main halo. Satellite halos with low concentration on less circular orbits are fragile to disruption, while satellite halos on wide orbits and high concentration preserve their identity. Mergers are preferred in correspondence of high concentration and close circular orbits. Courtesy of Taffoni et al. (2003)

3.2 Black Hole Paring in Unequal-Mass Mergers

3.2.1 Collisionless Unequal-Mass Mergers

Early works on collisionless mergers of unequal-mass spherical dark matter halos (Taffoni et al. 2003; Governato et al. 1994; Boylan-Kolchin et al. 2008) indicated that additional mechanisms are present, besides dynamical friction, that influence the structure and orbital evolution of the interacting halos (primarily the less massive, secondary): (i) progressive mass loss, or *tidal stripping*, induced by the tidal field of the main halo which reduces the mass of the secondary delaying the sinking by dynamical friction (the force scaling as the square of the satellite mass), and (ii) *tidal heating*, i.e. the effect of short impulses imparted to bound particles within the secondary satellite galaxy by the rapidly varying tidal force of the primary which heats the system causing its (partial) dissolution (Taffoni et al. 2003). Depending on the energy E of the orbit and its degree of circularity ε, on the relative mass concentration $c_{\mathrm{s}}/c_{\mathrm{h}}$ between satellite and main halo, and on the initial mass ratio of the primary to the satellite halos $M_{\mathrm{h}}/M_{\mathrm{s}}$, the encounter can lead either to rapid merging toward the centre of the primary halo (M), disruption (D), or survival (S) (when a residual mass remains bound and maintains its identity, orbiting in the main halo for a time longer than the Hubble time). Figure 3 illustrates the various outcomes of these experiments. In this context, merging times can be described by an empirical equation which accounts for the progressive mass loss of the secondary by tidal stripping and the progressive delay in the halo merging process

$$\frac{\tau_{\mathrm{df,tidal}}}{t_{\mathrm{dyn}}} \approx \frac{\Theta(E, \varepsilon, c_{\mathrm{s}}/c_{\mathrm{h}})}{\ln(1 + M_{\mathrm{h}}/M_{\mathrm{s}}(t))} \frac{M_{\mathrm{h}}}{M_{\mathrm{s}}(t)} \tag{11}$$

where Θ, function of the initial parameters, and satellite mass $M_{\mathrm{s}}(t)$ are computed from the numerical simulation. Figure 3 shows the fragility of less concentrated satellites to dispersal and disruption. These findings anticipate the possibility that unequal-mass mergers may release black holes on peripheral orbit inside the primary, due to tidal stripping of the less massive galaxy prior completion of the merger.

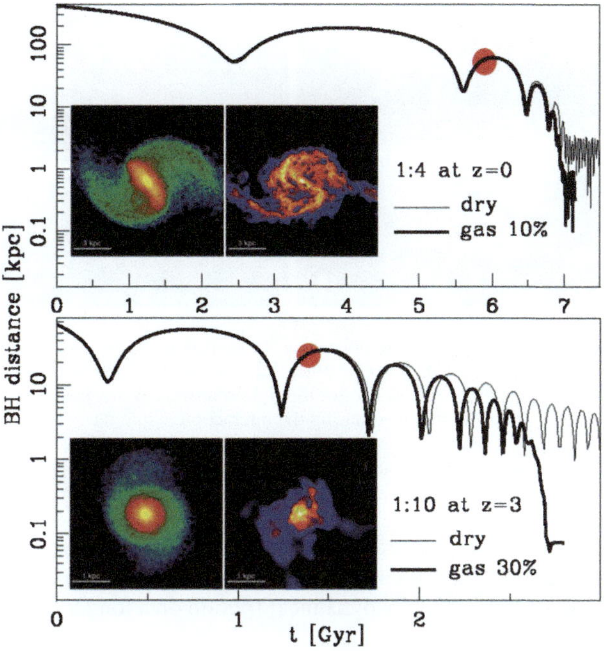

Fig. 4 *Upper panel*: black hole separation as a function of time for a 1:4 merger. The *thin* and *thick lines* refer to the dry (gas free) and wet (with gas fraction of 10 %) cases, respectively. The inset shows the color-coded density of stars (*left*) and gas (*right*) for the wet case at $t = 5.75$ Gyr (marked with a *red dot* on the curve); each image is 12 kpc on a side, and colors code the range $10^{-2} - 1\ M_\odot\,pc^{-3}$ for stars, and $10^{-3} - 10^{-1} M_\odot\,pc^{-3}$ for the gas. *Lower panel*: black hole separation as a function of time for a 1:10 merger (*upper panel*). The *thin* and *thick line* refer to the dry and wet (with gas fraction of 30 %) cases, respectively. The inset shows density maps at $t = 1.35$ Gyr for the wet merger: images are 4 kpc on a side (color coding as in upper panel). Courtesy of Callegari et al. (2009)

3.2.2 Gas-Rich Unequal-Mass Mergers

Recent suites of N-Body/SPH simulations of unequal-mass galaxy mergers have highlighted the occurrence of new key features in the dynamics of the discs and their embedded black holes that can be ascribed to differences in the central concentration of the interacting galaxies, and to the geometry of the encounter, but that go beyond the results inferred in Sect. 3.2.1. These new simulations illustrate the pivotal role played by gas which acts, through its cooling, to enhance the central mass concentration of the satellite and favours the sinking of the secondary black hole in the otherwise disrupted galaxy (Kazantzidis et al. 2005; Callegari et al. 2009, 2011; Van Wassenhove et al. 2012, 2014).

In unequal-mass mergers, the secondary, less massive galaxy undergoes major transformations. In particular, if the merger is wet, i.e. if the gas fraction in the disc of the secondary is relatively high ($\gtrsim 10$ %), tidal torques during the last peri-centre passage prior merging, trigger inflows which give rise to a nuclear starburst in the vicinity of the secondary black hole. This enhances the resilience of the galaxy's nucleus against tidal stripping due to the increased stellar density and degree of compactness of the nuclear bulge, at the time the secondary starts interacting with the disc of the primary. The denser stellar cusp surrounding the secondary black hole thus sinks rapidly toward the primary, dragging the black hole that reaches a separation of ~ 100 pc, close to the resolution limit of the simulation. This is illustrated in Fig. 4 for 1:4 and 1:10 wet mergers, where the relative separation of the black holes is plotted against time (heavy solid line). In Fig. 4 we also plot the stellar and gas distribution at the end of the nuclear starburst that created a denser stellar nucleus in the secondary. The disc of the secondary, rather turbulent and clumpy due to star formation, is later disrupted by ram pressure stripping by the gas of the primary. The secondary black hole continues its sinking toward the centre of the primary, being surrounded by the compact and massive star cluster. In Fig. 4, we also contrast the results from dry, i.e. gas free mergers. In

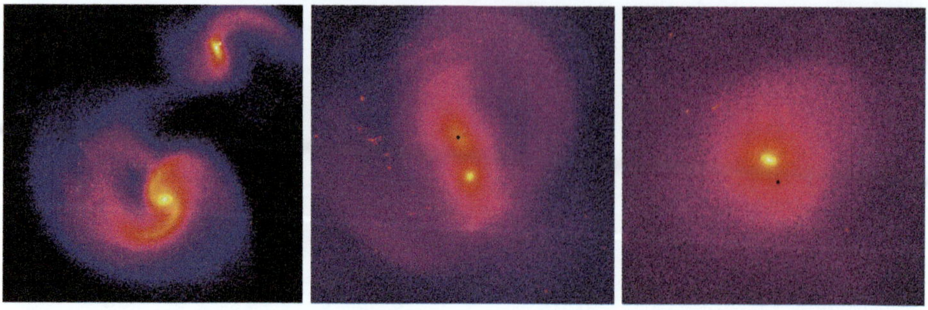

Fig. 5 Time sequence of stellar density snapshots in the 1:4 coplanar, prograde-prograde merger after the second peri-centre passage, at times 1.2, 1.43, and 1.48 Gyr, respectively. The scale of the left and central snapshots is 8 kpc, and 2 kpc for the right snapshot. A *black dot* marks the black hole in the primary galaxy nucleus which is dissolved during the interaction, while the secondary is at the centre of the highest density region of the secondary galaxy. In the last panel the secondary nucleus and black hole are near the centre of the mass distribution. Courtesy of Van Wassenhove et al. (2014)

the absence of the central starburst, dry mergers leave the secondary black hole wandering on a peripheral orbit at ~ 1 kpc away from the central, primary black hole. The naked black hole will then sink by dynamical friction on a longer timescale (Callegari et al. 2009, 2011; Khan et al. 2012).

Higher-resolution simulations of disc galaxies have recently revealed the occurrence of additional features, indicating how rich is the outcome of mergers under different initial conditions. Van Wassenhove et al. (2014) have shown that, as the gas-rich merger progresses, the newly formed stellar nucleus of the less massive galaxy, denser on small scales, is able to dissolve the less concentrated nucleus of the primary, via impulsive tidal heating. This is illustrated in Fig. 5 which shows how the denser nucleus of the secondary, at the end of the merger, finds itself in the midst of the mass distribution, having dissolved the nucleus of the main galaxy.

3.3 Black Hole Pairing in Minor Mergers: the Role of Mass Accretion

Minor mergers among galaxies with mass ratios 1:10 or less show behaviours that are extremes, along the sequence of unequal-mass mergers, and may lead to wandering black holes, even in presence of a sizeable fraction of cold gas (Callegari et al. 2009, 2011). The fate of black holes in minor mergers depends not only on the gas content but on the orbital parameters, such as the degree of co-planarity, and in addition, on new input physics (neglected for seek of simplicity), i.e. accretion. During the encounter the secondary black hole as well as the primary can accrete from the surrounding gas and increase their mass. A mass increase can influence the dynamics of the secondary black hole, as a larger mass implies a more rapid sinking by dynamical friction. This correlation has been found in a number of simulations by Callegari et al. (2011) who showed that the black hole mass ratio q is not conserved during the merger. The secondary black hole is subjected to episodes of accretion which enhance the mass by an order of magnitude when interacting with the gas of the primary galaxy. Thus, the black hole mass ratio does not mirror that of the galaxies, and can be much higher than the initial value indicating that black holes in unequal-mass mergers may carry comparable masses at the time they form a close pair.

Figure 6 summarises these findings, i.e. the correlation between the ability of pairing (measured evaluating the black hole relative separation) and the mass ratio q, evaluated at the

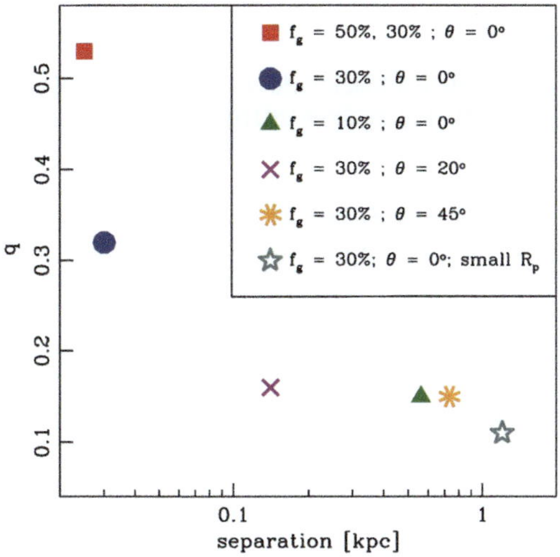

Fig. 6 Black hole mass ratio q versus relative separation, at the end of the simulation (when either a pair forms on the scale of the force resolution (10 pc), or the secondary black hole wanders at the periphery of the main galaxy), for the 1:10 mergers explored in Callegari et al. (2011), labelled according to their initial gas fractions f_g, orbital inclination θ and initial peri-centre R_p. Courtesy of Callegari et al. (2011)

end of the simulation. Coplanar prograde mergers with higher fractions of gas lead to higher q and smaller black hole separations. Inclined mergers with large gas fractions can instead fail in bringing the black holes to a small separation. Torques acting on the satellite during the early phases of the merger are weaker for higher inclinations, and for this reason the increase in mass ratio q during the first three orbits is milder than in the coplanar case with the same gas fraction. Moreover, a higher inclination corresponds to a slower orbital decay so that the satellite galaxy undergoes a larger number of tidal shocks before being disrupted, preventing further episodes of substantial accretion onto the secondary black hole. Finally, gas-rich mergers on closer orbits (i.e. with smaller peri-centre) are less effective in pairing contrary to what expected. Because of the smaller distance of approach and higher relative velocities between the satellite and the surroundings, ram pressure strips gas effectively, reducing the importance of the starburst that made the satellite less susceptible to stripping, and the accretion process onto the black hole. The joint action of these effects is therefore conducive to weak pairing, irrespective of the large amount of gas present initially.

In summary, minor mergers appear to fail in forming close black hole pairs in a number of cases, as the less massive galaxy is disrupted by tidal and ram pressure stripping at earlier times during the encounter so that dynamical friction is unable to deliver the secondary black hole to the centre of the main galaxy, within a Hubble time. The boundary between failure and success, i.e. between coalescence and wandering, is still poorly determined as it depends on the geometry, gas content and internal structure of the galaxies, and on the follow-up black hold dynamics on smaller scales (Khan et al. 2012). Direct N-body simulations of gas-free minor mergers have shown that black hole coalescences can occur on timescales of one to a few Gyrs, regardless the mass ratio provided that its value $q \gtrsim q_{\rm crit} \sim 0.05$–$0.1$ (Khan et al. 2012). The rather abrupt transition at $q_{\rm crit}$ appears to result from the monotonic decrease of merger-induced triaxiality in the main galaxy with decreasing mass ratio. The secondary galaxy is too small and light to significantly perturb the massive primary, slowing down the rate of binary single star interactions and hardening. Judging from the results of simulations of galaxy minor mergers from a limited sample comprising gas-poor and gas-rich cases, a very rough boundary between coalescence and wandering appears to be at $q_{\rm crit} \lesssim 0.1$. Along parallel lines, N-Body/SPH cosmological simulations of massive disc galaxies, inclusive of black hole seed formation and growth, have shown that satellite galaxies containing black hole seeds are often tidally stripped as they merge with the primary while building the main galaxy disc. This creates naturally a population of wandering middleweight black holes in the massive spiral (from 5–10 wanderers), remnants of satellite cores (Bellovary et al. 2010).

4 Black Hole Dynamics in Gaseous Nuclear Discs

SPH simulations of black hole dynamics in massive, rotationally supported nuclear discs represent a benchmark for studying the process of binary formation and coalescence, in gas-rich environments (Escala et al. 2005; Dotti et al. 2006, 2007, 2009; Fiacconi et al. 2013). These are *not ab initio* simulations, since the disc, in rotational equilibrium, is already in place as part of the remnant galaxy, or of the main galaxy, in case a minor merger has delivered the secondary black hole inside the disc of the massive host.

At present, there is no analytical model nor simulation that can trace the black hole dynamics in nuclear discs from the disc's periphery (at $\lesssim 100$ pc) down to the $\lesssim 10^{-3}$ pc scale (i.e. close and below $a_{\rm gw}$) where gravitational waves drive the inspiral, due to the complexity of the gas thermodynamics and to its susceptibility to undergo gravitational instabilities conducive to star formation episodes. In this context, key elements are the *rotation* of the underlying background, its *self-gravity*, the degree of *gas dissipation*, and the nature of *turbulence and viscosity*. Large scale gas discs can cool down, develop turbulence and inhomogeneities in the form of massive clumps which become sites of star formation. Gas can also dissipate the kinetic energy of the moving black holes via radiative cooling in a disc on smaller scales when the disc around the binary is nearly Keplerian. Thus, black hole inspiral in gaseous discs is mainly governed by processes of angular momentum exchange and radiative cooling. A compelling question to pose is whether angular momentum transport resulting from the gravitational interaction of the black holes with the gas is faster than that from the slingshot of stars.

In gaseous discs, we are led to distinguish three phases. There exists an early phase I-g of *nuclear-disc-driven migration* during which non axisymmetric perturbations in the density field excited by the gravitational field of the black hole(s) cause the braking of the orbit in regions where the disc dominates the gravitational potential (Escala et al. 2005; Dotti et al. 2006, 2007, 2009). The typical scale covered by I-g is between 100 pc down to ~ 0.1 pc.

With time, the gas mass enclosed in the orbit decreases below $m_{\bullet,{\rm t}}$ and the black hole dynamics is dominated by their own gravitational potential. The binary then forms a Keplerian system. This corresponds to the onset of phase II-g of *binary-disc-driven migration.* In phase II-g, the tidal torques exerted by the binary on the disc are sufficiently intense to repel gas away from the binary clearing a cavity, called *gap* (Farris et al. 2014; Rafikov 2013; Hayasaki et al. 2013; D'Orazio et al. 2013; Roedig et al. 2012; Kocsis et al. 2012; Shi et al. 2012; Noble et al. 2012; Roedig et al. 2011; Cuadra et al. 2009; Hayasaki 2009; MacFadyen and Milosavljević 2008; Hayasaki et al. 2008, 2007; Ivanov et al. 1999; Gould and Rix 2000). The binary is then surrounded by a *circum-binary* disc. Rotation in the circum-binary disc is nearly Keplerian but the disc's structure is affected by the binary, acting as a source of angular momentum. In phase II-g, black holes migrate under the combined action of viscous and gravitational torques which ultimately drive the binary into the third phase III of *gravitational-driven inspiral* where loss of orbital energy and angular momentum is due to the emission of gravitational waves. Not for all black hole masses this gas-assisted inspiral leads to coalescence in a Hubble time and more work is necessary along these lines (Cuadra et al. 2009).

Below, we explore phase I-g considering first a smooth nuclear disc and later a clumpy nuclear disc to study black hole migration on pc-scales. In a second step we will explore phase II-g when a circum-binary disc forms which controls the evolution of the binary on smaller scales.

4.1 Nuclear-Disc-Driven Migration

4.1.1 Smooth Circum-Nuclear Discs

In a number of targeted studies, the massive nuclear disc is described by a Mestel model: the disc, self-gravitating and axisymmetric, has rotation velocity $V_{\rm rot}$ independent of radius R. With constant $V_{\rm rot}$, fluid elements in the disc are in differential rotation with $\Omega = V_{\rm rot}/R$, and are distributed following a surface density profile $\Sigma(R) = \Sigma_{\rm d}[R_{\rm d}/R]$, where $R_{\rm d}$ is a scale radius. The disc mass within a radius R is then given by $M_{\rm Mestel}(R) = M_{\rm d}[R/R_{\rm d}]$, with $M_{\rm d} = 2\pi R_{\rm d}^2 \Sigma_{\rm d}$, and the circular velocity $V_{\rm rot}^2 = GM_{\rm d}/R_{\rm d}$. The disc is pressure supported vertically, with aspect ratio $h/R_{\rm d}$ of ~ 0.1–0.05, and isothermal sound speed $c_{\rm s}$ such that the Toomre parameter Q is $\gtrsim 3$ everywhere, to prevent the development of gravitational instabilities. The disc is embedded in a more massive Plummer stellar sphere, representing the innermost region of the galactic bulge: hereon we refer to this configuration as circum-nuclear disc (Escala et al. 2005; Dotti et al. 2006, 2007, 2009).

In this smooth background (guaranteed by the large Q) one can trace the black hole dynamics, assuming a primary black hole of mass $m_{\bullet,1}$ at rest in the centre of the circum-nuclear disc, and a secondary black hole of mass $m_{\bullet,2}$ initially moving on a wide eccentric co-planar orbit inside the disc. The mass ratio q, the initial binary orbital elements, and the disc mass $M_{\rm Mestel}$ enclosed in the binary orbit (in excess of the binary mass $m_{\bullet,\rm t}$, in the simulated volume), are free parameters to mimic different encounter geometries and mergers of galaxies with different stellar/gas mass contents.

Assisted by a series of N-Body/SPH simulations, these studies have highlighted key differences in the black hole dynamics compared to that in spherical collisionless backgrounds, the most remarkable being the dragging of the moving black hole into a co-planar co-rotating orbit with null eccentricity before the black holes form a binary (Dotti et al. 2006, 2007, 2009).

The simulations show that any orbit with large initial eccentricity is forced into circular rotation in the disc. In the different panels of Fig. 7, we show the over-density excited by the black hole along the orbital phase, for an initial value of the eccentricity e_0 equal to 0.9. The wake, that in a uniform medium, trails the motion of the perturber maintaining its shape, here changes both orientation and shape, due to the differential rotation of the underlying disc. The wake is trailing behind when the black hole is at pericentre, but is leading ahead when at apocentre, since there, the black hole is moving more slowly than the background, and this causes a temporary acceleration on $m_{\bullet,2}$. It is important to remark that this is a rapid change of e occurring on few orbital times. Circularisation is a fast process, and it is faster the cooler is the disc, i.e. the denser is the disc. Sinking times are found to depend on the equation of state adopted, i.e. on the polytropic index γ used to model the thermodynamic behaviour of the gas.

A further signature of a rotating background is the *angular momentum flip* of an initially counter-rotating orbit, if it exists (Dotti et al. 2009).[3] Initially, the gas opposes to the motion

[3] This is a possibility that may occur in the case of a minor merger where the incoming black hole in the satellite galaxy enters the main galaxy from a co-planar counter-rotating orbit.

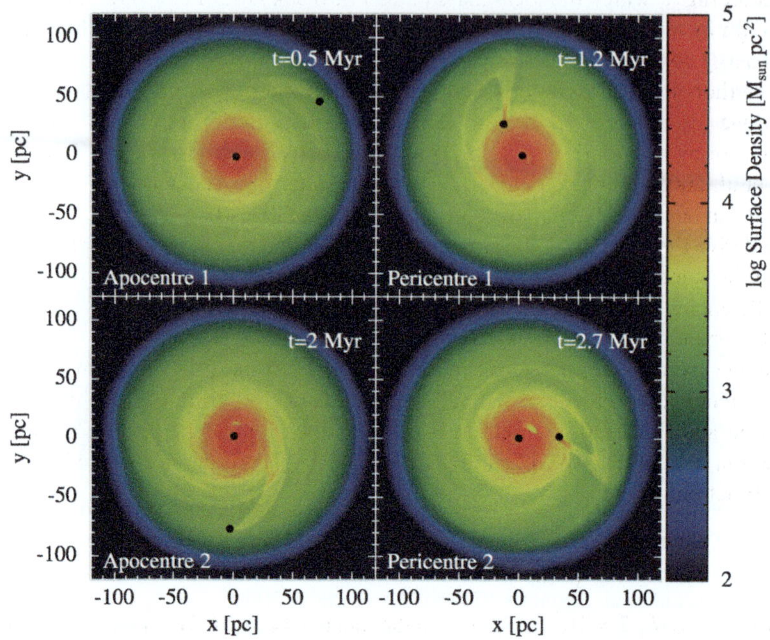

Fig. 7 Colour-coded gas surface density of a smooth disc of mass $M_d = 5 \times 10^8\ M_\odot$, for a black hole binary with $q = 0.1$ and a primary black hole of $10^7\ M_\odot$. The initial eccentricity is $e_0 = 0.7$. Snapshots refer to four different times, covering the process of circularisation lasting a few Myrs. The gas and the secondary black hole rotate counter-clockwise. The position of the black holes is marked by black dots. In the *top* and *bottom left* (*right*) panels the density wake excited by the secondary black hole is leading (trailing) the orbit, resulting in the circularisation of the relative orbit. Courtesy of D. Fiacconi

of the black hole as the density perturbation is always in the form of a trailing wake which causes an effective brake. The change of the orbital angular momentum from negative to nearly null values is further facilitated by the fact that, while the orbit decays, the black hole interacts with progressively denser regions of the disc. The orbit is nearly radial when the orbital angular momentum changes sign but the change is so rapid, relative to the orbital time, that the black hole is forced to co-rotate. When co-rotation establishes along an eccentric orbit, the orbital momentum increases under the circularising action of dynamical friction (non axisymmetric wake) in its co-rotating mode. Thus, a further prediction of black hole inspiral in rotating discs is that gas-dynamical friction conspires to turn counter-rotating orbits into co-rotatating ones, even before the formation of a Keplerian binary.

After circularisation, the secondary black hole continues to spiral in as it experiences a net negative torque, despite having reduced its relative velocity with respect to neighbouring fluid elements.[4] The black hole $m_{\bullet,2}$ is able to excite a non axisymmetric perturbation in the disc structure which produces a net negative torque on $m_{\bullet,2}$. This process is reminiscent

[4]In a uniform, isotropic gaseous background, gas-dynamical friction vanishes when the velocity of the perturber falls below the sound speed (Ostriker 1999). Dynamical friction is a non-local process and in a disc there is a residual velocity difference between the black hole and the more distant rotating fluid elements. One can view the migration process described in the text again as a manifestation of the large scale gravitational perturbation excited by the black hole mass, but this time the drag is inside a rotating inhomogeneous background. The net torque results from the sum of positive (inside the black hole orbit) and negative (outside) contributions as the perturbation is highly non axisymmetric due to differential rotation.

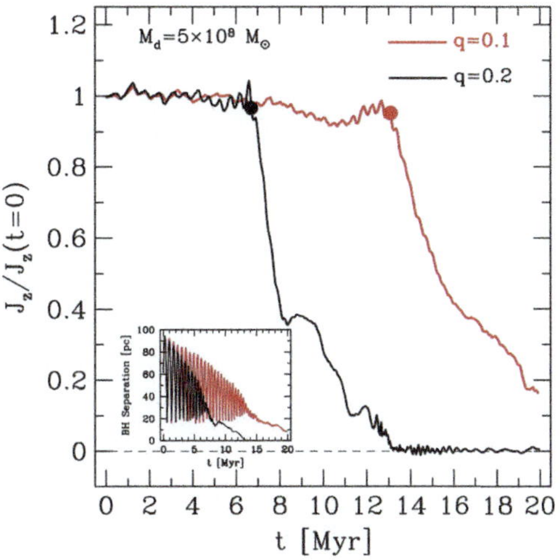

Fig. 8 The evolution of the angular momentum of the secondary black hole orbiting inside a smooth circum-nuclear disc of mass $M_{\rm d} = 5 \times 10^8\ {\rm M}_\odot$. The initial eccentricity is $e_0 = 0.9$. *Red* and *black* colours refer to $q = 0.1$ and $q = 0.2$, respectively ($m_{\bullet,1} = 10^7$ sun in the runs). The *dot* marks the time at which the circularisation of the orbit by dynamical friction is completed. The *inset* shows the black hole separation (in pc) versus time (in Myr). At circularisation, $M_{\rm Mestel}(a)/m_{\bullet,2} \sim 150$ and 75 for the two cases, respectively. Courtesy of D. Fiacconi

to Type I planet migration, but with key differences. While in planet migration, the central star dominates the gravitational potential and the disc's self-gravity is negligible, in disc-driven black hole migration (phase I-g) the disc is dominant, while the gravity of the central black hole is negligible. In Type I migration, the net torque on the planet is the sum of the Lindblad and co-rotating resonances, computed in the linear perturbation theory under the hypothesis that the planet migrates on a timescale much longer than the orbital time, so that the small-amplitude perturbation is periodic in the disc frame. By contrast, during phase I-g of black hole migration, the torque on the secondary black hole comes from the non-linear density perturbations that $m_{\bullet,2}$ excites in the disc. Figure 8 shows the fast orbital decay that the secondary black hole experiences after circularisation (Fiacconi et al. 2013).

To gain some insight into nuclear-driven-disc migration in a Mestel disc, we estimate the migration time following a simple argument to capture key dependences of $\tau_{\rm mig}$ on the disc properties and black hole mass (Armitage 2013). In the two-body approximation, a fluid particle, approaching $m_{\bullet,2}$ along a straight-line with impact parameter b and relative velocity ${\rm v}_{\rm rel} \sim a\delta\Omega \sim b\Omega$ experiences a velocity change parallel to $\mathbf{v}_{\rm rel}$ of the order of $\delta v_{||} \sim 2G^2 m_{\bullet,2}^2/(b^2 {\rm v}_{\rm rel}^3)$, where a denotes the black hole distance from the centre of the disc. As gas exterior to the secondary black hole moves more slowly than the black hole in the disc, it gains velocity parallel to $\mathbf{v}_{\rm rel}$ increasing its angular momentum. As angular momentum is conserved, this implies a *decrease* in the angular momentum per unit gas mass for the black hole equal to $\sim -a\delta v_{||}$. Note that gas interior to the black hole orbit exerts a torque of opposite sign, so that the net torque depends on a delicate balance. As simulations show inward migration and larger torques in the black hole vicinity, we compute the rate of change of angular momentum considering only neighbouring gas particles in the trailing side of the spiral density perturbation, contained in a cylinder of scale height $b \sim h$. Accordingly, the mass flux on $m_{\bullet,2}$ is $\delta m/\delta t \sim\sim 2\pi h \Sigma {\rm v}_{\rm rel} \sim 2\pi h^2 \Sigma \Omega$, where Σ and Ω are evaluated at a. The resulting torque on the black hole can thus be written as $T_I^{\rm Mestel} \sim -4\pi\zeta [m_{\bullet,2}/M_{\rm Mestel}(a)]^2 \Sigma a^4 \Omega^2$, where $\zeta = \zeta'(a/h)^3$ brackets uncertainties in the normalisation of the torque and its dependence on the aspect ratio. As in a Mestel disc the circular velocity is independent of radius a, the black hole sinks from an initial radius $a_{\rm i}$ to a much smaller radius $a_{\rm f}$ on a migration time scale $\tau^I_{\rm mig,Mestel} \sim C\Omega_{\rm i}^{-1}[M_{\rm Mestel}(a_{\rm i})/m_{\bullet,2}]$, where $C = (h/a)^3/(4\zeta')$.[5] We remark that the scaling of $\tau^I_{\rm mig,Mestel}$ with the disc and black hole mass holds true provided $M_{\rm Mestel}(a) > m_{\bullet,1} > m_{\bullet,2}$.

Nuclear discs, stable against fragmentation, are nevertheless ideal. The gas can not be treated as a simple one-phase fluid. Galactic discs are sites of local gravitational instabilities conducive to star formation episodes: massive stars inject energy in the form of winds and supernova blast waves, feeding back energy into the disc and the gas is multi-phase, and clumpy. Thus, it is of importance to understand how the black hole sinking is affected by the inhomogeneous substructure of star forming discs. To this aim, in the next subsection, we explore black hole dynamics in clumpy discs.

4.1.2 Clumpy Circum-Nuclear Discs and Stochastic Orbital Decay

In real astrophysical discs, massive gas clouds coexist with warmer phases and a polytropic equation of state, often used in SPH simulations, provides only an averaged representation of the real thermodynamical state. Cold self-gravitating discs are unstable to fragmentation and attain stability when stars, resulting from the collapse and/or collision of clouds, inject energy in the form of winds and supernova blast waves, feeding back energy into the disc now composed of stars and a multiphase gas. Due to the complexity of implementing this rich physics at the required level of accuracy, a first step ahead is to insert a phenomenological cooling prescription to allow the formation of clumps in a controlled way (Fiacconi et al. 2013). Clumps of size ~ 5 pc in the mass interval between 10^5 $M_\odot$ and 10^7 $M_\odot$ develop in the disc, and evolve as they mass segregate, collide with each other and interact with the secondary black hole. Acting as massive perturbers, they disturb the otherwise smooth black hole orbital decay due to the stochastic behaviour of their torques that are not coherent in time. Several close encounters between $m_{\bullet,2}$ and the massive clumps act as gravitational slingshots, causing an impulsive exchange of orbital energy and angular momentum. Thus, the black hole deviates from its original trajectory either outwards, or inwards or out of the disc plane (above the typical scale height of the disc). When moving on an inclined orbit the black hole experiences the weaker dynamical friction of the stellar background, resulting in a longer orbital decay timescale. The secondary black hole can also be captured by a massive clump forming a pair which segregates rapidly toward the centre. Figure 9 shows the evolution of the gas surface density of a selected run, and the position of the two black holes (marked as white dots) at four different times.

The *stochastic* behaviour of the black hole orbit, resulting from the incoherence of torques, emerges mainly when the clump to black hole mass ratio is $M_{\rm clump}/m_{\bullet,2} \gtrsim 1$. This enlarges the values of the decay time which now range from less than $\lesssim 1$ up to $\gtrsim 50$ Myr. This suggests that describing the cold clumpy phase of the interstellar medium in nuclear discs, albeit so far neglected, is important to predict the black hole dynamics. Ongoing simulations in a multi-phase star forming nuclear disc resulting from the collision of two discs following a merger produce results that are intermediate between the smooth and clumpy case (Lupi et al. in preparation).

[5] The coefficient ζ can be inferred from dedicated numerical experiments. In the case explored, the coefficient $\zeta' \sim 0.04$, to match the sinking time with a simulation. A systematic analysis is necessary to estimate ζ' in a Mestel disc (paper in preparation). Furthermore, the scaling of $\tau^I_{\rm mig,Mestel}$ with the aspect ratio h/a can not be derived from this elementary argument, as discussed in Armitage (2013).

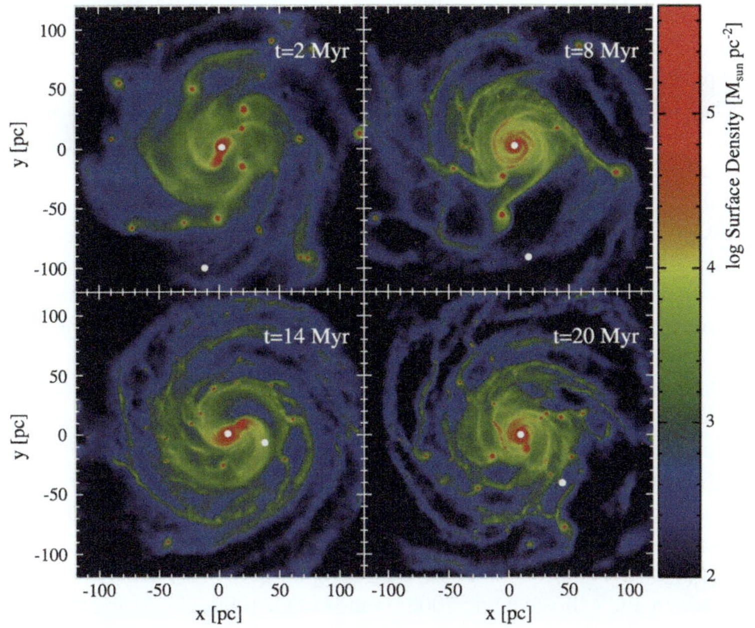

Fig. 9 Colour-coded face-on view of the gas surface density of a clumpy disc model with $M_d = 5 \times 10^8\,M_\odot$, $m_{\bullet,1} = 10^7\,M_\odot$, $q = 0.1$ and initial eccentricity $e = 0.7$, plotted at four different times. The position of the black holes is marked by *white dots*. Courtesy of D. Fiacconi

4.2 Binary-Disc-Driven Migration

In Sect. 4.1.1 we followed the black hole inspiral during phase I-g, in presence of a self-gravitating, rotationally supported disc much heavier than the binary, a condition leading to migration as described in Escala et al. (2005) and Dotti et al. (2006, 2007, 2009). However, with binary decay, the disc mass $M_{\rm Mestel}$ enclosed in the black hole orbit a decreases with time falling below $m_{\bullet,t}$. The black holes then form a Keplerian binary surrounded by a less massive disc, called *circum-binary* disc, dominated by the gravity of the binary and its quadrupolar field.

If we impose continuity in the physical processes, there might exist an intermediate phase whereby migration of the secondary black hole is controlled by *resonant* torques. In close resemblance to Type I planet migration, and for very small binary mass ratios $q \ll 1$, the resulting torque on $m_{\bullet,2}$ is $T_I^{\rm mig,Kep} = -\zeta_K [m_{\bullet,2}/m_{\bullet,1}]^2 \Sigma a^4 \Omega_{\rm K}^2$, where $\Omega_{\rm K}$ is the Keplerian rotational velocity in the gravitational field of $m_{\bullet,1}$ evaluated in a, Σ the disc surface density in the immediate vicinity of $m_{\bullet,2}$, $\zeta_K = (1.36 + 0.54\alpha)(a/h)^2$, and α the slope of the surface density profile $\Sigma \propto a^{-\alpha}$ of the underlying Keplerian disc (Tanaka et al. 2002; Armitage 2013). Notice that because of the natural scaling present in the problem, the torque $T_I^{\rm Kep}$ differs from the expression of $T_I^{\rm Mestel}$ having $m_{\bullet,1}$ in place of $M_{\rm Mestel}(a)$ as reference mass for the gravitational potential. In this case, the migration time reads as $\tau^I_{\rm mig,Kep} \propto [m_{\bullet,1}/m_{\bullet,2}][m_{\bullet,1}/M_{\rm disc}(a)]\Omega_K^{-1}$, under the condition that the disc mass enclosed in the black hole orbit $M_{\rm disc}(a) < m_{\bullet,1}$.

As described in Sect. 3, black hole binaries form preferentially in the aftermath of major mergers so that the binary mass ratio $q \lesssim 1$. Furthermore, accretion drives q to larger and larger values, in the case of minor mergers under specific circumstances (Callegari et al.

2011). Thus, it is quite likely that migration under the action of torques excited by resonances is a missing step, in the dynamical evolution of black hole binaries. At this time the binary is expected to alter profoundly the structure of the underlying disc.

Thus, a key question poses: will the black holes sink down to the domain of gravitational wave inspiral transferring their angular momentum to the disc, or would the binary stall? Is there a phase II-g of migration and under which conditions? This phase is in fact somewhat controversial, as the black hole fate depends on whether the disc is a one-time, short-lived excretion disc (Lodato et al. 2009; Pringle 1991), or an extended long-lived disc (Rafikov 2013). These are conditions that are not recoverable from realistic larger-scale simulations as it is difficult to model the transition from a disc dominated by self-gravity and gravito-turbulence to a disc dominated by magneto-hydrodynamical turbulence stirred by magneto-rotational instabilities in the conducting fluid (Shi et al. 2012).

Unless the binary is surrounded by a warm, geometrically thick disc or envelope and decays promptly (del Valle and Escala 2012, 2014), the tidal force exerted by the binary on the circum-binary disc is expected to eventually clear a cavity. The picture is that the binary transfers orbital angular momentum to the disc by exciting non-axisymmetric density perturbations in the disc body, causing the formation of a low-density, hollow region, called *gap* (Farris et al. 2014; Rafikov 2013; Hayasaki et al. 2013, 2008, 2007; Roedig et al. 2012, 2011; Kocsis et al. 2012; Shi et al. 2012; Cuadra et al. 2009; Hayasaki 2009; Haiman et al. 2009; MacFadyen and Milosavljević 2008; Ivanov et al. 1999; Gould and Rix 2000; Pringle 1991). Viscous torques in the disc oppose gas clearing by the tidal field of the binary and ensure strong binary-disc coupling. Under these conditions, the binary enters a regime of slow orbital decay (referred to as Type II migration in the case of planets; Gould and Rix 2000; Artymowicz and Lubow 1994, 1996; Armitage and Natarajan 2002; Armitage 2013) during which the inner edge of the circum-binary disc compresses in coordination with the hardening of the binary, so that the size $\delta(t)$ of the gap decays remaining close to twice the binary semi-major axis, $\delta \sim 2a(t)$.

Due to the tidal barrier offered by the binary, gas piles up at the inner rim of the disc. One can view the binary as acting as a dam, halting the gas inflow. Accordingly, the accretion rate in the circum-binary disc is not constant in radius and no strict steady state is ever attained in the disc body. In 1D modelling of circum-binary discs, the disc is seen to evolve into a state of constant angular momentum flux (Rafikov 2013), and binary decay leads to a secular and self-similar evolution of the disc as first suggested in Ivanov et al. (1999). This happens when the system loses memory of the natural scale a, set by the size of the cavity and binary orbit, soon after the angular momentum injected by the binary has been transmitted to the larger scale extended disc (Rafikov 2013). Gap opening implies longer hardening time scales compared to nuclear-disc-driven migration, now controlled by the viscous time at the inner disc edge.

The migration time can be estimated as $\tau^{II}_{\rm mig} \sim \tau_\nu[(m_{\bullet,2} + M_{\rm d}^{\rm edge})/M_{\rm d}^{\rm edge}]$ where $M_{\rm d}^{\rm edge} \sim \Sigma(R)R^2$ is the local mass near the inner edge of the disc, at $R \sim 2a$-$3a$ where the surface density has a peak, and τ_ν the disc viscous time there: $\tau_\nu \sim (2/3)R^2/\nu \sim 2\pi R^2\Sigma/\dot{M}$ where $\dot{M} \sim 3\pi\nu\Sigma$ is the mass accretion rate in an unperturbed reference disc (Shakura and Sunyaev 1973). When $M_{\rm d}^{\rm edge} > m_{\bullet,2}$, the secondary black hole behaves as a parcel in the viscous disc and migrates on the viscous timescale, whereas when $M_{\rm d}^{\rm edge} < m_{\bullet,2}$ migration slows down and occurs of a timescale longer than τ_ν. Condition $M_{\rm d}^{\rm edge} < m_{\bullet,2}$ is often referred to as secondary-dominated Type II migration (Haiman et al. 2009; Syer and Clarke 1995). The opposite regime is referred to as disc-dominated Type II migration.

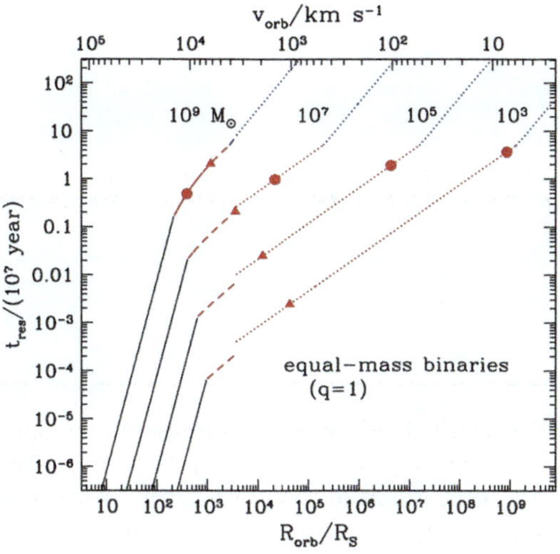

Fig. 10 Residence time $|a/\dot{a}|$ of equal-mass black hole binaries, embedded in a steady circum-binary disc, as a function of the black hole separation (in units of $2Gm_{\bullet,t}/c^2$), as computed in Haiman et al. (2009) for a reference disc model. The top x-axis label refers to the Keplerian relative orbital velocity of the black holes in the binaries. The *four curves* correspond to binaries with total masses of $m_{\bullet,t} = 10^3$, 10^5, 10^7 and 10^9 $M_\odot$ as labeled. The *large dots* denote the critical radius beyond which the assumed circum-binary Keplerian disc is unstable to fragmentation. Similarly, *triangles* denote radii beyond which the disc may be susceptible to ionisation instabilities (the gas temperature falls below 10^4 K). In each case, *blue/red* colors indicate whether the disc mass enclosed within the binary's orbit is larger/smaller than the black hole mass $m_{\bullet,2}$. The *dotted/dashed/solid portion* of each curve indicates the outer/middle/inner disc region. Note that in the disc-dominated regime (*blue* segments) the binary residence time is $\sim 10^9$ yrs, while it decreases below $\sim 10^7$ yrs for all binaries, i.e. independent of their mass, at the entrance in the stable region of a circum-binary disc (*red dots*). Courtesy of Haiman et al. (2009)

The timescale $\tau^{II}_{\rm mig}$ can be recovered if one assumes a torque on the binary of the form $T^{\rm mig}_{II} \sim -\xi j_o \dot{M} \sim \xi' j_o M^{\rm edge}_{\rm d} \Omega_K$ where $j_o = (\mu/m_{\bullet,t})(Gm_{\bullet,t}a)^{1/2}$ is the binary angular momentum per unit mass, and ξ or ξ' determined by numerical simulations, e.g. MacFadyen and Milosavljević (2008), Roedig et al. (2012) and Shi et al. (2012). The above expression for the torque relates the rate of binary orbital decay to the local disc mass $M^{\rm edge}_{\rm d}$ near the inner edge of the disc. Therefore, depending on how $M^{\rm edge}_{\rm d}$ varies over the relevant timescales, i.e. whether the disc is continuously re-filled of gas to keep $M^{\rm edge}_{\rm d}$ nearly stationary, or the disc mass is consumed before the binary has evolved substantially, orbital decay accelerates or decelerates, returning the problem to the old, outstanding issue on whether black holes are continuously fed in galactic nuclei or not, and on which timescale.

Semi-analytical expressions of the migration time have been derived in Haiman et al. (2009) considering orbital decay within a Shakura & Sunyaev accretion disc (Shakura and Sunyaev 1973). This enabled the authors to evaluate the disc surface density, opacity, viscosity and ultimately $M^{\rm edge}_{\rm d}$ as the binary transits through the outer/middle and inner zones of the disc. Under these simplifying assumptions (of a steady unperturbed 1D disc), Haiman et al. (2009) have shown that the sinking time of the binary is a *monotonic decreasing function of the binary orbital period* (or separation). The residence time $t_{\rm res} \sim |a/\dot{a}|$ for equal-mass binaries (which is in this context close to $\tau^{II}_{\rm mig}$) is plotted in Fig. 10 from Haiman et al. (2009) for a disc with α viscosity parameter equal to 0.3, a radiative efficiency of 0.1 and an accretion rate equal to 0.1 of the Eddington value. In the disc-dominated regime when

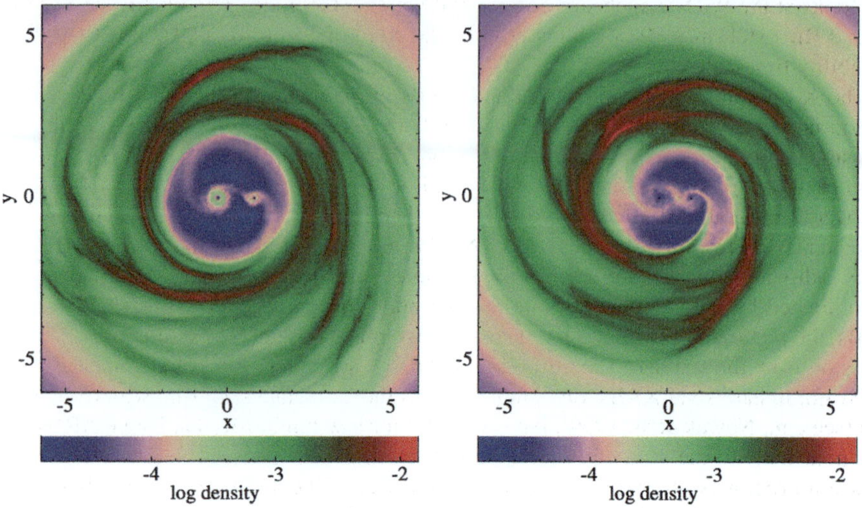

Fig. 11 Color-coded gas surface density of two Newtonian, self-gravitating circum-binary discs, showing the presence of a binary region with the two black holes and their mini-discs, a porous cavity filled with streams, the inner rim or edge working as a dam, and the body disc. *Left* (*right*) panel refers to a run with gas in the cavity treated with an isothermal (adiabatic) equation of state. Courtesy of Roedig et al. (2012)

$M_{\rm d}^{\rm edge} > m_{\bullet,2}$ the migration timescale is of the order of $\sim$ Gyr and when $M_{\rm d}^{\rm edge} < m_{\bullet,2}$ it drops below 10^7 yrs, showing weak dependence of the binary mass. Similar timescales have also been found in Rafikov (2013) when considering 1D disc models undergoing self-similar evolution (see also Fig. 6 of Haiman et al. 2009). Despite these studies, we are nonetheless far from having a reliable estimate of the migration timescale in circum-binary discs under a variety of conditions, given the rich physics involved.[6]

Gap opening and/or maintenence of the inner cavity around massive black holes have been seen in numerous numerical simulations of both Keplerian and self-gravitating circum-binary discs (MacFadyen and Milosavljević 2008; Shi et al. 2012; Cuadra et al. 2009; Roedig et al. 2011; del Valle and Escala 2012). But interestingly, recent 2D and 3D simulations have demonstrated that the binary+disc system contains as many as three discs and that these discs may persist being constantly fed by gas flowing through the gap. The three discs comprise the circum-binary disc plus two mini-discs around each member of the binary (Farris et al. 2014; Shi et al. 2012; Roedig et al. 2012, 2011). This is due to the fact that the disc inner edge is porous (for sufficiently high disc aspect ratios): high velocity, narrow streams of gas leak periodically through the dam into the inner cavity, modulated by the binary orbit (Roedig et al. 2012; Shi et al. 2012; Noble et al. 2012; D'Orazio et al. 2013; Farris et al. 2014).

Figure 11 shows the distribution of gas around the black hole binary after gap formation, from two SPH-3D simulations of Newtonian, massive circum-binary discs (Roedig et al. 2012) which differ from one anther due to a different thermodynamic modelling of the

[6]As an example, in recent studies of planet migration by Duffell et al. (2014) it has been shown, using highly accurate numerical calculations, that the actual migration rate is dependent on disc and planet parameters, and can be significantly larger or smaller than the viscous drift rate τ_ν^{-1}. In the case of disc-dominated migration the rate saturates to a constant value which is in excess of the viscous rate while in the opposite regime of a low-mass disc, the migration rate decreases linearly with disc mass.

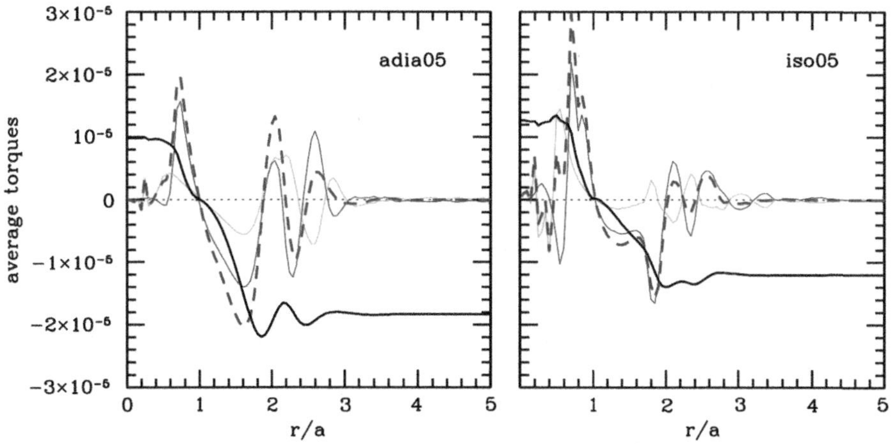

Fig. 12 Differential torques dT/dR and integrated torque T (averaged over the time span of the simulation, and in code units) exerted by the disc on the binary with mass ratio $q = 0.1$ as a function of the radial distance, in units of the binary separation a, for the adiabatic (*left*) and isothermal (*right*) run from Roedig et al. (2012). In each panel, the differential torque acting on the primary is plotted in *green*, on the secondary in *red*, and the sum of the two in *blue*. Notice that the torque density dT/dR shows different signs and starts oscillating around the zero point at distances far from the binary where the binary-disc coupling decreases sharply. The *black line* refers to the integrated torque T up to a distance R: T is positive inside a, and negative outside giving a total negative contribution. Courtesy of Roedig et al. (2012)

gaseous streams in the cavity: isothermal (on the right side) and adiabatic (on the left side). The figure highlights the occurrence of different domains in the disc (from outside in): the *disc body* $R > 2.5a$ where spiral patterns develop; the *cavity edge*, at radii $2 < R < 2.5a$, which is porous and leaky; the *cavity region* or gap, between $a < R < 2a$, which is almost devoid of gas except for the presence of tenuous streams; the *binary* region, at $0 < R < a$, with the two black holes and their mini-discs, fed by gas from the disc body flowing through the cavity across the porous dam. The mini-discs and the cavity are sharper in the isothermal case compared to the adiabatic case where the amount of gas impacting the gap is larger. Only a fraction of this gas is captured by the black holes to form the mini-discs, the remaining being swiftly ejected away. The different regions highlighted in Fig. 11 contribute to the differential torque dT/dR on the binary with different signs as illustrated in Fig. 12, for the case of a Newtonian self-gravitating disc (in the adiabatic and isothermal model, respectively). One can notice that the differential torque shows an oscillatory behaviour with a sharp maximum at the location of the secondary black hole ($R \sim 0.75a$), and a deep minimum in the cavity region. Positive and negative peaks alternate in the disc body that almost cancel out, giving a negligible contribution to the total torque. Torques on the secondary black hole are always larger than on the primary, due to its proximity to the inner rim of the disc, resulting in a stronger interaction.

In summary, simulations now indicate that clearing a cavity in the disc does not prevent the inflow of gas through streams across the cavity's edge. Thus accretion of a fraction of this gas on the black holes, and preferentially onto the secondary (nearer to the disc's edge) may be a persistent feature (Farris et al. 2014). Thus binary evolution becomes more complex than outlined in the first part of this section. All binary elements evolve over time and in some cases inward migration can turn into outward migration (Hayasaki 2009; Roedig et al. 2012). The evolution equation for the semi-major axis of a binary depends on changes in the eccentricity, mass, reduced mass and on the exchange of angular momentum between the binary and the three discs through a generalised T. All these contribute to $\dot{a}/a = 2T/J_\bullet - \dot{m}_{\bullet,t}/m_{\bullet,t} - 2\dot{\mu}/\mu + 2e\dot{e}/(1-e^2)$ where $J_\bullet$ is the binary angular momentum. The sign of this derivative thus depends on different effects.

The binary eccentricity tends to increase during the binary-disc coupling (Armitage and Natarajan 2005; MacFadyen and Milosavljević 2008), and the growth of e has also been seen in 3D numerical simulations (Roedig et al. 2011). Progress in the analysis of this process has revealed that such excitation can not grow indefinitely, as *saturation* occurs due to the interaction of the secondary black hole with gas near the inner rim of the disc body, and to the accumulation of gas around the black holes in the mini-discs (Roedig et al. 2011). The initial rise of e can be understood, as in Sect. 4.1.1, using dynamical friction in a differentially rotating background as leading argument. The secondary black hole, closer to the circum-binary disc, induces a trailing density wave near the inner rim which reduce its tangential velocity, causing a loss of orbital angular momentum. The eccentricity e grows and continues to grow as long as the gas at the inner edge of the circum-binary disc moves with a lower angular velocity. However the progressive decay of the black hole tangential velocity with increasing e leads eventually to a reversal of the sign of the relative velocities, the gas moving faster than the black holes, thus developing a wake heading in front which leads to an acceleration of the black hole. The process reaches saturation, and this is found to occur about $e \sim 0.6$–0.8. Furthermore, when the binary becomes very eccentric, the secondary, less massive black hole passes through the mini-disc of the primary suffering a deceleration at peri-centre which in turn decreases e, which then attains a saturation value.

The mass of the two black holes tend to increase as well, and the increase of the mass of the secondary black hole is even higher, being closer to the disc, thus driving q toward unity (Farris et al. 2014). The mass accretion rate is not severely limited (compared to the case of a single isolated black hole) and $\dot{M}$ is found to be modulated at the binary orbital period and higher harmonics (Farris et al. 2014; Roedig et al. 2011). Interestingly, modulated accretion suggests a promising avenue for producing a modulated electromagnetic signal permitting the identification of binaries during migration in circum-binary discs at different orbital phases along the path to coalescence (Eracleous et al. 2011; Decarli et al. 2013; Montuori et al. 2012).

5 Timescales: an Overview

Galaxy interactions and mergers are the sites of formation of dual, binary, coalescing and recoiling black holes. Associated to these different dynamical phases there is a zoo of sources, the dual, binary and recoiling candidate AGN that are now being discovered (Komossa 2012, 2006; Colpi and Dotti 2011; Schnittman 2011; Centrella et al. 2010; Bode et al. 2010; Eracleous et al. 2011; Decarli et al. 2013; Lusso et al. 2014). Their persistence in all these different phases is determined by their residence time, given by one of these scales: during phase I, the dynamical friction timescale $\tau_{\rm df}$, and/or the dynamical friction timescale which accounts for tidal mass loss $\tau_{\rm df,tidal}$; in phase II, the hardening time in a stellar background $\tau_{\rm H}$ (which falls in the interval between $\tau^*_{\rm hard}$ and $\tau_{\rm rel}$), and/or the timescales in the two regimes of nuclear-disc-driven (phase I-g) and binary-disc-driven migration (phase II-g) $\tau^{I}_{\rm mig,Mestel}$, and $\tau^{II}_{\rm mig}$; and ultimately the gravitational wave timescale $\tau_{\rm gw}$.

There is no simple recipe to calculate the residence times in terms of fundamental parameters such as the black hole mass and mass ratio since these timescales depend on the morphology of the interacting galaxies, the geometry of the encounter, the gas fraction, and most importantly the complex input physics of difficult implementation even in current state-of-the-art simulations.

The characteristic coalescence time $\tau_{\rm coal}$ would be the sum of the timescales associated to the different phases (I, II or I-g, II-g, and III), calculated along each individual pathway. Their value depends, even in the minimal model, on whether the merger is gas-poor (dry) or gas-rich (wet), and major or minor. As no unique pathway exists for a pair, $\tau_{\rm coal}$ can be estimated simply considering the maximum of all residence times. This timescale should then be compared with the Hubble time or better with the running age of the universe at the time of coalescence, given that eLISA sources are typically at high redshifts (Amaro-Seoane et al. 2013).

Here is a tentative summary of the timescales inferred from the whole body of works, in the black hole mass range $\lesssim 10^7$ M$_\odot$ and for values of the initial black hole mass ratio q (which indicates the mass ratio between the two interacting galaxies). If the "zero" time is calculated when the merger of the baryonic components (bulge and disc) is completed, i.e. when the black holes behave as individual objects moving in the relic galaxy, the relevant timescales in different environments and conditions are expected to cluster approximately around these values:

- In dry major mergers ($q > q_{\rm crit} \sim 0.1$): (i) dynamical friction time $\tau_{\rm df} \lesssim$ [10 Myr–100 Myr] (Yu 2002)—(ii) hardening timescale $\tau^*_{\rm hard} \sim 1$ Gyr to a few Gyr (Khan et al. 2011).
- In wet major mergers ($q > q_{\rm crit} \sim 0.1$): (i) gas-dynamical friction time $\tau_{\rm df} \lesssim$ [10 Myr–100 Myr] (Mayer et al. 2007; Chapon et al. 2013; Roškar et al. 2014)—(ii) nuclear-disc-driven migration time $\tau^{I}_{\rm mig,Mestel} \sim$ (5 Myr–50 Myr) (Escala et al. 2005; Dotti et al. 2006, 2007, 2009; Fiacconi et al. 2013)—(iii) binary-disc-driven migration $\tau^{II}_{\rm mig} \sim 10$ Myr (Haiman et al. 2009).
- In dry minor mergers ($q < q_{\rm crit} \sim 0.1$): (i) dynamical friction time $\tau_{\rm df} \lesssim$ [10 Myr–100 Myr] (Yu 2002)—(ii) hardening timescale $\tau^*_{\rm hard} \sim 1$ Gyr up to a few Gyr (Yu 2002; Khan et al. 2012).
- In wet minor mergers ($q < q_{\rm crit} \sim 0.1$): (i) dynamical friction time with corrections due to tidal stripping $\tau_{\rm df,tidal} \lesssim 100$ Myr or wandering (Callegari et al. 2009, 2011). The fate is uncertain. Depending on the geometry of the encounter and gas fraction, the secondary black hole may wander in the primary galaxy.

6 Summary and Future Prospects

The study of the dynamics of black holes, with masses from 10^4 M$_\odot$ up to 10^9 M$_\odot$, inside galaxies displaying a large variety of morphologies and masses, is not a side problem: it is central if we want to search for or recognise signs of their duality and/or coalescence at electromagnetic level, and if we want to detect the gravitational waves emitted at the time of black hole coalescence. There has been some advances in the study of the dynamics of black holes in merging galaxies, over the last years, and the points to remember and to take away for future reference are:

1. Black holes in binaries can reach coalescence under the emission of gravitational waves. But, for this to happen, the black holes have to be driven to separations as small as $\sim 10^{-3}$ pc or less, as gravity is a weak force and gravitational waves are a manifestation of the strong field regime (Sathyaprakash and Schutz 2009). This is a minuscule distance, compared to galaxy's sizes, and merging galaxies are the sites where these events can occur. Nature has thus to provide a series of mechanisms able to extract energy and

angular momentum, from the large scale of the merger (at least a few kpc) to the micro-parsec scale, i.e. the scale at which the black hole horizons touch. The path to coalescence is long and complex, and stalling of the binary at some scale is a possibility.

2. Three phases accompany the path to coalescence: the pairing, hardening, and gravitational-wave driven inspiral phases. Stars or gas, or stars and gas drive the black hole inspiral, depending on whether galaxies are gas-rich or gas-poor. Bottlenecks can appear at various scales and a major effort is to identify possible obstacles. The last parsec problem, i.e. the stalling of a massive black hole binary at the centre of a large spherical collisionless galaxy was highlighted as a critical step.
3. Thanks to recent advances in numerical computing, the last parsec problem appears to be an artefact of oversimplifying assumptions. Galaxies, relic of mergers, are not spherical systems and can retain a high degree of triaxiality or asymmetry. Under these circumstances the hardening of the binary via binary-single stellar encounters has no halt, at least in the cases explored and coalescence timescales are close to 1 to a few Gyr.
4. It is now possible to track the black hole dynamics during galaxy collisions using state-of-the-art simulations. The dynamics of the interacting galaxies is followed *ab initio*, from the large scale (several kpc) down to the central few parsecs, considering all components—the dark halo, the stellar and gaseous disc, and bulge. This enables us to trace the rise of asymmetries and instabilities in both the stellar and gas components which play a pivotal role in determining whether there is stalling or rapid sinking of the binary.
5. Major gas-rich (wet) mergers are conducive to the formation of close Keplerian binaries. The gas, thanks to its high degree of dissipation, controls the black hole inspiral inside the massive circum-nuclear disc that forms in the end-galaxy. When described as a single phase medium, the gas promotes rapid inspiral before the disc fragments into stars, and before stellar dynamical friction becomes effective. When the gas is multi-phase and clumpy on various mass scales, the black hole orbit shows a stochastic behaviour. The black holes in this case form a binary on timescales typically between 1 Myr and 100 Myr.
6. Coalescences of middleweight black holes of $\sim 10^4\ M_\odot$ at high redshifts $z \sim 10$–15 require either very dense, low velocity dispersion stellar environments, or large not yet quantified amounts of gas in unstable forming galaxies.
7. Minor mergers can release the less massive black hole on peripheral orbits in the main galaxy, due to the disruptive action of tidal torques on the less massive galaxy, rising a new problem: the last kpc problem. Gas plays a key role in the process of pairing in minor merger, as it makes the satellite galaxy more resilient against tidal stripping because central gas inflows, triggered during the interaction, steepen the stellar cusp. Due to the fragility of the satellite galaxy, the fate of black holes in minor mergers is uncertain: encounter geometry, gas fraction, degree of gas dissipation are key elements for establishing whether the black hole is a *sinking* or a *wandering* black hole inside the primary galaxy. The prediction is that there is a large scatter in the outcomes.
8. At sub-parsec scales, the Keplerian binary is likely surrounded by a circum-binary disc. When present, gas-assisted inspiral takes place which can be faster than star-driven inspiral. Both processes likely co-exist but have never been treated jointly. The black holes are expected to migrate on a timescale controlled by the interplay between the binary tidal torque which tends to clear a cavity (repelling the disc's gas in the immediate vicinity of the binary) and viscous torques in the disc which tend to fill and even overfill the cavity. Gas leaks through the gap and the black holes are surrounded by mini accretion discs. All these processes modify the orbital elements in a complex way. Theoretical

models indicate that if there is a sufficiently long-lived inflow of gas at the inner edge of the circum-binary disc, the binary hardens on timescales $\lesssim 10^9$ Gyr or even much less (depending on the previous history).
9. The path to coalescence still remains a complex problem to solve and there is no clear-cut answer. To be conservative, coalescence times range between several Myrs to about a Gyr.

The field needs to evolve farther along different lines and directions. Here are a few hints:

1. There is need to continue to study not only the growth of black holes along cosmic history, but their dynamics as they are inter-connected. Attempts to follow the dynamics of black holes during the cosmic assembly of galactic halos have been carried on, albeit at much lower revolution than required to track their accretion history and fate (Bellovary et al. 2010). Any effort along this line is central in order to understand the coevolution of black holes with galaxies.
2. Intriguingly enough, the fate of black hole binaries in galaxies takes us back to the unsolved problem of the *feeding* of black holes in galactic nuclei over cosmic ages, of the angular momentum barrier present on parsec scales, and on as to whether star formation goes hand in hand with black hole feeding or it anticipates or lags. Dynamical decay, star formation, accretion and their back-reactions are coupled. Star formation makes the ISM multi-phase and turbulent. Supernovae and AGN feed-back may heat/remove gas and the consequences of these effects on black hole migration have not been quantified yet during orbital evolution.
3. Dual, binary and recoiling AGN and also triple AGN are important observational targets. There is the need to improve upon observational strategies for identifying binary and recoiling AGN in large surveys, assisted by tailored and coordinated hydro-dynamical simulations.
4. Black hole migration in circum-binary disc is a challenging problem which deserves constant attention. Binary eccentricity growth, accretion and outflows, are processes that affect the dynamics and stability of the system as a whole. Understanding the nature of torques will become central in order to extrapolate the black hole migration timescale down to the domain controlled by gravitational wave inspiral, and to asses the observability of sub-parsec binaries during the phases which anticipate the merging.

Acknowledgements I would like to thank my collaborators Simone Callegari, Massimo Dotti, Davide Fiacconi, Lucio Mayer, Constanze Roedig, Alberto Sesana and Marta Volonteri for many useful and illuminating discussions over the years. I would like also to thank the International Space Science Institute for kind hospitality.

References

P. Amaro-Seoane, S. Aoudia, S. Babak, P. Binétruy, E. Berti, A. Bohé, C. Caprini, M. Colpi, N.J. Cornish, K. Danzmann, J.-F. Dufaux, J. Gair, I. Hinder, O. Jennrich, P. Jetzer, A. Klein, R.N. Lang, A. Lobo, T. Littenberg, S.T. McWilliams, G. Nelemans, A. Petiteau, E.K. Porter, B.F. Schutz, A. Sesana, R. Stebbins, T. Sumner, M. Vallisneri, S. Vitale, M. Volonteri, H. Ward, B. Wardell, eLISA: Astrophysics and cosmology in the millihertz regime. GW Notes, vol. 6, (2013) pp. 4–110, 6:4–110

P.J. Armitage, *Astrophysics of Planet Formation* (2013)

P.J. Armitage, P. Natarajan, Accretion during the merger of supermassive black holes. Astrophys. J. Lett. **567**, L9–L12 (2002)

P.J. Armitage, P. Natarajan, Eccentricity of supermassive black hole binaries coalescing from gas-rich mergers. Astrophys. J. **634**, 921–927 (2005)

P. Artymowicz, S.H. Lubow, Dynamics of binary-disk interaction. 1: Resonances and disk gap sizes. Astrophys. J. **421**, 651–667 (1994)
P. Artymowicz, S.H. Lubow, Mass flow through gaps in circumbinary disks. Astrophys. J. Lett. **467**, L77 (1996)
M.C. Begelman, R.D. Blandford, M.J. Rees, Massive black hole binaries in active galactic nuclei. Nature **287**, 307–309 (1980)
J.M. Bellovary, F. Governato, T.R. Quinn, J. Wadsley, S. Shen, M. Volonteri, Wandering black holes in bright disk galaxy halos. Astrophys. J. Lett. **721**, L148–L152 (2010)
P. Berczik, D. Merritt, R. Spurzem, H.-P. Bischof, Efficient merger of binary supermassive black holes in nonaxisymmetric galaxies. Astrophys. J. Lett. **642**, L21–L24 (2006)
O. Blaes, M.H. Lee, A. Socrates, The Kozai mechanism and the evolution of binary supermassive black holes. Astrophys. J. **578**, 775–786 (2002)
T. Bode, R. Haas, T. Bogdanović, P. Laguna, D. Shoemaker, Relativistic mergers of supermassive black holes and their electromagnetic signatures. Astrophys. J. **715**, 1117–1131 (2010)
T. Bogdanović, C.S. Reynolds, M.C. Miller, Alignment of the spins of supermassive black holes prior to coalescence. Astrophys. J. Lett. **661**, L147–L150 (2007)
M. Boylan-Kolchin, C.-P. Ma, E. Quataert, Dynamical friction and galaxy merging time-scales. Mon. Not. R. Astron. Soc. **383**, 93–101 (2008)
S. Callegari, S. Kazantzidis, L. Mayer, M. Colpi, J.M. Bellovary, T. Quinn, J. Wadsley, Growing massive black hole pairs in minor mergers of disk galaxies. Astrophys. J. **729**, 85 (2011)
S. Callegari, L. Mayer, S. Kazantzidis, M. Colpi, F. Governato, T. Quinn, J. Wadsley, Pairing of supermassive black holes in unequal-mass galaxy mergers. Astrophys. J. Lett. **696**, L89–L92 (2009)
J. Centrella, J.G. Baker, B.J. Kelly, J.R. van Meter, Black-hole binaries, gravitational waves, and numerical relativity. Rev. Mod. Phys. **82**, 3069–3119 (2010)
S. Chandrasekhar, Dynamical friction. I. General considerations: the coefficient of dynamical friction. Astrophys. J. **97**, 255 (1943)
D. Chapon, L. Mayer, R. Teyssier, Hydrodynamics of galaxy mergers with supermassive black holes: is there a last parsec problem? Mon. Not. R. Astron. Soc. **429**, 3114–3122 (2013)
M. Colpi, S. Callegari, M. Dotti, L. Mayer, Massive black hole binary evolution in gas-rich mergers. Class. Quantum Gravity **26**(9), 094029 (2009)
M. Colpi, M. Dotti, Massive binary black holes in the cosmic landscape. Adv. Sci. Lett. **4**, 181–203 (2011)
M. Colpi, L. Mayer, F. Governato, Dynamical friction and the evolution of satellites in virialized halos: the theory of linear response. Astrophys. J. **525**, 720–733 (1999)
J. Cuadra, P.J. Armitage, R.D. Alexander, M.C. Begelman, Massive black hole binary mergers within subparsec scale gas discs. Mon. Not. R. Astron. Soc. **393**, 1423–1432 (2009)
R. Decarli, M. Dotti, M. Fumagalli, P. Tsalmantza, C. Montuori, E. Lusso, D.W. Hogg, J.X. Prochaska, The nature of massive black hole binary candidates—I. Spectral properties and evolution. Mon. Not. R. Astron. Soc. **433**, 1492–1504 (2013)
L. del Valle, A. Escala, Binary-disk interaction: gap-opening criteria. Astrophys. J. **761**, 31 (2012)
L. del Valle, A. Escala, Binary-disk interaction. II. Gap-opening criteria for unequal-mass binaries. Astrophys. J. **780**, 84 (2014)
B. Devecchi, E. Rasia, M. Dotti, M. Volonteri, M. Colpi, Imprints of recoiling massive black holes on the hot gas of early-type galaxies. Mon. Not. R. Astron. Soc. **394**, 633–640 (2009)
T. Di Matteo, V. Springel, L. Hernquist, Energy input from quasars regulates the growth and activity of black holes and their host galaxies. Nature **433**, 604–607 (2005)
D.J. D'Orazio, Z. Haiman, A. MacFadyen, Accretion into the central cavity of a circumbinary disc. Mon. Not. R. Astron. Soc. **436**, 2997–3020 (2013)
M. Dotti, M. Colpi, F. Haardt, Laser interferometer space antenna double black holes: dynamics in gaseous nuclear discs. Mon. Not. R. Astron. Soc. **367**, 103–112 (2006)
M. Dotti, M. Colpi, F. Haardt, L. Mayer, Supermassive black hole binaries in gaseous and stellar circumnuclear discs: orbital dynamics and gas accretion. Mon. Not. R. Astron. Soc. **379**, 956–962 (2007)
M. Dotti, M. Ruszkowski, L. Paredi, M. Colpi, M. Volonteri, F. Haardt, Dual black holes in merger remnants—I. Linking accretion to dynamics. Mon. Not. R. Astron. Soc. **396**, 1640–1646 (2009)
M. Dotti, A. Sesana, R. Decarli, Massive Black Hole Binaries: Dynamical Evolution and Observational Signatures. Adv. Astron. **2012** (2012)
M. Dotti, M. Volonteri, A. Perego, M. Colpi, M. Ruszkowski, F. Haardt, Dual black holes in merger remnants—II. Spin evolution and gravitational recoil. Mon. Not. R. Astron. Soc. **402**, 682–690 (2010)
P.C. Duffell, Z. Haiman, A.I. MacFadyen, D.J. D'Orazio, B.D. Farris, Type II Migration is not Locked to Viscous Disk Evolution. ArXiv e-prints (2014)
eLISA Consortium, The Gravitational Universe, the science theme selected by ESA as L3 mission. ArXiv e-prints (2013)

M. Eracleous, T.A. Boroson, J.P. Halpern, J. Liu, A Large Systematic Search for Recoiling and Close Supermassive Binary Black Holes. ArXiv e-prints (2011)

A. Escala, R.B. Larson, P.S. Coppi, D. Mardones, The role of gas in the merging of massive black holes in galactic nuclei. II. Black hole merging in a nuclear gas disk. Astrophys. J. **630**, 152–166 (2005)

B.D. Farris, P. Duffell, A.I. MacFadyen, Z. Haiman, Binary black hole accretion from a circumbinary disk: gas dynamics inside the central cavity. Astrophys. J. **783**, 134 (2014)

L. Ferrarese, P. Côté, E. Dalla Bontà, E.W. Peng, D. Merritt, A. Jordán, J.P. Blakeslee, M. Haşegan, S. Mei, S. Piatek, J.L. Tonry, M.J. West, A fundamental relation between compact stellar nuclei, supermassive black holes, and their host galaxies. Astrophys. J. Lett. **644**, L21–L24 (2006)

L. Ferrarese, H. Ford, Supermassive black holes in galactic nuclei: past, present and future research. Space Sci. Rev. **116**, 523–624 (2005)

D. Fiacconi, L. Mayer, R. Roškar, M. Colpi, Massive black hole pairs in clumpy, self-gravitating circumnuclear disks: stochastic orbital decay. Astrophys. J. Lett. **777**, L14 (2013)

D. Gerosa, A. Sesana, Missing black holes in brightest cluster galaxies as evidence for the occurrence of superkicks in nature. ArXiv e-prints (2014)

A.M. Ghez, S. Salim, N.N. Weinberg, J.R. Lu, T. Do, J.K. Dunn, K. Matthews, M.R. Morris, S. Yelda, E.E. Becklin, T. Kremenek, M. Milosavljevic, J. Naiman, Measuring distance and properties of the Milky Way's central supermassive black hole with stellar orbits. Astrophys. J. **689**, 1044–1062 (2008)

S. Gillessen, F. Eisenhauer, S. Trippe, T. Alexander, R. Genzel, F. Martins, T. Ott, Monitoring stellar orbits around the massive black hole in the galactic center. Astrophys. J. **692**, 1075–1109 (2009)

A. Gould, H.-W. Rix, Binary black hole mergers from planet-like migrations. Astrophys. J. Lett. **532**, L29–L32 (2000)

F. Governato, M. Colpi, L. Maraschi, The fate of central black holes in merging galaxies. Mon. Not. R. Astron. Soc. **271**, 317 (1994)

A. Gualandris, D. Merritt, Ejection of supermassive black holes from galaxy cores. Astrophys. J. **678**, 780–797 (2008)

K. Gültekin, D.O. Richstone, K. Gebhardt, T.R. Lauer, S. Tremaine, M.C. Aller, R. Bender, A. Dressler, S.M. Faber, A.V. Filippenko, R. Green, L.C. Ho, J. Kormendy, J. Magorrian, J. Pinkney, C. Siopis, The M–σ and M–L relations in galactic bulges, and determinations of their intrinsic scatter. Astrophys. J. **698**, 198–221 (2009)

Z. Haiman, B. Kocsis, K. Menou, The population of viscosity- and gravitational wave-driven supermassive black hole binaries among luminous active galactic nuclei. Astrophys. J. **700**, 1952–1969 (2009)

N. Häring, H.-W. Rix, On the black hole mass-bulge mass relation. Astrophys. J. Lett. **604**, L89–L92 (2004)

K. Hayasaki, A new mechanism for massive binary black-hole evolution. Publ. Astron. Soc. Jpn. **61**, 65 (2009)

K. Hayasaki, S. Mineshige, L.C. Ho, A supermassive binary black hole with triple disks. Astrophys. J. **682**, 1134–1140 (2008)

K. Hayasaki, S. Mineshige, H. Sudou, Binary black hole accretion flows in merged galactic nuclei. Publ. Astron. Soc. Jpn. **59**, 427–441 (2007)

K. Hayasaki, H. Saito, S. Mineshige, Binary black hole accretion flows from a misaligned circumbinary disk. Publ. Astron. Soc. Jpn. **65**, 86 (2013)

A. Heger, C.L. Fryer, S.E. Woosley, N. Langer, D.H. Hartmann, How massive single stars end their life. Astrophys. J. **591**, 288–300 (2003)

L. Hoffman, A. Loeb, Dynamics of triple black hole systems in hierarchically merging massive galaxies. Mon. Not. R. Astron. Soc. **377**, 957–976 (2007)

P.F. Hopkins, T.J. Cox, L. Hernquist, D. Narayanan, C.C. Hayward, N. Murray, Star formation in galaxy mergers with realistic models of stellar feedback and the interstellar medium. Mon. Not. R. Astron. Soc. **430**, 1901–1927 (2013)

P.F. Hopkins, L. Hernquist, T.J. Cox, T. Di Matteo, B. Robertson, V.Springel.A. Unified, Merger-driven model of the origin of starbursts, quasars, the cosmic X-Ray background, supermassive black holes, and galaxy spheroids. Astrophys. J. Suppl. Ser. **163**, 1–49 (2006)

P.B. Ivanov, J.C.B. Papaloizou, A.G. Polnarev, The evolution of a supermassive binary caused by an accretion disc. Mon. Not. R. Astron. Soc. **307**, 79–90 (1999)

S. Kazantzidis, L. Mayer, M. Colpi, P. Madau, V.P. Debattista, J. Wadsley, J. Stadel, T. Quinn, B. Moore, The fate of supermassive black holes and the evolution of the M_{BH}–σ relation in merging galaxies: the effect of gaseous dissipation. Astrophys. J. Lett. **623**, L67–L70 (2005)

F.M. Khan, I. Berentzen, P. Berczik, A. Just, L. Mayer, K. Nitadori, S. Callegari, Formation and hardening of supermassive black hole binaries in minor mergers of disk galaxies. Astrophys. J. **756**, 30 (2012)

F.M. Khan, K. Holley-Bockelmann, P. Berczik, A. Just, Supermassive black hole binary evolution in axisymmetric galaxies: the final parsec problem is not a problem. Astrophys. J. **773**, 100 (2013)

F.M. Khan, A. Just, D. Merritt, Efficient merger of binary supermassive black holes in merging galaxies. Astrophys. J. **732**, 89 (2011)
F.M. Khan, M. Preto, P. Berczik, I. Berentzen, A. Just, R. Spurzem, Mergers of unequal-mass galaxies: supermassive black hole binary evolution and structure of merger remnants. Astrophys. J. **749**, 147 (2012)
B. Kocsis, Z. Haiman, A. Loeb, Gas pile-up, gap overflow and type 1.5 migration in circumbinary discs: application to supermassive black hole binaries. Mon. Not. R. Astron. Soc. **427**, 2680–2700 (2012)
S. Komossa, Observational evidence for binary black holes and active double nuclei. Mem. Soc. Astron. Ital. **77**, 733 (2006)
S. Komossa, Recoiling Black Holes: Electromagnetic Signatures, Candidates, and Astrophysical Implications. Adv. Astron. **2012** (2012)
J. Kormendy, L.C. Ho, Coevolution (or not) of supermassive black holes and host galaxies. Annu. Rev. Astron. Astrophys. **51**, 511–653 (2013)
G. Kulkarni, A. Loeb, Formation of galactic nuclei with multiple supermassive black holes at high redshifts. Mon. Not. R. Astron. Soc. **422**, 1306–1323 (2012)
G. Lodato, S. Nayakshin, A.R. King, J.E. Pringle, Black hole mergers: can gas discs solve the 'final parsec' problem? Mon. Not. R. Astron. Soc. **398**, 1392–1402 (2009)
C.O. Lousto, Y. Zlochower, Black hole binary remnant mass and spin: A new phenomenological formula. ArXiv e-prints (2013)
E. Lusso, R. Decarli, M. Dotti, C. Montuori, D.W. Hogg, P. Tsalmantza, M. Fumagalli, J.X. Prochaska, The nature of massive black hole binary candidates—II. Spectral energy distribution atlas. Mon. Not. R. Astron. Soc. **441**, 316–332 (2014)
A.I. MacFadyen, M. Milosavljević, An eccentric circumbinary accretion disk and the detection of binary massive black holes. Astrophys. J. **672**, 83–93 (2008)
A. Marconi, L.K. Hunt, The relation between black hole mass, bulge mass, and near-infrared luminosity. Astrophys. J. Lett. **589**, L21–L24 (2003)
A. Marconi, G. Risaliti, R. Gilli, L.K. Hunt, R. Maiolino, M. Salvati, Local supermassive black holes, relics of active galactic nuclei and the X-ray background. Mon. Not. R. Astron. Soc. **351**, 169–185 (2004)
L. Mayer, Massive black hole binaries in gas-rich galaxy mergers; multiple regimes of orbital decay and interplay with gas inflows. Class. Quantum Gravity **30**(24), 244008 (2013)
L. Mayer, S. Kazantzidis, P. Madau, M. Colpi, T. Quinn, J. Wadsley, Rapid formation of supermassive black hole binaries in galaxy mergers with gas. Science **316**, 1874 (2007)
A. Merloni, S. Heinz, *Evolution of Active Galactic Nuclei* (2013), p. 503
D. Merritt, Dynamics and Evolution of Galactic Nuclei (2013a)
D. Merritt, Loss-cone dynamics. Class. Quantum Gravity **30**(24), 244005 (2013b)
D. Merritt, M. Milosavljević, Massive black hole binary evolution. Living Rev. Relativ. **8**, 8 (2005)
D. Merritt, M.Y. Poon, Chaotic loss cones and black hole fueling. Astrophys. J. **606**, 788–798 (2004)
D. Merritt, J.D. Schnittman, S. Komossa, Hypercompact stellar systems around recoiling supermassive black holes. Astrophys. J. **699**, 1690–1710 (2009)
J.C. Mihos, L. Hernquist, Gasdynamics and starbursts in major mergers. Astrophys. J. **464**, 641 (1996)
M. Milosavljević, D. Merritt, Formation of galactic nuclei. Astrophys. J. **563**, 34–62 (2001)
M. Milosavljević, D. Merritt, Long-term evolution of massive black hole binaries. Astrophys. J. **596**, 860–878 (2003)
C. Montuori, M. Dotti, F. Haardt, M. Colpi, R. Decarli, Search for sub-parsec massive binary black holes through line diagnosis—II. Mon. Not. R. Astron. Soc. **425**, 1633–1639 (2012)
J.F. Navarro, C.S. Frenk, S.D.M. White, The structure of cold dark matter halos. Astrophys. J. **462**, 563 (1996)
S.C. Noble, B.C. Mundim, H. Nakano, J.H. Krolik, M. Campanelli, Y. Zlochower, N. Yunes, Circumbinary magnetohydrodynamic accretion into inspiraling binary black holes. Astrophys. J. **755**, 51 (2012)
K. Omukai, F. Palla, On the formation of massive primordial stars. Astrophys. J. Lett. **561**, L55–L58 (2001)
E.C. Ostriker, Dynamical friction in a gaseous medium. Astrophys. J. **513**, 252–258 (1999)
F. Özel, D. Psaltis, R. Narayan, J.E. McClintock, The black hole mass distribution in the galaxy. Astrophys. J. **725**, 1918–1927 (2010)
H.B. Perets, C. Hopman, T. Alexander, Massive perturber-driven interactions between stars and a massive black hole. Astrophys. J. **656**, 709–720 (2007)
M. Preto, I. Berentzen, P. Berczik, R. Spurzem, Fast coalescence of massive black hole binaries from mergers of galactic nuclei: implications for low-frequency gravitational-wave astrophysics. Astrophys. J. Lett. **732**, L26 (2011)
J.E. Pringle, The properties of external accretion discs. Mon. Not. R. Astron. Soc. **248**, 754–759 (1991)
G.D. Quinlan, The time-scale for core collapse in spherical star clusters. New Astron. **1**, 255–270 (1996)

R.R. Rafikov, Structure and evolution of circumbinary disks around supermassive black hole binaries. Astrophys. J. **774**, 144 (2013)
A.E. Reines, J.E. Greene, M. Geha, Dwarf galaxies with optical signatures of active massive black holes. Astrophys. J. **775**, 116 (2013)
L. Rezzolla, E. Barausse, E.N. Dorband, D. Pollney, C. Reisswig, J. Seiler, S. Husa, Final spin from the coalescence of two black holes. Phys. Rev. D **78**(4), 044002 (2008)
C. Roedig, M. Dotti, A. Sesana, J. Cuadra, M. Colpi, Limiting eccentricity of subparsec massive black hole binaries surrounded by self-gravitating gas discs. Mon. Not. R. Astron. Soc. **415**, 3033–3041 (2011)
C. Roedig, A. Sesana, M. Dotti, J. Cuadra, P. Amaro-Seoane, F. Haardt, Evolution of binary black holes in self gravitating discs. Dissecting the torques. Astron. Astrophys. **545**, A127 (2012)
R. Roškar, L. Mayer, D. Fiacconi, S. Kazantzidis, T.R. Quinn, J. Wadsley, Orbital Decay of Supermassive Black Hole Binaries in Clumpy Multiphase Merger Remnants. ArXiv e-prints (2014)
B.S. Sathyaprakash, B.F. Schutz, Physics, astrophysics and cosmology with gravitational waves. Living Rev. Relativ. **12**, 2 (2009)
D.R.G. Schleicher, F. Palla, A. Ferrara, D. Galli, M. Latif, Massive black hole factories: supermassive and quasi-star formation in primordial halos. Astron. Astrophys. **558**, A59 (2013)
J.D. Schnittman, Electromagnetic counterparts to black hole mergers. Class. Quantum Gravity **28**(9), 094021 (2011)
N. Scott, A.W. Graham, Updated mass scaling relations for nuclear star clusters and a comparison to supermassive black holes. Astrophys. J. **763**, 76 (2013)
A. Sesana, Self consistent model for the evolution of eccentric massive black hole binaries in stellar environments: implications for gravitational wave observations. Astrophys. J. **719**, 851–864 (2010)
A. Seth, M. Agüeros, D. Lee, A. Basu-Zych, The coincidence of nuclear star clusters and active galactic nuclei. Astrophys. J. **678**, 116–130 (2008)
N.I. Shakura, R.A. Sunyaev, Black holes in binary systems. Observational appearance. Astron. Astrophys. **24**, 337–355 (1973)
J.-M. Shi, J.H. Krolik, S.H. Lubow, J.F. Hawley, Three-dimensional magnetohydrodynamic simulations of circumbinary accretion disks: disk structures and angular momentum transport. Astrophys. J. **749**, 118 (2012)
D. Syer, C.J. Clarke, Satellites in discs: regulating the accretion luminosity. Mon. Not. R. Astron. Soc. **277**, 758–766 (1995)
G. Taffoni, L. Mayer, M. Colpi, F. Governato, On the life and death of satellite haloes. Mon. Not. R. Astron. Soc. **341**, 434–448 (2003)
H. Tanaka, T. Takeuchi, W.R. Ward, Three-dimensional interaction between a planet and an isothermal gaseous disk. I. Corotation and lindblad torques and planet migration. Astrophys. J. **565**, 1257–1274 (2002)
S. Van Wassenhove, P.R. Capelo, M. Volonteri, M. Dotti, J.M. Bellovary, L. Mayer, F. Governato, Nuclear coups: dynamics of black holes in galaxy mergers. Mon. Not. R. Astron. Soc. (2014)
S. Van Wassenhove, M. Volonteri, L. Mayer, M. Dotti, J. Bellovary, S. Callegari, Observability of dual active galactic nuclei in merging galaxies. Astrophys. J. Lett. **748**, L7 (2012)
E. Vasiliev, F. Antonini, D. Merritt, The final-parsec problem in non-spherical galaxies revisited. ArXiv e-prints (2013)
M. Vestergaard, X. Fan, C.A. Tremonti, P.S. Osmer, G.T. Richards, Mass functions of the active black holes in distant quasars from the sloan digital sky survey data release 3. Astrophys. J. Lett. **674**, L1–L4 (2008)
M. Volonteri, Formation of supermassive black holes. Astron. Astrophys. Rev. **18**, 279–315 (2010)
M. Volonteri, F. Haardt, P. Madau, The assembly and merging history of supermassive black holes in hierarchical models of galaxy formation. Astrophys. J. **582**, 559–573 (2003)
M. Volonteri, P. Natarajan, Journey to the M_{BH}–σ relation: the fate of low-mass black holes in the Universe. Mon. Not. R. Astron. Soc. **400**, 1911–1918 (2009)
L. Wang, P. Berczik, R. Spurzem, M.B.N. Kouwenhoven, The link between ejected stars, hardening and eccentricity growth of super massive black holes in galactic nuclei. Astrophys. J. **780**, 164 (2014)
S.D.M. White, M.J. Rees, Core condensation in heavy halos—a two-stage theory for galaxy formation and clustering. Mon. Not. R. Astron. Soc. **183**, 341–358 (1978)
Q. Yu, Evolution of massive binary black holes. Mon. Not. R. Astron. Soc. **331**, 935–958 (2002)

DOI 10.1007/978-1-4939-2227-7_12
Reprinted from *Space Science Reviews* Journal, DOI 10.1007/s11214-013-0030-6

Mass Measurements of Stellar and Intermediate-Mass Black Holes

J. Casares · P.G. Jonker

Received: 27 August 2013 / Accepted: 8 November 2013 / Published online: 10 December 2013

Abstract We discuss the method, and potential systematic effects therein, used for measuring the mass of stellar-mass black holes in X-ray binaries. We restrict our discussion to the method that relies on the validity of Kepler's laws; we refer to this method as the dynamical method. We briefly discuss the implications of the mass distribution of stellar-mass black holes and provide an outlook for future measurements. Further, we investigate the evidence for the existence of intermediate-mass black holes i.e. black holes with masses above 100 $M_{\odot}$, the limit to the black hole mass that can be produced by stellar evolution in the current Universe.

Keywords Black holes · X-ray binaries · Accretion disks

1 Introduction

Black holes (BH) are the pinnacle of extreme gravity. They provide astronomers with unique laboratories for observing some of the most fundamental and intriguing astrophysical phenomena, such as accretion, the ejection of relativistic outflows or the production of gamma-ray bursts. Consequently, BHs play an essential role in a variety of astrophysical phenomena

J. Casares
Instituto de Astrofísica de Canarias, 38205 La Laguna, S/C de Tenerife, Spain

J. Casares (✉)
Departamento de Astrofísica, Universidad de La Laguna, 38206 La Laguna, S/C de Tenerife, Spain
e-mail: jorge.casares@iac.es

P.G. Jonker
SRON, Netherlands Institute for Space Research, Sorbonnelaan 2, 3584 CA Utrecht, The Netherlands

P.G. Jonker
Department of Astrophysics/IMAPP, Radboud University Nijmegen, P.O. Box 9010,
6500 GL Nijmegen, The Netherlands

P.G. Jonker
Harvard–Smithsonian Center for Astrophysics, 60 Garden Street, Cambridge, MA 02138, USA

on various scales, ranging from binaries to ultra-luminous X-ray sources (ULXs), galaxies and quasars, the most powerful accretion engines in the Universe. However, it is *stellar-mass* BHs that offer us the best opportunity to study these objects in detail. Their proximity and variability time scales allow in-depth studies of their properties through a range of accretion regimes on time-scales convenient for study by humans. Thorough reviews on BH accretion and outflows are given in several other articles of this issue.

Astrophysical BHs are characterized by only two parameters, mass and spin, and their knowledge is key to probe space-time in the strong gravity regime (see articles by McClintock et al. and Reynolds in this issue). Accurate knowledge of BH masses is also critical to test models of massive progenitors, SNe Ibc explosions and compact binary evolution (e.g. Fryer and Kalogera 2001; Fryer et al. 2012; Belczynski et al. 2012). The current article presents an up-to-date overview of dynamical mass determinations in stellar-mass BHs. The main methods of analysis are summarized, together with a critical assessment on their limitations and possible systematics involved. In a second part of the article, new approaches and techniques are reviewed from which a significant advance in the precision of mass measurements is expected. Finally, a section on prospects for mass determination in ULXs is presented. Previous reviews on observational properties and mass determination in BH binaries can be found in van Paradijs and McClintock (1995), Tanaka and Shibazaki (1996), Orosz (2003), Charles and Coe (2006), Remillard and McClintock (2006), McClintock and Remillard (2006), Casares (2007) and Belloni et al. (2011).

2 Dynamical BHs in X-ray Transients

Stellar evolution predicts $\gtrsim 10^8$ BH remnants in the Galaxy (van den Heuvel 1992) but only BHs in compact binaries can be easily detected through accretion. X-ray binaries thus provide currently the best way to measure the mass of BHs. A large number of these X-ray binaries are found as X-ray transients (XRTs, for a comprehensive review see McClintock and Remillard 2006). XRTs are singled out by episodic outbursts caused by mass transfer instabilities in an accretion disc which is fed by a low-mass (donor) star (Mineshige and Wheeler 1989; Lasota 2001). The large fraction of BH systems among XRTs agrees with the predictions of the Disc Instability Model, modified by irradiation effects. The absence of a solid surface in the compact star inhibits disc stabilization through X-ray irradiation at the low accretion rates characteristic of overflowing low-mass stars (King et al. 1997; Coriat et al. 2012). XRTs may increase the integrated X-ray luminosity of the Milky Way by a factor ~ 2 and thus are promptly spotted by X-ray satellites. Between outbursts, they remain in a "quiescent" state, with typical X-ray luminosities below $\sim 10^{32}$ erg s^{-1}, allowing the optical detection of the faint low-mass donor star. This opens-up the possibility to perform radial velocity studies, probe the nature of the compact star and determine its mass.

The most robust method of measuring stellar masses relies on Kepler's Third law of motion. However, BH X-ray binaries are akin to single-lined spectroscopic binaries and hence only the radial velocity curve of the mass-losing star is available (Fig. 1). This yields the orbital period $P_{\rm orb}$ and the radial velocity semi-amplitude of the companion star $K_{\rm c}$. The two quantities combine in the mass function equation $f(M)$, a non-linear expression relating the masses of the compact object $M_{\rm x}$ and the companion star $M_{\rm c}$ (or binary mass ratio $q = M_{\rm c}/M_{\rm x}$) with the binary inclination angle (i):

$$f(M) = \frac{K_{\rm c}^3 P_{\rm orb}}{2\pi G} = \frac{M_{\rm x}^3 \sin^3 i}{(M_{\rm x} + M_{\rm c})^2} = \frac{M_{\rm x} \sin^3 i}{(1+q)^2} \quad (1)$$

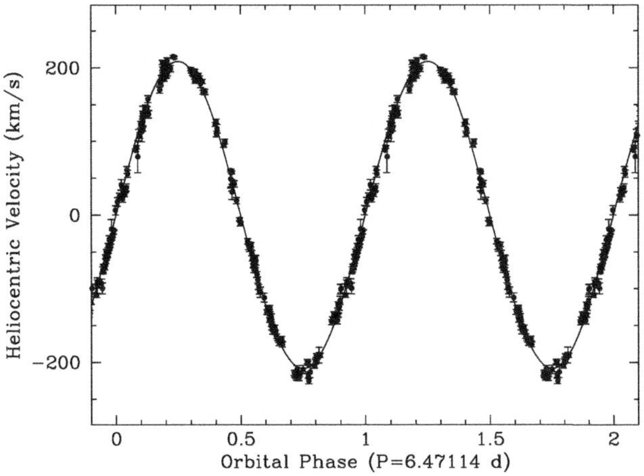

Fig. 1 Radial velocity curve of the K0 donor star in the XRT V404 Cyg

Note that this expression implicitly assumes orbits are circular, which is a fair assumption given the long lifetimes and short circularization timescales expected in X-ray binaries (Witte and Savonije 2001). The mass function provides a solid lower limit to the mass of the compact star for $M_c = 0$ and an edge-on geometry ($i = 90°$). A large mass function is widely considered as the best signature for a BH since observations and theoretical calculations of dense matter indicate that neutron stars cannot be more massive than $\sim$2.5 $M_{\odot}$ (Lattimer 2012). Besides, main sequence stars with masses above $\sim$3 $M_{\odot}$ would be B-type stars, and would be seen in the spectra. The fact that only a K/M-type stellar spectrum is detected together with the presence of an occasionally very bright X-ray source leads to the conclusion that the binary contains a BH.

The radial velocity curve of the donor star is best obtained through cross-correlation of the photospheric absorption lines with a stellar template of similar spectral type. Mass functions are gathered routinely to a few percent accuracy and this requires resolving powers better than about $\lambda/\Delta\lambda = 1500$. The use of 10 m class telescopes has allowed one to measure $f(M)$ for objects down to R$\sim$22 mag, as shown by the works on XTE J1859+226 (Filippenko and Chornock 2001; Corral-Santana et al. 2011). BH mass measurements require also the measurement of the mass ratio q and the binary inclination which, in the absence of eclipses, can only be obtained through indirect methods. These are based on information to be extracted from the *light curve* and the *spectrum* of the optical star, resulting in a full solution to the binary parameters with minimum assumptions. This procedure will be discussed in turn.

The best way to determine the mass ratio is through measuring the rotational broadening ($V \sin i$) of the photospheric lines from the companion star. This technique exploits the fact that the star fills its Roche lobe and it is tidally locked. This makes the absorption lines significantly broader than in single stars that are slowly rotating. Under the approximation of sphericity, the rotational broadening scales with the velocity of the donor star according to $V \sin i / K_c \simeq 0.462\, q^{(1/3)}(1+q)^{(2/3)}$ (Wade and Horne 1988) and hence q can be constrained. The rotational broadening is usually measured by comparing the target spectrum with a slowly rotating template, convolved with a limb-darkened rotational profile (e.g. Gray 1992). A χ^2 minimization yields the optimum broadening required by the template to match the spectrum of the companion to the BH (see Fig. 2). Typical rotational broadenings in BH transients range between 30 and 120 $\mathrm{km\,s^{-1}}$ and thereby moderately high spectral resolu-

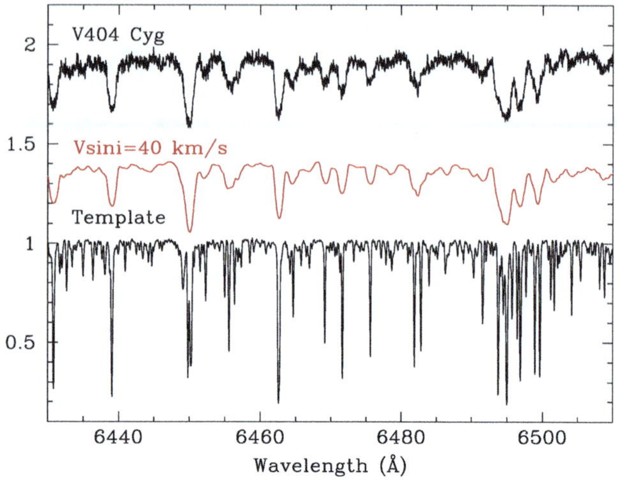

Fig. 2 Rotational broadening analysis. A K0IV template star (*bottom*) is broadened by $V \sin i = 40\ \mathrm{km\,s^{-1}}$ (*middle*) in order to reproduce the spectrum of the donor star in the BH transient V404 Cyg (*top*). The latter has been produced after coadding individual spectra in the rest frame of the companion star

tions ($\lambda/\Delta\lambda \gtrsim 5000$) are required for these measurements otherwise erroneous values for $V \sin i$ are likely obtained.

It should be noted that there are systematic errors involved in the calculation of $V \sin i$. First of all, Roche-lobe-filling stars are obviously non-spherical, with tidal distortion causing orbital variations of the broadening kernel. Fitting the phase-averaged spectrum with a template broadened using a spherical convolution profile yields $V \sin i$ values which underestimate q (Marsh et al. 1994). In addition, the use of a continuum limb-darkening approximation also leads to underestimates of the true broadening and thus a decreased mass ratio (Shahbaz 2003). Another potential source of systematics is introduced by assuming a gravity darkening law described by von Zeipel's theorem with exponent $\beta = 0.08$ (Lucy 1967, but see Sarna 1989). In any case, statistical uncertainties in the computation of $V \sin i$ are typically larger than systematic errors and hence mass ratios obtained through measuring rotational broadenings are, in most cases, robust. Furthermore, given the extreme mass ratios ($q \ll 1$) typical of BH XRTs, the impact of q uncertainties in the final BH mass is modest.

The binary inclination is commonly obtained through fitting optical/NIR light curves with synthetic ellipsoidal models. Light curves in XRTs show a characteristic double-humped modulation produced by the tidal distortion of the Roche lobe filling donor star and a non-uniform distribution of surface brightness. The amplitude of the modulation is a strong function of the inclination angle. Synthetic models are computed integrating the local flux intensity, modified by limb and gravity darkening effects, over the Roche geometry. The best results are obtained using Kurucz and NEXTGEN model atmospheres (see Orosz and Hauschildt 2000 for a critical comparison of several approaches). For example, Fig. 3 presents a textbook example of the ellipsoidal modulation from GRO J1655-40, an XRT with a F6IV intermediate-mass donor star. Synthetic model fits performed by different groups have resulted in very accurate inclination values distributed over a narrow range between 64–71° (Orosz and Bailyn 1997; van der Hooft et al. 1998; Greene et al. 2001; Beer and Podsiadlowsky 2002). However, the vast majority of XRTs possess faint K-M donor stars and, therefore, light curves can be seriously contaminated by other non-stellar sources of light. The impact of these on the determination of inclination angles and BH masses can be critical and will be discussed in Sect. 2.1.

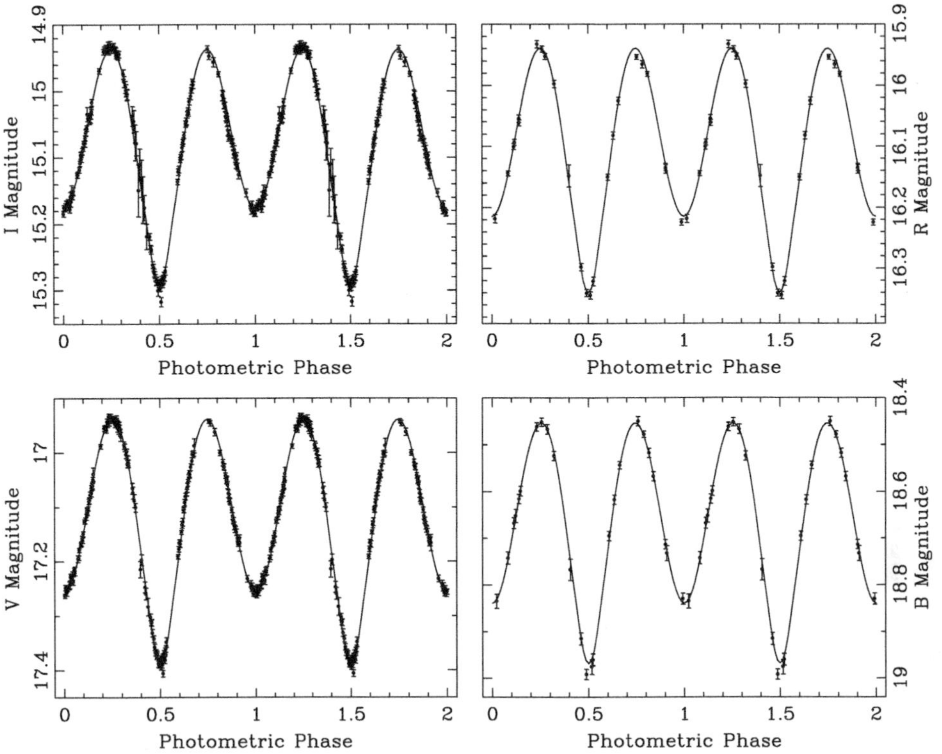

Fig. 3 Ellipsoidal fits to light curves of the XRT GRO J1655-40 in four colour bands simultaneously. Synthetic models were computed using Kurucz model atmospheres. From Beer and Podsiadlowsky (2002)

Table 1 presents a compilation of fundamental parameters and BH mass determinations for the 17 BH XRTs with dynamical confirmation currently known. Uncertainties are typically 1-σ except for errors in the inclination angle where a 90 or 95 percent confidence level is sometimes provided. The mass functions listed in the table have been obtained from radial velocity curves of the companion stars in quiescence, with the exception of GX 339-4 (see also footnotes on GRS 1915+105 and GRO J1655-40). In the case of GX 339-4, the donor star has not been detected in quiescence yet, although a lower limit to the mass function was derived using emission lines, excited on its irradiated hemisphere during an outburst episode (Hynes et al. 2003a; see Sect. 4.1). Table 1 also quotes mass ratios obtained exclusively through the $V \sin i$ technique, besides that from GRO J1655-40 where its high mass ratio also allows one to constrain q from model fits to the ellipsoidal light curves. Finally, binary inclinations refer to values derived from modelling ellipsoidal light curves in quiescence, except for GRS 1915+105, where it has been inferred from the orientation of the radio jets (Fender et al. 1999). In some cases, upper limits to the inclinations are given, based on the lack of X-ray eclipses and mass ratio constraints. Numbers highlighted in boldface indicate our recommended set of fundamental parameters to be adopted. In our view, these provide the best determinations currently available i.e. those least affected by possible systematic effects and having the lowest statistical uncertainties.

Table 1 Mass measurements of BHs in XRTs (continues on next page). We mark in bold face the measurements we consider the most reliable currently available in the literature

System	Donor Spect. Type	P_{orb} [days]	$f(M)$ [$M_\odot$]	q	i [deg]	M_X [$M_\odot$]	Ref.
GRS 1915+105	K1/5 III	33.5(1.5)	9.5 ± 3.0	0.058 ± 0.033	66 ± 2	14.0 ± 4.4	Greiner et al. (2001a), Harlaftis and Greiner (2004)
		33.8(1)	8.0 ± 0.6		66 ± 2	12.0 ± 1.4	Hurley et al. (2013)
		33.85(16)	$\mathbf{7.02 \pm 0.17}$	$\mathbf{0.042 \pm 0.024}$	$\mathbf{66 \pm 2}$	$\mathbf{10.1 \pm 0.6}$	Steeghs et al. (2013)
V404 Cyg	K0 IV	**6.47129(7)**	$\mathbf{6.07 \pm 0.05}$	$\mathbf{0.067 \pm 0.005}$			Casares et al. (1992), Casares and Charles (1994), Casares (1996)
					60–80	8–12	Wagner et al. (1992)
					56 ± 4	12^{+3}_{-2}	Shahbaz et al. (1994b)
					56 ± 2		Pavlenko et al. (1996)
					>62	$\lesssim 12.5$	Sanwal et al. (1996)
	K3 IV				67^{+3}_{-1}	$9.0^{+0.2}_{-0.6}$	Khargharia et al. (2010)
XTE J1819.3-2525	B9 III	**2.81730(1)**	$\mathbf{2.74 \pm 0.12}$	**0.63–0.70**	60–71	8.7–11.7	Orosz et al. (2001)
GRO J1655-40	~F5 IV	2.601(27)	3.16 ± 0.15				Bailyn et al. (1995)
	F3/6 IV	2.62157(15)	3.24 ± 0.09	0.33 ± 0.01	69.5 ± 0.1	7.0 ± 0.2	Orosz and Bailyn (1997)
		2.62168(14)		0.24–0.41	63.7–70.7	6.3–7.6	van der Hooft et al. (1998)
		2.62191(20)		0.34–0.4	70.2 ± 1.9	6.3 ± 0.5	Greene et al. (2001)
	F5/G0 IV			0.26 ± 0.04	$\mathbf{68.7 \pm 1.5}$	5.4 ± 0.3	Beer and Podsiadlowsky (2002)
	F6 IV		$\mathbf{2.73 \pm 0.09}$	0.34–0.44		5.5–7.9	Shahbaz et al. (1999)
				0.31 ± 0.08		7.9 ± 3.8	Buxton and Vennes (1999)
				$\mathbf{0.42 \pm 0.03}$		$\mathbf{6.6 \pm 0.5}$	Shahbaz (2003)
		2.62120(14)	3.16 ± 0.15	0.33 ± 0.05			González Hernández et al. (2008b)
BW Cir	~G5 IV	**2.54451(8)**	$\mathbf{5.73 \pm 0.29}$	$\mathbf{0.12^{+0.03}_{-0.04}}$	<79	>7.0	Casares et al. (2004, 2009)

Table 1 (*Continued*)

System	Donor Spect. Type	P_{orb} [days]	$f(M)$ [$M_\odot$]	q	i [deg]	M_X [$M_\odot$]	Ref.
GX 339-4	–	**1.7557(4)**	**5.8 ± 0.5**			>6.0	Hynes et al. (2003a), Muñoz-Darias et al. (2008a)
XTE J1550-564	K2/4 IV	**1.5420333(24)**	**7.65 ± 0.38**	≈0.03	74.7 ± 3.8	7.8–15.6	Orosz et al. (2002, 2011a)
4U 1543-475	A2 V	**1.123(8)**	**0.22 ± 0.02**		20–40	2.7–7.5	Orosz et al. (1998)
H1705-250	K3/M0 V	**0.5213(13)**	**4.86 ± 0.13**	≤0.053	60–80	4.9–7.9	Remillard et al. (1996), Filippenko et al. (1997), Harlaftis et al. (1997)
					48–51		Martin et al. (1995)
GS 1124-684	K3/5 V	**0.4326058(31)**	**3.01 ± 0.15**	**0.13 ± 0.04**			Remillard et al. (1992), Orosz et al. (1996), Casares et al. (1997)
					60^{+5}_{-6}	5.0–7.5	Orosz et al. (1996)
					54^{+20}_{-15}	$5.8^{+4.7}_{-2.0}$	Shahbaz et al. (1997)
					54 ± 2	7.0 ± 0.6	Gelino et al. (2001a)
GS 2000+250	K3/7 V	0.3440915(9)	**5.01 ± 0.15**	**0.042 ± 0.012**			Casares et al. (1995b), Filippenko et al. (1995b), Harlaftis et al. (1996)
					65 ± 9	8.5 ± 1.5	Callanan et al. (1996b)
					43–69	4.8–14	Beekman et al. (1996)
		0.3440873(2)			**58–74**	**5.5–8.8**	Ioannou et al. (2004)
A0620-00	K2/7 V	0.323014(1)	2.91 ± 0.08		<50	>7.3	McClintock and Remillard (1986), Orosz et al. (1994)
			2.72 ± 0.06	**0.067 ± 0.010**			Marsh et al. (1994)
			2.76 ± 0.04	0.060 ± 0.004			Neilsen et al. (2008)

Table 1 (*Continued*)

System	Donor Spect. Type	P_{orb} [days]	$f(M)$ [$M_{\odot}$]	q	i [deg]	M_X [$M_{\odot}$]	Ref.
		0.32301405(1)	$\mathbf{2.76 \pm 0.01}$				González Hernández and Casares (2010)
					63–74	4.1–5.4	Haswell et al. (1993)
					31–54	10^{+7}_{-5}	Shahbaz et al. (1994a)
	K3/7 V				38–75	3.3–13.6	Froning and Robinson (2001)
					41 ± 3	11.0 ± 1.9	Gelino et al. (2001b)
					$\mathbf{51 \pm 0.9}$	$\mathbf{6.6 \pm 0.3}$	Cantrell et al. (2010)
XTE J1650-500	≈K4 V	**0.3205(7)**	$\mathbf{2.73 \pm 0.56}$		>47		Orosz et al. (2004)
GRS 1009-45	K7/M0 V	**0.285206(14)**	$\mathbf{3.17 \pm 0.12}$				Filippenko et al. (1999)
					37–80		Shahbaz et al. (1996)
XTE J1859+226	≈K5 V	**0.274(2)**	$\mathbf{4.5 \pm 0.6}$		<70	>5.42	Filippenko and Chornock (2001), Corral-Santana et al. (2011)
GRO J0422+32	M0/4 V	0.21159(57)	1.21 ± 0.06	$0.12^{+0.08}_{-0.07}$			Orosz and Bailyn (1995), Casares et al. (1995a), Filippenko et al. (1995a), Harlaftis et al. (1999)
	M4/5 V	**0.2121600(2)**	$\mathbf{1.19 \pm 0.02}$	$\mathbf{0.11^{+0.05}_{-0.02}}$	<45	>2.2	Webb et al. (2000)
	M0/4 V				35–49	~2.5–5.0	Casares et al. (1995a)
					≤45		Callanan et al. (1996a)
					10–26	≳15	Beekman et al. (1997)
					45 ± 2	3.97 ± 0.95	Gelino and Harrison (2003)
				<0.04	<30	≳10.4	Reynolds et al. (2007)

Table 1 (*Continued*)

System	Donor Spect. Type	P_{orb} [days]	$f(M)$ [$M_\odot$]	q	i [deg]	M_X [$M_\odot$]	Ref.
XTE J1118+480	K5/M1 V	**0.1699339(2)**	**6.27 ± 0.04**	**0.024 ± 0.009**			Wagner et al. (2001), McClintock et al. (2001), Torres et al. (2004), González Hernández et al. (2008a), Calvelo et al. (2009)
				0.037 ± 0.007			Orosz (2001)
					68 ± 2	$8.30^{+0.28}_{-0.14}$	Gelino et al. (2006)
	K7/M1 V				**68–79**	**6.9–8.2**	Khargharia et al. (2013)

Notes to Table 1: GRS 1915+105: the inclination angle is derived from the orientation of radio jets (Fender et al. 1999). Note that the radial velocity curve, and thus the BH mass, might be affected by irradiation because the binary has remained active since its discovery, although there are currently no signs for such effects

V404 Cyg: Pavlenko et al. (1996) model the lower envelope of an I-band light curve to minimize flickering contribution, but do not account for dilution from non-variable disc light. Optical spectroscopy indicates veiling is very small at RI wavelengths but the impact in ellipsoidal modeling should be tested through simultaneous photometric and spectroscopic observations. Shahbaz et al. (1994b) model a K-band light curve assuming no disc contribution while Sanwal et al. (1996) demonstrate significant flickering is present in the H-band. Khargharia et al. (2010) fit the H-band light curve of Sanwal et al. (1996) after correcting for the disc contribution which is, however, estimated from non-simultaneous NIR spectroscopy

XTE J1819.3-2525: limits to the inclination angle are derived from ellipsoidal fits to a low quality archival photographic light curve

GRO J1655-40: the mass functions reported in Bailyn et al. (1995) and Orosz and Bailyn (1997) are likely biased by irradiation effects. Shahbaz et al. (1999) and González Hernández et al. (2008b) quote mass functions in true quiescence but they disagree at $>3\sigma$ level. The latter is obtained from an orbital solution with sparse phase coverage, resulting in a significantly different period and systemic velocity with respect to previous studies. Therefore, we tentatively favour the former solution although this issue needs to be investigated further. Orosz and Bailyn (1997), van der Hooft et al. (1998), Greene et al. (2001), Beer and Podsiadlowsky (2002) report q values derived from fits to the ellipsoidal modulations, the others are obtained from $V \sin i$. Shahbaz (2003) determine q by fitting line profiles with synthetic spectra computed using NEXTGEN model atmospheres in a Roche geometry, accounting for variations in temperature and gravity. We favour this q determination. Beer and Podsiadlowsky (2002) fit BVRI light curves from Orosz and Bailyn (1997) simultaneously, using Kurucz model atmospheres. The distance and colour excess are not fixed but included in the fit as free parameters, and the best (adopted) solution assumes the distance from Hjellming and Rupen (1995). We tentatively favour the inclination reported in Beer and Podsiadlowsky (2002) although it should be noted that the q value implied by their ellipsoidal fits is significantly lower than our favoured value. In any event, there is an excellent agreement between all inclination values reported by four different groups. This is the only case where the error on the BH mass is dominated by uncertainties in q and the mass function rather than in the inclination. We tentatively favour the BH mass reported in Shahbaz (2003) because it is based on both our favoured q and mass function from Shahbaz et al. (1999), which is free from irradiation effects. The BH mass quoted in the abstract of Shahbaz (2003) is wrong due to a typo (Shahbaz, private communication)

BW Cir: the light curves are dominated by strong flickering which disguise the ellipsoidal modulation

GX 339-4: the mass function is based on radial velocities of Bowen emission lines emanating on the inner hemisphere of the irradiated donor

XTE J1550-564: Orosz et al. (2011a) derive inclination constraints by simultaneously fitting optical and NIR lightcurves. Conservative BH masses are quoted after allowing for different values of disc contribution obtained through non-simultaneous spectroscopy

4U 1543-475: low inclinations are implied by the small amplitude of some low quality VI-band ellipsoidal light curves. However, quoted values should be treated with caution because of possible systematics derived from model assumptions, namely disc light is neglected and the early type donor is assumed to be synchronized and filling its Roche lobe

H1705-250: the limited spectral resolution in the spectra reported in Harlaftis et al. (1997) only allow one to set an upper limit to q. Remillard et al. (1996) derive broad limits to the inclination based on the absence of eclipses ($i < 80°$) and assuming no disc contribution in a V-band lightcurve ($i > 60°$)

GS 1124-684: Shahbaz et al. (1997) and Gelino et al. (2001a) derive inclinations assuming no disc contribution to the NIR light curves. The latter authors also quote uncertainties which seem unrealistically small given potential systematic effects. While Orosz et al. (1996) tries to account for disc veiling in their B + V & I-band lightcurves, large deviations from ellipsoidal morphology make the derived inclinations suspect. Note also that this XRT is strongly affected by large amplitude flickering (Hynes et al. 2003b; Shahbaz et al. 2010)

GS 2000+250: the mass ratio reported in Harlaftis et al. (1996) should be treated with some caution because of the limited spectral resolution of the spectroscopic observations. Callanan et al. (1996b) and Beekman et al. (1996) neglect any disc contribution in their ellipsoidal fits to NIR light curves. Inclinations reported in Ioannou et al. (2004) are tentatively favoured because modeling accounts for a hot-spot component and the maximum disc light contribution (<32 percent in the R-band) allowed by the absence of X-ray eclipses

A0620-00: the q value reported in Marsh et al. (1994) is slightly preferred over that in Neilsen et al. (2008) because the former accounts for the non-sphericity of the Roche lobe, although this correction may not be statistically significant. In any case both determinations are fully consistent

GRS 1009-45: wide limits to the inclination are obtained from the absence of X-ray eclipses and ellipsoidal fits to an R-band light curve, assuming no disc contribution

XTE J1859+226: light curves are dominated by strong episodic flickering

GRO J0422+32: the reported mass ratios should be treated with some caution because of the limited spectral resolution of the spectroscopic data. All inclinations reported are suspected because of very large flickering amplitude in optical and NIR light curves

XTE J1118+480: the q value reported in Calvelo et al. (2009) has been corrected for orbital smearing of the line due to the motion of the mass donor during the exposure. Inclination and BH masses given in Khargharia et al. (2013) are slightly preferred over those in Gelino et al. (2006) because the latter are based on multiwavelength ellipsoidal fits to BVRJHK lightcurves which are non-simultaneous. Khargharia et al. (2013) only fits H-band light curves but the disc contribution is accounted for via contemporaneous NIR spectroscopy. It also provides a wider range of inclinations which is seen as more realistic. A period derivative for the orbital period of $\dot{P} = -1.83 \pm 0.66$ ms yr^{-1} was derived by González Hernández et al. (2012)

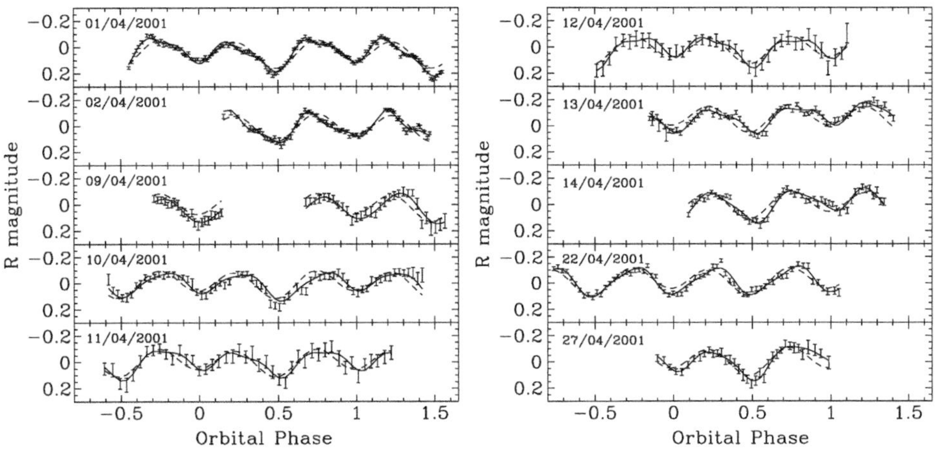

Fig. 4 Ellipsoidal light curves distorted by a superhump in XTE J1118+480. A combined model of an ellipsoidal plus superhump waves (*continuous line*) provides a better description of the data than a pure ellipsoidal fit (*dashed line*). From Zurita et al. (2002)

2.1 Systematic Errors and Biases in BH Mass Determinations

The error bars on BH mass measurements are dominated by uncertainties in the inclination angle because of its cubic dependence in Eq. (1). But most worryingly, Table 1 indicates that ellipsoidal fits performed by independent groups on the same binary often lead to a wide spread of inclinations and thereby BH masses. This is mainly thought to be due to systematic effects caused by contamination from non-stellar sources of light rather than statistical errors. There are two main sources of systematics affecting light curve analysis. The first one is the presence of a *superhump* modulation, a distorting wave produced by an eccentric disc precessing with a timescale a few percent longer than the orbital period. Superhumps are typically seen in outburst, when the accretion disc exceeds the 3:1 resonance radius (O'Donoghue and Charles 1996), but can also be detected near quiescence. When this happens, the ellipsoidal light curve is distorted by secular changes in shape and relative height of the maxima and minima (Fig. 4). Intensive monitoring over several orbital cycles is thus important to disentangle potential superhump waves from the true ellipsoidal modulation. Sometimes asymmetries in ellipsoidal light curves are interpreted as contamination by a hot spot (when extra flux is located at phase ~0.75, e.g. Khargharia et al. 2013) or stellar spots. Observations of sharp asymmetries in the light curves can also be mistaken for eclipse features, leading to overestimates in the inclination angle (Haswell et al. 1993).

The second source of systematics is caused by contamination from rapid aperiodic variability. The presence of optical flares in V404 Cyg, with a timescale of ~6 h, was already noticed long ago but thought to be peculiar to this system (Wagner et al. 1992; Pavlenko et al. 1996). However, subsequent high-time resolution (1–5 min) light curves have revealed that all quiescent XRTs display the same type of variability, with typical amplitudes ranging from 0.06 to 0.6 mag (Zurita et al. 2003; Hynes et al. 2003b; see Fig. 5 but also Shahbaz et al. 2013 for a record ~1.5 mag amplitude flaring activity). The variability seems stronger for systems with cooler companions and its characteristic time-scale appears to increase with orbital period, both properties suggesting an accretion disc origin. Magnetic reconnection events (Zurita et al. 2003), X-ray reprocessing (Hynes et al. 2004), instabilities in the transition between the thin and advective disc (Shahbaz et al.

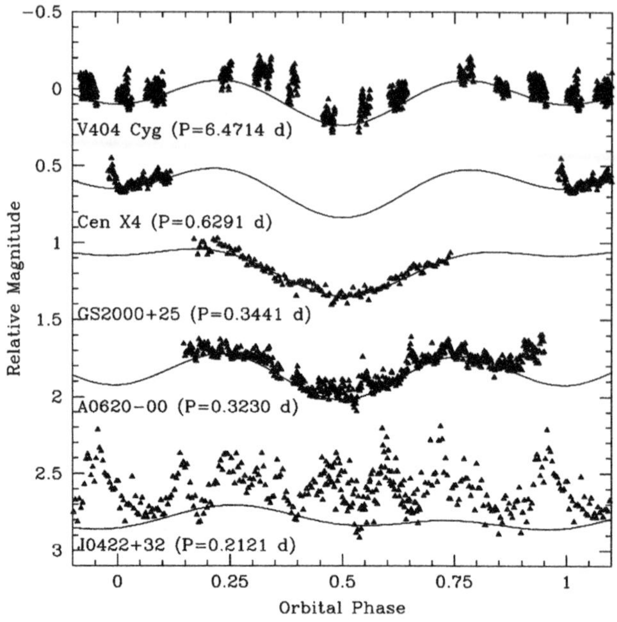

Fig. 5 High-time resolution light curves of quiescent XRTs showing the presence of fast variability. From Zurita et al. (2003)

2003) and variable synchrotron emission from a disc jet/outflow (Shahbaz et al. 2013) have been proposed but the physical mechanism responsible for the rapid variability is still unknown.

It is commonly assumed that the accretion disc light follows a negative power law with λ (see Table 1 in Garcia et al. 1996) and thereby most ellipsoidal fits tend to be performed at NIR wavelengths to minimize contamination. However, strong flaring activity is also observed in several NIR light curves, questioning this strategy. The most dramatic case is presented by K-band observations of GRO J0422+320, where the ellipsoidal modulation becomes completely diluted by the flaring variability (Reynolds et al. 2007). Further, flickering is not a white noise process (the PDS is described by a negative power-law with index ~ -1.3, see Shahbaz et al. 2003) and thus it is not cancelled out after binning or averaging light curves over many orbital cycles. The aperiodic variability can also vary with time for any particular system as shown by several cases in the literature (e.g. BW Cir: Casares et al. 2009; XTE J1859+226: Corral-Santana et al. 2011).

Using a decade-long data set of multicolour quiescent photometry, Cantrell et al. (2008) identify two main states of variability in A0620-00. The so-called *passive* and *active* states fall in separate parts of a colour-magnitude diagram. The system becomes redder in the *passive* state and displays minimum flaring activity (Fig. 6). Ellipsoidal fits to a subset of *passive VIH* light curves separately yield an inclination which is $\sim 10°$ higher than previous works have suggested (Cantrell et al. 2010). The latter were based on pure ellipsoidal fits to NIR light curves where the non-stellar contribution was neglected. Consequently, the mass of the BH in A0620-00 can be overestimated by a factor ≈ 2 if the disc contamination is ignored. This work has demonstrated that it is critical to employ light curves with minimum flickering activity during *passive* states to measure unbiased binary inclinations. Further discussion on the impact of the rapid variability in mass determinations of other BH XRTs is presented in Kreidberg et al. (2012).

One of the peculiar properties of the sample of BH XRTs listed in Table 1 is the absence of binaries with inclinations $i > 75°$. Furthermore, none of the other ~ 33 BH candidates

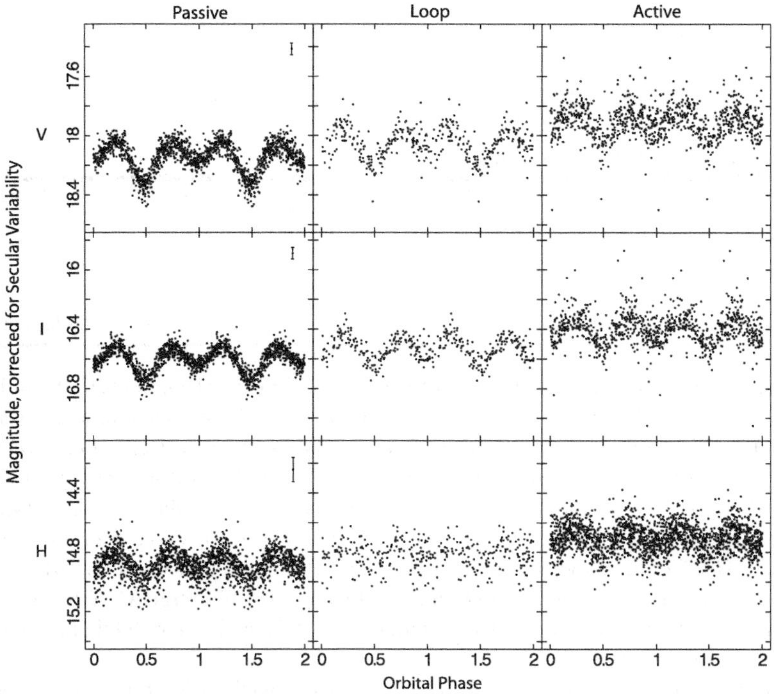

Fig. 6 Ellipsoidal light curves of A0620-00 in three different quiescent states: *passive* (*left panels*), *loop* (*middle panels*) and *active* (*right*). The strength of the aperiodic variability increases from the *passive* to the *active* state. From Cantrell et al. (2008)

(i.e. those XRTs with similar X-ray properties to dynamical BHs) shows eclipses, whereas ~20 percent are expected for a random distribution of inclinations (although this number is dependent on the assumed mass ratio, q). The lack of eclipsing BH XRTs is intriguing and strongly suggests that an observational bias is at play. It has been proposed that high inclination systems are hidden from view due to obscuration of the central X-ray source by a flared accretion disc (Narayan and McClintock 2005). The recent discovery of optical dips in the XRT Swift J1357.2-0933 suggests that the first extreme inclination BH transient may have been detected (Fig. 7). Aside from the unusual optical dips, the system is remarkable because of its extremely broad H_α emission profile and very low peak X-ray luminosity, properties which can be explained by orientation effects in an edge-on geometry (Corral-Santana et al. 2013; but see Armas Padilla et al. 2013a, 2013b for a different interpretation which proposes an intrinsically faint XRT). Based on the double-peak separation and radial velocities of the H_α profile indirect prove for a BH in a 2.8 h orbit is presented. In addition, evidence is provided for the presence of an obscuring torus in the inner disc. This brings a new ingredient to theoretical modelling of inner disc flows and jet collimation mechanisms in stellar-mass BHs. The discovery of edge-on BH XRTs is important because these systems will likely deliver the most precise BH mass determinations. Therefore, they will be key in the construction of the mass distribution of compact remnants.

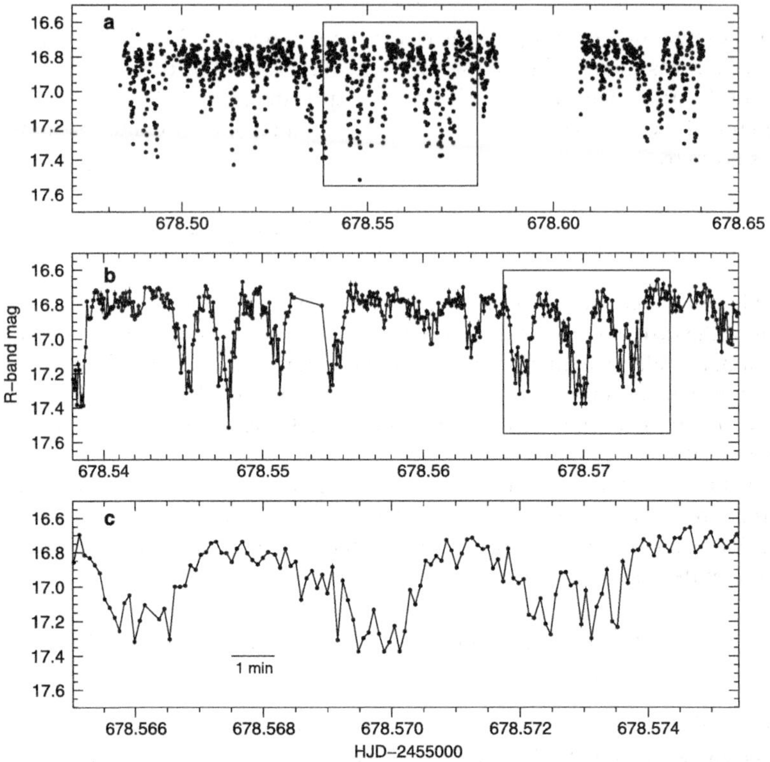

Fig. 7 Optical light curves of Swift J1357.2-0933 during outburst at 7 s time resolution. Regular dips repeating every 2 min cause a drop in brightness of up to $\simeq$0.8 mag. Close ups with lengths of 1 h and 15 min are displayed in the *middle* and *bottom panels*, respectively. From Corral-Santana et al. (2013)

3 Dynamical Black Holes in High-Mass X-ray Binaries (HMXBs)

In addition to the 17 transients listed in Table 1, dynamical mass measurements of BHs have also been possible for four X-ray binaries with high-mass donor stars: Cyg X-1 and the extragalactic sources LMC X-1, LMC X-3 and M33 X-7. HMXBs are persistent X-ray sources, with typical X-ray luminosities $\sim 10^{37}$ $\mathrm{ergs\,s^{-1}}$ powered by massive stellar winds or incipient Roche lobe overflow. Despite being persistent X-ray sources, irradiation effects are mostly negligible because the X-ray luminosity is smaller than or comparable to the bolometric luminosity of the mass-losing star. Further, the contribution of the accretion disc to the total optical flux can be ignored and thus ellipsoidal light curves are not affected by systematic effects as discussed in the previous section. There are, however, two important limitations regarding mass determination in HMXBs. First, the BH mass is very sensitive to uncertainties in the mass of the optical star. The latter is highly uncertain because donor stars in HMXBs are typically undermassive for their spectral types due to secular mass transfer and binary evolution (Rappaport and Joss 1983; Podsiadlowski et al. 2003; also compare spectral types and donor masses implied by Table 2). Second, mass transfer in HMXBs is mostly powered by stellar winds rather than Roche lobe overflow and this has a two-fold effect. On the one hand, wind emission can contaminate radial velocities from the massive star, specially if these are obtained from lower excitation Balmer lines (e.g. Ninkov et al.

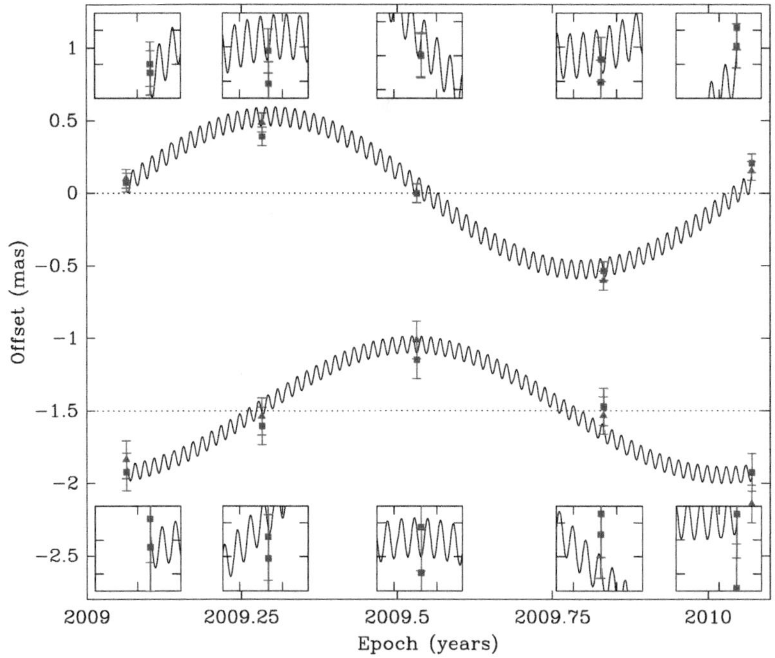

Fig. 8 Astrometric parallax of the compact radio source in Cyg X-1. The *long sinusoid* indicates the annual parallax after removing the proper motion of the system. The *short period sinusoid* reflects the 5.6 day orbit of the BH around the center of mass of the binary. From Reid et al. (2011)

1987). On the other hand, if the companion star is not filling its Roche lobe then one of the previous assumptions breaks down and q and i can be underestimated if derived from $V \sin i$ and ellipsoidal models as before. To circumvent this problem, extra parameters need to be included when modeling the observations, namely the Roche lobe filling factor and the degree of synchronization of the companion star (e.g. Gies and Bolton 1986; Orosz et al. 2007).

Despite these caveats, accurate masses can still be obtained if dynamical constraints are combined with a determination of the radius of the optical star. Accurate knowledge of the luminosity and hence the distance is required which is often difficult. For example, BH masses reported for Cyg X-1 over the last 3 decades show a large dispersion, with values between 7–29 $M_\odot$, mainly owing to distance uncertainties. Fortunately, a major advance has been allowed by the determination of the trigonometric parallax distance of Cyg X-1 with VLBA. Reid et al. (2011) have used emission from the compact radio jet to trace the position of the BH over 1 year. A parallax of 0.539 ± 0.033 mas has been derived which in turn implies a distance of 1.86 ± 0.12 kpc. The astrometric measurements are even sensitive to the size of the BH orbit allowing to constrain its radius to 0.18 ± 0.09 AU (Fig. 8). Using the accurate distance as an extra constraint to the dynamical model, a BH mass of $14.8 \pm 1.0\ M_\odot$ is derived, one of the most precise determinations to date (Orosz et al. 2011b). The best model solution also proves that the binary orbit is slightly eccentric ($e \simeq 0.02$) and the companion star rotates faster than the synchronization value at periastron.

Another remarkable result has been reported for M33 X-7, the first eclipsing stellar-mass BH. M33 X-7 is located in the nearby spiral galaxy M33 and hosts a late O companion which eclipses the X-ray source every 3.45 days (Larson and Schulman 1997). Ellipsoidal fits to

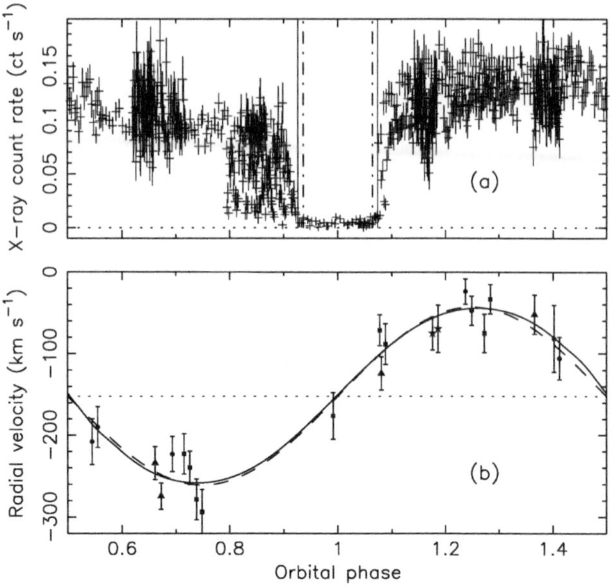

Fig. 9 (**a**) Chandra X-ray light curve of M33 X-7 showing the eclipse of the compact source by the optical companion. (**b**) Radial velocity curve of the O7-8III companion star. From Orosz et al. (2007)

the optical light curves alone have suggested the presence of a BH companion (Pietsch et al. 2006) but confirmation through radial velocities became challenging because of crowding and contamination by nebular lines. A radial velocity curve of the O7-8III companion was finally reported in Orosz et al. (2007), together with updated ellipsoidal fits (Fig. 9). The duration of the eclipse and the distance to M33 provides new restrictions which tightly constrain the parameter space of the dynamical model, in particular the binary inclination, the companion radius and its filling factor. As in Cyg X-1, the orbit is found to be slightly eccentric, while the radius of the donor star extends up to 78 percent of its Roche lobe. The best fit yields a BH mass of $15.7 \pm 1.5\ M_{\odot}$, one of the largest accurately known. The companion star is underluminous for its spectral type and, with $70.0 \pm 6.9\ M_{\odot}$, it is also one of the most massive stars known with high accuracy. M33 X-7 strongly impacts theories of BH formation and binary evolution since it is very hard to produce this massive HMXB in such a tight orbit (Valsecchi et al. 2010).

Table 2 lists a summary of fundamental parameters in the four HMXBs with dynamical BH mass measurements. In some cases, early mass estimates were derived after assuming that the optical star is synchronized and/or filling its Roche lobe. However, the most accurate results were subsequently obtained through fitting the ensemble of light curves, radial velocities and projected rotational velocities with a full model parametrization. These models account for non-synchronous rotation, Roche lobe filling fraction and (sometimes) eccentricity. As mentioned above, a determination of the distance, which is constrained to better than 6 percent in the four HMXBs, is key to deliver accurate BH masses. Despite this, a precise BH mass for LMC X-3 is not yet available because of the large impact of X-ray heating in the analysis. Note that we only quoted BH masses derived using the dynamical mass measurement method explained above. Mass estimates based on other techniques, such as fitting model atmospheres (Herrero et al. 1995; Caballero-Nieves et al. 2009), stellar evolutionary models (Ziółkowski 2005) or scaling X-ray spectral/timing properties (Shaposhnikov and Titarchuk 2007), are not considered here. Albeit with low number statistics, Table 2 shows that BHs in HMXBs have typically

Table 2 Mass measurements of BH in HMXBs. We mark in bold face the measurements we consider the most reliable currently available in the literature

System	Donor Spect. Type	P_{orb} [days]	$f(M)$ [$M_\odot$]	q	i [deg]	M_x [$M_\odot$]	Ref.
Cyg X-1	~B0Ib	~5.607	~0.16			>3	Webster and Murdin (1972), Bolton (1972a)
	09.7 Iab	5.5995(9)	0.182	≈2.1	≈30	≈14	Bolton (1972b)
		5.59974(8)	0.25 ± 0.01	1.95–2.31	28–38	16 ± 5	Gies and Bolton (1982, 1986)
		5.599829(16)	**0.244 ± 0.006**	**1.29 ± 0.15**	**27.1 ± 0.8**	**14.8 ± 1.0**	Orosz et al. (2011b)
LMC X-1	07 III	4.2288(6)	0.14 ± 0.05	$\gtrsim$2	40–63	≈6	Hutchings et al. (1983, 1987)
		3.90917(5)	**0.149 ± 0.007**	**4.91 ± 0.53**	**36.4 ± 1.9**	**10.9 ± 1.4**	Orosz et al. (2009)
M33 X-7	07-8 III	**3.453014(20)**	**0.46 ± 0.08**	**4.47 ± 0.60**	**74.6 ± 1.0**	**15.7 ± 1.5**	Orosz et al. (2007)
LMC X-3	B3 V	1.70491(7)	2.3 ± 0.3		50–70	7–14	Cowley et al. (1983)
	B3-5V	1.70479(4)	2.99±0.17		50–70	9.5–13.6	Val-Baker et al. (2007)
		1.7048089(11)	**2.77 ± 0.04**				Song et al. (2010)

Notes to Table 2: Cyg X-1: Bolton (1972b) uses ellipsoidal light curves from Cherepashchuk et al. (1973) to constrain q and i, assuming the star fills the Roche lobe. Also assumes a 30 $M_\odot$ companion to estimate the BH mass. Gies and Bolton (1986) assume synchronization and fill-out factor in the range 0.9–1. Orosz et al. (2011b) uses the orbital period determination, radial velocity data and light curves from Brocksopp et al. (1999). Also adopts a rotational broadening determination from Caballero-Nieves et al. (2009)

LMC X-1: Hutchings et al. (1983, 1987) use a lower limit to q from radial velocities of the NIII λ4640 emission. Synchronization is also assumed to derive a lower limit to the inclination

M33 X-7: Orosz et al. (2007) adopts the orbital period determination from X-ray eclipses (Pietsch et al. 2006)

LMC X-3: Cowley et al. (1983) assumes synchronization to derive a lower limit to the inclination. Val-Baker et al. (2007) finds evidence of irradiation in the donor star and the BH mass has been corrected for it a mass >10 $M_\odot$ and hence they tend to be more massive than BHs in XRTs. This seems to hint at an evolutionary difference, with BHs in HMXBs descending from massive progenitors which may have experienced comparatively lower stellar wind mass-loss rates, possibly through case C mass transfer (Wellstein and Langer 1999). Low metallicity environments, which strongly influence Wolf-Rayet mass-loss rates and the radius evolution of progenitor stars, is another important factor in the three extragalactic HMXBs (Crowther et al. 2010). It has also been proposed that BHs more massive than ~10 $M_\odot$ may form through direct implosion of the core remnant (Mirabel and Rodrigues 2003) although it remains unclear why prompt collapse should have a larger incidence in HMXBs.

We did not include the mass measurements for the two extra-Galactic HMXBs NGC 300 X-1 and IC 10 X-1 by Crowther et al. (2010) and Silverman and Filippenko (2008), respectively, as we deem these mass measurements less secure than those of the four sources we list in Table 2. The main reason for this is that the BH mass is strongly dependent on the assumed mass of the Wolf-Rayet star. The latter depends on the assumed stellar luminosity of the Wolf-Rayet star which in turn can be influenced by contamination from unresolved stars in the galaxies under study. These two sources are nevertheless very interesting as they may harbor BHs with the largest masses measured so far.

Several authors have examined the mass distribution of stellar mass BHs in order to gain insight into BH formation models. Bayesian analysis of the observed distribution suggests the presence of a mass gap or dearth of compact objects in the interval ~2–5 $M_\odot$ (Özel et al. 2010; Farr et al. 2011). This has been interpreted in the context of supernova models by a rapidly evolving explosion (within 0.2 s after the core bounce) through a Rayleigh-Taylor instability (Belczynski et al. 2012). On the other hand, the absence of low mass BHs has been attributed to a potential observational artefact caused by the systematic uncertainties affecting ellipsoidal fits and hence inclination measurements (see Sect. 2.1; Kreidberg et al. 2012). This has a strong impact on BH formation scenarios since confirmation of a mass gap would rule out accretion induced collapse while it would support delayed supernova explosion models. Future accurate mass measurements of BHs in high-inclination systems will allow testing whether the mass gap is caused by systematics in the binary inclination or it is a signature of the BH formation mechanism. This involves discovering new favourable XRTs and also exploiting novel techniques for mass measurements.

4 Future Perspectives

The best prospects to enlarge the current sample of stellar-mass BHs are offered by new discoveries of XRTs. Figure 10 presents the cumulative number of BH XRTs discovered in the X-ray astronomy era, starting with the historic detection of Cen X-2 by a rocket flight in 1966. A linear increase is apparent since the late 80's, when X-ray satellites with increased sensitivity and All-Sky-Monitors became operational. This sets a discovery rate of ~1.7 XRTs yr^{-1}. Extrapolation of the number of XRTs detected to date suggests that several thousand "dormant" BHs remain to be discovered (Romani 1998). And this estimate is most likely biased low because of sample incompleteness and complex selection effects. In particular a likely population of long period XRTs with very long outburst duty cycles is often ignored (c.f. Ritter and King 2002). In addition, there is mounting evidence for the existence of a significant number of intrinsically faint or obscured XRTs (Degenaar and Wijnands 2010; Corral-Santana et al. 2013). Incidentally, the latest population-synthesis models predict ~10^3–10^4 XRTs in the Galaxy (Yungelson et al. 2006;

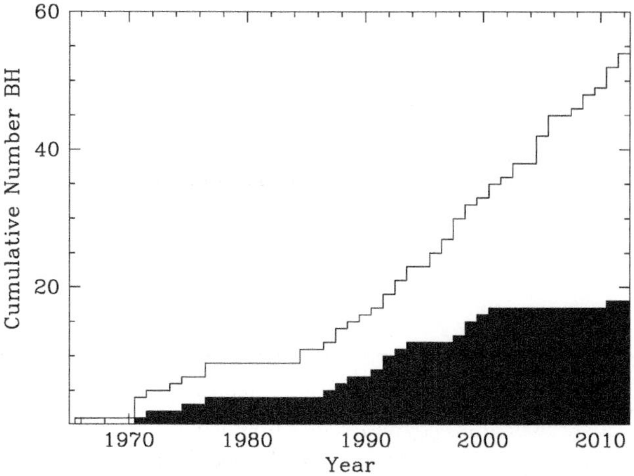

Fig. 10 Cumulative distribution of BH XRTs discovered during the era of X-ray astronomy. The *black histogram* indicates XRTs with BHs proven dynamically. Note, however, that the *black histogram* indicates the times when dynamical BHs were first discovered by X-ray satellites and not when the actual mass functions were measured/reported. The latest dynamical mass measurement BH corresponds to Swift J1357.2-0933 although the evidence is indirect (see Corral-Santana et al. 2013). Updated after Casares (2010)

Kiel and Hurley 2006). In either case, the observed sample is just the tip of the iceberg of a large "hibernating" population which becomes slowly unveiled through outburst episodes (van den Heuvel 1992). Figure 10 also shows that only 17 BH XRTs have been dynamically confirmed, representing about 30 percent of the total number discovered. The remaining ones are virtually lost during decay to quiescence because they become too faint, even for 10 m-class telescopes. Therefore, improving on the statistics of BH masses requires not only a new generation of ELT telescopes to tackle fainter targets but also new strategies aimed at unveiling a large fraction of the "hibernating" population of quiescent XRTs. A promising approach is presented by the *Galactic Bulge Survey (GBS*), where potential quiescent BH XRTs are selected in Chandra observations of regions in the Galactic Bulge 1 degree away from the Galactic plane (Jonker et al. 2011; Torres et al. 2013).

4.1 Reprocessed Bowen Emission

New opportunities for the study of BH XRTs are also opened by the analysis of emission lines and timing properties of the X-ray reprocessed radiation. To start with, the discovery of sharp high-excitation emission lines in active X-ray binaries have demonstrated that dynamical information can also be extracted from XRTs while in outburst (Steeghs and Casares 2002). These lines, first detected in the neutron star binary Sco X-1, arise from reprocessing in the irradiated companion. The most prominent are the NIII λ4634-40 and CIII λ4647-50 triplets in the core of the Bowen blend (Fig. 11). In particular, the NIII lines are powered by fluorescence resonance which requires seed photons of HeII Lyα. Since the velocities of the Bowen lines trace the orbit of the illuminated hemisphere, a K-correction (which chiefly depends on the mass ratio and the disc flaring angle) needs to be applied in order to obtain the true radial velocity curve of the donor star center of mass (Muñoz-Darias et al. 2005). The discovery of sharp Bowen lines during the 2002 outburst of GX 339-4 has allowed the first determination of the mass function in this Rosetta stone XRT and therefore the dynamical proof of a BH (Hynes et al. 2003a; Muñoz-Darias et al. 2008a; see Fig. 11). This work illustrates the power of the technique for systems which otherwise cannot be studied in quiescence because they become too faint for current instrumentation.

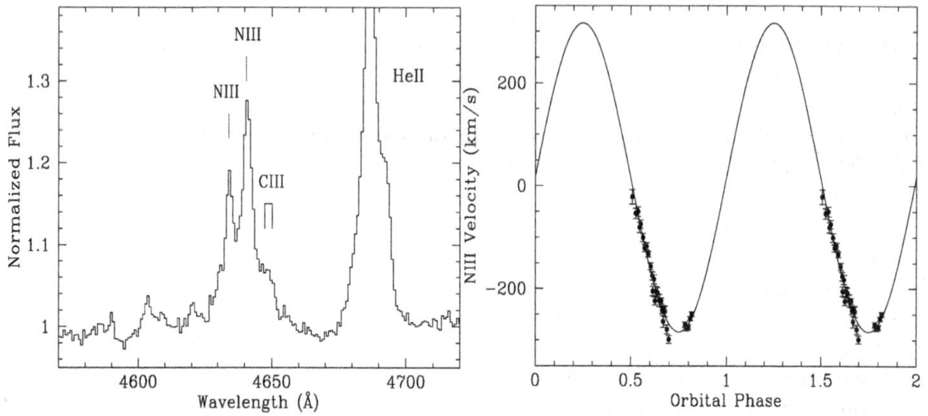

Fig. 11 Detecting companion stars in persistent X-ray binaries and XRTs in outburst. *Left*: high excitation emission lines from the irradiated donor in GX 339-4. *Right*: radial velocity motion of the Bowen CIII/NIII emission lines in GX 339-4 as a function of time. Adapted from Hynes et al. (2003a)

Second, the timing properties of the reprocessed light allows one to perform Echo Tomography experiments. Echo Tomography exploits time delays between X-ray and UV/optical variability as a function of orbital phase to map the illuminated sites in a binary (O'Brien et al. 2002). In particular, radiation reprocessed in the companion star gives rise to a sinusoidal modulation of time lag versus orbital phase. The shape of the modulation encodes information on the most fundamental parameters such as the binary inclination, mass ratio and stellar separation. Therefore, Echo Tomography offers an alternative route to derive accurate inclinations, which is critical for measuring BH masses. Unfortunately, attempts to measure correlated optical/X-ray variability using broad-band filters have resulted in little evidence for orbital modulation, with time lags pointing to reprocessing in the outer disc (Hynes 2005). Alternatively, the use of emission line light curves allows one to amplify the response of the donor's contribution by suppressing most of the background continuum light, dominated by the disc. This requires special instrumentation such as *ULTRACAM* (Dhillon et al. 2007), a high-speed triple-beam CCD camera, equipped with customized narrow-band filters centered in the Bowen blend and a nearby continuum. Following this strategy, time lags associated with the donor star have been finally presented for the neutron star X-ray binaries Sco X-1 and 4U 1636-536 (Muñoz-Darias et al. 2007, 2008b). In the latter case, three X-ray/optical bursts were observed at different orbital phases and their time lags were found to be consistent with those simulated for a plausible range of binary masses and inclinations. A careful subtraction of the underlying continuum seems critical to deliver unbiased inclinations and this requires further investigation. Echo Tomography has not been attempted on BH XRTs yet because it is very difficult to coordinate space facilities with adequate ground-based instruments, such as *ULTRACAM*, in ToO mode. The most efficient use of the technique would require implementing a fast read-out optical/UV camera onboard of an X-ray satellite. The camera should be provided with frame transfer EM3CCDs and a dual-channel, optimized for the Bowen lines (either in the optical or the higher energy OIII $\lambda3133/\lambda3444$ transitions) and an adjacent continuum. Such an instrument would guarantee both simultaneity and an optimal continuum subtraction.

Table 3 Best BH candidates for astrometric orbit determination

System	P_{orb} [days]	d [kpc]	a/d [μas]	V mag	K mag
V404 Cyg	6.5	2.4	61	18.4	12.4
Cyg X-1	5.6	1.9	45	8.9	6.5
GRS 1915+105	33.5	11.5	40	>25	13.0
A0620-00	0.33	1.1	20	18.3	14.5
GRO J1655-40	2.6	3.2	18	17.2	13.2

4.2 Optical/NIR Interferometry

A different avenue to obtain accurate inclinations relies on measuring the astrometric orbits of the companion stars to the BHs. Table 3 lists the angular size (a/d) of the companion's orbits for the best candidates, with values ranging between 20–60 μas. Resolving the projected orbits thus requires astrometry at micro-arcsec level which is technically challenging. Fortunately, prospects are bright thanks to new developments on optical/NIR interferometry in ground-based and space instrumentation. Three main facilities will be capable of measuring the orbits of some of the targets listed in Table 3: *GRAVITY*, *GAIA* and the proposed *Space Interferometry Mission SIM Lite* (Unwin et al. 2008). *GRAVITY*, a second generation instrument on VLTI, is scheduled to see first light in 2014 (Kendrew et al. 2012). With a goal of 10 μas precision astrometry for targets brighter than $K = 15$ it can potentially resolve the companion's orbit in GRS 1915+105. V404 Cyg and Cyg X-1 are also within reach, although pushing the observations to high airmasses. On the other hand, *SIM Lite* is designed to deliver micro-arcsecond astrometry in the optical band down to $V = 19$. While Cyg X-1 can be easily tackled, V404 Cyg is at the limit because of its optical faintness. Finally, *GAIA* will allow to measure the orbital motion of the companion star in Cyg X-1. The first direct measurements of binary orbits will clearly represent a major step forward in the knowledge of the fundamental parameters of these systems. Not only BH masses could be determined to better than 10 percent accuracy, but also precise inclinations for a sample of BHs will allow to quantify the impact of systematic errors in ellipsoidal light curve modeling. In addition, interferometry will deliver accurate distances and proper motions, key quantities to determine the BH natal kick and thus constrain the formation and evolution history of accreting BH X-ray binaries (e.g. Wong et al. 2012). Currently, parallax distances are only available for Cyg X-1 (Lestrade et al. 1999; Reid et al. 2011) and V404 Cyg (Miller-Jones et al. 2009) through radio Very Long Baseline Interferometry observations. The advent of the optical/NIR interferometry era will allow one to extend these measurements to any (sufficiently bright) BH binary in the Galaxy.

5 Intermediate-Mass Black Holes?

As one has seen above, stellar black holes with masses up to $\sim$16 $M_{\odot}$ (see Tables 1 and 2) have been found. Theoretically, the mass of a stellar mass black hole depends on the initial mass of the progenitor, how much mass is lost during the progenitor's evolution and on the supernova explosion mechanism (Belczynski et al. 2010; Fryer et al. 2012). Mass is lost through stellar winds, the amount of mass lost strongly depends on the metallicity of the star. For a low-metallicity star ($\sim$0.01 of the solar metallicity) it is possible to leave a black hole of $\lesssim$100 $M_{\odot}$ (Belczynski et al. 2010).

Thus, these models do allow for more massive stellar-mass black holes than have been found so far in our Galaxy. In current stellar evolution models the mass distribution of black hole remnants from massive stars peaks around 10 $M_\odot$, with a tail up to the said $\lesssim$100 $M_\odot$. In very massive stars ($\gtrsim$130 $M_\odot$) the production of free electrons and positrons, due to the increased gamma ray radiation, reduces the thermal pressure inside the core. This eventually leads to a runaway thermonuclear explosion that completely disrupts the star without leaving a black hole, causing the upper limit for a stellar black hole of $\sim$100 $M_\odot$.

Supermassive black holes (SMBHs) in AGN with masses of $>10^5$ $M_\odot$ (Greene and Ho 2007) have been identified. However, black holes with masses of several hundred to a few thousand solar masses remain elusive. Such intermediate-mass black holes (IMBHs) may be remnants from the first population of zero-metallicity massive stars. These Population III stars could have had masses above the pair-instability limit of around 130 $M_\odot$ and they may have collapsed into IMBHs (Madau and Rees 2001). It has also been suggested that IMBHs may form in the centers of dense stellar clusters via the merging of stellar-mass black holes (e.g., Miller and Hamilton 2002), or from the collapse of merged supermassive stars in very dense star clusters (e.g., Portegies Zwart and McMillan 2002). Primordial formation of a population of IMBHs is not ruled out (Afshordi et al. 2003), although they do not form a significant fraction of the dark matter (Ricotti et al. 2008).

IMBHs could allow for the assembly of supermassive black holes early in the Universe (e.g. Volonteri 2010, 2012). IMBHs may contain a similar or even a larger fraction of the baryonic matter than SMBH and they are potential sources of gravitational waves when they spiral into SMBHs (Madau and Rees 2001). Those IMBHs that were produced early in the life of the Universe but that did not evolve into SMBHs should still be wandering in the (outer) halos of galaxies, such as our Milky way (O'Leary and Loeb 2012; Rashkov and Madau 2013).

So, there is significant interest in finding these IMBHs and observationally determine their properties. A comprehensive review of constraints on the existence of IMBHs can be found in van der Marel (2003). The main problem in finding definitive proof for the existence of IMBHs is that the radius of their sphere of influence is small ($r_{infl} = \frac{GM_{BH}}{v^2}$). The meaning of v depends on the case under consideration: v can be the velocity of a recoiling BH with respect to that of surrounding stars, or for BHs in the centre of a stellar association/cluster it is the velocity dispersion of the stars). Below, we will revisit the possible observational evidence for the existence of IMBHs.

5.1 IMBHs in Globular Clusters?

Given that several formation mechanisms for IMBHs seem to require dense stellar environments (Miller and Hamilton 2002; Portegies Zwart and McMillan 2002), much effort has gone into investigating whether IMBHs are present in the cores of globular clusters.

5.1.1 Photometric and Kinematic Evidence

The number of stars as a function of distance (R) to the centre of a globular cluster has been used to investigate whether a central BH is present or not: this function has the form of N(R) $\propto R^{-0.75}$ (Bahcall and Wolf 1976, 1977). However, a globular cluster that has gone through core collapse (Djorgovski and King 1986) will show a similar distribution of the number of stars as a function of radius from the centre of the globular cluster (e.g. the case of M 15 see Grabhorn et al. 1992).

Instead, using both spectroscopic as well as photometric data and comparing that to a model of stellar orbits in the globular cluster potential one can derive whether a central

IMBH is present or not. The observed surface brightness profile together with an assumed mass-to-light ratio and a central IMBH of a given mass provides a model for the kinematics of the stars. This model is compared with the actual kinematics data such as the radial velocity profile and the velocity dispersion as a function of the distance to the cluster centre. The data is fit for the mass-to-light ratio and the mass of the IMBH in the cluster core (which can be zero). IMBHs, if present in the centre of a globular cluster, will increase the velocity dispersion in the core. This method has been used to argue for the presence of IMBHs in a sample of globular clusters (most notably G1 near M 31; Gebhardt et al. 2002, 2005). The presence of compact objects (white dwarfs, neutron stars and stellar-mass BHs that sink to the centre of the cluster due to dynamical friction) does not significantly alter the conclusion about the presence of an IMBH in the centre of the cluster (Gebhardt et al. 2005). However, conflicting reports on modeling the observable data exist: Baumgardt et al. (2003) obtain good fits to the spectroscopic and the photometric data using N-body calculations without the need of an IMBH (although the work of Gebhardt et al. 2005 challenges this). Lützgendorf et al. (2013) compile the globular clusters for which evidence exists that they harbor IMBHs and they further show that these IMBHs do not follow the M–σ correlation as found to SMBHs (Ferrarese and Merritt 2000).

5.1.2 Radio–X-ray Correlation

For G1, besides the dynamical evidence for the presence of an IMBH, Pooley and Rappaport (2006) showed that there was X-ray emission of a source associated with the globular cluster, however, the spatial scale of their XMM-*Newton* observation did not allow the authors to claim that the source resides in the core of G1. The source luminosity is both consistent with that expected for an IMBH accreting at a low (Bondi-Hoyle) rate as well as with the source being a low-mass X-ray binary (LMXB) accreting via Roche lobe overflow from a companion star. Kong et al. (2010) improved the X-ray position of the source using a *Chandra* observation. The source position is consistent with being equal to the centre of the globular cluster.

Ulvestad et al. (2007) provided evidence for the detection of radio emission from the same location as the X-ray emission. Given that the relation between BH mass, X-ray and radio luminosity appears to follow a "fundamental plane," in which the ratio of radio to X-ray luminosity increases as the ~0.8 power of the BH mass, an IMBH is more radio loud at a given X-ray luminosity than a stellar-mass BH (Merloni et al. 2003; Falcke et al. 2004). The result of Ulvestad et al. (2007) is consistent with an IMBH scenario but not with an LMXB scenario. However, the radio and X-ray observations were not simultaneous and the significance of the radio detection warranted further investigation. Miller-Jones et al. (2012) conducted the experiment where simultaneous X-ray (*Chandra*) and radio (VLA) observations of G1 were obtained. Whereas the X-ray luminosity was consistent with that found before, these authors find no evidence for radio emission at the position of the X-ray source down to a limit of 4.7 μJy per beam. Using the fundamental plane of BH activity this yields an upper limit on the mass of the IMBH in G1 of $<9.7 \times 10^3\ M_\odot$. Note that this upper limit is only marginally consistent with the mass of the IMBH derived dynamically (Gebhardt et al. 2005). Strader et al. (2012a) used deep VLA observations of the globular clusters M 15, M 19 and M 22 to search for radio emission associated with low-level accretion onto an IMBH. No radio sources were detected, putting additional constraints on any IMBH in these globular clusters. Overall, it is probably fair to state that IMBHs have not yet been detected beyond doubt in globular clusters although the evidence for their existence seems to be growing.

Incidentally, in the process of searching for IMBHs in globular clusters, Strader et al. (2012a) *did* find evidence for radio sources in globular clusters albeit not in their cores. The properties of some of these radio sources (flat spectrum radio emission and the limits on/detection of X-ray emission) did provide evidence for the presence of stellar-mass BHs in globular clusters (Strader et al. 2012b; Chomiuk et al. 2013).

5.2 Ultraluminous X-ray Sources

Ultraluminous X-ray sources (ULXs) are off-nuclear X-ray point sources in nearby galaxies with X–ray luminosities, $L_X \gtrsim 1 \times 10^{39}$–$10^{42}$ erg s^{-1} (e.g. Colbert and Mushotzky 1999). Their X-ray luminosities are suggestive of IMBHs if they radiate isotropically at sub-Eddington levels. Hence, ULXs could harbor IMBHs. Alternatively, the radiation in ULXs is not emitted isotropically or the Eddington limit is breached. So called slim disc models could potentially allow for the latter. Below we discuss these possible explanations for the high luminosity in ULXs in more detail.

5.2.1 Beaming and Super-Eddington Accretion

Relativistic beaming has been proposed as the cause for the high apparent luminosity of ULXs (Körding et al. 2002). This model predicts that for every high-luminosity source, there should be a larger number of lower luminosity sources; around 30 sources of $L_x \sim 10^{39}$ erg s^{-1} for every source of $L_x \sim 10^{40}$ erg s^{-1}. However, approximately 5–10 sources at 10^{39} erg s^{-1} are found for each source with luminosity 10^{40} erg s^{-1} (Walton et al. 2011). As this scenario requires a large number of jet sources beamed in other directions, for which there is no observational evidence it cannot explain the high luminosity for the majority of sources.

Geometrical beaming, where the emission is non-isotropic, provides, in combination with super-Eddington accretion, a viable explanation of the high ULX luminosity. The theory of super-Eddington black hole accretion was developed in the 1980s with the slim disk model of Abramowicz et al. (1980). Here, narrow funnels along the rotation axis of the accretion disk, in part caused by massive radiation-driven winds from the accretion disk, collimate radiation into beams, resulting in an apparent high luminosity as a combination of collimation and super-Eddington accretion rates. Recent simulations support this idea and even with a moderately super-critical mass supply an apparent luminosity $\approx 20 L_{Edd}$ can be reached (Ohsuga and Mineshige 2011). It has also been suggested that strong density inhomogeneities in the accretion disk could cause the escaping flux to exceed the Eddington limit by a factor of ~10–100 (Begelman 2002).

In these models ordinary stellar mass black holes can reach the luminosity observed for most ULXs. However, these scenarios do mean that the Eddington limit is violated and, so far, the Eddington limit works well for almost all known Galactic black holes and AGN (Raimundo et al. 2010). Some Galactic black holes do seem to approach or even breach the Eddington limit, but if so, they do so for only a brief period in time. The conclusion on this issue for Galactic black hole binaries is hampered by the uncertain distance for many sources (Jonker and Nelemans 2004). Of course, perhaps the limited inflow rate is causing the AGNs and most X-ray binaries to remain below the Eddington rate instead of there being a physical barrier at the Eddington limit (Rappaport et al. 2005).

The ULXs with the largest luminosities cannot easily be explained by the geometrical beaming model with super-Eddington accretion rates onto stellar mass black holes unless the mass of these stellar mass black holes is larger than those found in our own Galaxy. The

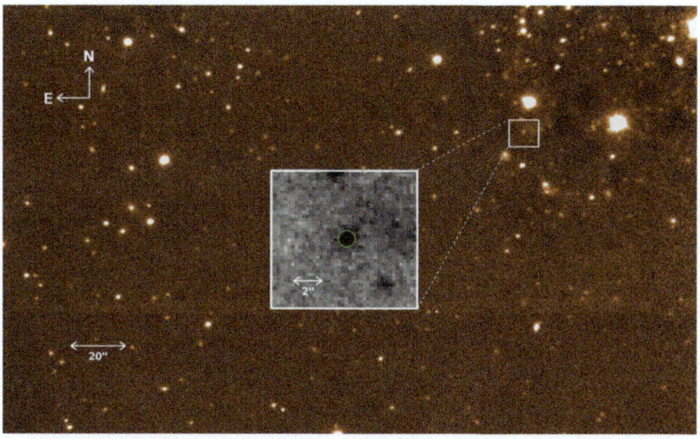

Fig. 12 William Herschel Telescope observations in the Ks-band of the ULX Holmberg II X-1. The *small circle* indicates the *Chandra* position of the ULX. The *inset* shows the unique Ks-band counterpart to the ULX (from Heida et al. in prep.)

higher the luminosity, the less likely it is that they can be explained as the high-luminosity end of the X-ray binary stellar-mass BH distribution.

A strong case for an IMBH is provided by the variable, very bright ULX ESO 243-49 X-1 (Farrell et al. 2009). Its X-ray luminosity is too high for a stellar-mass black hole even in the presence of some beaming. Given the evidence for the detection of a redshifted H_α emission line in the optical counterpart to ESO 243-49 X-1 (Wiersema et al. 2010), the uncertainty on its distance is reduced with respect to other bright ULXs. Another case for an IMBH is M82 X41.4+60, whereas it does not reach peak luminosities as high as ESO 243-49 X-1, the maximum luminosity is still uncomfortably high for a stellar-mass black hole (Strohmayer and Mushotzky 2003).

Are ULXs really IMBHs or stellar–mass black holes under peculiar accretion conditions? The answer to this question relies on dynamical mass measurements for the black holes in these systems similar to those available for stellar-mass BHs (see Sects. 2 and 3). Given the faintness of the optical counterparts (typically $V \geq 22$ mag; see for instance Liu et al. 2004 and Roberts et al. 2008), radial velocity studies of ULXs have concentrated on strong emission lines in the optical spectrum. However, these attempts to provide ULX dynamical masses have not met with success because the emission lines are originating in the accretion disc or a wind, and not in the donor star itself (cf. Liu et al. 2012; Roberts et al. 2011). One potential exception has been mentioned by Liu (2009) who interpret several emission lines as coming from a Wolf Rayet mass donor star to the ULX X-1 in M 101, although no follow-up work has been presented so far.

Searches for photospheric lines have so far concentrated on the blue part of the spectrum as many work on the hypothesis that the donor stars are blue supergiants. This is based on the fact that some ULXs are near a young star cluster and by the blue colours of ULX optical counterparts. However, the blue colours are also consistent with emission from accretion disks as recently has been confirmed by Soria et al. (2012) for a ULX (see also Jonker et al. 2012 and Grisé et al. 2012 for discussions). In fact, it could well be that a significant fraction of the donor stars are red supergiants (Copperwheat et al. 2005, 2007) which are intrinsically bright in the infrared, ($M_V \sim -6$, V–H ~ 4, H–K ~ 0) in contrast with the blue supergiants ($M_V \sim -6$, V–H ~ 0, H–K ~ 0). Therefore, some ULX systems may resemble the bright Galactic X-ray binary GRS 1915+105 that has a red giant donor star (Greiner et al. 2001b) and thereby radial velocity measurements from infrared photospheric lines will be possible.

From a near-infrared survey of ULX sources within 10 Mpc, Heida et al. (in prep.) found evidence (e.g. see Fig. 12) that about 10–15 percent of ULXs have a bright near-infrared counterpart which are consistent with a red-supergiant companion star. However, future near-infrared spectroscopy of these targets should be done to investigate the nature of these near-infrared counterparts.

Acknowledgements We thanks Cristina Zurita, Andrew Cantrell and Rob Hynes for providing us with Figs. 5, 6 and 11, respectively. J.C. acknowledges support by the Spanish Ministerio de Economía y Competitividad (MINECO) under grant AYA2010-18080.

References

M.A. Abramowicz, M. Calvani, L. Nobili, Astrophys. J. **242**, 772 (1980)
N. Afshordi, P. McDonald, D.N. Spergel, Astrophys. J. **594**, L71 (2003)
M. Armas Padilla, P.N. Degenaar, D.M. Russell, R. Wijnands, Mon. Not. R. Astron. Soc. **428**, 3083 (2013a)
M. Armas Padilla, R. Wijnands, D. Altamirano, M. Méndez, J.M. Miller, N. Degenaar, Mon. Not. R. Astron. Soc., arXiv:1308.4326 (2013b, submitted)
J.N. Bahcall, R.A. Wolf, Astrophys. J. **209**, 214–232 (1976)
J.N. Bahcall, R.A. Wolf, Astrophys. J. **216**, 883–907 (1977)
C.D. Bailyn, J.A. Orosz, J.E. McClintock, R.A. Remillard, Nature **378**, 157 (1995)
H. Baumgardt, J. Makino, P. Hut, S. McMillan, S. Portegies Zwart, Astrophys. J. Lett. **589**, 25–28 (2003)
G. Beekman, T. Shahbaz, T. Naylor, P.A. Charles, Mon. Not. R. Astron. Soc. **281**, L1 (1996)
G. Beekman, T. Shahbaz, T. Naylor, P.A. Charles, R.M. Wagner, P. Martini, Mon. Not. R. Astron. Soc. **290**, 303 (1997)
M.E. Beer, P. Podsiadlowsky, Mon. Not. R. Astron. Soc. **331**, 351 (2002)
M.C. Begelman, Astrophys. J. Lett. **568**, 97–100 (2002)
K. Belczynski, T. Bulik, C.L. Fryer, A. Ruiter, F. Valsecchi, J.S. Vink, J.R. Hurley, Astrophys. J. **714**, 1217 (2010)
K. Belczynski, G. Wiktorowicz, C.L. Fryer, D.E. Holz, V. Kalogera, Astrophys. J. **757**, 91 (2012)
T.M. Belloni, S.E. Motta, T. Muñoz-Darias, Bull. Astron. Soc. India **39**, 409 (2011)
C.T. Bolton, Nature **235**, 271 (1972a)
C.T. Bolton, Nature **240**, 124 (1972b)
C. Brocksopp, A.E. Tarasov, V.M. Lyuty, P. Roche, Astron. Astrophys. **343**, 861 (1999)
M. Buxton, S. Vennes, Publ. Astron. Soc. Pac. **18**, 91 (1999)
S.M. Caballero-Nieves et al., Astrophys. J. **701**, 1895 (2009)
P.J. Callanan, M.R. Garcia, J.E. McClintock, P. Zhao, R. Remillard, F. Haberl, Astrophys. J. **461**, 351 (1996a)
P.J. Callanan, M.R. Garcia, A.V. Filippenko, I. McLean, H. Teplitz, Astrophys. J. **470**, L57 (1996b)
D.E. Calvelo, S.D. Vrtilek, D. Steeghs, M.A.P. Torres, J. Neilsen, A.V. Filippenko, J.I. González Hernández, Mon. Not. R. Astron. Soc. **399**, 539 (2009)
A.G. Cantrell, C.D. Bailyn, J.E. McClintock, J.A. Orosz, Astrophys. J. **673**, L159 (2008)
A.G. Cantrell et al., Astrophys. J. **710**, 1127 (2010)
J. Casares, in *IAU Colloq.*, vol. 158 (1996), p. 395
J. Casares, in *IAU Symp.*, vol. 238 (2007), p. 3
J. Casares, in *Ap&SS Proc.* (Springer, Berlin, 2010), p. 3
J. Casares, P.A. Charles, Mon. Not. R. Astron. Soc. **271**, l5 (1994)
J. Casares, P.A. Charles, T. Naylor, Nature **355**, 614 (1992)
J. Casares, A.C. Martin, P.A. Charles, E.L. Martín, R. Rebolo, E.T. Harlaftis, A.J. Castro-Tirado, Mon. Not. R. Astron. Soc. **276**, L35 (1995a)
J. Casares, P.A. Charles, T.R. Marsh, Mon. Not. R. Astron. Soc. **277**, L45 (1995b)
J. Casares, E.L. Martín, P.A. Charles, P. Molaro, R. Rebolo, New Astron. **1**, 299 (1997)
J. Casares, C. Zurita, T. Shahbaz, P.A. Charles, R.P. Fender, Astrophys. J. **613**, L133 (2004)
J. Casares et al., Astrophys. J. Suppl. Ser. **181**, 238 (2009)
P.A. Charles, M.J. Coe, in *Compact Stellar X-Ray Sources*, vol. 39 (2006), p. 215
A.M. Cherepashchuk, V.M. Lyutiy, R.A. Sunyaev, Astron. Ž. **50**, 3 (1973)
L. Chomiuk, J. Strader, T.J. Maccarone, J.C.A. Miller-Jones, C. Heinke, E. Noyola, A.C. Seth, S. Ransom, arXiv:1306.6624 (2013)
E.J.M. Colbert, R.F. Mushotzky, Astrophys. J. **519**, 89 (1999)
C. Copperwheat, M. Cropper, R. Soria, K. Wu, Mon. Not. R. Astron. Soc. **362**, 79–88 (2005)
C. Copperwheat, M. Cropper, R. Soria, K. Wu, Mon. Not. R. Astron. Soc. **376**, 1407–1423 (2007)
M. Coriat, R.P. Fender, G. Dubus, Mon. Not. R. Astron. Soc. **424**, 1991 (2012)
J.M. Corral-Santana, J. Casares, T. Shabaz, C. Zurita, I.G. Martínez-Pais, P. Rodríguez-Gil, Mon. Not. R. Astron. Soc. **413**, L15 (2011)

J.M. Corral-Santana, J. Casares, T. Muñoz-Darias, P. Rodríguez-Gil, T. Shabaz, C. Zurita, M.A.P. Torres, A. Tyndall, Science **339**, 1048 (2013)
A. Cowley, D. Crampton, G.J.B. Hutchings, R. Remillard, J.E. Penfold, Astrophys. J. **272**, 118 (1983)
P.A. Crowther, R. Barnard, G.S. Carpano, J.S. Clark, V.S. Dhillon, A.M.T. Pollock, Mon. Not. R. Astron. Soc. **403**, L41 (2010)
N. Degenaar, R. Wijnands, Astron. Astrophys. **524**, 69 (2010)
V.S. Dhillon et al., Mon. Not. R. Astron. Soc. **378**, 825 (2007)
S. Djorgovski, I.R. King, Astrophys. J. Lett. **305**, 61–65 (1986)
H. Falcke, E. Körding, S. Markoff, Astron. Astrophys. **414**, 895–903 (2004)
W.M. Farr, N. Sravan, A. Cantrell, L. Kreidberg, C.D. Bailyn, I. Mandel, V. Kalogera, Astrophys. J. **741**, 103 (2011)
S.A. Farrell, N.A. Webb, D. Barret, O. Godet, J.M. Rodrigues, Nature **460**, 73–75 (2009)
R.P. Fender et al., Mon. Not. R. Astron. Soc. **304**, 865 (1999)
L. Ferrarese, D. Merritt, Astrophys. J. Lett. **539**, 9–12 (2000)
A.V. Filippenko, R. Chornock, IAUC 7644, 1 (2001)
A.V. Filippenko, T. Matheson, L.C. Ho, Astrophys. J. **455**, 614 (1995a)
A.V. Filippenko, T. Matheson, A.J. Barth, Astrophys. J. **455**, L139 (1995b)
A.V. Filippenko, T. Matheson, D.C. Leonard, A.J. Barth, S.D. van Dyk, Publ. Astron. Soc. Pac. **109**, 461 (1997)
A.V. Filippenko, D.C. Leonard, T. Matheson, W. Li, E.C. Moran, A.G. Riess, Publ. Astron. Soc. Pac. **111**, 969 (1999)
C.S. Froning, E.L. Robinson, Astron. J. **121**, 2212 (2001)
C.L. Fryer, V. Kalogera, Astrophys. J. **554**, 548 (2001)
C.L. Fryer, K. Belczynski, G. Wiktorowicz, M. Dominik, V. Kalogera, D.E. Holz, Astrophys. J. **749**, 91 (2012)
M.R. Garcia, P.J. Callanan, J.E. McClintock, P. Zhao, Astrophys. J. **460**, 932 (1996)
K. Gebhardt, R.M. Rich, L.C. Ho, Astrophys. J. Lett. **578**, 41–45 (2002)
K. Gebhardt, R.M. Rich, L.C. Ho, Astrophys. J. **634**, 1093–1102 (2005)
D.M. Gelino, T.E. Harrison, Astrophys. J. **599**, 1254 (2003)
D.M. Gelino, T.E. Harrison, B.J. McNamara, Astrophys. J. **122**, 971 (2001a)
D.M. Gelino, T.E. Harrison, J.A. Orosz, Astron. J. **122**, 2668 (2001b)
D.M. Gelino, S. Balman, U. Kiziloglu, A. Yilmaz, E. Kalemci, J.A. Tomsick, Astrophys. J. **642**, 438 (2006)
D.R. Gies, C.T. Bolton, Astrophys. J. **260**, 240 (1982)
D.R. Gies, C.T. Bolton, Astrophys. J. **304**, 371 (1986)
J.I. González Hernández, J. Casares, Astron. Astrophys. **516**, A58 (2010)
J.I. González Hernández et al., Astrophys. J. **679**, 732 (2008a)
J.I. González Hernández, R. Rebolo, G. Israelian, Astron. Astrophys. **478**, 203 (2008b)
J.I. González Hernández, R. Rebolo, J. Casares, Astrophys. J. **744**, L25 (2012)
R.P. Grabhorn, H.N. Cohn, P.M. Lugger, B.W. Murphy, Astrophys. J. **392**, 86–98 (1992)
D.F. Gray, *The Observation and Analysis of Stellar Photospheres* (Cambridge University Press, Cambridge, 1992)
J.E. Greene, L.C. Ho, Astrophys. J. **670**, 92 (2007)
J. Greene, C.D. Bailyn, J.A. Orosz, Astrophys. J. **554**, 1290 (2001)
J. Greiner, J.G. Cuby, M.J. McCaughrean, Nature **414**, 522 (2001a)
J. Greiner, J.G. Cuby, M.J. McCaughrean, A.J. Castro-Tirado, R.E. Mennickent, Astron. Astrophys. **373**, 37–40 (2001b)
F. Grisé, P. Kaaret, S. Corbel, H. Feng, D. Cseh, L. Tao, Astrophys. J. **745**, 123 (2012)
E.T. Harlaftis, J. Greiner, Astron. Astrophys. **414**, L13 (2004)
E.T. Harlaftis, K. Horne, A.V. Filippenko, Publ. Astron. Soc. Pac. **108**, 762 (1996)
E.T. Harlaftis, D. Steeghs, K. Horne, A.V. Filippenko, Astron. J. **114**, 1170 (1997)
E.T. Harlaftis, S. Collier, K. Horne, A.V. Filippenko, Astron. Astrophys. **341**, 491 (1999)
C.A. Haswell, E.L. Robinson, K. Horne, R.F. Stiening, T.M.C. Abbott, Astrophys. J. **411**, 802 (1993)
A. Herrero, R.P. Kudritzki, R. Gabler, J.M. Vilchez, A. Gabler, Astron. Astrophys. **297**, 556 (1995)
R.M. Hjellming, M.P. Rupen, Nature **375**, 464 (1995)
D.J. Hurley, P.J. Callanan, P. Elebert, M.T. Reynolds, Mon. Not. R. Astron. Soc. **430**, 1832 (2013)
J.B. Hutchings, D. Crampton, A. Cowley, Astrophys. J. **275**, L43 (1983)
J.B. Hutchings, D. Crampton, A. Cowley, L. Bianchi, I.B. Thompson, Astron. J. **94**, 340 (1987)
R.I. Hynes, in *ASP Conf. Ser.*, vol. 330 (2005), p. 237
R.I. Hynes, D. Steeghs, J. Casares, P.A. Charles, K. O'Brien, Astrophys. J. **583**, L95 (2003a)
R.I. Hynes, P.A. Charles, J. Casares, C.A. Haswell, C. Zurita, T. Shahbaz, Mon. Not. R. Astron. Soc. **340**, 447 (2003b)

R.I. Hynes et al., Astrophys. J. **611**, L125 (2004)
Z. Ioannou, E.L. Robinson, W.F. Welsh, C.A. Haswell, Astrophys. J. **127**, 481 (2004)
P.G. Jonker, G. Nelemans, Mon. Not. R. Astron. Soc. **354**, 355–366 (2004)
P.G. Jonker et al., Astrophys. J. Suppl. Ser. **194**, 18 (2011)
P.G. Jonker, M. Heida, M.A.P. Torres, J.C.A. Miller-Jones, A.C. Fabian, E.M. Ratti, G. Miniutti, D.J. Walton, T.P. Roberts, Astrophys. J. **758**, 28 (2012)
S. Kendrew et al., SPIE J. **8446**, 7 (2012)
J. Khargharia, C.S. Froning, E.L. Robinson, Astrophys. J. **716**, 1105 (2010)
J. Khargharia, C.S. Froning, E.L. Robinson, D.M. Gelino, Astron. J. **145**, 21 (2013)
P.D. Kiel, J.R. Hurley, Mon. Not. R. Astron. Soc. **369**, 1152 (2006)
A.R. King, U. Kolb, E. Szuszkiewicz, Astrophys. J. **488**, 89 (1997)
A.K.H. Kong, C.O. Heinke, R. di Stefano, H.N. Cohn, P.M. Lugger, P. Barmby, W.H.G. Lewin, F.A. Primini, Mon. Not. R. Astron. Soc. **407**, 84–88 (2010)
E. Körding, H. Falcke, S. Markoff, Astron. Astrophys. **382**, L31 (2002)
L. Kreidberg, C.D. Bailyn, W. Farr, V. Kalogera, Astrophys. J. **757**, 36 (2012)
D.T. Larson, E. Schulman, Astron. J. **113**, 618 (1997)
J.-P. Lasota, New Astron. Rev. **45**, 449 (2001)
J.M. Lattimer, Annu. Rev. Nucl. Part. Sci. **62**, 485 (2012)
J.-F. Lestrade et al., Astron. Astrophys. **344**, 1014 (1999)
J. Liu, Astrophys. J. **704**, 1628–1639 (2009)
J.-F. Liu, J.N. Bregman, P. Seitzer, Astrophys. J. **602**, 249–256 (2004)
J. Liu, J. Orosz, J.N. Bregman, Astrophys. J. **745**, 89 (2012)
L.B. Lucy, Z. Astrophys. **65**, 89 (1967)
N. Lützgendorf, M. Kissler-Patig, N. Neumayer, H. Baumgardt, E. Noyola, P.T. de Zeeuw, K. Gebhardt, B. Jalali, A. Feldmeier, Astron. Astrophys. **555**, 26 (2013)
P. Madau, M.J. Rees, Astrophys. J. **551**, L27 (2001)
T.R. Marsh, E.L. Robinson, J.H. Wood, Mon. Not. R. Astron. Soc. **266**, 137 (1994)
A.C. Martin, J. Casares, P.A. Charles, F. van dr Hooft, J. van Paradijs, Mon. Not. R. Astron. Soc. **274**, L46 (1995)
J.E. McClintock, R.A. Remillard, Astrophys. J. **308**, 110 (1986)
J.E. McClintock, R.A. Remillard, *Black Hole Binaries* (2006), pp. 157–213. Chap. 4
J.E. McClintock, M.R. Garcia, N. Caldwell, E.E. Falco, P.M. Garnavich, P. Zhao, Astrophys. J. **551**, L147 (2001)
A. Merloni, S. Heinz, T. di Matteo, Mon. Not. R. Astron. Soc. **345**, 1057–1076 (2003)
M.C. Miller, D.P. Hamilton, Mon. Not. R. Astron. Soc. **330**, 232 (2002)
J.C.A. Miller-Jones, P.G. Jonker, V. Dhawan, W. Brisken, M.P. Rupen, G. Nelemans, E. Gallo, Astrophys. J. **706**, L230 (2009)
J.C.A. Miller-Jones, J.M. Wrobel, G.R. Sivakoff, C.O. Heinke, R.E. Miller, R.M. Plotkin, R. Di Stefano, J.E. Greene, L.C. Ho, T.D. Joseph, A.K.H. Kong, T.J. Maccarone, Astrophys. J. Lett. **755**, 1 (2012)
S. Mineshige, J.C. Wheeler, Astrophys. J. **343**, 241 (1989)
I.F. Mirabel, I. Rodrigues, Science **300**, 1119 (2003)
T. Muñoz-Darias, J. Casares, I.G. Martínez-Pais, Mon. Not. R. Astron. Soc. **635**, 502 (2005)
T. Muñoz-Darias, I.G. Martínez-Pais, J. Casares, V.S. Dhillon, T.R. Marsh, R. Cornelisse, D. Steeghs, P.A. Charles, Mon. Not. R. Astron. Soc. **379**, 1673 (2007)
T. Muñoz-Darias, J. Casares, I.G. Martínez-Pais, Mon. Not. R. Astron. Soc. **385**, 2205 (2008a)
T. Muñoz-Darias et al., in *AIP Conf Proc.*, vol. 984 (2008b), p. 15
R. Narayan, J.E. McClintock, Astrophys. J. **623**, 1017 (2005)
J. Neilsen, D. Steeghs, S.D. Vrtilek, Mon. Not. R. Astron. Soc. **384**, 849 (2008)
Z. Ninkov, G.A.H. Walker, S. Yang, Astrophys. J. **321**, 425 (1987)
K. O'Brien, K. Horne, R.I. Hynes, W. Chen, C.A. Haswell, M.D. Still, Mon. Not. R. Astron. Soc. **334**, 4260 (2002)
D. O'Donoghue, P.A. Charles, Mon. Not. R. Astron. Soc. **282**, 191 (1996)
K. Ohsuga, S. Mineshige, Astrophys. J. **736**, 2 (2011)
R.M. O'Leary, A. Loeb, Mon. Not. R. Astron. Soc. **421**, 2737–2750 (2012)
J.A. Orosz, ATel **67** (2001)
J.A. Orosz, in *IAU Symp.*, vol. 212 (2003), p. 365
J.A. Orosz, C.D. Bailyn, Astrophys. J. **446**, L59 (1995)
J.A. Orosz, C.D. Bailyn, Astrophys. J. **477**, 876 (1997)
J.A. Orosz, P.H. Hauschildt, Astron. Astrophys. **364**, 265 (2000)
J.A. Orosz, C.D. Bailyn, J.E. McClintock, R.A. Remillard, C.B. Foltz, Astrophys. J. **436**, 848 (1994)
J.A. Orosz, C.D. Bailyn, J.E. McClintock, R.A. Remillard, Astrophys. J. **468**, 380 (1996)

J.A. Orosz, R.K. Jain, C.D. Bailyn, J.E. McClintock, R.A. Remillard, Astrophys. J. **499**, 375 (1998)
J.A. Orosz et al., Astrophys. J. **555**, 489 (2001)
J.A. Orosz et al., Astrophys. J. **568**, 845 (2002)
J.A. Orosz, J.E. McClintock, R.A. Remillard, S. Corbel, Astrophys. J. **616**, 376 (2004)
J.A. Orosz et al., Nature **449**, 872 (2007)
J.A. Orosz et al., Astrophys. J. **697**, 573 (2009)
J.A. Orosz, J.F. Steiner, J.E. McClintock, M.A.P. Torres, R.A. Remillard, C.D. Bailyn, J.M. Miller, Astrophys. J. **730**, 75 (2011a)
J.A. Orosz, J.E. McClintock, J.P. Aufdenberg, R.A. Remillard, M. Reid, R. Narayan, L. Gou, Astrophys. J. **742**, 840 (2011b)
F. Özel, D. Psaltis, R. Narayan, J.E. McClintock, Astrophys. J. **725**, 1918 (2010)
E.P. Pavlenko, A.C. Martin, J. Casares, P.A. Charles, N.A. Ketsaris, Mon. Not. R. Astron. Soc. **281**, 1094 (1996)
W. Pietsch et al., Astrophys. J. **646**, 420 (2006)
Ph. Podsiadlowski, S. Rappaport, Z. Han, Mon. Not. R. Astron. Soc. **341**, 385 (2003)
D. Pooley, S. Rappaport, Astrophys. J. Lett. **644**, 45–48 (2006)
S.F. Portegies Zwart, S.L.W. McMillan, Astrophys. J. **576**, 899–907 (2002)
S.I. Raimundo, A.C. Fabian, F.E. Bauer, D.M. Alexander, W.N. Brandt, B. Luo, R.V. Vasudevan, Y.Q. Xue, Mon. Not. R. Astron. Soc. **408**, 1714–1720 (2010)
S.A. Rappaport, P.C. Joss, in *Accretion Driven X-Ray Sources* (Cambridge University Press, Cambridge, 1983), p. 33
S.A. Rappaport, P. Podsiadlowski, E. Pfahl, Mon. Not. R. Astron. Soc. **356**, 401–414 (2005)
V. Rashkov, P. Madau, Astrophys. J., arXiv:1303.3929 (2013, submitted)
M.J. Reid, J.E. McClintock, R. Narayan, L. Gou, R.A. Remillard, J.A. Orosz, Astrophys. J. **742**, 83 (2011)
R.A. Remillard, J.E. McClintock, Annu. Rev. Astron. Astrophys. **44**, 49 (2006)
R.A. Remillard, J.E. McClintock, C.D. Bailyn, Astrophys. J. **399**, L145 (1992)
R.A. Remillard, J.A. Orosz, J.E. McClintock, C.D. Bailyn, Astrophys. J. **459**, 226 (1996)
M.T. Reynolds, P.J. Callanan, A.V. Filippenko, Mon. Not. R. Astron. Soc. **374**, 657 (2007)
M. Ricotti, J.P. Ostriker, K.J. Mack, Astrophys. J. **680**, 829 (2008)
H. Ritter, A.R. King, in *ASP Conference Series*, vol. 261 (2002), p. 531
T.P. Roberts, A.J. Levan, M.R. Goad, Mon. Not. R. Astron. Soc. **387**, 73–78 (2008)
T.P. Roberts, J.C. Gladstone, A.D. Goulding, A.M. Swinbank, M.J. Ward, M.R. Goad, A.J. Levan, Astron. Nachr. **332**, 398 (2011)
R.W. Romani, Astron. Astrophys. **333**, 583 (1998)
D. Sanwal et al., Astrophys. J. **460**, 337 (1996)
M.J. Sarna, Astron. Astrophys. **224**, 98 (1989)
T. Shahbaz, Mon. Not. R. Astron. Soc. **339**, 1031 (2003)
T. Shahbaz, T. Naylor, P.A. Charles, Mon. Not. R. Astron. Soc. **268**, 756 (1994a)
T. Shahbaz, F.A. Ringwald, J.C. Bunn, T. Naylor, P.A. Charles, J. Casares, Mon. Not. R. Astron. Soc. **271**, L10 (1994b)
T. Shahbaz, F. van der Hooft, P.A. Charles, J. Casares, J. van Paradijs, Mon. Not. R. Astron. Soc. **282**, L47 (1996)
T. Shahbaz, T. Naylor, P.A. Charles, Mon. Not. R. Astron. Soc. **285**, 607 (1997)
T. Shahbaz, F. van der Hooft, J. Casares, P.A. Charles, J. van Paradijs, Mon. Not. R. Astron. Soc. **306**, 889 (1999)
T. Shahbaz, V.S. Dhillon, T.R. Marsh, C. Zurita, C.A. Haswell, P.A. Charles, R.I. Hynes, J. Casares, Mon. Not. R. Astron. Soc. **346**, 1116 (2003)
T. Shahbaz, V.S. Dhillon, T.R. Marsh, J. Casares, C. Zurita, P.A. Charles, Mon. Not. R. Astron. Soc. **403**, 2167 (2010)
T. Shahbaz, D.M. Russell, C. Zurita, J. Casares, J.M. Corral-Santana, V.S. Dhillon, T.R. Marsh, Mon. Not. R. Astron. Soc. **434**, 2696 (2013)
N. Shaposhnikov, L. Titarchuk, Astrophys. J. **663**, 445 (2007)
J.M. Silverman, A.V. Filippenko, Astrophys. J. **678**, L17 (2008)
L. Song et al., Astron. J. **140**, 794 (2010)
R. Soria, K.D. Kuntz, P.F. Winkler, W.P. Blair, K.S. Long, P.P. Plucinsky, B.C. Whitmore, Astrophys. J. **750**, 152 (2012)
D. Steeghs, J. Casares, Astrophys. J. **568**, 273 (2002)
D. Steeghs, J.E. McClintock, S.G. Parsons, M.J. Reid, S. Littlefair, V.S. Dhillon, Astrophys. J. **768**, 185 (2013)
J. Strader, L. Chomiuk, T.J. Maccarone, J.C.A. Miller-Jones, A.C. Seth, C.O. Heinke, G.R. Sivakoff, Astrophys. J. Lett. **750**, 27 (2012a)

J. Strader, L. Chomiuk, T.J. Maccarone, J.C.A. Miller-Jones, A.C. Seth, Nature **490**, 71–73 (2012b)
T.E. Strohmayer, R.F. Mushotzky, Astrophys. J. Lett. **586**, 61–64 (2003)
Y. Tanaka, N. Shibazaki, Annu. Rev. Astron. Astrophys. **34**, 607 (1996)
M.A.P. Torres, P.J. Callanan, M.R. Garcia, P. Zhao, S. Laycock, A.K.H. Kong, Astrophys. J. **612**, 1026 (2004)
M.A.P. Torres et al., Mon. Not. R. Astron. Soc., arXiv:1310.0224v1 (2013, submitted)
J.S. Ulvestad, J.E. Greene, L.C. Ho, Astrophys. J. Lett. **661**, 151–154 (2007)
S.C. Unwin, M. Shao, S.J. Edberg, SPIE J. **7013**, 78 (2008)
A.K.F. Val-Baker, A.J. Norton, I. Negueruela, AIP Conf. Proc. **924**, 530 (2007)
F. Valsecchi et al., Nature **468**, 7320 (2010)
E.P.J. van den Heuvel, in *Proc. Inter. Space Year Conf. ESA ISY-3* (1992), p. 29
F. van der Hooft, M.H.M. Heemskerk, F. Alberts, J. van Paradijs, Astron. Astrophys. **329**, 538 (1998)
R.P. van der Marel, in *The Astrophysics of Gravitational Wave Sources*, ed. by J.M. Centrella. American Institute of Physics Conference Series, vol. 686 (2003), pp. 115–124
J. van Paradijs, J.E. McClintock, in *X-Ray Binaries*. Cambridge Astrophysics Series, vol. 26 (1995), p. 58
M. Volonteri, Nature **466**, 1049 (2010)
M. Volonteri, Science **337**, 544 (2012)
R.A. Wade, K. Horne, Astrophys. J. **324**, 411 (1988)
R.M. Wagner, T.J. Kreidl, S.B. Howell, S.G. Starrfield, Astrophys. J. **401**, L25 (1992)
R.M. Wagner, C.B. Foltz, T. Shahbaz, J. Casares, P.A. Charles, S.G. Starrfield, P. Hewett, Astrophys. J. **556**, 42 (2001)
D.J. Walton, T.P. Roberts, S. Mateos, V. Heard, Mon. Not. R. Astron. Soc. **416**, 1844 (2011)
N.A. Webb, T. Naylor, Z. Ioannou, P.A. Charles, T. Shahbaz, Mon. Not. R. Astron. Soc. **317**, 528 (2000)
B.L. Webster, P. Murdin, Nature **235**, 37 (1972)
S. Wellstein, N. Langer, Astron. Astrophys. **350**, 148 (1999)
K. Wiersema, S.A. Farrell, N.A. Webb, M. Servillat, T.J. Maccarone, D. Barret, O. Godet, Astrophys. J. Lett. **721**, 102–106 (2010)
M.G. Witte, G.J. Savonije, Astron. Astrophys. **366**, 840 (2001)
T.-W. Wong, F. Valsecchi, T. Fragos, V. Kalogera, Astrophys. J. **747**, 111 (2012)
L.R. Yungelson, J.P. Lasota, G. Nelemans, G. Dubus, E.P.J. van den Heuvel, J. Dewi, S. Portegies Zwart, Astron. Astrophys. **454**, 559 (2006)
J. Ziółkowski, Mon. Not. R. Astron. Soc. **358**, 851 (2005)
C. Zurita et al., Mon. Not. R. Astron. Soc. **333**, 791 (2002)
C. Zurita, J. Casares, T. Shahbaz, Astrophys. J. **582**, 369 (2003)

DOI 10.1007/978-1-4939-2227-7_13
Reprinted from *Space Science Reviews* Journal, DOI 10.1007/s11214-013-9987-4

Measuring the Masses of Supermassive Black Holes

Bradley M. Peterson

Received: 17 February 2013 / Accepted: 22 April 2013 / Published online: 8 May 2013

Abstract Supermassive black holes reside at the centers of most, if not all, massive galaxies: the difference between active and quiescent galaxies is due to differences in mass accretion rate and radiative efficiency rather than whether or not they have nuclear black holes. In this contribution, methods for measuring the masses of supermassive black holes are discussed, with emphasis on reverberation mapping which is most generally applicable to accreting supermassive black holes and, in particular, to distant quasars where time resolution can be used as a surrogate for angular resolution. Indirect methods based on scaling relationships from reverberation mapping studies are also discussed, along with their current limitations.

Keywords Active galactic nuclei · Black hole · Reverberation mapping

1 Introduction

As recently as 20 years ago, whether or not active galactic nuclei (AGNs) were powered by accretion onto supermassive black holes was widely regarded as an open question. The theoretical arguments supporting gravitational accretion as the primary source of power in AGNs were in place within a few years of the discovery of quasars (Zel'dovich and Novikov 1964; Salpeter 1964; Lynden-Bell 1969) and some two decades later Rees (1984) convincingly argued that supermassive black holes were the inevitable endpoint of *any* of the scenarios proposed to account for activity in galactic nuclei. But definitive observational proof remained illusive.

Now, however, it is generally accepted that black holes reside at the center of most, if not all, massive galaxies, both quiescent and active. And, perhaps ironically, the first convincing proof of the existence of supermassive black holes was not in AGNs, but in quiescent galaxies. It was primarily the high angular resolution afforded by *Hubble Space Telescope* that enabled determination of the masses of nuclear black holes; for the first time, the dynamics

B.M. Peterson (✉)
Department of Astronomy, The Ohio State University, Columbus, OH 43210, USA
e-mail: peterson.12@osu.edu

of stars and gas in the nuclei of nearby galaxies could be studied on scales smaller than the black hole radius of influence, $R_{\rm BH} = GM_{\rm BH}/\sigma_*^2$, where $M_{\rm BH}$ is the black hole mass, σ_* is the velocity dispersion of the stars in the host-galaxy bulge, and G is the gravitational constant. Detection of supermassive black holes in quiescent galaxies showed that AGNs are different from other galaxies not because they harbor supermassive black holes in their nuclei, but because their supermassive black holes are actively accreting mass at fairly high rates, typically more than $\sim$0.1 % of the Eddington rate.

The realization that supermassive black holes are ubiquitous led to improved understanding of quasar evolution. The AGN population at the present epoch is a small fraction of what it was at its peak at $2 < z < 3$ when the comoving space density of quasars was at its highest. The quiescent supermassive black holes in most galaxies are clearly the remnants of the quasars of the distant past. The demographics of black holes (e.g., Shankar 2009; Vestergaard and Osmer 2009; Shen and Kelly 2012, and references therein) are thus of keen interest for understanding the accretion history of the universe.

2 Measuring the Masses of Supermassive Black Holes

2.1 Direct versus Indirect Methods

A distinction must first be drawn between *direct* and *indirect* methods of measuring black hole masses.

- Direct measurements are those where the mass is derived from the dynamics of stars or gas accelerated by the black hole itself. Direct methods include stellar and gas dynamical modeling and reverberation mapping.
- Indirect methods are those where the black hole mass is inferred from observables that are correlated with the black hole mass. This includes masses based on correlations between black hole masses and host-galaxy properties, such as the velocity dispersion of bulge stars, i.e., the $M_{\rm BH}$–σ_* relationship (Ferrarese and Merritt 2000; Gebhardt et al. 2000a; Tremaine et al. 2002), or the bulge luminosity, i.e., the $M_{\rm BH}$–$L_{\rm bulge}$ relationship (Kormendy and Richstone 1995; Magorrian et al. 1998), and masses based on AGN scaling relationships, such as the R–L relationship discussed in Sect. 5.1.

It is also common to distinguish among "primary," "secondary," and even "tertiary" methods, based on the number of assumptions and model dependence. Reverberation mapping (Sect. 3) is an interesting example: it is a *direct* method as it is based on observations of gas that is accelerated by the gravitational potential of the central black hole, but, as generally practiced, it is also a *secondary* method because absolute calibration of the mass scale depends on reference to another method (see Sect. 4.1). This is, of course, a temporary condition since modeling of the dynamics of the emission-line gas in AGNs (Sect. 4.2) is beginning to lead to masses that are independent of other methods.

2.2 A Brief Summary of Primary Methods

2.2.1 Dynamics of Individual Sources

The most accurate and reliable mass measurements are based on studying the motions of individual sources that are accelerated by the gravity of the black hole. One of the most accurately measured masses, not surprisingly, is that of Sgr A* at the Galactic Center.[1] Two decades of observations of the proper motions and radial velocities of individual stars (e.g., Genzel et al. 2010; Meyer et al. 2012) enabled by the combination of advanced infrared detectors and adaptive optics on large telescopes, has led to a measurement of a black hole mass of $4.1(\pm 0.4) \times 10^6\, M_\odot$.

A second measurement of remarkable quality is that of the mass of the black hole in the galaxy NGC 4258 (Miyoshi et al. 1995), which is a Type 2 AGN. In AGN unified models, Type 2 AGNs are those which are viewed at high inclination and are heavily obscured by the dusty molecular "torus" in the AGN midplane. In Type 2 sources, the broad lines and continuum that characterize the spectra of Type 1 AGNs are thus not directly observed. A consequence of observing the nucleus through a large column of molecular gas is that under the right circumstances masers are formed, so bright that in this case they are called "megamasers." The proper motions and radial velocities of the individual megamaser sources in NGC 4258 (a.k.a. M 106) show that they arise in a warped rotating disk around a black hole of mass $M_{\rm BH} = 3.82\,(\pm 0.01) \times 10^7\, M_\odot$ (Herrnstein et al. 2005).

Both of these cases are special in that they are, relatively speaking, very close, ~8 kpc for the Galactic Center and ~7.2 Mpc for NGC 4258. The black hole radius of influence is better resolved in these two galaxies than in any other (see Gültekin et al. 2009), in the former case simply because of its proximity and in the latter case also because the observations are made at ~20 MHz, where it is possible to attain very high angular resolution with VLBI techniques.

2.2.2 Collective Motions

Black hole masses can also be determined by their gravitational effects on systems of stars or gas. The collective dynamics of stars or gas on scales of $R_{\rm BH}$ can be modeled, with the central $M_{\rm BH}$ a free parameter. An excellent review of these methods is provided by Ferrarese and Ford (2005).

Using stellar dynamics to determine black hole masses has the advantage that stars, unlike gas, respond to gravitational forces only. On the other hand, high angular resolution is required to resolve, or at least nearly resolve, $R_{\rm BH}$. Using gas dynamics has the advantage of being somewhat simpler, fundamentally because gas is viscous and settles into a rotating disk-like structure fairly quickly compared to the relaxation time for stars in a galactic nucleus. Also, at least in the case of reverberation mapping (Sect. 3), high angular resolution is not required.

Stellar dynamical modeling is based on the superpositioning of individual stellar orbits from a large orbit library to obtain a best fit to the observables, mainly the surface brightness profiles and line-of-sight velocity distributions. Several sophisticated computer codes for this have been developed (e.g., Gebhardt et al. 2003; van der Marel et al. 1998; Cretton et al. 1999; Thomas et al. 2004; Valluri et al. 2004).

A small, but non-negligible, fraction (<20 %; Tran et al. 2001) of early-type galaxies have small nuclear dusty disks. These are often found to be emission-line sources, so their Doppler motions can be detected. Their dynamics can be modeled with the central mass as a free parameter. This has enabled measurement of some of the largest known supermassive

[1] The phrase "not surprisingly" in this context is used because Sgr A* is 100 times closer than any other supermassive black hole so it is possible to observe the motions of individual stars around the black hole in its vicinity. But the very fact that individual stars at the Galactic Center can be observed, despite 30 mag of visual extinction, *is* surprising, one of the things the author least expected to see in his own scientific lifetime.

black holes (Macchetto et al. 1997; Bower et al. 1998; de Francisco et al. 2008; Dalla Bontà et al. 2009).

All together, the number of supermassive black holes whose masses have been determined by modeling stellar or gas dynamics numbers over 70 (McConnell and Ma 2013). This number is not likely to increase dramatically in the near term on account of the difficulty in resolving $R_{\rm BH}$ beyond the Virgo Cluster. However, reverberation mapping presents a viable alternative for measuring black hole masses at large cosmological distances via gas dynamics by substituting time resolution for angular resolution. The major limitation is that it is only applicable to Type 1 (broad emission-line) AGNs; while quasars are currently a trace population, they were more predominant at higher redshift. Since the focus of this discussion is on accreting black holes, emphasis on reverberation mapping of AGNs is not misplaced.

3 Reverbation Mapping of AGNs

3.1 Beginnings

That the continuum emission from AGNs varies on quite short time scales (as short as days) has been known since nearly the time of the discovery of quasars. Indeed, within a few years of their discovery optical variability was regarded as a defining characteristic of quasi-stellar objects (Burbidge 1967). There were also some early scattered reports of *emission-line* variability (e.g., Andrillat and Souffrin 1968; Pastoriza and Gerola 1970; Collin-Souffrin et al. 1973; Tohline and Osterbrock 1976), which were in each case quite extreme changes; this is not surprising, given that only dramatic changes (e.g., apparent change from Type 1 to Type 2) could be detected with the technology of the times. With the proliferation of sensitive electronic detectors in the 1980s, a strong connection between continuum and emission-line variability was established (Antonucci and Cohen 1983; Peterson et al. 1983; Ulrich et al. 1991, and references therein), though the initial results were met with skepticism as the broad-line region (BLR) appeared to be an order of magnitude smaller than predicted by photoionization theory (Peterson et al. 1985).[2] Nevertheless, an older idea (Bahcall et al. 1972) about how emission-line region structure could be probed by variability was cast into a mathematical formalism by Blandford and McKee (1982) and given the name "reverberation mapping."

3.2 Theory of Reverberation Mapping

3.2.1 Assumptions

In this section, the basic theory of reverberation mapping is briefly outlined. More thorough discussions are available elsewhere (e.g., Peterson 1993, 2001).

Some very basic observations allow us to make several simplifying assumptions:

[2]The discrepancy between theory and observation was due to an oversimplified theory; it was implicitly assumed in photoionization equilibrium modeling that all BLR "clouds" are intrinsically identical. A successful photoionization model was one that correctly predicts the emission-line intensity ratios in the emitted spectrum of some "standard" cloud. A key intensity ratio is C IV λ1549/C III] λ1909. Moreover, the very presence of C III] λ1909 set an upper limit to the density as this line is collisionally suppressed above $10^{9.5}$ cm^{-3}. The implicit assumption that C IV λ1549 and C III] λ1909 are produced cospatially was incorrect. The first high-sampling rate reverberation program (Clavel et al. 1991) showed that C IV λ1549 and C III] λ1909 arise at different distances from the central source, thus obviating the earlier arguments. The C IV-emitting zone is now believed to have a higher density, $\sim 10^{11}$ cm^{-3} (Ferland et al. 1992).

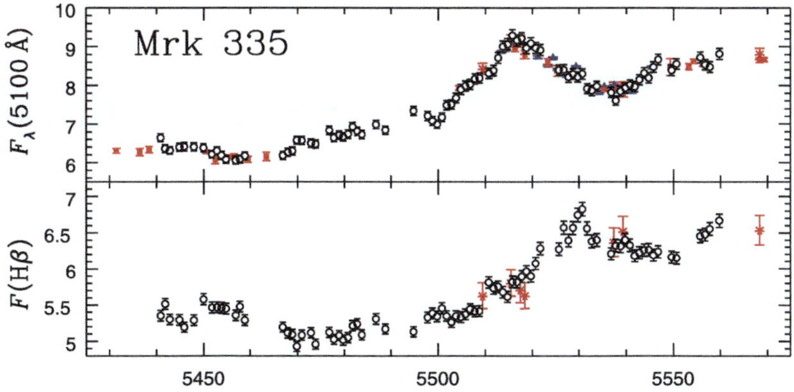

Fig. 1 Optical continuum (*top*) and broad Hβ emission-line (*bottom*) light curves for Mrk 335. The variations in Hβ follow those in the continuum by 13.9 ± 0.9 days. The continuum fluxes are in units of 10^{-15} ergs s^{-1} cm^{-2} Å^{-1} and the emission-line fluxes are in units of 10^{-13} ergs s^{-1} cm^{-2}. *Black points* are from MDM, *blue points* are from Wise Observatory, and *red points* are from Crimean Astrophysical Observatory. Data are from Grier et al. (2012a, 2012b)

1. The emission lines respond rapidly to continuum changes (Fig. 1), showing that the BLR is small (because the light-travel time is short) and the gas density in the BLR is high (so the recombination time is much shorter than the light-travel time). It is also noted that the dynamical timescale (of order $R_{\text{BLR}}/\Delta V$) of the BLR is much longer than the reverberation timescale (of order R_{BLR}/c), so the BLR is essentially stationary over a reverberation monitoring program.
2. The continuum-emitting region is so small compared to the BLR it can be considered to be a point source. It does *not* have to be assumed that the continuum emits isotropically, though that is often a useful starting point.
3. There is a simple, though not necessarily linear or instantaneous, relationship between variations of the ionizing continuum (at $\lambda < 912$ Å) and the observed continuum (typically at $\lambda \sim 5100$ Å). The fact that reverberation works at all justifies this at some level of confidence.

3.2.2 The Transfer Equation

Over the duration of a reverberation monitoring program, the continuum behavior over time can be written as $C(t) = \langle C \rangle + \Delta C(t)$ and the emission-line response as a function of line-of-sight velocity V_{LOS} is $L(V_{\text{LOS}}, t) = \langle L(V_{\text{LOS}}) \rangle + \Delta L(V_{\text{LOS}}, t)$ where $\langle C \rangle$ and $\langle L(V_{\text{LOS}}) \rangle$ represent mean values and $\Delta C(t)$ and $\Delta L(V_{\text{LOS}}, t)$ are the variations around the mean values. On a reverberation timescale, both continuum and emission-line variations are usually rather small (typically ∼10–20 %) so even if their relationship is non-linear, it can be modeled as linear on short timescales. In this case, the relationship between the continuum and emission-line variations can be written in terms of the time delay τ as

$$\Delta L(V_{\text{LOS}}, t) = \int \Psi(V_{\text{LOS}}, \tau) \Delta C(t - \tau)\, d\tau, \tag{1}$$

which is usually known as the "transfer equation" and $\Psi(V_{\text{LOS}}, \tau)$ is the "transfer function." Inspection of Eq. (1) shows that $\Psi(V_{\text{LOS}}, \tau)$ is the observed response to a δ-function continuum outburst.

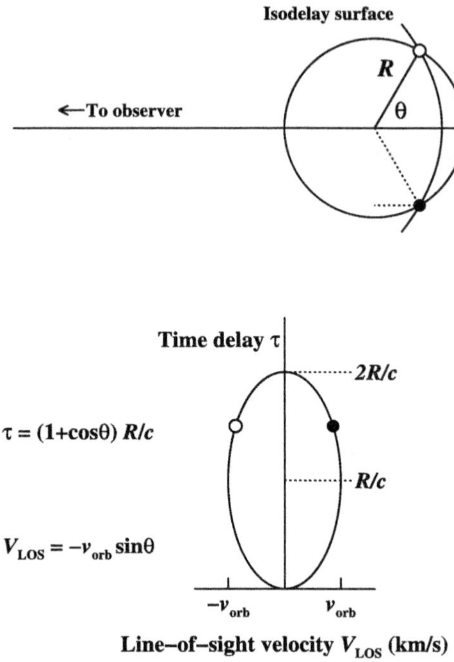

Fig. 2 *Top*: A notional BLR comprised of clouds orbiting the central source counterclockwise in a circular orbit of radius R. The *dotted line* shows the path of an ionizing photon to a BLR cloud at coordinate $(R, -\theta)$ plus the path of an emission-line photon until it is at the same distance from the observer as the continuum source. The light travel time along the dotted path is $\tau = (1 + \cos\theta)R/c$, which is the time lag the observer sees between a continuum outburst and the response of the cloud. All points on the "isodelay surface" have the same delay relative to the continuum. *Bottom*: the same circular BLR projected into the observable quantities of Doppler velocity and time-delay; this is a very simple velocity–delay map

3.2.3 Construction of a Velocity–Delay Map

The transfer function can be constructed geometrically as it is simply the six-dimensional phase space of the BLR projected into the observable coordinates, line-of-sight velocity (i.e., Doppler shift) and time delay relative to the continuum variations. It is therefore common to refer to the transfer function $\Psi(V_{\rm LOS}, \tau)$ as a "velocity–delay map." It should be clear that each emission line has a different velocity–delay map because the combination of emissivity and responsivity is optimized at different locations of the BLR for different lines. To map out *all* of the BLR gas would require velocity–delay maps for multiple emission lines with different response timescales.

Consider for illustrative purposes a very simple BLR model, a circular ring of gas orbiting counterclockwise at speed $v_{\rm orb}$ around the central source at a distance R. Suppose that a distant observer sees this system edge-on, as shown in the upper part of Fig. 2, and define a polar coordinate system centered on the continuum source with the angle θ measured from the observer's line of sight. Two clouds are shown at positions $(R, \pm\theta)$. Photons from a δ-function continuum outburst travel toward the observer along the $-x$ axis. The dotted line shows that the path taken by an ionizing photon from the same outburst will reach the BLR cloud pictured in the lower half of the figure after travel time R/c; an emission-line photon produced in response by the cloud and, by chance, directed toward the distant observer travels an additional distance $R\cos\theta/c$, where it is now at the same distance from the observer as the continuum source. So relative to the ionizing photons headed directly toward the observer from the continuum source, the emission-line photons are delayed by the sum of these two dotted segments, i.e., by

$$\tau = (1 + \cos\theta)R/c. \qquad (2)$$

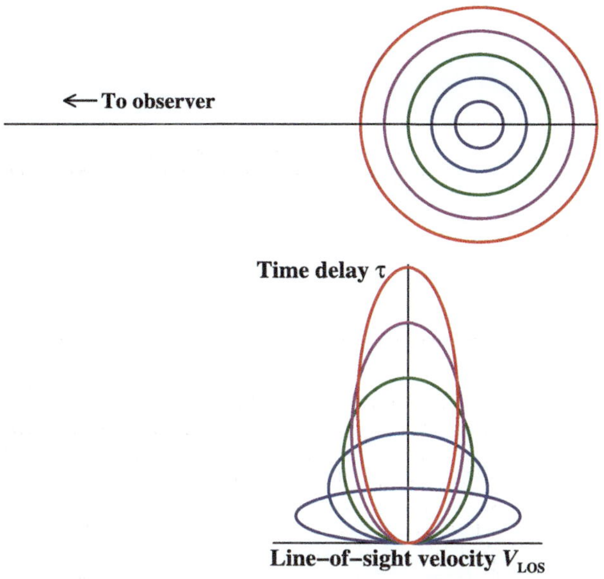

Fig. 3 Similar to Fig. 2, except for a series of rings in circular Keplerian orbits. Note in particular the "Keplerian taper" of the velocity–delay map at increasing time delay

The locus of points that all have the same time delay to the observer is labeled as an "isodelay surface" in the top part of the figure; a moment's reflection will convince the reader that the isodelay surface is a paraboloid. The corresponding Doppler shifts of the clouds at coordinate $(R, \pm\theta)$ are $\mp v_{\rm orb} \sin\theta$. These transformations are general, and a ring of radius R and orbital speed $v_{\rm orb}$ projects in velocity–delay space to an ellipse with axes $2v_{\rm orb}$ centered on $V_{\rm LOS} = 0$ and $2R/c$ centered on R/c. Here the ring is pictured edge-on, at inclination 90°; at any other inclination i, the projected axes of the ellipse in velocity–delay space are correspondingly reduced to $2v_{\rm orb} \sin i$ and $2R \sin i/c$. Thus, a face-on ($i = 0°$) disk projects in velocity–delay space to a single point at $(0, R/c)$; all of the ring responds simultaneously and no Doppler shift is detected.

Generalization of this structure to a Keplerian disk is straightforward by simply adding more rings such that $v_{\rm orb} \propto R^{-1/2}$. This is illustrated for a system of several rings in Fig. 3.

At this point, an assumption about how the individual clouds re-emit line radiation can be introduced. The simplest assumption is that the line emission is isotropic, i.e., $\Psi(\theta) = \epsilon$, a constant. To transform this to the observable velocity–time-delay coordinates,

$$\Psi(\tau)\,d\tau = \Psi(\theta)\frac{d\theta}{d\tau}\,d\tau. \tag{3}$$

From Eq. (2)

$$\frac{d\tau}{d\theta} = -\frac{R}{c}\sin\theta. \tag{4}$$

It is simple to show (Peterson 2009) then that

$$\Psi(\tau)\,d\tau = \frac{c\epsilon}{R(2c\tau/R)^{1/2}(1 - c\tau/2R)^{1/2}}\,d\tau \tag{5}$$

and that the mean response time for the ring is

$$\langle\tau\rangle = \frac{\int \tau\Psi(\tau)\,d\tau}{\int \Psi(\tau)\,d\tau} = \frac{R}{c}, \tag{6}$$

as is intuitively obvious. A more realistic assumption is that much of the line emission is directed back toward the ionizing source because the BLR clouds are very optically thick even in the lines. A simple parameterization is that $\Psi(\theta) = (1 + A\cos\theta)/2$. Isotropy is the case $A = 0$ and complete anisotropy (which is, incidentally, perhaps appropriate for Ly$\alpha\,\lambda$1215) corresponds to $A = 1$ (Ferland et al. 1992).

Using a similar transformation for the case of isotropic re-emission, it is easy to show that (Peterson 2009)

$$\Psi(V_{\rm LOS})\,dV_{\rm LOS} = \frac{\epsilon}{v_{\rm orb}(1 - V_{\rm LOS}^2/v_{\rm orb}^2)^{1/2}}\,dV_{\rm LOS}. \tag{7}$$

A useful measure of the line width is the line dispersion $\sigma_{\rm line}$; for a ring, the line dispersion is

$$\sigma_{\rm line} = \left(\langle V_{\rm LOS}^2\rangle - \langle V_{\rm LOS}\rangle^2\right)^{1/2} = \frac{v_{\rm orb}}{2^{1/2}}. \tag{8}$$

For comparison, for such a ring, FWHM $= 2v_{\rm orb}$.

A velocity–delay map for any other geometry and velocity field can be constructed similarly.

3.2.4 Reverberation Mapping Results: Velocity–Delay Maps

As noted above, both continuum and emission-line variations are usually not very large ($<$20 %) on a reverberation timescale. This alone makes it very difficult to recover a velocity–delay map from spectrophotometric monitoring data. Indeed, the relative photometric accuracy must be extremely good: for ground-based spectra, for which absolute spectrophotometry at even the 5 % level is notoriously difficult to achieve, this is usually accomplished by using the [O III] $\lambda\lambda$4959, 5007 narrow lines as an internal flux calibrator. These lines arise in the spatially extended low-density narrow-line region, so both the light-travel time and recombination time are sufficiently long to ensure that the flux in these lines is constant on reverberation timescales. An alternative calibration strategy is to rotate the spectrograph slit to a position where a nearby non-variable star can be observed simultaneously. An even better strategy that is gaining popularity is to use simultaneous imaging data to define the continuum light curve and calibrate the spectra.

Time sampling is critical (Horne et al. 2004). As a rule of thumb the duration of the monitoring campaign has to be at least $\sim$3 times the longest timescale of interest ($2R/c$) for an accurate measurement of the mean response time and rather longer than this to recover a velocity–delay map. The sampling rate ultimately transforms into the spatial resolution of the velocity–delay map and thus needs to be high.

When reverberation studies began in earnest in the late 1980s, most programs were designed with the modest goal of measuring mean response times for various emission lines and characterizing the variability characteristics of AGNs as a function of luminosity, as discussed in Sect. 3.2.5 below. There were early attempts to recover velocity–delay maps (e.g., Ulrich and Horne 1996; Wanders et al. 1997; Kollatschny 2003a); these revealed some structure, but little detail. It is only relatively recently that reliable velocity–delay maps have begun to appear in the literature (Bentz et al. 2010b;

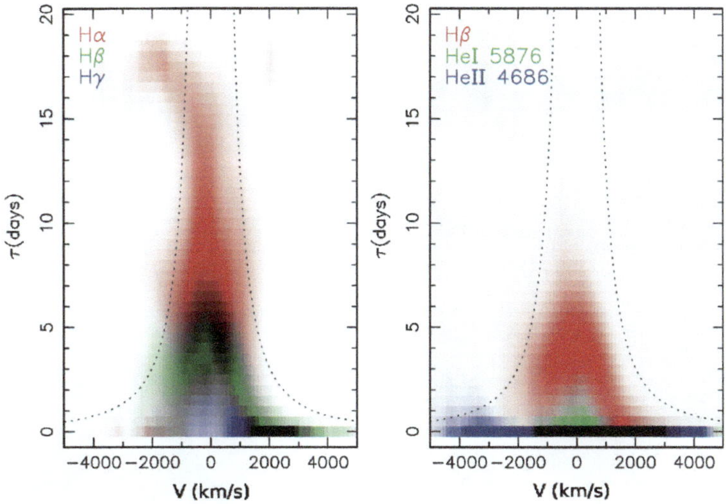

Fig. 4 *Left*: Velocity–delay maps for the Balmer lines in Arp 151. Note the clear inflow signature for time delays above 15 light days. *Right*: Velocity–delay maps for Hβ, He I λ5876, and He II λ4686 in Arp 151. From Bentz et al. (2010b)

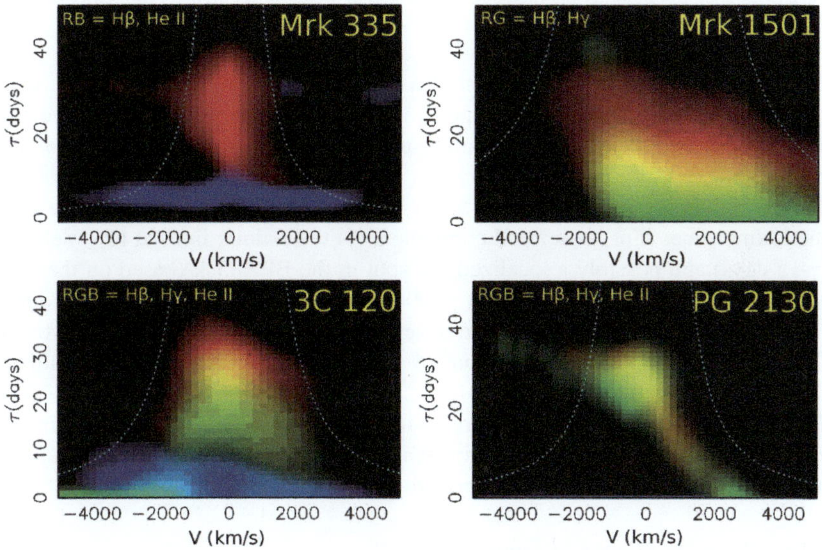

Fig. 5 Velocity–delay maps for four AGNs. 3C 120 has a disk-like structure and evidence for infall is apparent in each of these. From Grier et al. (2013)

Grier et al. 2013), with examples shown in Figs. 4 and 5. To provide some insight into interpreting these maps, Fig. 6 shows examples of velocity–delay maps for some simple "toy models;" these demonstrate how inflow models (spherical infall, in the top two panels of Fig. 6) are characterized by the earlier response of the red wing of the line (closer to us, but receding) followed by the response of the blue side of the line (farther away for us and approaching). BLRs in which the cloud motions are in Keplerian orbits, either in a disk as in

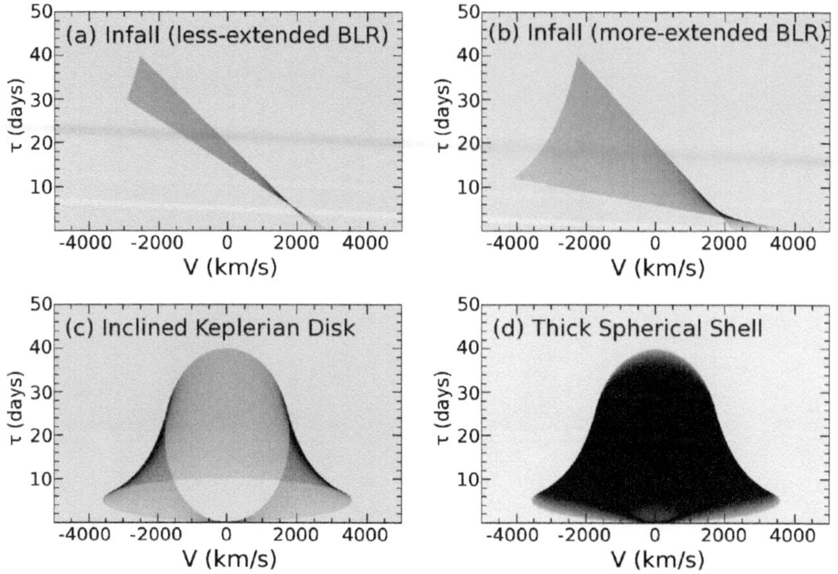

Fig. 6 Toy models of velocity–delay maps for spherical infall (*top two panels*) and a Keplerian disk (*lower left*) and a thick shell of randomly inclined circular Keplerian orbits (*lower right*). From Grier et al. (2013)

the lower left of Fig. 6 or in a spherical shell as in the lower right, have symmetric structures that show a characteristic Keplerian "taper:" all the gas with large line-of-sight velocities is at small lags, close to the continuum source, and at large lags, only small values of $V_{\rm LOS}$ are observed.

The velocity–delay maps for Arp 151 in Fig. 4 strongly suggest a disk-like structure, and the Balmer lines hint at an infall component (in the Balmer lines at time delays larger than ~15 days) and possibly a localized hot spot in the BLR disk (based on the enhanced emission at velocity–delay coordinates (+2000 km s^{-1}, 0 days). More detailed modeling by Brewer et al. (2011) based on formalism developed by Pancoast et al. (2011) favors a thick disk-like BLR geometry at an inclination[3] of $\sim 22°$, with infall favored, and a value for the central mass $M_{\rm BH} = 3.2\ (\pm 2.1) \times 10^6\ M_{\odot}$.

The velocity–delay maps in Fig. 5 are so recent that detailed physical modeling has not yet been done. Some of these strongly hint at a disk-like geometry (e.g., 3C 120) and in each case there is evidence for an infall component as well.

3.2.5 Reverberation Mapping Results: Lags

As already noted, the primary goal of most reverberation monitoring campaigns that have been undertaken to date has been to determine the mean response time of the integrated emission line, i.e., to measure the "lag" between continuum and emission-line variations. The most up-to-date methodology for making these measurements is that described by Zu et al. (2011).

[3]Brewer et al. (2011) define the inclination angle to be the complement of the usual astronomical convention where $i = 0$ is face-on. The value given here is corrected to the standard astronomical convention.

The first high-sampling rate multiwavelength reverberation monitoring program was undertaken in 1988–89 by the International AGN Watch (Clavel et al. 1991; Peterson et al. 1991; Maoz et al. 1993; Dietrich et al. 1993; Alloin et al. 1994). Since then, emission-line lags have been measured for about 50 AGNs, mostly for the Hβ emission line, and for multiple emission lines in only a few cases. Peterson et al. (2004) provide a homogeneous compilation of most of the high-quality reverberation results available as of a decade ago. More recent large-scale campaigns have been undertaken by the Lick AGN Monitoring Program (LAMP), described by Bentz et al. (2009c), Barth et al. (2011a, 2011b), by Crimean Astrophysical Observatory (e.g., Sergeev et al. 2011; Doroshenko et al. 2012) by a consortium of astronomers primarily in eastern Europe, Mexico, and Germany (e.g., Shapovalova et al. 2010, 2012; Popović et al. 2011) and by the author's group of collaborators (Bentz et al. 2006b, 2007; Denney et al. 2006, 2010; Grier et al. 2012b). Programs to measure the C IV λ1549 lag in a very low-luminosity AGN (Peterson et al. 2005) and a very high-luminosity AGN (Kaspi et al. 2007) have also been reported. All of these efforts are continuing, and new results appear regularly.

These studies have led to several important findings:

- Within a given AGN, the higher-ionization lines have smaller lags than the lower-ionization lags, demonstrating ionization stratification of the BLR. This also shows that the BLR gas is distributed over a range of radii from the central source, and the lag for a particular emission line represents the radius at which the combination of emissivity and responsivity is optimized for that particular emission line.
- The variable part of the emission line can be isolated by constructing the rms residual spectrum from all monitoring data. In the rms residual spectrum, the higher ionization lines are broader than the low ionization lines, in such a way that the product $\Delta V^2 \tau$ is constant within a given source, supporting the view that the BLR is virialized; this is necessary, though not sufficient, evidence for a BLR with gravitationally dominated dynamics.
- There is a relationship between the size of the BLR R and the AGN luminosity L of the approximate form $R \propto L^{1/2}$.

These results underpin our efforts to measure the masses of the central sources in these objects, as discussed in the next session.

4 Reverberation-Based Black Hole Masses

4.1 Virial Mass Estimates

For every AGN for which emission-line lags and line widths have been measured, consistency with the "virial relationship" is found (Peterson and Wandel 1999, 2000; Kollatschny 2003a; Peterson et al. 2004; Bentz et al. 2010a). This also appears to be true when the lag and line width are measured for the same emission line when the AGN is in very different luminosity states (e.g., Peterson 2008). This strongly suggests that the BLR dynamics are dominated by the central mass, which is then

$$M_{\mathrm{BH}} = f\left(\frac{\Delta V^2 R}{G}\right), \tag{9}$$

where ΔV is the line width and R is the reverberation radius $c\tau$. The quantity in parentheses that contains the two directly observable parameters has units of mass and is sometimes referred to as the "virial product." The effects of everything unknown—the BLR geometry, kinematics, and inclination—are then subsumed into the dimensionless factor f, which will be different for each AGN, but is expected to be of order unity. This is one of the main reasons for obtaining velocity–delay maps: one can then model the dynamics of the BLR which then yields the black hole mass, or equivalently, f. At this time this has been done for only two AGNs, as discussed further below (Brewer et al. 2011; Pancoast et al. 2012).

Individual values of f can also be determined if there is some other way of determining the black hole mass. In the absence of a second direct measurement, it has been common practice to use the $M_{\rm BH}$–σ_* relationship for this purpose. The relationship between central black hole mass and bulge velocity dispersion that is seen in quiescent galaxies (Gebhardt et al. 2000a; Ferrarese and Merritt 2000) is also seen in AGNs (Gebhardt et al. 2000b; Ferrarese et al. 2001; Nelson et al. 2004) although of course, the host-galaxy velocity dispersions are much more difficult to measure in AGNs because of the bright active nucleus and because even the nearest AGNs are typically quite distant (Dasyra et al. 2007; Watson et al. 2008). By assuming that the $M_{\rm BH}$–σ_* relationship is the same in quiescent and active galaxies, it becomes possible to compute a mean value for the scaling factor, which turns out to be $\langle f \rangle \sim 5$ (Onken et al. 2004; Woo et al. 2010; Park et al. 2012), although it is noted that Graham et al. (2011) argue that in practice this process has been oversimplified. Figure 7 shows the $M_{\rm BH}$–σ_* relationship for quiescent galaxies and AGNs using the assumption that $\langle f \rangle = 5.25$. The scatter around this relationship amounts to about ~0.4 dex, which is a reasonable estimate of the accuracy of the "virial method" of estimating black hole masses. It is important to understand that while this calibration of the reverberation-based mass scale seems to be statistically robust, it cannot be applied with any accuracy to individual sources (i.e., f cannot be inferred by comparing the virial product with the $M_{\rm BH}$–σ_* prediction of the mass) without knowing the geometry and inclination of the source.

Sometimes concern is expressed that the empirical value of $\langle f \rangle$ seems uncomfortably large for a truly virialized system. However, it must be kept in mind that AGN unification stipulates that Type 1 AGNs are generally observed at low values of inclination, much closer to face-on than edge-on. Actually, the fact that $\langle f \rangle$ is as *small* as it is tells us that the BLR must have a fairly significant velocity component in the polar direction; it is surely not a flat disk.

To return to a point made earlier, reverberation mapping is a *direct* measure of black hole mass, but it is a *secondary* method because, at the present time, it relies on an independent method, the $M_{\rm BH}$–σ_* relationship, to calibrate the mass scale through determination of $\langle f \rangle$.

4.2 Testing Reverberation-Based Masses

In addition to the virial relationship $\Delta V \propto R^{-1/2}$ seen in the broad emission lines and the similarity of the $M_{\rm BH}$–σ_* relationships for active and quiescent galaxies, there is other evidence that supports the general veracity of reverberation-based masses:

- The well-known correlation between central black hole mass and host-galaxy bulge luminosity (Kormendy and Richstone 1995; Magorrian et al. 1998) is also seen in active galaxies (Bentz et al. 2009a). There is very good agreement between these when the AGN black hole masses are calibrated by using the $M_{\rm BH}$–σ_* relationship.
- In a small number of nearby AGNs, it is possible to measure or at least constrain the black hole mass using more than one direct method of mass measurement, as shown in Table 1. To within the factor of two or three uncertainties in each of these methods[4]

[4]Note that the errors quoted in Table 1 are statistical. Systematic errors have not been included.

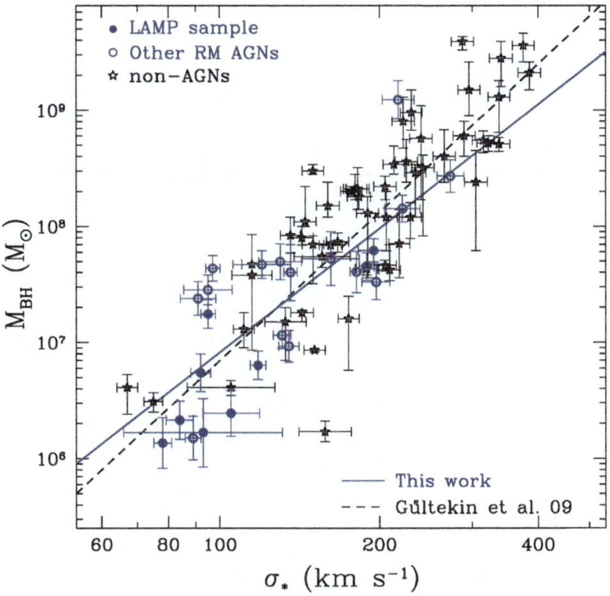

Fig. 7 The M_{BH}–σ_* relationship. Quiescent galaxies are shown in *black* and AGNs are shown in *blue*. From Woo et al. (2010)

Table 1 Comparison of black hole mass measurements

Method	NGC 4258 (Units $10^6\ M_\odot$)	NGC 3227	NGC 4151
Direct methods:			
Megamasers	38.2 ± 0.1[1]	N/A	N/A
Stellar dynamics	33 ± 2[2]	7–20[3]	≤ 70[4]
Gas dynamics	25–260[5]	20^{+10}_{-4}[6]	$30^{+7.5}_{-22}$[6]
Reverberation	N/A	$7.63^{+1.62}_{-1.72}$[7]	46 ± 5[8]
Indirect methods:			
M_{BH}–σ_*[9]	13	25.0	6.1
R–L scaling[10]	N/A	15	65

[1]Herrnstein et al. (2005). [2]Siopsis et al. (2009). [3]Davies et al. (2006). [4]Onken et al. (2007). [5]Pastorini et al. (2007). [6]Hicks and Malkan (2008). [7]Denney et al. (2010). [8]Bentz et al. (2006b). [9]Gültekin et al. (2009). [10]McGill et al. (2008)

(except for megamasers), the multiple measurements are in generally good agreement. Table 1 also underscores the important point that reverberation mapping and megamasers motions cannot be used in the same source: the former can be employed in Type 1 AGNs and the latter are likely to be found only in Type 2 AGNs.
- It was noted in Sect. 3.2.4 that the black hole mass is a free parameter in modeling velocity–delay maps. So far detailed modeling has been done in only two cases, Arp 151 (Brewer et al. 2011) and Mrk 50 (Pancoast et al. 2012). In both cases, the masses obtained from modeling are in reassuringly good agreement with those based on Eq. (9) using the nominal value $f = 5.25$.

In closing this section, it is noted that an unresolved issue is whether or not radiation pressure plays a sufficiently important role in the BLR dynamics that it affects black hole mass measurement (Marconi et al. 2008; Netzer and Marziani 2010). Since radiation pressure is a $1/r^2$ force, it is not easily distinguished from gravity: ignoring radiation pressure simply leads to systematic underestimation of AGN black hole masses. Radiation pressure is likely to be important only in high Eddington-rate sources, such as narrow-line Seyfert 1 galaxies.

5 Indirect Mass Estimates Anchored by Reverberation Results

As noted earlier, the distinct advantage of reverberation mapping compared to other methods of measuring the masses of supermassive black holes in galaxies is that it substitutes time resolution for angular resolution. In principle, it can be used to measure the masses of black holes even in the most distant quasars. It does, however, have the corresponding disadvantage of being resource-intensive: a reliable reverberation measurement requires a large number of observations, typically 30–50 at minimum. This drives a search for shortcuts that will allow mass estimates from simpler or fewer observations. Fortunately, and remarkably, reverberation mapping provides a suitable scaling relationship that allows estimation of AGN black hole masses from single spectra.

5.1 The Radius–Luminosity Relationship

In Sect. 3.2.5, it was noted that reverberation mapping has revealed an empirical relationship between the radius of the BLR and the luminosity of the AGN. This was not in any way unexpected. Prior to the advent of reverberation mapping, the principal means of studying the BLR was through photoionization equilibrium modeling (e.g., Ferland and Mushotzky 1982), where one tries to adjust the model parameters to produce relative emission-line intensities that match the observations. Photoionization equilibrium models are characterized by the shape of the ionizing continuum that irradiates the emission-line clouds, the chemical composition of the gas, and an ionization parameter U that is the ratio of the ionizing photon density to the particle density at the irradiated face of the clouds,

$$U = \frac{Q(H)}{4\pi R^2 n_e c}, \tag{10}$$

where n_e is the particle density and $Q(H)$ is the rate at which hydrogen-ionizing photons are produced by the central source, which is at a distance R from the emission-line clouds. The similarity of AGN spectra over many orders of magnitude in luminosity (e.g., Dietrich et al. 2002; Vanden Berk et al. 2004) suggests that U and n_e are approximately the same for all AGNs. By assuming that $L \propto Q(H)$,

$$R \propto L^{1/2}. \tag{11}$$

The great utility of the ionization parameter formulation was first recognized by Davidson (1972), although he is not often credited with this insight, possibly because instead of expressing U in terms of L and R, he used the ionizing flux $F = L/4\pi R^2$, so the distance did not appear explicitly.

The physics of the BLR is grossly oversimplfied in the discussion above. While there is currently no self-consistent model of the BLR, it is probably not far from the truth to suppose that at any given distance from the central source, there are clouds or filaments of various densities, but that most of the reverberation response comes from gas that has an optimal combination of emissivity and responsivity for a particular emission line (e.g., Baldwin et al. 1995; Korista et al. 1997).

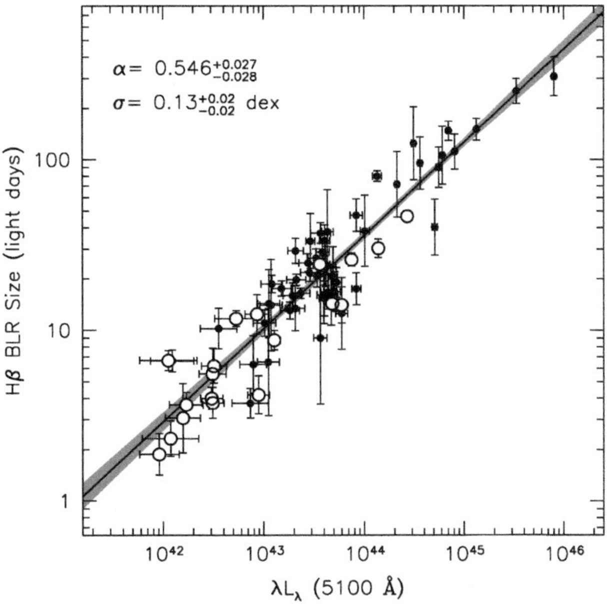

Fig. 8 The *R–L* relationship between the size of the BLR measured by the broad Hβ response as a function of the host-corrected AGN luminosity. *Open circles* are new or improved data since Bentz et al. (2009b). Based on Fig. 11 from Bentz et al. (2013)

Thus from the beginning of reverberation investigations, a relationship like Eq. (11) was expected, and it was searched for even with the first marginally sampled and undersampled reverberation data (e.g., Koratkar and Gaskell 1991; Peterson 1993). The first observationally well-defined version of the *R–L* relationship was by Kaspi et al. (2000), who found $R \propto L^{0.7}$. A refinement of the database led to a somewhat shallower slope (Kaspi et al. 2005), but it was only when contamination of the luminosity by host-galaxy starlight was accounted for (Bentz et al. 2006a, 2009b, 2013) that the slope was close to the naïve photoionization prediction. The most recent version of the *R–L* relationship for the broad Hβ emission line and the optical continuum is shown in Fig. 8. It is also worth noting that the existence of the *R–L* relationship has been confirmed independently by gravitational microlensing studies (Guerras et al. 2013).

While the existence of the *R–L* relationship is not surprising, the tightness of the relationship is. The intrinsic scatter in the relationship seems to be ∼0.13 dex (Fig. 8). Typical *good* individual reverberation lag measurements are accurate to about 0.09 dex, so, at least for Hβ, the *R–L* relationship does almost as well as an actual reverberation measurement, provided one has an accurate, host-corrected AGN luminosity. And, it must be said, it is surprising that the empirically determined slope of the *R–L* relationship is so close to 0.5 given the naïve assumptions that went into the theoretical prediction.

Unfortunately, Hβ is the *only* emission line for which the *R–L* relationship is empirically well-calibrated. There are, however, a limited number of observations that seem to indicate that a relationship similar to that between *Hβ* and the AGN optical continuum holds for C IV λ1549 and the UV continuum as well (Kaspi et al. 2007).

5.2 Indirect Masses from *R–L*

The great utility of Eq. (11) is that it provides a quick way to estimate the BLR size and thus black hole mass from a single spectrum, limited only by the accuracy of the flux calibration and the starlight removal. Of course, since many AGN luminosity measures are highly correlated, it is entirely possible to use other, sometimes easier-to-measure, luminosities—e.g., X-ray, broad Balmer and Paschen lines, narrow [O III] λ5007, narrow [O IV] λ25.8 μm (Wu et al. 2004; Greene et al. 2010; Landt et al. 2011)—as surrogates for the host-corrected optical luminosity, although none of these is yet very well calibrated.

The first attempts to use the R–L relationship in an informed way to estimate masses (variously called the "photoionization method" or "single-epoch spectrum method") were by Laor (1998) and Wandel et al. (1999), in both cases using the broad Hβ line. This was subsequently extended to the UV emission lines Mg II λ2798 and C IV λ1549 (McLure and Jarvis 2002; Vestergaard 2002, 2004; Vestergaard and Peterson 2006; Kollmeier et al. 2006) to allow applications at higher redshift.

For the sake of completeness, indirect measures of black hole mass are included in the bottom two lines of Table 1. These underscore the fact that indirect mass estimates are no more accurate than a factor of several. The principal value of the indirect methods is to produce large samples of mass estimates that are statistically, rather than individually, accurate.

5.3 Problems with Indirect Mass Measurements

The first thing one needs to keep in mind is that at the present time there is no consensus on the exact prescription one needs to follow to compute a black hole mass from a single spectrum of a quasar. This is a field that is still in a developmental stage.

Something that is generally not appreciated is that estimating the BLR size from the R–L relationship is the easy part (again assuming that L is determined accurately). The more ambiguous part is characterizing the line widths. In Sect. 3.2.5, it was noted that in determining a reverberation-based mass it is most common to use the line widths measured in the rms residual spectrum, which isolates the part of the emission line that is actually varying, and to characterize the line width by the line dispersion (second moment) of the profile. Characterizing the line width in "single-epoch" spectra presents some problems:

- The spectrum contains contaminating features that do not appear in the rms residual spectrum. The narrow-line components of the broad emission lines and the constant host-galaxy continuum disappear in the rms residual spectrum; in a "single-epoch" or mean spectrum, they need to be modeled out. In the vicinity of Hβ, there are also blended Fe II features—these vary slowly, but do not appear to "reverberate" as cleanly as other broad lines (Vestergaard and Peterson 2005; Kuehn et al. 2008), so they are usually not seen in the rms residual spectra. In the case of the C IV λ1549, He II λ1640, O III] λ1663 complex there is an unidentified emission feature at ~1600 Å that is problematic (Fine et al. 2010). Mg II λ2798 is also difficult because of contamination by broad blends of Fe II (Vestergaard and Wilkes 2001).
- There are intrinsic differences between the widths of certain emission lines in mean (or single-epoch) and rms residual spectra. As an example, Ferland et al. (1990) find that the far wings of Hβ do not vary in Mrk 590 and suggest that this might be because the highest velocity BLR gas, presumably closest to the continuum source, is optically thin in the ionizing continuum. As another example, Denney (2012) finds dramatic differences in the C IV λ1549 profiles between mean and rms residual spectra, but that these are correlated with the emission-line profile. In general, it seems that it should be possible to calibrate out the differences between mean and rms residual spectra one way or another.
- The most common, and intended, use of indirect methods is for surveys. Survey spectra, however, are generally quite noisy, typically below a threshold where there are not

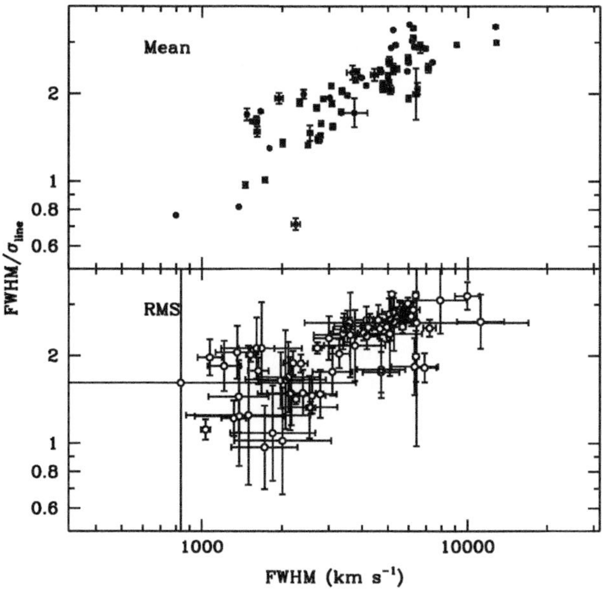

Fig. 9 Ratio of two measures of the emission line width for Hβ in reverberation-mapped AGNs, full-width at half maximum (FWHM) and line dispersion $\sigma_{\rm line}$ as a function of line width for the compilation of Peterson et al. (2004). *Top panel* is for mean spectra, *bottom panel* is for rms residual spectra. Values in top panel are very approximate since no deblending of contaminants (except narrow emission components) was undertaken

systematic errors of various types introduced by low signal-to-noise ratio (S/N) (Denney et al. 2009; Denney et al. 2013). Again, in principle, these effects are statistically correctable, though the uncertainties for individual sources can be quite large.

There are two other continuing problems in determining masses.

- Many studies continue to use FWHM to characterize the line widths. This is, of course, a very tempting thing to do because FWHM is generally easily measured and less affected by blending than the line dispersion $\sigma_{\rm line}$. However, the reverberation masses are calibrated using $\sigma_{\rm line}$, not FWHM. This actually matters a great deal because the ratio FWHM/$\sigma_{\rm line}$ is *a strong function of line width*, as shown in Fig. 9.
- A related problem is that many studies, particulary those based on lower S/N survey data, fit single Gaussians to the emission lines. For a Gaussian, as is well known, FWHM/$\sigma_{\rm line} \approx 2.35$ which, as can be seen in Fig. 9 is *on average* not a bad approximation for a quasar emission line, but it is terrible at both the high-width and low-width ends of the distribution. The problem again is that *the line profile is a strong function of line width.*

The impact of using FWHM rather than $\sigma_{\rm line}$ is, at fixed luminosity L (or, equivalently, BLR radius R), to stretch out the distribution in mass, overestimating the highest masses (with the broadest lines) and underestimating the lowest masses (with the narrowest lines). This introduces a bias into the mass functions derived from these measurements. Of course, it is possible to remove this bias through the clear relationship between $\sigma_{\rm line}$ and FWHM shown in Fig. 9; in principle, $\sigma_{\rm line}$ can be inferred, with some loss of precision, from FWHM (e.g., Collin et al. 2006).

A fair question to ask at this point is why not use FWHM to characterize the line widths in the original reverberation data instead of $\sigma_{\rm line}$? There are a number of reasons for preferring $\sigma_{\rm line}$ over FWHM as the line-width measure:

1. Multiple reverberation measurements of the same source should always yield the same mass. NGC 5548 is our ideal test case here, as there are over a dozen reverberation

measurements of the lag and line width for Hβ alone, with lags ranging from a few days to nearly a month. For these measurements, σ_{line} yields a more consistent mass than does FWHM (Collin et al. 2006).
2. For AGNs in which multiple emission lines have been measured, the virial relationship is tighter for σ_{line} than for FWHM (Peterson et al. 2004). This is essentially the same test as above.
3. The agreement between black hole masses based on C IV λ1549 with those based on Hβ is much better when σ_{line} is used as the line-width measure instead of FWHM (Denney et al. 2013).
4. Steinhardt and Elvis (2010) find that when they plot AGN luminosity as a function of black hole mass for individual redshift bins, the quasars at the high-mass end of the bin fall farther and farther below the Eddington limit, an effect they refer to as the "sub-Eddington boundary." This is, in fact, exactly the bias that would be expected by using FWHM (and/or Gaussian fits) instead of σ_{line}, which was indeed the case for the line-width measurements used in this study (Shen et al. 2008); the highest masses are overestimated and the lowest masses are underestimated, which would appear to rotate a distribution parallel to the Eddington limit (i.e., at fixed Eddington rate) to something shallower. Rafiee and Hall (2011) measured σ_{line} directly from the line profiles for essentially the same sample of AGNs and the "sub-Eddington boundary" effect either vanishes or is greatly reduced.

None of these arguments is iron-clad, particularly the argument based on the sub-Eddington boundary, which is only circumstantial evidence; it does, however, explain an observed phenomenon that is difficult to account for physically. In any case, a critical examination of how to characterize line widths deserves consideration.

6 Alternative Paths to Black Hole Masses

There are other methods of black hole mass measurement that are used less frequently, either because they require special cirumstances or lead to lower-confidence estimates of black hole mass.

- The black hole mass can be measured from the gravitational redshift of the broad emission lines. While the BLR, at several hundred gravitational radii from the black hole, is far enough from the black hole that relativistic effects can usually be ignored, the gravitational redshift

$$c\Delta z = \frac{GM_{\text{BH}}}{cR_{\text{BLR}}}, \tag{12}$$

can be of order 100 km s^{-1} for many of the reverberation-mapped AGNs and is therefore detectable at sufficiently high spectral resolution. A very credible detection has been reported by Kollatschny (2003b).
- The temperature structure of an accretion disk depends on the mass and mass accretion rate. It is therefore possible to obtain the mass of the black hole by fitting the observed continuum with an accretion disk model. While this is notoriously difficult, some current models (Gliozzi et al. 2011) yield masses that seem to be in reasonable agreement with reverberation-based masses.
- X-ray observations have been used in some cases to estimate black hole masses. For example, Iwasawa et al. (2004) report a tentative detection of modulations of the X-ray

Fe Kα emission line which they associate with orbital motion of an accretion disk. From this they make model-dependent estimate of the black hole mass in NGC 3516. In another case, a reverberation lag between the directly emitted X-ray continuum and its reflection off the accretion disk has also been used to estimate the mass of the central black hole in an AGN (Fabian et al. 2009, see also McHardy et al. 2007).
- X-ray observations also reveal that the power spectral density function (PSD) of accreting black holes shows a characteristic "break frequency," above which the slope of the PSD becomes much steeper (Markowitz et al. 2003; Papadakis 2012; McHardy et al. 2006). The break frequency is a function of both mass, moving to lower frequencies for higher masses, and mass accretion rate and thus can be used to estimate these quantities. The origin of the break is not well-understood, although it has been suggested recently that it is associated with the cooling timescale for Comptonization of electrons (Ishibashi and Courvoisier 2012). The amplitude of the high frequency part of the PSD may also depend inversely on black hole mass (e.g., Gierliński et al. 2008; McHardy 2013).
- By assuming that X-ray warm absorbers in AGNs are radiatively accelerated, an upper limit on the central black hole mass can be calculated (Morales and Fabian 2002). Given the model uncertainties, these are in surprisingly good agreement with the reverberation measurements.

7 The Future

In addition to trying to capture the current state of the art, a review article should also attempt to identify particularly important directions that might be undertaken in the future. Based on this discussion, here are a few investigations that might be pursued to improve measurement of the masses of black holes in AGNs:

- **Velocity–delay maps for high-ionization lines.** As discussed in Sect. 3.2.4, the Balmer lines are emitted by lower-ionization gas that seems to be infalling and at least sometimes in a disk-like structure. Absorption-line studies suggest, by contrast, that the strong high-ionization lines in the UV arise in outflows. A complete picture of gas flows in active nuclei will therefore require contemporaneous velocity–delay maps for both high-ionization and low-ionization lines.
- **Modeling BLR kinematics.** Velocity–delay maps provide entirely new constraints that should allow tremendous improvements in forward modeling of the BLR and, consequently, determination of the masses of the central black holes.
- **The radius–luminosity relationship at high luminosity.** This is especially important for the C IV line because of the potential importance of this line in determining black hole masses at high redshift. Even the Hβ R–L relationship is sparsely populated at high luminosity (Fig. 8) and by sources for which the light curves are not exceptionally well-sampled. It is important to quantify the systematic effects and intrinsic scatter because of the potential importance of using this relationship for cosmological applications (e.g., Watson et al. 2011).
- **Characterization of line widths.** At the present time, there is no consensus "best practice" for characterization of broad emission-line widths for black hole mass determination. How this is done for single-epoch spectra is even more controversial and, indeed, the case that C IV in particular can be used to estimate black hole masses with any confidence at all needs to be made.

- **Direct comparisons of methods of mass measurement.** The opportunities for this are limited, of course, by the necessity of resolving R_{BH} (Sect. 1) to relatively nearby systems. As the precision of mass measurement improves, even in the immediately foreseeable future the efficacy of the comparisons will be limited by uncertainties in the distances to the nearest AGNs on account of their peculiar motions relative to the Hubble flow (Bentz et al. 2013). Obtaining accurate distances to the nearest AGNs is a matter of some urgency since it will get a lot harder following the inevitable retirement of *Hubble Space Telescope*.
- **Expanding the reverberation-mapped population.** The sample of some 50 or so AGNs that have been the targets of reverberation-mapping programs is certainly plagued by selection effects: for the most part, the AGNs that have been the targets of reverberation monitoring campaigns have been selected by some combination of their apparent brightness, known variability, or favorable position in the sky. Richards et al. (2011) note, for example, the absence in the reverberation sample of quasars with blueshifted C IV emission. This is, of course, not surprising given that such objects are found at high redshift and the reverberation sample is mostly at $z < 0.2$. Certainly this is an area of concern that must be addressed in the future.

Acknowledgements The author is grateful for support by the US National Science Foundation through grant AST-1008882 to The Ohio State University and for the kind hospitality of the International Space Science Institute in Bern, where this work was first presented. The author thanks Misty Bentz for providing Figs. 4 and 8, Kate Grier for Figs. 1, 5 and 6, and Jong-Hak Woo for Fig. 7. The author also thanks Misty Bentz, Kelly Denney, Stephan Frank, Kate Grier, Peter Jonker, Smita Mathur, Ian McHardy, Yue Shen, Marianne Vestergaard, and an anonymous referee for advice, criticism, and comments on the manuscript.

References

D. Alloin, J. Clavel, B.M. Peterson, G.A. Reichert, G.M. Stirpe, in *Frontiers of Space and Ground-Based Astronomy: The Astrophysics of the 21st Century*, ed. by W. Wamsteker, M.S. Longair, Y. Kondo (Kluwer, Dordrecht, 1994)
Y. Andrillat, S. Souffrin, Astrophys. Lett. **1**, 111 (1968)
R.R.J. Antonucci, R.D. Cohen, Astrophys. J. **271**, 564 (1983)
J.N. Bahcall, B.Z. Kozlovsky, E.E. Salpeter, Astrophys. J. **171**, 467 (1972)
J.A. Baldwin, G.J. Ferland, K.T. Korista, D. Verner, Astrophys. J. **455**, L119 (1995)
A.J. Barth et al., Astrophys. J. **732**, 121 (2011a)
A.J. Barth et al., Astrophys. J. **743**, L4 (2011b)
M.C. Bentz, B.M. Peterson, R.W. Pogge, M. Vestergaard, C.A. Onken, Astrophys. J. **644**, 133 (2006a)
M.C. Bentz et al., Astrophys. J. **651**, 775 (2006b)
M.C. Bentz et al., Astrophys. J. **662**, 205 (2007)
M.C. Bentz, B.M. Peterson, R.W. Pogge, M. Vestergaard, Astrophys. J. **694**, L166 (2009a)
M.C. Bentz, B.M. Peterson, H. Netzer, R.W. Pogge, M. Vestergaard, Astrophys. J. **697**, 160 (2009b)
M.C. Bentz et al., Astrophys. J. **705**, 199 (2009c)
M.C. Bentz et al., Astrophys. J. **716**, 993 (2010a)
M.C. Bentz et al., Astrophys. J. **720**, L46 (2010b)
M.C. Bentz et al., Astrophys. J. **767**, 149 (2013)
R.D. Blandford, C.F. McKee, Astrophys. J. **255**, 419 (1982)
G. Bower et al., Astrophys. J. **492**, L111 (1998)
B.J. Brewer et al., Astrophys. J. **733**, L33 (2011)
E.M. Burbidge, Annu. Rev. Astron. Astrophys. **5**, 399 (1967)
J. Clavel et al., Astrophys. J. **366**, 64 (1991)
S. Collin, T. Kawaguchi, B.M. Peterson, M. Vestergaard, Astron. Astrophys. **456**, 75 (2006)
S. Collin-Souffrin, D. Alloin, Y. Andrillat, Astron. Astrophys. **22**, 343 (1973)
N. Cretton, P.T. de Zeeuw, R.P. van der Marel, H.-W. Rix, Astrophys. J. Suppl. **124**, 383 (1999)
E. Dalla Bontà, L. Ferrarese, E.M. Corsini, J. Miralde Escudé, L. Coccato, M. Sarzi, A. Pizzella, A. Beifiori, Astrophys. J. **690**, 537 (2009)

K.M. Dasyra, L.J. Tacconi, R.I. Davies, R. Genzel, D. Lutz, B.M. Peterson, S. Veilleux, A.J. Baker, M. Schweitzer, E. Sturm, Astrophys. J. **657**, 102 (2007)
K. Davidson, Astrophys. J. **171**, 213 (1972)
R.I. Davies et al., Astrophys. J. **646**, 754 (2006)
G. de Francisco, A. Capetti, A. Marconi, Astron. Astrophys. **479**, 355 (2008)
K.D. Denney, Astrophys. J. **759**, 44 (2012)
K.D. Denney et al., Astrophys. J. **653**, 152 (2006)
K.D. Denney, B.M. Peterson, M. Dietrich, M. Vestergaard, M.C. Bentz, Astrophys. J. **692**, 246 (2009)
K.D. Denney et al., Astrophys. J. **721**, 715 (2010)
K.D. Denney, R.W. Pogge, R.J. Assef, C.S. Kochanek, B.M. Peterson, M. Vestergaard. Astrophys. J. (2013, submitted). arXiv:1303.3889
M. Dietrich et al., Astrophys. J. **408**, 146 (1993)
M. Dietrich, F. Hamann, J.C. Shields, A. Constantin, M. Vestergaard, F. Chaffee, C.B. Foltz, V.T. Junkkarinen, Astrophys. J. **581**, 912 (2002)
V.T. Doroshenko, S.G. Sergeev, S.A. Klimanov, V.I. Pronik, Yu.S. Efimov, Mon. Not. R. Astron. Soc. **426**, 416 (2012)
A.C. Fabian et al., Nature **459**, 540 (2009)
G.J. Ferland, R.F. Mushotzky, Astrophys. J. **262**, 564 (1982)
G.J. Ferland, K.T. Korista, B.M. Peterson, Astrophys. J. **363**, L21 (1990)
G.J. Ferland, B.M. Peterson, K. Horne, W.F. Welsh, S.N. Nahar, Astrophys. J. **387**, 95 (1992)
L. Ferrarese, H. Ford, Space Sci. Rev. **116**, 523 (2005)
L. Ferrarese, D. Merritt, Astrophys. J. **539**, L9 (2000)
L. Ferrarese, R.W. Pogge, B.M. Peterson, D. Merritt, A. Wandel, C.L. Joseph, Astrophys. J. **555**, L79 (2001)
S. Fine, S.M. Croom, J. Bland-Hawthorn, K.A. Pimbblet, N.P. Ross, D.P. Schneider, T. Shanks, Mon. Not. R. Astron. Soc. **409**, 591 (2010)
K. Gebhardt et al., Astrophys. J. **539**, L13 (2000a)
K. Gebhardt et al., Astrophys. J. **543**, L5 (2000b)
K. Gebhardt et al., Astrophys. J. **583**, 92 (2003)
R. Genzel, F. Eisenhauer, S. Gillessen, Rev. Mod. Phys. **82**, 3121 (2010)
M. Gierliński, M. Nikołajuk, B. Czerny, Mon. Not. R. Astron. Soc. **383**, 741 (2008)
M. Gliozzi, L. Titarchuk, S. Satyapal, D. Prince, I. Jang, Astrophys. J. **735**, 16 (2011)
A.W. Graham, C.A. Onken, E. Athanassoula, F. Combes, Mon. Not. R. Astron. Soc. **412**, 2211 (2011)
J.E. Greene et al., Astrophys. J. **723**, 409 (2010)
C.J. Grier et al., Astrophys. J. **744**, L4 (2012a)
C.J. Grier et al., Astrophys. J. **755**, 60 (2012b)
C.J. Grier et al., Astrophys. J. **764**, 47 (2013)
E. Guerras, E. Mediavilla, J. Jimenez-Vicente, C.S. Kochanek, J.A. Muñoz, E. Falco, V. Motta, Astrophys. J. **764**, 160 (2013)
K. Gültekin et al., Astrophys. J. **698**, 198 (2009)
J.R. Herrnstein, J.M. Moran, L.J. Greenhill, A.S. Trotter, Astrophys. J. **629**, 719 (2005)
E.K.S. Hicks, M.A. Malkan, Astrophys. J. Suppl. **174**, 31 (2008)
K. Horne, B.M. Peterson, S.J. Collier, H. Netzer, Publ. Astron. Soc. Pac. **116**, 465 (2004)
W. Ishibashi, T.J.-L. Courvoisier, Astron. Astrophys. **540**, L2 (2012)
K. Iwasawa, G. Miniutti, A.C. Fabian, Mon. Not. R. Astron. Soc. **355**, 1073 (2004)
S. Kaspi, P.S. Smith, H. Netzer, D. Maoz, B.T. Jannuzi, U. Giveon, Astrophys. J. **533**, 631 (2000)
S. Kaspi, D. Maoz, H. Netzer, B.M. Peterson, M. Vestergaard, B.T. Jannuzi, Astrophys. J. **629**, 61 (2005)
S. Kaspi, W.N. Brandt, D. Maoz, H. Netzer, D.P. Schneider, O. Shemmer, Astrophys. J. **659**, 997 (2007)
W. Kollatschny, Astron. Astrophys. **407**, 461 (2003a)
W. Kollatschny, Astron. Astrophys. **412**, L61 (2003b)
J.A. Kollmeier et al., Astrophys. J. **648**, 128 (2006)
A.P. Koratkar, C.M. Gaskell, Astrophys. J. **370**, L61 (1991)
K.T. Korista, J.A. Baldwin, G.J. Ferland, D. Verner, Astrophys. J. Suppl. **108**, 401 (1997)
J. Kormendy, D. Richstone, Annu. Rev. Astron. Astrophys. **33**, 581 (1995)
C.A. Kuehn, J.A. Baldwin, B.M. Peterson, K.T. Korista, Astrophys. J. **673**, 69 (2008)
H. Landt, M.C. Bentz, B.M. Peterson, M. Elvis, M.J. Ward, K.T. Korista, M. Karovska, Mon. Not. R. Astron. Soc. **413**, L106 (2011)
A. Laor, Astrophys. J. **505**, L83 (1998)
D. Lynden-Bell, Nature **223**, 690 (1969)
F. Macchetto, A. Marconi, D.J. Axon, A. Capetti, W. Sparks, P. Crane, Astrophys. J. **489**, 579 (1997)
J. Magorrian et al., Astron. J. **115**, 2285 (1998)
D. Maoz et al., Astrophys. J. **404**, 576 (1993)

A. Marconi, D.J. Axon, R. Maiolino, T. Nagao, G. Pastorini, P. Pietrini, A. Robinson, G. Torricelli, Astrophys. J. **678**, 693 (2008)
A. Markowitz et al., Astrophys. J. **593**, 96 (2003)
N.J. McConnell, C.-P. Ma, Astrophys. J. **764**, 184 (2013)
K.L. McGill, J.-H. Woo, T. Treu, M.A. Malkan, Astrophys. J. **673**, 703 (2008)
I.M. McHardy, Mon. Not. R. Astron. Soc. **430**, L49 (2013)
I.M. McHardy, E. Koerding, C. Knigge, P. Uttley, R.P. Fender, Nature **444**, 730 (2006)
I.M. McHardy, P. Arévalo, P. Uttley, I.E. Papadakis, D.P. Summons, W. Brinkmann, M.J. Page, Mon. Not. R. Astron. Soc. **382**, 985 (2007)
R.J. McLure, M.J. Jarvis, Mon. Not. R. Astron. Soc. **337**, 109 (2002)
L. Meyer, A.M. Ghez, B. Schödel, S. Yelda, A. Boehle, J.R. Lu, T. Do, M.R. Morris, E.E. Becklin, K. Matthews, Science **338**, 84 (2012)
M. Miyoshi, J. Moran, J. Herrnstein, L. Greenhill, N. Nakai, P. Diamond, M. Inoue, Nature **373**, 127 (1995)
R. Morales, A.C. Fabian, Mon. Not. R. Astron. Soc. **329**, 209 (2002)
C.J. Nelson, R.F. Green, G. Bower, K. Gebhardt, D. Weistrop, Astrophys. J. **615**, 652 (2004)
H. Netzer, P. Marziani, Astrophys. J. **724**, 318 (2010)
C.A. Onken, L. Ferrarese, D. Merritt, B.M. Peterson, R.W. Pogge, M. Vestergaard, A. Wandel, Astrophys. J. **615**, 645 (2004)
C.A. Onken et al., Astrophys. J. **670**, 105 (2007)
A. Pancoast, B.J. Brewer, T. Treu, Astrophys. J. **730**, 139 (2011)
A. Pancoast et al., Astrophys. J. **754**, 49 (2012)
I.E. Papadakis, Mon. Not. R. Astron. Soc. **348**, 207 (2012)
D. Park, B.C. Kelly, J.-H. Woo, T. Treu, Astrophys. J. Suppl. **203**, 6 (2012)
G. Pastorini et al., Astron. Astrophys. **469**, 405 (2007)
M. Pastoriza, H. Gerola, Astrophys. Lett. **6**, 155 (1970)
B.M. Peterson, Publ. Astron. Soc. Pac. **105**, 247 (1993)
B.M. Peterson, in *Advanced Lectures on the Starburst–AGN Connection*, ed. by I. Aretxaga, D. Kunth, R. Mújica (World Scientific, Singapore, 2001)
B.M. Peterson, New Astron. Rev. **52**, 240 (2008)
B.M. Peterson, in *The Emission-Line Universe*, ed. by J. Cepa (Cambridge University Press, Cambridge, 2009)
B.M. Peterson, A. Wandel, Astrophys. J. **521**, L95 (1999)
B.M. Peterson, A. Wandel, Astrophys. J. **540**, L13 (2000)
B.M. Peterson, R.M. Wagner, D.M. Crenshaw, K.A. Meyers, P.L. Byard, C.B. Foltz, H.R. Miller, Astron. J. **88**, 926 (1983)
B.M. Peterson, K.A. Meyers, E.R. Capriotti, C.B. Foltz, B.J. Wilkes, H.R. Miller, Astrophys. J. **292**, 164 (1985)
B.M. Peterson et al., Astrophys. J. **368**, 119 (1991)
B.M. Peterson, L. Ferrarese, K.M. Gilbert, S. Kaspi, M.A. Malkan, D. Maoz, D. Merritt, C.A. Onken, R.W. Pogge, M. Vestergaard, A. Wandel, Astrophys. J. **613**, 682 (2004)
B.M. Peterson et al., Astrophys. J. **632**, 799 (2005)
L. Popović et al., Astron. Astrophys. **529**, A130 (2011)
A. Rafiee, P. Hall, Astrophys. J. Suppl. **194**, 42 (2011)
M.J. Rees, Annu. Rev. Astron. Astrophys. **22**, 471 (1984)
G.T. Richards et al., Astron. J. **141**, 167 (2011)
E.E. Salpeter, Astrophys. J. **140**, 796 (1964)
S.G. Sergeev, S.A. Klimanov, V.T. Doroshenko, Yu.S. Efimov, S.V. Nazarov, V.I. Pronik, Mon. Not. R. Astron. Soc. **410**, 1877 (2011)
F. Shankar, New Astron. Rev. **53**, 57 (2009)
A.I. Shapovalova et al., Astron. Astrophys. **509**, 106 (2010)
A.I. Shapovalova et al., Astrophys. J. Suppl. **202**, 19 (2012)
Y. Shen, B.C. Kelly, Astrophys. J. **746**, 169 (2012)
Y. Shen, J.E. Greene, M.A. Strauss, G.T. Richards, D.P. Schneider, Astrophys. J. **680**, 169 (2008)
C. Siopsis et al., Astrophys. J. **693**, 946 (2009)
C.L. Steinhardt, M. Elvis, Mon. Not. R. Astron. Soc. **402**, 2637 (2010)
J. Thomas, R.P. Saglia, R. Bender, D. Thomas, K. Gebhardt, J. Magorrian, D. Richstone, Mon. Not. R. Astron. Soc. **353**, 391 (2004)
J. Tohline, D.E. Osterbrock, Astrophys. J. **210**, L117 (1976)
H.D. Tran, Z. Tsvetanov, H.D. Ford, J. Davies, W. Jaffe, F.C. van den Bosch, A. Rest, Astron. J. **121**, 2928 (2001)
S. Tremaine et al., Astrophys. J. **574**, 740 (2002)

M.-H. Ulrich, K. Horne, Mon. Not. R. Astron. Soc. **283**, 748 (1996)
M.-H. Ulrich, A. Boksenberg, M.V. Penston, G.E. Bromage, J. Clavel, A. Elvius, G.C. Perola, M.A.J. Snijders, Astrophys. J. **382**, 483 (1991)
M. Valluri, D. Merritt, E. Emsellem, Astrophys. J. **602**, 66 (2004)
R.P. van der Marel, N. Cretton, P.T. de Zeeuw, H.-W. Rix, Astrophys. J. **493**, 613 (1998)
D.E. Vanden Berk, C. Yip, A. Connolly, S. Jester, C. Stoughton, in *AGN Physics with the Sloan Digital Sky Survey*, ed. by G.T. Richards, P.B. Hall (Astronomical Society of the Pacific, San Francisco, 2004)
M. Vestergaard, Astrophys. J. **571**, 733 (2002)
M. Vestergaard, Astrophys. J. **601**, 676 (2004)
M. Vestergaard, P.S. Osmer, Astrophys. J. **699**, 800 (2009)
M. Vestergaard, B.M. Peterson, Astrophys. J. **625**, 688 (2005)
M. Vestergaard, B.M. Peterson, Astrophys. J. **641**, 676 (2006)
M. Vestergaard, B.J. Wilkes, Astrophys. J. Suppl. **134**, 1 (2001)
A. Wandel, B.M. Peterson, M.A. Malkan, Astrophys. J. **526**, 579 (1999)
I. Wanders et al., Astrophys. J. **453**, L87 (1997)
L.C. Watson, P. Martini, K.M. Dasyra, M.C. Bentz, L. Ferrarese, B.M. Peterson, R.W. Pogge, L. Tacconi, Astrophys. J. **682**, L21 (2008)
D. Watson, K.D. Denney, M. Vestergaard, T.M. Davis, Astrophys. J. **740**, L49 (2011)
J. Woo et al., Astrophys. J. **716**, 269 (2010)
X.-B. Wu, R. Wang, M.Z. Kong, F.K. Liu, J.L. Han, Astron. Astrophys. **424**, 793 (2004)
Ya.B. Zel'dovich, I.D. Novikov, Sov. Phys. Dokl. **158**, 811 (1964)
Y. Zu, C.S. Kochanek, B.M. Peterson, Astrophys. J. **735**, 80 (2011)

DOI 10.1007/978-1-4939-2227-7_14
Reprinted from *Space Science Reviews* Journal, DOI 10.1007/s11214-013-0006-6

Measuring Black Hole Spin Using X-Ray Reflection Spectroscopy

Christopher S. Reynolds

Received: 24 February 2013 / Accepted: 15 July 2013 / Published online: 31 August 2013

Abstract I review the current status of X-ray reflection (a.k.a. broad iron line) based black hole spin measurements. This is a powerful technique that allows us to measure robust black hole spins across the mass range, from the stellar-mass black holes in X-ray binaries to the supermassive black holes in active galactic nuclei. After describing the basic assumptions of this approach, I lay out the detailed methodology focusing on "best practices" that have been found necessary to obtain robust results. Reflecting my own biases, this review is slanted towards a discussion of supermassive black hole (SMBH) spin in active galactic nuclei (AGN). Pulling together all of the available *XMM-Newton* and *Suzaku* results from the literature that satisfy objective quality control criteria, it is clear that a large fraction of SMBHs are rapidly-spinning, although there are tentative hints of a more slowly spinning population at high ($M > 5 \times 10^7\ M_\odot$) and low ($M < 2 \times 10^6\ M_\odot$) mass. I also engage in a brief review of the spins of stellar-mass black holes in X-ray binaries. In general, reflection-based and continuum-fitting based spin measures are in agreement, although there remain two objects (GRO J1655–40 and 4U 1543–475) for which that is not true. I end this review by discussing the exciting frontier of relativistic reverberation, particularly the discovery of broad iron line reverberation in *XMM-Newton* data for the Seyfert galaxies NGC 4151, NGC 7314 and MCG–5-23-16. As well as confirming the basic paradigm of relativistic disk reflection, this detection of reverberation demonstrates that future large-area X-ray observatories such as *LOFT* will make tremendous progress in studies of strong gravity using relativistic reverberation in AGN.

Keywords Accretion disks · Black holes · X-ray astronomy

1 Introduction

Despite their exotic nature, General Relativity (GR) tells us that black holes are remarkably simple objects, at least as far as the external Universe is concerned. In astrophysical settings

C.S. Reynolds (✉)
Dept. of Astronomy, University of Maryland, College Park, MD 20742, USA
e-mail: chris@astro.umd.edu

(where charge neutrality is essentially guaranteed and hence the spacetime is described by the Kerr metric; Kerr 1963), an isolated black hole is fully characterized by just its mass (M) and angular momentum (usually specified in terms of the dimensionless spin parameter $a = Jc/GM^2$, where J is the angular momentum and the cosmic censorship hypothesis ensures that $|a| \leq 1$). For gravitational processes, the mass acts as a trivial scaling factor for distances, timescales, and energies. This directly underlies the fact that accreting black holes across the entire mass scale (from those in X-ray binaries [BH-XRBs] to active galactic nuclei [AGN]) have very similar spectral and temporal characteristics, once one scales the energies and timescales appropriately.

Spin, however, is a much more interesting quantity. Depending upon the spin, the Kerr metric can possess qualitatively different properties that have astrophysical significance. For example, for slowly rotating black holes, the innermost stable circular orbit (ISCO) is far from the ergosphere; accreting matter will be well into its ballistic "death plunge" before encountering the ergosphere and the strong frame dragging effects. But for $a > 0.93$, the ISCO moves into the ergosphere raising the possibility that the accretion disk can tap into and radiate the rotational energy of the black hole (Agol and Krolik 2000). Indeed, it is possible that some that relativistic jets from black holes are powered by the magnetic extraction of rotational energy from a central Kerr black hole (Blandford and Znajek 1977).

Clearly, measuring black hole spin is of great interest. It is a necessary step in any Kerr-metric based tests of strong-field GR. Spin measurements are also vital if we are to understand the astrophysical consequences of spin and, particularly, give observational grounding to theoretical notions such as the spin-driving of relativistic jets. Measuring spin is a challenging endeavour, however—to find the signatures of spin, we must search for and characterize (with high precision) signals from within a few gravitational radii of the black hole.

In this chapter, I review the progress that has been made in measuring black hole spins using relativistic X-ray reflection spectroscopy (a.k.a. broad iron line spectroscopy). With this one technique, we can study black hole spin across the mass range, from the stellar-mass black holes in X-ray binaries to the supermassive black holes in active galactic nuclei (AGN). A major advantage of this method is that we require no knowledge of the black hole mass or its distance in order to derive the black hole spin. Neither do we require knowledge of the accretion disk inclination—indeed, X-ray reflection spectroscopy allows us to measure the inclination concurrently with the black hole spin.

In contrast to the continuum fitting (CF) method for measuring spins (McClintock and Narayan, this volume), the relativistic reflection method has been developed, evolved and debated by several groups over the past few years, producing a literature that can be rather bewildering. Hence, in this chapter, I review the assumptions on which the technique rests, summarize the current "best practices" for employing this technique, and describe how careful consideration of the astrophysical consistency (and, in some cases, instrumental issues compromising the data) can resolve discrepancies in published spins. Reflecting my own biases, much of this discussion will be dedicated to supermassive black holes (SMBHs), but I shall also describe recent work on stellar-mass black holes in BH-XRBs and address the concordance between relativistic reflection and CF spin measurements. I shall end by discussing recent results on relativistic reverberation.

This chapter is organized as follows. Section 2 lays out the assumptions and methodology of the relativistic reflection spin measurements. The results for SMBHs are discussed in Sect. 3, using two case studies to highlight important systematic issues that can affect spin measurements. Stellar-mass black holes in BH-XRBs are discussed in Sect. 4. The dramatic progress that has been made recently in characterizing relativistic reverberation is summarized in Sect. 5. The casual reader who is interested more in the current status of spin results rather than details of the methodology may wish to skip Sects. 2.2 and 3.2. Unless otherwise stated, all spins quoted in this review are given with their 90 % error ranges.

2 Assumptions and Methodology of Relativistic X-Ray Reflection Spectroscopy

2.1 Basic Assumptions and Geometry

In both spectroscopic methods for determining black hole spin (CF and relativistic reflection), the central assumption is that the accretion disk remains geometrically-thin, optically-thick, and radiatively efficient down to the ISCO. As the accretion flow crosses over the ISCO, the inward radial velocity rapidly increases, the flow quickly becomes super-sonic and super-Alfvenic (hence ballistic) and, due to conservation of mass flux, the density of the flow rapidly drops. This basic behavior is confirmed by magnetohydrodynamic simulations of thin accretion disks (Reynolds and Fabian 2008; Penna et al. 2010). As discussed below, the result is that the ISCO serves as the "inner-edge" of the various observables, including the X-ray reflection features that are our focus here. The ability to measure spin then follows directly from the fact that the location of the ISCO (in units of gravitational radii $r_g \equiv GM/c^2$) is a simple function of spin (Bardeen et al. 1972).

This central assumption is likely to be valid for sources accreting at modest rates, with bolometric luminosities in the range $10^{-2} - 0.3L_{\rm Edd}$ (where $L_{\rm Edd}$ is the usual Eddington luminosity). This includes the thermal dominant (a.k.a. high/soft) state the of BH-XRBs, as well as the more luminous broad-optical-line Seyfert galaxies and radio-galaxies. Thermal emission from the accretion disk is clearly visible in these classes of sources, in the soft X-ray band for BH-XRBs and the optical/UV/EUV bands for AGN. Generically, however, the spectra of these sources also show a hard X-ray power-law tail which must originate from some structure external to the accretion disk, either a disk corona or the base of a jet.

This geometry is key to the reflection method for spin determination. The X-ray source irradiates the underlying accretion disk with a hard X-ray continuum. Outside of the ISCO, where the accretion disk is optically-thick and still contains combined metal ions, the irradiated surface emits a complex spectrum consisting of atomic emission features sitting on a Compton-backscattered continuum (Fig. 1, left). This is the *X-ray reflection spectrum*, and provided that the disk photosphere is not too highly photoionized, the most prominent feature is the iron-Kα emission line (rest-frame energy 6.4–6.97 keV depending upon ionization state). Doppler shifts and gravitational redshifts, that become progressively stronger as one considers radii in the disk closer to the black hole, broaden and skew the observed features in a well-defined manner (Fig. 1, right).

Within the ISCO, the density of the accretion flow plummets and it will be fully ionized by any significant X-ray irradiation of the flow. Hence, the region within the ISCO will not contribute to the observed atomic signatures. There may still be some Compton-reflection from the region within the ISCO but, since there is no combined iron in this region and hence no iron absorption, it will not contribute to the "Compton-reflection hump" (which is sculpted on the low-energy side by iron-K bound-free absorption). In other words, the ISCO defines the inner edge to the region producing "spectrally structured" X-ray reflection.

2.2 Operationally Nuts and Bolts of Relativistic Reflection Spin Determinations

Given these basics, how do we proceed at an operational level to analyze a real spectrum and extract black hole spin? To isolate the signatures of relativistic reflection needed for

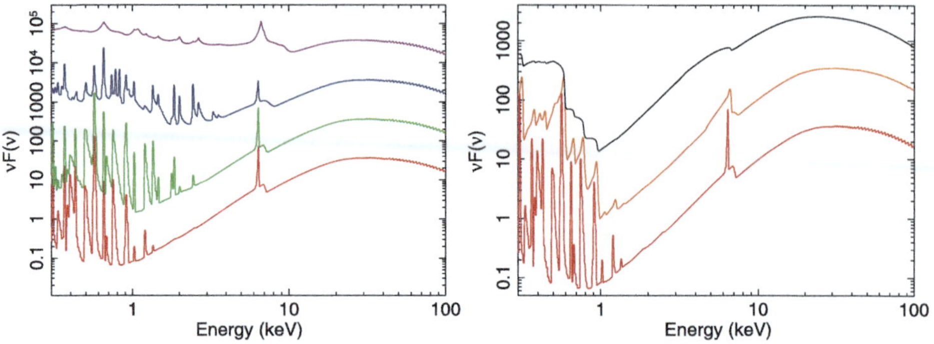

Fig. 1 *Left panel*: Rest-frame X-ray reflection spectrum for photospheric ionization parameters of $\log\xi = 0, 1, 2, 3$ (from bottom to top). In all cases, the irradiating source has a power-law spectrum with photon index $\Gamma = 2$. The broadening of the emission lines for the more highly-ionized cases is due to Compton scattering in the disk atmosphere. *Right panel*: Demonstration of the effects of relativistic Doppler/gravitational blurring on the reflection spectrum. *Curves show* the $\log\xi = 0$ rest-frame reflection spectrum (*bottom*), relativistic blurring with $r_{\rm in} = 10r_g$ and emissivity index $\beta = 3$ (*middle curve*), and extreme blurring with $r_{\rm in} = 2r_g$ and $\beta = 4$. In all cases, a disk inclination of $i = 30$ degrees is assumed

the spin measurement, we must understand the spectrum of the primary continuum, the effects of absorption, and the presence of other reprocessing signatures (e.g., emission from an optically-thin photoionized wind, or X-ray reflection from distant structures such as the companion star in a BH-XRB or the molecular torus in an AGN).

Here, I shall focus on the methodology developed for (non-blazar, Compton-thin) AGN, mentioning a few operational differences with BH-XRBs towards the end of the section. In a disk-dominated AGN, the primary X-ray continuum in the crucial 2–50 keV range is very well approximated by a power-law. This is readily understood in terms of a separation of energy-scales. The X-rays are believed to be produced by the thermal Comptonization of soft optical/UV photons ($h\nu \sim 1$–10 eV) by an electron population that is hot ($kT \sim 100$ keV). For photon energies E satisfying $h\nu \ll E \ll kT$, Comptonization generically produces a power-law. The most common and obvious complexities observed in AGN spectra are (i) a narrow iron-Kα emission line at 6.4 keV due to reflection/fluorescence by distant and low-velocity material, and (ii) absorption by cold and photoionized gas along the line-of-sight to the central X-ray source. Thus, to our power-law spectral model, we add a model of cold, distant X-ray reflection (e.g., using the `pexmon` model which self-consistently includes the iron-Kα, iron-Kβ, and nickel-Kα lines as well as the Compton reflected continuum; Nandra et al. 2007). To treat absorption, we add successive absorption components (each with its own ionization parameter and column density; computed using a photoionization code such as XSTAR; Kallman and Bautista 2001) until no further improvement in the fit to the data is found. The final step before dealing with the relativistic reflection is to add a phenomenological component (a blackbody or additional Comptonization component) to treat the soft excess that many AGN display.

In approximately 30–50 % of type-1 AGN, applying this procedure to data with sufficient signal-to-noise will leave unmodelled residuals indicative of a broad/redshifted iron line and excess Compton reflection, i.e. a hump at ~20 keV (Nandra et al. 2007). These are the signatures of relativistic reflection that we seek to characterize although, as discussed below, there are alternative explanations for the soft excess that can lead to important ambiguities. Thus, we add to the spectral model a component describing an ionized reflection spectrum (the current industry standard is `reflionx`; Ross and Fabian 2005) that has been convolved by a "disk-line transfer function" encoding the Doppler/gravitational redshifts from the accretion disk. The available and commonly employed disk-line convolution models (e.g., `relconv`; Dauser et al. 2010) are based on GR ray-tracing calculations within the Kerr metric, assuming a razor thin accretion disk lying in the equator of the rotating black hole. The rest-frame reflection spectrum depends upon the ionization parameter ξ of the plasma at the X-ray photosphere of the disk (see Fig. 1, left panel).

In addition to the black hole spin a and the disk-inclination i, the disk-line transfer function depends upon the intensity of the irradiating flux as a function of radius, $f_X(r)$, that, of course, depends upon the nature of the primary X-ray source. It is sometimes said that the lack of knowledge of the location and structure of the X-ray source is a major limitation for studies of relativistic reflection. In fact, from the point of view of spin measurements, all we require is a parameterized form of $f_X(r)$ that can then be constrained by the data themselves. Empirically, it is found that a broken-powerlaw is a good description of this irradiation profile, giving up to five additional fit parameters, the inner/break/outer radii $(r_{\rm in}, r_{\rm br}, r_{\rm out})$ and the two power-law indices (β_1, β_2) defined such that in the relevant patch we have $f_X(r) \propto r^{-\beta}$. As detailed above, we identify $r_{\rm in}$ with the ISCO. Furthermore, provided that $\beta_2 > 2$, the spectrum will be insensitive to a sufficiently large $r_{\rm out}$. Thus, the total set of parameters describing the relativistic reflection is $\{a, i, \xi, \beta_1, \beta_2, r_{\rm br}\}$. The final spectral model describing the continuum source, the distant reprocessing, absorption components, and relativistic reflection from the accretion disk, is then fitted to the data (with all interesting spectral parameters, including those describing the continuum form and the absorption components allowed to be free) and constraints on the spin can be obtained.

There are two aspects of this methodology that must be underscored. Firstly, in many situations, the error bar on the derived spin will be driven as much by uncertainties in the continuum/absorber parameters as by uncertainties in the form of the relativistic reflection itself. Thus, it is crucial to allow all relevant spectral parameters to vary in the final spectral analysis, including those describing the continuum and absorption. Secondly, when attempting to measure black hole spin, one must model the full ionized reflection spectrum. Phenomenological models in which isolated broad iron lines are added to the primary continuum are useful for initial explorations of spectra, but are inadequate for characterizing the subtle spectral signatures of spin.

A very similar procedure is adopted in the case of BH-XRBs, although there are a few differences compared with the case for AGN driven by the fact that XRB accretion disks are hotter, the astrophysical environments are cleaner and, typically, the signal-to-noise (s/n) in the datasets is much better. Absorption is much less of a concern in studies of the disk reflection in BH-XRBs. Strong outflows are confined to strongly thermally-dominated states (of potential concern for CF spin measurements) but are weak or absent during states that display prominent reflection (Miller et al. 2006, 2008; King et al. 2011, 2013). Even when present, the outflows tend to have significantly higher ionization states (such that iron has a fully stripped L-shell) and, hence, absorption tends not to introduce broad-band curvature that can confuse studies of the broad iron line. On the other hand, the thermal disk spectrum in BH-XRB is now in the X-ray band and must be modeled if one hopes to robustly probe the broad iron line and other disk reflection signatures. Furthermore, the disk reflection features (and especially the iron-K lines) have significant broadening even in the rest frame from Compton scattering in the hot disk atmosphere. While this additional broadening is captured by the ionized reflection models (see top model in left panel of Fig. 1), there is some uncertainty in the degree of Comptonization that can affect spin determinations (see discussion of XTE J1550–564 in Sect. 4). Of course, the s/n is often extremely high for BH-XRBs. Indeed, one must start to consider "bright source" instrumental issues such as errors in the calibration of the effective area and photon pile-up. Miller et al. (2010) has explicitly shown that pile-up tends to make a broad line narrower and, hence, always acts to decrease the inferred spin. We return to this point in the discussion of Cygnus X-1 in Sect. 4.

We end this section on methodology by highlighting the complementarities and strengths of the relativistic reflection method in comparison to the CF method. As discussed by McClintock and Narayan (this volume), the CF method, in its essence, uses the Stefan–Boltzamann law to measure the emitting area of the accretion disk (employing a Novikov and Thorne (1974) emission profile) and hence the size (in physical units) of the ISCO. To apply this technique requires accurate optical/IR measurements of mass (to convert the ISCO into units of gravitational radii, $r_g = GM/c^2$), distance (in order to derive luminosities), and inclination, as well as a trustworthy absolute X-ray flux calibration. By contrast, the relativistic transfer function underlying the reflection method is completely independent of mass, i.e., all radii are naturally scaled in units r_g, and one never needs to know the explicit size of r_g. Hence, mass uncertainties are irrelevant to reflection-based spins. In addition, since we are interested purely in spectral shape, the distance to the source and the absolute flux is also irrelevant (although, for AGN, the cosmological redshift is important to know). Finally, while the inclination of the inner disk is an important quantity, it can be directly measured along with spin from the X-ray spectrum.

3 Supermassive Black Holes

3.1 Early History

The first quantitative attempt to constrain black hole spin was based on modeling of the broad iron line profile in the Seyfert galaxy MCG–6-30-15. A long *ASCA* observation of this source had caught its transition to the so-called "Deep Minimum State" in which Iwasawa et al. (1996) noticed that the iron line was too broadened and redshifted to be explained by a non-rotating black hole. Subsequent modeling of this line profile by Dabrowski et al. (1997) concluded that $a > 0.95$ under the assumption that the irradiation profile of the reflection follows a Novikov-Thorne profile. In its original form, however, this analysis was not completely robust—Reynolds and Begelman (1997) showed that, in a lamp-post geometry[1], the combination of light-bending effects and a small penetration of the line-emitting region inside of the ISCO could render the Deep Minimum line profile consistent with Schwarzschild ($a = 0$) geometry. Young et al. (1998) showed that the need for rapid spin in MCG–6-30-15 was restored when one considered the full reflection continuum, showing that the Reynolds and Begelman (1997) scenario predicted a strong, and unobserved, absorption edge from ionized iron. This was the first realization that modeling the full reflection spectrum, not just the iron line, was crucial for spin determinations.

The first spin-analysis to follow the modern methodology (Sect. 2) was by Brenneman and Reynolds (2006), also for MCG–6-30-15. Using a long and high s/n *XMM-Newton* observation, they confirmed the rapid spin in MCG–6-30-15, obtaining a formal limit of $a > 0.98$ (although finite disk-thickness effects can relax this limit to $a > 0.92$; Reynolds and Fabian (2008)). Wider application of these techniques to other AGN had to await the availability of sufficiently deep observations. Over the past few years, the number of useful datasets has expanded, leading to an explosion in spin studies.

[1] In the lamp-post geometry, the X-ray source is situated at some height above the disk plane on the spin-axis of the black hole.

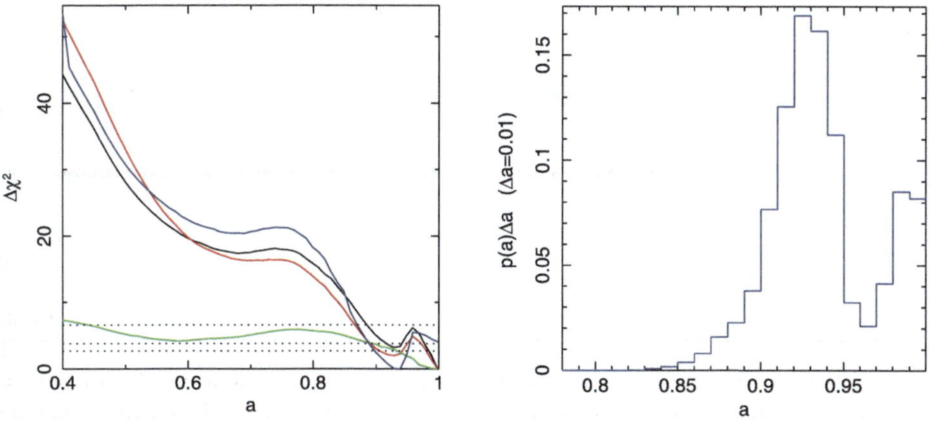

Fig. 2 Spin constraints on the SMBH in NGC3783 from the 2009 *Suzaku* observation. *Left panel*: Goodness of fit relative to the best fit, $\Delta\chi^2$, as a function of the spin parameter a. Different *lines* show the effects of different data analysis assumptions; a fiducial analysis (*black*), an analysis in which the warm absorber parameters are frozen at their best values (*red*), an analysis in which the XIS and PIN instrumental cross-normalizations are allowed to float (*blue*), and an analysis that ignores all data below 3 keV (*green*). From Brenneman et al. (2011). *Right panel*: Probability distribution for a as derived from a Monte Carlo Markov Chain (MCMC) analysis using the fiducial spectral model. From Reynolds et al. (2012)

3.2 Two Illustrative Case Studies

We now discuss two case studies that demonstrate the quality, as well as possible ambiguities, of reflection-based SMBH spin measurements. We will focus on results from one particular program, the *Suzaku AGN Spin Survey* (SASS), a Cycle 4–6 *Suzaku* Key Program (PI: Reynolds).

A posterchild for SASS is the bright Seyfert 1.5 galaxy NGC 3783 ($z = 0.0097$). Despite the complex (multi-zone) photoionized absorber and the presence of significant distant reflection, the relativistic reflection signatures from the inner accretion disk could be isolated and the spin constrained to be $a > 0.88$ (Fig. 2; Brenneman et al. 2011; Reis et al. 2012a, 2012b; Reynolds et al. 2012). Figure 3 shows the residuals when comparing the data with the best fitting spectral model ($a = 0.94$) and a model in which we have forced a non-spinning black hole (and, for physical consistency, $\beta_1 = \beta_2 = 3$) but allowed all other spectral parameters to fit. We see that the non-spinning model is over-predicting the spectrum in the 5–6 keV band. There is a clear way to understand this. In this spectrum, which includes hard-band data (up to 45 keV), the overall fractional amount of reflection is well determined. The question then arises whether the observed iron line is consistent with that reflection. Consistency of the expected line strength with the data demands that the iron line is strongly broadened and hence that the black hole is rapidly spinning. If a non-spinning black hole is imposed in the spectral fit, the iron line is "too narrow" and, given that its strength is set by the overall amount of reflection, one over-predicts the 5–6 keV flux.

The strength of the iron line also depends upon the iron abundance. Thus, from the same considerations described in the previous paragraph, we would expect the spectral fit to show a statistical correlation between the spin and the iron abundance. Our analysis of NGC 3783 finds exactly such a correlation (Fig. 4, left). In a free-fit, the spectral analysis of NGC 3783 clearly demands iron abundances that are 2–4× the cosmic average, in line with the metallicity enhancement often inferred within Seyfert nuclei on the basis of the optical/UV line

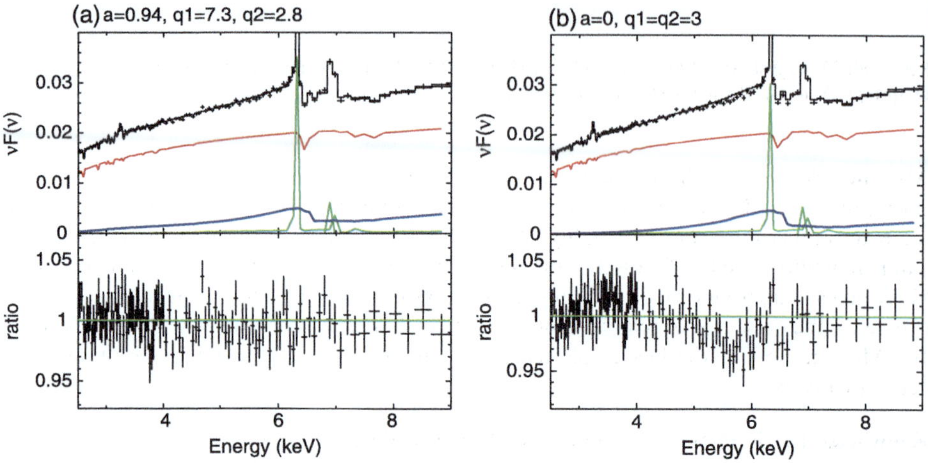

Fig. 3 Demonstration of the spectral signatures which, in practice, drive spin constraints using the *Suzaku* data for NGC 3783. *Left panel*: Unfolded XIS data overlaid with the best fitting model (*top*) and the associated data/model ratio (*bottom*). *Right panel*: Same, except that the spin parameter has been frozen at $a = 0$ and (for physical consistency) the irradiation indices have been frozen at $\beta_1 = \beta_2 = 3$. All other parameters (including those associated with the warm absorbers) have been allowed to fit freely. For both panels, the model components are colored as follows: absorbed power-law continuum (*red*), distant reflection (*green*), and relativistically smeared disk reflection (*blue*). Figure from Reynolds et al. (2012)

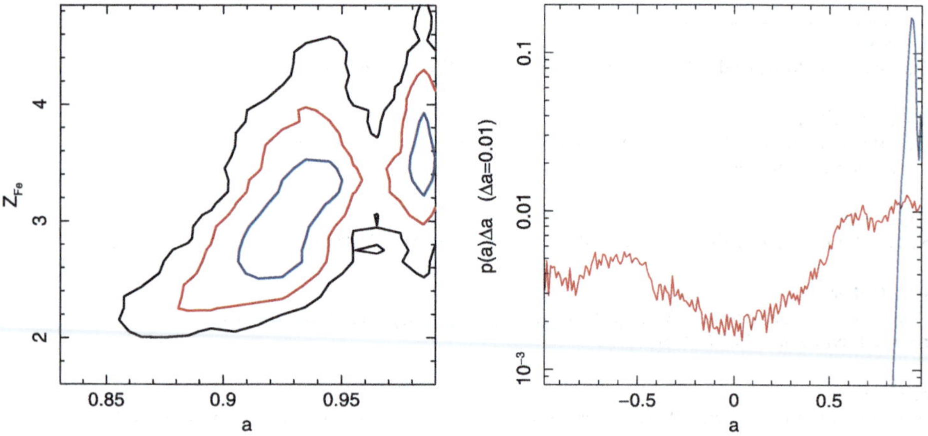

Fig. 4 Influence of iron abundance on the measured spin, illustrated using the 2009 *Suzaku* data for NGC 3783. *Left panel*: Two-dimensional probability distribution for iron abundance $Z_{\rm Fe}$ and spin a showing the existence of a statistical correlation between these two variables. *Right panel*: Probability distribution for spin a assuming a free-fitting iron abundance (*blue*) and an iron abundance fixed to solar values (*red*). Figures from Reynolds et al. (2012)

ratios (e.g., Warner et al. 2004; Nagao et al. 2006). However, if we simply impose solar iron abundances, the spectral model is a poorer fit to the data ($\Delta\chi^2 = 36$) and we essentially lose the ability to diagnose spin (Fig. 4, right). This exercise clearly demonstrates the need to permit non-solar iron abundance in the spectral fit if one is interested in measuring spin. We also note that, as discussed in detail in Reynolds et al. (2012), robust spin measurements

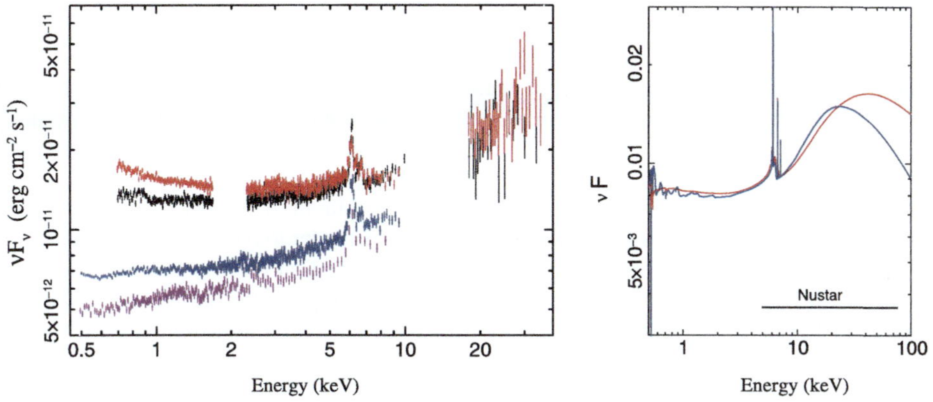

Fig. 5 *Left panel*: Unfolded spectra for the two *XMM-Newton* (*blue* and *magneta*) and two *Suzaku* pointings (*red* and *black*) of Fairall 9. The 1.5–2.5 keV band have been excluded from the *Suzaku* spectra due to the presence of known calibration artifacts. Note the presence of the soft excess that appears to become more prominent as the source brightens. *Right panel*: Extrapolation of the two spectral models for Fairall 9 to higher energies; the *red line* shows the model with a thermal Comptonization soft excess, and the *blue* shows the case where the soft excess is described by an additional highly ionized relativistic reflection component. Figures from Lohfink et al. (2012b)

in absorbed AGN such as NGC 3783 absolutely require tabulated absorption models with adequate resolution in the ξ-dimension.

While NGC 3783 is an example of an object in which we can obtain a clean and robust spin result, some objects can display spectral complexity that conspires with the medium-resolution of today's spectrometers in order to produce ambiguities in the measured spin. These ambiguities can be particularly severe if hard X-ray data is unavailable. We illustrate this with another of the SASS objects, the luminous Seyfert galaxy Fairall 9 ($z = 0.047$). This is an example of a "bare" Seyfert nucleus in the sense that every pointed X-ray observation shows it to be free of intrinsic absorption, either neutral or ionized (although *RXTE* monitoring does show unusual X-ray "dips" that may correspond to transient Compton-thick absorption events; Lohfink et al. 2012a). Analysis of earlier *Suzaku* data by Schmoll et al. (2009) found an obvious broad iron line with a red-wing extending down to 5 keV or possibly lower and, by imposing some restrictions on the irradiation profile, was able to estimate the black hole spin to be $a = 0.6 \pm 0.1$.

However, the broad line in Fairall 9 is weaker than in NGC 3783 or MCG–6-30-15 and, while it lacks absorption, there are some additional spectral complications. Due to the higher inclination of the accretion disk, the blue-peak of the broad line is blended with narrow Fe25/Fe26 emission lines from circumnuclear photoionized material. There is also a smooth and variable soft excess below 2 keV (Fig. 5, left). To deal with such complexity, Lohfink et al. (2012b) accumulated all available *XMM-Newton* and *Suzaku* data (taken over the course of several years) and applied the technique of multi-epoch fitting in which physical arguments are used to tie certain spectral parameters across epochs, but all other spectral parameters are allowed to fit freely to the spectrum from each epoch. Specifically, they tied the black hole spin, accretion disk inclination, iron abundance, and the strength of the narrow iron line components (which originate from parsec scales and hence will be stable between different observations). This procedure finds two acceptable solutions. In the first solution, the soft excess is described by an additional continuum component (modeled as thermal Comptonization from a moderately hot $kT \sim 20$ keV plasma) and the relativistic disk reflection, driven by the modestly broad iron line, implies a modestly spinning black hole ($a = 0.52^{+0.19}_{-0.15}$, completely consistent with the previous analysis of Schmoll et al. 2009). Interestingly, such a solution demands a very high iron abundance ($Z_{\rm Fe} \sim 10\ Z_\odot$) which would drive one to consider somewhat exotic phenomena such as radiative-levitation to enhance the photospheric iron content of the inner disk (Reynolds et al. 2012). However, a second solution exists which is a slightly worse description of the data but still statistically acceptable overall. Here, the soft excess is described by a highly-ionized ($\xi \sim 10^3$) relativistic reflection component from the inner accretion disk. In order to model the modest breadth of the broad iron line, the same disk must have low ionization patches that become more dominant at larger radii. This solution has an iron abundance that is slightly sub-solar, but the need to produce a very smooth soft excess from a structured reflection spectrum drives the fit to a rapidly-spinning black hole ($a > 0.93$).

These two solutions predict very different behavior above 20 keV (Fig. 5, right). In principle, the PIN instrument on *Suzaku* should have distinguished these models. Unfortunately, due to pointing and aspect problems during the observation (discussed in Lohfink et al. 2012b), the PIN data cannot be normalized with respect to the XIS data and, together with the low s/n of Fairall 9 in the PIN detector, provides no additional constraints. This experience underscores the importance of data above 10 keV even when one is attempting to extract spin from signatures below 10 keV. We look forward to *NuSTAR* and *Astro-H* studies of Fairall 9 to squash these ambiguities.

3.3 Summary of Current SMBH Spin Measurements

The past couple of years has seen an explosion in the number of published SMBH spin measurements based on relativistic reflection spectra, including two major works that use the public data archives to compile samples of SMBH spins (Patrick et al. 2012; Walton et al. 2013). Drawing on the published refereed literature at the time of writing (January 2013), Table 1 compiles the 20 SMBH/AGN spin measurements that satisfy the following "quality control" criteria:

1. Spin measurement are based on fits of the fully ionized reflection spectrum (not just an isolated broad iron line). In practice, all of these measurements have employed the Ross and Fabian (2005) model `reflionx`.
2. The iron abundance characterizing the disk reflection is allowed to be a free parameter in the spectral fit. This is important given the abundance-spin partial degeneracy noted in Sect. 3.2.
3. The inclination of the accretion disk is allowed to be a free parameter in the spectral fit and can be constrained. In cases where the inclination cannot be constrained, the spin constraint must be driven by some aspect of the spectrum other than the broad iron line (e.g., the soft excess) and one must be concerned about lack of uniqueness of the spectral model.
4. The irradiation index β is allowed to be a free parameter and converges on a value $\beta > 2$. In cases where the best-fit implies $\beta < 2$, the X-ray reflection is dominated by the *outer* accretion disk and so we would not expect to be sensitive to spin.

In addition to these quality-control criteria, we have rejected two sources from the Walton et al. (2013) sample; Mrk 509 has a *Suzaku*-based spin value that is not robust to the treatment of the PIN normalization, and PDS 456 has a spin solution that appears to be double-valued. We have included Fairall 9, listing its intermediate spin value from Lohfink et al. (2012b; based on the strict implementation of the methodology discussed in Sect. 2), but we note the ambiguity of this particular spin value (see Sect. 3.2).

Table 1 Summary of published AGN/SMBH spin measurements that pass the quality-criteria laid out in Sect. 3.3. All measurements are based upon *XMM-Newton* and/or *Suzaku* data. Reflecting the conventions in the primary literature, all masses are quoted with 1σ error bars whereas spins are quoted with 90 % error ranges. Column (1) lists the common name for the AGN, Columns (2) and (3) give the mass and spin, respectively, and Column (4) gives the references for these mass/spin measurements. Key to references: Be11 = Bennert et al. (2011); BR06 = Brenneman and Reynolds (2006); Br11 = Brenneman et al. (2011); Fa13 = Fabian et al. (2013); Ga11 = Gallo et al. (2011); Go12 = Gonzáles-Martin and Vaughan (2012); Lo12 = Lohfink et al. (2012b); Ma08 = Malizia et al. (2008); Mc05 = McHardy et al. (2005); Mi09 = Miniutti et al. (2009); Pe04 = Peterson et al. (2004); Pa12 = Patrick et al. (2012); Ta12 = Tan et al. (2012); Wa13 = Walton et al. (2013); Zo10 = Zoghbi et al. (2010); ZW05 = Zhou and Wang (2005)

Object	Mass ($\times 10^6\ M_\odot$)	Spin	Mass/Spin references
Mrk335	14.2 ± 3.7	$0.83^{+0.09}_{-0.13}$	Pe04/Wa13
IRAS 00521–7054	–	>0.84	–/Ta12
Tons180	~ 8.1	$0.92^{+0.03}_{-0.11}$	ZW05/Wa13
Fairall 9	255 ± 56	$0.52^{+0.19}_{-0.15}$	Pe04/Lo12
Mrk359	~ 1.1	$0.66^{+0.30}_{-0.54}$	ZW05/Wa13
Mrk1018	~ 140	$0.58^{+0.36}_{-0.74}$	Be11/Wa13
1H0419-577	~ 340	>0.89	ZW05/Wa13
Ark120	150 ± 19	$0.64^{+0.19}_{-0.11}$	Pe04/Wa13
Swift J0501.9-3239	–	>0.99	–/Wa13
1H0707-495	~ 2.3	>0.97	ZW05/Zo10
Mrk79	52.4 ± 14.4	0.7 ± 0.1	Pe04/Ga11
Mrk110	25.1 ± 6.1	>0.89	Pe04/Wa13
NGC3783	29.8 ± 5.4	>0.88[a]	Pe04/Br11
NGC4051	1.91 ± 0.78	>0.99	Pe04/Pa12
RBS1124	–	>0.97	–/Wa13
IRAS13224–3809	~ 6.3	>0.987	Go12/Fa13
MCG–6-30-15	$2.9^{+1.8}_{-1.6}$	$a>0.98$	Mc05/BR06
Mrk841	~ 79	>0.52	ZW05/Wa13
Swift J2127.4+5654	~ 1.5	0.6 ± 0.2	Ma08/Mi09
Ark564	~ 1.1	$0.96^{+0.01}_{-0.11}$	ZW05/Wa13

[a]Note that Patrick et al. (2012) find a spin of $a < -0.35$, clearly discrepant with the results of Brenneman et al. (2011) report here. However, this object has a high column density warm absorber. The modeling of this absorber, specifically the resolution of the table-models used to characterize the absorber, appears to be the root cause of the spin discrepancy

While this is far from a well-defined or complete sample of objects, we can already draw some interesting conclusions. Firstly, there are a large number of objects with rapidly spinning black holes. For the most rapid spinners ($a > 0.9$), the numbers must be taken with a pinch of salt—the reported spins are the formal results (with statistical errors) that arise from the application of our standard methodology which assumes a razor thin Keplerian accretion disk. Using the radiation-pressure disk theory of Shakura and Sunyaev (1973), we expect the inner disks of these Seyfert galaxies, most of which are between 1–30 % of their Eddington luminosities, to have thicknesses of $h \sim 0.1 - 1r_g$. Hence, especially for sources with rather large Eddington ratios, we may expect finite disk thickness effects at the ISCO to become relevant when the ISCO becomes small, i.e. for large spins. Further work is needed

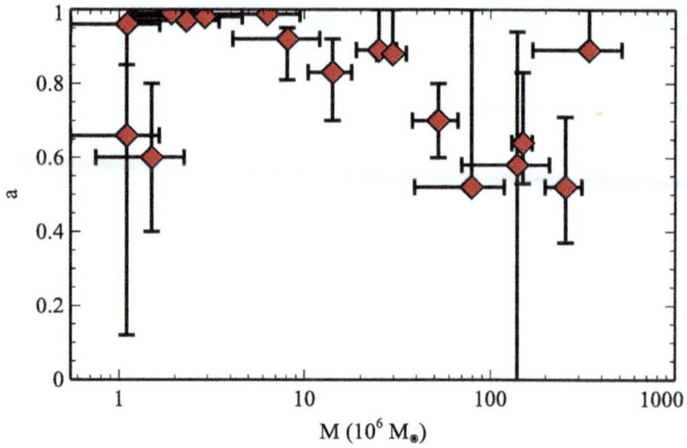

Fig. 6 Plot of the SMBH mass M and spin a from the sample listed in Table 1. Reflecting the conventions in the primary literature, all masses are marked with 1σ error bars whereas spins are marked with 90 % error ranges. When no error estimate is available for the mass, we have assumed an error of $\pm 0.5M$

to fully characterize this systematic error, but preliminary considerations by Reynolds and Fabian (2008) suggested that finite-thickness effects could relax the limits on the "true" spin value in these extreme cases to $a > 0.9$.

Still, it seems clear that a significant number of SMBHs in these AGN are rapidly spinning ($a > 0.8$), at least in the mass range 10^6 – few $\times$ 10^7 $M_\odot$. This suggests that the last mass doubling of these SMBHs occurred via coherent accretion and not chaotic accretion (King and Pringle 2006) or SMBH-SMBH major merger (Volonteri et al. 2005). Interestingly, as shown in Fig. 6 there are hints that the most massive black holes in this sample ($>10^8$ $M_\odot$) as well as the least massive black holes ($M < 2 \times 10^6$ $M_\odot$) may have more modest spins. If these trends hold up in more rigorous analyses, they would provide direct evidence for the increased role of chaotic accretion and/or major mergers at these two extreme ends of the SMBH mass spectrum.

The second noteworthy point is that all of the AGN in Table 1 are radio-quiet, and many of them are hosted by late-type (i.e., spiral disk) galaxies. This appears to conclusively reject the notion that black hole spin drives the radio-quiet/radio-loud dichotomy and its association with host galaxy type (Sikora et al. 2007).

4 Stellar Mass Black Holes

4.1 Early History

The first black hole broad iron line was discovered by Barr et al. (1985) in the *EXOSAT* spectrum of the BH-XRB Cygnus X-1. These authors interpreted the line broadening as the effects of Compton scattering of iron emission line photons in the hot disk atmosphere. It was several years before this result was re-interpreted as broadening due to Doppler/gravitational shifts in the innermost regions of the black hole accretion disk (Fabian et al. 1989).

Due to bright-source limitations by CCD spectrometers, relativistic reflection studies in BH-XRB lagged those of AGN during the *ASCA* era. However, this situation changed with the launch of *Chandra*, *XMM-Newton* at the turn of the millennium. The first robust sign of spin in the reflection spectrum of a BH-XRB was found in the *XMM-Newton*/EPIC-pn data of XTE J1650–500 (Miller et al. 2002), and quantitative constraints on spin using an early version of the methodology described in Sect. 2 were derived for the BH-XRB GX339–4 (Miller et al. 2004).

4.2 Challenges and Rewards

To measure the spin of a stellar mass black hole in a BH-XRB, one must confront several challenges. Typically, the crucial 2–10 keV part of the X-ray spectrum contains contributions from both the thermal (blackbody) disk emission and the Comptonized hard X-ray tail. When applying disk-reflection methods to determine spin, one must model the thermal disk emission and be cognizant of the fact that any mis-modeling of the continuum may affect the inferred spin. Similarly, when applying the CF method, one attempts to select spectral states that are overwhelmingly dominated by the thermal disk emission but any remaining Comptonization of that emission (beyond that already included in the disk atmosphere models) will skew the spin results.

Additional complications are caused, ironically, by the fact that these sources are so bright. In many of today's more capable X-ray (CCD) spectrometers, special instrumental modes are required in order to observe such sources while avoiding telemetry saturation as well as photon-pileup and other more subtle instrumental effects. Thus, unlike the case for AGN, one is often in the realm where systematic uncertainties related to both the spectral models as well as the calibration of the spectral data dominate statistical (photon) errors; there is the danger that one will end up with very precise but inaccurate spin measures. These concerns are valid for both the reflection- and CF-methods.

Once these challenges are overcome, however, BH-XRBs offer a tremendous reward—the possibility of measuring spin using multiple techniques and thereby validate the underlying assumptions of all spin measurements. For this reason, BH-XRBs have deservedly received significant attention.

4.3 Summary of Current Stellar-Mass Black Hole Spin Measurements

Table 2 draws together the BH-XRBs with published spins based on the relativistic reflection method. Of the six objects with both reflection- and CF-based spin measurements, there is general concordance for four. Clear discrepancies remain for two objects, 4U 1543–475 and GRO J1655–40, the origin of which remain unclear.

Two of the systems in which there are concordant spin measurements deserve additional discussion. For Cygnus X-1, there is now general agreement between reflection-based measurements (Fabian et al. 2012a, 2012b) and CF-measurements (Gou et al. 2011) that the black hole is a rapid spinner, $a > 0.95$. Again, this is the formal result and the finite disk thickness considerations discussed in Sect. 3.3 may weaken this limit to $a > 0.9$. However, it must be noted that early spin measurements using the reflection method suggested a small spin ($a < 0.05$). It now seems apparent that these early *XMM-Newton* data on which this measurement was based were inflicted by photon-pileup effects (J.M. Miller, private communication) that tend to narrow line features and hence decrease inferred spins (Miller et al. 2010). Analysis of more recent *XMM-Newton* data using a modified timing mode to handle the high count rate found two solutions for spin (Duro et al. 2011). Assuming an irradiation profile described by a single power-law, the Duro et al. analysis suggested either an intermediate spin with a steep irradiation profile ($\beta > 4$) or a rapid spin with a standard irradiation profile ($\beta \sim 3$). Guided by the realization that steep profiles correspond to either extreme light-bending effects or inner-disk torques, both of which require rapid spins, these authors rejected the intermediate spin solution. Finally, Fabian et al. (2012a, 2012b) examined *Suzaku* data and showed that rapid spin is favored unambiguously if one adopts a broken power-law form for the irradiation profile, a form that is verified by a non-parametric determination of the irradiation.

Table 2 BH-XRBs that have reflection-based spin determinations, together with the CF result when one exists. Spins are listed with their 90 % error ranges. Key to reference: Bl09 = Blum et al. (2009); Fa12 = Fabian et al. (2012a, 2012b); Go11 = Gou et al. (2011); Hi = Hiemstra et al. (2011); Mc06 = McClintock et al. (2006); Mi09 = Miller et al. (2009); Rei09 = Reis et al. (2009); Rei11 = Reis et al. (2011); Rei12 = Reis et al. (2012b); Sh06 = Shafee et al. (2008); St11 = Steiner et al. (2011); St12 = Steiner et al. (2012)

Object	Spin from reflection	Spin from CF	References
4U 1543–475	0.3 ± 0.1	0.8 ± 0.1	Mi09/Sh06
Cygnus X-1	>0.95[a]	>0.95	Fa12/Go11
GX339–4	0.94 ± 0.02	–	Mi09/–
GRS1915+105	>0.97[b]	>0.95	Bl09/Mc06
GRO J1655–40	>0.9[a]	0.7 ± 0.1	Rei09/Sh06
LMC X 1	>0.55	>0.87	St12/Go09
MAXI J1836–194	0.88 ± 0.03	–	Rei12/–
SAX J1711.6–3808	$0.6^{+0.2}_{-0.4}$	–	Mi09/–
Swift J1753.5–0127	$0.76^{+0.11}_{-0.15}$	–	Rei09/–
XTE J1550–564	$0.33 - 0.77$[c]	$0.34^{+0.37}_{-0.45}$	Mi09/St11
XTE J1650–500	0.79 ± 0.01	–	Mi09/–
XTE J1652–453	0.45 ± 0.02	–	Hi11/–
XTE J1752–223	0.52 ± 0.11	–	Rei11/–
XTE J1908+094	0.75 ± 0.09	–	Mi09/–

[a]We note that Mi09 quote a spin value of $a < 0.05$. However, it appears likely that the *XMM-Newton* data on which this measurement rested were affected by photon-pileup (J.M. Miller, private communication). See further discussion in Sect. 4

[b]We quote the high-spin model of Bl09 which employs a broad-band fit of a broken power-law primary continuum and a corresponding reflection continuum modeled with a blurred `pexriv` (Magdziarz and Zdziarski 1995). The more self-consistent `reflionx` model (Ross and Fabian 2005) could not be adequately applied to the broad-band data due to the complex continuum

[c]Result combines the work of Mi09 and St11 to develop a composite error range that encompasses systematic uncertainties of the reflection model. See discussion in Sect. 4

XTE J1550–564 is another interesting case. The CF method yields a rather broad range of allowable spins, $-0.11 < a < 0.71$ at the 90 % confidence level (Steiner et al. 2011). Miller et al. (2009) apply the reflection method to *ASCA*/GIS data for this object, using the standard `reflionx` (Ross and Fabian 2005) ionized reflection model, and find a spin slightly above the upper end of the CF range, $a = 0.76 \pm 0.01$. However, motivated by the fact that, for the highly ionized disks found in BH-XRBs, the iron line can be appreciably Compton broadened, Steiner et al. (2011) re-examined this object using a very high-density, hot, variant of the reflection model (`refbhb`; Ross and Fabian 2007). They find that the additional Compton broadening of the iron line present in `refbhb` leads to a slightly lower best fit spin, $a = 0.55^{+0.15}_{-0.22}$. The fact that the spin does not change dramatically implies that the broadening of the reflection spectrum is dominated by the disk dynamics (Doppler/gravitational), with Compton-broadening having a noticeable but sub-dominant effect. Given that the true density and vertical structure of the X-ray photosphere is still uncertain, we take these two reflection results as bracketing the possible systematic error introduced by the reflection model, giving us the range of $a = 0.33 - 0.77$ that we quote in Table 2.

The observed broadened reflection spectrum depends upon the disk inclination. Thus, there is the opportunity to test the CF assumption that the inner disk inclination is the same as the measured binary orbital inclination. In practice, the uncertainties in the reflection-determined inclinations can be significant given that the disks are typically strongly ionized and hence the sharp blue edge to the iron line (the cleanest indication of inclination in AGN spectra) is Compton broadened. Thus, even in many current reflection-based analyzes, the inclination is fixed to (or constrained to lie within the error bounds of) the binary orbital plane or (when superluminal motion is observed) the plane orthogonal to the inferred jet-axis. In cases where independent inclinations have been obtained from reflection spectroscopy, they are found to be in agreement with those assumed in CF.

We end this section with a brief discussion of the astrophysical implications. Given that it is difficult to conceive of an astrophysical scenario whereby a stellar-mass black hole in a binary system can accrete a substantial fraction of its mass, we conclude that the measured spin must be close to the birth spin of the black hole. Furthermore, irrespective of whether we use the reflection- or CF-based spin measures, all of the stellar mass black holes examined to date have spin parameters that greatly exceed estimates for the natal spin parameters of neutron stars ($a_{\rm NS} \sim 0.01 - 0.03$, Miller et al. 2011). This constitutes direct evidence for differences in the core-collapse supernovae that have produced the black hole and neutron star populations. One possibility is that core collapse generically forms rapidly-spinning compact objects but, in the case of neutron stars, there is speedy reduction in the angular momentum from either gravitational wave loses due to R-mode instabilities (Andersson et al. 1999) or a magnetic propeller mechanism.

5 Beyond Spectroscopy: The Age of Relativistic Reverberation Mapping

We conclude this chapter with a brief discussion of an exciting frontier, the recent discovery of relativistic iron line reverberation.

A key aspect of the geometry underlying the X-ray reflection picture is that the hard X-ray source is external to the optically-thick flow—i.e., there is a finite distance between the emission site for the hard X-rays and the X-ray reflector. It is straightforward to see that variations in the luminosity of the primary X-ray source will drive variations in the strength of the observed X-ray reflection with a time-delay related to the light travel time from the source to the disk. The full energy-dependent transfer function that relates continuum fluctuations with changes in the broad iron line encodes a tremendous amount of information including the location of the X-ray source and the spin parameter of the black hole (Reynolds et al. 1999). For this reason, relativistic iron line reverberation has been a focus of proposals for future high-throughput X-ray observatories such as *LOFT* (Feroci et al. 2012).

While measurements of the full transfer function must await future observatories, recent analyses of *XMM-Newton* data have already found evidence for time delays between the continuum emission and relativistic reflection. The narrow-line Seyfert-1 galaxy 1H0707–495 has a strong soft excess that can be explained as soft reflection from an ionized relativistic disk. This interpretation is bolstered by spectral structure in the soft excess that resembles a relativistically-broadened iron-L line (Fabian et al. 2009). Fabian et al. (2009) discovered that, at intermediate timescales, the soft-excess lags the 2–4 keV continuum band by ~30 s (also see detailed follow-up work by Zoghbi et al. 2010). Since the continuum source itself always displays a pattern of hard-lags-soft ("hard lags"), this pattern of soft-lags-hard ("soft lags") is best interpreted as reverberation delays due to a displacement of the variable X-ray source from the disk (however, see alternative interpretation of these soft-lags by Miller et al. 2010, and a discussion of the problems with this alternative model by Zoghbi et al. 2011). Since the initial discovery of soft-lags in 1H0707–495, they have been found in a number of other AGN (Zoghbi and Fabian 2011; de Marco et al. 2011; Fabian et al. 2012b; Kara et al. 2013a, 2013b).

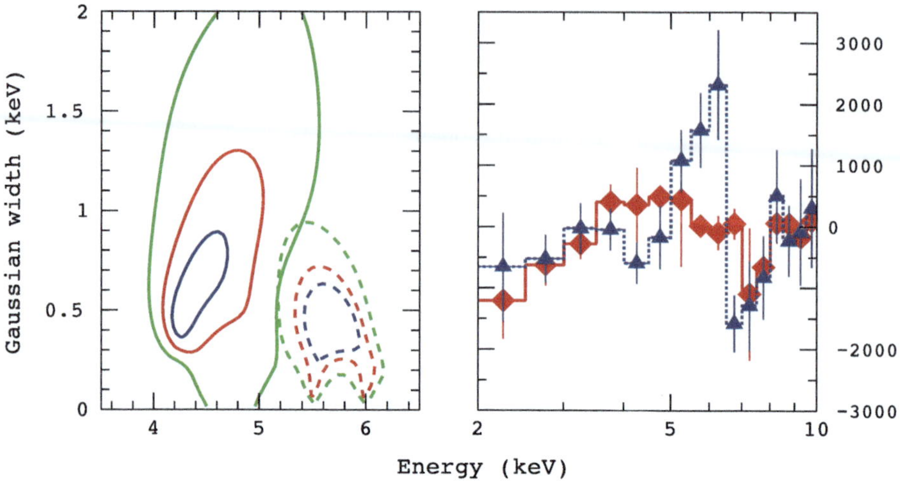

Fig. 7 Detection of lags in the broad iron line in the Seyfert galaxy NGC 4151. *Right panel*: Energy lags as a function of energy for two Fourier frequency ranges, $f < 2 \times 10^{-5}$ Hz (*blue triangles*) and $f = (5 - 50) \times 10^{-5}$ Hz (*red diamonds*). *Left panel*: Error contours from a Gaussian fit to the pag-spectra, showing the 68 %, 90 % and 95 % confidence levels. *Continuous lines* refer to the $(5 - 50) \times 10^{-5}$ Hz lags, and *dashed lines* refer to the $f < 2 \times 10^{-5}$ Hz lags. Figures from Zoghbi et al. (2012)

Very recently, Zoghbi et al. (2012, 2013) and Kara et al. (2013a, 2013b) have used similar techniques to find reverberation lags in the broad iron lines of seven AGN that are particularly bright in the iron-K band. Here, we briefly illustrate these results with a discussion NGC 4151. In this source, a skewed broad iron line can be clearly seen when the lags are examined as a function of energy (Fig. 7). Furthermore, the shape of this "lag spectrum" is a function of Fourier frequency, with lower-frequency variations showing a narrower line with longer time-lags. This is exactly as expected from reverberation theory, suggesting that we are gaining our first glimpse at the structure of the transfer function. Putting this together suggests that the variable X-ray source is located above the accretion disk, possibly on the black hole spin axis, at a height of a few $\times r_g$. This positive detection of iron line reverberation with *XMM-Newton* shows conclusively that future observatories, such as *LOFT*, will be able to make tremendous progress in studies of strong gravity using relativistic reverberation in AGN.

6 Conclusion

X-ray reflection signatures from the inner accretion disk of BH-XRBs and AGN give us one of the most direct probe of black hole spin. In this chapter, I have reviewed the assumptions that underlie this technique as well as the detailed methodology as applied to both stellar-mass black hole systems and supermassive black holes. The key points to take away from this discussion are:

1. Relativistic reflection spectroscopy can be applied to derive spin across the full mass range of astrophysical black holes. A major strength of this method is that one does not need to know the black hole mass or the distance of the system in order to derive spin. Furthermore, one does not need prior knowledge of the inclination of the inner accretion disk (it is a measured quantity along with the spin).

2. There has been an explosion of SMBH spin measurements in recent years. Applying some objective quality-control criteria to results in the literature, we obtain a list of 20 AGN with SMBH spin measurements. It is clear that a large fraction of SMBHs are rapidly-spinning, although there are tentative hints of a more slowly spinning population at high ($M > 5 \times 10^7\ M_\odot$) and low ($M < 2 \times 10^6\ M_\odot$) mass.
3. Stellar-mass black holes in BH-XRBs provide the invaluable opportunity to measure spin with multiple techniques, thereby allowing us to validate the basic assumptions of the various methodologies. Comparing relativistic reflection results to those derived from continuum fitting, there is generally concordance (exceptions being 4U 1543–475 and GRO J1655–40).
4. The recent discovery of relativistic reverberation, especially of time delays between continuum fluctuations and the broad iron-K line, has opened a new frontier. In addition to definitively confirming the broad iron line paradigm, this proves that future large-area X-ray observatories will make tremendous progress in studies of strong gravity using strong-field reverberation mapping.

Acknowledgements I thank Andy Fabian, Anne Lohfink, Jon Miller and Abdu Zoghbi for invaluable discussion during the writing of this review. I gratefully acknowledge support from NASA under ADAP grant NNX12AE13G.

References

E. Agol, J.H. Krolik, Astrophys. J. **528**, 161 (2000)
N. Andersson, K.D. Kokkotas, N. Stergioulas, Astrophys. J. **516**, 307 (1999)
J. Bardeen, W.H. Press, S.A. Teukolsky, Astrophys. J. **178**, 347 (1972)
P. Barr, N.E. White, C.G. Page, Mon. Not. R. Astron. Soc. **216**, 65 (1985)
V. Bennert, M.W. Auger, T. Treu, J.H. Woo, M.A. Malkan, Astrophys. J. **726**, 59 (2011)
R.D. Blandford, R.L. Znajek, Mon. Not. R. Astron. Soc. **179**, 433 (1977)
J.L. Blum et al., Astrophys. J. **706**, 60 (2009)
L.W. Brenneman, C.S. Reynolds, Astrophys. J. **652**, 1028 (2006)
L.W. Brenneman et al., Astrophys. J. **736**, 103 (2011)
Y. Dabrowski, A.C. Fabian, K. Iwasawa, A.N. Lasenby, C.S. Reynolds, Mon. Not. R. Astron. Soc. **288**, L11 (1997)
T. Dauser, J. Wilms, C.S. Reynolds, L.W. Brenneman, Mon. Not. R. Astron. Soc. **409**, 1534 (2010)
B. de Marco, G. Ponti, P. Uttley, M. Cappi, M. Dadina, A.C. Fabian, G. Miniutti, Mon. Not. R. Astron. Soc. **417**, L98 (2011)
R. Duro et al., Astron. Astrophys. **533**, L3 (2011)
A.C. Fabian, M.J. Rees, L. Stella, N.E. White, Mon. Not. R. Astron. Soc. **238**, 729 (1989)
A.C. Fabian et al., Nature **459**, 540 (2009)
A.C. Fabian et al., Mon. Not. R. Astron. Soc. **424**, 217 (2012a)
A.C. Fabian et al., Mon. Not. R. Astron. Soc. **419**, 116 (2012b)
A.C. Fabian et al., Mon. Not. R. Astron. Soc. **429**, 2917 (2013)
M. Feroci et al., Exp. Astron. **34**, 415 (2012)
L.C. Gallo, G. Miniutti, J.M. Miller, L.W. Brenneman, A.C. Fabian, M. Guainazzi, C.S. Reynolds, Mon. Not. R. Astron. Soc. **411**, 607 (2011)
O. Gonzáles-Martin, S. Vaughan, Astron. Astrophys. **544**, 80 (2012)
L. Gou et al., Astrophys. J. **742**, 85 (2011)
B. Hiemstra, M. Mendez, C. Done, M. Diaz Trigo, D. Altamirano, P. Casella, Mon. Not. R. Astron. Soc. **411**, 137 (2011)
K. Iwasawa et al., Mon. Not. R. Astron. Soc. **282**, 1038 (1996)
T. Kallman, M. Bautista, Astrophys. J. Suppl. Ser. **133**, 221 (2001)
E. Kara, A.C. Fabian, E.M. Cackett, G. Miniutti, P. Uttley, Mon. Not. R. Astron. Soc. **430**, 1408 (2013a)
E. Kara, A.C. Fabian, E.M. Cackett, J.F. Steiner, P. Uttley, D.R. Wilkins, A. Zoghbi, Mon. Not. R. Astron. Soc. (2013b, in press). arXiv:1306.2551
R.P. Kerr, Phys. Rev. Lett. **11**, 237 (1963)

A.R. King, J.E. Pringle, Mon. Not. R. Astron. Soc. **373**, L90 (2006)
A.L. King et al., Astrophys. J. **729**, 19 (2011)
A.L. King et al., Astrophys. J. **762**, 103 (2013)
A.M. Lohfink, C.S. Reynolds, J.M. Miller, L.W. Brenneman, R.F. Mushotzky, M.A. Nowak, A.C. Fabian, Astrophys. J. **758**, 67 (2012b)
A.M. Lohfink, C.S. Reynolds, R.F. Mushotzky, J. Wilms, Astrophys. J. **749**, L31 (2012a)
P. Magdziarz, A. Zdziarski, Mon. Not. R. Astron. Soc. **273**, 837 (1995)
A. Malizia et al., Mon. Not. R. Astron. Soc. **389**, 1360 (2008)
J.E. McClintock et al., Astrophys. J. **652**, 518 (2006)
I.M. McHardy, K.F. Gunn, P. Uttley, M.R. Goard, Mon. Not. R. Astron. Soc. **359**, 1469 (2005)
J.M. Miller et al., Astrophys. J. **570**, L69 (2002)
J.M. Miller et al., Astrophys. J. **606**, L131 (2004)
J.M. Miller et al., Nature **441**, 953 (2006)
J.M. Miller et al., Astrophys. J. **680**, 1359 (2008)
J.M. Miller, C.S. Reynolds, A.C. Fabian, G. Miniutti, L. Gallo, Astrophys. J. **697**, 900 (2009)
J.M. Miller et al., Astrophys. J. **724**, 1441 (2010)
L. Miller, T.J. Turner, J.N. Reeves, V. Braito, Mon. Not. R. Astron. Soc. **408**, 1928 (2010)
J.M. Miller, M.C. Miller, C.S. Reynolds, Astrophys. J. **731**, L5 (2011)
G. Miniutti et al., Mon. Not. R. Astron. Soc. **398**, 255 (2009)
T. Nagao, R. Maiolino, A. Marconi, Astron. Astrophys. **459**, 85 (2006)
K. Nandra, P.M. O'Neill, I.M. George, J.N. Reeves, Mon. Not. R. Astron. Soc. **382**, 194 (2007)
I.D. Novikov, K.S. Thorne, in *Black Holes*, ed by. C. Dewitt, B.S. Dewitt (Gordon & Breach, New York, 1974), p. 343
A.R. Patrick, J.N. Reeves, D. Porquet, A.G. Markowitz, V. Braito, A.P. Lobban, Mon. Not. R. Astron. Soc. **426**, 2522 (2012)
R.F. Penna, J.C. McKinney, R. Narayan, A. Tchekhovskoy, R. Shaffe, J.E. McClintock, Mon. Not. R. Astron. Soc. **408**, 752 (2010)
B.M. Peterson et al., Astrophys. J. **613**, 682 (2004)
R.C. Reis, A.C. Fabian, R.R. Ross, J.M. Miller, Mon. Not. R. Astron. Soc. **395**, 1257 (2009)
R.C. Reis et al., Mon. Not. R. Astron. Soc. **410**, 2497 (2011)
R.C. Reis et al., Astrophys. J. **745**, 93 (2012a)
R.C. Reis, J.M. Miller, M.T. Reynolds, A.C. Fabian, D.J. Walton, Astrophys. J. **751**, 34 (2012b)
C.S. Reynolds, M.C. Begelman, Astrophys. J. **488**, 109 (1997)
C.S. Reynolds, A.C. Fabian, Astrophys. J. **675**, 1048 (2008)
C.S. Reynolds, A.J. Young, M.C. Begelman, A.C. Fabian, Astrophys. J. **514**, 164 (1999)
C.S. Reynolds, L.W. Brenneman, A.M. Lohfink, M.L. Trippe, J.M. Miller, A.C. Fabian, M.A. Nowak, Astrophys. J. **755**, 88 (2012)
R.R. Ross, A.C. Fabian, Mon. Not. R. Astron. Soc. **358**, 211 (2005)
R.R. Ross, A.C. Fabian, Mon. Not. R. Astron. Soc. **381**, 1697 (2007)
S. Schmoll et al., Astrophys. J. **703**, 2171 (2009)
R. Shafee et al., Astrophys. J. **687**, L25 (2008)
N.I. Shakura, R.A. Sunyaev, Astron. Astrophys. **24**, 337 (1973)
M. Sikora, L. Stawarz, J. Lasota, Astrophys. J. **658**, 815 (2007)
J.F. Steiner et al., Mon. Not. R. Astron. Soc. **416**, 941 (2011)
J.F. Steiner et al., Mon. Not. R. Astron. Soc. **427**, 2552 (2012)
Y. Tan, J.X. Wang, W. Shu, Y. Zhou, Astrophys. J. **747**, L11 (2012)
M. Volonteri, P. Madau, E. Quataert, M.J. Rees, Astrophys. J. **620**, 69 (2005)
D.J. Walton, E. Nardini, A.C. Fabian, L.C. Gallo, R.C. Reis, Mon. Not. R. Astron. Soc. **428**, 2901 (2013)
C. Warner, F. Hamann, M. Dietrich, Astrophys. J. **608**, 136 (2004)
A.J. Young, R.R. Ross, A.C. Fabian, Mon. Not. R. Astron. Soc. **300**, L11 (1998)
X.-L. Zhou, J.-M. Wang, Astrophys. J. **618**, L83 (2005)
A. Zoghbi, A.C. Fabian, Mon. Not. R. Astron. Soc. **418**, 2642 (2011)
A. Zoghbi, A.C. Fabian, P. Uttley, G. Miniutti, L.C. Gallo, C.S. Reynolds, J.M. Miller, G. Ponti, Mon. Not. R. Astron. Soc. **401**, 2419 (2010)
A. Zoghbi, P. Uttley, A.C. Fabian, Mon. Not. R. Astron. Soc. **412**, 59 (2011)
A. Zoghbi, A.C. Fabian, C.S. Reynolds, E.M. Cackett, Mon. Not. R. Astron. Soc. **422**, 129 (2012)
A. Zoghbi, C.S. Reynolds, E.M. Cackett, G. Miniutti, E. Kara, A.C. Fabian, Astrophys. J. **767**, 121 (2013)

DOI 10.1007/978-1-4939-2227-7_15
Reprinted from *Space Science Reviews* Journal, DOI 10.1007/s11214-013-0003-9

Black Hole Spin via Continuum Fitting and the Role of Spin in Powering Transient Jets

Jeffrey E. McClintock · Ramesh Narayan · James F. Steiner

Received: 3 March 2013 / Accepted: 24 June 2013 / Published online: 9 July 2013

Abstract The spins of ten stellar black holes have been measured using the continuum-fitting method. These black holes are located in two distinct classes of X-ray binary systems, one that is persistently X-ray bright and another that is transient. Both the persistent and transient black holes remain for long periods in a state where their spectra are dominated by a thermal accretion disk component. The spin of a black hole of known mass and distance can be measured by fitting this thermal continuum spectrum to the thin-disk model of Novikov and Thorne; the key fit parameter is the radius of the inner edge of the black hole's accretion disk. Strong observational and theoretical evidence links the inner-disk radius to the radius of the innermost stable circular orbit, which is trivially related to the dimensionless spin parameter a_* of the black hole ($|a_*| < 1$). The ten spins that have so far been measured by this continuum-fitting method range widely from $a_* \approx 0$ to $a_* > 0.95$. The robustness of the method is demonstrated by the dozens or hundreds of independent and consistent measurements of spin that have been obtained for several black holes, and through careful consideration of many sources of systematic error. Among the results discussed is a dichotomy between the transient and persistent black holes; the latter have higher spins and larger masses. Also discussed is recently discovered evidence in the transient sources for a correlation between the power of ballistic jets and black hole spin.

Keywords Black hole physics · Accretion disks · X-Ray binaries · Stars: winds, outflows

1 Introduction

In his Ryerson Lecture, Chandrasekhar (1975) described the Kerr solution as the "most shattering experience" of his entire scientific life. He found himself "shuddering before the

J.E. McClintock (✉) · R. Narayan · J.F. Steiner
Harvard-Smithsonian Center for Astrophysics, 60 Garden Street, Cambridge, MA 02138, USA
e-mail: jem@cfa.harvard.edu

R. Narayan
e-mail: narayan@cfa.harvard.edu

J.F. Steiner
e-mail: jsteiner@cfa.harvard.edu

beautiful, the incredible fact" that each of the many trillions of black holes in the universe is completely described by a single pair of numbers that specify the black hole's mass and its spin.[1] In Chandrasekhar's time, 1910–1995, the masses of Cyg X-1 and three other stellar black holes had been estimated (Webster and Murdin 1972; Bolton 1972; Cowley et al. 1983; McClintock and Remillard 1986; Remillard et al. 1992; Shahbaz et al. 1994). Today, accurate dynamical mass measurements have been achieved for more than a dozen stellar black holes (McClintock and Remillard 2006; Özel et al. 2010; Orosz et al. 2011a; Steeghs et al. 2013), as well as for several supermassive black holes, e.g., Sgr A* (Ghez et al. 2008; Gillessen et al. 2009), NGC 4258 (Herrnstein et al. 2005), and others (Gültekin et al. 2009, and references therein).

In 1989, the first practical approach to measuring black hole spin was suggested by Fabian et al. (1989), namely, modeling the relativistically-broadened Fe K emission line emitted from the inner accretion disk. The first compelling observation of such a line was reported just two months before Chandrasekhar died (Tanaka et al. 1995). Presently, the spins of more than a dozen black holes have been estimated by modeling the "reflected" spectrum of an accretion disk, which includes as its most prominent feature the Fe K line. For a review of this method of measuring black hole spin, we refer the reader to Reynolds (2013).

It was not until 1997 that a new approach to measuring black hole spin—the continuum-fitting method—which is the subject of this chapter, was pioneered by Zhang et al. (1997). In brief, in applying this method one fits the thermal continuum spectrum of a black hole's accretion disk to the relativistic thin-disk model of Novikov and Thorne (1973) and thereby determines the radius of the inner edge of the disk. One then identifies this radius with the radius of the innermost stable circular orbit (R_{ISCO}), which is simply related to the spin parameter a_* (Bardeen et al. 1972). The method is simple: It is strictly analogous to measuring the radius of a star whose flux, temperature and distance are known. By this analogy, it is clear that it is essential to know the luminosity of the accretion disk; hence, one must have estimates of the source distance D as well as the disk inclination i. Additionally, one must know M in order to scale R_{ISCO} and thereby determine a_*.

In 2006, the continuum-fitting method was employed to estimate the spins of three stellar black holes (Shafee et al. 2006; McClintock et al. 2006). Presently, ten spins have been measured using this method (Sect. 6). Not only is the continuum-fitting method simple, it is also demonstrably robust. For example, there is strong observational and theoretical evidence (discussed in Sect. 4) that the disk is truncated quite sharply at R_{ISCO}. Furthermore, there is an abundance of suitable X-ray spectral data for many black holes; consequently, for a given black hole one can typically obtain tens or even hundreds of independent measurements of spin that agree to within a few percent (Sect. 4.1). The one open question for this method is whether the black hole's spin is aligned with the orbital angular momentum vector of the inner disk (Sect. 5.4). Meanwhile, a limitation of the continuum-fitting method is that it is only readily applicable to stellar black holes (but see Jolley et al. 2009; Czerny et al. 2011), while the Fe K method is applicable to both stellar and supermassive black holes (Reynolds 2013).

In order to obtain secure measurements of spin using the continuum-fitting method, and to establish the reliability of this method, substantial and comparable effort is required on three fronts: (1) The selection and fitting of X-ray spectral data to the Novikov–Thorne model (in conjunction with ancillary models); (2) testing and exploring extensions of the

[1] Spin is commonly expressed in terms of the dimensionless parameter $a_* \equiv \mathrm{cJ/GM^2}$, where J and M are respectively the angular momentum and mass of the black hole.

Novikov–Thorne model via general relativistic magnetohydrodynamic (GRMHD) simulations; and (3) obtaining accurate estimates of D, i and M. The first two topics are discussed in Sects. 3–5. Concerning the third topic, we refer the reader to recent papers on the measurements of these crucial parameters for M33 X-7 (Orosz et al. 2007); LMC X-1 (Orosz et al. 2009); A0620–00 (Cantrell et al. 2010); XTE J1550–564 (Orosz et al. 2011b); Cyg X-1 (Reid et al. 2011; Orosz et al. 2011a); H1743–322 (Steiner et al. 2012a); and GRS 1915+105 (Steeghs et al. 2013). The uncertainties in D, i and M are critically important because they dominate the error budget in the final determination of a_*, including the error incurred by reliance on the Novikov–Thorne disk model (Sect. 5).

Initial efforts are under way to use the available continuum-fitting spin data to investigate the formation and evolution of black holes, as well as their host systems (e.g., Lee et al. 2002; Wong et al. 2012), and to understand how a black hole interacts with its environment (e.g., Wang et al. 2003; Cooke et al. 2008). The most important application to date of spin data is the discovery of a long-predicted correlation between jet power and black hole spin, which is the subject of Sect. 7. Very recently, Russell et al. (2013) challenged the validity of this correlation; Sect. 7.4 answers this challenge.

2 Stellar Black Holes in X-Ray Binaries

There are 24 confirmed black hole binaries: the 23 listed in Table 1 in Özel et al. (2010) plus H1743–322 (Steiner et al. 2012a).[2] A schematic sketch to scale of 21 of these confirmed black-hole systems is shown in Fig. 1.

For decades, it has been customary to define two classes of X-ray binaries, commonly referred to as LMXBs (low-mass X-ray binaries) and HMXBs (high-mass X-ray binaries), based on whether the mass of the secondary star is relatively low or relatively high (e.g., Bradt and McClintock 1983). Here, we use a different classification scheme that differentiates two distinct classes of black hole binaries by the primary mode of mass transfer to the black hole and the effect that this has on the stability of the X-ray source (White et al. 1995).

The black holes in five of the 24 systems are steadily fed by the winds of massive O-supergiant or Wolf–Rayet companions, and consequently their bolometric X-ray luminosities are relatively stable. Sketches of three of these systems (M33 X-7, LMC X-1 and Cyg X-1) appear in the top-right corner of Fig. 1. We refer to these systems and their black holes as "persistent."

The black holes in the remaining systems are fed by Roche lobe overflow through the L1 point, and all of them have been observed to vary in luminosity by factors of >100 ($\sim 10^8$ in several extreme cases; e.g., see Narayan and McClintock 2008). We refer to these systems and their black holes as "transient."

2.1 Persistent Black Hole Binaries

These systems are distinguished by the large masses of their secondary stars ($20\ M_\odot$–$70\ M_\odot$) and by the extreme optical/UV luminosities of these stars, which exceed the X-ray luminosities of their black hole companions. Consequently, the effects of X-ray heating are minimal

[2] Apart from H1743–322, our selection is based on firm dynamical evidence, and we therefore exclude some important systems for which there is significant evidence that the primary is a black hole, e.g., Cyg X-3 (Zdziarski et al. 2013), or a strong presumption that it is, e.g., SS433 (Begelman et al. 2006) and 4U 1957+11 (Nowak et al. 2012).

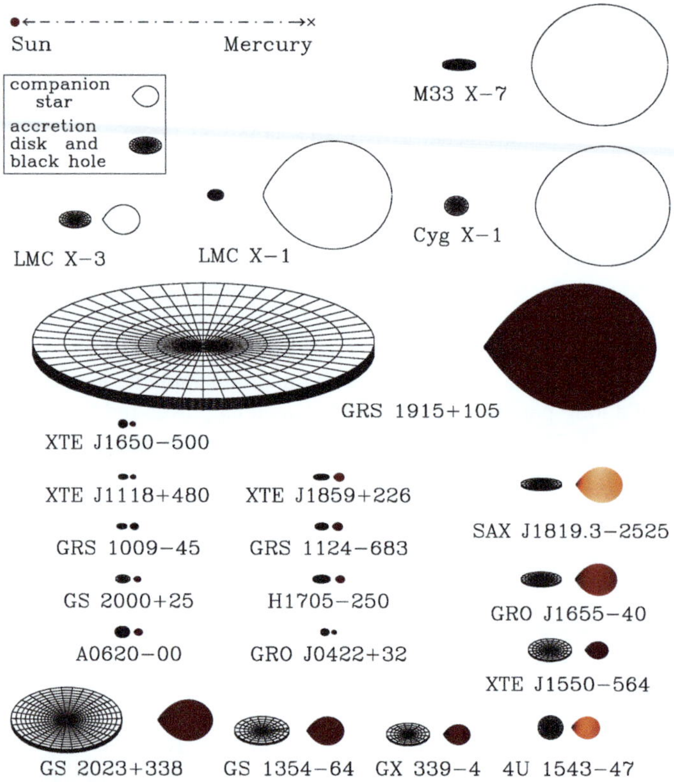

Fig. 1 Schematic sketch to scale of 21 black hole binaries (see scale and legend in the upper-left corner). The tidally-distorted shapes of the companion stars are accurately rendered in Roche geometry. The black holes are located in the center of the disks. A disk's tilt indicates the inclination angle i of the binary, where $i = 0$ corresponds to a system that is viewed face-on; e.g., $i = 21°$ for 4U 1543–47 (*bottom right*) and $i = 75°$ for M33 X-7 (*top right*). The size of a system is largely set by the orbital period, which ranges from 33.9 days for the giant system GRS 1915+105 to 0.2 days for tiny XTE J1118+480. Three well-studied persistent systems (M33 X-7, LMC X-1 and Cyg X-1) are located in the *upper-right corner*. The other 18 systems are transients. (Figure courtesy of J. Orosz.)

and the optical star dominates the optical properties of the system. The key distinguishing feature of these systems is their X-ray persistence, which is a consequence of the star's massive stellar wind ($\sim 10^{-5}$–$10^{-8} M_\odot$ yr^{-1}), a significant fraction of which is captured by the black hole.

Because the secondaries are massive these systems are obviously young ($\lesssim 10^7$ yr). They are also very rare: There is only one confirmed system in the Galaxy, Cyg X-1, and, despite many deep *Chandra* and *XMM-Newton* X-ray observations of Local Group galaxies, only four other such systems have been discovered, one each in the LMC, M33, IC 10 and NGC 300.

In this review, we do not consider further the two persistent systems that contain Wolf–Rayet secondaries, namely IC 10 X-1 and NGC 300 X-1, because the masses of their black holes depend strongly on the very uncertain masses of their secondaries, and also because no attempt has so far been made to estimate their spins. By contrast, the three remaining persistent systems—M33 X-7, LMC X-1 and Cyg X-1—have well-determined values of both mass and spin (see Sect. 6). Relative to the black hole primaries in the transient systems (apart from GRS 1915+105), the black holes in these three persistent systems have large masses, $M = 11$–$16\ M_{\odot}$, and high spins that range from $a_* = 0.84$ to $a_* > 0.95$, a point that we return to in Sect. 6.1.

2.2 Transient Black Hole Binaries

With few exceptions, the 18 transient black hole binaries (hereafter simply referred to as transients) manifest and then rise to maximum X-ray luminosity on a timescale of several days, thereafter returning to a quiescent state over a period of many tens or hundreds of days, as illustrated in Fig. 2. The masses of the black holes in these systems are relatively low, as are their spins (with the exception of GRS 1915+105), and their orbital periods range widely from 0.2–33.9 days. By comparison, the orbital periods of the persistent systems span a relatively narrow range. The transients are, on average, likely Gyrs old (White and Ghosh 1998; Fragos et al. 2013).

During a major outburst, the peak luminosities of transient sources approach the Eddington limit (Steiner et al. 2013), while in quiescence their luminosities are typically in the range $10^{-8.5}$ to 10^{-6} of Eddington (Narayan and McClintock 2008). Figure 2 shows X-ray, optical and radio light curves of a typical short-period transient. The optical emission is generated largely by reprocessing of X-rays in the accretion disk, and the radio outburst is primarily the result of synchrotron emission produced in a jet. The ballistic jets, which are the subject

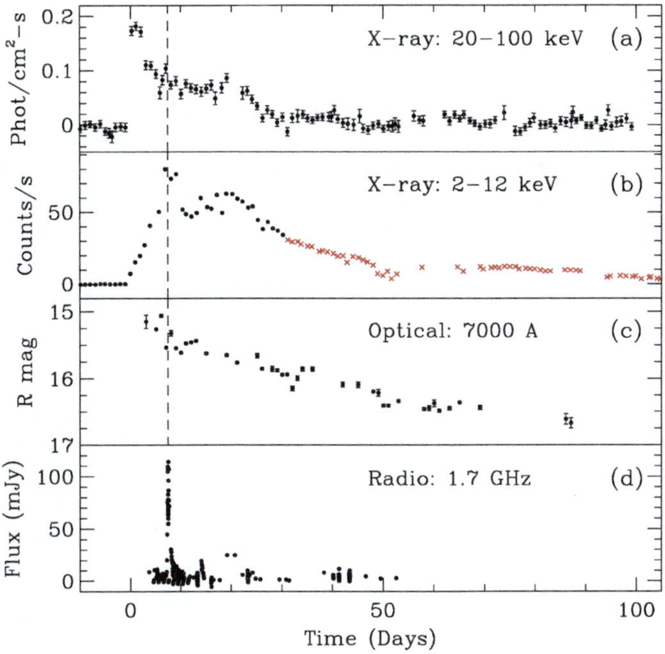

Fig. 2 Outburst cycle of XTE J1859+226 in 1999. The *dashed line* (*top three panels*) marks the time of peak radio flux (panel **d**). The ≈1-day radio spike (panel **d**) is shown fully resolved in Fig. 2 in Brocksopp et al. (2002). The *red crosses* (panel **b**) indicate times when the X-ray spectrum is dominated by the thermal component. These *BATSE* and *RXTE*/ASM X-ray, and Merlin (and other) radio data (panels **a**, **b** and **d**, respectively) appear in Fig. 1 in Brocksopp et al. (2002) and the optical data (panel **c**) appear in Fig. 2 in Sánchez-Fernández et al. (2001). For further details, consult the references

of Sect. 7, are launched very near the time of peak radio emission (panel d), which in the case of XTE J1859+226 occurred just 0.5 days after the X-ray luminosity (panel b) peaked. Spin can be reliably measured when the thermal component dominates the spectrum during the latter part of the outburst cycle (panel b). For a complete and state-coded version of the X-ray light curve of XTE J1859+226, see Fig. 8b in Remillard and McClintock (2006).

There are several oddballs among the transient systems: Four have relatively massive secondaries, $\sim$2–6 $M_\odot$, compared to the typical value of $\lesssim$1 $M_\odot$ (Charles and Coe 2006). GRS 1915+105 has remained very luminous continuously since its appearance in 1992, and GX 339–4 never reaches a deep quiescent state (McClintock and Remillard 2006). LMC X-3 is almost always active and highly variable (Sect. 4.1), although it does have extended low states (Smale and Boyd 2012).

3 The Continuum-Fitting Method

The two foundations of the continuum-fitting method are (1) the existence of an ISCO for a test particle orbiting a black hole and (2) the strong observational and theoretical evidence that—for a wide range of conditions—accretion disks in black hole binaries are truncated quite sharply at the ISCO radius. In this section, we first discuss the physics of these disks, and we close by describing the mechanics of continuum fitting.

3.1 Accretion Disk Theory

The basic physics of black hole accretion is straightforward (Frank et al. 2002; Kato et al. 2008; Abramowicz and Fragile 2013). Gas with angular momentum flows in from the outside and settles into a circular orbit stabilized by centrifugal force. The gas steadily loses angular momentum as a result of magnetic stresses from the magnetorotational instability (Balbus and Hawley 1998), whose effect is often approximated via the α-viscosity prescription of Shakura and Sunyaev (1973). As the gas loses angular momentum, it moves inward, occupying at each instant a circular orbit appropriate to its instantaneous angular momentum. The inward drift continues until the gas reaches the radius of the ISCO, $R_{\rm ISCO}$. Inside $R_{\rm ISCO}$, no stable circular orbits are available and the gas falls dynamically into the black hole.

As described above, the ISCO represents a major transition point in disk physics, where gas switches from slow viscous accretion on the outside to inviscid free-fall on the inside. The ISCO is thus effectively the inner edge of the disk. Correspondingly, information on the linear dimensions of the radius $R_{\rm ISCO}$ is imprinted on the emitted radiation. Since $R_{\rm ISCO}$ varies monotonically with the black hole spin parameter a_* (Bardeen et al. 1972), as illustrated in Fig. 3a, it is thus possible to measure a_* by modeling the disk emission.

The model of choice for this purpose is that described by Novikov and Thorne (1973), hereafter referred to as the NT model, which is the relativistic generalization of the thin accretion disk model of Shakura and Sunyaev (1973). Using nothing more than the Kerr metric, basic conservation laws of mass, momentum, angular momentum and energy, and assumptions of axisymmetry and steady state, the NT model (see also Page and Thorne 1974; Riffert and Herold 1995) derives an analytical formula for the differential luminosity $dL(R)/dR$ emitted by the disk as a function of radius R.

The solid lines in Fig. 3b show for three values of a_* the differential disk luminosity predicted by the NT model. The disk flux vanishes at $R_{\rm ISCO}$ because the model has, by assumption, no viscous stress inside this radius (see Sect. 4.2 for further discussion). More

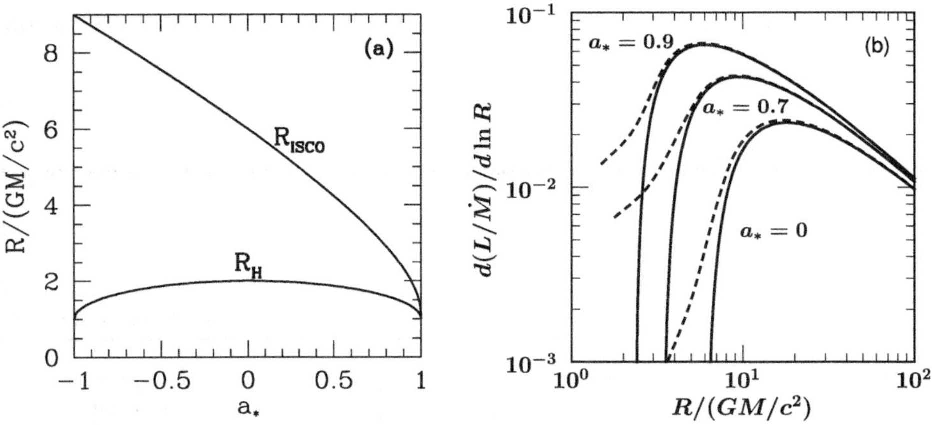

Fig. 3 (**a**) Radius of the ISCO $R_{\rm ISCO}$ and of the horizon $R_{\rm H}$ in units of GM/c^2 plotted as a function of the black hole spin parameter a_*. Negative values of a_* correspond to retrograde orbits. Note that $R_{\rm ISCO}$ decreases monotonically from 9 GM/c^2 for a retrograde orbit around a maximally spinning black hole, to 6 GM/c^2 for a non-spinning black hole, to GM/c^2 for a prograde orbit around a maximally spinning black hole. (**b**) Profiles of $d(L/\dot{M})/d\ln R$, the differential disk luminosity per logarithmic radius interval normalized by the mass accretion rate, versus radius $R/(\mathrm{GM/c^2})$ for three values of a_*. *Solid lines* are the predictions of the NT model. The *dashed curves* from Zhu et al. (2012), which show minor departures from the NT model, are discussed in Sect. 5.2

importantly, the peak emission occurs at a radius that tracks the ISCO (it is a factor of ~2–3 larger than $R_{\rm ISCO}$). This means that the radiation is emitted from a progressively smaller effective area, roughly $\propto R_{\rm ISCO}^2$, as the black hole spin increases. Therefore, for a given total disk luminosity, the temperature of the emitted radiation increases with increasing a_*. This is the key physical effect that underlies the continuum-fitting method. By measuring the characteristic temperature and luminosity of the disk emission, and applying the NT model, one is able to estimate both a_* and the mass accretion rate $\dot{M}$.

As should be clear from the above, the accuracy of the continuum-fitting method ultimately depends on the reliability of the NT model; this issue is discussed further in Sects. 4.2 and 5.2. It also depends on our ability to calculate the spectrum of the radiation, which would be trivial if the disk radiated as a perfect blackbody. Unfortunately, because electron scattering plays a prominent role at the X-ray temperatures found in black hole binaries, the emitted spectrum is substantially harder than a blackbody spectrum of the same flux. Hence it is necessary to employ detailed disk atmosphere models. Most of the work to date is based on the atmosphere model BHSPEC developed by Davis and Hubeny (2006), which is discussed in Sects. 3.2 and 5.3.

3.2 Continuum Fitting in Practice

In broad outline, one fits the X-ray continuum spectrum to the Novikov–Thorne model of a thin accretion disk with other spectral components as needed, principally a Compton component. As stressed in Sects. 1 and 5, in order to obtain useful constraints on a_*, one must inform the fitting process by inputting accurate values of the external parameters D, i and M. The spectral fit returns two output parameters: the spin a_* and the mass accretion rate $\dot{M}$. An important derived quantity is the Eddington-scaled luminosity of the disk component $L(a_*, \dot{M})/L_{\rm Edd}$.

In practice, one usually fits the thermal component using KERRBB2 (McClintock et al. 2006),[3] which is a hybrid code implemented in XSPEC (Arnaud 1996) that combines the capabilities of two relativistic disk models, BHSPEC (Davis et al. 2005) and KERRBB (Li et al. 2005). This latter model, KERRBB, which is a straightforward implementation of the analytic Novikov–Thorne model, has three principal fit parameters: a_*, $\dot{M}$, and the spectral hardening factor f, which relates the observed color temperature to the effective temperature, $f = T/T_{\rm eff}$.

In fitting the disk component with KERRBB, it is quite generally the case that one can only determine two parameters, a shape parameter (e.g., a_* or T) and a normalization constant (e.g., $\dot{M}$). That is, in practice one cannot additionally obtain a useful constraint on f. However, this limitation of KERRBB is handily overcome by pairing it with BHSPEC, which is based on non-LTE disk atmosphere models within an α-viscosity prescription. BHSPEC has just two principal fit parameters (spin and mass accretion rate), and it can be used to fit directly for a_*. However, it does not include the effects of self-illumination of the disk ("returning radiation"), which is a feature that is included in KERRBB.

The pairing of KERRBB and BHSPEC is achieved using KERRBB2, which is a modified version of KERRBB that contains a pair of look-up tables for f corresponding to two values of the viscosity parameter: $\alpha = 0.01$ and 0.1. The entries in the tables are computed using BHSPEC. The two tables give f versus $L/L_{\rm Edd}$ for a wide range of the spin parameter ($|a_*| \leq 0.9999$). The computations of f versus $L/L_{\rm Edd}$ are done using the appropriate, corresponding response matrices and energy ranges used in fitting the spectra with KERRBB. Thus, KERRBB and the subroutine/table computed using BHSPEC (which together constitute KERRBB2) allow one to fit directly for a_* and $L/L_{\rm Edd}$, while retaining the returning-radiation feature of KERRBB.

Depending on the quality of a particular spectrum, it may be necessary to include minor spectral components (e.g., line or edge features), but these cosmetic features do not significantly affect the spin results. Typically, three model components are fitted in conjunction with the thermal component: a low-energy cutoff, a "reflected" component (e.g., Ross and Fabian 2007), and a Compton component. The cutoff is straightforward to model (e.g., Wilms et al. 2000), and the reflected component is relatively weak in disk-dominated spectra, even in the most extreme circumstances (Gou et al. 2011). It is the modeling of the Compton component that has been of central concern in applying the continuum-fitting method, and we discuss this issue now.

All spectra of black hole binaries, even the most disk-dominated, show a high-energy tail component of emission, which is widely attributed to Compton upscattering of soft photons by coronal electrons (Remillard and McClintock 2006). In early continuum-fitting work (Shafee et al. 2006; McClintock et al. 2006), this component was modeled unsatisfactorily by adding a power-law component to the spectrum. All subsequent work has used a much-improved empirical model of Comptonization called SIMPL (Steiner et al. 2009b). This model self-consistently generates the Compton component from the thermal seed spectrum of photons. It allows reliable measurements of spin to be obtained even as the fraction of seed photons $f_{\rm SC}$ that are scattered into the power-law component approaches 25 % (Steiner et al. 2009a; Steiner et al. 2009b). The use of SIMPL in place of the standard power law has doubled the body of useful data for several sources (e.g., see Steiner et al. 2011; Steiner et al. 2012a), and it has enabled the measurement of the spins of black holes whose spectra are persistently quite strongly Comptonized such as LMC X-1 (Gou et al. 2009) and Cyg X-1 (Gou et al. 2011).

[3]For alternatives, see Gierliński et al. (2001), Kolehmainen and Done (2010), and Straub et al. (2011).

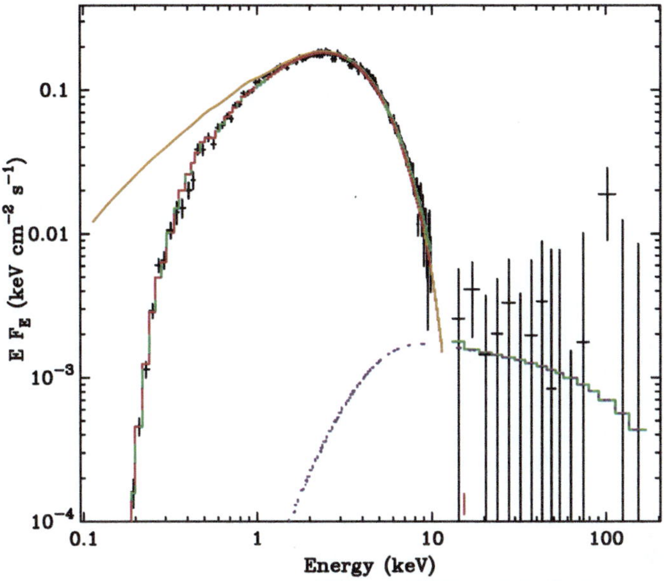

Fig. 4 Model fit to a disk-dominated spectrum of LMC X-3 obtained using detectors aboard the *BeppoSAX* satellite for $D = 52$ kpc, $i = 67°$ and $M = 10\ M_\odot$ (Davis et al. 2006). A *green solid curve*, which is difficult to discern because it hugs the data, is the total model. Also shown is the thermal component (*red long-dashed curve*) and the Compton component (*violet short-dashed curve*). The reflected component is negligible and was not included. The *orange solid curve* shows the total model with the effects of interstellar absorption removed. Note that the peak Compton flux is only 1 % of the peak thermal flux

Successful application of the continuum-fitting method requires the selection of spectra that are disk-dominated. For transient sources, such spectra are typically observed during the latter part of an outburst cycle (Fig. 2b). Figure 4 shows a spectrum with a peak flux in the Compton component that is only 1 % of the peak thermal flux. For spectra that are this disk-dominated, how one chooses to model the adulterating Compton component is obviously quite unimportant. Meanwhile, there is an abundance of spectra of comparable quality available for several sources, i.e., sources with $f_{SC} \sim 1$ % (e.g., see Fig. 1 in Steiner et al. 2009a and Table 1 in Steiner et al. 2011).

While it is essential to select spectra that have a substantial thermal component (i.e., $f_{SC} \lesssim 25$ %; Steiner et al. 2009a), it is equally important to select data of moderate luminosity, specifically spectra with Eddington-scaled disk luminosities $L/L_{Edd} < 0.3$. Otherwise, the disk scale-height grows and the thin disk model is invalidated (Sects. 4.2 and 5.2; McClintock et al. 2006). Fortunately, there is usually an abundance of such data because a typical transient source remains for months in a suitable disk-dominated state of moderate luminosity (see Fig. 2b). A very wide range of detectors are capable of providing suitable data (see example in Sect. 4.1). The principal requirements are that the data can be corrected for dead time, and that the detector have a dozen or more energy channels, an appropriate bandwidth, and be well calibrated (Sect. 5.1).

4 Truncation of the Disk at the ISCO

We review the large body of observational evidence that there exists a constant inner-disk radius in disk-dominated states of black hole binaries. We follow with theoretical evidence, based on GRMHD simulations, that this fixed radius can be identified with the radius of the ISCO.

4.1 Observational Evidence

It has been clear for decades that fitting the X-ray continuum might prove to be a promising approach to measuring black hole spin. The earliest indications came with the advent in the mid-1980s of a nonrelativistic disk model (Mitsuda et al. 1984; Makishima et al. 1986), now referred to as DISKBB, which returns the color temperature $T_{\rm in}$ at the inner-disk radius $R_{\rm in}$. In an important review paper, Tanaka and Lewin (1995) show the remarkable stability of $R_{\rm in}$ for three transients as the thermal flux of these sources steadily decays on a timescale of months by factors of 10–100 (see their Fig. 3.14). Tanaka & Lewin remark that the constancy of $R_{\rm in}$ suggests that this fit parameter is related to the radius of the ISCO. Subsequently, similar evidence for a constant inner-disk radius in disk-dominated states of black hole binaries has been demonstrated for many sources by showing that the bolometric luminosity of the thermal component is approximately proportional to $T_{\rm in}^4$ (Kubota et al. 2001; Kubota and Makishima 2004; Gierliński and Done 2004; Abe et al. 2005; McClintock et al. 2009).

A recent study of the persistent source LMC X-3 presents the most compelling evidence to date for a constant inner-disk radius (Steiner et al. 2010). This result is based on an analysis of a large sample of X-ray spectra collected during eight X-ray missions that span 26 years. As illustrated in Fig. 5 for a selected sample of 391 *RXTE* spectra, the radius of the accretion disk was found to be constant over time and unaffected by the gross variability of the source to within ≈2 percent. Even considering an ensemble of eight X-ray missions, the radius was observed to be stable to within ≈5 percent. These results provide compelling evidence for the existence of a fixed inner-disk radius and establish a firm empirical foundation

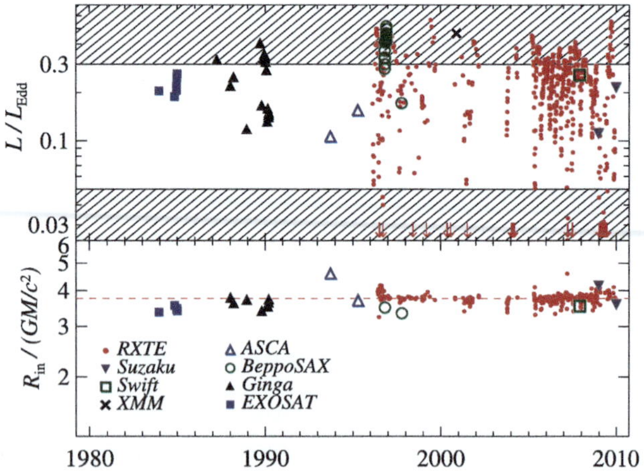

Fig. 5 (*top*) Accretion disk luminosity in Eddington-scaled units (for $M = 10\ M_\odot$) versus time for all the 766 spectra considered in a study of LMC X-3 by Steiner et al. (2010). (*Downward arrows* indicate data that are off scale.) Selected data in the unshaded region satisfy the thin-disk selection criterion $L/L_{\rm Edd} < 0.3$ and avoid confusion with strongly-Comptonized hard-state data with $f_{\rm SC} \gtrsim 25\ \%$ (Sect. 3.2; Remillard and McClintock 2006). (*bottom*) Fitted values of the inner-disk radius are shown for *thin-disk* data in the *top panel* that meet the selection criteria of the study (a total of 411 spectra). Despite large variations in luminosity, the inner-disk radius remains constant to within a few percent over time. The median value for just the 391 selected *RXTE* spectra is shown as a *red dashed line*

for the measurement of black hole spin. The only reasonable inference is that this radius is closely associated with the radius of the ISCO, as we show to be the case in the following section.

4.2 Theoretical Evidence

The NT model makes one key assumption: It assumes that the viscous torque vanishes inside the ISCO. While eminently reasonable (e.g., Paczyński 2000; Afshordi and Paczyński 2003; Shafee et al. 2008b), this "zero-torque" assumption does not follow directly from basic conservation laws but is applied as an extra ad hoc boundary condition. Furthermore, the luminosity profiles shown for the NT model in Fig. 3b depend critically on this boundary condition because this condition causes the luminosity profiles to vanish at the ISCO, which in turn fixes the radius of peak disk emission. Krolik (1999) argued that magnetic stresses can operate freely across the ISCO and will cause strong torques at the ISCO, as well as in the inner plunging region. Also, Gammie (1999) came up with a simple analytical MHD model of the plunging region with demonstrably non-zero torques. What do real disks do?

To answer this question, geometrically thin accretion disks around black holes have been simulated by a number of authors (Shafee et al. 2008a; Noble and Krolik 2009; Penna et al. 2010) using state-of-the-art GRMHD codes. The main advantage of simulations is that they do not require ad hoc assumptions. One simply introduces magnetized gas in a Kerr space-time and lets the system evolve to a quasi-steady state. Since the ISCO lies inside the simulation box, well away from computational boundaries, it is not treated differently from other regions of the system. In other words, no boundary condition is applied by hand at the ISCO, a great improvement over analytical models. On the other hand, for technical reasons, simulations to date have not treated radiation transfer self-consistently but instead have assumed local cooling. This is not considered serious for the purposes of testing the zero-torque condition.

The dashed lines in Fig. 3b show results from simulations of very thin disks ($H/R \sim 0.05$; Penna et al. 2010; Kulkarni et al. 2011; Zhu et al. 2012). The simulation-derived disk luminosity profiles show modest deviations from the NT model predictions; in particular, the disk flux does not vanish inside the ISCO. On the other hand, the deviations are minor, and we discuss and quantify these effects in Sect. 5.2. Importantly, the radius corresponding to the luminosity peak, which is the most relevant quantity for the continuum-fitting method, agrees quite well. As discussed in Sect. 5.2, the good agreement between model predictions and simulation results translates into modest uncertainties in spin estimates. GRMHD simulation results such as these shown in Fig. 3b are viewed as a strong validation of the NT model.

Interestingly, the magnetic stress in the simulations does not vanish in the plunging region. Indeed, Penna et al. (2010) found that the stress there agrees remarkably well with Gammie's (1999) model. However, there is little energy dissipation associated with this stress (Gammie's analytical model has zero dissipation), so it has little bearing on the continuum-fitting method. Another interesting result is that deviations from the NT model seem to increase as the luminosity and, concomitantly, the disk thickness increases[4] (Kulkarni et al. 2011), as anticipated in previous work (Paczyński 2000; Shafee et al. 2008b).

[4] In contrast, Noble et al. (2010) find that the stress profile is almost completely independent of disk thickness.

5 Uncertainties in Spin Estimates

The bottom line of this section is that the error in a_* is dominated by the observational errors in the external input parameters D, i and M. By comparison, the errors due to reliance on the NT model, as well as on the disk atmosphere model that is used to correct for the effects of spectral hardening, are less important. Meanwhile, the one significant question that hangs over most of the spin results is the assumption that the black hole's spin vector is aligned with the orbital angular momentum vector. We discuss these points in turn.

5.1 Observational Errors

In early work, the error in a_* attributable to the uncertainties in D, i and M was only crudely estimated (Shafee et al. 2006; Davis et al. 2006; McClintock et al. 2006). However, in all subsequent work, starting with Liu et al. (2008), the error in the spin due to the combined uncertainties in these three parameters has been computed in detail via Monte Carlo simulations. That the error budget for a_* is dominated by the uncertainties in D, i and M has been thoroughly demonstrated in recent work, which provides error estimates for a very wide range of statistical and systematic errors associated with (1) the details of the spectral models employed, (2) flux calibration uncertainties, (3) the effects of a warm absorber, etc. Instead of discussing such technical details here, we refer the reader to Sect. 5 and Appendix A in Steiner et al. (2011) and Sect. 5 in Gou et al. (2011).

5.2 Errors from the Novikov–Thorne Model

As described in Sect. 3.1, the NT model is robust and makes very few untested assumptions. It is true that some properties of the disk, e.g., the density and temperature of gas at the disk mid-plane, depend on the magnitude of the viscosity parameter α, but the all-important luminosity profiles shown in Fig. 3b do not. These profiles are a direct consequence of energy conservation—gas drifts inward, it converts gravitational potential energy into orbital kinetic and gas thermal energy, and the latter is radiated. This physics is independent of the value of α, or even the validity of the α prescription.

The NT model assumes that dissipated energy is radiated by the gas locally at the same radius. This is a very safe assumption. The cooling time of the gas in a thin accretion disk is approximately $(H/R)^2$ times the viscous radial advection time. For the disks of interest to the continuum-fitting method ($H/R < 0.1$), this means that cooling is about 100 times faster than energy advection and hence very local.

Another approximation in the NT model is the neglect of disk self-irradiation. This is acceptable near the peak of the luminosity profile, where local energy dissipation greatly exceeds irradiation. However, it is less safe at larger radii. The models KERRBB and KERRBB2 (Sect. 3.2) include self-irradiation consistently. In practice, self-irradiation seems to have a minor effect on spin estimates.

As already discussed, the NT model assumes zero torque at the ISCO. Although this approximation turns out to be less severe than one might have anticipated (Fig. 3b), we still expect it to have some effect on the continuum-fitting method. Several authors (Kulkarni et al. 2011; Noble et al. 2011; Zhu et al. 2012) have investigated this issue quantitatively. The general consensus is that the zero-torque approximation introduces uncertainties in spin estimates of around $\Delta a_* \sim 0.1$ for low spin values $a_* < 0.5$ and much smaller errors as $a_* \to 1$. For example, Kulkarni et al. (2011) estimate for $a_* = 0, 0.7, 0.9$ and 0.98 that the respective values of Δa_* are 0.11, 0.06, 0.014 and 0.007. (Noble et al. 2011 estimate

$\Delta a_* \sim 0.2 - 0.3$ for $a_* = 0$.) These results are for a disk inclination angle of $i = 60°$ and a disk thickness of $H/R = 0.05$, which corresponds to $L/L_{\rm Edd} \sim 0.35$ (see Table 1 in Zhu et al. 2012). The errors are more severe for thicker disks. Meanwhile, the results quoted here are for the thinnest, lowest-luminosity disks simulated to date; presently, it is not practical to resolve the MRI turbulence in thinner disks.

Not only are these estimates of the NT model errors significantly less than the observational errors presented in Sect. 6, they are overestimates of the model errors because the continuum-fitting method is applied only to very thin disks: A strict requirement of the method is $L/L_{\rm Edd} < 0.3$, while most measurements are based on spectral data with $L/L_{\rm Edd} \lesssim 0.1$. Looking to the future, it might be possible to do better by replacing the NT model with a more accurate simulation-based model (e.g., Penna et al. 2012), but this step is not presently warranted. In conclusion, all continuum-fitting spin measurements published to date (see Sect. 6) are based on the NT model which systematically overestimates the spin; however, this source of error is presently small compared to the observational errors.

5.3 Errors from the Disk Atmosphere Model

An essential cornerstone of the continuum-fitting method is a reliable model of the disk's atmosphere. Such a model is BHSPEC (Davis and Hubeny 2006), which can be used either alone or including the effect of self-irradiation via KERRBB2. BHSPEC, which is quite sophisticated and includes a wide range of physical effects, is based on the non-LTE radiative transfer code TLUSTY (Hubeny and Lanz 1995), which was originally developed for stellar atmospheres.

At a given location on an accretion disk, BHSPEC computes the emitted spectrum using three supplied parameters: the effective temperature $T_{\rm eff}$ defined such that radiative flux $F = \sigma T_{\rm eff}^4$, local vertical gravity parameter Q, and disk column density Σ. As discussed in previous sections, a robust estimate of $T_{\rm eff}$ can be obtained from the NT model, while the parameter Q is calculated directly from the Kerr metric. The main uncertainty is in the value of Σ.

In standard disk theory, Σ varies inversely as the viscosity parameter α and is thus quite uncertain. Fortunately, in the case of optically thick disks (which all thermal state disks are) Σ has only a weak effect on the emerging spectrum. This is analogous to the case of a star where the spectrum depends on the effective temperature and surface gravity, but not at all on the optical depth to the stellar core, which is effectively infinite. The optical depth through a disk is not quite infinite, hence there is some spectral dependence on Σ. However, this dependence is weak for models with $L/L_{\rm Edd} < 0.3$ (Davis and Hubeny 2006; Done and Davis 2008).

For the same reason, details of exactly how viscous heating is distributed vertically within the disk are unimportant. So long as energy dissipation occurs in the disk interior at optical depths greater than a few, the emerging spectrum depends only on $T_{\rm eff}$ and Q (Davis et al. 2005; Davis et al. 2009). This is not true if there is substantial energy dissipation close to or above the photosphere. Disks in the thermal state probably do not have such dissipation since their spectra show very little hard "coronal" emission (Remillard and McClintock 2006). Whatever little coronal emission is present is fitted for via a model for the Compton power law such as SIMPL (Sect. 3.2).

The standard BHSPEC model assumes hydrostatic equilibrium and does not include the force from magnetic fields. However, numerical simulations (e.g., Hirose et al. 2009) indicate that the photospheric surface regions show modest deviations from hydrostatic equilibrium and are primarily supported by magnetic forces. Including these effects in BHSPEC generally leads to a modest (<10 %) increase in the spectral hardness (Davis et al. 2009). The effects of irradiation (both self- and from a corona), which have not yet been rigorously explored, may also lead to a slight hardening of the spectrum. In summary, while there are uncertainties associated with the disk spectral model used in the continuum-fitting method, it appears unlikely that the resulting errors in R_{in} are more than 10 %, which for low values of spin implies $\Delta a_* \sim 0.1$, decreasing as $a_* \to 1$ (Fig. 3a).

5.4 Assumption of Spin-Orbit Alignment

In determining the spins of eight of the ten black holes (see Sect. 6), it is assumed that the plane of the inner X-ray-emitting portion of the disk is aligned with the binary orbital plane, whose inclination angle i is determined from optical observations (e.g., Orosz et al. 2011a). However, if a black hole's spin is misaligned with the orbital vector, this will warp a thin disk because the Bardeen–Petterson effect will force the inner disk to align with the black hole spin vector (Bardeen and Petterson 1975).[5] An error in estimating the inclination of the inner disk of $\sim 10°$ or more, resulting from an erroneous use of i as a proxy for the inclination of the inner disk, would substantially corrupt most continuum-fitting measurements of spin.[6]

There is evidence for gross spin-orbit misalignment for one transient system (SAX J1819.3-2525); however, this evidence is weak (Narayan and McClintock 2005). For the transients generally, more recent evidence, which is summarized in Sect. 1 of Steiner and McClintock (2012), argues in favor of alignment. Briefly, the timescale for accretion to torque the black hole into alignment is estimated to be $\sim 10^6$–10^8 years, which is short compared to the typical lifetime of a transient system (Sect. 2.2). In the case of the persistent supergiant systems, there is some evidence that their more massive black holes are formed by direct, kickless collapse (Mirabel and Rodrigues 2003; Reid et al. 2011). Finally, a population synthesis study based on a maximally conservative (i.e., minimum-torque) assumption indicates that the spin axes of most black hole primaries will be tilted less than 10° (Fragos et al. 2010).

In determining the spins of the remaining two black holes (see Sect. 6), the inclination of the inner disk is taken to be the inclination θ of the radio or X-ray jet axis, which is presumed to be aligned with the black hole's spin axis. The jet inclination angle for these microquasars, GRS 1915+105 and H1743–322, was determined by modeling proper-motion data derived from radio and X-ray observations (Mirabel and Rodríguez 1994; Fender et al. 1999; Steiner et al. 2012a). Fortunately, radio/X-ray jet data have also yielded a strong constraint on θ for a third microquasar, XTE J1550–564, thereby providing a rare opportunity to check directly the assumption of spin orbit alignment because its orbital inclination angle i has also been measured (Orosz et al. 2011b). In this case, Steiner and McClintock (2012) find no evidence for misalignment and place an upper limit on the difference between the spin and orbital inclinations of $|\theta - i| < 12$ deg (90 % confidence).

6 Results and Discussion

Table 1 lists the masses and spins of ten stellar black holes. By virtue of the no-hair theorem, this table provides complete descriptions of each of these ten black holes. The spins span

[5]While thin disks are subject to warping, thick disks are not (Dexter and Fragile 2011).

[6]Unfortunately the continuum-fitting method cannot fit for the inclination of the inner disk because there is a degeneracy between the inclination and spin parameter (Li et al. 2009).

Table 1 The masses and spins, measured via continuum-fitting, of ten stellar black holes[a]

System	a_*	$M/M_\odot$	References
Persistent			
Cyg X-1	>0.95	14.8 ± 1.0	Gou et al. 2011; Orosz et al. 2011a
LMC X-1	$0.92^{+0.05}_{-0.07}$	10.9 ± 1.4	Gou et al. 2009; Orosz et al. 2009
M33 X-7	0.84 ± 0.05	15.65 ± 1.45	Liu et al. 2008; Orosz et al. 2007
Transient			
GRS 1915+105	$>0.95^{b}$	10.1 ± 0.6	McClintock et al. 2006; Steeghs et al. 2013
4U 1543–47	0.80 ± 0.10^{b}	9.4 ± 1.0	Shafee et al. 2006; Orosz 2003
GRO J1655–40	0.70 ± 0.10^{b}	6.3 ± 0.5	Shafee et al. 2006; Greene et al. 2001
XTE J1550–564	$0.34^{+0.20}_{-0.28}$	9.1 ± 0.6	Steiner et al. 2011; Orosz et al. 2011b
H1743–322	0.2 ± 0.3	$\sim 8^{c}$	Steiner et al. 2012a
LMC X-3	$<0.3^{d}$	7.6 ± 1.6	Davis et al. 2006; Orosz 2003
A0620–00	0.12 ± 0.19	6.6 ± 0.25	Gou et al. 2010; Cantrell et al. 2010

[a]Errors are quoted at the 68 % level of confidence, except for the three spin limits, which are estimated to be at the 99.7 % level of confidence.

[b]Uncertainties greater than those in papers cited because early error estimates were crude.

[c]Mass estimated using an empirical mass distribution (Özel et al. 2010).

[d]Preliminary result pending improved measurements of M and i.

the full range of prograde values, and the masses range from 6 to 16 $M_\odot$. In addition to the continuum-fitting spin data in Table 1, Gierliński et al. (2001) provide preliminary estimates for the spins of LMC X-1 and GRO J1655–50, Kolehmainen and Done (2010) report a hard upper limit of $a_* < 0.9$ on the spin of GX 339–4, and Nowak et al. (2012) argue that the spin of 4U 1957+11 is extreme. Concerning 4U 1957+11, it is unclear if the compact object is a black hole, and the key parameters D and M are essentially unconstrained. Finally, Middleton et al. (2006) find an apparently moderate value of spin for GRS 1915+105, which is at odds with the extreme value in Table 1; Middleton et al. obtained a depressed value of spin because they relied on high-luminosity data, as explained in Sect. 5.3 in McClintock et al. (2006).

Caution is required in considering the errors for the values of spin quoted in Table 1 assuming that they are Gaussian, particularly for $a_* \gtrsim 0.7$. Note in Fig. 3a how insensitive a_* is to large changes in the observable $R_{\rm ISCO}$ as a_* approaches unity. As a consequence of this limiting behavior of a_*, doubling a 1σ error to approximate a 2σ error can lead to nonsense. For example, formally increasing the nominal spin of LMC X-1 ($a_* = 0.92$; Table 1) by doubling the 1σ error ($\Delta a_* = 0.05$) implies a 2σ upper limit of $a_* < 1.02$, whereas the correct 2σ upper limit is $a_* < 0.98$ (see Fig. 8 in Gou et al. 2009).

6.1 The Persistent Systems vs. the Transients

There is a dichotomy between the black holes in persistent systems and those in transients, both in their masses and their spins (Table 1). Considering spin first, the three persistent black holes all have high or extreme spins. In contrast, the spins of the transient black holes range widely: Four have spins consistent with zero, two have intermediate values of spin, and one is a near-extreme Kerr hole. The dichotomy is sharpened if one considers six additional transient black holes all of whose spins are predicted to be $a_* \lesssim 0.8$ (Steiner et al. 2013) based on a fitted correlation between radio power and spin (Sect. 7.2).

Not only are the persistent black holes rapidly spinning, they are also massive—11–16 $M_\odot$—compared to the transient black holes. The masses of the transients are significantly lower and, remarkably, their mass distribution is narrow: $7.8 \pm 1.2\ M_\odot$ (Özel et al. 2010; Farr et al. 2011).

6.2 Prograde Spins that Obey the Kerr Bound

The lack of negative spins in Table 1 may be the result, in a close binary system, of the expected alignment of the spin of the black hole progenitor with the orbital angular momentum, and it may also indicate that black hole kicks are not strong enough to flip the black hole into a retrograde configuration. While interesting that there are no negative spins, it is equally interesting that the spins of all ten black holes obey the Kerr bound $|a_*| < 1$. In particular, if the distances to either Cyg X-1 or GRS 1915+105 were ~30 % less than the best current estimates, then it would be impossible to fit the data with the KERRBB2 model, which only accommodates spin values $a_* < 1$. Because the observed values of each of the three external fit parameters (D, i and M) place hard constraints when fitting the data, a failure to fit a spectrum that requires $a_* > 1$ has the potential to falsify the spin model. For a discussion of this point in relation to the near-extreme Kerr hole GRS 1915+105, see Sect. 6.4 in McClintock et al. (2006).

6.3 The High Natal Spins of the Persistent Black Holes

It is reasonable to conclude that the black holes in the persistent systems were born with high spins because their host systems are too young for these black holes to have been spun up by accretion torques. Consider, for example, the persistent system Cyg X-1 (Gou et al. 2011): For its black hole to achieve its present spin of $a_* > 0.95$ via disk accretion, an initially nonspinning black hole would have had to accrete $>7.3\ M_\odot$ from its donor (Bardeen 1970; King and Kolb 1999) to become the $14.8\ M_\odot$ black hole we observe today. However, even at the maximum (Eddington-limited) accretion rate this would require >31 million years, while the age of the system is between 4.8 and 7.6 million years (Wong et al. 2012). Likewise for M33 X-7 and LMC X-1, the corresponding minimum spin-up timescales are >17 and >25 million years, respectively, while the respective ages of the systems are $\lesssim 3$ and $\lesssim 5$ million years (Gou et al. 2011). It therefore appears that the spins of these systems must be chiefly natal, although possibly such high spins could be achieved during a short-lived evolutionary phase of hypercritical accretion (Moreno Méndez et al. 2008).

6.4 Applications

The data in Table 1 have a number of applications to physics and astrophysics, both immediate and potential. In physics, a high goal is to use such data as a springboard to test the no-hair theorem (see Sect. 8), and the foundation for any such test is high-quality measurements of mass and spin for a good sample of black holes. In astrophysics, knowledge of the spins of stellar black holes is crucial for example in constraining models of gamma-ray burst sources (Woosley 1993; MacFadyen and Woosley 1999; Woosley and Heger 2006); supernovae and black hole formation (Lee et al. 2002; Wong et al. 2012); exotic black hole states and state transitions (Remillard and McClintock 2006); and in informing LIGO/VIRGO modelers who are computing gravitational-wave signals (Campanelli et al. 2006). A central question, which we turn to in the next section, is the role of spin in powering jets.

7 Jet Power and Black Hole Spin

Since the spin parameter a_* is one of only two numbers that completely characterize a black hole (mass M being the other), it stands to reason that it should influence at least some observational properties of the hole. The most widely discussed connection is to relativistic jets.

The story goes back to Penrose (1969) who showed that a spinning black hole has free energy that can in principle be tapped by specially prepared infalling particles. Although Penrose's specific proposal is not considered promising, the idea of extracting energy from spinning black holes has stuck and has become popular in astrophysics. Ruffini and Wilson (1975; see also Damour et al. 1978) and Blandford and Znajek (1977) suggested a specific mechanism whereby a force-free poloidal magnetic field around a spinning black hole is twisted by frame dragging, thereby producing outgoing Poynting flux along twin jets. We refer to this as the generalized Penrose process.

GRMHD simulations of accreting black holes have found MHD jets forming spontaneously from generic initial conditions (e.g., Koide et al. 2002; McKinney and Gammie 2004; McKinney 2005; Beckwith et al. 2008; McKinney and Blandford 2009). Moreover, in one particular simulation involving a rapidly spinning black hole and a strong poloidal field, Tchekhovskoy et al. (2011) showed that the power carried by the jet exceeded the total rest mass energy of accreted gas, meaning that the jet extracted energy from the spinning black hole.

On the observational front, until recently there was no empirical evidence for a connection between black hole spin and relativistic jets. We discuss here the first such evidence.

7.1 Two Kinds of Jets in Black Hole Binaries

Fender et al. (2004) identified a number of systematic properties in the radio emission of black hole binary jets. They showed that there are two kinds of jets, which we refer to as "steady jets" and "ballistic jets," each associated with a specific spectral state of the X-ray source. Although we discuss both kinds of jets for completeness, our focus here is on the ballistic jet.

The steady jet is observed as a continuous outflow of plasma in the hard spectral state (for a discussion of spectral states in black hole binaries, see Remillard and McClintock 2006). This jet is small-scale, being observable only out to a few tens of AU, and it appears not to be very relativistic. It is present at the very start of a transient's outburst cycle. Referring to the X-ray light curves for XTE J1859+226 in Fig. 2a, the jet is present during the first few days when the hard flux (panel a) is most intense and the 2–12 keV flux (panel b) is increasing rapidly. It then disappears, and it returns only near the end of the outburst cycle (beyond the right edge of the plot). The steady jet is seen in all transients at low values of $\dot{M}$.

The far more dramatic ballistic jet is launched when a transient goes into outburst (Fender et al. 2004). This powerful transient jet usually appears near (or soon after) the time of outburst maximum, as the source switches from its initial hard state to a soft state via the "steep power-law" state. Ballistic jets manifest themselves as blobs of radio (and occasionally X-ray) emitting plasma that move ballistically outward at relativistic speeds (Lorentz factor $\Gamma > 2$). They are often observed out to distances of order a parsec. Because ballistic jets resemble the kpc-scale jets seen in quasars, black hole binaries that produce them are called microquasars (Mirabel and Rodríguez 1999).

Ballistic jet ejection occurs at a very specific stage during the spectral evolution of a given system (Fender et al. 2004). In Fig. 2, the strong spike in the radio light curve (panel d), which is characteristically delayed relative to the corresponding spike in the X-ray luminosity (panel b), is associated with a ballistic jet. As most clearly demonstrated for the prototypical microquasar GRS 1915+105 (Fender and Belloni 2004), this ejection stage appears to correspond to the inward-moving inner edge of the accretion disk reaching the ISCO, which apparently results in some violent event that launches a large-scale relativistic blob.

On general principles, one expects jet power to depend on a black hole's mass M and spin a_*, and the mass accretion rate $\dot{M}$, plus other factors such as the strength and topology of the magnetic field. If one wishes to investigate the dependence of jet power on a_*, one needs first to eliminate the other variables.

For steady jets, $\dot{M}$ spans a wide range, and it is not straightforward to eliminate the effects of this variable. It is possible to do this, however, by using the disk X-ray luminosity as a proxy for $\dot{M}$ (e.g., Heinz and Sunyaev 2003; Merloni et al. 2003; Falcke et al. 2004; Fender et al. 2004; Fender and Belloni 2004), but one must have knowledge of the radiative efficiency of the disk, which is generally both low and variable in the hard state where the steady jet forms (see Narayan and McClintock 2008, for a discussion of radiatively inefficient accretion in the hard state). The procedure is relatively uncertain and it is difficult to obtain robust results. Nevertheless, Fender et al. (2004) have performed such a study and have claimed that there is no evidence in the data that the power of a steady jet depends on spin.

Ballistic jets on the other hand invariably occur near the peak of transient outbursts. Steiner et al. (2013) have shown that during major outbursts the peak disk luminosities in various transients are near the Eddington limit and are clustered within a factor of ~ 2 in luminosity, which means that these systems behave for all purposes like "standard candles." This crucially allows one to compare the power of ballistic jets observed for different black holes at the same $\dot{M}$, namely $\dot{M} \sim \dot{M}_{\rm Edd}$. In addition, all black holes in transients have similar masses to better than a factor of two (Özel et al. 2010); furthermore, it is easy to correct for any mass dependence (see below). This leaves a_* (with magnetic field as a wild card) as the sole remaining parameter that could have any influence on jet power.

Let us define the jet efficiency factor η of a ballistic jet,

$$\eta_{\rm jet}(a_*) = \langle L_{\rm jet}\rangle / \langle \dot{M}\rangle c^2, \tag{1}$$

where $\langle L_{\rm jet}\rangle$ is the time-average kinetic luminosity flowing out through the jet and $\langle \dot{M}\rangle c^2$ is the time-average rate at which rest-mass energy flows into the black hole. Using (1) radio luminosity as a proxy for $L_{\rm jet}$ (Narayan and McClintock 2012) and (2) observed values for the peak radio luminosities for five black holes that are all accreting at $\sim \dot{M}_{\rm Edd}$ (Steiner et al. 2013), one can infer directly how jet luminosity depends on spin, as we discuss in the following section.

7.2 Correlation Between Spin and Ballistic Jet Power

A typical ballistic jet blob is initially optically thick and has a low radio power. As the blob moves out and expands, the larger surface area causes its radio power to increase. This continues until the blob becomes optically thin, after which the flux declines rapidly. The overall behavior is generally consistent with an expanding conical jet (van der Laan 1966; Hjellming and Johnston 1988). Moreover, as discussed in Sect. 7.4.3, the peak radio luminosity is expected to scale more or less linearly with the jet kinetic energy or kinetic luminosity. Thus, peak radio luminosity is a good proxy for jet kinetic luminosity.

Narayan and McClintock (2012) considered the peak radio luminosities of ballistic jet blobs in four transients, A0620–00, XTE J1550–564, GRO J1655–40, GRS 1915+105, and

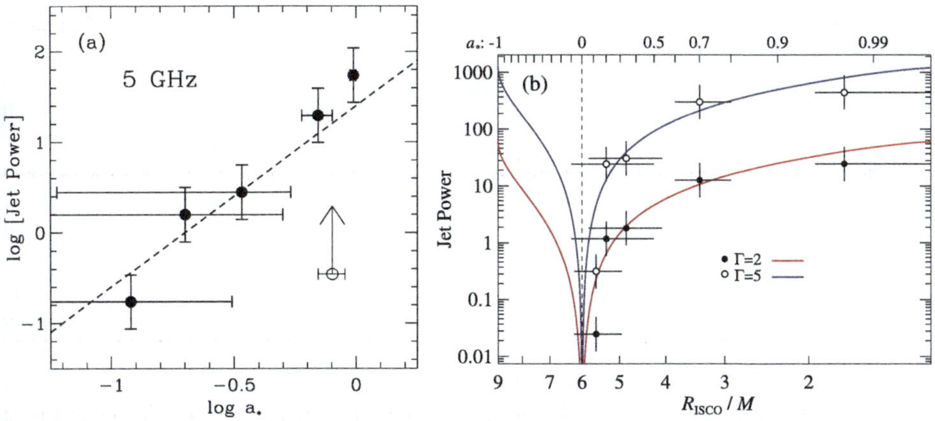

Fig. 6 (**a**) Plot of the quantity Jet Power, which measures the 5 GHz radio luminosity at light curve maximum, versus black hole spin, measured via the continuum-fitting method for five transients (Narayan and McClintock 2012; Steiner et al. 2013). The *dashed line* has slope equal to 2. (**b**) Plot of Jet Power versus $R_{ISCO}/(GM/c^2)$. Here the radio luminosity has been corrected for beaming assuming a bulk Lorentz factor $\Gamma = 2$ (*filled circles*) or $\Gamma = 5$ (*open circles*). The *solid lines* correspond to Jet Power $\propto \Omega_H^2$, where Ω_H is the angular frequency of the horizon (Steiner et al. 2013)

showed that they correlated well with the corresponding black hole spins measured via the continuum-fitting method.[7] Later, Steiner et al. (2013) included a fifth transient, H1743–322, whose spin had been just measured. Figure 6a shows a plot of the black hole spins of these five objects versus a measured quantity called "Jet Power," which refers to the radio luminosity $\nu L_\nu = (\nu S_\nu) D^2/M$ (here, not corrected for beaming), where $\nu = 5$ GHz is the radio frequency, S_ν is the flux density in Jy at the peak of the ballistic-jet radio light curve, D is the distance in kpc, and M is the black hole mass in solar units.[8] That is, the proxy adopted for jet kinetic luminosity is simply the peak radio luminosity at 5 GHz.[9] Figure 6a shows unmistakable evidence for a strong correlation between Jet Power and a_*. Note that Jet Power varies by nearly three orders of magnitude as the spin parameter varies from $\approx$0.1–1.

The uncertainty in the estimated values of Jet Power, which is difficult to assess, is arbitrarily and uniformly assumed to be a factor of two (Narayan and McClintock 2012). The very unequal horizontal error bars in Fig. 6a are a feature of the continuum-fitting method of measuring a_*. Recall that the method in effect measures R_{ISCO} and then deduces the value of a_* using the mapping shown in Fig. 3a. Since the mapping is highly non-linear, especially as $a_* \to 1$, comparable errors in R_{ISCO} correspond to vastly different uncertainties in a_*. In addition, the use of log a_* along the horizontal axis tends to stretch error bars excessively for low spin values. This point is clarified by considering Fig. 6b, based on Steiner et al. (2013). Here the horizontal axis tracks log R_{ISCO} rather than log a_*, and the horizontal error

[7] In the case of a fifth transient, 4U1543–47, radio observations did not include the peak of the light curve, so one could only deduce a lower limit to the jet power. Note that the radio peak can be very narrow in time, e.g., $\approx$1-day in the case of XTE J1859+226 (Fig. 2), so one requires dense radio monitoring to catch the peak.

[8] The scaling by mass is sensible because the sources are near the Eddington luminosity limit, which is proportional to mass. However, since the masses of the black holes differ little (Table 1), the results would be virtually identical if the mass scaling were eliminated.

[9] None of the results change if one chooses a different reference frequency, e.g., 1.4 GHz or 15 GHz.

bars are therefore more nearly equal. The key point is, regardless of how one plots the data, the correlation between Jet Power and black hole spin appears to be strong.

7.3 What Does it Mean?

Assuming the correlation shown in Fig. 6 is real, there are two immediate implications: (i) Ballistic jets in black hole binaries are highly sensitive to the spins of their underlying black holes, presumably because these jets derive their power directly from the spin energy of the hole (á la Penrose 1969; Ruffini and Wilson 1975; Blandford and Znajek 1977). (ii) Spin estimates of stellar black holes obtained via the continuum-fitting method are sufficiently reliable to reveal this long-sought connection between relativistic jets and black hole spin.

With respect to item (i), we note that the mere existence of a correlation does not necessarily imply that the Penrose process is at work. We know that the gravitational potential in the inner disk deepens with increasing black hole spin, since the inner radius of the disk ($R_{\rm ISCO}$) shrinks with increasing a_* (Fig. 3a).

Could this disk-related effect be the reason for the increasing jet power? It seems unlikely. The radiative efficiency of a Novikov–Thorne thin accretion disk increases only modestly with spin; for the five objects shown in Fig. 6, the radiative efficiencies are 0.061, 0.069, 0.072, 0.10 and 0.19, respectively, varying by a factor of $\approx$3. It seems implausible that a disk-powered jet could vary in radio Jet Power by the orders of magnitude seen in Fig. 6.

In contrast, any mechanism that taps directly into the black hole spin energy via something like the Penrose process can easily account for the observed large variation in Jet Power. For instance, the generalized Penrose effect predicts that the jet efficiency factor should vary as $\eta_{\rm jet} \propto a_*^2$ (Ruffini and Wilson 1975; Blandford and Znajek 1977) or, more precisely, as $\eta_{\rm jet} \propto \Omega_{\rm H}^2$ (Tchekhovskoy et al. 2010), where $\Omega_{\rm H}$ is the angular frequency of the black hole horizon,[10]

$$\Omega_{\rm H} = \frac{c^3}{2{\rm GM}}\left(\frac{a_*}{1+\sqrt{1-a_*^2}}\right). \tag{2}$$

The dashed line in Fig. 6a corresponds to Jet Power $\propto a_*^2$ and the solid lines in Fig. 6b to $\propto \Omega_{\rm H}^2$. The observational data agree remarkably well with these scalings, strongly suggesting that a Penrose-like process is in operation.

While the above conclusions are highly satisfying, one should not discount other possibilities. First, the correlation shown in Fig. 6 is based on only five objects. Although this is mitigated by the very wide range of Jet Powers, the correlation might become weaker with the next spin measurement. Second, the correlation might arise if Jet Power and spin are each correlated with some third parameter. For instance, it is intriguing that the binary orbital periods of the five transients under consideration increase systematically with Jet Power. One could imagine scenarios in which the energy of ejected blobs depends on the size of the accretion disk, which depends in turn on the orbital period. However, it is less easy to see why the values of a_* measured with the continuum-fitting method should correlate with the orbital period, while conspiring to produce the scaling predicted by Blandford and Znajek (1977).

Returning to the subject of steady jets, it is interesting to consider why there is apparently no correlation between jet radio luminosity and spin (Fender et al. 2010; Russell et al.

[10]The two scalings agree for small values of a_*, but differ as $a_* \to 1$.

2013). One likely answer is that steady jets and ballistic jets are produced via very different mechanisms. Perhaps ballistic jets are launched within a few gravitational radii, near R_{ISCO}, where the black hole spin could plausibly have a strong effect, whereas steady jets in the hard state originate much further out at radii ∼10–100 GM/c^2 (e.g., Markoff et al. 2005), where the effects of spin are relatively weak. In support of this explanation, ballistic jets are definitely relativistic, with Lorentz factors of up to several (Fender et al. 2004; Fender et al. 2006), whereas there is little evidence that steady jets are relativistic (Gallo et al. 2003; Narayan and McClintock 2005).

7.4 A Challenge

Based on an analysis of a heterogeneous data sample of uneven quality, Fender et al. (2010) claim that there is no evidence for a correlation between the power of ballistic jets and black hole spin. The substantial difference between the results obtained by these authors and Narayan and McClintock (2012) is, in the end, determined by the quantity used to represent jet power (which is discussed further in Sect. 7.4.3). Fender et. al compute jet power from the peak radio luminosity and rise time of a particular synchrotron event; adopt a formula relating jet power to X-ray luminosity, namely, $\log_{10} L_{\text{jet}} = c + 0.5(\log_{10} L_{\text{x}} - 34)$; and use the normalization constant c as their proxy for the jet power. Narayan and McClintock, on the other hand, use the model-independent proxy discussed above, namely, the maximum observed radio luminosity at 5 GHz.

Very recently, the three authors of Fender et al. (2010) have written a second paper (Russell et al. 2013) repeating their claim that there is no evidence for a correlation between the power of ballistic jets and black hole spin. Therein, they challenge the methodologies and findings of Narayan and McClintock (2012) and Steiner et al. (2013). A response to this challenge is being readied (J. Steiner et al., in preparation); meantime, the following is a preliminary sketch of the elements of this response. The following comments pertain only to ballistic jets and continuum-fitting spin measurements. (Russell et al. 2013; also discuss steady jets and Fe-line spin measurements.)

7.4.1 Significance of the Result

Russell et al. (2013) contend that the empirical correlation shown in Fig. 6a is only marginally statistically significant (≈90 % confidence). Their analysis is based on a Bayesian linear regression model (Kelly 2007) which incorporates an additional parameter that allows for intrinsic noise; this putative and undefined source of noise is distinct from measurement error. Adopting precisely the procedures of Russell et al., but using the Steiner et al. (2013) data set, we confirm the correlation at ≈95 % confidence.[11] In contrast, a traditional analysis (i.e., a linear fit using two-dimensional error bars) produces a correlation at ≈99.9 % confidence. A comparably strong result is achieved using the Kelly (2007) model if one applies the usual Jeffreys noninformative prior to the intrinsic noise term rather than the default flat prior, which preferentially selects solutions corresponding to large estimates of intrinsic noise.

[11] The lower level of confidence reported by Russell et al. (2013) is largely attributable to their use of radio data for a flare of H1743–322 that is unrelated to the ballistic jet, namely a flare event that occurred 28 days before the X-ray flux peaked (McClintock et al. 2009). We focus solely on post-Eddington radio flares. The relevant radio flare event, which we consider, occurred 2.6 days after X-ray maximum (Steiner et al. 2013).

Of greater importance, Russell et al. (2013) consider only the empirical correlation shown in Fig. 6a, and they disregard the physical model which takes into account beaming effects, namely, the model shown in Fig. 6b. The application of this physical model is based on four simple assumptions: (i) spin-orbit alignment (Steiner and McClintock 2012); (ii) Γ is the same for all five sources and bracketed between 2 and 5 (Fender 2006); (iii) jet power is proportional to $\Omega_{\rm H}^2$ (Blandford and Znajek 1977; Tchekhovskoy et al. 2010); and (iv) radio luminosity can be used as a proxy for jet power (Sects. 7.2 and 7.4.3). Fitting the Steiner et al. (2013) data set to this model, with the model normalization as its *sole* fit parameter, one obtains good fits (shown in Fig. 6b) with $\chi^2/\nu = 0.3$ and 0.5 for $\Gamma = 2$ and 5, respectively. This relationship is determined over a span of ~3 orders of magnitude in jet power, and over the full allowed range of prograde spins. It is reasonable to view this result as evidence that the Blandford–Znajek model successfully describes the behavior of a ballistic jet produced by a black hole transient as it approaches its Eddington limit.

7.4.2 Issue of Data Selection

Russell et al. (2013) furthermore argue that Narayan and McClintock (2012) and Steiner et al. (2013) omit data for several systems that should be included in the correlation plots shown in Fig. 6. These additional data, which are plotted in Fig. 1c in Russell et al. (2013), destroy the clean correlation shown in Fig. 6a. However, it is inappropriate to include these data, and we reject them for the following reasons.

Cygnus X-1 radiates persistently at a few percent of Eddington and, during its periods of jet ejection (Fender et al. 2006), its mass accretion rate is both very low and poorly constrained. At the same time, as stressed in Sect. 7.1, comparing sources at the same mass accretion rate, namely $\dot{M} \sim \dot{M}_{\rm Edd}$, is the essential methodological requirement that eliminates the otherwise unknown dependency of the jet efficiency $\eta_{\rm jet}$ on $\dot{M}$ (see Eq. (1)). In order to include Cyg X-1 in the Steiner et al. (2013) sample, it would be necessary to know the precise scaling of jet power with $\dot{M}$ and then to estimate $\dot{M}$ at the time of jet ejection; $\dot{M}$ is ~1–2 orders of magnitude below Eddington and highly uncertain. Russell et al. (2013) ignore this problem.

For **GRS 1124–68** and **GS 2000+25**, Russell et al. (2013) adopt continuum-fitting spin data that are completely unreliable. These data are from a pioneering, proof-of-concept paper on continuum fitting (Zhang et al. 1997) whose authors note that the crucial "system parameters [i.e., D, i and M] are mostly unknown." More importantly, the values of spin adopted by Russell et al. for these two sources were simply inferred by Zhang et al. from crude estimates of the inner-disk radius $R_{\rm in}$ taken from a review paper (Tanaka and Lewin 1995); these estimates of $R_{\rm in}$ were, moreover, computed using the nonrelativistic disk model DISKBB that assumes a grossly incorrect inner-boundary condition (Zimmerman et al. 2005), while neglecting the effects of spectral hardening.

GRS J1655–40 had major outbursts in both 1994 and 2005. Russell et al. (2013) plot in their Fig. 1c a data point for the 2005 outburst which corresponds to a very faint radio flare. Narayan and McClintock (2012) and Steiner et al. (2013) do not include data for the 2005 outburst in their sample because the radio coverage was too sparse: The proximate observations that bracket the peak 4.3 Crab flare (see RXTE/ASM plot in Fig. 1 in Brocksopp et al. 2006) occur 2.3 days before and 4.7 days after the 2–12 keV maximum (see observation log in the NRAO VLA archive and web link to a plot of the radio data in Rupen et al. 2005). We argue that the source produced a bright and brief radio flare (e.g., like the 1-day radio spike observed for XTE J1859+226 shown in Fig. 2d), which was not observed because of the week-long gap in radio coverage. For the 2002 outburst of 4U 1543–47, we likewise hold that the radio coverage was inadequate and the radio flare was missed, as discussed in Narayan and McClintock (2012).

7.4.3 Synchrotron Bubble Model

Another point of contention is how to relate the radio luminosity νL_ν observed at the peak of the light curve—this is the quantity "Jet Power" in Fig. 6—to physical conditions in the jet. Steiner et al. (2013) used synchrotron theory with the following standard assumptions: (1) The nonthermal radio-emitting electrons in the jet blob have an energy distribution $N(\gamma) \sim \gamma^{-p}$ where γ is the electron Lorentz factor in the blob frame; (2) $p = 5/2$ to be consistent with the synchrotron spectrum; (3) the magnetic energy in the blob is in equipartition with the energy in the nonthermal electrons; (4) there is one proton for each nonthermal electron and the total energy of the blob is dominated by the kinetic energy E of the protons; and (5) at light curve maximum, the jet blob transitions from optically thick to thin (van der Laan 1966). Steiner et al. showed that $E \sim (\nu L_\nu)^{1.2}$, i.e., the blob energy varies approximately linearly with νL_ν. They thus argued that the latter is a good proxy for the former.

Why consider blob energy E? In the synchrotron bubble model (van der Laan 1966), the jet ejection is some brief episode that is not observationally resolved. Hence the total ejected energy is all that one can measure. Russell et al. (2013) (and references therein) focus on the jet kinetic luminosity, $L_{\rm jet} = dE/dt \sim E/t_{\rm jet}$, where $t_{\rm jet}$ is the duration of the jet ejection. They further assume that the jet is ejected continuously, with a constant luminosity, until light curve maximum.[12] How does $L_{\rm jet}$ depend on νL_ν? Steiner et al. (2013) find that the radius of the jet blob scales as $R \sim (\nu L_\nu)^{0.5}$. Therefore, if the blob expands with some constant speed, say c, then $L_{\rm jet} \sim E/(R/c) \sim (\nu L_\nu)^{0.7}$, i.e., jet kinetic luminosity varies as νL_ν to a power somewhat less than unity.[13] The truth is probably somewhere in between this result and that obtained for the blob energy E, i.e., very close to a linear dependence. Note that Russell et al. (2013) ignore altogether the fact that the jet blob transitions from optically thick to thin at light curve maximum, thereby missing a key piece of information. As a result, they do not have an analytic estimate of R and need to estimate $t_{\rm jet}$ from the poorly constrained rise time to maximum of the radio flux.

7.4.4 On Resolving the Controversy

In our view, the significant challenge posed by Russell et al. (2013) is whether GRO J1655–40 did or did not produce a strong radio flare during its 2005 outburst (Sect. 7.4.2); a similar challenge is posed for the 2002 outburst of 4U 1543–47 (Narayan and McClintock 2012). We maintain that both sources produced bright radio flares, but that they were missed because of the sparse radio coverage. On the other hand, Russell et al. (2013) and D. Russell (private communication) contend that the radio coverage was adequate to detect the strong flares during their decay phase, and they conclude that neither source produced a strong flare. This controversy cannot be firmly decided because the radio data collected for these events are inadequate.

Fortunately, we can expect the controversy to be settled relatively soon via radio observations of transient outburst events using new facilities such as the MeerKAT array (assuming

[12]The ejection apparently shuts off, coincidentally, as the jet becomes optically thin.

[13]Using somewhat different assumptions, (Willott et al. 1999) estimated that kinetic jet kinetic luminosities of radio galaxies should vary as $(\nu L_\nu)^{6/7}$, i.e., a slope again close to but less than unity.

a continuing capability to monitor the X-ray sky), and by the continual progress in obtaining secure measurements of the spins of transient black holes. MeerKAT, a forerunner of the SKA, is an array of 64 dishes scheduled for commissioning in 2014–2015 that will have outstanding sensitivity. R. Fender and P. Woudt, the PIs of the science project ThunderKAT, will obtain definitive measurements of all bright black-hole transients at high cadence (R. Fender, private communication).

8 Conclusions and Future Prospects

The continuum-fitting method has a number of virtues. A principal one is the simple and elegant model upon which it is based, namely the model of a thin, viscous accretion disk. This model was anticipated and developed well before the existence of black holes was established (e.g., Lynden-Bell 1969). Shortly thereafter, the analytic theory of thin disks was fully developed, an effort that culminated in the workhorse NT model (Novikov and Thorne 1973). The most important predictions of the model have been validated recently via GRMHD simulations (Sects. 4.2 and 5.2). This venerable model, with the addition of a model of the disk's atmosphere (Sects. 3.1 and 5.3), measurements of three key parameters (D, i and M), and suitable X-ray data, allows one to estimate the inner-disk radius. Meanwhile, an abundance of strong observational and theoretical evidence allows one to identify the inner-disk radius with the radius of the ISCO, which is simply related to the spin of the black hole.

Another key virtue of the continuum-fitting method is an abundance of data for which the thermal disk component is strongly dominant. For most stellar black holes, one has many suitable archival spectra for which only a few percent or less of the thermal seed photons are upscattered by a hot corona into a Compton tail component of emission. In short: The continuum-fitting model is tried and true, and there is an abundance of suitable data for many stellar black holes obtained for the simplest and best understood state of an accreting black hole, namely, an optically-thick thermal disk.

The spins and masses of ten stellar black holes are given in Table 1. Their spins span the full range of prograde values, and their masses range from 6–16 $M_{\odot}$. Setting aside the extreme spin of the transient GRS 1915+105, the three persistent black holes have higher spins and larger masses than their transient cousins. Furthermore, the high spins of these persistent, young black holes are unlikely to have been achieved via accretion torques, which implies that their spins are natal. The spins and masses of the ten black holes in Table 1 provide their complete description and are the essential data for testing astrophysical models of how an accreting black hole interacts with its environment. They are likewise essential data for underpinning the physical theory of black holes, and for ultimately attempting tests of the no-hair theorem by, e.g., observing deviations from the multipoles predicted by the Kerr metric, all of which are functions of just the two parameters a_* and M (Vigeland and Hughes 2010; Johannsen and Psaltis 2010; Johannsen and Psaltis 2011; Bambi 2013).

To date, the most important application of the data in Table 1 is to one class of jets, namely, ballistic jets that are produced in major, Eddington-limited outbursts of black holes in transient systems. For such outbursts the peak radio luminosities of five of these microquasars correlate strongly with their spins, increasing by a factor of $\sim$1000 as spin increases from $\sim$0 to $>$0.95. Meanwhile, a simple synchrotron jet model shows radio luminosity to be a good proxy for jet power. As Narayan et al. (2013) discuss in detail, the fitted relationship between jet power and spin (Fig. 6b) is not only a validation of the classic model of Blandford and Znajek (1977), it was also anticipated by GRMHD simulations showing that a jet can extract energy directly from a spinning black hole.

During the next several years, one can hope to double the number of black holes with spins measured via the continuum-fitting method. It will be equally important to improve the quality of each measurement, largely by obtaining more accurate measurements of the parameters D, i and M, but also by making methodological advances and by pursuing more advanced GRMHD and atmosphere models of thin disks. The payoff for this effort will be the widening applications of these spin data to problems in astrophysics and physics.

Especially important will be the possibility of validating both the continuum-fitting and Fe-line methods by comparing spin results obtained for individual stellar black holes. The validation of the Fe-line method is particularly important because it is the most direct approach to measuring the spins of AGN. Presently, concordant results are being obtained using these two leading methods (e.g., Steiner et al. 2011; Steiner et al. 2012b; Gou et al. 2011; Fabian et al. 2012; Reynolds 2013).

Two other methods for measuring the spins of stellar black holes appear promising, namely, via X-ray polarimetry (Dovčiak et al. 2008; Li et al. 2009; Schnittman and Krolik 2009) and high-frequency quasiperiodic oscillations (Török et al. 2005; Remillard and McClintock 2006; Belloni et al. 2012). The former method is stymied by a lack of data, and the latter by the lack of an adequate model. It is reasonable to hope that the mass and spin data in Table 1 will assist in identifying the appropriate physical model for the high-frequency QPOs. Because spin is such a critical parameter, it is important to attempt to measure it by as many methods as possible, as this will arguably provide our best check on the results. Stellar black holes are central to this effort because all of the methods of measuring spin mentioned above can be applied to them.

Acknowledgements The authors thank S. W. Davis for important input on Sect. 5.3. We also thank C. Brocksopp, E. Kuulkers, M. L. McCollough, C. Sánchez-Fernández and C. Zurita for help in preparing Fig. 2; J. García and T. Fragos for their comments on a version of the manuscript; R. Fender for discussions on MeerKAT; and an anonymous referee for several important criticisms. JEM was supported in part by NASA grant NNX11AD08G and RN by NASA grant NNX11AE16G. JFS was supported by NASA Hubble Fellowship grant HST-HF-51315.01.

References

Y. Abe, Y. Fukazawa, A. Kubota, D. Kasama, K. Makishima, Publ. Astron. Soc. Jpn. **57**, 629 (2005)
M.A. Abramowicz, P.C. Fragile, Living Rev. Relativ. **16**, 1 (2013)
N. Afshordi, B. Paczyński, Astrophys. J. **592**, 354 (2003)
K.A. Arnaud, in *Astronomical Society of the Pacific Conference Series*, ed. by G.H. Jacoby, J. Barnes. Astronomical Data Analysis Software and Systems V, vol. 101 (1996), p. 17
S.A. Balbus, J.F. Hawley, Rev. Mod. Phys. **70**, 1 (1998)
C. Bambi, Astron. Rev. **8**, 010000 (2013)
J.M. Bardeen, Nature **226**, 64 (1970)
J.M. Bardeen, J.A. Petterson, Astrophys. J. **195**, L65 (1975)
J.M. Bardeen, W.H. Press, S.A. Teukolsky, Astrophys. J. **178**, 347 (1972)
K. Beckwith, J.F. Hawley, J.H. Krolik, Astrophys. J. **678**, 1180 (2008)
M.C. Begelman, A.R. King, J.E. Pringle, Mon. Not. R. Astron. Soc. **370**, 399 (2006)
T.M. Belloni, A. Sanna, M. Méndez, Mon. Not. R. Astron. Soc. **426**, 1701 (2012)
R.D. Blandford, R.L. Znajek, Mon. Not. R. Astron. Soc. **179**, 433 (1977)
C.T. Bolton, Nature **235**, 271 (1972)
H.V.D. Bradt, J.E. McClintock, Annu. Rev. Astron. Astrophys. **21**, 13 (1983)
C. Brocksopp et al., Mon. Not. R. Astron. Soc. **331**, 765 (2002)
C. Brocksopp et al., Mon. Not. R. Astron. Soc. **365**, 1203 (2006)
M. Campanelli, C.O. Lousto, Y. Zlochower, Phys. Rev. D **74**, 041501 (2006)
A.G. Cantrell et al., Astrophys. J. **710**, 1127 (2010)
S. Chandrasekhar, *Shakespeare, Newton and Beethoven: or Patterns of Creativity* (University of Chicago Press, Chicago, 1975)

P.A. Charles, M.J. Coe, in *Optical, ultraviolet and infrared observations of X-ray binaries*, ed. by W.H.G. Lewin, M. van der Klis (2006), pp. 215–265
R. Cooke, J. Bland-Hawthorn, R. Sharp, Z. Kuncic, Astrophys. J. **687**, L29 (2008)
A.P. Cowley, D. Crampton, J.B. Hutchings, R. Remillard, J.E. Penfold, Astrophys. J. **272**, 118 (1983)
B. Czerny, K. Hryniewicz, M. Nikołajuk, A. Sądowski, Mon. Not. R. Astron. Soc. **415**, 2942 (2011)
T. Damour, R. Ruffini, R.S. Hanni, J.R. Wilson, Phys. Rev. D **17**, 1518 (1978)
S.W. Davis, I. Hubeny, Astrophys. J. Suppl. Ser. **164**, 530 (2006)
S.W. Davis, O.M. Blaes, I. Hubeny, N.J. Turner, Astrophys. J. **621**, 372 (2005)
S.W. Davis, C. Done, O.M. Blaes, Astrophys. J. **647**, 525 (2006)
S.W. Davis, O.M. Blaes, S. Hirose, J.H. Krolik, Astrophys. J. **703**, 569 (2009)
J. Dexter, P.C. Fragile, Astrophys. J. **730**, 36 (2011)
C. Done, S.W. Davis, Astrophys. J. **683**, 389 (2008)
M. Dovčiak, F. Muleri, R.W. Goosmann, V. Karas, G. Matt, Mon. Not. R. Astron. Soc. **391**, 32 (2008)
A.C. Fabian, M.J. Rees, L. Stella, N.E. White, Mon. Not. R. Astron. Soc. **238**, 729 (1989)
A.C. Fabian et al., Mon. Not. R. Astron. Soc. **424**, 217 (2012)
H. Falcke, E. Körding, S. Markoff, Astron. Astrophys. **414**, 895 (2004)
W.M. Farr, N. Sravan, A. Cantrell, L. Kreidberg, C.D. Bailyn, I. Mandel, V. Kalogera, Astrophys. J. **741**, 103 (2011)
R. Fender, in *Jets from X-Ray Binaries*, ed. by W.H.G. Lewin, M. van der Klis (2006), pp. 381–419
R. Fender, T. Belloni, Annu. Rev. Astron. Astrophys. **42**, 317 (2004)
R.P. Fender, S.T. Garrington, D.J. McKay, T.W.B. Muxlow, G.G. Pooley, R.E. Spencer, A.M. Stirling, E.B. Waltman, Mon. Not. R. Astron. Soc. **304**, 865 (1999)
R.P. Fender, T.M. Belloni, E. Gallo, Mon. Not. R. Astron. Soc. **355**, 1105 (2004)
R.P. Fender, A.M. Stirling, R.E. Spencer, I. Brown, G.G. Pooley, T.W.B. Muxlow, J.C.A. Miller-Jones, Mon. Not. R. Astron. Soc. **369**, 603 (2006)
R.P. Fender, E. Gallo, D. Russell, Mon. Not. R. Astron. Soc. **406**, 1425 (2010)
T. Fragos, M. Tremmel, E. Rantsiou, K. Belczynski, Astrophys. J. **719**, L79 (2010)
T. Fragos et al., Astrophys. J. **764**, 41 (2013)
J. Frank, A. King, D.J. Raine, Accretion Power in Astrophysics: Third Edition (2002)
E. Gallo, R.P. Fender, G.G. Pooley, Mon. Not. R. Astron. Soc. **344**, 60 (2003)
C.F. Gammie, Astrophys. J. **522**, L57 (1999)
A.M. Ghez et al., Astrophys. J. **689**, 1044 (2008)
M. Gierliński, C. Done, Mon. Not. R. Astron. Soc. **347**, 885 (2004)
M. Gierliński, A. Maciołek-Niedźwiecki, K. Ebisawa, Mon. Not. R. Astron. Soc. **325**, 1253 (2001)
S. Gillessen, F. Eisenhauer, S. Trippe, T. Alexander, R. Genzel, F. Martins, T. Ott, Astrophys. J. **692**, 1075 (2009)
L. Gou et al., Astrophys. J. **701**, 1076 (2009)
L. Gou, J.E. McClintock, J.F. Steiner, R. Narayan, A.G. Cantrell, C.D. Bailyn, J.A. Orosz, Astrophys. J. **718**, L122 (2010)
L. Gou et al., Astrophys. J. **742**, 85 (2011)
J. Greene, C.D. Bailyn, J.A. Orosz, Astrophys. J. **554**, 1290 (2001)
K. Gültekin et al., Astrophys. J. **695**, 1577 (2009)
S. Heinz, R.A. Sunyaev, Mon. Not. R. Astron. Soc. **343**, L59 (2003)
J.R. Herrnstein, J.M. Moran, L.J. Greenhill, A.S. Trotter, Astrophys. J. **629**, 719 (2005)
S. Hirose, J.H. Krolik, O. Blaes, Astrophys. J. **691**, 16 (2009)
R.M. Hjellming, K.J. Johnston, Astrophys. J. **328**, 600 (1988)
I. Hubeny, T. Lanz, Astrophys. J. **439**, 875 (1995)
T. Johannsen, D. Psaltis, Astrophys. J. **716**, 187 (2010)
T. Johannsen, D. Psaltis, Phys. Rev. D **83**, 124015 (2011)
E.J.D. Jolley, Z. Kuncic, G.V. Bicknell, S. Wagner, Mon. Not. R. Astron. Soc. **400**, 1521 (2009)
S. Kato, J. Fukue, S. Mineshige, Black-Hole Accretion Disks. Towards a New Paradigm (2008)
B.C. Kelly, Astrophys. J. **665**, 1489 (2007)
A.R. King, U. Kolb, Mon. Not. R. Astron. Soc. **305**, 654 (1999)
S. Koide, K. Shibata, T. Kudoh, D.L. Meier, Science **295**, 1688 (2002)
M. Kolehmainen, C. Done, Mon. Not. R. Astron. Soc. **406**, 2206 (2010)
J.H. Krolik, Astrophys. J. **515**, L73 (1999)
A. Kubota, K. Makishima, Astrophys. J. **601**, 428 (2004)
A. Kubota, K. Makishima, K. Ebisawa, Astrophys. J. **560**, L147 (2001)
A.K. Kulkarni et al., Mon. Not. R. Astron. Soc. **414**, 1183 (2011)
C.-H. Lee, G.E. Brown, R.A.M.J. Wijers, Astrophys. J. **575**, 996 (2002)
L. Li, E.R. Zimmerman, R. Narayan, J.E. McClintock, Astrophys. J. Suppl. Ser. **157**, 335 (2005)

L. Li, R. Narayan, J.E. McClintock, Astrophys. J. **691**, 847 (2009)
J. Liu, J.E. McClintock, R. Narayan, S.W. Davis, J.A. Orosz, Astrophys. J. **679**, L37 (2008)
D. Lynden-Bell, Nature **223**, 690 (1969)
A.I. MacFadyen, S.E. Woosley, Astrophys. J. **524**, 262 (1999)
K. Makishima, Y. Maejima, K. Mitsuda, H.V. Bradt, R.A. Remillard, I.R. Tuohy, R. Hoshi, M. Nakagawa, Astrophys. J. **308**, 635 (1986)
S. Markoff, M.A. Nowak, J. Wilms, Astrophys. J. **635**, 1203 (2005)
J.E. McClintock, R.A. Remillard, Astrophys. J. **308**, 110 (1986)
J.E. McClintock, R.A. Remillard, in *Black Hole Binaries*, ed. by W.H.G. Lewin, M. van der Klis (2006) pp. 157–213
J.E. McClintock, R. Shafee, R. Narayan, R.A. Remillard, S.W. Davis, L. Li, Astrophys. J. **652**, 518 (2006)
J.E. McClintock, R.A. Remillard, M.P. Rupen, M.A.P. Torres, D. Steeghs, A.M. Levine, J.A. Orosz, Astrophys. J. **698**, 1398 (2009)
J.C. McKinney, Astrophys. J. **630**, L5 (2005)
J.C. McKinney, R.D. Blandford, Mon. Not. R. Astron. Soc. **394**, L126 (2009)
J.C. McKinney, C.F. Gammie, Astrophys. J. **611**, 977 (2004)
A. Merloni, S. Heinz, T. di Matteo, Mon. Not. R. Astron. Soc. **345**, 1057 (2003)
M. Middleton, C. Done, M. Gierliński, S.W. Davis, Mon. Not. R. Astron. Soc. **373**, 1004 (2006)
I.F. Mirabel, I. Rodrigues, Science **300**, 1119 (2003)
I.F. Mirabel, L.F. Rodríguez, Nature **371**, 46 (1994)
I.F. Mirabel, L.F. Rodríguez, Annu. Rev. Astron. Astrophys. **37**, 409 (1999)
K. Mitsuda et al., Publ. Astron. Soc. Jpn. **36**, 741 (1984)
E. Moreno Méndez, G.E. Brown, C. Lee, I.H. Park, Astrophys. J. **689**, L9 (2008)
R. Narayan, J.E. McClintock, Astrophys. J. **623**, 1017 (2005)
R. Narayan, J.E. McClintock, New Astron. Rev. **51**, 733 (2008)
R. Narayan, J.E. McClintock, Mon. Not. R. Astron. Soc. **419**, L69 (2012)
R. Narayan, J.E. McClintock, A. Tchekhovskoy, in *Relativity and Gravitation: 100 Years After Einstein in Prague*, ed. by J. Bicak, T. Ledvinka (2013, to appear)
S.C. Noble, J.H. Krolik, Astrophys. J. **703**, 964 (2009)
S.C. Noble, J.H. Krolik, J.F. Hawley, Astrophys. J. **711**, 959 (2010)
S.C. Noble, J.H. Krolik, J.D. Schnittman, J.F. Hawley, Astrophys. J. **743**, 115 (2011)
I.D. Novikov, K.S. Thorne, in *Black Holes (Les Astres Occlus)*, ed. by A. Giannaras (1973), pp. 343–450
M.A. Nowak, J. Wilms, K. Pottschmidt, N. Schulz, D. Maitra, J. Miller, Astrophys. J. **744**, 107 (2012)
J.A. Orosz, in *IAU Symposium*, ed. by K. van der Hucht, A. Herrero, C. Esteban. A Massive Star Odyssey: from Main Sequence to Supernova, vol. 212, (2003), p. 365
J.A. Orosz et al., Nature **449**, 872 (2007)
J.A. Orosz et al., Astrophys. J. **697**, 573 (2009)
J.A. Orosz, J.E. McClintock, J.P. Aufdenberg, R.A. Remillard, M.J. Reid, R. Narayan, L. Gou, Astrophys. J. **742**, 84 (2011a)
J.A. Orosz, J.F. Steiner, J.E. McClintock, M.A.P. Torres, R.A. Remillard, C.D. Bailyn, J.M. Miller, Astrophys. J. **730**, 75 (2011b)
F. Özel, D. Psaltis, R. Narayan, J.E. McClintock, Astrophys. J. **725**, 1918 (2010)
B. Paczyński (2000). arXiv:astro-ph/0004129v1
D.N. Page, K.S. Thorne, Astrophys. J. **191**, 499 (1974)
R.F. Penna, J.C. McKinney, R. Narayan, A. Tchekhovskoy, R. Shafee, J.E. McClintock, Mon. Not. R. Astron. Soc. **408**, 752 (2010)
R.F. Penna, A. Sąowski, J.C. McKinney, Mon. Not. R. Astron. Soc. **420**, 684 (2012)
R. Penrose, in *Nuovo Cimento Rivista Serie* vol. 1 (1969), p. 252
M.J. Reid, J.E. McClintock, R. Narayan, L. Gou, R.A. Remillard, J.A. Orosz, Astrophys. J. **742**, 83 (2011)
R.A. Remillard, J.E. McClintock, Annu. Rev. Astron. Astrophys. **44**, 49 (2006)
R.A. Remillard, J.E. McClintock, C.D. Bailyn, Astrophys. J. **399**, L145 (1992)
C.S. Reynolds (2013). arXiv:1302.3260 [astro-ph.HE]
H. Riffert, H. Herold, Astrophys. J. **450**, 508 (1995)
R.R. Ross, A.C. Fabian, Mon. Not. R. Astron. Soc. **381**, 1697 (2007)
R. Ruffini, J.R. Wilson, Phys. Rev. D **12**, 2959 (1975)
M.P. Rupen, A.J. Mioduszewski, V. Dhawan ATel #434 (2005)
D.M. Russell, E. Gallo, R.P. Fender, Mon. Not. R. Astron. Soc. **431**, 405 (2013)
C. Sánchez-Fernández, A.J. Castro-Tirado, A. Giménez, C. Zurita, J. Casares, N. Lund, Astrophys. Space Sci. Suppl. **276**, 51 (2001)
J.D. Schnittman, J.H. Krolik, Astrophys. J. **701**, 1175 (2009)
R. Shafee, J.E. McClintock, R. Narayan, S.W. Davis, L. Li, R.A. Remillard, Astrophys. J. **636**, L113 (2006)

R. Shafee, J.C. McKinney, R. Narayan, A. Tchekhovskoy, C.F. Gammie, J.E. McClintock, Astrophys. J. **687**, L25 (2008a)
R. Shafee, R. Narayan, J.E. McClintock, Astrophys. J. **676**, 549 (2008b)
T. Shahbaz, F.A. Ringwald, J.C. Bunn, T. Naylor, P.A. Charles, J. Casares, Mon. Not. R. Astron. Soc. **271**, L10 (1994)
N.I. Shakura, R.A. Sunyaev, Astron. Astrophys. **24**, 337 (1973)
A.P. Smale, P.T. Boyd, Astrophys. J. **756**, 146 (2012)
D. Steeghs, J.E. McClintock, S.G. Parsons, M.J. Reid, S. Littlefair, V.S. Dhillon, Astrophys. J. **768**, 185 (2013)
J.F. Steiner, J.E. McClintock, Astrophys. J. **745**, 136 (2012)
J.F. Steiner, J.E. McClintock, R.A. Remillard, R. Narayan, L. Gou, Astrophys. J. **701**, L83 (2009a)
J.F. Steiner, R. Narayan, J.E. McClintock, K. Ebisawa, Publ. Astron. Soc. Pac. **121**, 1279 (2009b)
J.F. Steiner, J.E. McClintock, R.A. Remillard, L. Gou, S. Yamada, R. Narayan, Astrophys. J. **718**, L117 (2010)
J.F. Steiner et al., Mon. Not. R. Astron. Soc. **416**, 941 (2011)
J.F. Steiner et al., Mon. Not. R. Astron. Soc. **427**, 2552 (2012b)
J.F. Steiner, J.E. McClintock, M.J. Reid, Astrophys. J. **745**, L7 (2012a)
J.F. Steiner, J.E. McClintock, R. Narayan, Astrophys. J. **762**, 104 (2013)
O. Straub et al., Astron. Astrophys. **533**, A67 (2011)
Y. Tanaka, W.H.G. Lewin, in *X-Ray Binaries*, ed. by W.H.G. Lewin, J. van Paradijs, E.P.J. van den Heuvel, (1995), pp. 126–174
Y. Tanaka et al., Nature **375**, 659 (1995)
A. Tchekhovskoy, R. Narayan, J.C. McKinney, Astrophys. J. **711**, 50 (2010)
A. Tchekhovskoy, R. Narayan, J.C. McKinney, Mon. Not. R. Astron. Soc. **418**, L79 (2011)
G. Török, M.A. Abramowicz, W. Kluźniak, Z. Stuchlík, Astron. Astrophys. **436**, 1 (2005)
H. van der Laan, Nature **211**, 1131 (1966)
S.J. Vigeland, S.A. Hughes, Phys. Rev. D **81**, 024030 (2010)
X.Y. Wang, Z.G. Dai, T. Lu, Astrophys. J. **592**, 347 (2003)
B.L. Webster, P. Murdin, Nature **235**, 37 (1972)
N.E. White, P. Ghosh, Astrophys. J. **504**, L31 (1998)
N.E. White, F. Nagase, A.N. Parmar, in *X-Ray Binaries*, ed. by W.H.G. Lewin, J. van Paradijs, E.P.J. van den Heuvel, (1995), p. 1
C.J. Willott, S. Rawlings, K.M. Blundell, M. Lacy, Mon. Not. R. Astron. Soc. **309**, 1017 (1999)
J. Wilms, A. Allen, R. McCray, Astrophys. J. **542**, 914 (2000)
T.-W. Wong, F. Valsecchi, T. Fragos, V. Kalogera, Astrophys. J. **747**, 111 (2012)
S.E. Woosley, Astrophys. J. **405**, 273 (1993)
S.E. Woosley, A. Heger, Astrophys. J. **637**, 914 (2006)
A.A. Zdziarski, J. Mikołajewska, K. Belczyński, Mon. Not. R. Astron. Soc. **429**, L104 (2013)
S.N. Zhang, W. Cui, W. Chen, Astrophys. J. **482**, L155 (1997)
Y. Zhu, S.W. Davis, R. Narayan, A.K. Kulkarni, R.F. Penna, J.E. McClintock, Mon. Not. R. Astron. Soc. **424**, 2504 (2012)
E.R. Zimmerman, R. Narayan, J.E. McClintock, J.M. Miller, Astrophys. J. **618**, 832 (2005)

DOI 10.1007/978-1-4939-2227-7_16
Reprinted from *Space Science Reviews* Journal, DOI 10.1007/s11214-014-0069-z

An Overview of Jets and Outflows in Stellar Mass Black Holes

Rob Fender · Elena Gallo

Received: 19 March 2014 / Accepted: 26 June 2014 / Published online: 7 August 2014

Abstract In this article, we will briefly review the current empirical understanding of the relation between accretion state and outflows in accreting stellar mass black holes. The focus will be on the empirical connections between X-ray states and relativistic ('radio') jets, although we are now also able to draw accretion disc winds into the picture in a systematic way. We will furthermore consider the latest attempts to measure/order jet power, and to compare it to other (potentially) measurable quantities, most importantly black hole spin.

Keywords Black hole physics · X-ray binaries · Jets

1 Introduction

Jets, collimated relativistic outflows carrying large amounts of energy away from the deepest parts of the gravitational potential well, are a relative latecomer to the overall picture of accretion around stellar mass black holes. Much of the basics of accretion theory, namely how the matter gets in while angular momentum and radiation get out, was developed in between the 1960s and 1980s, with roots in much earlier works, and is still applicable today (and much of it will have been covered in other articles of this issue). In contrast, it took until the 2000s for it be accepted that jets are a key part of the accretion process in stellar mass accretors, both in terms of their uniquity and importance in having very large kinetic powers. Furthermore, despite knowledge of their existence in active galactic nuclei for a century, the basics of how relativistic jets are launched are still a long way behind our understanding of the accretion process.

Nevertheless, once we discovered that jets in X-ray binaries follow certain patterns, revealed mainly (but not exclusively) by coordinated observing campaigns in the radio and

R. Fender (✉)
Astrophysics, Department of Physics, University of Oxford, Keble Road, OX1 3RH, Oxford, UK
e-mail: rob.fender@astro.ox.ac.uk

E. Gallo
Department of Astronomy, University of Michigan, 500 Church St., Ann Arbor, MI 48109, USA

X-ray bands, we realised that they offered perhaps our best opportunity to understand the connection between accretion and jet formation in relativistic objects.

In this paper we will describe the current state of play in our empirical understanding of the relation between accretion, jets and winds in stellar mass black holes (and neutron stars). We will focus on thoroughly summarising the observational evidence, and highlighting recent progress.

For a lengthier introduction to the basic observables of jets and what can be understood from them, the reader is directed towards Fender (2006) and references therein. For a very quick summary of the state of the field, see Fender and Belloni (2012). Other relevant historical reviews include Hjellming and Han (1995), Mirabel and Rodríguez (1999), Martí (2005).

2 The Empirical Picture

In this section we shall summarise the empirical evidence for a connection between accretion and outflow in X-ray binaries. We shall start with the black hole X-ray binaries, for which the evidence is strongest. Figure 1 summarises both the current state of our knowledge of the connection between accretion and outflows in black hole X-ray binaries, and how we most commonly choose to represent it. This representation of the relation between jets and X-ray state originated with Fender et al. (2004), and has since been greatly augmented by the work of Ponti et al. (2012) who found a clear relation between accretion disc winds and the soft X-ray state.

We are very fortunate that, on timescales fairly well tuned to PhD durations or grant cycles, we can observe a single stellar mass black hole in a binary system undergo accretion in a small number of different modes over a large range of luminosities. The cycle illustrated in Fig. 1 takes place over about one year, as is illustrated in Fig. 2. Noteworthy points to take from this figure are that the quasi-steady jet is present most of the time, although mainly at low luminosities, that the transient/flaring jet phase constitutes a very small fraction of the outburst, and that for most of the time at the highest luminosities there appears to be no core jet but a strong disc wind.

The following subsections refer to the (quasi-cyclic) phases labelled A through F in Figs. 1 and 2.

2.1 A to B (and F to A): Quiescence to Bright Hard States

Outbursts of black hole X-ray binaries (BHXRBs) seem to be caused by a switch in the viscosity of the accretion disc, associated with the ionisation of hydrogen. The most recent test of this theory, originally developed for white dwarf accretion in cataclysmic variables (Meyer and Meyer-Hofmeister 1981), continues to support the scenario (Coriat et al. 2012). Once the viscosity increases, the accretion rate at the inner edge of the accretion disc rises dramatically. Since this is where most of the accretion luminosity is produced, the source gets brighter on a short timescale (days).

In our experience to date, most BHXRBs spend most of their time outside of outbursts at 'quiescent' levels of accretion with corresponding luminosities of $\sim 10^{30-32}$ erg s^{-1}. This population is not very well sampled, however, and some sources are stable at higher luminosities (e.g. V404 Cyg which is 'quiescent' at $\sim 10^{34}$ erg s^{-1}) and selection effects suggests that there's likely to be a tail to lower luminosities. Once the outburst begins, however, the sources can rise to luminosities $\geq 10^{38}$ erg s^{-1} very rapidly, an extremely

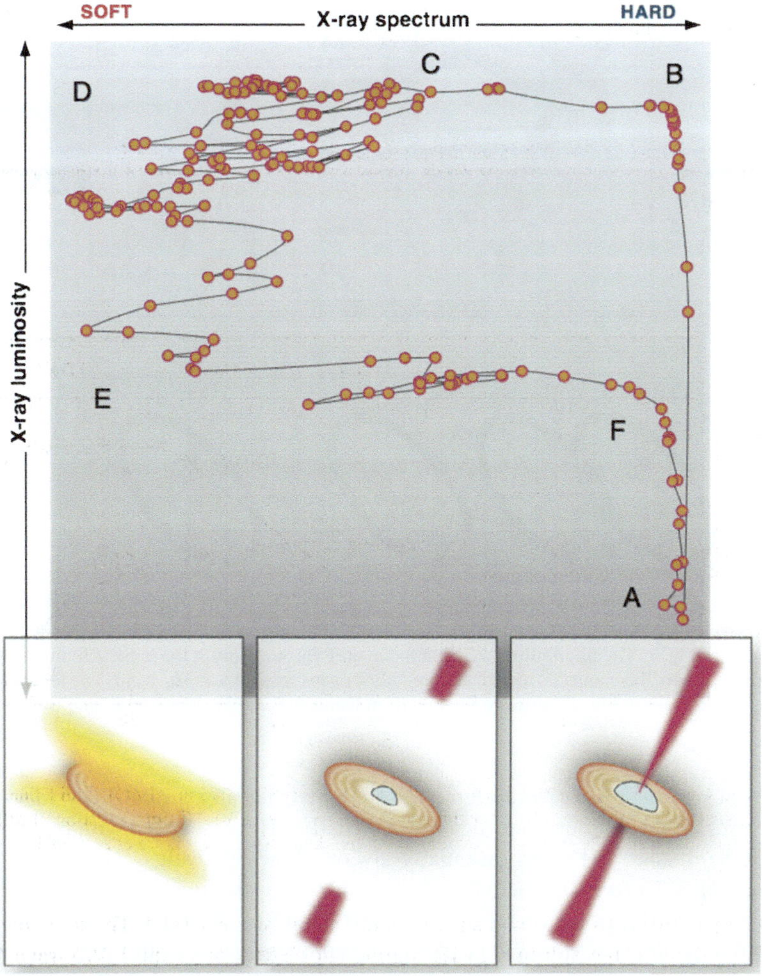

Fig. 1 A qualitative summary of the relation between accretion states and outflows in black hole X-ray binaries. The *upper panel* represents the relation between X-ray spectral hardness and luminosity, and the path A-B-C-D-E-F represents the path taken by the recurrently-outbursting binary GX 339-4 over the course of about one year. Hard states (*to the right*) are well-studied and show quasi-steady radio jets. Soft states (*to the left*) appear to have no radio jets but strong accretion disc winds. Transitions between the states, at least at high luminosity (e.g. B-C-D), appear to be associated with discrete jet ejection events. From Fender and Belloni (2012), where a concise summary of the cycle can be found

dramatic change. For most of this range in luminosity the X-ray spectrum remains in a 'hard' spectral state (however, it is likely that most of the action is going on in the ultraviolet, which is hard/impossible to observe in for most X-ray binaries, due to large dust extinction in the galactic plane). This state is characterised by a broad-band X-ray spectrum with photon index $\Gamma \sim 1.6$–2.1 (with the spectrum hardening as the luminosity increases; e.g. Plotkin et al. 2013) often observed to show a cutoff between ~50–130 keV (with the temperature decreasing the luminosity increases; e.g., Motta et al. 2009; Joinet et al. 2008), and generally ascribed to thermal Comptonisation in a hot plasma or

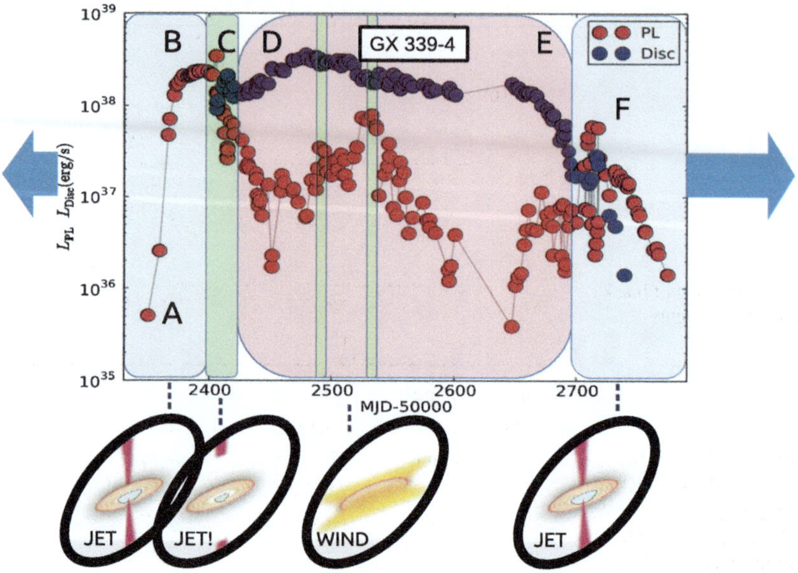

Fig. 2 An outburst of GX 339-4, this time shown as temporal evolution. The *red* and *blue curves* correspond to power-law and disc components of the X-ray spectrum: blue-dominated states are 'soft' and red are 'hard', in the context of Fig. 1. The figure illustrates that the majority of the time at the highest luminosities is spent in the soft state where there appears to be non core radio jet, but instead a strong wind. The letters correspond approximately to those in Fig. 1. Partially adapted from Dunn et al. (2010) and Fender and Belloni (2012)

'corona' above/around/near the accretion disc. There is usually strong X-ray variability (up to 40 % r.m.s. in the Fourier frequency range 0.01–100 Hz) associated with this state.

In this hard state, BHXRBs essentially always show relatively weak but steady radio emission, with a flat ($\alpha \sim 0$, where flux density $S_\nu \propto \nu^\alpha$) spectrum and low levels of polarisation (Fender 2001; but see Brocksopp et al. 2013 and Russell and Shahbaz 2014). The relative steadiness, flat spectrum and lack of polarisation suggest that this emission originates in a more or less continuously replenished, partially self-absorbed outflow, such as those originally conceived of to explain the flat spectrum cores of some quasars. The radio-X-ray correlation has been studied extensively, and is discussed in more detail in the following subsection.

2.1.1 Luminosity Correlations

Quasi-simultaneous radio and X-ray monitoring has become the standard tool of investigation for the so called "jet-accretion coupling" in hard state BHXRBs. A strong and repeating correlation has been established between the radio and X-ray luminosity for two systems, with L_X being proportional to $L_r^{0.6-0.7}$ (GX339−4: Corbel et al. 2003, 2013 and V404 Cyg: Gallo et al. 2003; 2005b; Corbel et al. 2008). However, the universality of this relation 10 years after its discovery is far from obvious. Figure 3 summarizes the current state of the problem by assembling what is likely the most complete data collection as of today (data from Gallo et al. 2012 plus Corbel et al. 2013, and references therein). Two main differences stand out with respect to the simple picture we used to draw a decade ago: firstly, there appears to be *two luminosity tracks* (see Gallo et al.

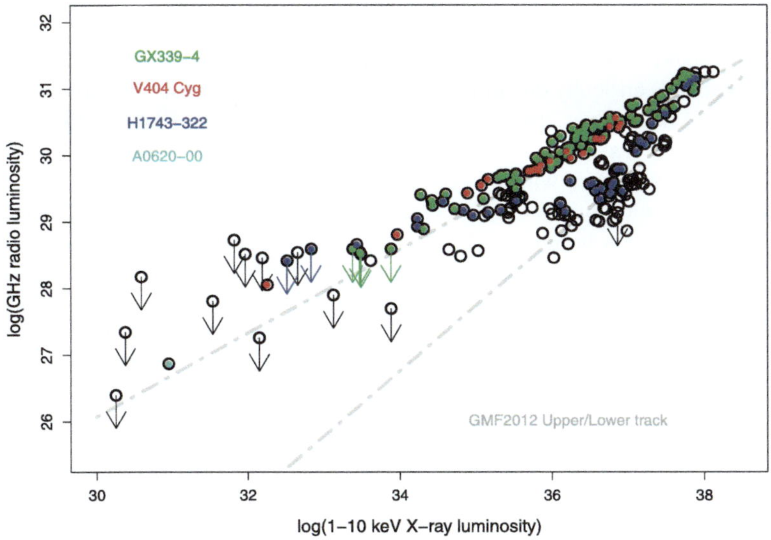

Fig. 3 The radio/X-ray luminosity plane of black hole X-ray binaries (c.a. 2014). Shown are quasi-simultaneous radio and X-ray luminosity measurements (in log CGS units) for 24 black hole X-ray binaries, ranging from quiescence (A0620−00; Gallo et al. 2006) up to bright hard states. The *dashed grey lines* indicate the best-fit relations to the *upper* and *lower tracks* as identified by Gallo et al. (2012) ($L_r \propto L_X^{0.63}$ and $L_r \propto L_X^{0.98}$, respectively). We also highlight the two sources for which a tight non-linear correlation of the form $L_X \propto L_r^{0.6-0.7}$ has been reported over a wide dynamic range (i.e. GX339–4; Corbel et al. 2003, and V404 Cyg; Gallo et al. 2003; Corbel et al. 2008), plus the enigmatic source H1743–322, which appeared to "jump" from the lower to the upper track as it faded into quiescence following its 2008 outburst (Jonker et al. 2010; Miller-Jones et al. 2012)

2012 for more details); secondly, the behavior of the BHXRB H1743–22 (in blue) as observed during the decline of its 2008 outburst (Jonker et al. 2010; Miller-Jones et al. 2012; Miller et al. 2012) is in stark contrast with that of GX339–4 and V404 Cyg (green and red, respectively): H1743–22 starts off as under-luminous in the radio band during the initial outburst decay phase ($10^{36} \lesssim L_X \lesssim 10^{38}$ erg s^{-1}), proceeds to make a nearly horizontal excursion toward lower X-ray luminosities (between $10^{36} \lesssim L_X < \lesssim 10^{36}$ erg s^{-1}), and finally reaches a comparable radio luminosity level (for the same L_X) as GX339–4 and V404 Cyg (this happens below $\simeq 10^{34}$ erg s^{-1}). Whatever drives the relative radio quietness/loudness of BHXRBs in hard and quiescent states, thus, cannot be linked to any physical property that remains constant over typical outburst time-scales (e.g., black hole spin, orbital inclination; see Soleri and Fender 2011).

Coriat et al. (2011) ascribe the erratic behavior of H1743−22, and by analogy that of other radio-quiet systems, to the onset of a radiatively efficient inflow (see also Körding et al. 2006a). Along the same lines, a more theoretical basis for the existence of a radio-quiet (or rather, X-ray-bright) track has been recently proposed by Meyer-Hofmeister and Meyer (2014), who argue that thermal photons from a weak, cool, inner disc (sustained by re-condensation of optically thin gas within the inner regions; Meyer et al. 2007; Liu et al. 2007) could be responsible for enhancing the seed photon field available for Comptonisation, and hence the hard X-ray flux. From a theoretical standpoint (as well as observational, see e.g. Miller et al. 2006; Reis et al. 2010; Reynolds and Miller 2013 and references therein), the inner cold disc would cease to exist at low accretion rate. Thus, *the radio-quiet track is not expected to extend down to the quiescent regime.* Though this exact behavior is just what was observed in the case H17143−22, whether the inner disc re-condensation scenario will stand the test of time (in terms of reproducing the X-ray and radio behavior of BHXRBs) depends critically on our ability to probe the truly quiescent domain (see Calvelo et al. 2010 and Miller-Jones et al. 2011 for a detailed discussion on how distance limitations are likely to hamper a systematic investigation in the radio band). To date, A0620−00 is the only truly quiescent system with a robust radio detection, and its radio luminosity seems to lie on the extrapolation of the GX339−4 and V404 Cyg radio/X-ray correlations down to Eddington-scaled X-ray luminosities as low as $\sim 10^{-9}$ (Gallo et al. 2006). However, with a sample of one, and considering selection effects, this is a long way from conclusive.

At the same time, a systematic study on the radio properties (most notably spectral index and polarization degree) of BHXRBs while on the radio quiet/X-ray bright track could provide complementary information on the physics that drives this apparent divide (there are hints that the "radio quiet" objects have slightly more optically thin radio spectra, Brocksopp et al. 2013). A good candidate for the origin of the flat spectrum, left unresolved in Blandford and Konigl (1979), is internal shocks driven at the frequencies of the X-ray variability (Malzac 2013).

Determining the compact jet contribution at mm, infrared (IR) and optical frequencies is admittedly challenging; much progress has been made in this direction over the last 10 years. A growing number of works (based both on spectral energy distribution analysis as well as variability studies) support the claim that, while in the hard state, the compact jet extends all the way up to IR and often optical frequencies (Fender 2001; Chaty et al. 2003; Bradley et al. 2007; Gallo et al. 2007; Russell et al. 2010, 2011b; Brocksopp et al. 2010; Malzac et al. 2004; Hynes et al. 2004, 2006, 2009; Markoff and Nowak 2007; Casella et al. 2010). Most relevant to this section, Russell et al. (2006) collected nearly-simultaneous IR/optical and X-ray observations of 33 X-ray binaries (black holes and neutron stars) to estimate the relative contributions of various IR/optical emission processes as a function of X-ray luminosity. They found evidence for a positive correlation between the IR/optical and X-ray luminosity, of the form $L_{\mathrm{opt-IR}} \propto L_{\mathrm{r}}^{0.6}$, extending all the way from the peak of the hard state down to the loses quiescent X-ray luminosities. Notwithstanding the large scatter, no strong outliers, exceptions, sub-clusters or bifurcation have been reported so far for this relation.

2.2 B-C-D: Transition from Hard to Soft

Once the rising source passes an X-ray luminosity of $\sim 10^{37}$ erg s^{-1} (albeit with substantial variations between sources and even between different outbursts of the same source; see, e.g., Dunn et al. 2010), at which level it will still be in the hard state, it is very likely to make a subsequent transition to a softer spectral state. The transition from the hard state to the soft state can occur with very little change (of order unity) in the broadband luminosity of the system (in spite of rather dramatic changes in the spectral shape), and has a number of very interesting characteristics. For a 'typical' BHXRB the spectral transition from hard to soft states can take days or weeks, but the evolution can be much more dramatic in terms of the X-ray power spectra. While the hard and soft states appear to be dominated by strong ($\geq$30 %) and very weak ($\leq$5 %) broad band noise respectively, some phases during the transition can be completely dominated by strong peaked and QPO-like components. Casella et al. (2005) classified these QPOs into three types: A, B and C, with strong type-Bs being the most indicative of the transition.

The hard to soft transition is also where the brightest radio flares occur, sometimes singly and sometimes in sequences. These radio flares can peak, in the 1–5 GHz band, at many times the preceding hard state flux density, and (mostly) evolve in a 'standard' (e.g. van der Laan 1966) way from optically thick to thin as they (presumably) expand. Such flares are occasionally spatially resolved by radio telescopes into individual radio knots which move away from the radio core (presumably the site of the black hole) at (at least) mildly relativistic speeds (e.g. Mirabel and Rodríguez 1994; Hjellming and Rupen 1995; Fender et al. 1999; Miller-Jones et al. 2012). It appears, although it is not at all proven, that these jets may move faster than those in the hard state, and so what we may be seeing is at least partially associated with internal shocks in a flow of variable (increasing) speed (Fender et al. 2004 and references therein; Fender et al. 2009). However, it is certainly not clear that they are significantly more *powerful* than the preceding, bright, hard state jets.

There have been a number of attempts to associate the 'moment of jet launch' with an associated change or event in the accretion flow. Fender, Homan and Belloni (2009) and Miller-Jones et al. (2012) both attempted to see if the moment of jet launch could be associated with changes in the timing (X-ray variability) properties of a small number of black hole binaries, in particular with the occurrence of the type-B QPOs (see also discussion in Soleri et al. 2008). While they are certainly broadly coincident, a one-to-one relation could not be firmly established and whether or not there exists a key signature of the moment of launch remains unclear. Seemingly at odds with this statement, in the bright black hole binary GRS 1915+105, which undergoes long cycles of accretion state changes on timescales as short as minutes, it is clear that each cycle is associated with an ejection event, and possibly even with a particular moment in that cycle (e.g. Mirabel et al. 1998; Klein-Wolt et al. 2001). However, these state changes appear to be compressed versions of what happens on longer timescales in other systems, and the blurring effect caused by the size scale of the radio emission makes direct association with some change difficult. Success is more likely in the infrared band, where much effort was made for GRS 1915+105 (e.g. Fender et al. 1997; Eikenberry et al. 1998; Mirabel et al. 1998) but it is extremely hard to maintain minutes-resolution monitoring of a more normal black hole binary across its entire week-long transition phase. Finally, it should be noted that there may be no clear 'moment of jet launch' at all: internal shock models are showing promise in reproducing some of the radio properties of X-ray binaries (e.g. Jamil et al. 2010; Malzac 2013) and it may be that a rapid, but not instantaneous, change in jet properties (such as injection cycle time or speed) are enough to reproduce what we see as bright flares in the radio band.

2.3 D to E: The Soft State

In the soft state the X-ray spectrum is dominated by a blackbody-like component which peaks around 1 keV, combined with a steeper and weaker power law. The origin of the thermal component is likely to be an optically thick accretion disc. The overall X-ray variability drops dramatically in this state (often below 5 %). The core radio, mm and near-infrared emission drop dramatically in the soft state (Fender et al. 1999; Russell et al. 2011a), the simplest interpretation of which is that the core jet has switched off.

A major breakthrough in understanding states appears to have arrived recently with Ponti et al. (2012) demonstrating that accretion disc winds, revealed in X-ray spectra, appear to be uniquely observed in *edge-on, soft-state* BHXRBs (caveat relatively poor sampling of late-time soft states, and some remaining small uncertainty about how evolving ionisation could affect this conclusion). Thus it seems that in moving from the hard to the soft state, we leave a regime of strong, quasi-steady jets and little, if any accretion disc wind and enter the converse regime, with strong winds and weak jets (see also Miller et al. 2006, 2008; Neilsen and Lee 2009). These two regimes are probably not simply a rebalancing of the same outflow power, however, with the wind probably carrying less kinetic power but more mass than the hard state jet (detailed calculations of kinetic energy, mass and momentum flux in these two modes have yet to be carried out, however).

2.4 X-Ray States: Other Views and Potential Origins

The X-ray states described above have been done so from a viewpoint which is very much biased towards the outflows. Other very useful reviews, usually more focussed on the X-ray emission and what it tells us about the workings of the inner accretion flow, can be found in Remillard and McClintock (2006), Done et al. (2007), Belloni et al. (2011). The general pattern of hysteresis can also be viewed in other ways than the *hardness-intensity* diagram used in Fig. 1. In particular Munoz-Darias et al. (2011) and Plant et al. (2014) utilised *rms-intensity* and *reflection-intensity* diagrams respectively to show broadly the same patterns but shedding new light on the detailed evolution of the accretion flow (see also Belloni and Stella, this issue).

It should be noted that while the root cause of the outburst may be agreed upon as being the hydrogen ionisation instability, the origin of the accretion state cycles and associated spectral hysteresis is not at all clear. There are published models which may be testable in the future (e.g. Petrucci et al. 2008; Begelman and Armitage 2014 and references therein), and very similar cycles of behaviour appear to be present in systems containing both neutron stars and white dwarfs (Körding et al. 2008b; Munoz-Darias et al. 2014).

There are good reasons to believe that the patterns of behaviour observed in stellar-mass black holes should be scalable to intermediate mass black holes and supermassive black holes in active galactic nuclei, in fact there is strong observational support for this view. Merloni et al. (2003) and Falcke et al. (2004) presented the first evidence for relatively neat scalings of X-ray/radio luminosities and black hole mass, across the whole black hole mass range. McHardy et al. (2006) and Körding et al. (2007) made similar analyses for timing frequencies. Körding et al. (2006b) made an argument that the overall patterns of coupling between states and jets might be the same in AGN, although this connection has really yet to be established convincingly (see also Körding, this issue).

3 Jet Power

Radiative and mechanical feedback from relativistic jets is thought to play a key role in regulating the growth of galactic bulges and their nuclear super-massive black holes. Empirical scaling relations between the mass of super-massive black holes in nearby galaxies and large scale properties of their host bulges, such as the well know "black-hole-mass/stellar velocity dispersion" relation (see Kormendy and Ho 2013 for a recent review) are interpreted as strong observational evidence for strong co-evolution. At the same time, state-of-the-art cosmological simulations (e.g. Khandai et al. 2014) that trace the assembly and merger history of galaxies and their nuclear black holes rely on some form of black hole- (plus supernova) -driven feedback in order to reproduce them. Nevertheless, even the most sophisticated models make use of over-simplified recipes for the relative efficiencies of accretion vs. jet power, and *posit* that the jet power, at least during the quasar phases, is set by the spin of the black hole. As we discuss in the following sections, BHXBRs can actually help shed light on some of these processes.

3.1 Jet Power and Black Hole Spin

The holy grail for relativistic jet astrophysics is to establish the jet formation mechanism. More specifically, we would like to be able to identify observational signatures pointing unequivocally to models that rely solely on differential rotation (Blandford and Payne 1982) vs. models that tap directly into the rotational energy of the black hole (Blandford and Znajek 1977; BZ). While super-massive black holes in AGN offer the advantage of vast demographics, BHXRBs in soft states provide us with a more favorable environment to systematically employ X-ray spectral fitting techniques to measure the temperature, and thus the extent, of the inner accretion disc (see Miller 2007 and Reynolds 2013 for recent reviews, as well as Reynolds, and McClintock, Narayan and Steiner, this issue). This translates into an estimate of the black hole spin parameter a, which can then be compared against the jet power P_j in order to (dis)prove a relation of the form $P_j \propto a^2$, predicted by the BZ model (see Tchekhovskoy et al. 2010 for a full relativistic treatment).

How to measure P_j? When it comes to BHXRBs, persistent, hard state, compact jets are a natural place to start. Unfortunately, that the face-value radio luminosity of compact jets is at best a poor indicator of jet power is apparent from a number of considerations; firstly, although compact jets seem to be persistently on during hard states, their flux density varies by orders of magnitude, with some non linear power of L_X (see Sect. 2.1.1). In addition, it should be kept in mind that any inference of radio luminosity that is based on a single-frequency flux density measurement relies on assuming a specific spectral shape at lower frequencies (often taken as flat). Further, extrapolating the integrated radio luminosity to total (i.e. radiative plus kinetic) jet power relies on even more crucial assumptions: (i) on the location of the optically-thin-to thick jet break (which is known to vary with overall luminosity, yet not necessarily as basic scaling relations would predict; see next section); (ii) on the location of the cooling break (possibly constrained in one source, see again next section) (iii) on the radiative efficiency of the synchrotron process.

Based on the above arguments, in attempt to test the presence of a correlation between jet power and spin parameter in hard state BHXRBs, Fender et al. (2010) adopted a phenomenological approach whereby the relative normalization of the (thought to be universal) radio/X-ray and/or infrared/X-ray correlation were taken as a proxy for total jet power. Perhaps not surprisingly, no evidence for a positive correlation between these normalisations and reported spin measurements emerged—a result that could be interpreted as due the large uncertainties in measuring jet power (and, to a second extent, spin). A different approach was taken by King et al. (2013a), who found a marginally significant positive correlation between the mass-scaled radio luminosity and spin parameter across a sample of 11 BHRBs and 37 Seyfert galaxies.

Ideally, one would like to calibrate the radio luminosity-to-total jet power relation conversion via robust observational constraints on the amount of work exerted by the jets on the surrounding inter-stellar medium, and use jet-inflated radio lobes and cavities as effective jet calorimeters, similarly to what has been done for super-massive black holes in radio galaxies for decades. Unfortunately, the number of jet-powered large-scale structures is very slim for BHXRBs, mainly as a result of the jets propagating though under-dense environments compared to AGN (see Heinz 2002). Albeit the small number statistics, these large scale structures point towards the jets carrying away a substantial fraction of the overall accretion energy budget (Heinz and Grimm 2005; Gallo et al. 2005a, 2005b; Fender et al. 2005; Körding et al. 2008a, 2008b; Heinz et al. 2007, King et al. 2013b).

More recently, much excitement has grown around the claim of a positive correlation—consistent with the BZ-predicted scaling—between spin parameter and jet power *for tran-*

sient jets (Narayan and McClintock 2012; Steiner et al. 2013). In these works, the mass-scaled peak radio luminosity of the giant radio flare that is often observed during hard-to-soft state transitions is taken as proxy for jet power (while the spin measurements rely on fitting the broadband X-ray spectrum in pure thermal states). In a subsequent work, Russell et al. (2013a) argued against there being a significant correlation. While the controversy is still open, much of the uncertainty depends on the relative scarcity of observations of Eddington-limited BHXRBs, which are supposed to act as "standard candles" (cf. Steiner et al. 2013). The debate will likely settle over the next few years, as more and better data become available; in the meantime, the number of possible exceptions to the $P_{\rm j} \propto a^2$ scaling relation for transient jets may be growing; the extragalactic BHXRB in M31 (Middleton et al. 2014), and the Galactic system 4U 1630–472 (King et al. 2014) seem to display too high/low (respectively) flare radio luminosity for their reported spin values (unless exceptionally high values of jet boosting/de-boosting factor are at play). For the "microquasar" in M31 (Middleton et al. 2013), however, the significant uncertainty on the system inclination also allows for a reconciliation with the best-fit scaling relation reported by Steiner et al. (2013). For 4U 1630–472, the high spin parameter reported by King et al. (2014) relies on fitting the X-ray reflection spectrum rather than thermal continuum spectrum. While the number of discrepancies between the two spin-measuring methods is reported to be diminishing (see discussions in Miller et al. 2009; Fabian et al. 2012), the consensus is that any convincing relation between jet power and spin parameter ought to be verified by both (or, at least, that either spin-fitting method leading to the existence of a positive relation with jet power does not constitute a legitimate argument for invalidating the other).

An entirely new approach to measuring black hole spin has been recently devised by Motta et al. (2014a, 2014b); rather than fitting X-ray energy spectra, this technique relies on the relativistic precession model quasi-period oscillations (QPOs) in the power density spectra of BHXRBs (Stella and Vietri 1999). This method has been successfully applied to the BHXRB GRO J1655-40 (Motta et al. 2014a) and XTE J1550-564 (Motta et al. 2014b). In the former case, the recovered black hole mass value matches the known system's mass function, while the spin value is not consistent with X-ray spectroscopy. In the latter, the spin value estimated from the relativistic precession model is consistent with X-ray spectroscopy (while the known black hole mass was used to solve for spin). Going forward, this could prove an exceptionally powerful method to infer black hole spin and mass values, in that it is not affected by many of the complexities of radiative models.

3.2 Jet Power and Spectral Breaks

As briefly discussed in the previous section, the location where the optically thick (partially self-absorbed) synchrotron spectrum breaks and becomes optically thin also determines the peak of the jet flux density. Standard synchrotron theory predicts this jet break frequency (ν_b) to scale with the mass accretion rate and black hole mass (Heinz and Sunyaev 2003; Markoff et al. 2001, 2003, 2005). For BHXRBs, assuming that the magnetic field and jet acceleration region size do not vary dramatically, a positive relation should exist between ν_b and mass accretion rate. In order to test for such a correlation (in the form of $\nu_b \propto L_{\rm X}^{1/3}$) Russell et al. (2013b) performed a comprehensive literature search for nearly simultaneous multi-wavelength (including radio, IR and optical) data of BHXRBs in quiescent and hard states. This work represents the most accurate and complete study on BHXBR jet spectral breaks to date, with 12 BHRXBs having constraints on ν_b. No global relation was found with $L_{\rm X}$, from 10^{-8} up to the Eddington limit.

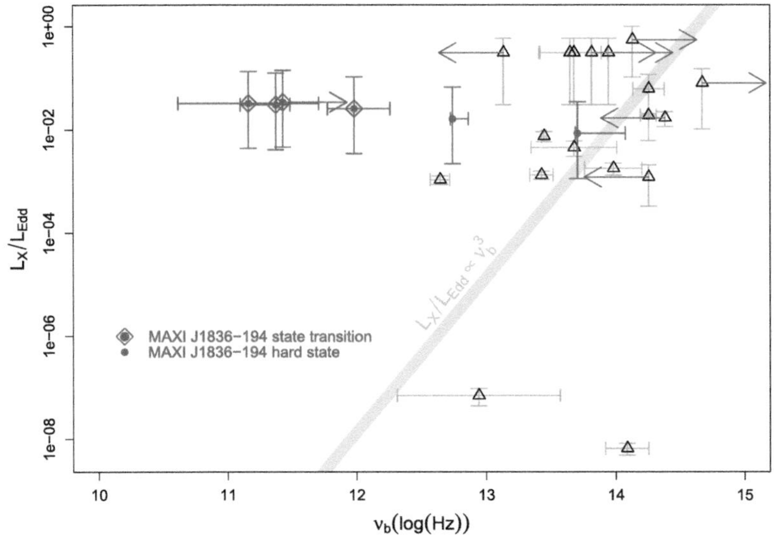

Fig. 4 Inferred values of jet spectral break frequencies as a function of Eddington scaled luminosity for a sample of 12 BHXRBs (Russell et al. 2013b, 2014). The evolution of MAXI J1836–194, *highlighted in green*, is in particularly stark contrast with the simple relation expected by basic synchrotron theory plus accretion scaling arguments. Data courtesy of Dave Russell and Tom Russell

Prior to this work, a *shifting* jet spectral break frequency (that is, within the duration of a single outburst) had been observed for one BHXRB only; GX339–4 (Gandhi et al. 2011; Corbel et al. 2013). Russell et al. (2013b) reported on the same behavior for XTE J1118+480 , whose jet break frequency varied by more than one order of magnitude while the X-ray luminosity change was negligible. More recently, high cadence, multi-wavelength monitoring of the 2011 outburst of MAXI J1836–194 (Russell et al. 2014), showed even more severe discrepancies between the data and the expected scaling between ν_b and L_X; during the early days of the outburst, ν_b for MAXI J1836–194 moved almost perpendicular to theoretical line (see Fig. 4), strongly indicating a much higher degree of complexity in the processes that regulate the jet special energy distribution, particularly during the initial phases of the compact jet recovery.

A second, major finding of this multi-wavelength campaign was the observational inference, for the first time, of the high energy synchrotron cooling break: $3.2 \times 10^{14} \lesssim \nu_c \lesssim 4.5 \times 10^{14}$ Hz. Though the cooling break was not directly visible in the broad-band spectral energy distribution data (see Fig. 7 of Russell et al. 2014 for a zoom into the mid-IR to X-ray spectrum of MAXI J1836–194 as measured on 2011 Sept 03, along with the inset showing the Ag 31 and Sept 01 VLT spectrum), the presence of a high energy break in the above frequency range is necessary in order to reconcile the near-IR synchrotron power-law spectrum with the strength of the Balmer absorption lines seen in the nearly simultaneous (within a day) VLT optical spectrum.[1] It is important to stress that, for X-ray binaries, the cooling break is expected to shift from the ultraviolet to the X-ray band (above 10 keV) as

[1] As discussed by Russell et al. (2014), strong Balmer absorption lines (with relative depths as high as 20 per cent of the continuum flux) are inconsistent with a synchrotron-dominated continuum; their presence demands an optically thick disc that, at the time of the observations, was contributing to about 50 per cent of the optical emission.

L_X increases from quiescence up to bright, hard X-ray states (Pe' er and Markoff 2012), while the reported value for MAXI J1836–194 falls into the optical band. This has been interpreted by Russell and collaborators as evidence for the jet "already evolving early in the outburst, at the same time as the system is brightening and the disc begins filling in".

The works discussed above serve to illustrate how high cadence, simultaneous, multi-wavelength monitoring of BHXBRs in outburst—particularly when the more traditional radio/X-ray approach is integrated with sensitive spectroscopy at IR and optical frequencies—hold the key to making further progress in this field.

A new avenue is also close to opening up for the study of X-ray binary outbursts: radio selected sources from wide-field searches. By the end of the current decade (i.e. 2020) the first phase of the SKA will have been constructed, alongside/incorporating the MeerKAT and ASKAP telescopes. For the first time large fractions of the sky (of order unity steradians) will be surveyed regularly (daily) and we should begin to find such events first by their radio emission. The mid-frequency component of the first phase of the SKA, 'SKA_1 Mid', should be able to easily detect most galactic XRBs during the hard state rising phase, providing plenty of alert for new outbursts, as well as tracking sources all the way to quiescence, and monitoring XRBs in other galaxies (Fender 2004; see www.skatelescope.org for more information on the SKA project).

3.3 Baryons?

One of the most important uncertainties in the study of relativistic jets ever since their discovery has been the question of their composition. In nearly all cases the radiation from jets is synchrotron emission, which only requires the presence of leptons, leaving us little the wiser as to whether the jets are primarily electron:proton or electron:positron in composition. The only exception to this case is SS433, which is itself an highly unusual system which is hard to fit into existing classification schemes, and in which the jet may become baryon-loaded after launch by interactions with a massive and dense environment (see Begelman et al. 2006 and references therein).

In the past year, Diaz-Trigo et al. (2013) have reported the first evidence for strong atomic emission lines associated with relativistic jets in a (low-mass) black hole X-ray binary, 4U 1630-47. They report evidence for Doppler-shifted emission lines originating in gas moving at $\sim 0.7c$, at the same time as optically thin radio emission is observed. This is a very exciting discovery, although it seems unlikely this can be ubiquitous for all strong-jet states and it may be only associated with certain rare phases of the outburst cycle (Neilsen et al. 2014). Regardless of the duration of the active baryon-loaded jet phase, its presence gives us very important new clues into the power of the jets and their role in the mass flow close to the black hole.

4 Summary

In this brief review we've attempted to provide a picture of the current empirical understanding of the coupling between accretion and outflow in stellar mass black holes. A key development in the past few years have been a clearer picture of the role of baryons in outflows, both in the systematic coupling of winds to accretion states reported in Ponti et al. (2012), and the recent, very exciting, discovery of relativistically moving baryons in a low-mass X-ray binary (Diaz-Trigo et al. 2013). Understanding these phenomena more deeply should allow us to really being to understand the flow of mass and energy near accreting black holes in different accretion states. The investigation of whether or not black hole spin really powers relativistic jets (part or all of the time) continues, and the results remain tantalising but unclear (to us). Without doubt this is one of the most important avenues of research in high-energy astrophysics research right now. What drives the 'radio loud' and 'radio quiet' branches in the hard X-ray states is a mystery (although its not spin!), and whether or not everything (anything?) we learn from black hole X-ray binaries is really applicable to AGN remains to be proven. Much, much, more remains to be done, observationally and theoretically, and this will be in part driven by new facilities such as the SKA, and by renewed interests in the origin of X-ray binary states and accretion cycles.

Acknowledgements R.F. would like to acknowledge an almost uncountable number of conversations with many collaborators over the past two decades. During this particular writing period he has learnt from Teo Munoz-Darias, Gabriele Ponti, Mickael Coriat, Dan Plant and Joey Neilsen. E.G. would like to thank Stephane Corbel for sharing his radio and X-ray luminosity data set, Dave and Tom Russell for sharing their data on spectral break frequencies.

References

M.C. Begelman, P.J. Armitage, Mon. Not. R. Astron. Soc. **782**, L18 (2014)
M.C. Begelman, A.R. King, J.E. Pringle, Mon. Not. R. Astron. Soc. **370**, 399 (2006)
T.M. Belloni, S.E. Motta, T. Munoz-Darias, Bull. Astron. Soc. India **39**, 409 (2011)
R. Blandford, A. Konigl, Astrophys. J. **232**, 34 (1979)
R.D. Blandford, D.G. Payne, Mon. Not. R. Astron. Soc. **199**, 883 (1982)
R.D. Blandford, R.L. Znajek, Mon. Not. R. Astron. Soc. **179**, 433 (1977)
C.K. Bradley et al., Astrophys. J. **667**, 427 (2007)
C. Brocksopp et al., Mon. Not. R. Astron. Soc. **404**, 908 (2010)
C. Brocksopp et al., Mon. Not. R. Astron. Soc. **432**, 931 (2013)
D.E. Calvelo et al., Mon. Not. R. Astron. Soc. **409**, 839 (2010)
P. Casella, T. Belloni, L. Stella, Astrophys. J. **629**, 403 (2005)
P.G. Casella et al., Mon. Not. R. Astron. Soc. **404**, L21 (2010)
S. Chaty et al., Mon. Not. R. Astron. Soc. **346**, 689 (2003)
S. Corbel et al., Astron. Astrophys. **400**, 1007 (2003)
S. Corbel, E. Körding, P. Kaaret, Mon. Not. R. Astron. Soc. **389**, 1697 (2008)
S. Corbel et al., Mon. Not. R. Astron. Soc. **431**, L107 (2013)
M. Coriat et al., Mon. Not. R. Astron. Soc. **414**, 677 (2011)
M. Coriat, R.P. Fender, G. Dubus, Mon. Not. R. Astron. Soc. **424**, 1991 (2012)
M. Diaz-Trigo et al., Nature **504**, 260 (2013)
C. Done, M. Gierlinski, A. Kubota, Astron. Astrophys. Rev. **15**, 1 (2007)
R.J.H. Dunn et al., Mon. Not. R. Astron. Soc. **403**, 61 (2010)
S. Eikenberry et al., Astrophys. J. **494**, L61 (1998)
A.C. Fabian et al., Mon. Not. R. Astron. Soc. **424**, 217 (2012)
H. Falcke, E. Körding, S. Markoff, Astron. Astrophys. **414**, 895 (2004)
R. Fender, Mon. Not. R. Astron. Soc. **322**, 31 (2001)
R. Fender, New Astron. Rev. **48**, 1399 (2004)
R.P. Fender, *Jets from X-ray binaries* (Cambridge University Press, Cambridge, 2006), p. 381
R.P. Fender, T. Belloni, Science **337**, 540 (2012)
R. Fender, G. Pooley, C. Brocksopp, S. Newell, Mon. Not. R. Astron. Soc. **290**, L65 (1997)
R. Fender et al., Astrophys. J. **519**, L165 (1999)
R.P. Fender, T. Belloni, E. Gallo, Mon. Not. R. Astron. Soc. **355**, 1105 (2004)
R.P. Fender, T.J. Maccarone, Z. van Kesteren, Mon. Not. R. Astron. Soc. **360**, 1085 (2005)
R. Fender, J. Homan, T. Belloni, Mon. Not. R. Astron. Soc. **396**, 1370 (2009)
R.P. Fender, E. Gallo, D.M. Russell, Mon. Not. R. Astron. Soc. **406**, 1425 (2010)
E. Gallo, R.P. Fender, G.G. Pooley, Mon. Not. R. Astron. Soc. **344**, 60 (2003)
E. Gallo et al., Nature **436**, 819 (2005a)
E. Gallo, R.P. Fender, R. Hynes, Mon. Not. R. Astron. Soc. **356**, 1017 (2005b)
E. Gallo et al., Mon. Not. R. Astron. Soc. **370**, 1351 (2006)
E. Gallo et al., Astrophys. J. **670**, 600 (2007)

E. Gallo, B.P. Miller, R.P. Fender, Mon. Not. R. Astron. Soc. **423**, 590 (2012)
P. Gandhi et al., Astrophys. J. **740**, L13 (2011)
S. Heinz, Astron. Astrophys. **388**, L40 (2002)
S. Heinz, H. Grimm, Astrophys. J. **633**, 384 (2005)
S. Heinz, R. Sunyaev, Mon. Not. R. Astron. Soc. **343**, L59 (2003)
S. Heinz, A. Merloni, J. Schwab, Astrophys. J. **658**, L9 (2007)
R.M. Hjellming, X. Han, *Radio Properties of X-Ray Binaries* (1995), p. 308
R.M. Hjellming, M. Rupen, Nature **6531**, 464 (1995)
R. Hynes et al., Astrophys. J. **611**, L125 (2004)
R. Hynes et al., Astrophys. J. **651**, 401 (2006)
R. Hynes et al., Mon. Not. R. Astron. Soc. **399**, 2239 (2009)
O. Jamil, R.P. Fender, C.R. Kaiser, Mon. Not. R. Astron. Soc. **401**, 394 (2010)
A. Joinet, E. Kalemci, F. Senziani, Astrophys. J. **679**, 655 (2008)
P.G. Jonker et al., Mon. Not. R. Astron. Soc. **401**, 1255 (2010)
N. Khandai et al., Mon. Not. R. Astron. Soc. arXiv:1402.0888 (2014, submitted)
A.L. King et al., Astrophys. J. **771**, 84 (2013a), 12
A.L. King et al., Astrophys. J. **762**, 18 (2013b)
A.L. King et al., Astrophys. J. **784**, L2 (2014), 6
M. Klein-Wolt et al., Astrophys. Space Sci. **276**, 291 (2001)
E.G. Körding, R.P. Fender, S. Migliari, Mon. Not. R. Astron. Soc. **369**, 1451 (2006a)
E.G. Körding, S. Jester, R.P. Fender, Mon. Not. R. Astron. Soc. **372**, 1366 (2006b)
E.G. Körding et al., Mon. Not. R. Astron. Soc. **380**, 301 (2007)
E.G. Körding, S. Jester, R.P. Fender, Mon. Not. R. Astron. Soc. **383**, 277 (2008a)
E.G. Körding et al., Science **320**, 1318 (2008b)
J. Kormendy, L. Ho, Annu. Rev. Astron. Astrophys. **51**, 511 (2013)
B.F. Liu et al., Astrophys. J. **671**, 695 (2007)
J. Malzac, Mon. Not. R. Astron. Soc. **429**, L20 (2013)
J. Malzac, A. Merloni, A.C. Fabian, Mon. Not. R. Astron. Soc. **351**, 253 (2004)
S. Markoff, M. Nowak, Astrophys. J. **609**, 972 (2007)
S. Markoff, H. Falcke, R.P. Fender, Astron. Astrophys. **372**, L25 (2001)
S. Markoff et al., Astron. Astrophys. **397**, 645 (2003)
S. Markoff, M. Nowak, J. Wilms, Astrophys. J. **635**, 1203 (2005)
J. Martí, Mem. Soc. Astron. Ital. **76**, 592 (2005)
I.M. McHardy et al., Nature **444**, 730 (2006)
A. Merloni, S. Heinz, T. Di Matteo, Mon. Not. R. Astron. Soc. **345**, 1057 (2003)
F. Meyer, E. Meyer-Hofmeister, Astron. Astrophys. **104**, L10 (1981)
F. Meyer, B.F. Liu, E. Meyer-Hofmeister, Astron. Astrophys. **463**, 1 (2007)
E. Meyer-Hofmeister, F. Meyer, Astron. Astrophys. **562**, 142 (2014)
M.J. Middleton et al., Nature **7431**, 187 (2013)
M.J. Middleton, J.C.A. Miller-Jones, R.P. Fender, Mon. Not. R. Astron. Soc. **439**, 1740 (2014)
J.M. Miller, Annu. Rev. Astron. Astrophys. **45**, 441 (2007)
J.M. Miller, J. Homan, G. Miniutti, Astrophys. J. **652**, L113 (2006)
J.M. Miller et al., Astrophys. J. **680**, 1359 (2008)
J.M. Miller et al., Astrophys. J. **697**, 900 (2009)
J.M. Miller et al., Astrophys. J. **759**, L6 (2012)
J.C.A. Miller-Jones et al., Astrophys. J. **739**, L18 (2011)
J.C.A. Miller-Jones et al., Mon. Not. R. Astron. Soc. **421**, 468 (2012)
I.F. Mirabel, L.F. Rodríguez, Nature **371**, 46 (1994)
I.F. Mirabel, L.F. Rodríguez, Annu. Rev. Astron. Astrophys. **37**, 409 (1999)
I.F. Mirabel et al., Astron. Astrophys. **330**, L9 (1998)
S.E. Motta, T. Belloni, J. Homan, Mon. Not. R. Astron. Soc. **400**, 1603 (2009)
S.E. Motta et al., Mon. Not. R. Astron. Soc. **437**, 2554 (2014a)
S.E. Motta et al., Mon. Not. R. Astron. Soc. **439**, L65 (2014b)
T. Munoz-Darias, S. Motta, T. Belloni, Mon. Not. R. Astron. Soc. **410**, 679 (2011)
T. Munoz-Darias, R. Fender, S. Motta, T. Belloni, Mon. Not. R. Astron. Soc. (2014, in press). arXiv:1407.1318
R. Narayan, J.E. McClintock, Mon. Not. R. Astron. Soc. **419**, L69 (2012)
J. Neilsen, J.C. Lee, Nature **458**, 481 (2009)
J. Neilsen et al., Astrophys. J. Lett. **784**, L5 (2014)
A. Pe' er, S. Markoff, Astrophys. J. **753**, 177 (2012)
P.O. Petrucci et al., Mon. Not. R. Astron. Soc. **385**, L88 (2008)

D.S. Plant et al., Mon. Not. R. Astron. Soc. (2014, submitted)
R. Plotkin, E. Gallo, P. Jonker, Astrophys. J. **773**, 59 (2013)
G. Ponti et al., Mon. Not. R. Astron. Soc. **422**, 11 (2012)
R.C. Reis, A.C. Fabian, J.M. Miller, Mon. Not. R. Astron. Soc. **402**, 836 (2010)
R. Remillard, J.E. McClintock, Annu. Rev. Astron. Astrophys. **44**, 49 (2006)
C.S. Reynolds, Space Sci. Rev. arXiv:1302.3260 (2013)
M.T. Reynolds, J.M. Miller, Astrophys. J. **769**, 16 (2013)
D.M. Russell et al., Mon. Not. R. Astron. Soc. **371**, 1334 (2006)
D.M. Russell, D. Maitra, R.J. Dunn, S. Markoff, Mon. Not. R. Astron. Soc. **405**, 1759 (2010)
D.M. Russell et al., Astrophys. J. **739**, L19 (2011a)
D.M. Russell et al., Mon. Not. R. Astron. Soc. **416**, 2311 (2011b)
D.M. Russell, E. Gallo, R. Fender, Mon. Not. R. Astron. Soc. **431**, 405 (2013a)
D.M. Russell et al., Astrophys. J. **768**, L35 (2013b)
T.D. Russell et al., Mon. Not. R. Astron. Soc. **439**, 1390 (2014)
D.M. Russell, T. Shahbaz, Mon. Not. R. Astron. Soc. **438**, 2083 (2014)
P. Soleri, R. Fender, Mon. Not. R. Astron. Soc. **413**, 2268 (2011)
P. Soleri, T. Belloni, P. Casella, Mon. Not. R. Astron. Soc. **383**, 1089 (2008)
J.F. Steiner, J.E. McClintock, R. Narayan, Astrophys. J. **762**, 104 (2013)
L. Stella, M. Vietri, Phys. Rev. Lett. **82**, 17 (1999)
A. Tchekhovskoy et al., Astrophys. J. **711**, 50 (2010)
H. van der Laan, Nature **211**, 1131 (1966)

DOI 10.1007/978-1-4939-2227-7_17
Reprinted from *Space Science Reviews* Journal, DOI 10.1007/s11214-013-0008-4

X-Ray Observations of Powerful AGN Outflows

Implications for Feedback

Ken Pounds

Received: 18 February 2013 / Accepted: 18 July 2013 / Published online: 31 August 2013

Abstract Highly ionised winds with velocities ~0.1–0.2c were first detected in X-ray spectra of non-BAL AGN a decade ago. Subsequent observations and archival searches have shown such winds to be a common feature of luminous AGN, increasing the belief that powerful ionised winds have a wider importance in galaxy feedback models. Paradoxically, for the best-quantified high velocity outflow (the luminous Seyfert PG1211+143) the wind appears *too powerful* to be compatible with the observed stellar bulge and black hole masses, suggesting the energy coupling of wind to bulge gas must be inefficient. A recent *XMM-Newton* observation of the narrow line Seyfert NGC 4051 offers an explanation of this apparent paradox, finding evidence for the fast ionised wind to lose most of its kinetic energy after shocking against the ISM. Importantly, the wind momentum is maintained through such a shock, supporting the view that a momentum-driven flow provides the critical link between black hole and stellar bulge growth implied by the observed M–σ relationship.

Keywords AGN · Feedback · X-Ray astronomy

1 Introduction

Early X-ray observations of AGN yielded soft X-ray spectra frequently showing the imprint of absorption from ionised gas, the 'warm absorber' (Halpern 1984; Reynolds and Fabian 1995). However, the limited spectral resolution of the *Einstein Observatory* and *ASCA* observations prevented important parameters of the warm absorbers, in particular the outflow velocity and mass rate, to be determined with useful precision. The higher resolution and high throughput afforded by contemporary X-ray observatories, *Chandra*, *XMM-Newton* and *Suzaku* has transformed that situation over the past decade, with the warm absorber being shown, typically, to be dominated by K-shell ions of the lighter metals (C, N, O, etc.)

K. Pounds (✉)
X-ray and Observational Astrophysics Group, Dept of Physics & Astronomy, University of Leicester, Leicester LE1 7RH, UK
e-mail: kap@le.ac.uk

and Fe-L, with outflow velocities of several hundred km s^{-1}. More dramatic was the detection of blue-shifted X-ray absorption lines in the iron K band, indicating the presence of highly ionised outflows with velocities $\sim$0.1–0.25c (Chartas et al. 2002; Pounds et al. 2003; Reeves et al. 2003). In addition to adding an important dimension to AGN accretion studies, the mechanical power of such winds was quickly recognised to have a wider significance in galaxy feedback models.

Additional detections of high velocity AGN winds were delayed by a combination of the low absorption cross section of highly ionised gas, and strongly blue-shifted lines in low redshift objects coinciding with falling telescope sensitivity. However, extended observations, particularly with *XMM-Newton*, found evidence in 7 additional AGN for outflow velocities of $\sim$0.1–0.2c, with 5 further claims of similar red-shifted components (Cappi 2006). Residual doubts remained, however, as the majority of detections were of a single absorption line (with consequent uncertainty of identification), and had moderate statistical significance, raising concerns of 'publication bias' (Vaughan and Uttley 2008).

Those doubts were finally removed following a blind search of AGN observations in the *XMM-Newton* archive (Tombesi et al. 2010), finding compelling evidence in 13 (of 42) radio quiet objects for blue-shifted iron K absorption lines, with implied outflow velocities of $\sim$0.05–0.25c. A more recent search of the *Suzaku* data archive has yielded a further group of strong detections, with a median outflow velocity again $\sim$0.1c (Gofford et al. 2013). In addition to confirming that high velocity, highly ionised AGN winds are common, the high yield from these archival searches shows the flows must typically have a large covering factor, and therefore are likely to involve substantial mass and energy fluxes.

Paradoxically, for the best-quantified high velocity outflow (the luminous Seyfert PG1211+143), in which a wide-angle flow was directly measured (Pounds and Reeves 2007, 2009), the wind appears *too energetic* to be compatible with the observed stellar bulge and black hole masses, suggesting that the energy coupling of wind to bulge gas must be inefficient.

A recent 600 ks *XMM-Newton* observation of the narrow line Seyfert NGC 4051 has provided a possible explanation of that paradox, finding evidence that the fast ionised wind loses much of its kinetic energy after shocking against the ISM, at a sufficiently small radius for Compton cooling to be strong (Pounds and King 2013). Crucially, the wind momentum is maintained through such a shock, with repeated events eventually leading to a momentum-driven thrust as shown (King 2003, 2005) to correctly predict the observed M–σ relationship (Ferrarese and Merritt 2000; Gebhardt et al. 2000).

In the following section, the energetic outflow of PG1211+143 is reviewed. Section 3 then summarises the essential features of a simple physical model, based on the PG1211+143 observation, offering a natural explanation for the existence of highly ionised winds in AGN with velocities $\sim$0.1c. Section 4 summarises the key results of archival searches showing that such winds are indeed common. Section 5 reviews the recent observation of NGC 4051, with evidence for the shocking of such a high velocity wind on collision with the ISM. Finally, Sect. 6 outlines a self-consistent model of the full outflow in NGC 4051, where the fast wind is shocked at $\sim$0.1 pc, with the post-shock flow rapidly cooling to yield the characteristic soft X-ray features of the 'warm absorber'.

2 The Fast Outflow in PG1211+143

PG1211+143, at a redshift of 0.0809 (Marziani et al. 1996), is one of the brightest AGN at soft X-ray energies (Elvis 2006). It was classified (Kaspi et al. 2000) as a Narrow Line

Seyfert 1 galaxy (FWHM Hβ 1800 km s^{-1}), with black hole mass ($\sim 4 \times 10^7 M_\odot$) and bolometric luminosity 4×10^{45} erg s^{-1}, indicating the mean accretion rate is probably Eddington-limited.

EPIC spectra from an *XMM-Newton* observation of PG1211+143 in 2001 provided the first evidence for a high velocity ionised outflow in a non-BAL AGN, with the identification of a blue-shifted Fe Lyman-α absorption line corresponding to an outflow velocity of $\sim 0.09c$ (Pounds et al. 2003). That observation, closely followed by the detection of a still higher outflow velocity from the luminous QSO PDS 456 (Reeves et al. 2003), attracted wide attention in potentially involving a significant fraction of the bolometric luminosity, and which might be typical of AGN accreting near the Eddington rate (King and Pounds 2003).

Initial questions were raised about the validity of the high velocity in PG1211+143. The near-coincidence of the observed absorption line blueshift and the redshift of the host galaxy was a concern, notwithstanding the uncomfortably high column density of heavy metals implied by a local origin. Then, in a detailed modelling of the soft X-ray RGS data, Kaspi and Behar (2006) found only a much lower velocity. Any doubts relating to the absorption being local were removed, however, by a revised velocity of 0.13–0.15c (Pounds and Page 2006), and when subsequent observations demonstrated the Fe K absorption to be variable (Reeves et al. 2008). Furthermore, it now appears that the highest velocity outflows are in general too highly ionised to have significant opacity in the soft X-ray band (Sect. 4).

Figure 1 illustrates the initial *XMM-Newton* observation of PG1211+143 in 2001. The upper panel is a ratio plot of EPIC pn data to a power law continuum, showing 3 significant 'narrow' absorption lines. Identifying the lines at $\sim$7.1 keV, $\sim$2.7 and $\sim$1.5 keV with the resonance Ly-α transitions of FeXXVI, SXVI and MgXII yields the outflow velocity of $0.09 \pm 0.01c$ reported in Pounds et al. (2003). A subsequent analysis, taking advantage of the higher energy resolution (albeit lower sensitivity) of the EPIC MOS camera, found the absorption features at $\sim$1.5 keV and $\sim$2.7 keV to be resolved as line pairs, with the correct energy spacing of the respective He-α and Ly-α resonance lines of Mg and S. Additional narrow absorption features match the same K-shell resonance line pairs of Ne, Si and possibly Ar, where—crucially—all the line energies (Fig. 1, mid panel) scale from the 7.1 keV line (if now identified with the FeXXV resonance line) with the same observed blueshift (with respect to the observer) of $\sim$0.05, yielding a revised outflow velocity in the PG1211+143 rest frame of $v \sim 0.13$–$0.15c$ (Pounds and Page 2006).

The lower panel of Fig. 1 illustrates absorption in a photoionised gas (derived from the XSTAR code of Kallman et al. 1996), modelling the MOS spectrum with column density $N_H \sim 2 \times 10^{22}$ cm^{-2} and ionisation parameter of $\log \xi = 2.9 \pm 0.4$ erg cm s^{-1}. The strongest absorption lines (in order of increasing energy) correspond to K-shell resonance transitions of Ne, Mg, Si, S, Ar and Fe. The apparent blueshift read from the XSTAR model of $4.9 \pm 0.3 \times 10^{-2}$ was consistent with the individual line fitting in corresponding to an outflow velocity (in the AGN rest frame) of $v \sim 0.14 \pm 0.01c$.

2.1 Mass Rate and Mechanical Energy in the PG1211+143 Outflow

Although the detection of high speed winds in a substantial fraction of bright AGN (Sect. 4) suggests most such flows have a large covering factor, a wide angle flow has only been demonstrated directly to date for PG1211+143.

Using stacked data from 4 *XMM-Newton* observations between 2001 and 2007, Pounds and Reeves (2007, 2009) examined the relative strength of ionised emission and absorption features to estimate the flow collimation. Analysis of the FeXXV PCygni profile (Fig. 2) indicated a collimation factor b ($= \Omega/4\pi$) of 0.75 ± 0.25. The emission line component in

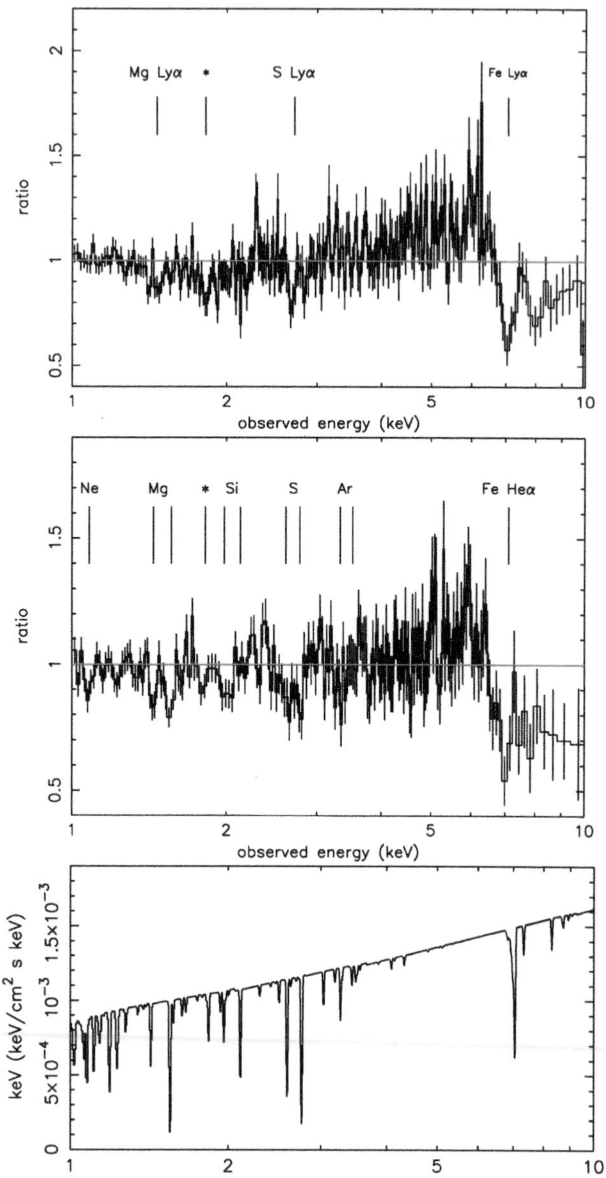

Fig. 1 (*top panel*) Ratio of EPIC pn spectrum of PG1211+143 to a power law continuum. Absorption lines are identified with Ly-α of Mg, S and Fe, with Fe Ly-β at ~8 keV. (*mid panel*) EPIC MOS data resolving H- and He-like ion pairs for a revised blue shift. (*lower panel*) Photoionised absorber model fit to MOS spectrum showing principal K-shell absorption lines of Ne, Mg, Si, S and Ar, together with K- and L-shell lines of Fe

the Fe K profile in a 2005 Suzaku observation appears very similar to that in the stacked *XMM-Newton* data, with a mean energy of ~6.5 keV and width of $\sigma \sim 250$ eV (Reeves et al. 2008)). The assumption of velocity broadening in a radial flow corresponds to a flow cone of half angle ~50 deg and $b \sim 0.3$.

A further estimate, independent of an individual line profile, compared the total continuum energy absorbed and re-emitted by the highly ionised gas components from modelling the stacked EPIC data (Pounds and Reeves 2009), finding $b \sim 0.5$, while a good fit to the broad band EPIC spectrum also *required* the X-ray emission lines to be broad. Taken to-

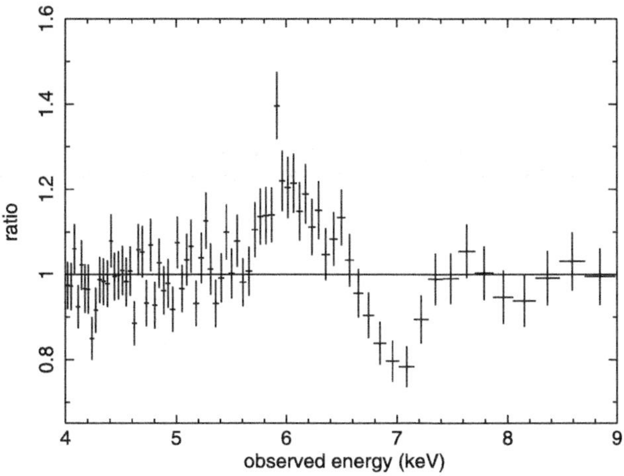

Fig. 2 The PCygni profile of Fe XXV from the stacked *XMM-Newton* pn observations of PG1211+143 shows the emission and blue-shifted absorption characteristic of a wide angle outflow

gether, the above indicators confirmed the highly ionised outflow in PG1211+143 is *not* highly collimated. It must therefore involve a significant mass flux, as shown below.

For a uniform radial outflow of velocity v the mass rate is

$$\dot{M}_{\text{out}} \simeq 4\pi bnr^2 m_p v, \tag{1}$$

where n is the gas density at a radial distance r, and m_p is the proton mass.

Modelling the broad band X-ray spectrum of PG1211+143 provides estimates of the wind ionisation parameter ($\xi = L/nr^2$) and the ionising luminosity, yielding $nr^2 \sim 3 \times 10^{39}\ \text{cm}^{-1}$. For a wind velocity of $0.12c$, and assuming $b = 0.3$, the mass loss rate $\dot{M}_{\text{out}} \sim 7 \times 10^{25}\ \text{gm}\,\text{s}^{-1}$ ($\sim 2.5 M_\odot\ \text{yr}^{-1}$), and mechanical energy $\sim 4.5 \times 10^{44}\ \text{erg}\,\text{s}^{-1}$.

The mass loss rate is comparable to the Eddington accretion rate $\dot{M}_{\text{Edd}} = 1.3 M_\odot\ \text{yr}^{-1}$ for a supermassive black hole of mass $\sim 4 \times 10^7 M_\odot$ accreting at an efficiency of 10 %. The outflow mechanical energy is ~6 % of the Eddington luminosity, close to that predicted by continuum driving (Eq. (5) in Sect. 3 below). A simple calculation shows that such a wind is more than sufficient to disrupt star formation in the host galaxy.

Consider a current episode of Eddington-limited accretion during which the black hole and bulge masses of PG1211+143 doubles over $\sim 3 \times 10^7$ yr. Assuming a mass ratio of $M_{\text{bulge}} \sim 10^3 M_{\text{BH}}$, the binding energy of the bulge gas increases by $E_{\text{bind}} \sim M_{\text{bulge}}\sigma^2 \sim 2 \times 10^{58}$ ergs (taking $\sigma = 160\ \text{km}\,\text{s}^{-1}$ from the M–σ relation), while the high velocity wind injects mechanical energy of $\sim 4 \times 10^{59} \eta_{\text{mech}}$ erg, where η_{mech} is the fraction of the wind energy transferred to the bulge gas. It appears that $\eta_{\text{mech}} \le 0.1$ for the continued growth of PG1211+143 not to be disrupted.

A potential explanation for such low efficiency coupling is presented in Sect. 5, with new observational evidence for the fast ionised wind in NGC 4051 shocking with the ISM at a sufficiently small radial distance from the AGN for strong Compton cooling to result in most of the mechanical energy being lost long before reaching the stellar bulge.

3 Continuum Driving of a Highly Ionised Wind

The Black Hole Winds (BHW) model (King and Pounds 2003) provides a simple physical basis for high velocity outflows in AGN accreting at or above the Eddington limit, and has the bonus of useful predictive power.

The BHW model was inspired by noting that a radial outflow similar to that of PG1211+143 will be optically thick at a sufficiently small radius. Unit electron scattering optical depth near the base of such a wind will cause each photon to be scattered once on average, before escaping, with the resulting wind momentum being of the order of the photon momentum

$$\dot{M}_{\rm out} v \simeq \frac{L_{\rm Edd}}{c}. \tag{2}$$

Since

$$\dot{M}_{\rm Edd} = \frac{L_{\rm Edd}}{\eta c^2} \tag{3}$$

where η is the accretion efficiency, the wind velocity

$$v \simeq \frac{\eta}{\dot{m}} c \sim 0.1c \tag{4}$$

with mechanical energy

$$1/2 \dot{M}_{\rm out} v^2 \simeq \frac{\eta}{2} L_{\rm Edd} \tag{5}$$

The v^3 dependency of the mechanical energy flux underlines the importance of high velocity for wind feedback, while equation 4 offers a physical basis on which to compare the observed range of outflow velocities (Sect. 4).

Assuming the wind is launched and then coasts, the observed outflow velocity will be of order the escape velocity at the launch radius. For PG1211+143 the wind velocity $v \sim 0.12c$ corresponds to $R_{\rm launch} \sim 70 R_{\rm s}$ (where $R_{\rm s} = 2GM/c^2$), or $\sim 8 \times 10^{14}$ cm. It is interesting to note that EPIC data show significant flux variability in the harder (2–10 keV) band of PG1211+143 on timescales of 2–3 hours (Fig. 1 in Pounds et al. 2003), compatible with the above scale size relating to the primary (disc/corona) X-ray emission region.

The results from archival searches presented in Sect. 4 show that AGN winds with $v \sim 0.1c$ are indeed common, perhaps indicating that continuum driving is also common. As only a minority of local AGN appear to be radiating continuously at the Eddington limit (however, see King 2010a), high velocity winds in such AGN may be intermittent, on timescales perhaps related to the inner accretion disc. The measured absorption column will be an integral of such a fluctuating wind in the line of sight, while its ionisation state will depend on the current luminosity.

4 High Speed Winds Become Common

As noted earlier, the general acceptance of high speed winds in AGN was delayed by concerns relating to the prototype case of PG1211+143. High velocity outflows in two BAL AGN (Chartas et al. 2002) and in the most luminous low redshift QSO PDS 456 (Reeves et al. 2003; O'Brien et al. 2005) might also be considered rare objects. However, a significant detection of an outflow with velocity $\sim 0.1c$ was reported for IC4329A (Markowitz et al. 2006), and repeated detections in the range ~ 0.14–$0.2c$ were found in multiple observations of Mkn 509 (Dadina et al. 2005). A review in 2006 (Cappi 2006) listed 7 non-BAL objects with blue shifts $\sim 0.1c$ and several with red-shifted absorption lines.

A major step forward was achieved with the results of the *XMM-Newton* archival search by Tombesi et al. (2010), finding strong statistical evidence in 13 of 42 radio quiet objects of

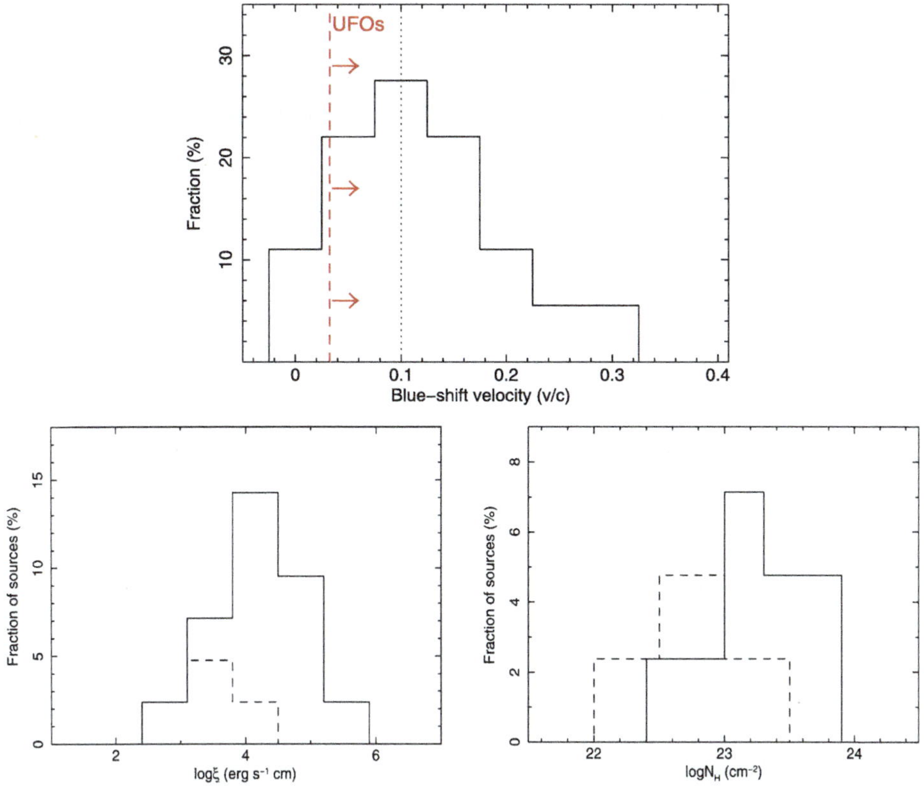

Fig. 3 (*top*) Distribution of outflow velocities in the Tombesi et al. sample obtained from examination of extended observations of radio quiet AGN in a blind search of the *XMM-Newton* data archive. The *vertical dotted line* indicates the mean value of ~0.1*c*. (*lower*) Distribution of ionisation parameter and column density from modelling the individual spectra (from Tombesi et al. 2011). *Solid lines* refer to the ultra fast outflows, with velocity above 10^4 km s^{-1}, and *dashed lines* to lower velocity Fe XXV/XXVI absorption

blue-shifted iron K absorption lines, with implied outflow velocities up to ~0.3*c*, and clustering near $v \sim 0.1c$ (Fig. 3, top). Subsequent modelling with XSTAR photoionised grids (Tombesi et al. 2011) showed the outflows were highly ionised, with $\log \xi \sim 3$–6 erg cm s^{-1}, and had large column densities in the range $N_H \sim 10^{22}$–10^{24} cm^{-2} (Fig. 3, lower). Among the strongest positive detections was that of PG1211+143, with an outflow velocity ~0.15*c* and an F-test probability of 99.99 %.

The high ionisation parameter typical of the fast outflows explains why their detection (in non-BAL AGN) has been restricted to X-ray observations in the Fe K band, leaving open the additional possibility that fully ionised outflows (quite likely at launch with continuum driving) will only be detectable when the AGN luminosity falls. As noted above, the observed column density will be a line-of-sight integration over the flow, albeit dominated by the higher density at small radii, while the ionisation parameter will be governed by the current AGN luminosity.

A more recent search of the *Suzaku* data archive has yielded a further group of detections, finding significant absorption in the Fe K band in 20 (of 51) AGN (Gofford et al. 2013), with velocities up to ~0.3*c* and a flatter distribution than the *XMM-Newton* sample, but a median value again ~0.1*c*. In terms of the BHW model the higher velocities could relate to a higher value of η, although it might be premature to suggest such observations as a reliable measure of black hole spin! Reference to equation 4 also suggests the low velocity tail in both Tombesi et al. and Gofford et al. distributions could indicate a higher accretion ratio, or possibly confusion with a secondary wind that has slowed after being shocked (Sect. 5).

5 Evidence for a Shocked Flow

The mechanical energy in a fast wind, such as that in PG1211+143, was noted in Sect. 3 to be incompatible with the continued growth of the black hole and stellar bulge in that object, unless the coupling of wind energy to bulge gas is highly inefficient. A recent *XMM-Newton* observation of the narrow line Seyfert galaxy NGC 4051 has provided the first evidence that much of the energy in such a wind may be lost after collision with the ISM.

NGC 4051 was found in the *XMM-Newton* archival search to have a high velocity wind, the initial identification with Fe XXVI Lyman-α in Tombesi et al. (2010) indicating a velocity of $\sim$0.15c. Detailed photoionisation modelling subsequently preferred an identification with Fe XXV, for a velocity $v \sim 0.2c$ (Tombesi et al. 2011), the absorption being detected at high significance (F-test probability 99.95 %) during an observation in 2002 when the source was in an unusually low state. However, in a 2001 observation, when the X-ray flux was much higher, no fast wind was seen.

A 600 ks *XMM-Newton* observation of the Seyfert 1 galaxy NGC 4051 in 2009, extending over 6 weeks and 15 spacecraft orbits, revealed a rich absorption spectrum (Fig. 4) with outflow velocities, in both RGS and EPIC spectra, up to $\sim$9000 km s^{-1} (Pounds and Vaughan 2011a). Evidence for a higher velocity wind ($v \sim 0.13c$) was again stronger during periods when the ionising continuum was low (Pounds and Vaughan 2012), suggesting a fast wind that may be fully ionised at higher continuum levels. The low redshift ($z = 0.00234$) of NGC 4051 also makes a high velocity wind more difficult to detect as the sensitivity of current X-ray telescopes is falling sharply above $\sim$7 keV.

6 A Self-Consistent Model for the Shocked Wind in NGC 4051

More complete modelling of both RGS and EPIC pn absorption spectra of NGC 4051 confirmed a general correlation of outflow velocity and ionisation state (Fig. 5), as expected by mass conservation in the post-shock flow (King 2010; Pounds and Vaughan 2011a). Importantly, the new analysis (Pounds and King 2013) found a large range of column densities to be required by the individual XSTAR absorption components, suggesting an inhomogeneous shocked flow, perhaps with lower ionisation gas clumps or filaments embedded in a more extended, lower density and more highly ionised flow. However, a more important factor in determining the structure of the post-shock flow is likely due to a variable cooling time, as discussed below.

The key factor in determining the impact of a fast wind is likely to be the distance it travels before colliding with the ISM or previous ejecta. Compton cooling will dominate for a shock occurring close to the AGN continuum source, with the Compton cooling time and mean flow speed then determining the shell thickness of the hotter, more highly ionised flow component. However, in the later stages of the flow the density is likely to increase to the point where two-body cooling becomes important, and then perhaps dominant.

At the (adiabatic) shock the free–free and Compton cooling times are

$$t_{\rm ff} \cong 3 \times 10^{11} \frac{T^{1/2}}{N}\ {\rm s} = 2 \frac{R_{16}^2}{M_8 \dot{m}}\ {\rm yr} \tag{6}$$

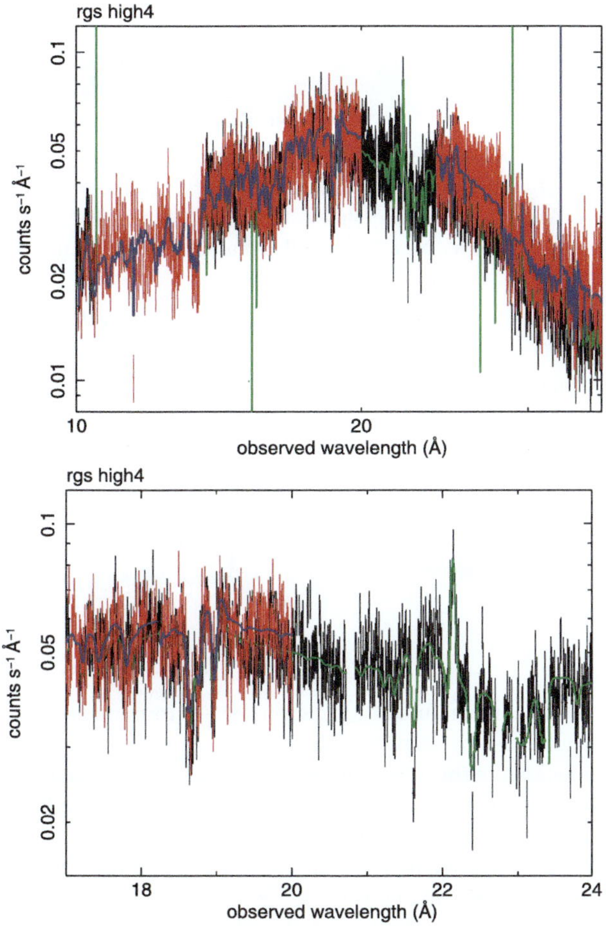

Fig. 4 XSTAR model fit to the combined RGS1 and RGS2 spectra summed over four successive high-flux orbits from the 2009 *XMM-Newton* observation of NGC 4051, for the full 10–36 Å waveband in the upper panel, with the section covering strong absorption in OV, VI, VII and VIII highlighted in the lower panel

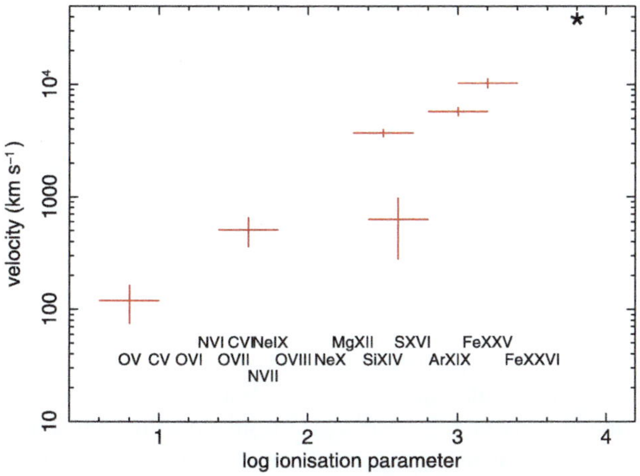

Fig. 5 Outflow velocity and ionisation parameter for each of the XSTAR photoionised absorbers derived from fitting to the RGS and EPIC spectra, together with a high point representative of the pre-shock wind, show the linear correlation expected for a mass conserved cooling flow (from Pounds and King 2013)

and

$$t_C = 10^{-4} \frac{R_{16}^2}{M_8} \text{ yr} \tag{7}$$

respectively (see King et al. 2011: here T, N are the postshock temperature and number density, R_{16} is the shock radius in units of 10^{16} cm, M_8 is the black hole mass in units of $10^8 M_\odot$, and $\dot{m} \sim 1$ is the Eddington factor of the mass outflow rate).

After the adiabatic shock, the gas cools rapidly via inverse Compton cooling, while its density rises as $N \propto T^{-1}$ (isothermal shock—pressure almost constant). So

$$t_{\rm ff} \propto \frac{T^{1/2}}{N} \propto T^{3/2}, \tag{8}$$

which means that the free–free cooling time decreases sharply while the Compton time does not change. Eventually free–free (and other atomic 2-body processes) become faster than Compton when T has decreased sufficiently below the original shock temperature $T_s \sim 1.6 \times 10^{10}$ K. From (6), (7) above this requires

$$\left(\frac{T}{T_s}\right)^{3/2} < 5 \times 10^{-5} \tag{9}$$

or

$$T < 2 \times 10^7 \text{ K}. \tag{10}$$

The temperature of ionization species forming around a few keV is therefore likely to be determined by atomic cooling processes rather than Compton cooling. The strong recombination continua in NGC 4051 (Pounds and Vaughan 2011b; Pounds and King 2013) are direct evidence for that additional cooling, with the RRC flux yielding an emission measure for the related flow component.

In particular, the onset of strong 2-body cooling will result in the lower ionisation, lower velocity gas being constrained in a relatively narrow region in the later stages of the post-shock flow. The structure and scale of both high and low ionisation flow regions can be deduced from the observations and modelling parameters.

For the highly ionised post-shock flow, the Fe Ly-α to He-α ratio will be governed by the ionising continuum and recombination time. Significant variations in this ratio are found on inter-orbit timescales (Pounds and Vaughan 2012), with an example shown in Fig. 6. For a mean temperature of $\sim$1 keV, and recombination coefficient of 4.6×10^{-12} (Verner and Ferland 1996), the observed recombination timescale of $\sim 2 \times 10^5$ s corresponds to an average particle density of $\sim 4 \times 10^6$ cm^{-3}. Comparison with a relevant absorption column $N_H \sim 4 \times 10^{22}$ cm^{-2} from the XSTAR modelling indicates a column length scale of $\sim 10^{16}$ cm. Assuming a mean velocity of the highly ionised post-shock flow of 6000 km s^{-1}, the observed absorption length corresponds to a flow time $\sim 1.7 \times 10^7$ s (0.6 yr). Equation (7) finds a comparable cooling time for NGC 4051 at a shock radius $R \sim 10^{17}$ cm.

For the low ionisation flow component, decay of strong radiative recombination continua (RRC) of NVII, CVI and CV (Pounds and Vaughan 2011b; Pounds and King 2013), occurs over $\sim$2–6 days. With an electron temperature from the mean RRC profile of $\sim$5 eV, and recombination coefficient for CVI of $\sim 10^{-11}$ (Verner and Ferland 1996), the observed RRC decay timescale corresponds to a (minimum) electron density of $\sim 2 \times 10^6$ cm^{-3}. A column density of 1.5×10^{21} cm^{-2} from modelling absorption in the main low ionisation flow component then corresponds to an absorbing path length of 7×10^{14} cm.

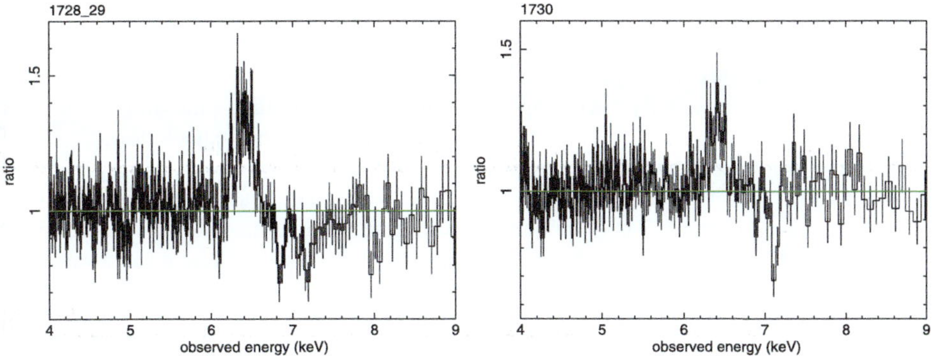

Fig. 6 Fe K profiles for the composite of orbits 6 and 7 and for orbit 8, the latter coinciding with a hard X-ray flare (from Pounds and Vaughan 2012). The ratio of resonance absorption lines of Fe XXV and Fe XXVI is a sensitive measure of the ionisation state of the absorbing gas

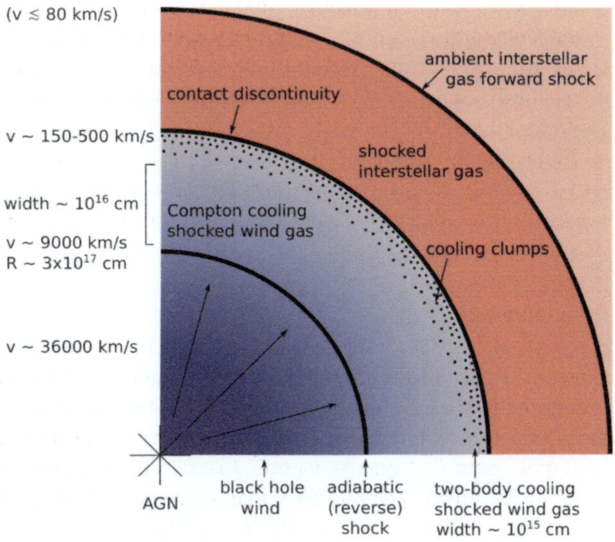

Fig. 7 Structure of the NGC 4051 outflow, not to scale, showing the highly ionised wind colliding with the ISM at ~0.1 pc radius, the strong shock causing a factor 4 drop in velocity. Strong Compton cooling of the shocked gas defines a thin shell where the velocity continues to fall but the ionisation state remains sufficiently high for strong Fe K absorption. Further along, when 2-body cooling becomes important, the flow rapidly cools and slows over a narrower region where absorption (and emission) are dominated by the lighter metals seen in the soft X-ray spectrum (from Pounds and King 2013)

Finally, the RRC emission flux provides a consistency check on the above scaling. Assuming a solar abundance, and 30 percent of recombinations direct to the ground state, a CVI RRC flux of $\sim 10^{-5}$ photons $\text{cm}^{-2}\,\text{s}^{-1}$ corresponds to an emission measure of $\sim 2 \times 10^{62}$ cm^{-3}, for a Tully-Fisher distance to NGC 4051 of 15.2 Mpc.

With a mean particle density of $\sim 2 \times 10^{6}$ cm^{-3} the emission volume $(4\pi.R^2.\Delta R)$ is $\sim 5 \times 10^{49}$ cm^3. Assuming a spherical shell geometry of the flow, with fractional solid angle b, shell thickness $\Delta R \sim 7 \times 10^{14}$ cm, and shell radius $R \sim 10^{17}$ cm, the measured RRC flux is reproduced for $b \sim 0.5$. While this excellent agreement might be fortuitous given the approximate nature and averaging of several observed and modelled parameters, the mutual consistency of absorption and emission of the photoionised flow is encouraging. Given that only blue-shifted RRC emission is seen, $b \sim 0.5$ is consistent with a wide angle flow, visible only on the near side of the accretion disk.

Figure 7 illustrates the main features of the overall NGC 4051 outflow, with a fast primary wind being shocked at a radial distance of order 0.1 pc, within the zone of influence of a

SMBH of $1.7 \times 10^6 M_\odot$. The initially hot gas then cools in the strong radiation field of the AGN, with a Compton cooling length determining the absorption columns of Fe and the other heavy metal ions. Two-body recombination provides additional cooling as the density rises downstream, eventually becoming dominant. Absorption (and emission) in the soft X-ray band is located primarily in this thinner, outer layer of the post-shock flow.

It is interesting to note that similar shocking of fast outflows provides a natural link between UFOs (Tombesi et al. 2010) and the equally common 'warm absorbers' in AGN. While the onset of strong 2-body cooling, resulting in the intermediate column densities being small, might explain why evidence for intermediate flow velocities has awaited an unusually long observation of a low mass AGN, the accumulated 'debris' of shocked wind and ISM could be a major component of the 'warm absorber'.

7 Summary

The requirement of X-ray observations with high sensitivity and high spectral resolution delayed the discovery of powerful, highly ionised winds from AGN until the launch of *Chandra* and *XMM-Newton*. A decade after the initial reports, high velocity ($v \sim 0.1c$) ionised winds are now established to be common in low redshift AGN.

The observed distributions of velocity, ionisation parameter and column density are all compatible with winds launched from close to the black hole, where the optical depth $\tau_{es} \sim 1$, and carrying the local escape velocity. As the mean luminosity in most low redshift AGN is on average sub-Eddington, such winds are likely to be intermittent, a view supported by the range of observed column densities.

Recent evidence of a fast wind in NGC 4051 being shocked at a distance of ~0.1 pc from the black hole offers an explanation of why such powerful winds remain compatible with the continued growth of such systems, strong Compton cooling in the AGN radiation field causing most of the wind energy to be lost before reaching the main star-forming region. Conversely, momentum is conserved through the shock, suggesting a momentum-driven flow is the likely mechanism to eventually curtail growth of the stellar bulge and black hole, as the hole reaches the mass implied by the M–σ relation.

Acknowledgements Many thanks to ISSI for the opportunity to take part in a stimulating Workshop; also to Andrew King, Simon Vaughan, Kim Page and James Reeves for helpful discussions over the period covered by this research.

References

M. Cappi, Astron. Nachr. **327**, 1012 (2006)
G. Chartas, W.N. Brandt, S.C. Gallagher, G. Garmire, Astrophys. J. **569**, 179 (2002)
M. Dadina, M. Cappi, G. Malaguti, G. Ponti, A. de Rosa, Astron. Astrophys. **442**, 461 (2005)
M. Elvis, Mem. Soc. Astron. Ital. **77**, 573 (2006)
L. Ferrarese, D. Merritt, Astrophys. J. **539**, L9 (2000)
K. Gebhardt et al., Astrophys. J. **539**, L13 (2000)
J. Gofford, J.N. Reeves, T. Tombesi, V. Braito, T.J. Turner, L. Miller, M. Cappi, Mon. Not. R. Astron. Soc. **430**, 60 (2013)
J.P. Halpern, Astrophys. J. **1984**, 90 (1984)
T. Kallman, D. Liedahl, A. Osterheld, W. Goldstein, S. Kahn, Astrophys. J. **465**, 994 (1996)
S. Kaspi, E. Behar, Astrophys. J. **636**, 674 (2006)
S. Kaspi, P.S. Smith, H. Netzer, D. Maoz, B.T. Jannuzi, U. Giveon, Astrophys. J. **533**, 631 (2000)
A.R. King, Astrophys. J. **596**, L27 (2003)

A.R. King, Astrophys. J. **635**, L121 (2005)
A.R. King, Mon. Not. R. Astron. Soc. **402**, 1516 (2010)
A.R. King, Mon. Not. R. Astron. Soc. **408**, L95 (2010a)
A.R. King, K.A. Pounds, Mon. Not. R. Astron. Soc. **345**, 657 (2003)
A.R. King, K. Zubovas, C. Power, Mon. Not. R. Astron. Soc. **415**, L6 (2011)
A. Markowitz, J.N. Reeves, V. Braito, Astrophys. J. **646**, 783 (2006)
P. Marziani, J.W. Sulentic, D. Dultzin-Hacyan, M. Clavani, M. Moles, Astrophys. J. Suppl. Ser. **104**, 37 (1996)
P.T. O'Brien, J.N. Reeves, C. Simpson, M.J. Ward, Mon. Not. R. Astron. Soc. **360**, 25 (2005)
K.A. Pounds, A.R. King, Mon. Not. R. Astron. Soc. **433**, 1369 (2013)
K.A. Pounds, K.L. Page, Mon. Not. R. Astron. Soc. **372**, 1275 (2006)
K.A. Pounds, J.N. Reeves, Mon. Not. R. Astron. Soc. **374**, 823 (2007)
K.A. Pounds, J.N. Reeves, Mon. Not. R. Astron. Soc. **397**, 249 (2009)
K.A. Pounds, S. Vaughan, Mon. Not. R. Astron. Soc. **413**, 1251 (2011a)
K.A. Pounds, S. Vaughan, Mon. Not. R. Astron. Soc. **415**, 2379 (2011b)
K.A. Pounds, S. Vaughan, Mon. Not. R. Astron. Soc. **423**, 165 (2012)
K.A. Pounds, J.N. Reeves, A.R. King, K.L. Page, P.T. O'Brien, M.J.L. Turner, Mon. Not. R. Astron. Soc. **345**, 705 (2003)
J.N. Reeves, P.T. O'Brien, M.J. Ward, Astrophys. J. **593**, 65 (2003)
J.N. Reeves, C. Done, K.A. Pounds, Y. Tereshima, K. Hayashida, N. Anabuki, M. Uchino, M.J.L. Turner, Mon. Not. R. Astron. Soc. **385**, L108 (2008)
C. Reynolds, A. Fabian, Mon. Not. R. Astron. Soc. **273**, 1167 (1995)
F. Tombesi, M. Cappi, J.N. Reeves, G.C. Palumbo, T. Yaqoob, V. Braito, M. Dadina, Astron. Astrophys. **521**, A57 (2010)
F. Tombesi, M. Cappi, J.N. Reeves, G.C. Palumbo, V. Braito, M. Dadina, Astrophys. J. **742**, 44 (2011)
S. Vaughan, P. Uttley, Mon. Not. R. Astron. Soc. **390**, 421 (2008)
D.A. Verner, G.J. Ferland, Astrophys. J. Suppl. Ser. **103**, 467 (1996)

DOI 10.1007/978-1-4939-2227-7_18
Reprinted from *Space Science Reviews* Journal, DOI 10.1007/s11214-013-0017-3

Outflow Launching Mechanisms

Ken Ohsuga · Shin Mineshige

Received: 1 April 2013 / Accepted: 12 August 2013 / Published online: 9 October 2013

Abstract Outflow from accretion flow seems to be quite general features of accretion processes. A number of launching mechanisms are known for accretion flow onto black holes: thermal wind, (continuum and line) radiation-pressure driven wind, and magnetically driven wind. To investigate the physics of outflow it is essential first to build good models for underlying accretion flow. Historically, various accretion disk models have been constructed under radially one-zone approximations, but they are not appropriate for exploring the outflow mechanisms, since the multi-dimensional motion of gas in accretion flow, including outflows, is totally neglected. Another limitation comes from that the disk viscosity, the most important ingredient, is described by the phenomenological α-viscosity model. We, here, elucidate the theory of accretion flow and associated outflow based on global, two-dimensional radiation-magnetohydrodynamic (radiation MHD) simulations, not relying on the α-viscosity prescription. We have succeeded in producing three distinct states of accretion flow by controlling only one parameter, a density normalization, and confirmed the occurrence of ubiquitous outflow from all the three states of accretion flow: supercritical, standard, and low-luminosity states. Especially strong outflow is confirmed from the supercritical and low-luminosity accretion flow. Several noteworthy features of the supercritical (or super-Eddington) accretion flows are found; that is, relativistic, collimated outflows (jets), and low-velocity, uncollimated outflows with clumpy structure.

Keywords Black hole · Accretion · Outflow · Magnetohydrodynamics · Radiation hydrodynamics

K. Ohsuga (✉)
Theory Division, National Astronomical Observatory, Mitaka, Tokyo 181-8588, Japan
e-mail: ken.ohsuga@nao.ac.jp

S. Mineshige
Department of Astronomy, Graduate School of Sciences, Kyoto University, Kyoto 606-8502, Japan

1 Introduction

Accretion onto a self-gravitating body is the most efficient energy production mechanism in the Universe. It can actually generate and release huge energy in forms of not only radiation but also gas outflow. Any kinds of astrophysical objects which have been formed via accretion, such as galaxies and stars, seem to have made enormous impacts on their environments by generated outflow. Thus deep understanding of outflow from accretion flow is one of the most important issues in modern astronomy.

There are a number of outstanding issues addressed regarding the nature of outflow. What is the outflow mechanism and on what condition can outflow emerge? How much material, momentum, and energy can be taken away to which direction? What is the relationship with the accretion modes (and spectral states)? It has been suggested that outflow rate is occasionally comparable to accretion rate so that it can significantly modify the accretion flow and its spectra. Instabilities may also be triggered (Begelman et al. 1983). Outflow may take away angular momentum from the accretion flow, thus promoting accretion motion (Blandford and Payne 1982). They can carry enormous energy and momentum, thus producing large impacts on its environments. The co-evolution of supermassive black holes and their host galaxies may also have been created by outflow.

What is obvious is that outflow itself cannot normally produce its energy; it should be powered by accretion processes. To understand how outflow emerges, therefore, it is important to have good knowledge about the nature of accretion flow and its state. This requires to build reliable accretion disk (or flow) models. Various types of accretion disk models have been proposed and discussed for over four decades (see Kato et al. 2008 for an extensive review).

Historically, disk models have been constructed under radially one-zone approximations. Good examples are the standard disk model (Shakura and Sunyaev 1973), the slim disk model (Abramowicz et al. 1988), and the ADAF (advection-dominated accretion flow) or RIAF (radiatively inefficient accretion flow) models (Ichimaru 1977; Narayan and Yi 1994; Abramowicz et al. 1995) proposed for the accretion disk (or flow) with moderate, high, and low luminosities, respectively. Those classical models are really useful for understanding the basic flow structure and for predicting emergent spectra, which have good correspondence with the observational ones. Our knowledge on the accretion flow has been remarkably enriched with these one-zone models.

Those classical models, however, possess a number of limitations and difficulties which have recently been made clear. The most problematic fact in our context is that they cannot obviously treat multi-dimensional motion, such as outflow, convection, and global circulation. Adiabatic inflow-outflow solution proposed by Blandford and Begelman (1999) takes the outflow into consideration, but it is also basically one-dimensional model. Further, viscosity, the most important key ingredient in the accretion disk theory, is prescribed by the phenomenological α-viscosity model, although its physical basis is not so clear. In addition, complex coupling between radiation, magnetic fields, and matter is not properly solved. All these drawbacks are fatal when discussing dynamics of gas outflow.

Since the disk viscosity is thought to be of magnetic origin (Balbus and Hawley 1991), global, multi-dimensional magneto-hydrodynamical (MHD) simulations are being performed rather extensively in this century as a model for the disks with low luminosities (see Kato et al. 2008, Chap. 9 and references therein). There are also attempts to incorporate radiative cooling effects in global/local MHD simulations.

However, MHD simulations are not enough for adequately describing the supercritical case, in which radiation fields are directly coupled with gas dynamics. It is well known that there is a stringent limit to the luminosities of spherically accreting objects, the Eddington luminosity, L_E. This is because above the Eddington luminosity strong radiation- pressure force does not allow material to accrete towards a central object. It may be possible to exceed the Eddington luminosity, however, in the case of disk accretion, since the main direction of the radiation output and mass input differ in the disk geometry (Shakura and Sunyaev 1973; Ohsuga and Mineshige 2007). In this particular context we address the following key questions: How high luminosities can be achieved by supercritical accretion? How much gas can go out as outflow and how fast is the outflow? What are their observational signatures? These are fundamental questions addressed already over several decades, yet no clear answers have been obtained, mainly because for considering complex matter-radiation interactions we need to rely on radiation-hydrodynamical (RHD) simulations.

Since the 1980's several groups performed global, two-dimensional (2-D) RHD simulations of luminous accretion flow (Eggum et al. 1988; Okuda and Fujita 2000; Ohsuga et al. 2005). Those simulations did not consider-MHD processes, however. Hence, they were obliged to utilize the phenomenological α-viscosity model. In the 2000's multi-dimensional radiation-MHD simulations were started (e.g. Turner et al. 2003) but under the shearing-box approximations. The global coupling of magnetic fields is artificially quenched there.

More recently, Ohsuga et al. (2009) presented a new type of accretion flow simulations; that is, global, 2-D radiation-MHD simulations. With such simulations one can for the first time discuss the multi-dimensional outflow properties from different types of accretion flow. In fact, one of the most important finding by these simulations is the existence of ubiquitous outflow from any types of accretion flow.

In this chapter, we discuss launching mechanisms mainly based on our global radiation-MHD simulations and provide basic information specifically on the observable signatures of supercritical accretion flow. For this purpose we first overview the various outflow mechanisms in the next section, and then discuss what we know from the simulation results.

2 Various Outflow Mechanisms

In this section we summarize the very basics of several outflow mechanisms, starting with general introduction to this problem.

2.1 Basic Considerations

Let us first consider why and how outflows can arise from underlying accretion flow. This is not an obvious question, since gas in accretion flow usually possesses negative total energy in the Newtonian sense (i.e., sum of the internal, kinetic, and gravitational energy). For gas on the disk surface to escape to form outflow, pressure force asserted on the gas should dominate over the gravitational force (in the perpendicular direction to the disk plane). That is, there must be some additional energy source other than viscous dissipation within the disk and/or mechanism to selectively inject energy to a fraction of gas. A number of such processes which seem to operate in different situations have been so far discussed.

We need to make remark on the relationship between the disk states and possible onset of outflow. In this respect it is interesting to introduce the argument made by Narayan and Yi (1994). In a steady. inviscid, adiabatic (non-radiative) flow, the quantity (called Bernoulli parameter)

$$\mathrm{Be} \equiv \frac{1}{2}v_r^2 + \frac{1}{2}\Omega^2 R^2 - \Omega_{\mathrm{K}}^2 R^2 + \frac{\gamma}{\gamma - 1}c_s^2$$

should be constant along a stream line (where v_r, Ω, Ω_K,and c_s are radial component of the velocity, angular momentum, and the Keplerian angular momentum, and sound velocity, respectively). They demonstrated that this value can be generally positive in ADAFs by using their self-similar solutions and claim that winds, jets, and other outflows may easily originate in ADAFs (than standard-type disks). This is due to a work made by the torque asserted by the inner parts of the disk.

Their argument has a profound meaning, despite the facts that a positive Bernouilli parameter does not necessarily mean the occurrence of outflow and that the Bernoulli parameter is mostly positive if we properly consider the boundary conditions (Nakamura 1998). We can, at least, say that the gas in a radiatively inefficient state is easier to escape than that in radiative cooling-dominated regimes of accretion (see Blandford and Begelman 1999). (Note $\mathrm{Be} \sim -(1/2)\Omega_K^2 R^2 < 0$ in standard-type disks, since radial velocity and thermal pressure are negligible and Ω–Ω_K.) This seems consistent with the observations that indicate strong (radio) jet emission observed in low/hard state and very high state. But we should keep in mind that the emergence of (low-velocity) outflow from a standard-type disk is not totally prohibited. We thus need more careful analysis to settle down this issue.

2.2 Thermal Wind (Gas-Pressure Driven Wind)

One of the plausible additional energy sources is the heating by the intense radiation from the luminous innermost part of the accretion flow. It is likely that the flux which is generated at the center and which is absorbed by the outer parts may exceed the local heating by viscous processes at large radii (R), since the former rate decays as R^{-2}, whereas the latter rate decays as R^{-3}. The outer parts gets Compton-heated up to the temperature of

$$T_{\mathrm{IC}} \equiv \frac{\langle h\nu\rangle}{4k} = \frac{1}{4kL}\int_0^\infty h\nu L_\nu d\nu \sim 5\times 10^6\ \mathrm{K}$$

for stellar-mass black holes, where k and h denote the Boltzmann constant and the Planck constant, respectively (Begelman et al. 1983; Woods et al. 1996). Accordingly gas pressure is increased and outflow arises when the local sound velocity (roughly equal to wind velocity) exceeds the escape velocity. We can calculate Compton radius, R_{IC}, where the sound velocity with T_{IC} is equal to the escape velocity,

$$R_{\mathrm{IC}} \equiv \frac{GM\mu}{kT_{\mathrm{IC}}} \sim 3\times 10^5 \left(\frac{T_{\mathrm{IC}}}{10^7\ \mathrm{K}}\right) R_{\mathrm{S}}.$$

In reality, launching radius of the wind is about (0.1–0.2) R_{IC}; that is, this process mainly works at large radii. For the cases of higher accretion rates, see Proga and Kallman (2002, 2004, see also next subsection).

2.3 Radiation-Pressure Driven Wind

As disk luminosity increases, so does the fraction of radiation pressure to total pressure. When the disk luminosity approaches the Eddington luminosity,

$$L_{\mathrm{E}} \equiv \frac{4\pi cGM}{\kappa} \sim 1.4\times 10^{38}\left(\frac{M}{10M_{\mathrm{sun}}}\right)\ \mathrm{erg\ s^{-1}}$$

radiation-pressure force which works against gravity force eventually drives powerful outflow at luminosities near the Eddington. It is well known that the luminosity of spherical accretion system cannot exceeds the Eddington, but disk accretion system may be able to

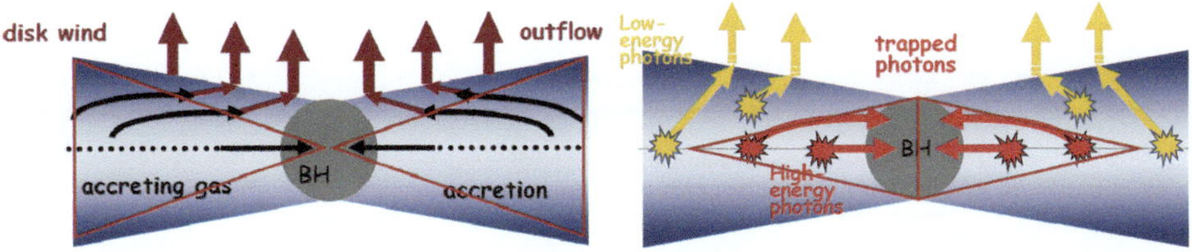

Fig. 1 Schematic figures explaining two important processes within supercritical accretion flow onto a black hole (BH): disk wind (outflow) which occurs from disk surface (*left*) and photon trapping which is effective deep inside the disk (*right*). The critical radii for these two processes are approximately the same; that is, they occur simultaneously

do, since the directions of the inflow motion of gas and of out-going radiation could be different. The critical radius is the spherization radius,

$$R_{\rm sph} \equiv \left(\dot{M}c^2/L_{\rm E}\right)R_{\rm S}$$

inside which quasi-spherical outflow emerges (see the left panel of Fig. 1). There is another critical radius, the photon trapping radius,

$$R_{\rm trap} \equiv \left(\dot{M}c^2/L_{\rm E}\right)(H/R)R_{\rm S}$$

inside which the photon diffusion time in the vertical direction exceeds the accretion time (i.e., photons are "trapped," see the right panel of figures, see also Chap. 10 of Kato et al. 2008).

We here wish to point out that significant photon trapping effects may lead to suppression of radiation-pressure driven outflow, since some fraction of photons are trapped and directly fall onto black holes without exerting strong radiation-pressure force.

Then, a question arises, which wins?

Since the pressure scale-height (H) is comparable to the radius, $H–R$, the trapping radius is on the same order of the spherization radius. That is, both effects work simultaneously but in slightly different ways: the former process starts to work from the disk surface where radiation-pressure force reaches its maximum, while the latter from the mid-plane of the disk where the photon diffusion time reaches its maximum. To understand such complex matter-radiation interactions we need to perform radiation-hydrodynamic and/or radiation-MHD simulations. We also need long-term simulations to follow accretion motion of gas on the viscous diffusion timescale.

Such long-term simulations have become possible only recently and have presented a number of unique features, such as significant internal circulation, emergence of high velocity ($\sim 0.1c$) outflow, which carries large mass and momentum, anisotropic radiation, and so on (Ohsuga et al. 2005; Ohsuga et al. 2009).

Radiation-induced outflow is realized even below the Eddington luminosity. Since line opacity is greater than the electron-scattering opacity, matters can be pushed upward by line-driven force. This process is extensively studied by Proga and his group (e.g., Proga et al. 2000b; Proga and Kallman 2002). More details will be discussed in Sect. 3.4.

2.4 Magnetically Driven Wind

Finally, we introduce the most ubiquitous mechanism; magnetically driven wind. We do not require any special conditions for the occurrence of this type of wind unlike the previous cases. In any situations where magnetic fields exist there is a possibility of magnetically driven wind.

Table 1 Line-driven wind and continuum-driven wind

Wind type	Line-driven wind	Continuum-driven wind
Relevant disk luminosities	Near and above $L_E(\kappa_{es}/\kappa_{line})$	Near and above L_E
Conditions for gas	Not for fully ionized gas	For ionized gas

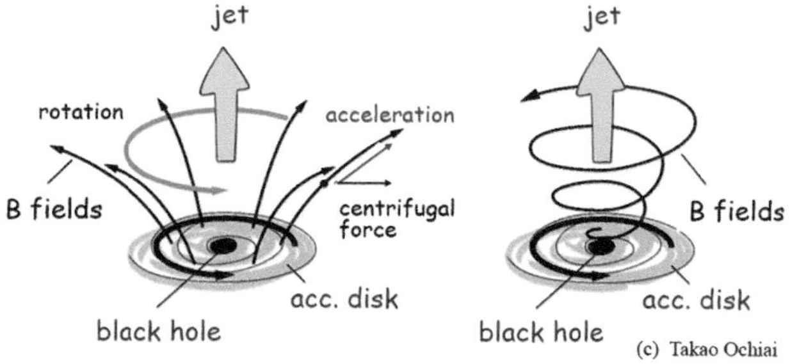

Fig. 2 Two types of MHD jet (or outflow): centrifugal-force driven jet (outflow, *left*) and magnetic-pressure driven jet (outflow, *right*)

Good analogy of this type of winds are the solar wind. It is ubiquitous but may not be so powerful. It is not yet clear how fast and how much can be material carried away under magnetic action. Their properties seem to rely on the strengths of magnetic fields, or the plasma-β, defined by the ratio of the gas pressure to the magnetic pressure.

Blandford and Payne (1982) proposed magnetohydrodynamical outflow. On the basis of self-similar solutions of axially symmetric, cold flow from a Keplerian accretion disk, they conclude that a centrifugally driven outflow is possible, if the poloidal component of the magnetic field makes an angle of less than 60 degree with the disk surface (see the left panel of Fig. 2). This is because for larger angles of the field lines centrifugal force is no longer strong enough to drive outflow. Alternative ideas have been pointed out by MHD simulations by Uchida and Shibata (1985), who demonstrated the occurrence of magnetic-pressure driven jets. Such possibilities are now confirmed by recent MHD simulations under a variety of initial conditions. Kato et al. (2004), for example, have shown the formation of magnetic-tower structure (made by accumulation by toroidal magnetic fields), and jet and outflow emerge driven by its magnetic-pressure force (see the right panel of Fig. 2).

These classical models of magnetically driven outflow describes those arising from the vicinity of black holes and are more or less collimated. Recent MHD simulations, however, revealed the emergence of uncollimated weak outflow from accretion surface at every radius driven by MRI turbulence (see, e.g., Hawley and Krolik 2001; Machida et al. 2004). The remarkable features of these outflows obtained by these MHD simulations are in their violent spatial and temporal variations on local dynamical timescales. But, in overall, quasi-steady outflows realize.

The most essential feature of these MHD models is that the occurrence of outflow does not depend on specific physical conditions of accretion disks. What is required is that disk materials are ionized so that they can strongly couple with magnetic fields, condition that is quite generally satisfied in black hole accretion flows.

3 Outflow by Global Radiation-MHD Simulations

In this section, we discuss the main features of outflow based on our radiation-MHD simulations. The precise numerical procedures and more detailed results have been presented by Ohsuga et al. (2009) and Ohsuga and Mineshige (2011).

3.1 Overview

Ohsuga and Mineshige (2011) solved the accretion flow and outflow structure from the first principle, thus not relying on the α-viscosity prescription. In these simulations the following processes are considered: transport of angular momentum via magnetic torque, leading to the accreting motion, the conversion of the mechanical energy to the thermal energy via MHD processes (MRI and Jule heating), dissipation of thermal energy, radiative transfer, and radiation-pressure and Lorentz forces which play important roles in launching outflows and supporting the disks in the vertical direction.

Our radiation-MHD simulations are performed by employing the flux-limited diffusion approximation for the radiation field. The calculations are performed in Newtonian dynamics.

We start simulations from a torus threaded with weak poloidal magnetic fields (with plasma $\beta = 100$) around a non-spinning black hole of $10M_{\rm sun}$. There is only one model parameter, a density normalization (density at the center of the initial torus). We calculate three models, in total (see Table 1). Since radiation emissivity sensitively depends on the density of gas, three distinct regimes of accretion flow can be reproduced by controlling this density normalization. When it is small, radiative cooling is not efficient (Model C). When it is moderate, radiative cooling turns on but radiation pressure is never dominant (Model B). When it is large enough, radiation pressure dominates over gas pressure (Model A).

Figure 3 is a perspective views of three simulated accretion flows. The (color) contours in the upper panels, which indicate the distributions of normalized density, clearly visualizes distinct flow patterns among three models. Especially, we find that accretion flows are geometrically thick in Models A and C, whereas it is geometrically thin in Model B. Therefore, the flows at high and low luminosities look apparently similar, though it is optically very thick in the former, while very thin in the latter.

The reasons for these differences can be understood in the following way: In Model A, gas density is so high that large amount of radiation must be produced, asserting strong radiation-pressure force to thicken the accretion flow. In Model B, density is moderately high so that radiative cooling can be effective. The disk is supported by gas pressure. The temperature is moderately low due to radiative loss, and so is the pressure scale-height small. A standard-type thin disk thus forms. In Model C, density is too low for radiative cooling to be effective. The flow temperature is highest among the three models, so is the pressure scale-height large. Hence, a geometrically thick, optically thin disk forms. Table 2 gives a summary of the three calculated models.

The outflow regions are also indicated in the upper panels of Fig. 3. We find ubiquitous outflow from every mode of accretion flow, which was not always anticipated. Among the three models Model A records the largest outflow rate and Model C produces the second largest one. Such strong outflow will inevitably affect the environments of the flow (discussed later).

It is also important to note that the r–ϕ component of the shear-stress tensor is shown by our simulations to be roughly proportional to the local pressure in the time averaged sense. Although such a proportionality relation was assumed in the α-viscosity model and has been

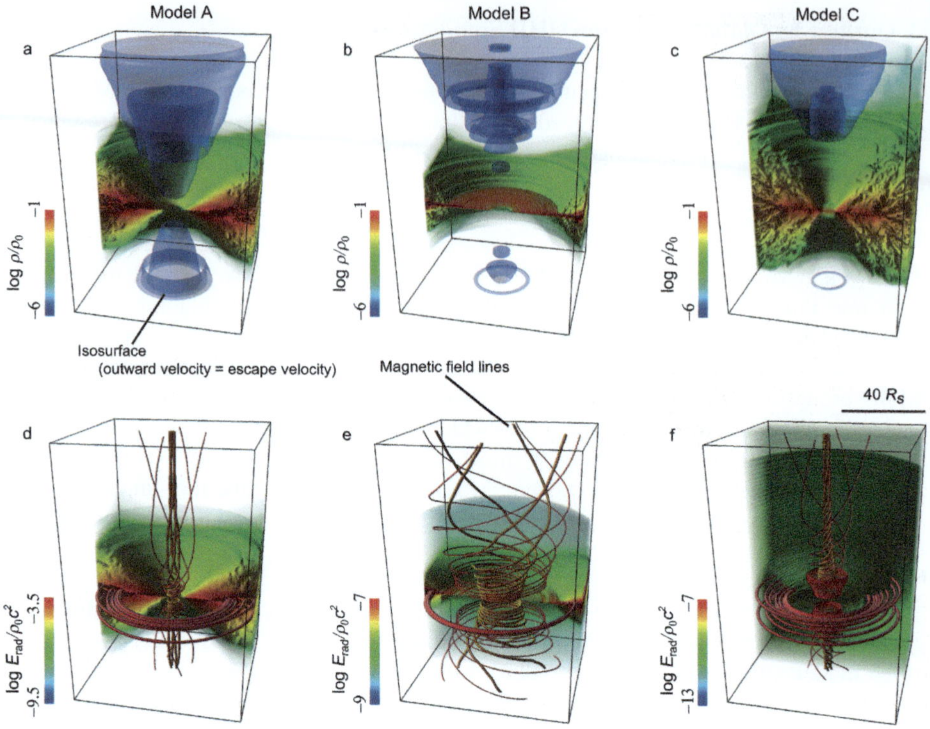

Fig. 3 Perspective view of inflow and outflow patterns for Models A, B, and C, from left to right via 2D radiation-MHD simulations. Normalized density distributions (*center*) are overlaid with isosurfaces (*above and below*), at which the outward velocity equals to the escape velocity in the upper panels, while distribution of radiation energy density is overlaid with magnetic field lines in the lower panels. For more details, see Ohsuga et al. (2009)

Table 2 Various accretion modes calculated by global radiation-MHD simulations

Model/states	Typical density and temperature	Accretion flow	Radiation & kinetic luminosities	Outflow mechanisms
Model A/ supercritical state	10^{-2} g cm^{-3} 10^{8} K	Geometrically thick, optically thick disk	$\sim L_E$ $\sim 0.1 L_E$	Strong radiation-pressure (continuum) driven outflow
Model B/ standard state	10^{-5} g cm^{-3} 10^{6} K	Geometrically thin, optically thick disk	$\sim 10^{-2} L_E$ $\sim 10^{-5} L_E$	Weak, magnetically driven outflow
Model C/RIAF state	10^{-9} g cm^{-3} 10^{10} K	Geometrically thick, optically thin disk	$\sim 10^{-8} L_E$ $\sim 10^{-6} L_E$	Strong, gas-pressure and magnetically driven outflow

widely used since the 1970's, there was no proof for this. We for the first time have proven it in the standard, RIAF, and slim accretion modes by the global radiation-MHD simulations, while non-radiative MHD simulations did only for RIAF mode (see Hawley et al. 2003).

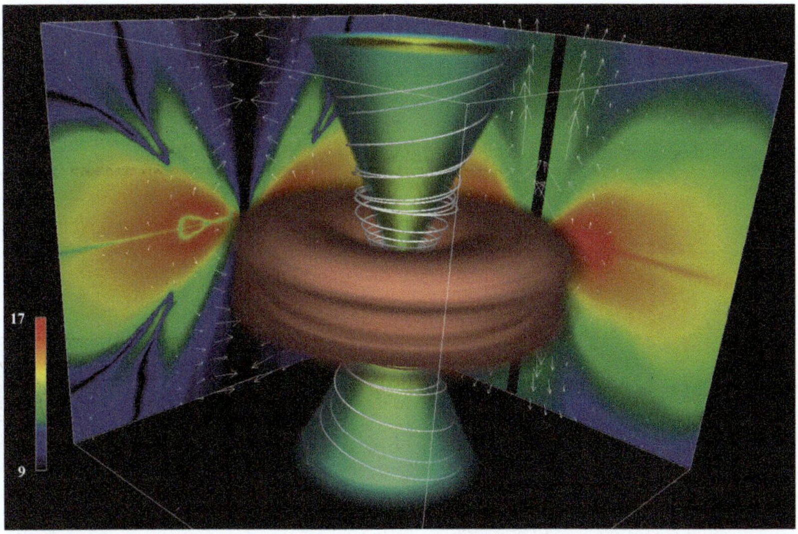

Fig. 4 Bird's-eye view of the luminous accretion flow and the associated radiation-MHD jet obtained by 2D radiation-MHD simulations. The accretion flow (the gas mass density, *brown*) and the radiation-MHD jet in which velocities exceed the escape velocity (the velocity, *white*, *blue*) are plotted. The high-speed jet (0.6–0.7c) is represented by *blue*. *White lines* indicate the magnetic field lines. The zz-component of the radiation-pressure tensor (color), overlaid with the radiation-pressure force vectors (*arrows*) on the meridional plane is projected on the left wall surface, while the magnetic pressure from the azimuthal component of the magnetic field (color), overlaid with the Lorentz force vectors (*arrows*) on the meridional plane is projected on the right wall surface. For more detailed explanation, see Takeuchi et al. (2010)

3.2 Radiation-MHD Jet

From now on, we focus on Model A, the 2-D radiation-MHD version of the supercritical flow. The accretion disk in Model A is geometrically thick, as shown in Fig. 3, and is supported by the radiation-pressure force.

The accretion rate onto the black hole is $\sim 50 L_{\rm E}/c^2$, much larger than the critical value, and the luminosity exceeds the Eddington luminosity, $L_{\rm ph} \sim 1.7 L_{\rm E}$. Note that in the calculations of the photon luminosity of Model A we only consider the radiative energy released from the inner disk surface ($R < 25 R_{\rm S}$) per unit time to avoid possible influence from the initial torus. [Here, we used the cylindrical coordinates (R, θ, z).] So the actual photon luminosity might be higher, if we add the radiative flux at the outer disk surface ($R > 25 R_{\rm S}$).

Figure 3 also shows that the matter goes outward above and below the disk. The mass-outflow rate is a few percent of the mass accretion rate. We also find kinetic luminosity of $L_{\rm kin} \sim 0.1 L_{\rm ph}$; that is, the supercritical flows release energy via radiation rather than via outflows.

A part of gas elements emerges and accelerates close to the photosphere in the inner region of accretion flows in which the radiation pressure dominates over gas pressure in the vertical direction. Radiation-pressure driven outflows thus arise. Within supercritical accretion flows, by contrast, rough force balance is achieved between the gravity force and the radiation force. The outflow is composed of the inner, relativistic, collimated (<10 degree) jet along the disk rotation axis and the outer, uncollimated, low-speed ($\sim 0.1c$) outflow emerging over a wide angle (see Fig. 4). We evaluated the mass accretion rate is $\dot{M}_{\rm acc} \sim 50 L_{\rm E}/c^2$, the mass outflow rate is $\dot{M}_{\rm out} \sim 10 L_{\rm E}/c^2$, and the photon luminosity is

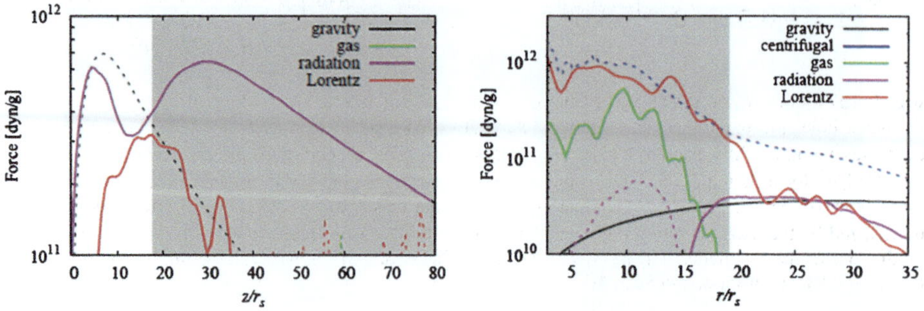

Fig. 5 Acceleration and collimation mechanism of the radiation-MHD jet. (*Left panel*) The vertical profiles of the gravitational force (*black*), the gas-pressure force (green), the radiation-pressure force (*red*), and the Lorentz force (*blue*) at $r = 10R_S$. The *solid lines* and the *dashed lines* indicate the outward force and the inward force, respectively. The radiation-pressure force is responsible for the jet acceleration. (*Right panel*) The radial profiles of the gravitational force (*black*), the centrifugal force (*magenta*), the gas-pressure force (*green*), the radiation-pressure force (*red*), and the Lorentz force (*blue*), at $z = 40R_S$. The *solid lines* and the *dashed lines* indicate the inward force and the outward force. We understand that the Lorentz force is responsible for the collimation of the jet. In both panels each quantity is time-averaged over 1 s, and the shadowed areas indicate the jet region in which the outflow velocity exceeds the escape velocity. For more detailed explanation, see Takeuchi et al. (2010)

$L_{\rm rad} \sim 2L_{\rm E}$. Here, the mass accretion rate and the mass outflow rate were calculated by summing up the mass passing through the inner boundary and the upper boundary per unit time with higher velocities than the escape velocity. The photon luminosity were calculated based on the radiative flux at the upper boundary.

To see what force is responsible for the acceleration and collimation of the jet, we evaluate the time-averaged strengths of the vertical components of the gravitational force, the gas-pressure force, the radiation-pressure force, and the Lorentz force (Takeuchi et al. 2010). The results shown in Fig. 5 indicate that the jet is accelerated by the radiation-pressure force, while the Lorentz force contributes much to the collimation. The latter was not expected, since the radiation energy density greatly exceeds the magnetic energy density, typically, by more than one order of magnitude. The key to understanding this magnetic collimation is the formation of a magnetic tower structure (Lynden-Bell 1996).

The magnetic tower structure is created by the inflation of the toroidal component of magnetic fields accumulating around the black hole (Kato et al. 2004). There should be something which prevents the magnetic tower structure from expansion. The agent that works could be an uncollimated outflow surrounding the jet (see Sect. 3.3).

The radiative flux from the supercritical flow in our simulations is highly dependent on the viewing angle. For a face-on observer, the maximum apparent (isotropic) luminosities of $L_{\rm iso} \sim 22L_{\rm E}$ is achieved for the mass accretion rate of $\sim 50L_{\rm E}/c^2$. (The maximum apparent luminosity is higher than that found by the radiation-MHD simulations, indicating that the beaming effects are enhanced by magnetic collimation.) Even higher luminosities are also feasible, if we increase mass accretion rate. Note, however, Compton cooling should be very effective for the cases with higher mass accretion rates, which may lead to a suppression of the photon luminosities (Kawashima et al. 2012).

For an edge-on observer, by contrast, the apparent luminosity will be much less because emission is mildly beamed and the innermost bright part of the disk is obscured by the outer part (see Begelman et al. 2006). We can thus expect large ratios of the kinetic luminosity ($L_{\rm kin} \sim 0.1L_{\rm E}$) to the isotropic luminosity ($L_{\rm iso} \ll L_{\rm E}$) for edge-on observers. The large

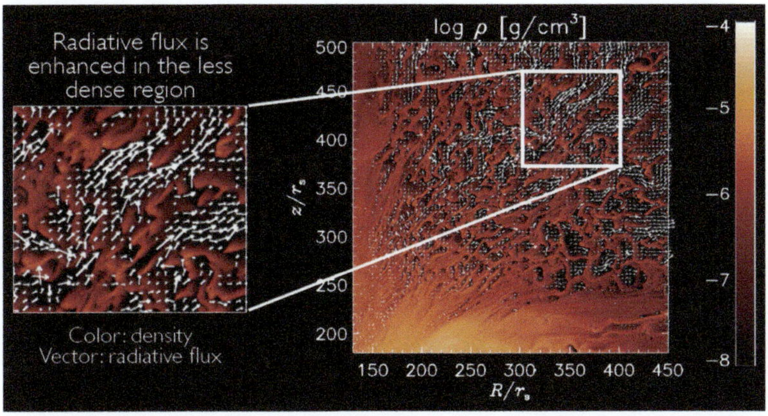

Fig. 6 Density structure (color) and radiative flux (vectors) of the clumpy outflow simulated by 2D radiation-MHD simulations. The clumps appear in the distant regions from the central *black* hole. The clump size is typically $\sim 10R_S$ or one optical depth. More details are shown in Takeuchi et al. (2013)

kinetic luminosity may account for the large (100–500 pc) ionizing nebulae around ultraluminous X-ray sources (ULXs) and/or microquasars (e.g. Pakull et al. 2010).

3.3 Clumpy Outflow

In order to examine large-scale behavior of the outflow, we run another simulation by significantly expanding the simulation box to be $(R_{\max}, z_{\max}) = (500R_s, 500R_s)$. As a result, we found clumpy structure in the distant outflow region at $R > 200R_s$, and $z > 250R_s$ (see Fig. 6, Takeuchi et al. 2013). Gas clumps with gas mass density of $\sim 10^{-6}$ g cm^{-3} and velocity of $\sim 0.1c$ are blown away over wide angles.

To quantify the clumpy structure, we have performed auto-correlation function (ACF) analyses for the density distribution. We confirm a sharp peak in the ACF for the density distribution above $z > 250R_s$. The width of the central peak (i.e, the typical clump size) is $\sim 5R_s$, which roughly corresponds to one optical depth. We should note that the layer where clumps are formed (at $z \sim 250R_s$) approximately coincides with the 'photosphere' of the flow; i.e., the average optical depth to this layer is unity. These facts indicate some kind of radiation-related instability being responsible for clump formation.

We, next, performed cross-correlation function (CCF) analysis, finding a clear anti-correlation between the gas mass density and the perpendicular component of the radiation force to the velocity vector in the clumpy outflow region. That is, the radiative flux avoids dense regions (i.e., clumps) and instead selectively passes through low-density channels between the clumps, pushing aside gas in the channel towards the clumps. Clumps thus grow until their optical depth becomes unity.

What is then a mechanism creating clumpy features of the outflow? There are a number of unique features confirmed by the simulations: (1) clumpy structure appears in the layer where upward radiation force overcomes downward gravity force, (2) a clump size is about one optical depth, (3) there is a clear anti-correlation between the gas density and radiation force, (4) temperature variations of some clumps are not monotonic increase nor monotonic decrease.

On the basis of these facts let us discuss plausible physical mechanism underlying clump formation. The fact (4) obviously indicates that clump formation cannot be explained by

Fig. 7 Schematic picture of supercritical accretion flow and associated outflow composed of a collimated, high-velocity jet and surrounding uncollimated, low-velocity outflows. The latter exhibit clumpy structure at distant regions

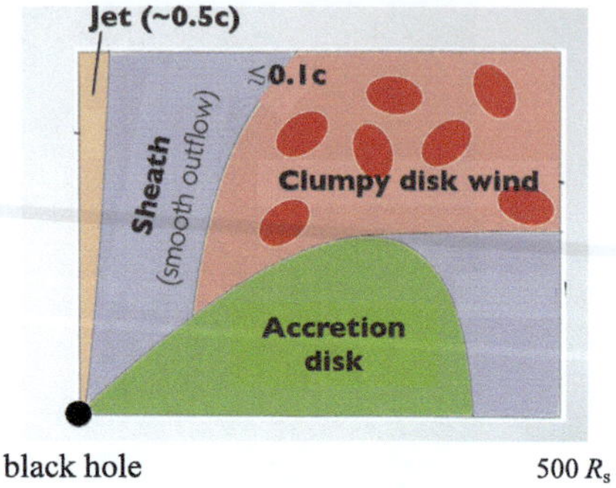

a thermal instability, which cause that the gas temperature monotonically decreases (increases) with a monotonic increase (decrease) of the matter density (Balbus and Soker 1989).

The fact (1) means that the clumpy outflow is *Rayleigh-Taylor unstable*. In luminous black hole inflow-outflow system, radiation field acts like an effective gravitational field. Since the matter density decreases in the direction of the acceleration, the condition for Rayleigh-Taylor instability is well satisfied. The overturning motion of the gases would form the seed of clumpy density pattern. We thus conclude that Rayleigh-Taylor instability is the primary cause of the clump formation. Jacquet and Krumholz (2011) performed linear stability analysis of a plane-parallel superposition of two media immersed in radiation field. They examined the stability in the optically thin and thick limit, finding that both case could be Rayleigh-Taylor unstable. Such radiation-induced Rayleigh-Taylor instability also appears in massive star formation system and H II region. The Rayleigh-Taylor instability, however, cannot explain the facts (2) and (3), which would imply radiation processes being responsible. We thus further examine radiation-related instabilities. Radiation force could work as the positive feedback to grow the initial perturbation of the matter density, which forms the inhomogeneous density pattern.

Shaviv (2001) made a linear stability analysis for the optically thick, radiation dominated atmospheres of the super-Eddington wind from stars. He has clearly shown that perturbations with wavelengths on the order of one scale-height grow on dynamical timescales under a fixed temperature condition at the bottom. This is a radiation hydrodynamic instability and is expected to create inhomogeneous density structure in the atmosphere. This is exactly the situation which we encounter here.

The directivity of radiation force (or radiation flux) is fundamental in this radiation hydrodynamic instability. The instability does not occur in the optically thick limit due to synchronization between matter and radiation field. It does not occur in the optically thin limit, either, since radiation cannot push material effectively but just passes through it. Further, this instability only occurs in the optically thick outflow from supercritical flows.

The outflow will produce several observable features. Photons from the underlying disk will be Compton up-scattered by hot plasmas in the outflow, which will create Compton-dominated spectra. Further, clumpy outflow structure will produce significant time variability in the photon luminosities. Those features are actually observed in ULXs (e.g., Middleton et al. 2011).

Finally, Fig. 7 summarizes the inflow and outflow structure obtained by our radiation-MHD simulations. It is important to note the smooth outflow (or "sheath") works to confine magnetic tower from expansion. We should also note that we have made a number of approximations in our simulations; axisymmetry, Newtonian dynamics, and the employment of the flux-limited diffusion. These points will be improved in future simulations.

Here, we wish to note that the properties obtained by our simulations share plenty of similarities with the observational ones (see chapters by Pounds and by King).

3.4 Line-Driven Outflow

The radiation force due to the spectral lines (line force) works to accelerate the matter, since metals in lower ionization state absorbs ultra-violet (UV) radiation via the bound-bound transition. The line opacity is much larger than the electron-scattering opacity so that outflow powered by the line force (line-driven outflow) appears even if the luminosity is less than the Eddington luminosity, in contrast to the continuum-driven outflow. The line-driven outflow was studied so as to explain the stellar wind (Castor et al. 1975), and was later applied to the disk wind (Proga et al. 1998, 2000a).

Here, we note that X-ray radiation works to prevent launching the line-driven outflow. Since the metals are fully photoionized by the strong X-ray irradiation, the line-force is made powerless and fails to form the outflow. Thus the line driven outflow would not appear in the case of the stellar mass black holes, since the standard-type thin disks are very hot $\sim 10^7$ K, thereby effectively emitting X-rays. Even in cases of the supermassive black holes, the matter is thought to be highly ionized in the vicinity of the black holes ($r \ll 100R_s$.). Thus, the line driven outflows are ejected only from the distant regions at $> \sim 100R_s$. Although the outflow fails to launch (failed wind) near the black hole, such failed wind plays an important role for forming the line driven outflow. The X-ray radiation is attenuated by the absorption/scattering by the failed wind, so that the ionization degree decreases outward and line force can blow away the matter behind the failed wind. Such structure was mentioned by Murray et al. (1995), and clearly shown by hydrodynamic simulations (Proga et al. 2000a, Proga and Kallman 2004) and by non-hydrodynamics method (Risaliti and Elvis 2010, Nomura et al. 2013).

Figure 8 illustrates the density profile of the line driven outflow, which is obtained by two-dimensional hydrodynamic simulations (Proga et al. 2000b). Here, the line force is simply assessed with using the force multiplier (Stevens and Kallman 1990). The geometrically thin disk with sub-Eddington luminosity (i.e., a standard disk) is assumed to be located at the equatorial plane. The disk structure is not solved and matter is assumed to be supplied from the disk surface. This figure shows the outflow is ejected around $r/r_* = 200$ (which corresponds to $r = 600R_s$), and its opening angle is 60–80 degree. The outflow is not smooth and a cloud sometimes appears in the outflow (see the red region around $r/r_* = 800$).

The line driven outflow is one of the most plausible models for explaining the blueshifted absorption features found in spectra of active galactic nuclei (AGNs). Since the line force accelerates the gas containing the metals in low or moderate ionization state, the line driven wind can explain both outflowing motion and ionization state. Schurch et al. (2009) calculated the spectra in X-ray band based on simulated line driven outflow (Proga and Kallman 2004).

Figure 9 shows the emergent spectra. The left panel shows the resulting spectra with the viewing angle of $\theta = 50$ (red), 57 (green), 62 (blue), 64 (cyan), 67 (pink) degrees. It is found that the observer with $\theta \sim >60$ detects strong absorption features, since the outflow mainly interrupt the lines of sight. In contrast, the absorption lines are not noticeable if we observe the AGN from the angle of $\theta < 50$. The system is Compton-thick for the observers with $\theta > 70$.

The right panel represents the time evolution of the spectra with $\theta = 62$. The red line indicates a spectrum at a given time, in which we find some absorption lines. After 4.46×10^6 s, the nucleus is heavily obscured, the X-ray radiation is attenuated (green). The blue

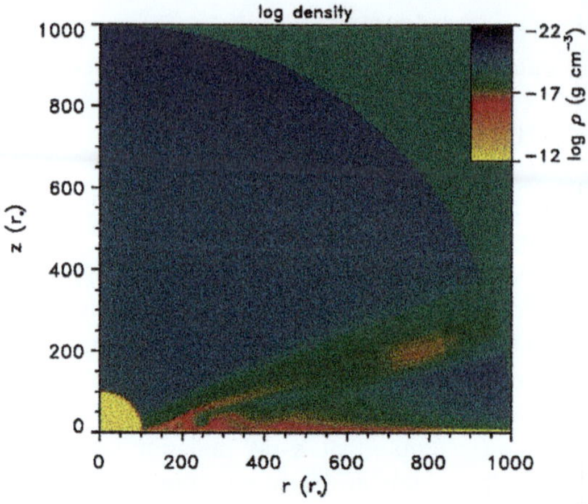

Fig. 8 Structure of the line driven outflow launched form the disks around the supermassive black holes, obtained by 2D hydrodynamic simulations (Proga et al. 2000b). The outflow, of which the opening angle is 60–80 degree, is blown away (see the *green* and *red* regions)

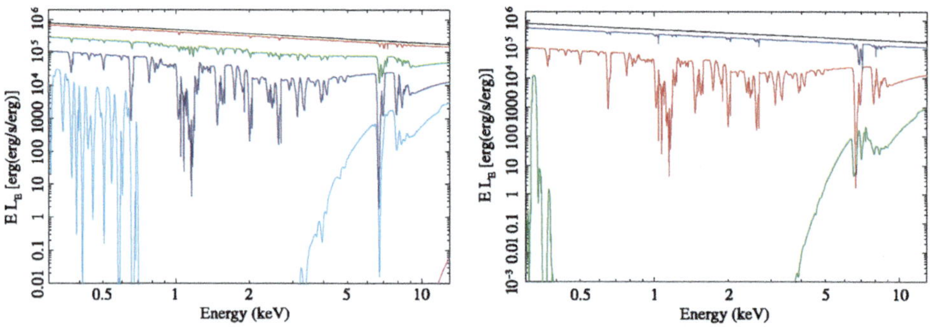

Fig. 9 Simulated X-ray spectra (Schurch et al. 2009). (*Left*) Spectra from five different line of singts, $\theta = 50$ (*red*), 57 (*green*), 62 (*blue*), 64 (*cyan*), 67 (*pink*). (*Right*) Spectra from $\theta = 62$. *Red line* indicates an initial spectrum. The *green* and *blue lines* show the spectra after 4.46×10^6 and 9.5×10^6 s, respectively

line shows the spectrum at 9.5×10^6 s, in which the absorption features are unnoticeable. Such results imply that the absorption features sensitively depends on the viewing angle and changes with time.

4 Outstanding Issues

4.1 Inverse-Compton Effects

When soft photons are abundantly supplied by underlying supercritical accretion flow, hot outflow should undergoes remarkable inverse-Compton cooling, thus not only outflow temperature but also emergent spectra being significantly modified, Such effects have not been taken into account in the simulation results presented in the previous section. It is still difficult to incorporate Compton scattering processes in global radiation-MHD simulations, though there are some attempts in (non-MHD) radiation hydrodynamical simulations.

Kawashima et al. (2012) have demonstrate Compton effects lead to low temperatures and Compton dominated spectra, as are observed in ULXs (e.g., Gladstone et al. 2009). More

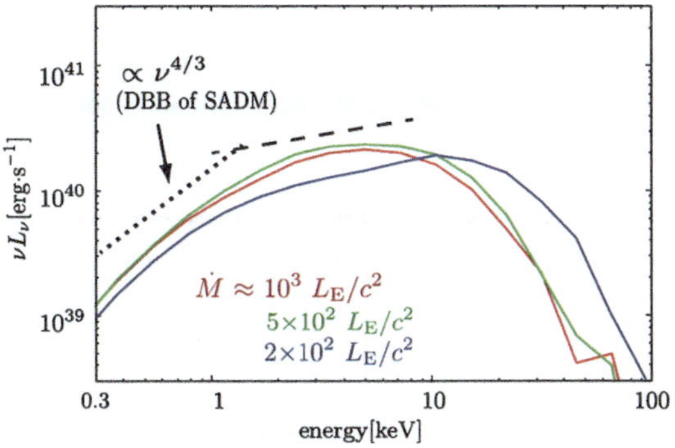

Fig. 10 Spectral energy distribution of supercritical accretion flow obtained by global 2D radiation hydrodynamical simulations with inverse Compton scattering (adapted from Kawashima et al. 2012). Because of Compton cooling the total radiation luminosity is saturated

importantly in the context of outflow, they found that the total radiation luminosities no longer grow in proportion to mass supply rate, when it reaches some critical value of several tens of L_E/c^2 (see Fig. 10). Proper treatment of MHD processes to this kind of simulations are one of the most important future issues.

4.2 How Much Fraction of Gas Energy Can Be Extracted?

One of the naive questions regarding the outflow from super/subcritical accretion flow will be: how much mass, momentum, and energy can be finally carried away to leave a system to reach infinity? This is a fundamental question and is also important when considering the feedback effects, but yet an unanswered question at this moment. This is because we need really long-term (grossly over $10^6 R_s/c$) large scale (over $10^5 R_s$) radiation-MHD simulations, which are impossible because of a very limited computer performance. Again, however, (non-MHD) radiation hydrodynamical simulations can be performed with existing supercomputers. Such simulations have recently performed and have shown considerable fraction of outflow can actually go out.

5 Conclusion

We here summarize our current theoretical understanding supercritical accretion flow.

1. A number of launching mechanisms are known from accretion flow onto black holes: thermal wind, radiation (continuum and line) pressure driven wind, and magnetically driven wind. Thermal wind originates from large radius of the disk, while radiation-pressure driven wind occurs only at high luminosities around the Eddington luminosity. Magnetically driven wind, in contrast, can be launched from every radii at any luminosity state, although it does not always dominate everywhere (Proga 2003 see also Ohsuga et al. 2009, Ohsuga and Mineshige 2011).
2. We elucidate the theory of accretion flow and associated outflow based on global, two-dimensional radiation-magnetohydrodynamic (radiation MHD) simulations, not relying on the α-viscosity prescription. We have succeeded in producing three distinct states of accretion flow by controlling a density parameter, and confirmed the occurrence of ubiquitous outflow from all the three states of accretion flow: supercritical, standard, and low-luminosity states.
3. Especially strong outflow is confirmed from the supercritical and low-luminosity accretion flow. Supercritical accretion flow accompanies large-scale outflows at rates over

$\sim 10L_{\mathrm{E}}/c^2$. The outflows are composed of the inner, relativistic, collimated (<10 degree) jets and the outer, relatively low-speed ($\sim 0.1c$) uncollimated outflows emerging over a wide angle. The jets are accelerated by the radiation-pressure force and is collimated by the Lorentz force.
4. We expect a number of observational features in the supercritical outflow. Hot materials in the outflow will Compton up-scatter soft photons emitted from the underlying disk surface, thus producing Compton-dominated spectra. The clumpy structure, which is created by a radiation-hydrodynamic instability, will produce significant time variability.

Acknowledgements This work is supported in part by the Grant-in-Aid of MEXT (22340045, S.M.; 20740115, K.O.), and the global COE programs on The Next Generation of Physics, Spun from Diversity and Emergence from MEXT (SM). Numerical computations were in part carried out on Cray XT4 at CfCA of NAOJ.

References

M.A. Abramowicz, B. Czerny, J.-P. Lasota, E. Szuszkiewicz, Astrophys. J. **332**, 646 (1988)
M.A. Abramowicz, X. Chen, S. Kato, J.P. Lasota, O. Regev, Astrophys. J. **438**, L37 (1995)
S.A. Balbus, J.F. Hawley, Astrophys. J. **376**, 214 (1991)
S. Balbus, N. Soker, Astrophys. J. **341**, 611 (1989)
M.C. Begelman, C.F. McKee, G.A. Shields, Astrophys. J. **271**, 70 (1983)
M.C. Begelman, A.R. King, J.E. Pringle, Mon. Not. R. Astron. Soc. **370**, 399 (2006)
R.D. Blandford, D.G. Payne, Mon. Not. R. Astron. Soc. **199**, 883 (1982)
R.D. Blandford, M.C. Begelman, Mon. Not. R. Astron. Soc. **303**, L1 (1999)
J.I. Castor, D.C. Abbott, I. Klein, Astrophys. J. **195**, 157 (1975)
G.E. Eggum, F.V. Coroniti, J.I. Katz, Astrophys. J. **330**, 142 (1988)
J.C. Gladstone, T.P. Roberts, C. Done, Mon. Not. R. Astron. Soc. **397**, 1836 (2009)
J.F. Hawley, J.H. Krolik, Astrophys. J. **548**, 348 (2001)
J.F. Hawley, S.S. Balbus, J.M. Stone, Astrophys. J. **554**, L49 (2003)
S. Ichimaru, Astrophys. J. **214**, 840 (1977)
E. Jacquet, M.R. Krumholz, Astrophys. J. **730**, 116 (2011)
Y. Kato, S. Mineshige, K. Shibata, Astrophys. J. **605**, 307 (2004)
S. Kato, J. Fukue, S. Mineshige, *Black-Hole Accretion Disks—Towards a New Paradigm* (Kyoto Univ. Press, Kyoto, 2008)
T. Kawashima, K. Ohsuga, S. Mineshige, T. Yoshida, D. Heinzeller, R. Matsumoto, Astrophys. J. **752**, 18 (2012)
D. Lynden-Bell, Mon. Not. R. Astron. Soc. **279**, 389 (1996)
M. Machida, K. Nakamura, R. Matsumoto, Publ. Astron. Soc. Jpn. **56**, 671 (2004)
M.J. Middleton, T.P. Roberts, C. Done, F.E. Jackson, Mon. Not. R. Astron. Soc. **411**, 644 (2011)
N. Murray, J. Chiang, S.A. Grossman, G.M. Voit, Astrophys. J. **451**, 498 (1995)
R. Narayan, I. Yi, Astrophys. J. **428**, L13 (1994)
K. Nakamura, Publ. Astron. Soc. Jpn. **50**, L11 (1998)
M. Nomura et al., Publ. Astron. Soc. Jpn. **65**, 40 (2013)
K. Ohsuga, S. Mineshige, M. Mori, Y. Kato, Publ. Astron. Soc. Jpn. **61**, L7 (2009)
K. Ohsuga, S. Mineshige, Astrophys. J. **736**, 2 (2011)
K. Ohsuga, S. Mineshige, Astrophys. J. **670**, 1283 (2007)
K. Ohsuga, T. Mori, S. Nakamoto, S. Mineshige, Astrophys. J. **628**, 368 (2005)
T. Okuda, M. Fujita, Publ. Astron. Soc. Jpn. **52**, L5 (2000)
M.W. Pakull, R. Soria, C. Motch, Nature **466**, 209 (2010)
D. Proga, Astrophys. J. **585**, 406 (2003)
D. Proga, T.R. Kallman, Astrophys. J. **565**, 455 (2002)
D. Proga, T.R. Kallman, Astrophys. J. **616**, 688 (2004)
D. Proga, J.M. Stone, J.E. Drew, Mon. Not. R. Astron. Soc. **295**, 595 (1998)
D. Proga, J.M. Stone, T.R. Kallman, Astrophys. J. **543**, 686 (2000a)
D. Proga, J.M. Stone, T.R. Kallman, Astrophys. J. **543**, 686 (2000b)
G. Risaliti, M. Elvis, Astron. Astrophys. **516**, 89 (2010)
N.J. Schurch, C. Done, D. Proga, Astrophys. J. **694**, 1 (2009)

N.I. Shakura, R.A. Sunyaev, Astron. Astrophys. **24**, 337 (1973)
N. Shaviv, Astrophys. J. **549**, 1093 (2001)
I.R. Stevens, T.R. Kallman, Astrophys. J. **436**, 599 (1990)
S. Takeuchi, K. Ohsuga, S. Mineshige, Publ. Astron. Soc. Jpn. **62**, L43 (2010)
S. Takeuchi, K. Ohsuga, S. Mineshige, Publ. Astron. Soc. Jpn. (2013, in press)
N.J. Turner, J.M. Stone, J.H. Krolik, T. Sano, Astrophys. J. **593**, 992 (2003)
Y. Uchida, K. Shibata, Publ. Astron. Soc. Jpn. **37**, 515 (1985)
D.T. Woods, R.I. Klein, J.I. Castor, C.F. McKee, J.B. Bell, Astrophys. J. **461**, 767 (1996)

DOI 10.1007/978-1-4939-2227-7_19
Reprinted from *Space Science Reviews* Journal, DOI 10.1007/s11214-013-0001-y

Energetic and Broad Band Spectral Distribution of Emission from Astronomical Jets

Asaf Pe'er

Received: 15 March 2013 / Accepted: 5 June 2013 / Published online: 4 July 2013

Abstract Emission from astronomical jets extend over the entire spectral band: from radio to the TeV γ-rays. This implies that various radiative processes are taking place in different regions along jets. Understanding the origin of the emission is crucial in understanding the physical conditions inside jets, as well as basic physical questions such as jet launching mechanism, particle acceleration and jet composition. In this chapter I discuss various radiative mechanisms, focusing on jets in active galactic nuclei (AGN) and X-ray binaries (XRB) environment. I discuss various models in use in interpreting the data, and the insights they provide.

Keywords Jets · Radiation mechanism: non-thermal · Galaxies: active · Gamma-ray bursts · Microquasars

1 Introduction

Jets and outflows are very ubiquitous in astrophysics. They are observed in both galactic objects such as X-ray emitting binaries (XRBs) (for reviews, see, e.g., Fender 2006, 2010; Gallo 2010; Markoff 2010; Maccarone 2012), as well as extra-galactic sources, such as active galactic nuclei (AGNs) (Begelman et al. 1984; Urry and Padovani 1995; Harris and Krawczynski 2006; Marscher 2009; Ghisellini 2012), and on a much smaller scale, gamma-ray bursts (GRBs) (e.g., Levinson and Eichler 1993; Piran 2004; Mészáros 2006). Recently, the existence of jet was inferred in a tidal disruption event (TDE) of a stray star passing near a massive black hole (Burrows et al. 2011; Levan et al. 2011). Emission from jets in both galactic and extra-galactic sources is observed over the entire spectral range: from radio to the highest γ-rays, at TeV energies. In addition to the spectral information, in galactic sources (X-ray binaries) as well as in jets from GRBs and TDEs, a wealth of temporal information exists. Similarly, in the high energy (GeV–TeV) emission from AGNs, flaring activities on time scales shorter than $\sim$1 h has been observed (Kniffen et al. 1993; Buckley

A. Pe'er (✉)
Physics Department, University College Cork, Cork, Ireland
e-mail: a.peer@ucc.ie

et al. 1996; Aharonian et al. 2007; Albert et al. 2007a; Aleksić et al. 2011). Moreover, in nearby jets from AGNs, such as Cen A, spatial information about emission from different region along the jet exist (Hardcastle et al. 2009 and references therein).

While a wealth of data has existed for several decades now, a detailed theoretical understanding of emission from jets is still lacking. One potential explanation is that most work to date has focused on emission from the accreting (inflow) material, and only in the past decade or so have more advanced models of emission from the outflow (jets) emerged. One notable exception is the emission from GRB jets: as our understanding of these objects relies nearly entirely on studying emission from their jets, the theory of emission from GRB jets is likely the most advanced one to date. A second reason is the enormous complexity of these systems. As will be discussed here, although the nature of the radiative processes is well understood, as the physical conditions inside and in the vicinity of the jets are poorly constrained, the data can be interpreted in more than one way. As a result, a plethora of models exist, and a conclusive picture is still lacking.

Although different objects share the common property of having jets, there is a huge difference in scales of the observed objects. While XRBs and GRBs are stellar-size objects, with a typical mass of the central BH of $\sim$ few $M_\odot$, the black holes in the center of AGNs have masses of 10^6–$10^9 M_\odot$. This difference in scaling results in very different scales of the resulting jets. In galactic XRBs the inferred size of the observed jets is typically 100's of AU's ($\gtrsim 10^{14}$–10^{15} cm) (Miller-Jones et al. 2012) while radio 'blobs' are seen on much larger, sub-pc scales ($\sim 10^{17}$ cm). In GRBs, the jet does not deposit most of its energy in the environment before reaching $\sim 10^{18}$ cm (Meszaros and Rees 1997; Wijers et al. 1997), although analysis shows that emission exists from the photosphere at $\sim 10^{12}$ cm (Axelsson et al. 2012). Sizes of AGN jets extend to much larger scales, with giant radio lobes extending to hundreds of kpc, $\sim 10^{23}$ cm (Alvarez et al. 2000 and references therein).

This difference in scaling implies that the physical conditions, and hence the leading radiative mechanisms inside the jets, vary with distance. Nonetheless, the basic emission processes are the same in all sources. The leading radiative processes include synchrotron emission, synchrotron self-Compton and Compton scattering of photons external to the jet—either photons originating from the accretion disk, companion star (in XRBs) or from the cosmic microwave (or infra-red) background (CMB). If hadrons (mainly protons) are accelerated to high energies in jets, they can also make a significant contribution to the emission, particularly at high energies (X and γ-ray bands). This contribution is both by direct emission (e.g., synchrotron), and indirectly, by interacting with photons and protons to produce secondaries (pions, Kaons and electron-positron pairs) which contribute to the emitted spectra. In addition, emission from the photosphere, defined here as emission that originates from regions in space in which the optical depth of photons to reach the observer is larger than unity, may play an important role.

In addition to the spectral analysis, two very important sources of information exist. The first is temporal variability which provides strong constraints on the physical conditions inside the jets, and hence on the emission processes. This played a crucial role in the development of the leading theory of emission from GRB jets (the "fireball" model). Studying the correlated variability seen in the emission at different wavelengths (radio/IR/optic and X/γ-rays) in XRBs is likely the key to understanding of the emission from these objects (see Uttley et al. 2011, and the chapters by Gallo and Casella in this book). The second source of information is spatial analysis, which is particularly useful when studying the largest-scale jets in AGNs. The "hot spots" frequently seen imply that the physical conditions and the radiative processes vary along the jet. Thus, a full physical picture must take into account first the *dynamics* of the outflow and second the *radiation*. Clearly, both parts are connected, and, in addition, give information about the jet launching process and the properties of the inner accretion disk.

Another important factor that needs to be considered in analyzing the emission is *geometry*. There are several aspects to this issue. First, as astronomical jets are mildly relativistic in XRBs ($\Gamma \gtrsim few$), often relativistic in AGNs ($\Gamma \gtrsim 10$) and highly relativistic in GRBs ($\Gamma \gtrsim 100$–1000), relativistic Doppler effect is important when analyzing the spectrum, as the jets will rarely point directly towards us. Emission from relativistically expanding blobs can lead to an apparent motion faster than the speed of light (a phenomenon known as "superluminal motion" Rees 1966; Mirabel and Rodríguez 1994). Second, jets, by definition, have spatial structure (often referred to as "structured jets"): a velocity profile exists, namely $v = v(r, \theta, \phi)$, where the angles θ, ϕ are measured relative to the jet axis. Thus, a velocity gradient in the transverse direction (perpendicular to the jet propagation direction) exists, with an obvious effect on the scattering between electrons and photons, and hence on the observed spectra.

Finally, the velocity structure in the radial direction can lead to confusing definition of jets. One possibility is that the outflow is continuous (generating a smooth velocity gradient in the radial direction) in which case it will be seen as a continuous jet. Alternatively, the outflow may be fragmented: in this case, the outflow will be observed as 'blobs' that propagate outward, while expanding (possibly, but not necessarily, adiabatically). Of course, the observed emission from these blobs imply that the conditions inside the blobs are different than those outside. Thus, when studying emission from these blobs one needs to consider the conditions both inside the blobs and in the surrounding material. In this chapter, we will treat both emission from continuous jets as well as from the blobs.

Thus, a full description of the emission requires understanding of (1) the dynamics, (2) the geometry and (3) the various radiative processes. Clearly, I cannot possibly cover the entire physics of jet emission in one chapter. I will thus focus on key radiative processes. I will show how the basic, well-known radiative processes can lead to the wealth of spectra observed. I will also try to point to basic, unsolved questions which naturally arise when analyzing the emission. The discussion will be focused on the jet emission from XRB and AGN environments, which show several similar key properties, although having different scales. Clearly, many of the results are relevant to jets in GRBs and TDEs as well.

2 Basic Radiative Processes: Synchrotron Emission

Variable radio emission in AGNs and XRBs is conventionally interpreted as synchrotron radiation from a non-thermal distribution of relativistic electrons. Indeed, synchrotron emission, being perhaps the most straightforward emission mechanism for explaining non-thermal radiation has been extensively studied since the 1960's (Ginzburg and Syrovatskii 1965; Blumenthal and Gould 1970). Two basic ingredients are needed: energetic particles and a strong magnetic field.

Consider a source at redshift z which is moving at velocity $\beta \equiv v/c$ (corresponding Lorentz factor $\Gamma = (1-\beta)^{-1/2}$) at angle θ with respect to the observer. The emitted photons are thus seen with a Doppler boost $\mathcal{D} = [\Gamma(1-\beta\cos\theta)]^{-1}$. Synchrotron emission from electrons having random Lorentz factor $\gamma_{\rm el}$ in a magnetic field B (all in the comoving frame) is observed at a typical energy

$$\varepsilon_m^{ob} = \frac{3}{2}\hbar\frac{qB}{m_e c}\gamma_{\rm el}^2\frac{\mathcal{D}}{(1+z)} = 1.75\times10^{-19}B\gamma_{\rm el}^2\frac{\mathcal{D}}{(1+z)}\ \text{erg}. \tag{1}$$

Thus, when studying this emission, the basic physical questions are:

1. What is the origin of the magnetic field?
2. What is the mechanism that accelerates particles to high energies? Does this mechanism accelerate only electrons? Are protons being accelerated similarly, thereby contributing to the emission? What is the resulting energy distribution of the energetic particles, $n(E)dE$?
3. What is the spatial/temporal evolution of the magnetic field and particle distribution in different regions along the jet?

While significant progress has been made in the last few decades, proper understanding of any of these issues remains elusive. These questions are deeply related to the physics of the jet launch mechanism, and jet composition. While there is no direct observational test that addresses any of these phenomena, very considerable theoretical effort supported by state of the art numerical simulations, as well as indirect interpretation of existing data, all suggest that these questions are likely intimately related to each other. We discuss these questions in what follows.

Origin of Magnetic Field Although the question of magnetic field generation is a fundamental one, little is known about the exact mechanism at work in these objects. Roughly speaking, there are two main (separated) sources of strong magnetic fields: the first is related to the accretion flow and the jet launching process. Possibly, even if the magnetic fields do not carry a large fraction of the kinetic energy, they may still play a key role in jets collimation. The second, independent source is magnetic field generation in shock waves that exist inside the outflow itself (assuming it is irregular), or when the outflow interacts with its surroundings—the interstellar medium (ISM) or intergalactic medium (IGM). A third possible source is amplification of ISM or IGM magnetic fields when compressed by the expanding shock waves, but in this case the amplified fields can at most explain the observed emission in the interaction of the outflow with its surroundings—they are much too weak to be consistent with the ones required to explain the observed properties in the inner jet regions. For a recent review on magnetic fields in astrophysical jets, see Pudritz et al. (2012).

The two leading mechanisms believed to operate for jet launching are the Blandford and Znajek (1977) and Blandford and Payne (1982) mechanisms. In the Blandford and Znajek (1977) mechanism, the source of energy is the rotational energy extracted from a rotating black hole, embedded in a strong magnetic field. The field itself must be anchored into the accretion flow (Livio et al. 1999; Meier 2001). In the Blandford and Payne (1982) mechanism, energy is extracted from the accretion disk by magnetic field lines that leave the disk surface and extend to large distances. This is accompanied by centrifugally-driven outflow of material from the inner parts of the disk, that is attached to the field lines (for further explanation see Spruit 2010, as well as the chapter on jet acceleration in this book). Both mechanisms require a strong magnetic field attached to the disk. At larger distances along the jet, the magnetic field decays as Poynting flux is conserved.

These ideas have been recently tested and validated with state of the art numerical GR-MHD simulations (Meier et al. 2001; McKinney and Gammie 2004; McKinney 2005, 2006; Tchekhovskoy et al. 2010, 2011). These models imply that the magnetic field originates in the inner parts of the disk. The inner parts of the jets are strongly magnetized (Poynting-flux dominated), and the magnetic energy is gradually dissipated along the jet. The dissipated energy is then used to accelerate the particles along the jet (see, e.g., Vlahakis and Königl 2004; Komissarov et al. 2007). Recent observations on parsec-scale in AGNs indicate magnetic field strengths consistent with those expected from theoretical models of magnetically powered jets (O'Sullivan and Gabuzda 2009). However, the picture is far more complicated, since modeling the broad-band emission (radio–X-rays) on a $\gtrsim$ parsec scale from several AGNs show that the magnetic field must be sub-dominant, and most of the kinetic energy is carried by protons (particle-dominated jets) (Celotti and Fabian 1993; Krawczynski et al. 2002; Kino et al. 2002; Celotti and Ghisellini 2008). The mechanism in which magnetic-dominated outflow at the core becomes particle dominated at larger distances is far from being clear.

Independent of the question of jet launching, a second source of strong magnetic fields are shock waves that develop as a result of instabilities within the outflow. These shock waves can result, e.g., from fluctuations in the ejection process itself: if a slower moving plasma shell (or "blob") is followed by a faster moving one, the two shells will eventually collide, producing a pair of forward and reverse shock waves propagating into each of these blobs. These shock waves may generate strong magnetic fields by two-stream instabilities (Weibel 1959; Medvedev and Loeb 1999). In recent years, advances in particle-in cell (PIC) simulations enabled to study this process by tracing the instability growth modes (Silva et al. 2003; Frederiksen et al. 2004; Nishikawa et al. 2005; Spitkovsky 2008a). The results of these works have demonstrated that strong magnetic fields are indeed created in collisionless shock waves.

The key question though, is the decay length of the magnetic field: the results discussed above also show that the generated field decays on a very short length scale, of the order of few hundred skin depths (Spitkovsky 2008a). As observations imply that the emitting region is many orders of magnitude larger than this scale, there must be a mechanism that maintains a strong magnetic field extending to much larger scales. One suggestion is that the amplification of the magnetic field may be closely related to the acceleration of particles to high energies (Keshet et al. 2009). Thus, while initially the magnetic field may occupy only a small region close to the shock front, over time, as particles are accelerated to increasingly higher energies, the magnetized region expands. This suggestion is difficult to directly test, due to the numerical complexity of the problem. Another suggestion is that, due to the turbulent nature of the post-shock outflow, the magnetic field is maintained over a long distance behind the shock front (Zrake and MacFadyen 2012). Thus, while it is clear that magnetic fields can be generated in shock waves, the exact scaling (strength and decay length) of these fields is still a matter for debate.

Particle Acceleration It should be emphasized that the existence of cosmic rays, (charged particles that are observed at energies as high as $\gtrsim 10^{20}$ eV; for a recent review, see Kotera and Olinto 2011), is a direct evidence that particle acceleration to ultra-high energies takes place in astronomical objects. However, there is no direct information on the exact nature of the cosmic ray sources, nor on the nature of the acceleration process itself. Hence the question of lepton (electrons and positrons) acceleration is inferred indirectly, by fitting the observed spectra from various objects. It is most likely that acceleration takes place in several different locations: in the nucleus, in the hot spots and possibly additional locations along the jet axis.

The most widely discussed mechanism for acceleration of particles is the *Fermi mechanism* (Fermi 1949, 1954), which requires the particles to cross back and forth a shock wave. A basic explanation of this mechanism can be found in the textbook by Longair (2011). For reviews see Bell (1978), Blandford and Ostriker (1978), Blandford and Eichler (1987), Jones and Ellison (1991). In this process, the accelerated particle crosses the shock multiple times, and in each crossing its energy increases by a (nearly) constant fraction, $\Delta E/E \sim 1$. This results in a power law distribution of the accelerated particles, $N(E) \propto E^{-S}$ with power law index $S \approx 2.0$–2.4 (Kirk et al. 1998, 2000; Ellison et al. 1990; Achterberg et al. 2001; Ellison and Double 2004). Recent developments in particle-in-cell (PIC) simulations have allowed to model this process from first principles, and study it in more detail (Silva et al. 2003; Nishikawa et al. 2003; Spitkovsky 2008b; Sironi and Spitkovsky 2009; Haugbølle 2011). However, due to the numerical complexity of the problem, these simulations can only cover a tiny fraction ($\sim 10^{-8}$) of the actual emitting region in which energetic particles exist. Thus, these simulations can only serve as guidelines, and the problem is still far from being fully resolved. Regardless of the exact details, it is clear that particle acceleration via the Fermi mechanism requires the existence of shock waves, and is thus directly related to the internal dynamics of the gas inside the jet, and possibly to the generation of magnetic fields, as mentioned above.

An alternative model for particle acceleration is magnetic reconnection. The basic idea is that when magnetic field lines change their topology and form a reconnection layer, magnetic energy is released. Part of the generated energy may be used to accelerate particles to high energies (Romanova and Lovelace 1992; Lyutikov 2003; Lyubarsky and Liverts 2008; Lazarian et al. 2011; McKinney and Uzdensky 2012). This idea is very appealing if jets are highly magnetized (at least close to the core), as is suggested by the leading theories of jet launching. In fact, it is not clear that the conditions that enable particle acceleration to high energies in shock waves exist at all in highly magnetized outflows (Sironi and Spitkovsky 2009, 2011), in which case the Fermi mechanism may not be at work. However, theoretical understanding of this process, and its details (e.g., what fraction of the reconnected energy is being used in accelerating particles, or the energy distribution of the accelerated particles) is still very limited.

Although the power law distribution of particles resulting from Fermi-type, or perhaps magnetic-reconnection acceleration is the most widely discussed, we point out that alternative models exist. One such model involves particle acceleration by a strong electromagnetic potential, which can exceed 10^{20} eV close to the jet core (Lovelace 1976; Blandford 1976; Neronov et al. 2009). The accelerated particles may produce a high energy cascade of electron-positron pairs. Additional model involves stochastic acceleration of particles due to resonant interactions with plasma waves in the black hole magnetosphere (Dermer et al. 1996).

Several authors have also considered the possibility that particles in fact have a relativistic quasi-Maxwellian distribution (Jones and Hardee 1979; Cioffi and Jones 1980; Wardziński and Zdziarski 2000; Pe'er and Casella 2009). Such a distribution, with the required temperature ($\sim 10^{11}$–10^{12} K) may be generated if particles are roughly thermalized behind a relativistic strong shock wave (e.g., Blandford and McKee 1977). Interestingly, this model is consistent with several key observations, as will be discussed below.

Spatial and Temporal Distributions The uncertainty that exists in both the origin of the magnetic field as well as the nature of the particle acceleration process is directly translated to an uncertainty in the spatial and temporal distributions of these two quantities, and hence on the emission pattern. If the magnetic field originates in the disk, then as the jet expands the magnetic field must decay. For example, if the cross sectional radius of the jet is $r = r(z)$ where z is the direction along the jet axis, then Poynting flux conservation implies $B \propto r^{-1}$. If, on the other hand, the dominant process for magnetic field generation is two stream instability in shock waves, the magnetic field then traces the shock wave location. Thus, strong magnetic fields are expected only above a certain radius, where plasma shells collide. This is very likely the case in the spatially resolved "radio blobs" seen in XRBs, as well as in "knots" observed along AGN jets.

The magnetic field strength may also be different in the two possibilities discussed. Lacking a complete theory, it is commonly assumed that the generated magnetic field carries a constant fraction, ϵ_B of the kinetic energy dissipated by the shock wave, $B^2/8\pi = \epsilon_B U$. Here, U is the (post-shock) energy density of the plasma. Estimated values for ϵ_B based on fitting the data vary from equipartition ($\epsilon_B = 1/3$; Miller-Jones et al. 2005; Cerruti et al. 2013) to $\epsilon_B \sim 10^{-2}$, possibly even lower (Celotti and Ghisellini 2008; Santana et al. 2013).

The spatial and temporal distribution of the energetic particles is determined by several factors. Once accelerated to high energies, the radiating particles lose their energy both adiabatically as the jet expands, and radiatively, as they radiate their energy. Thus, in order to understand their spatial distribution, one needs to know (1) the initial distribution of the energetic particles accelerated by the acceleration process (determined by the unknown nature of this process). (2) The dynamics of the plasma; and (3) the physical conditions inside the plasma, that govern the energy loss rate.

2.1 Spectral Shape: Basic Considerations and Maxwellian Distribution of Electrons

The discussion above points towards high uncertainty in our knowledge of the initial energy distribution of particles produced by the acceleration process. It is commonly believed that the acceleration process produces a power law distribution $n(E)dE \propto E^{-S}$, with $S \approx 2.0$–2.4. This is based on (1) theoretical expectations from Fermi acceleration, and (2) interpretation of broad band synchrotron emission. However, a few words of caution are necessary here. First, as discussed above, it is not clear that the Fermi process is necessarily the acceleration mechanism at work in these objects. Second, as was recently shown (Pe'er and Casella 2009) and will be discussed below, the observed data can be interpreted in a way that does not require a power law distribution of the radiating particles. Thus, evidence for the existence of a power law distribution is inconclusive.

Even if the acceleration process is indeed Fermi-type in shock waves, then the resulting power law distribution is expected to be limited to a certain region in energy space. As particles cross the shock front, they thermalize. Strong shock waves propagating at Lorentz factor Γ into a cold material of density n compress the material so that its density in the downstream region is $4\Gamma n$. The material is being heated: the energy density in the downstream region is $4\Gamma^2 n m_p c^2$. Thus, the average energy per particle in the downstream region is $\approx \Gamma m_p c^2$ (Blandford and McKee 1976). If a fraction $\epsilon_e \leq 1$ of this energy is carried by the energetic electrons, then (neglecting a possible contribution from pairs) the expected Lorentz factor of the electrons is $\gamma_{\rm el} \approx \Gamma \epsilon_e m_p/m_e$. Note that this is the Lorentz factor associated with the random motion of the electrons as they cross the shock front, and should not be confused with the Lorentz factor associated with the bulk motion of the flow, which is of the order of Γ. Thus, even in mildly-relativistic outflows ($\Gamma \gtrsim 1$), the electrons in the downstream region may still have (random) Lorentz factor of few hundreds, provided that ϵ_e is close to equipartition.

One can thus conclude that regardless of the question of whether electrons are accelerated to a power law distribution, they are still expected to be heated to high energies (high Lorentz factors) when shock waves exist. Hence, if no further acceleration process are present, the electrons will have a Maxwellian distribution with typical Lorentz factor $\gamma_{\rm el} \lesssim 10^3$ (assuming ϵ_e close to equipartition). In the vicinity of a magnetic field B, which could naturally be generated by the same shock wave, electrons at the peak of the Maxwellian distribution will emit at a characteristic energy given by Eq. (1). For typical value $\gamma_{\rm el} \sim 10^3$ and $B \sim 1$ G, Eq. (1) implies a characteristic observed frequency in the optical band.

This result implies that in order to explain the observed flat radio spectra seen in many objects, there is no need to invoke a power law distribution of the accelerated particles. It is enough to consider a power law decay of the magnetic field along the jet, $B(r) \propto r^{-\alpha}$ to obtain a power law decay of the peak synchrotron frequency below the optical band, in accordance to Eq. (1). A power law spectrum would be observed if the emission is not spatially resolved, but is integrated over some distance along the jet along which the magnetic field decays. This is a typical scenario for the inner parts of AGN jets, as well as for jets in XRBs.

2.2 Power Law Distribution of the Accelerated Particles

It is possible to envision a different model, in which the energy distribution of the accelerated electrons is a power law. In fact, historically this model was the first to be suggested in explaining the observed spectra (van der Laan 1966; Blandford and Konigl 1979), and is still the most widely-discussed one.

An uncertainty lies in the fraction of particles that are being accelerated: as the electrons cross the shock wave, they have a thermal distribution with typical Lorentz factor $\gamma_{\rm el}$. As some fraction continues to cross the shock front multiple times, this fraction obtain a power law distribution. What fraction of particles are accelerated to a power law distribution above the Maxwellian is unclear. Recent PIC simulations suggest that only a small fraction of the population, $\epsilon_{\rm pl} \approx 1~\%$–$10~\%$ form a power law tail at higher energies (Spitkovsky 2008b). However, as discussed above, these conclusions are far from being certain, and it is possible that the fraction is much higher, perhaps even closer to 100 %.

A theoretical limit on the maximum energy is obtained by the requirement that the acceleration time must be shorter than the minimum energy loss time (e.g., due to synchrotron radiation or Compton scattering) and the time in which the accelerated particle is confined to the accelerated region. In a plasma which moves relativistically with Lorentz factor Γ, the acceleration time in Fermi-type acceleration is (e.g., Norman et al. 1995)

$$t_{\rm acc} = \frac{\eta E^{\rm ob}}{\Gamma Z q B c}. \tag{2}$$

Here, Zq is the charge of the particle (the same equation holds for electrons, protons as well as heavy nuclei with Z protons), and $E^{\rm ob}$ is the energy of the energetic particle in the observer's frame. The exact value of the dimensionless factor $\eta \geq 1$ depends on the uncertain details of the acceleration process: for example, in non-relativistic diffusive shock acceleration, this factor corresponds to $\eta = (20/3)\beta^{-2}$ in the Bohm limit for parallel shocks (e.g., Blandford and Eichler 1987).

The second requirement constraints the size of the acceleration region. For typical values of parameters that govern the emission in XRBs and AGNs, it is not very restrictive. On the other hand, the requirement that the acceleration time is shorter than the radiative cooling time puts a stronger constraint on the maximum energy of the accelerated particles. The radiative cooling time of energetic electrons due to synchrotron emission and Compton scattering is

$$t_{\rm cool} = \frac{E}{P} = \frac{\gamma_{\rm el} m_e c^2}{(4/3) c \sigma_T \gamma_{\rm el}^2 u_B (1+Y)} = \frac{6\pi m_e c}{\sigma_T B^2 \gamma_{\rm el} (1+Y)}, \tag{3}$$

where $u_B \equiv B^2/8\pi$ is the energy density in the magnetic field, σ_T is Thomson's cross section and Y is Compton parameter. Comparing the radiative cooling time in Eq. (3) to the acceleration time in Eq. (2) gives a theoretical upper limit on the energy of the accelerated electrons,

$$\gamma_{\max} = \left(\frac{6\pi q}{\eta B \sigma_T (1+Y)} \right). \tag{4}$$

Using the derived value of $\gamma_{\max}$ from Eq. (4) in Eq. (1) gives a very interesting result: the characteristic energy of photons emitted by these electrons,

$$\varepsilon_{\max}^{\mathrm{ob}} = 240 \frac{\mathcal{D}}{\eta(1+Y)(1+z)} \,\mathrm{MeV} \tag{5}$$

is *independent on the strength of the magnetic field*. This result implies that regardless of the value of the magnetic field, if indeed particles are accelerated by a Fermi-type mechanism in shock waves, synchrotron emission is expected to be observed at all energies up to the γ-ray band. Thus, it is possible, at least from a theoretical perspective, that synchrotron photons have a significant contribution to the emission at X and γ-ray energies.

2.3 Broad Band, Spatially Resolved Synchrotron Spectrum

The discussion above implies that even if synchrotron emission is the only source of radiation, and even if the magnetic field is constant, the complex distribution of the energetic particles leads to a complex observed spectrum. In addition to the two frequencies discussed, $\nu_m = \varepsilon_m/h$ (see Eq. (1)) and $\nu_{\max} = \varepsilon_{\max}/h$ (the later exists only if the acceleration process produces a power law), there are additional two inherent characteristic frequencies. The first is the synchrotron self absorption frequency, ν_{SSA}, below which synchrotron photons are absorbed. This frequency can be either above or below ν_m. The exact value of ν_{SSA} depends on the magnetic field strength and the distribution of the radiating particles (for discussion see Rybicki and Lightman 1979). For typical parameters, $\nu_{\mathrm{SSA}} < \nu_m$. However, given the uncertainty that exists in the acceleration process and the strength of the magnetic fields, it is possible to envision scenarios in which $\nu_m < \nu_{\mathrm{SSA}}$.

The fourth frequency is the cooling frequency, ν_c. This is the characteristic emission frequency from particles whose radiative cooling time (given by Eq. (3)) is equal to the characteristic plasma expansion time, $t_{\mathrm{dyn}} \simeq r/\Gamma c$, where r is the radius of the expanding plasma. Since the cooling time is inversely proportional to the particle's Lorentz factor, $t_{\mathrm{cool}} \propto \gamma^{-1}$ (see Eq. (3)), energetic particles cool faster than low energy ones. Above a certain Lorentz factor, denoted by γ_c, particles cool faster than the dynamical time. Thus, if particles are accelerated only once, $\gamma_c = \gamma_{\max}$. However, if the acceleration continuously produces a power law distribution of energetic particles, $n(\gamma)d\gamma \propto \gamma^{-S}$, then γ_c marks a transition in the steady state distribution. By solving the continuity equation, it is easy to show (see, e.g., Longair 2011) that for $\gamma \gg \max(\gamma_c, \gamma_m)$, the steady particle distribution is $n(\gamma)d\gamma \propto \gamma^{-(S+1)}$. If $\gamma_c < \gamma_{\mathrm{el}}$, then in the region $\gamma_c \ll \gamma \ll \gamma_{\mathrm{el}}$ the steady distribution is $n(\gamma)d\gamma \propto \gamma^{-2}$. This break in the particle distribution is directly translated to a break in the emitted spectrum. As the synchrotron spectrum from a power law distribution of particles with power law index S is $F_\nu \propto \nu^{-(S-1)/2}$, at frequencies above ν_c, this power law changes to $F_\nu \propto \nu^{-S/2}$.

Thus, in Fermi-type acceleration, four breaks in the spectrum are expected (see Table 1). Even if the acceleration mechanism produces only a Maxwellian distribution of hot particles, at least two break frequencies (ν_{SSA} and ν_m) are unavoidable.

In the scenario where $\nu_{\mathrm{SSA}} < \nu_m < \nu_c$ the peak of the spectrum occurs at $\nu = \nu_m$. Denoting by $F_{\nu,\max}$ the observed peak flux, the broad band synchrotron spectrum is (e.g., Meszaros

Table 1 Key break frequencies

ν_m	Synchrotron emission frequency from electrons at typical Lorentz factor γ_{el} (given in Eq. (1))
ν_{ssa}	Self absorption frequency; Photons at $\nu < \nu_{ssa}$ are absorbed
ν_c	Cooling frequency; A break in the spectrum caused by rapid cooling of electrons at high energies
ν_{max}	Maximum emission frequency from Fermi-accelerated electrons; No synchrotron emission is expected at higher frequencies

and Rees 1997; Sari et al. 1998)

$$F_\nu = F_{\nu,\max} \times \begin{cases} (\nu/\nu_{SSA})^2(\nu_{SSA}/\nu_m)^{1/3}, & \nu < \nu_{SSA} \\ (\nu/\nu_m)^{1/3}, & \nu_{SSA} < \nu < \nu_m \\ (\nu/\nu_m)^{-(S-1)/2}, & \nu_m < \nu < \nu_c \\ (\nu_c/\nu_m)^{-(S-1)/2}(\nu/\nu_c)^{-S/2}, & \nu_c < \nu < \nu_{\max} \end{cases} \tag{6}$$

If, on the other hand $\nu_{SSA} < \nu_c < \nu_m$, the peak of the emission is at ν_c, and the spectral shape is

$$F_\nu = F_{\nu,\max} \times \begin{cases} (\nu/\nu_{SSA})^2(\nu_{SSA}/\nu_c)^{1/3}, & \nu < \nu_{SSA} \\ (\nu/\nu_c)^{1/3}, & \nu_{SSA} < \nu < \nu_c \\ (\nu/\nu_c)^{-1/2}, & \nu_c < \nu < \nu_m \\ (\nu_m/\nu_c)^{-1/2}(\nu/\nu_m)^{-S/2}, & \nu_m < \nu < \nu_{\max} \end{cases} \tag{7}$$

Finally, the model of Blandford and Konigl (1979) can be viewed as a model in which $\nu_m < \nu_{ssa} \ll \nu_c$. In this case, the peak of the emission is at ν_{SSA}, and the broad band spectrum is

$$F_\nu = F_{\nu,\max} \times \begin{cases} (\nu/\nu_m)^2(\nu_m/\nu_{SSA})^{5/2}, & \nu < \nu_m \\ (\nu/\nu_{SSA})^{5/2}, & \nu_m < \nu < \nu_{SSA} \\ (\nu/\nu_{SSA})^{-(S-1)/2}, & \nu_{SSA} < \nu < \nu_c \\ (\nu/\nu_c)^{-S/2}(\nu_c/\nu_{SSA})^{-(S-1)/2}, & \nu_c < \nu < \nu_{\max} \end{cases} \tag{8}$$

These spectra are shown in Fig. 1.

2.4 Integrated Spectrum: Flat Radio Emission

The broad band spectrum considered above is developed under the assumptions that the radiating particles have a power law distribution and that the magnetic field is steady. In reality, once accelerated, or even during the acceleration the energetic particles propagate along the jet. As the magnetic field strength varies along the jet, the break frequencies—ν_{SSA}, ν_m and ν_c (but not $\nu_{\max}$!) are different in different regions along the jet. If the jet is spatially resolved, this implies that different regions along the jet are characterized by different spectra. If the jet is spatially unresolved, as is the case in XRBs and the inner parts of jets in AGNs, then the observed spectrum is obtained by integrating over different emission regions, each characterized by different break frequencies. This integration naturally leads to the observed power law spectra, such as the flat spectra frequently observed at radio frequencies, in both AGNs (the so called "flat spectrum radio quasars", or FSRQ), and more recently in XRBs (Hynes et al. 2000; Fender 2001, 2006, and chapters in this book by Fender, Gallo and Casella).

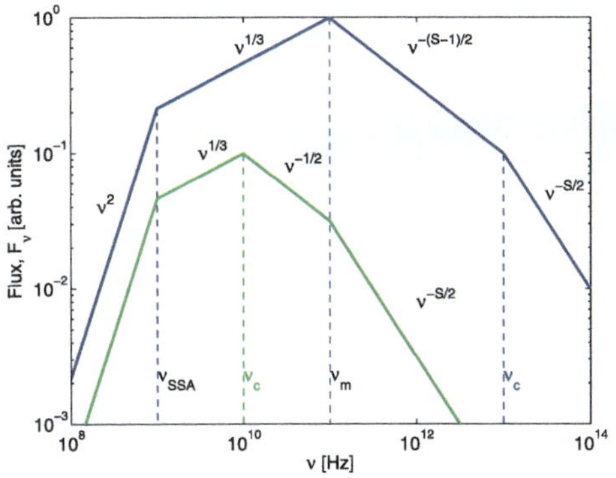

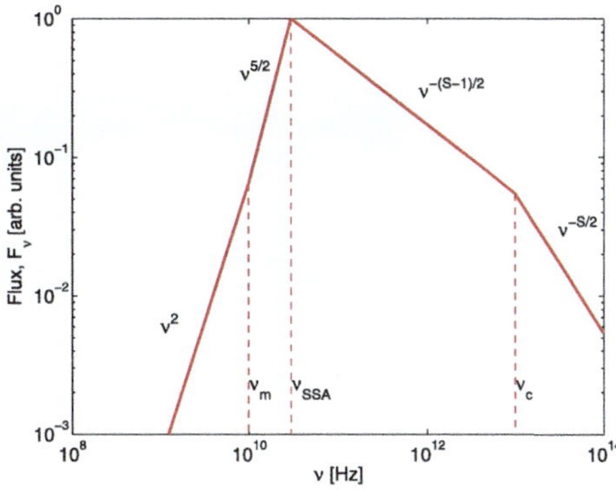

Fig. 1 Broad band synchrotron spectra from a power law distribution of energetic particles in a steady magnetic field, as expected from a small jet slab. The flux and the chosen values of the break frequencies are arbitrary, and depend on the exact values of the magnetic field, particles energies and number of radiating particles. (*Left*) The spectra expected when $\nu_{\rm SSA} < \nu_m < \nu_c$ (*blue*) peaks at ν_m. In the case $\nu_{\rm SSA} < \nu_c < \nu_m$ (*green*), the spectrum peaks at ν_c. These spectra were considered in the model of Pe'er and Casella (2009). (*Right*) The broad band spectra in the scenario $\nu_m < \nu_{\rm SSA} < \nu_c$ peaks at $\nu_{\rm SSA}$. This is the scenario considered in the model of Blandford and Konigl (1979), and is commonly considered in the literature thereafter

If the origin of the magnetic field is in the disk, it is expected to decay along the jet as a power law with distance. This decay, which results in a corresponding decrease of the break frequencies, is all that is needed to produce a power law spectrum in the radio band, in particular the flat spectrum that is typically observed. This was first noted by Blandford and Konigl (1979). In this model, a conical jet with $B(r) \propto r^{-1}$ and steady outflow velocity resulting in a number density variation along the jet $n(r) \propto r^{-2}$ was analyzed. Only the evolution of the self absorption frequency, $\nu_{\rm SSA}$ along the jet was considered. In the more general framework considered here, this is equivalent to a model in which particles are accelerated to a power law energy distribution with $\nu_m < \nu_{\rm SSA} \ll \nu_c$ as is presented in Fig. 1 (right). In this scenario, the emission from a jet slab (in which the magnetic field is constant), peaks at $\nu_{\rm SSA}$. It is straightforward to show (see, e.g., Rybicki and Lightman 1979; Blandford and Konigl 1979) that these conditions lead to a decay of the self absorption frequency along the jet, $\nu_{\rm SSA} \propto r^{-1}$, while the flux from a slab along the jet axis at the self absorption frequency is constant, $dF_\nu|_{\nu_{\rm SSA}} \propto \nu^0$. Thus, when integrated over an unresolved distance along the jet a flat radio spectrum is obtained. This is demonstrated in Fig. 2, taken from Markoff (2010).

This basic idea was extended by several authors in various aspects. For example, Marscher (1980) considered different viewing angles, while Reynolds (1982) considered different dynamical models for the outflow. In another work, Hjellming and Johnston (1988) considered a more refined jet geometry, as well as adiabatic (though, not radiative) energy losses. Several works (Falcke and Biermann 1995, 1999; Levinson and Blandford 1996; Heinz and Sunyaev 2003; Bosch-Ramon et al. 2006) have connected the jet properties to the disk properties, and refined the inner jet dynamics. This dynamics was used by several authors (Markoff et al. 2001, 2003, 2005; Yuan et al. 2005; Yuan and Cui 2005; Maitra et al. 2009) to model the broad band spectra of XTE J1118+480 and GX 339-4. These works included a self consistent modeling of the emission from the radio all the way to the X-ray band. In modeling the X-ray emission, the power law assumption was used to fit both the synchrotron emission and the synchrotron self Compton (SSC) emission (Markoff et al. 2001; Gallo et al. 2007; Migliari et al. 2007; Maitra et al. 2009; see further discussion below).

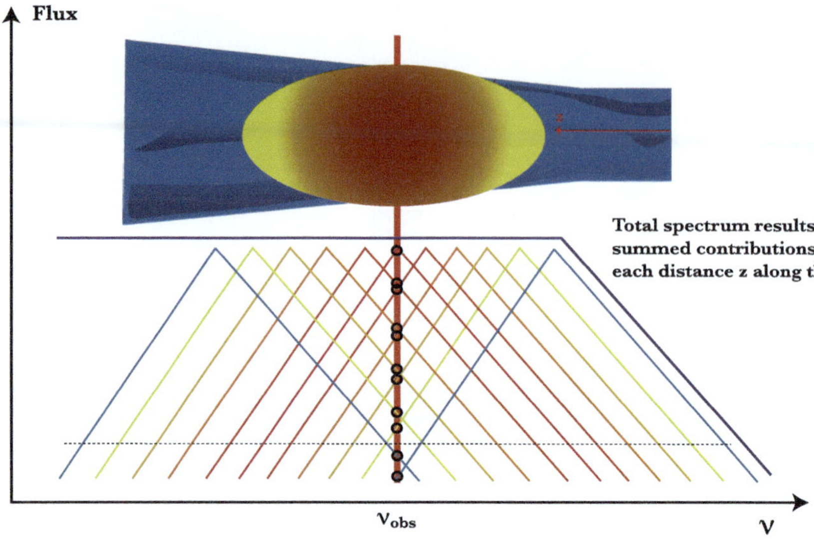

Fig. 2 Along the jet, the magnetic field decays. Thus, as the radiating particles propagate along the jet, the break frequencies presented in Fig. 1 decay. While the emission from each slab has the same spectrum as presented in Fig. 1 (*right*) peaking at ν_{SSA} (*thin lines*), when integrated over a spatially unresolved region along the jet, the observed radio spectrum is flat. The cartoon here is taken from Markoff (2010)

While these models show significant improvement in treating the dynamical properties of jets, the basic radiative mechanism discussed by Blandford and Konigl (1979) remains key to all of them. The radiative particles are assumed to have a power law distribution, and the peak of the emission is at ν_{SSA}. The decrease of ν_{SSA} along the jet due to the decay of the magnetic field is the origin of the flat radio spectra.

An alternative approach was suggested by Pe'er and Casella (2009). Based on the idea of a single acceleration episode and the inclusion of particle cooling first proposed by Kaiser (2006), this model considers a scenario in which $\nu_{SSA} < \nu_m$. Thus, the peak of the emission from a given 'slab' is at ν_m (or ν_c) rather than at ν_{SSA}. As was shown in this work, as a result of the decaying magnetic field along the jet, the decay law of ν_m is identical to the decay law of ν_{SSA}. Thus, a flat radio spectrum is naturally obtained in an analogous way to the Blandford and Konigl (1979) model, by integrating over emitting regions inside the jet (see Fig. 3).

A conceptual difference between this model and the Blandford and Konigl (1979) model is that the former does *not* require a power law distribution of the accelerating electrons. As ν_m is one of the natural frequencies obtained from a Maxwellian distribution of radiating particles, a flat radio spectrum is obtained even if the acceleration process does not accelerate particles to a power law distribution. A second difference is that the change in particle distribution due to cooling is inherently considered. Thus, in a region of strong magnetic field the radiating particles rapidly cool, $\nu_c < \nu_m$, and the flux $F_\nu \propto \nu^{-1/2}$ (see Eq. (7)). This result is independent of the details of the acceleration process and the magnetic field structure. Finally, if the magnetic field is very strong, rapid cooling of the particles as they propagate along the jet leads to absorption of the emission peak, $\nu_m < \nu_{SSA}$ for emission that occurs above a certain radius. This, in turn, results in a suppression of the radio emission, as seen in several objects (Casella and Pe'er 2009).

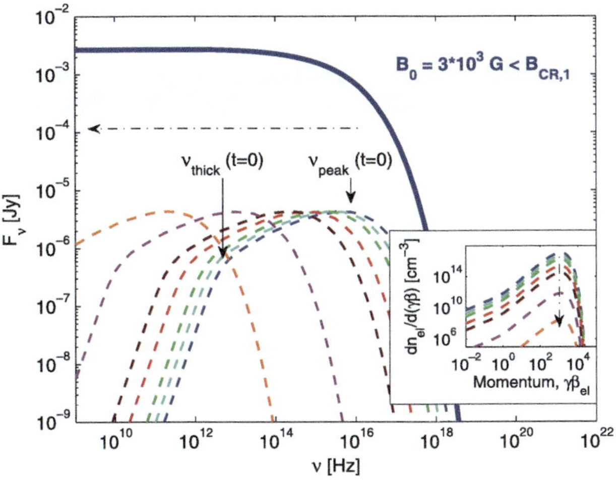

Fig. 3 The model of Pe'er and Casella (2009) provides an alternative way to explain flat radio spectra without the need for a power law distribution of the energetic particles. Emission from a Maxwellian distribution (*dashed lines*) is characterized by two break frequencies, ν_m (described in the figure as $\nu_{\rm peak}$) and $\nu_{\rm SSA}$ (or $\nu_{\rm thick}$). In the model considered, $\nu_m > \nu_{\rm SSA}$, and thus the emission from a single slab peaks at ν_m. As the particles propagate along the jet ν_m decays, and thus the integrated spectra resulting from a Maxwellian distribution of particles (inner set) is flat (*thick blue line*). The value of the magnetic field, $B = 10^{3.5}$ G is arbitrarily chosen for demonstration purposes only

One assumption common to all jet models is that the decay of the magnetic field leads to a decay of the characteristic frequencies along the jet. Thus, although unresolved, emission at low radio frequencies is expected from distant regions along the jet. Conversely, emission at higher energies—microwave, optical X- and γ-rays—must originate from the inner parts of the jet, where the discrimination between the outflow (jet) and inflow (accretion) may be very difficult. As shown in Eq. (5), at least from a theoretical perspective synchrotron photons can be observed nearly to the GeV-range, irrespective of the strength of the magnetic field.

2.5 Fragmented Outflow: Emission from Radio Blobs

The flat radio emission is a natural outcome of emission in a power law decay of the magnetic field along the jet. Thus, the above discussed models are relevant when the outflow is continuous. However, fragmented ejection of material, in the form of radio "blobs" are frequently observed in both XRBs, such as GRS1915+15 (Mirabel and Rodríguez 1994; Rodríguez and Mirabel 1999; Fender et al. 1999a; Miller-Jones et al. 2005) as well as AGNs.

The different emission observed from these blobs with respect to their environment indicates different physical conditions inside the blobs. Particularly, it implies that the magnetic field and/or the distribution of radiating particles inside the blobs is different than outside of them. This is a natural consequence if the magnetic field and particle acceleration originate in shock waves, as discussed above. As the blobs propagate outward they expand. The expansion can be adiabatic, but not necessarily (it could be confined by, say an external magnetic field). Thus, one can deduce scaling laws for the evolution of the magnetic field and the particle distribution inside the expanding blobs. The basic model was suggested by van der Laan (1966), and extended by Hjellming and Johnston (1988), Atoyan and Aharonian (1999).

The key radiative model is similar to the Blandford and Konigl (1979) model, namely that particles are accelerated to a power law distribution, and the emission peaks at ν_{SSA}. However, the scaling laws are different. The basic assumption is that particles do not enter or leave a blob, which is adiabatically expanding. Conservation of magnetic flux implies a decay of the magnetic field $B \propto L^{-2}$, and adiabatic cooling implies a decline in the particles' energy, $\gamma \propto L^{-1}$, where L is the comoving size of the expanding blob. Since the emitted frequency $\nu \propto B\gamma_{\rm el}^2$ (see Eq. (1)), one can derive the scaling law of the Lorentz factor of the radiating electrons at observed frequency ν to be $\gamma_{\rm el} \propto \nu^{1/2}B^{-1/2} \propto L$. For power law distribution $N(\gamma)d\gamma = k\gamma^{-S}$, the synchrotron flux then scales as $F_\nu \propto kB^{(S+1)/2}L^3$, where $k \propto L^{(-S+2)}$ (see van der Laan 1966; Rybicki and Lightman 1979).

These scaling laws thus give a testable prediction, $F_\nu \propto L^{-2S} \sim t^{-2S}$. When confronted with observations (Rodriguez et al. 1995), the observed decline is not as steep as the theoretical prediction. Thus, the simplified version of the theory needs to be adjusted. One natural possibility is that the expansion is not adiabatic. For example, reverse shock may play a significant role in determining the evolution of these blobs (R. Narayan, private communication).

3 Compton Scattering and the Origin of the X-Ray Spectrum

While there is a consensus that the radio spectrum originates from synchrotron radiation (although the full details of the process are uncertain), the origin of the X-ray and γ-ray emission is far more debatable. As shown above, synchrotron emission can extend up to hundreds of MeV. However, at these energies, there are alternative sources of emission. In particular, Compton scattering of low energy photons by energetic electrons is a natural, alternative way to produce emission at these bands. Due to the larger cross section, even if hadrons (protons) contribute to the emission, their contribution to IC scattering is expected to be negligible compared to the electrons contribution.

Energetic electrons radiate their energy via both synchrotron radiation and IC scattering. The total power emitted by IC process is (Rybicki and Lightman 1979)

$$\frac{P_{\rm IC}}{P_{\rm syn}} = \frac{U_{\rm ph}}{U_B}, \tag{9}$$

where $P_{\rm syn}$ is the synchrotron power, U_B and $U_{\rm ph}$ are the energy densities in the magnetic and photon fields, respectively. Thus, if $U_{\rm ph} > U_B$, most of the electrons' energy is radiated by IC scattering rather than synchrotron. However, even if $U_B > U_{\rm ph}$, it is still possible that IC scattering is the main source of emission at a given frequency band.

In understanding Compton scattering, the basic questions are therefore:

1. What is the origin and spectral distribution of the energetic electrons? Obviously, this is a similar question to the one that lies in the heart of understanding synchrotron emission, as the same electrons radiate both synchrotron photons and IC photons.
2. What is the origin of the upscattered photon field? Do these photons originate inside the jet (e.g., by synchrotron emission), or are they external to the jet (e.g., originating in the accretion flow or CMB)?
3. Since electrons in the inner parts of the accretion flow are hot enough to emit in the X-ray band, is there a simple way to discriminate disk and jet emission by observing at this band?

The third question is particularly puzzling, and is the source of an intense debate. As discussed above, synchrotron emission from the inner parts of the jet, where the magnetic field is strongest, are expected to contribute to the observed flux at the X- and possibly also γ-ray frequencies. These regions are close to the inner parts of the inflow. Thus, discriminating between the inflow and jet as the sources of X-ray radiation is very challenging.

While IC emission from particles in AGN jets is well established, most works on X-ray emission in XRBs are focused on IC emission from the inner parts of the accretion flow. A few notable works are by Sunyaev and Titarchuk (1980), Haardt and Maraschi (1993), Titarchuk (1994), Magdziarz and Zdziarski (1995), Esin et al. (1997), Poutanen (1998), Cadolle Bel et al. (2006), Yuan et al. (2007), to name only a handful. As this is the subject of a separate chapter in this book, we will not discuss it here. I will point though, that there are various reasons to consider IC scattering from electrons inside the jets in XRBs. These include:

1. As the radio spectrum originates from synchrotron photons, energetic electrons exist in the jet. These electrons must upscatter low energy photons.
2. From a theoretical perspective, models in which the dominant contribution is IC emission from the inflow do not well connect to the need for strong magnetic fields required in leading jet-launching models discussed above (though a few recent accretion models may overcame this problem; see Ferreira et al. 2006; Fragile and Meier 2009; Bu et al. 2009; Oda et al. 2010; Petrucci et al. 2010).
3. Detection of X-ray emitting blobs propagating outward in the inner regions of jets in several microquasars (Corbel et al. 2002) indicate that part of the radiation in these objects is from the jet (or interaction of the jet with the ambient medium), and not all of it originates from the accretion flow.
4. Finally, the very high energy emission ($\gtrsim$100 GeV) observed in several microquasars (or microquasar candidates) (Aharonian et al. 2005; Albert et al. 2006, 2007b) is difficult to explain in disk models.

3.1 Origin of the Seed Photons

As particles acceleration were discussed in Sect. 2 above, let us focus on the origin of the seed photons for IC scattering.

Synchrotron Self Compton A natural source of seed photons are the synchrotron photons emitted by the energetic electrons, namely SSC. As long as the scattering is in the Thomson regime, namely the energy of the upscattered photons is much less than the energy of the incoming electron, the outgoing photon energy is $\varepsilon_{\rm out} \simeq 4\gamma_{\rm el}^2 \varepsilon_{\rm in}$, where $\gamma_{\rm el}$ is the Lorentz factor of the electron (e.g., Rybicki and Lightman 1979). In this case, the spectral shape of SSC emission from a power law distribution of energetic electrons is *similar* to that of synchrotron emission discussed above (see, e.g., Sari and Esin 2001). It is characterized by four break frequencies. If $\nu_m < \nu_c$, the values of these break frequencies are $\nu_{\rm SSA}^{\rm IC} \simeq 4\gamma_m^2 \nu_{\rm SSA}$, $\nu_m^{\rm IC} \simeq 4\gamma_m^2 \nu_m$, and $\nu_c^{\rm IC} \simeq 4\gamma_c^2 \nu_c$. If $\nu_c < \nu_m$, spectral break corresponding to the self absorption occurs at $\nu_{\rm SSA}^{\rm IC} \simeq 4\gamma_c^2 \nu_{\rm SSA}$, while the energy of the other breaks is not changed.

Thus, for $\nu_m < \nu_c$ the peak of the IC spectrum is at $\nu_m^{\rm IC}$ while for $\nu_c < \nu_m$ it is at $\nu_c^{\rm IC}$, both naturally extend up and above the MeV range. The ratio of the IC and synchrotron energy fluxes is given by $(\nu F_\nu)_{\rm peak,IC}/(\nu F_\nu)_{\rm peak,syn} = Y$, where $Y = (4/3)\gamma_{\rm el}^2 \tau$ is the Compton Y parameter and τ is the optical depth.

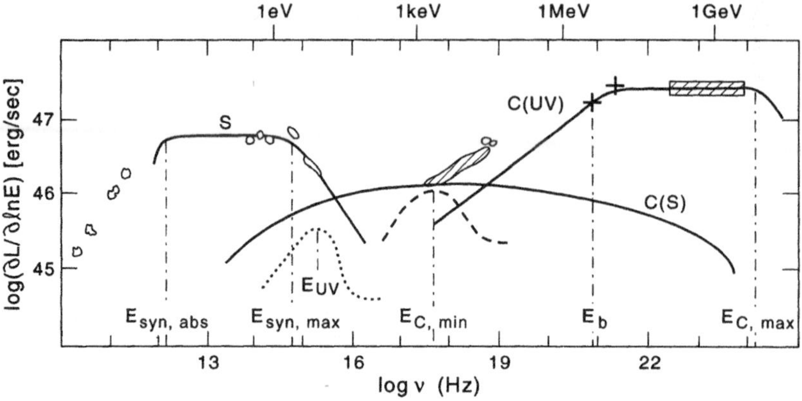

Fig. 4 Broad band spectrum of the AGN 3C 279, taken from Sikora et al. (1994). Solid lines mark the fitted model spectral components: (S)—synchrotron, C(S)—SSC, C(UV)—Comptonization of diffusive UV radiation. For further details as well as references about the data points, see Sikora et al. (1994)

In AGN jets two distinct, broad spectra components are observed. The low energy component is peaking at the sub-mm to IR regime (in FSRQ) or in the UV/X-rays (in high-frequency peaked BL Lacs, or HBLs). The high energy component peaks at the MeV energies in FSRQs and GeV energies in HBLs (e.g., Fossati et al. 1998; Donato et al. 2001; Sambruna et al. 2004; Levinson 2006 and references therein). An example of this spectra is presented in Fig. 4, taken from Sikora et al. (1994). It is therefore natural to attribute the peak at the radio band to synchrotron emission, while that at the X-ray band to IC, as is done by many authors (Konigl 1981; Marscher and Gear 1985; Maraschi et al. 1992; Hartman et al. 2001; Finke et al. 2008; Ghisellini et al. 2009, 2010 and many more). In fact, lacking good theoretical knowledge of the electron number density, hence of the optical depth, very often it is being determined by fitting the ratio of IC peak flux to the synchrotron peak flux. In recent years, similar fitting was done to the X-ray spectra in XRBs (e.g., Gupta et al. 2006; Gallo et al. 2007; Migliari et al. 2007; Maitra et al. 2009).

External Seed Photons In addition to SSC, there are other sources of seed photons. As the spectral shape of the resulting IC emission depends on the spectral shape of the incoming photons, broad band fitting of the spectra are required to determine which is the dominant field. One natural source is the photon field created by the accretion disk (Begelman and Sikora 1987; Dermer et al. 1992; Dermer and Schlickeiser 1993; Blandford and Levinson 1995). In XRBs, photons from the companion star can also serve as seed photons for IC scattering (Dermer and Böttcher 2006).

Alternative source of photons is reprocessing of disk emission by the surrounding material, such as the broad emission line region in AGNs (Sikora et al. 1994; Dermer et al. 1997; see Fig. 4). Additional suggestions for seed photons include reprocessing of the synchrotron emitted photons from the jet itself by the surrounding medium before being IC scattered (Ghisellini and Madau 1996), infrared emission from circumnuclear dust (Błażejowski et al. 2000) or synchrotron radiation from other regions along the jet itself (Georganopoulos and Kazanas 2003). Clearly, these models require additional assumptions about the environment and/or the material that acts to reprocess the original emission. The addition of degrees of freedom with respect to the synchrotron-SSC model enables much better fits to existing broad band data at the price of more complex modeling of the environment.

In nearby objects such as Cen A, the extended giant radio lobes can be spatially resolved. Two distinguished spectral components, one at the radio band (Hardcastle et al. 2009) and one at the X-ray and/or γ-rays (Feigelson et al. 1995; Kataoka and Stawarz 2005; Croston et al. 2005; Abdo et al. 2010) are detected in these lobes. While the radio spectrum is naturally attributed to synchrotron emission, at such large distances from the core (typically hundreds of kpc) it is insufficient to provide enough photons to explain the X- and γ-ray flux observed. Instead, this is attributed to IC emission of the cosmic microwave background (CMB) or extra-galactic background (EBL) light (Tavecchio et al. 2000; Celotti et al. 2001), who's' spectra are well known (Georganopoulos et al. 2008; Finke et al. 2010). This scenario has a great advantage, as it enables decoupling of the electrons distribution and the magnetic field. First, the electron distribution is inferred from the IC spectra and the known seed photon field, and at a second step the magnetic field strength is inferred from the synchrotron spectrum. This separation thus enables to infer the values of the magnetic fields in these regimes, which is found to be close to equipartition. Since these values are much larger than can be achieved by Poynting-flux conservations from the core, as well as higher than the (compressed) external field, these results point towards magnetic field generation in shock waves, as discussed above. Moreover, analyzing the spectra enables to show that the conditions in these lobes enable acceleration of particles to ultra-high energies (Pe'er and Loeb 2012).

3.2 Separation Between Disk and Jet Photons

As much as inferring the origin of seed photons at large distances along the jet is not easy, close to the jet base the situation is far more complicated. Separating jet-based emission models from disk-based emission models is a very difficult task, as both disk-based and jet-based models can produce good fits to the data (Markoff et al. 2005).

On the one hand, there are some indirect evidence based on correlation between emission at the X-ray band and lower energy bands (radio, IR and optic) for jet-dominated X-ray emission (Yuan et al. 2009; Russell et al. 2010). This interpretation is strengthen by extrapolation of the spectral energy distribution (SED) above the turn over at mid- IR bands (Gandhi et al. 2011; Rahoui et al. 2011, 2012; Russell et al. 2013; Corbel et al. 2013). An independent support comes from polarization measurements by *INTEGRAL* satellite (Laurent et al. 2011; Jourdain et al. 2012), which show strong polarization above 400 keV in Cyg X-1, hinting towards jet origin in these energies.

In spite of these indications, it should be stressed that as of today, disk-based models for the X-ray emission are much more developed, and are favored by a central part of the community. As these models are thoroughly discussed in other chapters of this book, I will only briefly mention some aspects here. In these models, many of the X-ray properties are explained by Comptonization in hot accretion flows, including detailed X-ray spectral shape (e.g., Sobolewska et al. 2012; Qiao and Liu 2013), spectral evolution during state transition (Del Santo et al. 2013), and many timing properties of X-ray variability (Kotov et al. 2001; Ingram and Done 2011, 2012). Moreover, detailed fitting of X-ray spectra with the jet models require, in some cases, optical depth of $\tau \sim 2$–3 (Malzac and Belmont 2009; Poutanen and Vurm 2009; Droulans et al. 2010), which put strong constraints on the jet kinetic power (Malzac et al. 2009). Further critical discussions about jet vs. disk models can be found in Poutanen and Zdziarski (2003), Zdziarski et al. (2003), Maccarone (2005), Veledina et al. (2013) as well in other chapters of this book.

One difference between the two scenarios is that while electrons in the inflow are expected to be continuously heated as they spiral in, it is possible that once they enter the jet

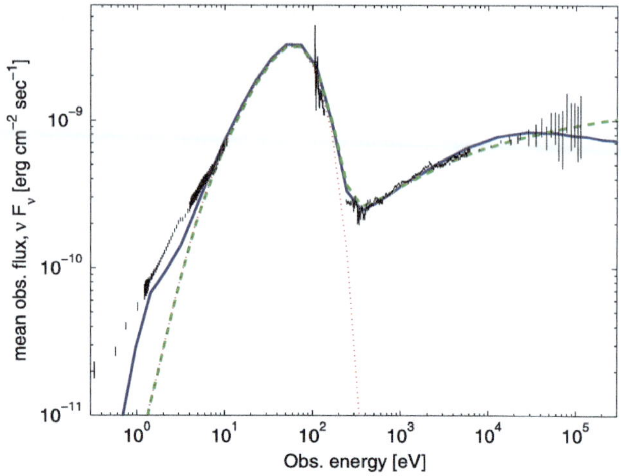

Fig. 5 Fitting the 2000 outburst of XTE J1118+480 (taken from Pe'er and Markoff 2012). The solid (*blue*) curve represent a model in which synchrotron emission is the main source of radiation, while the *dashed* (*green*) represents an IC-dominated model. Both models provide good fits to the data, and are consistent with emission from electrons in the inner part of the jet. For further details see Pe'er and Markoff (2012)

region they are no longer heated. As they propagate outwards inside the jet, both the radiation field and the magnetic field decay, and thus cooling of the electrons is suppressed. During their initial propagation outward, they do though radiatively cool very rapidly. For rapidly cooling electrons, both synchrotron emission discussed above and Compton scattering produce the same spectrum: $F_\nu \propto \nu^{-1/2}$ in the range $\nu_c < \nu < \nu_m$. This spectrum is consistent with the X-ray spectra observed in many outbursts in XRBs (e.g., Hynes et al. 2000; Esin et al. 2001; Homan et al. 2005; Joinet et al. 2008).[1] Thus, by identifying the break frequencies seen in the spectra with ν_m and ν_c, it is possible to constraint the physical parameters from the emitting region. In particular, this analysis may enable to discriminate between emission from the inner parts of the jet (in which the electrons reside only a short time), and emission from the inflow, which is expected to last over a longer period, during the spiral-in. Such an analysis was carried by Pe'er and Markoff (2012), and one of its results is presented in Fig. 5.[2]

4 Hadronic Contribution to the High Energy Spectra

The uncertainty in the nature of the acceleration process implies that protons may be accelerated as well inside the jets. Once accelerated to high energies, protons contribute to the observed spectrum. Although synchrotron emission and Compton scattering are suppressed with respect to emission from leptons due to the much smaller cross section, protons may still have a significant contribution to the high energy emission. First, if the acceleration process acts in such a way that most of the energy is deposited in accelerated protons, it is possible that synchrotron emission from these protons have a significant contribution to the high energy (X- and γ-ray) flux (Aharonian 2000). Second, energetic protons can deposit their

[1]Jet dominated models are expected in XRBs during the low/hard state, where $L \sim 1\ \% L_{\rm Edd}$. At higher luminosities, disk contribution is expected, and the spectral slope varies; the luminosity-dependence of the spectral index can be found in Wu and Gu (2008).

[2]Data taken from McClintock et al. (2001); Note, though, that a different analysis of BeppoSAX (Frontera et al. 2001) and Chandra (Reis et al. 2009) data done in the context of disk models, resulted in a somewhat softer slope below 2 keV.

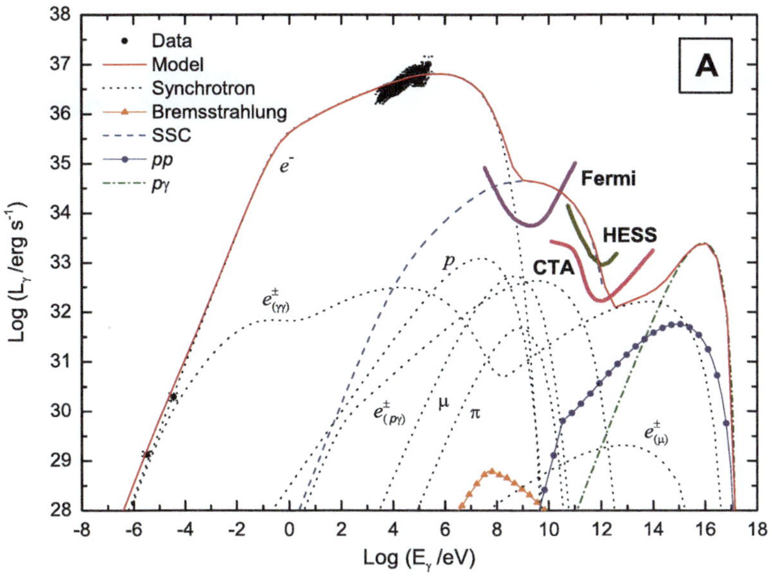

Fig. 6 Fitting the broad band spectra of the 1997 outburst of GX 339-4 by a leptonic/hadronic model of Vila and Romero (2010). Contribution from the various processes are marked: generally, protons contribution is expected mainly at high energies, X- and γ-rays. These fits are used to infer the uncertain values of the magnetic field as well as the electron/proton distribution. For details see Vila and Romero (2010)

energy by photo-meson production, $p\gamma \rightarrow n + \pi^+, p + \pi^0$ (Mannheim and Biermann 1992; Mannheim 1993). The created π mesons are unstable; the π^+ can radiate synchrotron emission before decaying into $\mu^+ + \nu_\mu \rightarrow e^+ + \nu_e + \bar{\nu}_e + \nu_\mu$, while the π^0 decays into a pair of energetic photons. These particles thus produce a high energy electromagnetic cascade, as the created photons are energetic enough to produce a pair of electron-positron, $\gamma + \gamma \rightarrow e^\pm$. Emission from these secondaries may thus be responsible for the high energy (up to TeV) emission seen in AGNs (blazars) (Rachen and Mészáros 1998; Mücke et al. 2003; Murase et al. 2012), as well as in GRBs (Pe'er and Waxman 2004, 2005a) and XRBs (Romero et al. 2005).

In addition to photomeson production, protons can interact with photons by photopair production ($p + \gamma \rightarrow p + e^\pm$), and with other protons through proton-proton (pp) collisions, producing pions and Kaons (Koers et al. 2006). The rate of these interactions depend of course on the ambient photon field, as well as the uncertain distribution of the energetic protons. Combined leptonic/hadronic models that explain the radio emission in AGNs and XRBs as due to synchrotron radiation from electrons and the high energy (up to TeV) emission as due to hadronic-originated cascade exist (e.g., Vila and Romero 2010; see Fig. 6). Although, as can be seen from Fig. 6, these models are still lagging behind disk models, and can thus currently can still only provide approximate fits to the observed spectra. These models suffer two main drawback: First, the inherent uncertainty in the knowledge of the accelerated proton distribution. Second, calculating the evolution of the high energy electromagnetic cascade is extremely difficult, due to the non-linearity of the process, and the fact that it is very rapid, namely, many orders of magnitude shorter than the dynamical time. Thus, it is numerically challenging. While in recent years models of cascade evolution in GRB environment exist (Pe'er and Waxman 2005b), this field is still at its infancy.

An interesting consequence of hadronic models, is that if indeed protons are accelerated to high energies, among the secondaries produced are high energy neutrinos (for a review about neutrino production in AGN jets see Gaisser et al. 1995; for neutrino production in XRB jets, see Levinson and Waxman 2001; Christiansen et al. 2006; Zhang et al. 2010). Thus, such neutrinos—if detected—would be a direct proof of proton acceleration in these environments.

5 Temporal Information

5.1 XRBs: Temporal Correlation Between Different Spectral Bands

Although the basic radiative processes are well known, the emitted spectra from jets are very complex, due to the complex nature of these systems. Emission originates from both the accretion flow, different regions along the jet where the physical conditions vary, as well as external photons that can be reprocessed (IC scattered) from particles along the jet. The physical conditions along the jet, such as the magnetic field and particle distribution, and their connection to the physical conditions in the inner parts of the inflow are uncertain. It is thus not surprising that the observed spectra can be interpreted in more than one way, and that plethora of models aimed at explaining the broad band spectra exist. An in depth discussion in some of the models appear in the chapters authored by Fender, Gallo, Casella and Koërding in this book.

Thus, in order to obtain a full picture additional information is needed. In XRBs, a natural source of information is temporal analysis, since the emission pattern conveniently changes over time scale of $\lesssim$ months. As the emission changes with time, correlation between emission from the inflow and the jet at different times (the different "states") is established. Such correlation is the switch off of jet radio emission in the high/soft state (Tananbaum et al. 1972; Fender et al. 1999b). Others are the correlation found between the X-ray luminosity and the radio luminosity (Hannikainen et al. 1998; Corbel et al. 2000, 2003; Gallo et al. 2003), which are found to scale as $L_R \propto L_X^{0.7}$. However, recently, it was shown that the system H1743-322 follow a different correlation, $L_R \propto L_X^{1.4}$ (Coriat et al. 2011; Gallo et al. 2012). Other correlations are found at different wavebands: between the X and near infrared (NIR) (Russell et al. 2006; Coriat et al. 2009; Casella et al. 2010), and radio—optic and X-rays (Kanbach et al. 2001; Gandhi et al. 2010; Cadolle Bel et al. 2011).

The wealth of emitting zones and radiative processes enables to interpret the observed correlations in various ways. One type of models explore the obvious (yet uncertain) connection between the properties of the inner parts of the accretion flow and the jet (e.g., Markoff et al. 2003; Heinz and Sunyaev 2003). Other ideas include the obvious connection between the synchrotron radiation and IC scattering by the same population of electrons [(Giannios 2005; Veledina et al. 2011); see details in the chapter by Poutannen in this book], as well as correlation between (synchrotron) emission by the same electrons as they propagate along the jet thereby occupying different regions in the jet at different times (Casella et al. 2010).

Existence of different emission zones reflects the complex internal dynamics of the outflow. For non-steady outflow, shock waves naturally develop when two "blobs", or shells of plasma collide. This happens once the ejection of a slower plasma blob is followed by ejection of faster moving one. Once the blobs collide, two shock waves are formed, propagating into both plasmas. By heating (and possibly accelerating to high energies) the particles, these shock waves are the initial source of radiation. This scenario was invoked to explain the complex lightcurve seen during GRB prompt emission (Rees and Meszaros 1994;

Daigne and Mochkovitch 1998). In recent years, similar ideas were studied in the context of emission from XRBs (Kaiser et al. 2000; Jamil et al. 2010; Malzac 2013) and TDEs (Giannios and Metzger 2011; although a structured jet model was suggested by Liu et al. 2012).

This idea, though, is incomplete: currently, the internal shock model is lacking a predictive power about the radii at which the collisions, hence the energy dissipation takes place—these are determined by the initial conditions. Thus, overall, my personal opinion is that understanding the nature of the correlations observed is at its infancy, and that this field is a very promising path to take. Future models will inevitably combine both dynamical models and radiative models, which will mature in the coming years.

5.2 Flaring Activities in AGNs

In AGNs (blazars), flaring activity is observed in the X- and γ-rays up to the highest energies, at the TeV band. This is often observed on a very short time scales, of the order of hours and in some cases even minutes (Kniffen et al. 1993; Buckley et al. 1996; Aharonian et al. 2007; Albert et al. 2007a; Aleksić et al. 2011). Radio observations showed that radio outburst seem to follow the γ-ray flares (Reich et al. 1993; Zhang et al. 1994). While significant variability in the optical band is observed as well, its correlation with the variability in the γ-band is not fully clear (Wehrle et al. 1998; Palma et al. 2011).

The main implication of this rapid variability in the flux is constraining the size of the emitting region and the bulk motion Lorentz factor. An observed variability time Δt^{ob} implies that the size of the emitting region cannot exceed

$$r \leq r_{\mathrm{var}} \approx \frac{\Gamma c \Delta t^{\mathrm{ob}} \mathcal{D}}{1+z}, \tag{10}$$

where z is the redshift, and Γ is the Lorentz factor associated with the bulk motion (Ghisellini and Madau 1996). On the other hand, the fact that TeV photons are observed implies that the optical depth to pair production with the low energy photons in the plasma cannot exceed unity. Thus, the emitting region cannot be too compact. Combined together, these two constraints imply high bulk Lorentz factor (e.g., in PKS 2155-304, $\Gamma \gtrsim 50$ was inferred by Begelman et al. 2008). The exact value of the constraint on the emitting region thus depend on the variability time, as well as the photon field. The variability itself reflects changing conditions within the outflow, e.g., due to the existence of internal shock waves (Spada et al. 2001). The fact that the constraints found on Γ in PKS 2155-40 were found to be inconsistent with direct measurements, have led Giannios et al. (2009) to suggest a jet within a jet model for the high energy emission. In a more general form, this can be viewed as an indication for an internal structure within the jets.

6 Jet Power

Estimating the total deposited energy (or power) in astronomical jets is a very tricky task. The complexity of the problem is most easily understood if one considers the different episodes of energy transfer in these systems. First, there is the *kinetic* energy associated with the bulk motion of particles inside the jet. Naturally, this is some fraction of the gravitational energy of the inflowing material in the accretion disk. Theoretical determination of this fraction is possible only after the theory of jet production is fully understood. Until then, it can only be estimated from observations.

The observed radiation, in turn, reveals only a small fraction of this energy. Following jet launching, the second energy transfer occurs at a certain location(s) along the jet, where particles are accelerated to high energies. This acceleration must occur on the expense of (part of) the bulk motion kinetic energy, but possibly also due to magnetic reconnection—in which case it is at the expense of magnetic energy. Finally, the accelerated particles radiate *some fraction* of their energy as photons, producing the observed signal. Thus, direct observation of the photon signal reveals only an unknown fraction—likely a small fraction, of the kinetic energy initially given to the particles inside the jets.

Estimating the kinetic jet power is thus difficult, and rely on several assumptions. For example, Rawlings and Saunders (1991) estimated the average kinetic power of jets in blazars by dividing the total energy stored in the form of electrons and the magnetic field energy in the radio lobes (as deduced from synchrotron theory and the equipartition assumption) by the lobe age, which was computed from spectral aging or expansion velocity arguments. Similarly, Celotti and Fabian (1993) estimated the jet power of blazer jets using the framework of the standard synchrotron self-Compton theory. As explained above, these works suffer from substantial uncertainties, due to the various underlying assumptions needed.

In an alternative approach, Allen et al. (2006) and Balmaverde et al. (2008) estimated the jet kinetic power by estimating the mechanical work, PdV required to inflate the observed giant X-ray cavities. Since here too there are uncertainties in estimating the size of these cavities, these translate into uncertainties in the jet power. This will further be discussed in the chapter of this book authored by Heinz.

These works found a strong correlation between the estimated jet power and the disk luminosity. Such a correlation is expected in the leading mechanisms for jet production. As material from the jet originates from the disk, such correlations are of no surprise. Additional clue may come from a correlation between the jet power (as estimated from the radio flux) and spin of the black hole, as recently reported (Narayan and McClintock 2012; Steiner et al. 2013). While this result is still debatable (see Russell et al. 2013), if confirmed it may serve as a strong clue for the mechanism that launches jets in nature. See further discussions in the chapters by McClintock, Narayan, Fender and Gallo in this book.

Thus, while various models that estimate the kinetic power of jets exist, they all suffer from uncertainties, caused both by uncertainties in the measurements, and also by the need to rely on uncertain emission models. I thus view this subject as one which is far from being matured, and will be further developed in the near future.

7 Summary and Conclusions

In this chapter, I reviewed some of the basic radiative mechanisms that produce the broad band emission seen in astronomical jets. Due to the broad nature of this subject, I focused on XRBs and AGNs (mainly blazars). The main radiative processes considered are synchrotron emission, SSC, Compton scattering of external, or reprocessed photons, and hadronic contribution, via proton-synchrotron emission and electromagnetic cascade caused by secondaries produced by proton-photon (and to a lesser extent, proton-proton) interactions.

Although each of these processes is well understood, the changing conditions inside the jets lead to complex observed spectra. This leads to the fact that inspite a wealth of broadband data, no single model is commonly accepted. On the contrary, as discussed here, the same data can be interpreted in more than one way. Thus, the main "take away" massage from this chapter, is that modeling emission from jets is one of the most challenging tasks.

The questions that need to be addressed when studying emission from jets extend far beyond the realm of the radiative processes involved, and require addressing questions in basic physics and astronomy. Broadly speaking, in order to fully understand the emission, one needs to understand:

1. The connection between disk and jet, and the mechanism that leads to jet launching.
2. The varying physical conditions in different regions inside the jet, such as the magnetic field along the jet.
3. The jet composition that governs the contribution of leptons and hadrons to the observed spectra.
4. The nature and details of the acceleration mechanisms that determine the energy distribution of energetic particles in different parts of the jet.
5. The internal (synchrotron) and external (accretion disk, companion star, CMB, etc.) photon fields that serve as seed photons to scattering by jet material.
6. The geometry of the jets, including velocity profile and its angle towards the observer, that determine the different scattered field, as well as the Doppler boost.
7. The dynamics of material inside the jet, that determines the spatial distribution of the radiating particles and their temporal evolution.

Addressing each of these questions is a task so challenging by itself, that despite decades of research and numerous works (unfortunately, only very few could be mentioned here) we still have only clues, but no definite answer to any of them. Moreover, these questions, while can be addressed separately, should be addressed in the context of the different environments in which jets are observed—XRBs, AGNs, GRBs and recently also TDEs. Thus, full answer to all these questions is not expected any time in the near future. However, the wealth of current and future data—both spectral, temporal and spatial data, ensures that there is plenty of room for new ideas in the coming years.

Acknowledgements I would like to thank Paul Callanan, Piergiorgio Casella, Amir Levinson and Abraham Loeb for providing useful comments on this manuscript.

References

A.A. Abdo, M. Ackermann, M. Ajello, W.B. Atwood, L. Baldini, J. Ballet, G. Barbiellini, D. Bastieri, B.M. Baughman, K. Bechtol, et al., Fermi gamma-ray imaging of a radio galaxy. Science **328**, 725 (2010). doi:10.1126/science.1184656

A. Achterberg, Y.A. Gallant, J.G. Kirk, A.W. Guthmann, Particle acceleration by ultrarelativistic shocks: theory and simulations. Mon. Not. R. Astron. Soc. **328**, 393–408 (2001). doi:10.1046/j.1365-8711.2001.04851.x

F.A. Aharonian, TeV gamma rays from BL Lac objects due to synchrotron radiation of extremely high energy protons. New Astron. **5**, 377–395 (2000). doi:10.1016/S1384-1076(00)00039-7

F. Aharonian, A.G. Akhperjanian, K.-M. Aye, A.R. Bazer-Bachi, M. Beilicke, W. Benbow, D. Berge, P. Berghaus, et al., Discovery of very high energy gamma rays associated with an X-ray binary. Science **309**, 746–749 (2005). doi:10.1126/science.1113764

F. Aharonian, A.G. Akhperjanian, A.R. Bazer-Bachi, B. Behera, M. Beilicke, W. Benbow, D. Berge, K. Bernlöhr, et al., An exceptional very high energy gamma-ray flare of PKS 2155-304. Astrophys. J. Lett. **664**, 71–74 (2007). doi:10.1086/520635

J. Albert, E. Aliu, H. Anderhub, P. Antoranz, A. Armada, M. Asensio, C. Baixeras, J.A. Barrio, et al., Variable very-high-energy gamma-ray emission from the microquasar LS I +61 303. Science **312**, 1771–1773 (2006). doi:10.1126/science.1128177

J. Albert, E. Aliu, H. Anderhub, P. Antoranz, A. Armada, C. Baixeras, J.A. Barrio, H. Bartko, et al., Variable very high energy γ-ray emission from Markarian 501. Astrophys. J. **669**, 862–883 (2007a). doi:10.1086/521382

J. Albert, E. Aliu, H. Anderhub, P. Antoranz, A. Armada, C. Baixeras, J.A. Barrio, H. Bartko, et al., Very high energy gamma-ray radiation from the stellar mass black hole binary Cygnus X-1. Astrophys. J. Lett. **665**, 51–54 (2007b). doi:10.1086/521145

J. Aleksić, L.A. Antonelli, P. Antoranz, M. Backes, J.A. Barrio, D. Bastieri, J. Becerra González, W. Bednarek, et al., MAGIC discovery of very high energy emission from the FSRQ PKS 1222+21. Astrophys. J. Lett. **730**, 8 (2011). doi:10.1088/2041-8205/730/1/L8
S.W. Allen, R.J.H. Dunn, A.C. Fabian, G.B. Taylor, C.S. Reynolds, The relation between accretion rate and jet power in X-ray luminous elliptical galaxies. Mon. Not. R. Astron. Soc. **372**, 21–30 (2006). doi:10.1111/j.1365-2966.2006.10778.x
H. Alvarez, J. Aparici, J. May, P. Reich, The radio continuum spectrum of Centaurus A's large-scale components. Astron. Astrophys. **355**, 863–872 (2000)
A.M. Atoyan, F.A. Aharonian, Modelling of the non-thermal flares in the galactic microquasar GRS 1915+105. Mon. Not. R. Astron. Soc. **302**, 253–276 (1999). doi:10.1046/j.1365-8711.1999.02172.x
M. Axelsson, L. Baldini, G. Barbiellini, M.G. Baring, R. Bellazzini, J. Bregeon, M. Brigida, P. Bruel, R. Buehler, G.A. Caliandro, R.A. Cameron, P.A. Caraveo, C. Cecchi, R.C.G. Chaves, et al., GRB110721A: an extreme peak energy and signatures of the photosphere. Astrophys. J. Lett. **757**, 31 (2012). doi:10.1088/2041-8205/757/2/L31
B. Balmaverde, R.D. Baldi, A. Capetti, The accretion mechanism in low-power radio galaxies. Astron. Astrophys. **486**, 119–130 (2008). doi:10.1051/0004-6361:200809810
M.C. Begelman, M. Sikora, Inverse Compton scattering of ambient radiation by a cold relativistic jet. a source of beamed, polarized continuum in blazars? Astrophys. J. **322**, 650–661 (1987). doi:10.1086/165760
M.C. Begelman, R.D. Blandford, M.J. Rees, Theory of extragalactic radio sources. Rev. Mod. Phys. **56**, 255–351 (1984). doi:10.1103/RevModPhys.56.255
M.C. Begelman, A.C. Fabian, M.J. Rees, Implications of very rapid TeV variability in blazars. Mon. Not. R. Astron. Soc. **384**, 19–23 (2008). doi:10.1111/j.1745-3933.2007.00413.x
A.R. Bell, The acceleration of cosmic rays in shock fronts I. Mon. Not. R. Astron. Soc. **182**, 147–156 (1978)
R. Blandford, D. Eichler, Particle acceleration at astrophysical shocks: a theory of cosmic ray origin. Phys. Rep. **154**, 1–75 (1987). doi:10.1016/0370-1573(87)90134-7
R.D. Blandford, Accretion disc electrodynamics—a model for double radio sources. Mon. Not. R. Astron. Soc. **176**, 465–481 (1976)
R.D. Blandford, A. Konigl, Relativistic jets as compact radio sources. Astrophys. J. **232**, 34–48 (1979). doi:10.1086/157262
R.D. Blandford, A. Levinson, Pair cascades in extragalactic jets. 1. Gamma rays. Astrophys. J. **441**, 79–95 (1995). doi:10.1086/175338
R.D. Blandford, C.F. McKee, Fluid dynamics of relativistic blast waves. Phys. Fluids **19**, 1130–1138 (1976). doi:10.1063/1.861619
R.D. Blandford, C.F. McKee, Radiation from relativistic blast waves in quasars and active galactic nuclei. Mon. Not. R. Astron. Soc. **180**, 343–371 (1977)
R.D. Blandford, J.P. Ostriker, Particle acceleration by astrophysical shocks. Astrophys. J. Lett. **221**, 29–32 (1978). doi:10.1086/182658
R.D. Blandford, D.G. Payne, Hydromagnetic flows from accretion discs and the production of radio jets. Mon. Not. R. Astron. Soc. **199**, 883–903 (1982)
R.D. Blandford, R.L. Znajek, Electromagnetic extraction of energy from Kerr black holes. Mon. Not. R. Astron. Soc. **179**, 433–456 (1977)
M. Błażejowski, M. Sikora, R. Moderski, G.M. Madejski, Comptonization of infrared radiation from hot dust by relativistic jets in quasars. Astrophys. J. **545**, 107–116 (2000). doi:10.1086/317791
G.R. Blumenthal, R.J. Gould, Bremsstrahlung, synchrotron radiation, and Compton scattering of high-energy electrons traversing dilute gases. Rev. Mod. Phys. **42**, 237–271 (1970). doi:10.1103/RevModPhys.42.237
V. Bosch-Ramon, G.E. Romero, J.M. Paredes, A broadband leptonic model for gamma-ray emitting microquasars. Astron. Astrophys. **447**, 263–276 (2006). doi:10.1051/0004-6361:20053633
D.-F. Bu, F. Yuan, F.-G. Xie, Self-similar solution of hot accretion flows with ordered magnetic field and outflow. Mon. Not. R. Astron. Soc. **392**, 325–331 (2009). doi:10.1111/j.1365-2966.2008.14047.x
J.H. Buckley, C.W. Akerlof, S. Biller, D.A. Carter-Lewis, M. Catanese, M.F. Cawley, V. Connaughton, D.J. Fegan, J.P. Finley, J. Gaidos, A.M. Hillas, J.F. Kartje, A. Koenigl, F. Krennrich, R.C. Lamb, R. Lessard, D.J. Macomb, J.R. Mattox, J.E. McEnery, G. Mohanty, J. Quinn, A.J. Rodgers, H.J. Rose, M.S. Schubnel, G. Sembroski, P.S. Smith, T.C. Weekes, C. Wilson, J. Zweerink, Gamma-ray variability of the BL Lacertae object Markarian 421. Astrophys. J. Lett. **472**, 9 (1996). doi:10.1086/310352
D.N. Burrows, J.A. Kennea, G. Ghisellini, V. Mangano, B. Zhang, K.L. Page, M. Eracleous, P. Romano, T. Sakamoto, A.D. Falcone, J.P. Osborne, S. Campana, A.P. Beardmore, A.A. Breeveld, M.M. Chester, R. Corbet, S. Covino, J.R. Cummings, P. D'Avanzo, V. D'Elia, P. Esposito, P.A. Evans, D. Fugazza, J.M. Gelbord, K. Hiroi, S.T. Holland, K.Y. Huang, M. Im, G. Israel, Y. Jeon, Y.-B. Jeon, H.D. Jun, N. Kawai, J.H. Kim, H.A. Krimm, F.E. Marshall, P. Mészáros, H. Negoro, N. Omodei, W.-K. Park, J.S. Perkins, M. Sugizaki, H.-I. Sung, G. Tagliaferri, E. Troja, Y. Ueda, Y. Urata, R. Usui, L.A. Antonelli,

S.D. Barthelmy, G. Cusumano, P. Giommi, A. Melandri, M. Perri, J.L. Racusin, B. Sbarufatti, M.H. Siegel, N. Gehrels, Relativistic jet activity from the tidal disruption of a star by a massive black hole. Nature **476**, 421–424 (2011). doi:10.1038/nature10374

M. Cadolle Bel, P. Sizun, A. Goldwurm, J. Rodriguez, P. Laurent, A.A. Zdziarski, L. Foschini, P. Goldoni, C. Gouiffès, J. Malzac, E. Jourdain, J.-P. Roques, The broad-band spectrum of Cygnus X-1 measured by INTEGRAL. Astron. Astrophys. **446**, 591–602 (2006). doi:10.1051/0004-6361:20053068

M. Cadolle Bel, J. Rodriguez, P. D'Avanzo, D.M. Russell, J. Tomsick, S. Corbel, F.W. Lewis, F. Rahoui, M. Buxton, P. Goldoni, E. Kuulkers, Overview of an extensive multi-wavelength study of GX 339-4 during the 2010 outburst. Astron. Astrophys. **534**, 119 (2011). doi:10.1051/0004-6361/201117684

P. Casella, A. Pe'er, On the role of the magnetic field on jet emission in X-ray binaries. Astrophys. J. Lett. **703**, 63–66 (2009). doi:10.1088/0004-637X/703/1/L63

P. Casella, T.J. Maccarone, K. O'Brien, R.P. Fender, D.M. Russell, M. van der Klis, A. Pe'er, D. Maitra, D. Altamirano, T. Belloni, G. Kanbach, M. Klein-Wolt, E. Mason, P. Soleri, A. Stefanescu, K. Wiersema, R. Wijnands, Fast infrared variability from a relativistic jet in GX 339-4. Mon. Not. R. Astron. Soc. **404**, 21–25 (2010). doi:10.1111/j.1745-3933.2010.00826.x

A. Celotti, A.C. Fabian, The kinetic power and luminosity of parsecscale radio jets—an argument for heavy jets. Mon. Not. R. Astron. Soc. **264**, 228 (1993)

A. Celotti, G. Ghisellini, The power of blazar jets. Mon. Not. R. Astron. Soc. **385**, 283–300 (2008). doi:10.1111/j.1365-2966.2007.12758.x

A. Celotti, G. Ghisellini, M. Chiaberge, Large-scale jets in active galactic nuclei: multiwavelength mapping. Mon. Not. R. Astron. Soc. **321**, 1–5 (2001). doi:10.1046/j.1365-8711.2001.04160.x

M. Cerruti, C.D. Dermer, B. Lott, C. Boisson, A. Zech, Gamma-ray blazars near equipartition and the origin of the GeV spectral break in 3C 454.3. ArXiv e-prints (2013)

H.R. Christiansen, M. Orellana, G.E. Romero, High-energy neutrino emission from X-ray binaries. Phys. Rev. D **73**(6), 063012 (2006). doi:10.1103/PhysRevD.73.063012

D.F. Cioffi, T.W. Jones, Internal Faraday rotation effects in transparent synchrotron sources. Astron. J. **85**, 368–375 (1980). doi:10.1086/112685

S. Corbel, R.P. Fender, A.K. Tzioumis, M. Nowak, V. McIntyre, P. Durouchoux, R. Sood, Coupling of the X-ray and radio emission in the black hole candidate and compact jet source GX 339-4. Astron. Astrophys. **359**, 251–268 (2000)

S. Corbel, R.P. Fender, A.K. Tzioumis, J.A. Tomsick, J.A. Orosz, J.M. Miller, R. Wijnands, P. Kaaret, Large-scale, decelerating, relativistic X-ray jets from the microquasar XTE J1550-564. Science **298**, 196–199 (2002). doi:10.1126/science.1075857

S. Corbel, M.A. Nowak, R.P. Fender, A.K. Tzioumis, S. Markoff, Radio/X-ray correlation in the low/hard state of GX 339-4. Astron. Astrophys. **400**, 1007–1012 (2003). doi:10.1051/0004-6361:20030090

S. Corbel, H. Aussel, J.W. Broderick, P. Chanial, M. Coriat, A.J. Maury, M.M. Buxton, J.A. Tomsick, A.K. Tzioumis, S. Markoff, J. Rodriguez, C.D. Bailyn, C. Brocksopp, R.P. Fender, P.O. Petrucci, M. Cadolle-Bel, D. Calvelo, L. Harvey-Smith, Formation of the compact jets in the black hole GX 339-4. Mon. Not. R. Astron. Soc. **431**, 107–111 (2013). doi:10.1093/mnrasl/slt018

M. Coriat, S. Corbel, M.M. Buxton, C.D. Bailyn, J.A. Tomsick, E. Körding, E. Kalemci, The infrared/X-ray correlation of GX 339-4: probing hard X-ray emission in accreting black holes. Mon. Not. R. Astron. Soc. **400**, 123–133 (2009). doi:10.1111/j.1365-2966.2009.15461.x

M. Coriat, S. Corbel, L. Prat, J.C.A. Miller-Jones, D. Cseh, A.K. Tzioumis, C. Brocksopp, J. Rodriguez, R.P. Fender, G.R. Sivakoff, Radiatively efficient accreting black holes in the hard state: the case study of H1743-322. Mon. Not. R. Astron. Soc. **414**, 677–690 (2011). doi:10.1111/j.1365-2966.2011.18433.x

J.H. Croston, M.J. Hardcastle, D.E. Harris, E. Belsole, M. Birkinshaw, D.M. Worrall, An X-ray study of magnetic field strengths and particle content in the lobes of FR II radio sources. Astrophys. J. **626**, 733–747 (2005). doi:10.1086/430170

F. Daigne, R. Mochkovitch, Gamma-ray bursts from internal shocks in a relativistic wind: temporal and spectral properties. Mon. Not. R. Astron. Soc. **296**, 275–286 (1998). doi:10.1046/j.1365-8711.1998.01305.x

M. Del Santo, J. Malzac, R. Belmont, L. Bouchet, G. De Cesare, The magnetic field in the X-ray corona of Cygnus X-1. Mon. Not. R. Astron. Soc. **430**, 209–220 (2013). doi:10.1093/mnras/sts574

C.D. Dermer, M. Böttcher, Gamma rays from Compton scattering in the jets of microquasars: application to LS 5039. Astrophys. J. **643**, 1081–1097 (2006). doi:10.1086/502966

C.D. Dermer, R. Schlickeiser, Model for the high-energy emission from blazars. Astrophys. J. **416**, 458 (1993). doi:10.1086/173251

C.D. Dermer, J.A. Miller, H. Li, Stochastic particle acceleration near accreting black holes. Astrophys. J. **456**, 106 (1996). doi:10.1086/176631

C.D. Dermer, R. Schlickeiser, A. Mastichiadis, High-energy gamma radiation from extragalactic radio sources. Astron. Astrophys. **256**, 27–30 (1992)

C.D. Dermer, S.J. Sturner, R. Schlickeiser, Nonthermal Compton and synchrotron processes in the jets of active galactic nuclei. Astrophys. J. Suppl. Ser. **109**, 103 (1997). doi:10.1086/312972

D. Donato, G. Ghisellini, G. Tagliaferri, G. Fossati, Hard X-ray properties of blazars. Astron. Astrophys. **375**, 739–751 (2001). doi:10.1051/0004-6361:20010675

R. Droulans, R. Belmont, J. Malzac, E. Jourdain, Variability and spectral modeling of the hard X-ray emission of GX 339-4 in a bright Low/Hard state. Astrophys. J. **717**, 1022–1036 (2010). doi:10.1088/0004-637X/717/2/1022

D.C. Ellison, G.P. Double, Diffusive shock acceleration in unmodified relativistic, oblique shocks. Astropart. Phys. **22**, 323–338 (2004). doi:10.1016/j.astropartphys.2004.08.005

D.C. Ellison, S.P. Reynolds, F.C. Jones, First-order Fermi particle acceleration by relativistic shocks. Astrophys. J. **360**, 702–714 (1990). doi:10.1086/169156

A.A. Esin, J.E. McClintock, R. Narayan, Advection-dominated accretion and the spectral states of black hole X-ray binaries: application to nova MUSCAE 1991. Astrophys. J. **489**, 865–889 (1997). doi:10.1086/304829

A.A. Esin, J.E. McClintock, J.J. Drake, M.R. Garcia, C.A. Haswell, R.I. Hynes, M.P. Muno, Modeling the low-state spectrum of the X-ray nova XTE J1118+480. Astrophys. J. **555**, 483–488 (2001). doi:10.1086/321450

H. Falcke, P.L. Biermann, The jet-disk symbiosis. I. Radio to X-ray emission models for quasars. Astron. Astrophys. **293**, 665–682 (1995)

H. Falcke, P.L. Biermann, The jet/disk symbiosis. III. What the radio cores in GRS 1915+105, NGC 4258, M 81 and SGR A* tell us about accreting black holes. Astron. Astrophys. **342**, 49–56 (1999)

E.D. Feigelson, S.A. Laurent-Muehleisen, R.I. Kollgaard, E.B. Fomalont, Discovery of inverse-Compton X-rays in radio lobes. Astrophys. J. Lett. **449**, 149 (1995). doi:10.1086/309642

R. Fender, Jets from X-Ray Binaries, ed. by W.H.G. Lewin, M. van der Klis (2006), pp. 381–419

R. Fender, 'Disc-jet' coupling in black hole X-ray binaries and active galactic nuclei, in *Lecture Notes in Physics*, vol. 794, ed. by T. Belloni (Springer, Berlin, 2010), p. 115

R.P. Fender, Powerful jets from black hole X-ray binaries in low/hard X-ray states. Mon. Not. R. Astron. Soc. **322**, 31–42 (2001). doi:10.1046/j.1365-8711.2001.04080.x

R.P. Fender, S.T. Garrington, D.J. McKay, T.W.B. Muxlow, G.G. Pooley, R.E. Spencer, A.M. Stirling, E.B. Waltman, MERLIN observations of relativistic ejections from GRS 1915+105. Mon. Not. R. Astron. Soc. **304**, 865–876 (1999a). doi:10.1046/j.1365-8711.1999.02364.x

R. Fender, S. Corbel, T. Tzioumis, V. McIntyre, D. Campbell-Wilson, M. Nowak, R. Sood, R. Hunstead, A. Harmon, P. Durouchoux, W. Heindl, Quenching of the radio jet during the X-ray high state of GX 339-4. Astrophys. J. Lett. **519**, 165–168 (1999b). doi:10.1086/312128

E. Fermi, On the origin of the cosmic radiation. Phys. Rev. **75**, 1169–1174 (1949). doi:10.1103/PhysRev.75.1169

E. Fermi, Galactic magnetic fields and the origin of cosmic radiation. Astrophys. J. **119**, 1 (1954). doi:10.1086/145789

J. Ferreira, C. Dougados, S. Cabrit, Which jet launching mechanism(s) in T Tauri stars? Astron. Astrophys. **453**, 785–796 (2006). doi:10.1051/0004-6361:20054231

J.D. Finke, C.D. Dermer, M. Böttcher, Synchrotron self-Compton analysis of TeV X-ray-selected BL Lacertae objects. Astrophys. J. **686**, 181–194 (2008). doi:10.1086/590900

J.D. Finke, S. Razzaque, C.D. Dermer, Modeling the extragalactic background light from stars and dust. Astrophys. J. **712**, 238–249 (2010). doi:10.1088/0004-637X/712/1/238

G. Fossati, L. Maraschi, A. Celotti, A. Comastri, G. Ghisellini, A unifying view of the spectral energy distributions of blazars. Mon. Not. R. Astron. Soc. **299**, 433–448 (1998). doi:10.1046/j.1365-8711.1998.01828.x

P.C. Fragile, D.L. Meier, General relativistic magnetohydrodynamic simulations of the hard state as a magnetically dominated accretion flow. Astrophys. J. **693**, 771–783 (2009). doi:10.1088/0004-637X/693/1/771

J.T. Frederiksen, C.B. Hededal, T. Haugbølle, Å. Nordlund, Magnetic field generation in collisionless shocks: pattern growth and transport. Astrophys. J. Lett. **608**, 13–16 (2004). doi:10.1086/421262

F. Frontera, A.A. Zdziarski, L. Amati, J. Mikołajewska, T. Belloni, S. Del Sordo, F. Haardt, E. Kuulkers, N. Masetti, M. Orlandini, E. Palazzi, A.N. Parmar, R.A. Remillard, A. Santangelo, L. Stella, A measurement of the broadband spectrum of XTE J1118+480 with BeppoSAX and its astrophysical implications. Astrophys. J. **561**, 1006–1015 (2001). doi:10.1086/323258

T.K. Gaisser, F. Halzen, T. Stanev, Particle astrophysics with high energy neutrinos. Phys. Rep. **258**, 173–236 (1995). doi:10.1016/0370-1573(95)00003-Y

E. Gallo, Radio emission and jets from microquasars, in *Lecture Notes in Physics*, vol. 794, ed. by T. Belloni (Springer, Berlin, 2010), p. 85

E. Gallo, R.P. Fender, G.G. Pooley, A universal radio-X-ray correlation in low/hard state black hole binaries. Mon. Not. R. Astron. Soc. **344**, 60–72 (2003). doi:10.1046/j.1365-8711.2003.06791.x

E. Gallo, B.P. Miller, R. Fender, Assessing luminosity correlations via cluster analysis: evidence for dual tracks in the radio/X-ray domain of black hole X-ray binaries. Mon. Not. R. Astron. Soc. **423**, 590–599 (2012). doi:10.1111/j.1365-2966.2012.20899.x

E. Gallo, S. Migliari, S. Markoff, J.A. Tomsick, C.D. Bailyn, S. Berta, R. Fender, J.C.A. Miller-Jones, The spectral energy distribution of quiescent black hole X-ray binaries: new constraints from Spitzer. Astrophys. J. **670**, 600–609 (2007). doi:10.1086/521524

P. Gandhi, V.S. Dhillon, M. Durant, A.C. Fabian, A. Kubota, K. Makishima, J. Malzac, T.R. Marsh, J.M. Miller, T. Shahbaz, H.C. Spruit, P. Casella, Rapid optical and X-ray timing observations of GX339-4: multicomponent optical variability in the low/hard state. Mon. Not. R. Astron. Soc. **407**, 2166–2192 (2010). doi:10.1111/j.1365-2966.2010.17083.x

P. Gandhi, A.W. Blain, D.M. Russell, P. Casella, J. Malzac, S. Corbel, P. D'Avanzo, F.W. Lewis, S. Markoff, M. Cadolle Bel, P. Goldoni, S. Wachter, D. Khangulyan, A. Mainzer, A variable mid-infrared synchrotron break associated with the compact jet in GX 339-4. Astrophys. J. Lett. **740**, 13 (2011). doi:10.1088/2041-8205/740/1/L13

M. Georganopoulos, D. Kazanas, Decelerating flows in TeV blazars: a resolution to the BL Lacertae-FR I unification problem. Astrophys. J. Lett. **594**, 27–30 (2003). doi:10.1086/378557

M. Georganopoulos, R.M. Sambruna, D. Kazanas, A.N. Cillis, C.C. Cheung, E.S. Perlman, K.M. Blundell, D.S. Davis, A novel method for measuring the extragalactic background light: Fermi application to the lobes of Fornax A. Astrophys. J. Lett. **686**, 5–8 (2008). doi:10.1086/592833

G. Ghisellini, Jetted active galactic nuclei. Int. J. Mod. Phys. Conf. Ser. **8**, 1 (2012). doi:10.1142/S2010194512004345

G. Ghisellini, P. Madau, On the origin of the gamma-ray emission in blazars. Mon. Not. R. Astron. Soc. **280**, 67–76 (1996)

G. Ghisellini, F. Tavecchio, G. Ghirlanda, Jet and accretion power in the most powerful Fermi blazars. Mon. Not. R. Astron. Soc. **399**, 2041–2054 (2009). doi:10.1111/j.1365-2966.2009.15397.x

G. Ghisellini, F. Tavecchio, L. Foschini, G. Ghirlanda, L. Maraschi, A. Celotti, General physical properties of bright Fermi blazars. Mon. Not. R. Astron. Soc. **402**, 497–518 (2010). doi:10.1111/j.1365-2966.2009.15898.x

D. Giannios, Spectra of black-hole binaries in the low/hard state: from radio to X-rays. Astron. Astrophys. **437**, 1007–1015 (2005). doi:10.1051/0004-6361:20041491

D. Giannios, B.D. Metzger, Radio transients from stellar tidal disruption by massive black holes. Mon. Not. R. Astron. Soc. **416**, 2102–2107 (2011). doi:10.1111/j.1365-2966.2011.19188.x

D. Giannios, D.A. Uzdensky, M.C. Begelman, Fast TeV variability in blazars: jets in a jet. Mon. Not. R. Astron. Soc. **395**, 29–33 (2009). doi:10.1111/j.1745-3933.2009.00635.x

V.L. Ginzburg, S.I. Syrovatskii, Cosmic magnetobremsstrahlung (synchrotron radiation). Annu. Rev. Astron. Astrophys. **3**, 297 (1965). doi:10.1146/annurev.aa.03.090165.001501

S. Gupta, M. Böttcher, C.D. Dermer, Time-dependent synchrotron and Compton spectra from jets of microquasars. Astrophys. J. **644**, 409–423 (2006). doi:10.1086/503552

F. Haardt, L. Maraschi, X-Ray spectra from two-phase accretion disks. Astrophys. J. **413**, 507–517 (1993). doi:10.1086/173020

D.C. Hannikainen, R.W. Hunstead, D. Campbell-Wilson, R.K. Sood, MOST radio monitoring of GX 339-4. Astron. Astrophys. **337**, 460–464 (1998)

M.J. Hardcastle, C.C. Cheung, I.J. Feain, L. Stawarz, High-energy particle acceleration and production of ultra-high-energy cosmic rays in the giant lobes of Centaurus A. Mon. Not. R. Astron. Soc. **393**, 1041–1053 (2009). doi:10.1111/j.1365-2966.2008.14265.x

D.E. Harris, H. Krawczynski, X-Ray emission from extragalactic jets. Annu. Rev. Astron. Astrophys. **44**, 463–506 (2006). doi:10.1146/annurev.astro.44.051905.092446

R.C. Hartman, M. Böttcher, G. Aldering, H. Aller, M. Aller, D.E. Backman, T.J. Balonek, D.L. Bertsch, et al., Multiepoch multiwavelength spectra and models for blazar 3C 279. Astrophys. J. **553**, 683–694 (2001). doi:10.1086/320970

T. Haugbølle, Three-dimensional modeling of relativistic collisionless ion-electron shocks. Astrophys. J. Lett. **739**, 42 (2011). doi:10.1088/2041-8205/739/2/L42

S. Heinz, R.A. Sunyaev, The non-linear dependence of flux on black hole mass and accretion rate in core-dominated jets. Mon. Not. R. Astron. Soc. **343**, 59–64 (2003). doi:10.1046/j.1365-8711.2003.06918.x

R.M. Hjellming, K.J. Johnston, Radio emission from conical jets associated with X-ray binaries. Astrophys. J. **328**, 600–609 (1988). doi:10.1086/166318

J. Homan, M. Buxton, S. Markoff, C.D. Bailyn, E. Nespoli, T. Belloni, Multiwavelength observations of the 2002 outburst of GX 339-4: two patterns of X-ray-Optical/Near-infrared behavior. Astrophys. J. **624**, 295–306 (2005). doi:10.1086/428722

R.I. Hynes, C.W. Mauche, C.A. Haswell, C.R. Shrader, W. Cui, S. Chaty, The X-ray transient XTE J1118+480: multiwavelength observations of a low-state minioutburst. Astrophys. J. Lett. **539**, 37–40 (2000). doi:10.1086/312836
A. Ingram, C. Done, A physical model for the continuum variability and quasi-periodic oscillation in accreting black holes. Mon. Not. R. Astron. Soc. **415**, 2323–2335 (2011). doi:10.1111/j.1365-2966.2011.18860.x
A. Ingram, C. Done, Modelling variability in black hole binaries: linking simulations to observations. Mon. Not. R. Astron. Soc. **419**, 2369–2378 (2012). doi:10.1111/j.1365-2966.2011.19885.x
O. Jamil, R.P. Fender, C.R. Kaiser, iShocks: X-ray binary jets with an internal shocks model. Mon. Not. R. Astron. Soc. **401**, 394–404 (2010). doi:10.1111/j.1365-2966.2009.15652.x
A. Joinet, E. Kalemci, F. Senziani, Hard X-ray emission of the microquasar GRO J1655-40 during the rise of its 2005 outburst. Astrophys. J. **679**, 655–663 (2008). doi:10.1086/533512
F.C. Jones, D.C. Ellison, The plasma physics of shock acceleration. Space Sci. Rev. **58**, 259–346 (1991). doi:10.1007/BF01206003
T.W. Jones, P.E. Hardee, Maxwellian synchrotron sources. Astrophys. J. **228**, 268–278 (1979). doi:10.1086/156843
E. Jourdain, J.P. Roques, M. Chauvin, D.J. Clark, Separation of two contributions to the high energy emission of Cygnus X-1: polarization measurements with INTEGRAL SPI. Astrophys. J. **761**, 27 (2012). doi:10.1088/0004-637X/761/1/27
C.R. Kaiser, The flat synchrotron spectra of partially self-absorbed jets revisited. Mon. Not. R. Astron. Soc. **367**, 1083–1094 (2006). doi:10.1111/j.1365-2966.2006.10030.x
C.R. Kaiser, R. Sunyaev, H.C. Spruit, Internal shock model for microquasars. Astron. Astrophys. **356**, 975–988 (2000)
G. Kanbach, C. Straubmeier, H.C. Spruit, T. Belloni, Correlated fast X-ray and optical variability in the black-hole candidate XTE J1118+480. Nature **414**, 180–182 (2001)
J. Kataoka, Ł. Stawarz, X-Ray emission properties of large-scale jets, hot spots, and lobes in active galactic nuclei. Astrophys. J. **622**, 797–810 (2005). doi:10.1086/428083
U. Keshet, B. Katz, A. Spitkovsky, E. Waxman, Magnetic field evolution in relativistic unmagnetized collisionless shocks. Astrophys. J. Lett. **693**, 127–130 (2009). doi:10.1088/0004-637X/693/2/L127
M. Kino, F. Takahara, M. Kusunose, Energetics of TeV blazars and physical constraints on their emission regions. Astrophys. J. **564**, 97–107 (2002). doi:10.1086/323363
J.G. Kirk, F.M. Rieger, A. Mastichiadis, Particle acceleration and synchrotron emission in blazar jets. Astron. Astrophys. **333**, 452–458 (1998)
J.G. Kirk, A.W. Guthmann, Y.A. Gallant, A. Achterberg, Particle acceleration at ultrarelativistic shocks: an eigenfunction method. Astrophys. J. **542**, 235–242 (2000). doi:10.1086/309533
D.A. Kniffen, D.L. Bertsch, C.E. Fichtel, R.C. Hartman, S.D. Hunter, G. Kanbach, P.W. Kwok, Y.C. Lin, J.R. Mattox, H.A. Mayer-Hasselwander, P.F. Michelson, C. von Montigny, P.L. Nolan, K. Pinkau, E. Schneid, P. Sreekumar, D.J. Thompson, Time variability in the gamma-ray emission of 3C 279. Astrophys. J. **411**, 133–136 (1993). doi:10.1086/172813
H.B.J. Koers, A. Pe'er, R.A.M.J. Wijers, Parameterization of the energy and rapidity distributions of secondary pions and kaons produced in energetic proton-proton collisions. ArXiv High Energy Physics. Phenomenology e-prints (2006)
S.S. Komissarov, M.V. Barkov, N. Vlahakis, A. Königl, Magnetic acceleration of relativistic active galactic nucleus jets. Mon. Not. R. Astron. Soc. **380**, 51–70 (2007). doi:10.1111/j.1365-2966.2007.12050.x
A. Konigl, Relativistic jets as X-ray and gamma-ray sources. Astrophys. J. **243**, 700–709 (1981). doi:10.1086/158638
K. Kotera, A.V. Olinto, The astrophysics of ultrahigh-energy cosmic rays. Annu. Rev. Astron. Astrophys. **49**, 119–153 (2011). doi:10.1146/annurev-astro-081710-102620
O. Kotov, E. Churazov, M. Gilfanov, On the X-ray time-lags in the black hole candidates. Mon. Not. R. Astron. Soc. **327**, 799–807 (2001). doi:10.1046/j.1365-8711.2001.04769.x
H. Krawczynski, P.S. Coppi, F. Aharonian, Time-dependent modelling of the Markarian 501 X-ray and TeV gamma-ray data taken during 1997 March and April. Mon. Not. R. Astron. Soc. **336**, 721–735 (2002). doi:10.1046/j.1365-8711.2002.05750.x
P. Laurent, J. Rodriguez, J. Wilms, M. Cadolle Bel, K. Pottschmidt, V. Grinberg, Polarized gamma-ray emission from the galactic black hole Cygnus X-1. Science **332**, 438 (2011). doi:10.1126/science.1200848
A. Lazarian, G. Kowal, E. Vishniac, E. de Gouveia Dal Pino, Fast magnetic reconnection and energetic particle acceleration. Planet. Space Sci. **59**, 537–546 (2011). doi:10.1016/j.pss.2010.07.020
A.J. Levan, N.R. Tanvir, S.B. Cenko, D.A. Perley, K. Wiersema, J.S. Bloom, A.S. Fruchter, A.d.U. Postigo, P.T. O'Brien, N. Butler, A.J. van der Horst, G. Leloudas, A.N. Morgan, K. Misra, G.C. Bower, J. Farihi, R.L. Tunnicliffe, M. Modjaz, J.M. Silverman, J. Hjorth, C. Thöne, A. Cucchiara, J.M.C. Cerón, A.J. Castro-Tirado, J.A. Arnold, M. Bremer, J.P. Brodie, T. Carroll, M.C. Cooper, P.A. Curran, R.M. Cutri, J. Ehle, D. Forbes, J. Fynbo, J. Gorosabel, J. Graham, D.I. Hoffman, S. Guziy, P. Jakobsson, A. Kamble,

T. Kerr, M.M. Kasliwal, C. Kouveliotou, D. Kocevski, N.M. Law, P.E. Nugent, E.O. Ofek, D. Poznanski, R.M. Quimby, E. Rol, A.J. Romanowsky, R. Sánchez-Ramírez, S. Schulze, N. Singh, L. van Spaandonk, R.L.C. Starling, R.G. Strom, J.C. Tello, O. Vaduvescu, P.J. Wheatley, R.A.M.J. Wijers, J.M. Winters, D. Xu, An extremely luminous panchromatic outburst from the nucleus of a distant galaxy. Science **333**, 199 (2011). doi:10.1126/science.1207143

A. Levinson, High-energy aspects of astrophysical jets. Int. J. Mod. Phys. A **21**, 6015–6054 (2006). doi:10.1142/S0217751X06035063

A. Levinson, R. Blandford, On the jets associated with galactic superluminal sources. Astrophys. J. Lett. **456**, 29 (1996). doi:10.1086/309851

A. Levinson, D. Eichler, Baryon purity in cosmological gamma-ray bursts as a manifestation of event horizons. Astrophys. J. **418**, 386 (1993). doi:10.1086/173397

A. Levinson, E. Waxman, Probing microquasars with TeV neutrinos. Phys. Rev. Lett. **87**(17), 171101 (2001). doi:10.1103/PhysRevLett.87.171101

D. Liu, A. Pe'er, A. Loeb, A two-component jet model for the tidal disruption event Swift J164449.3+573451. ArXiv e-prints (2012)

M. Livio, G.I. Ogilvie, J.E. Pringle, Extracting energy from black holes: the relative importance of the Blandford-Znajek mechanism. Astrophys. J. **512**, 100–104 (1999). doi:10.1086/306777

M.S. Longair, *High Energy Astrophysics* (Cambridge University Press, Cambridge, 2011).

R.V.E. Lovelace, Dynamo model of double radio sources. Nature **262**, 649–652 (1976). doi:10.1038/262649a0

Y. Lyubarsky, M. Liverts, Particle acceleration in the driven relativistic reconnection. Astrophys. J. **682**, 1436–1442 (2008). doi:10.1086/589640

M. Lyutikov, Role of reconnection in AGN jets. New Astron. Rev. **47**, 513–515 (2003). doi:10.1016/S1387-6473(03)00083-6

T.J. Maccarone, Constraints on jet X-ray emission in low/hard-state X-ray binaries. Mon. Not. R. Astron. Soc. **360**, 68–72 (2005). doi:10.1111/j.1745-3933.2005.00047.x

T.J. Maccarone, Jets from Galactic Binaries. ArXiv e-prints (2012)

P. Magdziarz, A.A. Zdziarski, Angle-dependent Compton reflection of X-rays and gamma-rays. Mon. Not. R. Astron. Soc. **273**, 837–848 (1995)

D. Maitra, S. Markoff, C. Brocksopp, M. Noble, M. Nowak, J. Wilms, Constraining jet/disc geometry and radiative processes in stellar black holes XTE J1118+480 and GX 339-4. Mon. Not. R. Astron. Soc. **398**, 1638–1650 (2009). doi:10.1111/j.1365-2966.2009.14896.x

J. Malzac, Internal shocks at the origin of the flat spectral energy distribution of compact jets. Mon. Not. R. Astron. Soc. **429**, 20–24 (2013). doi:10.1093/mnrasl/sls017

J. Malzac, R. Belmont, The synchrotron boiler and the spectral states of black hole binaries. Mon. Not. R. Astron. Soc. **392**, 570–589 (2009). doi:10.1111/j.1365-2966.2008.14142.x

J. Malzac, R. Belmont, A.C. Fabian, Energetics of a black hole: constraints on the jet velocity and the nature of the X-ray emitting region in Cyg X-1. Mon. Not. R. Astron. Soc. **400**, 1512–1520 (2009). doi:10.1111/j.1365-2966.2009.15553.x

K. Mannheim, The proton blazar. Astron. Astrophys. **269**, 67–76 (1993)

K. Mannheim, P.L. Biermann, Gamma-ray flaring of 3C 279—A proton-initiated cascade in the jet? Astron. Astrophys. **253**, 21–24 (1992)

L. Maraschi, G. Ghisellini, A. Celotti, A jet model for the gamma-ray emitting blazar 3C 279. Astrophys. J. Lett. **397**, 5–9 (1992). doi:10.1086/186531

S. Markoff, From multiwavelength to mass scaling: accretion and ejection in microquasars and Agn, in *Lecture Notes in Physics*, vol. 794, ed. by T. Belloni (Springer, Berlin, 2010), p. 143

S. Markoff, H. Falcke, R. Fender, A jet model for the broadband spectrum of XTE J1118+480. Synchrotron emission from radio to X-rays in the Low/Hard spectral state. Astron. Astrophys. **372**, 25–28 (2001). doi:10.1051/0004-6361:20010420

S. Markoff, M.A. Nowak, J. Wilms, Going with the flow: can the base of jets subsume the role of compact accretion disk coronae? Astrophys. J. **635**, 1203–1216 (2005). doi:10.1086/497628

S. Markoff, M. Nowak, S. Corbel, R. Fender, H. Falcke, Exploring the role of jets in the radio/X-ray correlations of GX 339-4. Astron. Astrophys. **397**, 645–658 (2003). doi:10.1051/0004-6361:20021497

A.P. Marscher, Relativistic jets and the continuum emission in QSOs. Astrophys. J. **235**, 386–391 (1980). doi:10.1086/157642

A.P. Marscher, Jets in Active Galactic Nuclei. ArXiv e-prints (2009)

A.P. Marscher, W.K. Gear, Models for high-frequency radio outbursts in extragalactic sources, with application to the early 1983 millimeter-to-infrared flare of 3C 273. Astrophys. J. **298**, 114–127 (1985). doi:10.1086/163592

J.E. McClintock, C.A. Haswell, M.R. Garcia, J.J. Drake, R.I. Hynes, H.L. Marshall, M.P. Muno, S. Chaty, P.M. Garnavich, P.J. Groot, W.H.G. Lewin, C.W. Mauche, J.M. Miller, G.G. Pooley, C.R. Shrader,

S.D. Vrtilek, Complete and simultaneous spectral observations of the black hole X-ray Nova XTE J1118+480. Astrophys. J. **555**, 477–482 (2001). doi:10.1086/321449

J.C. McKinney, Total and jet Blandford-Znajek power in the presence of an accretion disk. Astrophys. J. Lett. **630**, 5–8 (2005). doi:10.1086/468184

J.C. McKinney, General relativistic magnetohydrodynamic simulations of the jet formation and large-scale propagation from black hole accretion systems. Mon. Not. R. Astron. Soc. **368**, 1561–1582 (2006). doi:10.1111/j.1365-2966.2006.10256.x

J.C. McKinney, C.F. Gammie, A measurement of the electromagnetic luminosity of a Kerr black hole. Astrophys. J. **611**, 977–995 (2004). doi:10.1086/422244

J.C. McKinney, D.A. Uzdensky, A reconnection switch to trigger gamma-ray burst jet dissipation. Mon. Not. R. Astron. Soc. **419**, 573–607 (2012). doi:10.1111/j.1365-2966.2011.19721.x

M.V. Medvedev, A. Loeb, Generation of magnetic fields in the relativistic shock of gamma-ray burst sources. Astrophys. J. **526**, 697–706 (1999). doi:10.1086/308038

D.L. Meier, The association of jet production with geometrically thick accretion flows and black hole rotation. Astrophys. J. Lett. **548**, 9–12 (2001). doi:10.1086/318921

D.L. Meier, S. Koide, Y. Uchida, Magnetohydrodynamic production of relativistic jets. Science **291**, 84–92 (2001). doi:10.1126/science.291.5501.84

P. Mészáros, Gamma-ray bursts. Rep. Prog. Phys. **69**, 2259–2321 (2006). doi:10.1088/0034-4885/69/8/R01

P. Meszaros, M.J. Rees, Optical and long-wavelength afterglow from gamma-ray bursts. Astrophys. J. **476**, 232 (1997). doi:10.1086/303625

S. Migliari, J.A. Tomsick, S. Markoff, E. Kalemci, C.D. Bailyn, M. Buxton, S. Corbel, R.P. Fender, P. Kaaret, Tracing the jet contribution to the mid-IR over the 2005 outburst of GRO J1655-40 via broadband spectral modeling. Astrophys. J. **670**, 610–623 (2007). doi:10.1086/522023

J.C.A. Miller-Jones, D.G. McCormick, R.P. Fender, R.E. Spencer, T.W.B. Muxlow, G.G. Pooley, Multiple relativistic outbursts of GRS1915+105: radio emission and internal shocks. Mon. Not. R. Astron. Soc. **363**, 867–881 (2005). doi:10.1111/j.1365-2966.2005.09488.x

J.C.A. Miller-Jones, G.R. Sivakoff, D. Altamirano, M. Coriat, S. Corbel, V. Dhawan, H.A. Krimm, R.A. Remillard, M.P. Rupen, D.M. Russell, R.P. Fender, S. Heinz, E.G. Körding, D. Maitra, S. Markoff, S. Migliari, C.L. Sarazin, V. Tudose, Disc-jet coupling in the 2009 outburst of the black hole candidate H1743-322. Mon. Not. R. Astron. Soc. **421**, 468–485 (2012). doi:10.1111/j.1365-2966.2011.20326.x

I.F. Mirabel, L.F. Rodríguez, A superluminal source in the galaxy. Nature **371**, 46–48 (1994). doi:10.1038/371046a0

A. Mücke, R.J. Protheroe, R. Engel, J.P. Rachen, T. Stanev, BL Lac objects in the synchrotron proton blazar model. Astropart. Phys. **18**, 593–613 (2003). doi:10.1016/S0927-6505(02)00185-8

K. Murase, C.D. Dermer, H. Takami, G. Migliori, Blazars as ultra-high-energy cosmic-ray sources: implications for TeV gamma-ray observations. Astrophys. J. **749**, 63 (2012). doi:10.1088/0004-637X/749/1/63

R. Narayan, J.E. McClintock, Observational evidence for a correlation between jet power and black hole spin. Mon. Not. R. Astron. Soc. **419**, 69–73 (2012). doi:10.1111/j.1745-3933.2011.01181.x

A.Y. Neronov, D.V. Semikoz, I.I. Tkachev, Ultra-high energy cosmic ray production in the polar cap regions of black hole magnetospheres. New J. Phys. **11**(6), 065015 (2009). doi:10.1088/1367-2630/11/6/065015

K.-I. Nishikawa, P. Hardee, G. Richardson, R. Preece, H. Sol, G.J. Fishman, Particle acceleration in relativistic jets due to Weibel instability. Astrophys. J. **595**, 555–563 (2003). doi:10.1086/377260

K.-I. Nishikawa, P. Hardee, G. Richardson, R. Preece, H. Sol, G.J. Fishman, Particle acceleration and magnetic field generation in electron-positron relativistic shocks. Astrophys. J. **622**, 927–937 (2005). doi:10.1086/428394

C.A. Norman, D.B. Melrose, A. Achterberg, The origin of cosmic rays above 10 18.5 eV. Astrophys. J. **454**, 60 (1995). doi:10.1086/176465

H. Oda, M. Machida, K.E. Nakamura, R. Matsumoto, Thermal equilibria of optically thin, magnetically supported, two-temperature, black hole accretion disks. Astrophys. J. **712**, 639–652 (2010). doi:10.1088/0004-637X/712/1/639

S.P. O'Sullivan, D.C. Gabuzda, Magnetic field strength and spectral distribution of six parsec-scale active galactic nuclei jets. Mon. Not. R. Astron. Soc. **400**, 26–42 (2009). doi:10.1111/j.1365-2966.2009.15428.x

N.I. Palma, M. Böttcher, I. de la Calle, I. Agudo, M. Aller, H. Aller, U. Bach, E. Benítez, C.S. Buemi, L. Escande, J.L. Gómez, M.A. Gurwell, J. Heidt, D. Hiriart, S.G. Jorstad, M. Joshi, A. Lähteenmäki, V.M. Larionov, P. Leto, Y. Li, J.M. López, B. Lott, G. Madejski, A.P. Marscher, D.A. Morozova, C.M. Raiteri, V. Roberts, M. Tornikoski, C. Trigilio, G. Umana, M. Villata, D. Wylezalek, Multiwavelength observations of the gamma-ray blazar PKS 0528+134 in quiescence. Astrophys. J. **735**, 60 (2011). doi:10.1088/0004-637X/735/1/60

A. Pe'er, P. Casella, A model for emission from jets in X-ray binaries: consequences of a single acceleration episode. Astrophys. J. **699**, 1919–1937 (2009). doi:10.1088/0004-637X/699/2/1919
A. Pe'er, A. Loeb, Constraining sources of ultra high energy cosmic rays using high energy observations with the Fermi satellite. J. Cosmol. Astropart. Phys. **3**, 7 (2012). doi:10.1088/1475-7516/2012/03/007
A. Pe'er, S. Markoff, X-Ray emission from transient jet model in black hole binaries. Astrophys. J. **753**, 177 (2012). doi:10.1088/0004-637X/753/2/177
A. Pe'er, E. Waxman, Prompt gamma-ray burst spectra: detailed calculations and the effect of pair production. Astrophys. J. **613**, 448–459 (2004). doi:10.1086/422989
A. Pe'er, E. Waxman, High-energy photon emission in the early afterglow of GRBs. Astrophys. J. **633**, 1018–1026 (2005a). doi:10.1086/468175
A. Pe'er, E. Waxman, Time-dependent numerical model for the emission of radiation from relativistic plasma. Astrophys. J. **628**, 857–866 (2005b). doi:10.1086/431139
P.O. Petrucci, J. Ferreira, G. Henri, J. Malzac, C. Foellmi, Relevance of jet emitting disc physics to microquasars: application to Cygnus X-1. Astron. Astrophys. **522**, 38 (2010). doi:10.1051/0004-6361/201014753
T. Piran, The physics of gamma-ray bursts. Rev. Mod. Phys. **76**, 1143–1210 (2004). doi:10.1103/RevModPhys.76.1143
J. Poutanen, Accretion disc-corona models and X/γ-ray spectra of accreting black holes, in *Theory of Black Hole Accretion Disks*, ed. by M.A. Abramowicz, G. Bjornsson, J.E. Pringle (1998), p. 100
J. Poutanen, I. Vurm, On the origin of spectral states in accreting black holes. Astrophys. J. Lett. **690**, 97–100 (2009). doi:10.1088/0004-637X/690/2/L97
J. Poutanen, A.A. Zdziarski, Radiative processes in microquasars, in *New Views on Microquasars*, ed. by P. Durouchoux, Y. Fuchs, J. Rodriguez (2003), p. 95
R.E. Pudritz, M.J. Hardcastle, D.C. Gabuzda, Magnetic fields in astrophysical jets: from launch to termination. Space Sci. Rev. **169**, 27–72 (2012). doi:10.1007/s11214-012-9895-z
E. Qiao, B.F. Liu, A model for the correlation of hard X-ray index with Eddington ratio in black hole X-ray binaries. Astrophys. J. **764**, 2 (2013). doi:10.1088/0004-637X/764/1/2
J.P. Rachen, P. Mészáros, Photohadronic neutrinos from transients in astrophysical sources. Phys. Rev. D **58**(12), 123005 (1998). doi:10.1103/PhysRevD.58.123005
F. Rahoui, J.C. Lee, S. Heinz, D.C. Hines, K. Pottschmidt, J. Wilms, V. Grinberg, A multiwavelength study of Cygnus X-1: the first mid-infrared spectroscopic detection of compact jets. Astrophys. J. **736**, 63 (2011). doi:10.1088/0004-637X/736/1/63
F. Rahoui, M. Coriat, S. Corbel, M. Cadolle Bel, J.A. Tomsick, J.C. Lee, J. Rodriguez, D.M. Russell, S. Migliari, Optical and near-infrared spectroscopy of the black hole GX 339-4. I. A focus on the continuum in the low/hard and high/soft states. Mon. Not. R. Astron. Soc. **422**, 2202–2212 (2012). doi:10.1111/j.1365-2966.2012.20763.x
S. Rawlings, R. Saunders, Evidence for a common central-engine mechanism in all extragalactic radio sources. Nature **349**, 138–140 (1991). doi:10.1038/349138a0
M.J. Rees, Appearance of relativistically expanding radio sources. Nature **211**, 468–470 (1966). doi:10.1038/211468a0
M.J. Rees, P. Meszaros, Unsteady outflow models for cosmological gamma-ray bursts. Astrophys. J. Lett. **430**, 93–96 (1994). doi:10.1086/187446
W. Reich, H. Steppe, R. Schlickeiser, P. Reich, M. Pohl, H.P. Reuter, G. Kanbach, V. Schonfelder, The radio state of extragalactic gamma-ray sources detected by CGRO. Astron. Astrophys. **273**, 65 (1993)
R.C. Reis, J.M. Miller, A.C. Fabian, Thermal emission from the stellar-mass black hole binary XTE J1118+480 in the low/hard state. Mon. Not. R. Astron. Soc. **395**, 52–56 (2009). doi:10.1111/j.1745-3933.2009.00640.x
S.P. Reynolds, Theoretical studies of compact radio sources. I. Synchrotron radiation from relativistic flows. Astrophys. J. **256**, 13–37 (1982). doi:10.1086/159881
L.F. Rodríguez, I.F. Mirabel, Repeated relativistic ejections in GRS 1915+105. Astrophys. J. **511**, 398–404 (1999). doi:10.1086/306642
L.F. Rodriguez, E. Gerard, I.F. Mirabel, Y. Gomez, A. Velazquez, Radio monitoring of GRS 1915+105. Astrophys. J. Suppl. Ser. **101**, 173 (1995). doi:10.1086/192236
M.M. Romanova, R.V.E. Lovelace, Magnetic field, reconnection, and particle acceleration in extragalactic jets. Astron. Astrophys. **262**, 26–36 (1992)
G.E. Romero, H.R. Christiansen, M. Orellana, Hadronic high-energy gamma-ray emission from the microquasar LS I+61 303. Astrophys. J. **632**, 1093–1098 (2005). doi:10.1086/444446
D.M. Russell, E. Gallo, R.P. Fender, Observational constraints on the powering mechanism of transient relativistic jets. Mon. Not. R. Astron. Soc. **431**, 405–414 (2013). doi:10.1093/mnras/stt176
D.M. Russell, R.P. Fender, R.I. Hynes, C. Brocksopp, J. Homan, P.G. Jonker, M.M. Buxton, Global optical/infrared-X-ray correlations in X-ray binaries: quantifying disc and jet contributions. Mon. Not. R. Astron. Soc. **371**, 1334–1350 (2006). doi:10.1111/j.1365-2966.2006.10756.x

D.M. Russell, F. Lewis, P. Roche, J.S. Clark, E. Breedt, R.P. Fender, A long-term optical-X-ray correlation in 4U 1957+11. Mon. Not. R. Astron. Soc. **402**, 2671–2681 (2010). doi:10.1111/j.1365-2966.2009.16098.x
D.M. Russell, T.D. Russell, J.C.A. Miller-Jones, K. O'Brien, R. Soria, G.R. Sivakoff, T. Slaven-Blair, F. Lewis, S. Markoff, J. Homan, D. Altamirano, P.A. Curran, M.P. Rupen, T.M. Belloni, M. Cadolle Bel, P. Casella, S. Corbel, V. Dhawan, R.P. Fender, E. Gallo, P. Gandhi, S. Heinz, E.G. Körding, H.A. Krimm, D. Maitra, S. Migliari, R.A. Remillard, C.L. Sarazin, T. Shahbaz, V. Tudose, An evolving compact jet in the black hole X-ray binary MAXI J1836-194. Astrophys. J. Lett. **768**, 35 (2013). doi:10.1088/2041-8205/768/2/L35
G.B. Rybicki, A.P. Lightman, *Radiative Processes in Astrophysics* (1979)
R.M. Sambruna, J.K. Gambill, L. Maraschi, F. Tavecchio, R. Cerutti, C.C. Cheung, C.M. Urry, G. Chartas, A survey of extended radio jets with Chandra and the Hubble space telescope. Astrophys. J. **608**, 698–720 (2004). doi:10.1086/383124
R. Santana, R. Barniol Duran, P. Kumar, Magnetic fields in relativistic collisionless shocks, in *American Astronomical Society Meeting Abstracts*, vol. 221 (2013), p. 438-01
R. Sari, A.A. Esin, On the synchrotron self-Compton emission from relativistic shocks and its implications for gamma-ray burst afterglows. Astrophys. J. **548**, 787–799 (2001). doi:10.1086/319003
R. Sari, T. Piran, R. Narayan, Spectra and light curves of gamma-ray burst afterglows. Astrophys. J. Lett. **497**, 17 (1998). doi:10.1086/311269
M. Sikora, M.C. Begelman, M.J. Rees, Comptonization of diffuse ambient radiation by a relativistic jet: the source of gamma rays from blazars? Astrophys. J. **421**, 153–162 (1994). doi:10.1086/173633
L.O. Silva, R.A. Fonseca, J.W. Tonge, J.M. Dawson, W.B. Mori, M.V. Medvedev, Interpenetrating plasma shells: near-equipartition magnetic field generation and nonthermal particle acceleration. Astrophys. J. Lett. **596**, 121–124 (2003). doi:10.1086/379156
L. Sironi, A. Spitkovsky, Particle acceleration in relativistic magnetized collisionless pair shocks: dependence of shock acceleration on magnetic obliquity. Astrophys. J. **698**, 1523–1549 (2009). doi:10.1088/0004-637X/698/2/1523
L. Sironi, A. Spitkovsky, Particle acceleration in relativistic magnetized collisionless electron-ion shocks. Astrophys. J. **726**, 75 (2011). doi:10.1088/0004-637X/726/2/75
M.A. Sobolewska, A. Siemiginowska, G. Migliori, Ł. Stawarz, M. Jamrozy, D. Evans, C.C. Cheung, Nuclear X-ray properties of the peculiar radio-loud hidden AGN 4C+29.30. Astrophys. J. **758**, 90 (2012). doi:10.1088/0004-637X/758/2/90
M. Spada, G. Ghisellini, D. Lazzati, A. Celotti, Internal shocks in the jets of radio-loud quasars. Mon. Not. R. Astron. Soc. **325**, 1559–1570 (2001). doi:10.1046/j.1365-8711.2001.04557.x
A. Spitkovsky, On the structure of relativistic collisionless shocks in electron-ion plasmas. Astrophys. J. Lett. **673**, 39–42 (2008a). doi:10.1086/527374
A. Spitkovsky, Particle acceleration in relativistic collisionless shocks: Fermi process at last? Astrophys. J. Lett. **682**, 5–8 (2008b). doi:10.1086/590248
H.C. Spruit, Theory of magnetically powered jets, in *Lecture Notes in Physics*, vol. 794, ed. by T. Belloni (Springer, Berlin, 2010), p. 233
J.F. Steiner, J.E. McClintock, R. Narayan, Jet power and black hole spin: testing an empirical relationship and using it to predict the spins of six black holes. Astrophys. J. **762**, 104 (2013). doi:10.1088/0004-637X/762/2/104
R.A. Sunyaev, L.G. Titarchuk, Comptonization of X-rays in plasma clouds—typical radiation spectra. Astron. Astrophys. **86**, 121–138 (1980)
H. Tananbaum, H. Gursky, E. Kellogg, R. Giacconi, C. Jones, Observation of a correlated X-ray transition in Cygnus X-1. Astrophys. J. Lett. **177**, 5 (1972). doi:10.1086/181042
F. Tavecchio, L. Maraschi, R.M. Sambruna, C.M. Urry, The X-ray jet of PKS 0637-752: inverse Compton radiation from the cosmic microwave background? Astrophys. J. Lett. **544**, 23–26 (2000). doi:10.1086/317292
A. Tchekhovskoy, R. Narayan, J.C. McKinney, Black hole spin and the radio loud/quiet dichotomy of active galactic nuclei. Astrophys. J. **711**, 50–63 (2010). doi:10.1088/0004-637X/711/1/50
A. Tchekhovskoy, R. Narayan, J.C. McKinney, Efficient generation of jets from magnetically arrested accretion on a rapidly spinning black hole. Mon. Not. R. Astron. Soc. **418**, 79–83 (2011). doi:10.1111/j.1745-3933.2011.01147.x
L. Titarchuk, Generalized comptonization models and application to the recent high-energy observations. Astrophys. J. **434**, 570–586 (1994). doi:10.1086/174760
C.M. Urry, P. Padovani, Unified schemes for radio-loud active galactic nuclei. Publ. Astron. Soc. Pac. **107**, 803–845 (1995). doi:10.1086/133630
P. Uttley, T. Wilkinson, P. Cassatella, J. Wilms, K. Pottschmidt, M. Hanke, M. Böck, The causal connection between disc and power-law variability in hard state black hole X-ray binaries. Mon. Not. R. Astron. Soc. **414**, 60–64 (2011). doi:10.1111/j.1745-3933.2011.01056.x

H. van der Laan, A model for variable extragalactic radio sources. Nature **211**, 1131 (1966). doi:10.1038/2111131a0

A. Veledina, J. Poutanen, I. Vurm, A synchrotron self-Compton-disk reprocessing model for optical/X-ray correlation in black hole X-ray binaries. Astrophys. J. Lett. **737**, 17 (2011). doi:10.1088/2041-8205/737/1/L17

A. Veledina, J. Poutanen, I. Vurm, Hot accretion flow in black hole binaries: a link connecting X-rays to the infrared. Mon. Not. R. Astron. Soc. **430**, 3196–3212 (2013). doi:10.1093/mnras/stt124

G.S. Vila, G.E. Romero, Leptonic/hadronic models for electromagnetic emission in microquasars: the case of GX 339-4. Mon. Not. R. Astron. Soc. **403**, 1457–1468 (2010). doi:10.1111/j.1365-2966.2010.16208.x

N. Vlahakis, A. Königl, Magnetic driving of relativistic outflows in active galactic nuclei. I. Interpretation of parsec-scale accelerations. Astrophys. J. **605**, 656–661 (2004). doi:10.1086/382670

G. Wardziński, A.A. Zdziarski, Thermal synchrotron radiation and its comptonization in compact X-ray sources. Mon. Not. R. Astron. Soc. **314**, 183–198 (2000). doi:10.1046/j.1365-8711.2000.03297.x

A.E. Wehrle, E. Pian, C.M. Urry, L. Maraschi, I.M. McHardy, A.J. Lawson, G. Ghisellini, R.C. Hartman, et al., Multiwavelength observations of a dramatic high-energy flare in the blazar 3C 279. Astrophys. J. **497**, 178 (1998). doi:10.1086/305461

E.S. Weibel, Spontaneously growing transverse waves in a plasma due to an anisotropic velocity distribution. Phys. Rev. Lett. **2**, 83–84 (1959). doi:10.1103/PhysRevLett.2.83

R.A.M.J. Wijers, M.J. Rees, P. Meszaros, Shocked by GRB 970228: the afterglow of a cosmological fireball. Mon. Not. R. Astron. Soc. **288**, 51–56 (1997)

Q. Wu, M. Gu, The X-ray spectral evolution in X-ray binaries and its application to constrain the black hole mass of ultraluminous X-ray sources. Astrophys. J. **682**, 212–217 (2008). doi:10.1086/588187

F. Yuan, W. Cui, Radio-X-ray correlation and the "quiescent state" of black hole sources. Astrophys. J. **629**, 408–413 (2005). doi:10.1086/431453

F. Yuan, W. Cui, R. Narayan, An accretion-jet model for black hole binaries: interpreting the spectral and timing features of XTE J1118+480. Astrophys. J. **620**, 905–914 (2005). doi:10.1086/427206

F. Yuan, Z. Yu, L.C. Ho, Revisiting the "fundamental plane" of black hole activity at extremely low luminosities. Astrophys. J. **703**, 1034–1043 (2009). doi:10.1088/0004-637X/703/1/1034

F. Yuan, A.A. Zdziarski, Y. Xue, X.-B. Wu, Modeling the hard states of XTE J1550-564 during its 2000 outburst. Astrophys. J. **659**, 541–548 (2007). doi:10.1086/512078

A.A. Zdziarski, P. Lubiński, M. Gilfanov, M. Revnivtsev, Correlations between X-ray and radio spectral properties of accreting black holes. Mon. Not. R. Astron. Soc. **342**, 355–372 (2003). doi:10.1046/j.1365-8711.2003.06556.x

J.F. Zhang, Y.G. Feng, M.C. Lei, Y.Y. Tang, Y.P. Tian, High-energy neutrino emission from low-mass microquasars. Mon. Not. R. Astron. Soc. **407**, 2468–2474 (2010). doi:10.1111/j.1365-2966.2010.17072.x

Y.F. Zhang, A.P. Marscher, H.D. Aller, M.F. Aller, H. Terasranta, E. Valtaoja, Radio and X-ray observations of the gamma-ray bright quasar PKS 0528+134. Astrophys. J. **432**, 91–102 (1994). doi:10.1086/174551

J. Zrake, A.I. MacFadyen, Numerical simulations of driven relativistic magnetohydrodynamic turbulence. Astrophys. J. **744**, 32 (2012). doi:10.1088/0004-637X/744/1/32

DOI 10.1007/978-1-4939-2227-7_20
Reprinted from *Space Science Reviews* Journal, DOI 10.1007/s11214-013-0016-4

Jet–Environment Interactions as Diagnostics of Jet Physics

Sebastian Heinz

Received: 16 April 2013 / Accepted: 12 August 2013 / Published online: 12 November 2013

Abstract In this chapter, we will explore the interaction of jets with their environments. Jets can transport a sizable fraction of accretion energy away from black holes and neutron stars. Because they are collimated, they can travel to distances far beyond the gravitational sphere of influence of the black hole. Yet, their interaction with the interstellar and intergalactic medium must eventually halt their advance and dissipate the energy they carry. The termination of the jet, and the inflation of large scale cavities of relativistic plasma offers one of the most powerful ways to constrain the physics of jets. In this chapter, we will review the inflation of radio lobes, the propagation of hot spots, the creation of shells and cavities, and the bending of jet by proper motion through their environment, both in the context of AGN jets and microquasars.

Keywords Black holes · Jets · Accretion · Interstellar medium

1 Introduction

The discussion in previous chapters underscores the importance of jets for a complete picture of accretion. Similarly, the understanding of jets as both a diagnostic and a foreground for our understanding of strong gravity is becoming clear as observational probes like the Event Horizon Telescope approach the fundamental scale of accretion. Thus, understanding jet properties is critical both to understand accretion and to understand the radio properties of black holes near the event horizon.

Yet, while computational models of jet formation are making rapid progress (Penna et al. 2010; Tchekhovskoy et al. 2011), we know frustratingly little about the properties of jets on intermediate scales: Jet velocities, power, and composition are extremely difficult to extract

This work was supported by NSF grant AST1109347.

S. Heinz (✉)
Astronomy Department, University of Wisconsin-Madison, 475 N. Charter St., Madison, WI 53706, USA
e-mail: heinzs@astro.wisc.edu

directly from observations of relativistic jets (with the notable exception of SS433 Dubner et al. 1998).

The most powerful way to study jet properties is by way of their interaction with the large scale environment. This is in many ways an analog to the way accelerators study energetic particles: The momentum and energy content of an unknown object can best be determined by its interaction with a target. In this case, the target is the interstellar or intergalactic medium (ISM and IGM, respectively; in the following, we will refer to the external medium as the ISM)

From the very beginning of radio astronomy, the large scale properties of radio sources have been used to constrain the power output from jets (see below). With the launch of Chandra and the discovery of numerous jet-driven X-ray cavities in the centers of galaxy clusters, the study of jets from their interaction with the environment has become a standard tool and has delivered important global calibrations of jet power for a large number of AGN.

In fact, the discovery of cavities in clusters has led to the realization that energy injected into the surrounding gas by jets from accreting black holes likely plays an important role in the evolution of cosmic structure, both in the context of heating cool core galaxy clusters and keeping early type galaxies on the red sequence. While this cosmic "feedback" is an important topic of current research, we will concentrate in this chapter on what we can learn about the properties of jets themselves, and what this implies about accretion and jet creation, from studying jet-environment interactions.

2 Jet Propagation and Lobe Generation

Before discussing the phenomenology of different manifestations of jet-environment interactions, it is worth reviewing the mechanisms important in this process. The discussion in this section will focus mostly on AGN jets, with appropriate references to differences and similarities in the case of relativistic jets from X-ray binaries, often called microquasars, which will be discussed in Sect. 4

2.1 Phases of Jet-ISM/IGM Interaction

A good first step is to look at the relevant dimensionless numbers that govern jet propagation. For simplicity, let us consider a jet as flux of momentum and energy confined to a narrow conical beam with opening angle θ_{jet}. Even if the jet is self-collimated and more appropriately described as cylindrical, small variations in jet orientation will effectively spread the jet flux over a conical volume, in an average sense. Let us, for simplicity, also assume that the jet is propagating into a uniform density medium.

Given that jet formation occurs very close to the event horizon of the black hole, and given that the Kerr metric is scale invariant (that is, all radii scale linearly with the black hole mass), it is commonly assumed that the process of jet formation is the same in black holes of different mass. Thus, for black holes with different mass but otherwise identical parameters, namely, identical spin parameter a and identical accretion rate in units of the Eddington rate, $\dot{m} \equiv \dot{M}/\dot{M}_{\text{Edd}}$, we may reasonably assume that the jet power P and all relevant dynamical size scales (e.g., the Alfvén radius) vary linearly with black hole mass; explicitly: $P \propto M$. Heinz and Sunyaev (2003) called this the "scale invariance hypothesis".

A reasonable question to ask is, then, whether scale invariance carries over to the interaction of jets with their environment, and if not, how badly it is broken in different environments.

2.1.1 Jet Stopping

For jet material to be affected meaningfully by its interaction with the environment, a significant fraction of its momentum must be transferred to environmental gas. Before this happens, the jet simply bores through the environment unhindered, creating a strong bow shock ahead of it but not slowing down. Let us first consider a magnetized jet fluid that is kept from mixing with the gas it is interacting with by flux freezing. In this case, the two fluids (jet and ISM) will occupy separate volumes. This approximation is appropriate for the discussion of lobe/cocoon formation, but jet propagation itself may be affected by mass loading (see Sect. 2.2).

The propagation of the leading edge of the jet, henceforth referred to as the jet "head", is governed by ram pressure balance in the bow shock. Roughly speaking, the momentum flux of the jet into the bow shock is $p_{\rm ram,jet} \approx P/(\pi l^2 \theta_{\rm jet}^2 c)$, where P is the power and l is the length of the jet. The ram pressure of the ISM that must balance this momentum flux is $p_{\rm ram,ism} \approx \rho_{\rm ISM} v_{\rm head}^2$. These non-relativistic expressions assume that the head moves slowly compared to the jet, i.e., that the jet is effectively slowed down by the ISM; before that, the jet deposits relativily little plasma along its path which slowly expands sideways into a cocoon that shrouds the jet (see cartoon in the left panel of Fig. 1). The propagation velocity of the head is then simply

$$v_{\rm head} \sim \sqrt{\frac{P}{\pi \rho_{\rm ISM} c} \frac{1}{l \theta_{\rm jet}}} \tag{1}$$

and the condition that the jet is stopped by the ISM is simply that $v_{\rm head} \ll c$. The jet head then propagates with a velocity $v_{\rm head} \propto t^{-1/2}$. The stopping length for a relativistic jet can then be defined as

$$l_{\rm s} \equiv \sqrt{\frac{P}{\pi \rho_{\rm ISM} c^3 \theta_{\rm jet}^2}} \tag{2}$$

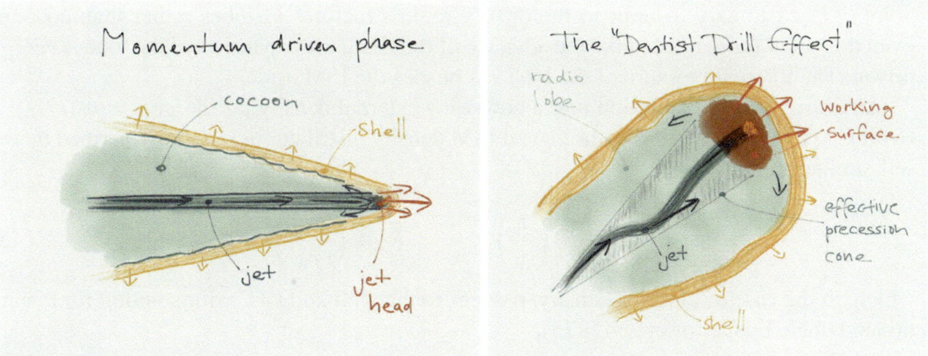

Fig. 1 *Left:* Cartoon of the early momentum-driven phase of jet head propagation (production of a narrow cocoon) at close to the jet velocity; *Right:* cartoon of the "dentist drill effect", which explains the observed aspect ratios of lobes (close to unity) by dynamical instabilities or precession spreading the jet thrust effectively over a much larger are than the jet cross section and leading to an earlier onset of the energy drive phase of lobe evolution

Setting the stopping length in relation to the fundamental scale of the accreting system, given by the gravitational radius $r_g = GM/c^2$, defines one of the fundamental dimensionless numbers for jet dynamics (Heinz 2002), which we shall call the thrust ratio $\eta_{\rm jet}$:

$$\eta_{\rm jet} \equiv \frac{l_s}{r_g} = \frac{P^{1/2}}{\rho_{\rm ISM}^{1/2} M} \frac{c^{1/2}}{\pi^{1/2} G \theta_{\rm jet}} \propto \sqrt{\frac{1}{\rho_{\rm ISM} M}} \tag{3}$$

where we may expect $\theta_{\rm jet}$ to be independent of black hole mass (though it may depend on accretion rate and spin).

Let us compare the thrust ratio for microquasars and typical radio galaxies. For microquasars, it is reasonable to assume ISM density of $n_{\rm ISM} \sim 1\ {\rm cm}^{-3}$, while the density in the intergalactic medium ranges from similar densities within the host galaxies of the AGN to $n_{\rm ISM} \sim 10^{-3}\ {\rm cm}^{-3}$ in clusters and lower densities yet in the environments of field galaxies. For representative black hole masses of $M_{\rm BH} \sim 10\ M_\odot$ and $M_{\rm BH} \sim 10^9\ M_\odot$ for microquasars and AGN jets, respectively, the thrust ratios for microquasars are much larger than those for AGN jets:

$$\eta_{\rm microquasar} \sim 10^3 \text{ to } 10^4 \eta_{\rm AGN} \tag{4}$$

Thus, the ISM provides a much weaker barrier to microquasar jets than it does to AGN jets.

One important consequence of this is that the structures generated by the interaction of microquasar jets with the ISM will appear on observable scales on the sky, despite the fact that the angular scales of Galactic X-ray Binary (XRB) accretion disks on the sky are many orders of magnitude smaller than those of nearby AGN.

Another important consequence is that the surface brightness of the observational signatures of this interaction is generally low, i.e., signatures of microquasar–ISM interaction should generally be hard to detect. This is consistent with the fact that such signatures have only been found in a handful of sources.

2.1.2 Lobe Formation

Once the jet is effectively stopped, the plasma transported along the jet is shed sideways at the head pressure and deposited alongside the slowly advancing head. This plasma then expands laterally to inflate lobes into the ISM (as the aspect ratio starts to increase at this point, it is customary to refer to the diffuse radio structures as lobes rather than cocoons from this point forward). Unlike the advance of the head, the expansion of the lobe is energy driven: The internal pressure of the lobe gas pushes the ISM aside.

An energy driven expansion into a powerlaw external density profile $\rho_{\rm ISM} = \rho_0 (r/r_0)^{-\beta}$ (with $\beta < 2$ for typical, non-pathological ISM density distributions) is well described by the self similar scaling

$$R_{\rm lobe} \sim R_0 \left(\frac{t}{t_0}\right)^{\frac{3}{5-\beta}} \qquad R_0 \equiv \left(\frac{L t_0^3}{\rho_0}\right)^{1/5} \tag{5}$$

which, in the case of uniform density, reduces to the well known Castor solution for a wind driven bubble by Castor et al. (1975),

$$R_{\rm wind} \sim \left(\frac{L t^3}{\rho_{ISM}}\right)^{1/5} \tag{6}$$

which itself is very similar to the self-similar Sedov–Taylor expansion of a blast wave. Because the expansion velocity of the lobe decreases with time for realistic distribution of $\rho_{\rm ISM}$, it is often appropriate to use Eq. (6) with the external density set to $\rho_{\rm ISM} = \rho_0 (R/R_0)^{-\beta}$ for the current size of R of the lobe.

From simple dimensional analysis it follows that the radius of the lobe will expand with velocity

$$v_{\rm lobe} = v_0 \left(\frac{t}{t_0}\right)^{\frac{\beta-2}{5-\beta}} \tag{7}$$

and, in the case of uniform density, $v \propto t^{-2/5}$. Because this is a shallower powerlaw in time than that for the head velocity, it follows that the aspect ratio of the lobe will tend towards unity, i.e., the cavity blown by the jet will become more spherical with time.

As the radio lobe expands into the surroundings, the gas that previously occupied the lobe volume is pushed aside. As long as the expansion is supersonic, this material will be piled up into a thin, shocked shell. The thickness of this shell depends on whether the gas inside is cooling radiatively. If not, it is of the order of 10 % of the size of the lobe.

As the pressure near the jet termination region (where the jet energy is dissipated) is high, a dynamical back flow may occur inside the lobe, where spent jet plasma flows backwards to contribute to the inflation of the lobe. This is confirmed in simulations of radio galaxies and often leads to the development of Kelvin–Helmholtz instability along the lobe boundary in these simulations (Reynolds et al. 2001). Observations of radio sources show rather stable lobe boundaries, suggestive of dynamical stabilization through magnetic fields.

2.1.3 Detachment

Once the expansion becomes sub-sonic, the further evolution will depend more strongly on the details of the environment.

If the jet is driven into a stratified atmosphere under the influence of gravity, the plasma filling the lobe will become buoyant as soon as the expansion velocity $v_{\rm lobe}$ dips below the buoyancy speed.

Let us consider a spherical buoyant plasma cavity with radius R at radial distance r from the center of a hydrostatic cluster. If the density inside the cavity is well below the density of the external medium, we may neglect it and write the buoyancy force acting on the cavity as

$$F_{\rm buoy} \sim -\frac{4\pi}{3} R^3 \rho_{\rm ISM} g = -\frac{4\pi}{3} R^3 \frac{dp}{dr} \sim \frac{4\beta \rho_{\rm ISM} c_{\rm s}^2 R^3}{r} \tag{8}$$

which will accelerate the cavity away from the center, inducing a buoyant velocity $v_{\rm buoy}$. Consequently, the cavity will experience a drag force of

$$F_{\rm drag} = \pi R^2 \rho_{\rm ISM} v_{\rm buoy}^2 C_{\rm W} \sim \frac{\pi R^2 \rho_{\rm ISM} v_{\rm buoy}^2}{2} \tag{9}$$

where $C_{\rm W} \sim 1/2$ is the drag coefficient for a spherical cavity, giving a characteristic buoyancy speed of

$$v_{\rm buoy} \sim c_{\rm s} \sqrt{8\beta \frac{R}{r}} \lesssim c_{\rm s} \tag{10}$$

where the latter inequality derives from the facts that the radial distance of a cavity must be at least $r \geq R$ and that the drag coefficient increases for velocities approaching the sound speed—buoyancy is a sub-sonic phenomenon. Once the expansion velocity of a cavity decreases below the buoyancy speed, it will detach and float away from the cluster center.

Naturally, a lobe will also detach buoyantly when the expansion is already sub-sonic and the jet turns off (i.e., jet power is reduced significantly for a time scale comparable to the age of the lobe). In cases where the jet power fluctuates significantly while the source is still supersonic, the solution for lobe evolution can be expected to approach the Sedov–Taylor solution once the jet has been off for longer than the age of the cavity.

In contrast, if a jet from a stationary accreting source is driven into uniform medium, the expansion can, in principle, continue indefinitely to inflate a single structure. However, the effects of dynamical instabilities along the back flow inside the lobe will affect the long term evolution of such a source.

Finally, if the jet is driven by a source moving with respect to its environment, the lobe will eventually be dragged into a trail of radio plasma. This will happen once the expansion velocity of the lobe dips below the velocity of the source with respect to the environment, $v_{\rm lobe} < v_{\rm ISM}$ (Yoon et al. 2011).

2.1.4 *Slow Bias*

Because the expansion velocity of the lobe decreases with time, most sources found observationally will be in a slow expansion phase, with a velocity around the sound speed, i.e., there is a bias against finding young and thus rapidly expanding sources. In gravitationally stratified environments, this sets the characteristic age of radio lobes, as sub-sonic lobes will detach and be transported outward by buoyancy. This explains the observational trend to find attached cavities without signs of strong shocks (Fabian et al. 2000; McNamara and Nulsen 2007).

Let us suppose for simplicity that a source becomes undetectable as a radio source when it detaches from the AGN, thus no longer being fed with fresh radio plasma and also floating away from the cluster center, where detectability as an X-ray cavity is strongly favored (Enßlin and Heinz 2002). Given that the buoyancy speed is approximately $v_{\rm buoy} \sim c_{\rm s}$, we can roughly identify the on-time of a source as the time during which it is supersonic (within a factor of $\sim$2).

Based on the evolution described in Sect. 2.1.2, we can characterize the fractional cumulative time spent at or above a given Mach number, compared to the total detectable time as

$$f(>M) \equiv \frac{t(v > Mc_{\rm s})}{t(v > c_{\rm s})} = M^{-\frac{5-\beta}{2-\beta}} \tag{11}$$

which, in the most conservative case of a uniform density distribution, gives $f(>M) = M^{-2.5}$, indicating that we should expect a fraction of less than $f(>M)_{\beta=0} = 18\ \%$ of all supersonic sources to be above Mach 2. In more realistic cluster density distributions, with $\beta \sim 1.5$, this fraction drops to $f(>M)_{\beta=1.5} \sim 0.8\ \%$.

This is consistent with the fact that radio sources in the centers of clusters are expanding at roughly sonic speeds. However, it does not imply that radio sources are never strongly supersonic.

2.2 Radio Galaxies

The evolutionary sequence outlined above is reflected in the observations of radio sources: We observe radio jets surrounded by radio lobes/lobes of diffuse, synchrotron emitting plasma. In sufficiently dens environments (i.e., galaxy clusters), we also observe the swept up ISM surrounding the lobes.

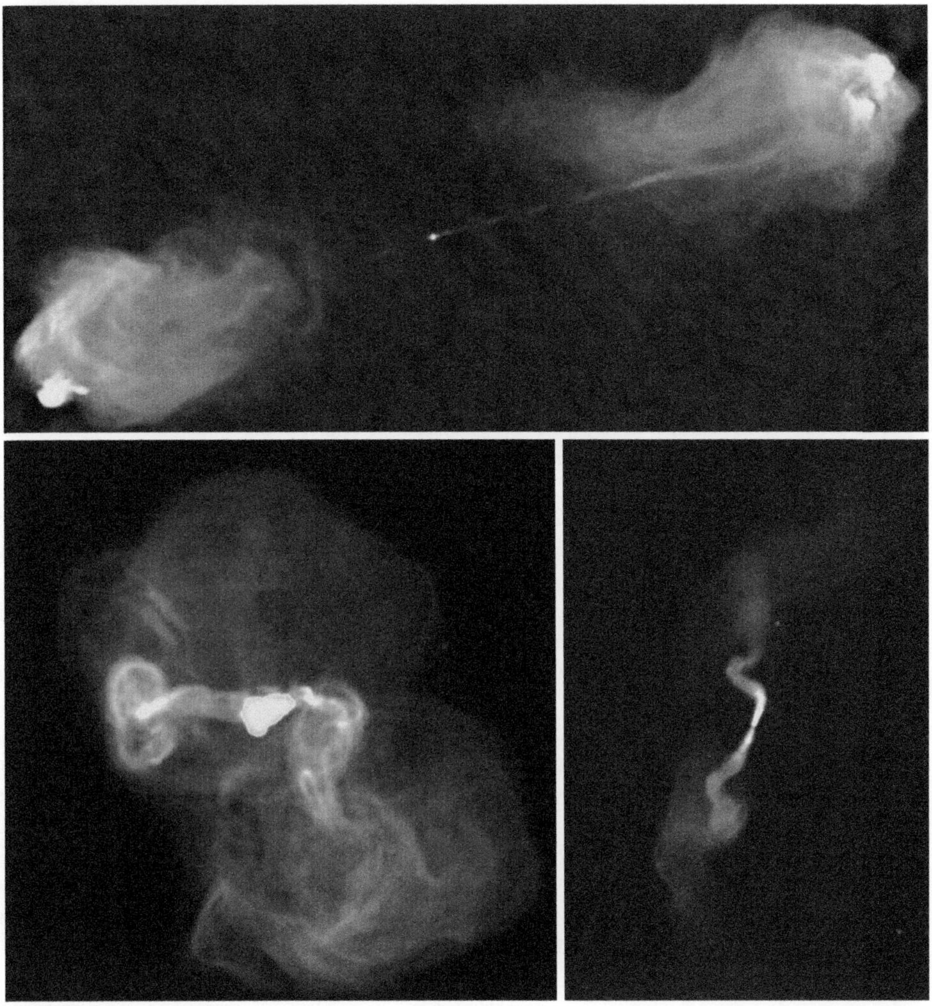

Fig. 2 *Top:* VLA image of the classic FR II radio galaxy Cygnus A (NRAO, Perley et al. 1984); *Bottom Left:* VLA image of the FR I radio galaxy Virgo A (NRAO, Owen et al. 2000); *Bottom Right:* VLA image of the classical FR I source 3C31 (NRAO, Laing and Bridle 2002)

We can identify different classes of radio sources with different evolutionary stages and different environmental conditions, with the understanding that bias will make observations of certain kinds of radio sources unlikely or impossible, such that certain stages in radio source evolution may not have observational equivalents.

Classic radio galaxies are defined following the Fanaroff–Riley scheme (Fanaroff and Riley 1974) into FR Is and FR IIs, based on their radio morphology: Class one sources exhibit core-brightened morphology, showing jets and radio lobes that decrease in brightness with distance from the black hole. The prototypical FR I source 3C31 is shown on the bottom-right of Fig. 2. The jets tend to de-collimate. Class two sources show bright extended structures, often with hot spots indicating the active working surface, with the brightest emission at the largest distance from the black hole. An image of the classic FR II source Cygnus A is shown in the top panel of Fig. 2.

The general consensus is that the difference between FR I and FR II sources are in large part due to differences in the interaction between jet and environment (other proposed differences include differences in jet composition). Generally, the two classes divide by radio power and, by inference, also by jet power, with FR I sources exhibiting lower radio luminosities and, where measurable, lower jet powers.

It has been proposed that the most important discriminator is, in fact, the amount of mass loading of the jet that occurs within the host galaxy, with FR I jets picking up a sufficient fraction of atomic matter from stellar mass loss to slow them down to sub-sonic speeds (Laing and Bridle 2002). This explains the lack of hot spots in FR I sources and detailed kinematic models require mass loading to explain the observed opening angles, Doppler ratios of jet to counter-jet, and synchrotron brightness distributions.

The clear distinction in radio power between class I and II sources strongly suggests that the reason for the increased mass loading in FR I sources is a result of lower intrinsic power and thus momentum flux.

Detached remnants of radio galaxies are generally hard to detect because the effects of synchrotron aging on the radio plasma are compounded by adiabatic expansion, shifting the cooling cutoff frequency below detectable bands. The best cases for detections of fossil radio lobes no longer inflated by jets are (a) the ghost cavities found in clusters of galaxies (Fabian et al. 2003; McNamara et al. 2001), and (b) radio relics in the outskirts of galaxy clusters, some of which are likely fossil radio lobes revived by the passage of merger shocks (Enßlin and Brüggen 2002).

The young end of the radio source spectrum is very likely represented by compact steep spectrum sources and possibly Gigahertz peaked spectrum sources (O'Dea 1998). These sources have been interpreted as very young radio sources still embedded within the dens gas in the center of their host galaxies, which explains their high brightness temperature, small size scales, and steep, absorbed low frequency spectral indices. Alternatively, GPS sources have been interpreted as frustrated old sources embedded within too dens an environment to generate a typical radio lobe. While the question is still not fully settled, the evidence points towards the former interpretation (Holt et al. 2008; Kunert-Bajraszewska et al. 2010).

3 Diagnostics

3.1 Cavities

The simple lobe size relations in Eqs. (5) and (10) suggest that observations of lobe sizes offer a very robust diagnostic of jet power. And indeed, one of the fundamental results of the past 15 years of *Chandra* observations has been the discovery of jet-driven cavities in the centers of cool core clusters (with a fractional incidence of nearly 100 %) and the subsequent calibration of jet power in a large number of clusters.

The most prominent example is the Perseus cluster, shown in Fig. 3, which was the fist cluster with clearly detected X-ray cavities.

3.1.1 Practical Application

In order to determine the jet power from observed cavities, a number of assumptions and simplifications have to be made: Rather than using Eq. (5), a simpler, more robust approach

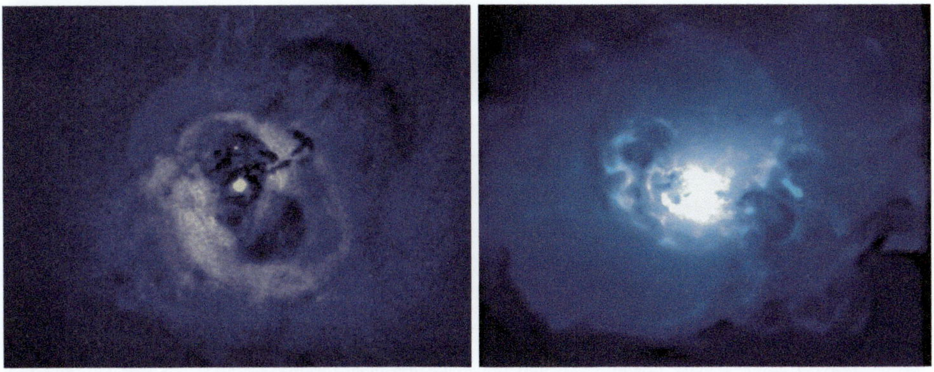

Fig. 3 *Left:* deep *Chandra* observation of the Perseus cluster, showing X-ray cavities, the swept up shell, and sound and shock waves (NASA/*Chandra X-Ray Observatory* Fabian et al. 2006); *Right:* Virtual *Chandra* X-ray observation of a hydrodynamic simulation of cavity evolution in dynamic cluster atmospheres, showing the formation of multiple cavities by the action of a continuously firing jet (Morsony et al. 2010)

of directly estimating the energy required for cavity inflation is most often used. In the case of a cavity with volume V, the total energy required for inflation is given by the total enthalpy of the cavity:

$$E_{\rm tot,cav} = H_{\rm cav} = \gamma_{\rm ad}\frac{p_{\rm cav}V_{\rm cav}}{\gamma_{\rm ad}-1} = 4p_{\rm cav}V_{\rm cav} \tag{12}$$

where $p_{\rm cav} = p_{\rm shock}$ is the pressure inside the cavity and the shocked ISM and we assumed a relativistic equation of state for the material inside the lobe with $\gamma_{\rm ad} = 4/3$. For cavities that are expanding at or below the sound speed, we may approximate $p_{\rm cav} \gtrsim p_{\rm ISM}$, technically giving a lower limit on the energy.

This energy is then divided by the expansion age of the cavity to derive the average jet power (averaged over the expansion history). Because we cannot yet determine the expansion velocity directly with current instrumentation, it is common to use the sound speed or the buoyancy speed (which is close to the sound speed) for the velocity, which may be assumed to be a lower limit to the velocity (as the source will have expanded more quickly in the past and may still be moderately supersonic in some cases), $v_{\rm cav} \gtrsim c_{\rm s}$.

Finally, it is typically assumed that the cavities detected in X-rays are roughly spherical. For a pair of equal size opposite cavities in a cluster of temperature $T_{\rm ISM}$ and total particle density $n_{\rm ISM}$ (ions plus electrons), the average jet power is thus

$$\langle P_{\rm jet}\rangle \gtrsim \frac{H_{\rm cav}c_{\rm s}}{R} = \frac{16\pi R^2 n_{\rm ISM}kT_{\rm ISM}c_{\rm s}}{3} \tag{13}$$

Figure 4 shows a recent collection of cavity powers derived from *Chandra* observations of nearby clusters, compared to the cooling luminosity of the host cluster. The implication for jet-driven feedback in clusters are obvious from the figure (i.e., the two quantities track each other and are comparable in magnitude). However, for the purpose of this chapter, we will restrict the discussion to the implications of this technique as a jet diagnostic.

3.1.2 Multiple/Ghost Cavities

A number of clusters exhibit multiple sets of cavities at different distances from the center, with outer sets of cavities showing steeper radio spectra, or, in some cases, no radio emission

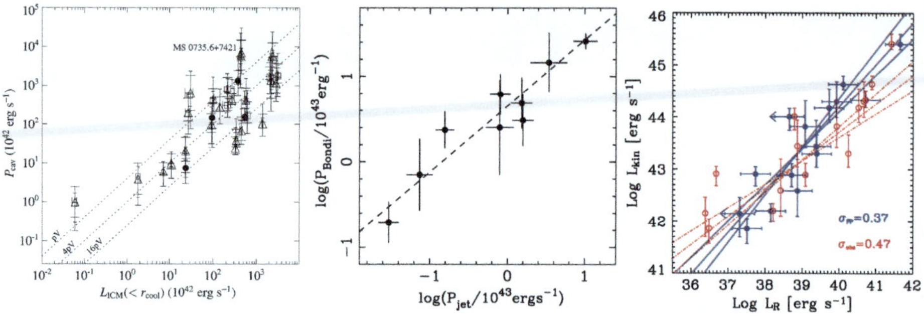

Fig. 4 *Left:* Jet powers derived from X-ray cavities in a sample of clusters observed with *Chandra* (Rafferty et al. 2006); *Center:* Cavity power vs. Bondi power (Allen et al. 2006); *Right:* Jet power vs. core radio luminosity for a sample of nearby clusters, using the observed core radio luminosity (*red*) and a beaming correction based on the fundamental plane relation (*blue*) (Merloni and Heinz 2007)

at all. This can safely be interpreted as an indication that the outer cavities are older and are currently rising buoyantly. The Perseus cluster once again is the most prominent example of this class (see Fig. 3).

Multiple sets of cavities can be generated in two ways. Jets are expected to have duty cycles of tens of millions of years (presumably set by the detailed feedback between gas supply for accretion and the heating of nearby gas by the jet). A jet switching on and off with a sufficiently long interval between active periods will generate distinct sets of cavities.

However, as argued above, lobes become buoyant when the expansion velocity becomes significantly sub-sonic. While numerical simulations of jets in clusters with stationary, fully hydrostatic gas (Vernaleo and Reynolds 2006) showed no clear separation of cavities, Heinz et al. (2006) showed that including the expected dynamic sub-structure of clusters expected from their formation through gas accretion and from turbulence can significantly affect the long term evolution of radio lobes. Morsony et al. (2010) showed that a continuously operating jet can, in fact, generate multiple well-separated X-ray cavities, as shown in the right panel of Fig. 3. Thus, the mere detection of multiple sets of cavities is not unequivocal evidence for jet power cycling.

3.1.3 Caveats

While this method is very robust it only provides the power averaged of typically millions of years. The approximation of using the buoyancy age introduces a factor of roughly two in uncertainty in the power.

A probably even larger error stems from the fact that cavity sizes are often estimated by visual inspection, which is appropriate in bright, well resolved clusters with deep observations like Perseus, where the cavity edges are well defined in the *Chandra* images. However, in the majority of clusters, the uncertainty in the radius of the cavities should be expected to be of order 50 %, introducing another factor of order 2 in the total error budget.

Another unknown factor in the determination of cavity size and ISM density is the line-of-sight angle of the cavity axis (though a bias favors detection of cavities in the plane of the sky Enßlin and Heinz 2002), introducing error of order of a few tens of percent (this factor enters into the density, cavity volume, and age, however, the latter two cancel each other to lowest order).

Finally, it is typically assumed that the cavities have a 100 % filling factor of relativistic gas. The limits on possible contamination by thermal gas are of the order of 10 %, entering directly into the error budget.

3.2 Radio Lobes

Cavity powers can also be estimated the old-fashioned way: From radio fluxes and estimates of the synchrotron age. If radio spectra are available and the effects of synchrotron aging are detectable either through a synchrotron cutoff or cooling break, a lower limit on the lobe energy can be derived from minimum energy arguments, and, coupled with the synchrotron age, a robust lower limit on the power can be derived, independent of the equipartition fraction of the magnetic field. The strongest constraints on the energetics of radio lobes comes from cases where inverse Compton X-ray emission is detectable from within the lobes, as in the case of Fornax A (Kaneda et al. 1995). In such a case, the magnetic field strength can be independently determined, and the uncertainty over the equipartition fraction is removed.

Because synchrotron emission cannot constrain the fraction of energy in relativistic protons, while cavity measurements of jet power are calorimetric in that they include all forms of energy in the plasma, the comparison between power estimates from both sources has been used to constrain the equipartition fraction or, alternatively, the presence of protons or other synchrotron-dark forms of energy inside the lobes (Fabian et al. 2002). Because of the uncertainties inherent in synchrotron power estimates, however, this method has not yet yielded a robust argument for a hadronic component in the plasma of radio lobes.

3.3 Shocks

The detection and proper analysis of a jet–drive shock removes some of the uncertainties inherent in cavity diagnostics, because a detected temperature jump implies, by the Rankine–Hugoniot shock jump conditions, the Mach number and thus expansion velocity of the shock. Detection of a shock requires the detection of a discontinuity in pressure, in addition to a discontinuity in density, which sets shocks apart from cold fronts.

Shocks have been detected in the Perseus cluster (Fabian et al. 2003), Virgo (Forman et al. 2007), Abell 2052 (Blanton et al. 2011), Hydra A (Nulsen et al. 2005a), Hercules A (Nulsen et al. 2005b), and MS0735.6+7421 (McNamara et al. 2005), among others. The implications especially in the latter two cases are significant: while these sources are old and thus very large compared to other nearby cool core cluster cavities, the implied power is very large, indicating that a set of cluster center black holes power jets at close to the Eddington rate for up to a Salpeter time (the e-folding time of a black hole's mass if accreting at Eddington). It has been suggested that such a large release of energy requires jet production efficiencies in excess of the canonical 10 %, given the estimates for their black hole masses, hinting at the possibility that black hole spin extraction might boost the efficiency of jet production (McNamara et al. 2011)

The case of the Mach 1.3 shock surrounding M87 in the Virgo cluster is clear demonstration that even in classically low power sources like M87, the residual signatures of the initial highly supersonic expansion of the lobes can be detected. Surface brightness profiles in this case are surprisingly similar to what one would expect for a simple 1D piston model for shock propagation (Forman et al. 2007).

3.4 Sound Waves

Unsharp masking of deep observations of the Perseus cluster reveal a set of concentric ripples that have been interpreted as sound waves, given their amplitude and lack of sharp, shock-like edges. They have been interpreted as signatures of variability in jet power (Ruszkowski et al. 2004), though Morsony et al. (2010) showed that similar features are expected due to simple dynamical interaction between the jet and the environment in the case that the jet runs into denser gas, resulting in an effective "pulsing" of the radio source.

Estimating the power going into sound waves in Perseus, Fabian et al. (2003) concluded that a significant fraction of jet power (up to the 25 % one would expect in p-dV work on the environment) could be deposited in the waves.

As a diagnostic of jet power, this technique is currently of limited use because of the required deep observations. However, the implications for jet dynamics are potentially important, as the amplitude of the sound waves indicates that the cause of the waves must affect the jet globally—either, the jet periodically shuts off, in which case the waves suggest a duty cycle of several million years, or it indicates that the jet is interrupted dynamically by interaction with dens material interior to the cavities, as suggested by Heinz et al. (2006)

3.5 Hot Spots

The termination shock in FR II radio galaxies and in microquasars like XTE J1550 and Circinus X-1 can also provide diagnostic constraints:

The propagation velocity of the hot spot can be used to measure the ratio of jet thrust to ISM ram pressure using Eq. (1) in cases where it is measurable, i.e., in microquasars like XTE J1550.

Most commonly, the synchrotron emission from the hot spot can be used to constrain the energy density inside the hot spot. Corrected for Doppler boosting (as some hot spots may move relativistically), and with a dynamical model for the flow velocity out of the hot spot (which may approach c), this measure can be translated into an estimate of the jet power.

A particularly interesting case is presented by Cygnus A, where it had been noted that the hot spots are not aligned with the current jet direction. I has been argued that this is evidence for either precession (Steenbrugge and Blundell 2008) or the dentist drill effect (initially proposed by Scheuer 1982; see cartoon in the right panel of Fig. 1). The relative straightness of the jet and the significant angular offset constrain the jet travel time and the radiative lifetime of the hot spot.

The dentist drill effect is critical in understanding the observed aspect ratios of powerful radio galaxies like Cygnus A, as the narrow jets observed in some of these sources would otherwise still be in the momentum dominated phase, contrary to the observed fairly wide radio lobes. This conclusion is supported by 3D jet simulations (Heinz et al. 2006).

3.6 Jet Bending

We have so far considered sources whose central engines are stationary in the center of the densest local halo (cluster or field galaxy). However, galaxy clusters often contain multiple radio galaxies, not all of which are thus central sources and must, therefore, imply that some of the radio galaxies are moving through the cluster gas at velocities comparable to the velocity dispersion of the cluster. The same is true for galaxy groups. And, as mentioned above, microquasar sources are all non-central, and a sub-class of these sources will be moving at substantial speeds relative to the ISM.

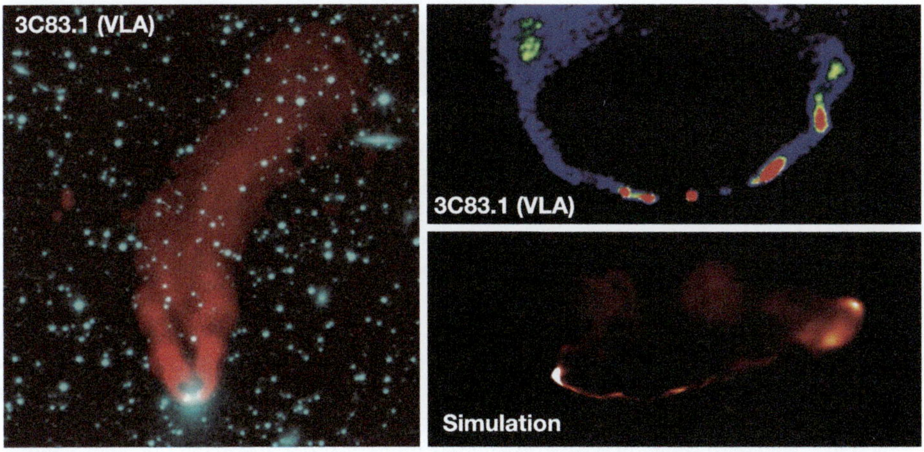

Fig. 5 *Left and Top Right:* VLA image of classical bent-double 3C83.1 (NRAO (Odea and Owen 1986)); *Bottom Right:* Simulation of a bent double in a similar environment as 3C83.1 (Morsony et al. 2013)

In this case, the ISM provides an effective head-wind. Lobe formation under these circumstances can be substantially different: As soon as the expansion velocity of the cocoon is comparable to the velocity of the central engine relative to the environment, the asymmetric ISM ram pressure will become important. The classic signs of this asymmetry are bent radio tails observed in bent-double radio sources (see Fig. 5).

The jets themselves will also be bent: The pressure gradient within the cocoon will impose a transverse momentum flux across the jet. Since the jets are traveling supersonically, this will induce a weak oblique shock that raises the jet pressure to the forward pressure of the cocoon, which itself is set by the ram pressure of the ISM, $\rho_{\rm ISM}v_{\rm ISM}^2$. The shock also alters the jet direction, bending the jet backwards in the direction of motion of the ISM, relative to the black hole. As shown by Morsony et al. (2013), the radius of curvature $R_{\rm curv}$ of such a bent jet can be used to constrain the jet power or environmental density:

$$\frac{P_{\rm jet}}{\rho_{\rm ISM}v_{\rm ISM}^2} = \frac{\pi R_{\rm curv}^2 (\sin\theta_{\rm jet})^2 v_{\rm jet}}{8} < \frac{\pi R_{\rm curv}^2 c}{8} \tag{14}$$

where $v_{\rm jet}$ and $\theta_{\rm jet}$ are the poorly known jet velocity and initial opening angle, respectively.

A reasonable guess for $v_{\rm ISM}$ in the case of group and cluster AGN will be given by the velocity dispersion, where one might use $v_{\rm ISM} \sim \sqrt{\sigma^2 - v_{\rm LOS}^2}$, since the radial velocity $v_{\rm LOS}$ of the system will not contribute to observable bending perpendicular to the line of sight. With density measurements from X-rays, bent-double radio sources can thus be used to constrain the jet power $P_{\rm jet}$ or, conversely, if estimates of the jet power can be derived from the observed synchrotron brightness, they can be used to provide lower limits on the density in group environments (Freeland et al. 2008).

3.7 Global Energetics & Efficiencies

The ability to measure the average kinetic power for large class of sources through cluster cavity measurements has cemented jets as the primary mechanisms of feedback in clusters. It has also inspired the suggestion that jet feedback is responsible for maintaining the red sequence in the process of galaxy formation and evolution through the so-called radio mode.

And it has allowed a number of important conclusions about the efficiency of accretion and jet formation:

First, it was noticed that low-luminosity AGN may, in fact, all be radio loud, relative to their bolometric luminosity (Ho and Ulvestad 2001; Falcke et al. 2000; Ho 2008). This realization was reached in parallel in the case of X-ray binaries, which, in the low-hard state, appear to universally exhibit jet radio emission (Fender 2001).

Chandra observations of M87 revealed that the Bondi accretion rate for this object is larger than what would be inferred for standard efficient accretion in terms of the radiative output. However, the inferred Bondi-accretion power is close to the jet power (Di Matteo et al. 2003).

Allen et al. (2006) expanded this analysis to a larger sample of clusters and found a clear relation between cavity power and Bondi accretion rate, as can be seen in Fig. 4. Given that the Bondi rate may be interpreted as a reasonable upper limit to the actual accretion rate, this indicates that the jet production efficiency of black holes in non-radiative states (called the "low/hard state" in X-ray binaries) is high, of order 10 %. On the other hand, an expanded sample and a different analysis shows that some jet powers are clearly in excess of the naively predicted Bondi upper limit (McNamara et al. 2011). This, and the very large energies implied by the cavities in Hydra A and MS0735, comparable to the total rest mass energy in black hole mass, have led the authors to suggest that an additional source of energy would be required to drive such large cavities, such as black hole spin.

Merloni and Heinz (2007) plotted the cavity power against the core radio power of the black hole. If the scale invariance hypothesis holds, the radio core power should be proportional to the jet power following (Heinz and Sunyaev 2003)

$$L_{\rm rad} = L_0 P_{\rm jet}^{17/12} \tag{15}$$

which is consistent with the data as shown in Fig. 4. Using the observed relation to calibrate L_0 in Eq. (15), combined with the observed core (flat spectrum) AGN radio luminosity function, one can estimate the integrated kinetic power density of jets as a function of red shift. Comparing the red-shift integrated energy density released by jets $\epsilon_{\rm jet}$ to the black hole mass density $\rho_{\rm BH}$ at redshift zero, one can then derive the average jet creation efficiency of supermassive black holes to be (Merloni and Heinz 2008)

$$\langle\eta_{\rm jet}\rangle \equiv \left\langle\frac{e_{\rm jet}}{\rho_{\rm BH}c^2}\right\rangle \sim 0.005 \tag{16}$$

Given that most of the black hole mass in SMBH was accreted during radiatively efficient phases, this is a very robust lower limit to the actual jet production efficiency during phases of active jet production. If a fraction ζ of the local mass density of black holes was accreted in the radio loud, inefficient mode, the jet production efficiency during this mode will be $\langle\eta_{\rm jet}\rangle \sim 0.005/\zeta$ which will be of the order of 10 %.

4 Microquasars

Just like AGN jets, jets launched by microquasars must interact with the ISM as they propagate outward from the compact object. As argued above, the thrust ratio for the class of microquasars if orders of magnitude larger than ANG jets, so the scales on which they interact with the ISM will be different, relative to the scale of the inner accretion flow. While this should make it easier to find signs of this interaction on resolvable angular scales, the observational search for such signs has been difficult and only a relatively small number of clear examples are known to date.

4.1 Radio Lobes—Why Microquasar-ISM Interactions are Hard to Observe

We can once again argue on energetic grounds why this should be so (Heinz 2002): The total energy one might expect to be ejected by microquasar jets should be of order $P_{\mathrm{jet}} \Delta t_{\mathrm{Salpeter}}$, where the Salpeter time $t_{\mathrm{Salpeter}} \equiv M/\dot{M}_{\mathrm{Eddington}}$ is the growth time for a Black hole accreting at the Eddington rate.

For X-ray binaries, this is certainly a reasonable upper limit on the total energy, as the compact object will likely not grow by more than a few solar masses of accretion over its accretion lifetime from stellar evolution calculations. In this case, the total energy within the lobe/cavity will be proportional to the black hole mass.

While detailed predictions for the observability of radio lobes, shells, and hot spots of microquasar jets require dedicated modeling, it is instructive to consider the approximate scaling in surface brightness one might expect under the simple consideration that Δt is similar in different mass black holes (we can certainly pick systems for this to be the case), and consider only differences in black hole mass and density of the ISM (as was done in deriving Eq. (3)). While the assumption of similar Δt is clearly an oversimplification, estimates of AGN duty cycles (of order $\Delta t \sim 10^7 - 10^8$ yrs) are, coincidentally, at least loosely comparable to the expected evolution time scales of X-ray binary companion stars (massive main sequence stars or evolved low mass stars).

The scale of the lobes inflated by such a jet is given by Eq. (5). A simplistic measure of how bright one might expect the signatures of the impact of a source on its environment to be is given by the surface energy density that it presents on the sky, Σ_{E}. If we write the jet power as $P \propto \dot{m} M$, where $\dot{m} = \dot{M}/\dot{M}_{\mathrm{Edd}}$ is the accretion rate in units of the Eddington rate, then

$$\Sigma_{\mathrm{E}} = \frac{E}{\pi R^2} \propto \frac{M \dot{m} \Delta t}{R^2} \propto M^{3/5} \rho_{\mathrm{ISM}}^{2/5} \left[\dot{m}^{3/5} \Delta t^{-1/5}\right] \tag{17}$$

which is, conservatively estimated, about three orders of magnitude smaller for microquasars than for typical radio galaxies (for similar $\dot{m}$ and Δt).

The synchrotron emissivity from within a radio lobe will be approximately proportional to the square of the pressure (assuming equipartition), which, from Eq. (5) follows

$$p_{\mathrm{lobe}} \sim \rho_{\mathrm{ISM}} v_{\mathrm{lobe}}^2 \propto \rho_{\mathrm{ISM}}^{3/5} M^{2/5} \left[\dot{m}^{2/5} \Delta t^{-4/5}\right] \tag{18}$$

such that the total synchrotron power will follow

$$L_{\mathrm{lobe}} \propto R_{\mathrm{lobe}}^3 p_{\mathrm{lobe}}^2 \propto M^{7/5} \rho_{\mathrm{ISM}}^{3/5} \left[\dot{m}^{7/5} \Delta t^{1/5}\right] \tag{19}$$

while the mean surface brightness will follow

$$I_{\mathrm{lobe}} \propto R_{\mathrm{lobe}} p_{\mathrm{lobe}}^2 \propto M \rho_{\mathrm{ISM}} \left[\dot{m} \Delta t^{-1}\right] \tag{20}$$

Thus, for similar $\dot{m}$ and Δt, the optically thin low-frequency radio synchrotron surface brightness of microquasar lobes should be orders of magnitude smaller than that found in radio galaxies.

It is clear that radio lobes from microquasars will be detectable only for very young sources, preferentially in dens environments. The only two XRBs with clearly detected radio lobes are SS433 and Circinus X-1, shown in Fig. 6, supporting the notion that such detections should be rare. On the basis of the total energy required to inflate the asymmetric extensions of the radio lobes (which also contain the supernova remnant W50 which gave birth to the compact object in SS433), it was estimated that the average jet power of SS433 is in excess of $P_{\mathrm{SS433}} \gtrsim 10^{39}$ ergs s^{-1}.

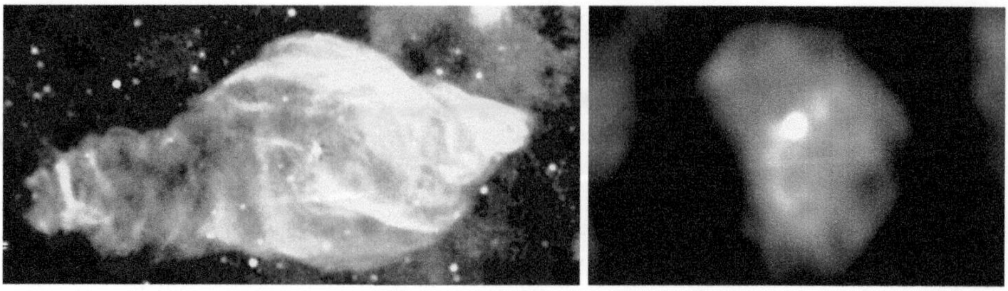

Fig. 6 *Left:* VLA observation of SS433 (NRAO Dubner et al. 1998); *Right:* ATCA radio observation of the Circinus X-1 radio nebula (see also Tudose et al. 2006)

While the evolution time scale for radio lobes is likely set by Δt, thus depending on the time averaged jet power, it is also clear that transient events of increased jet power (such as radio flares and optically thin jet ejections observed in microquasars) could lead to observable short term increases in emission from parts of the radio nebula (namely, the working surface). These events will evolve on time scales much shorter than Δt, of order months to years in the case of microquasars, which we will discuss next.

4.2 X-ray Ejections

The saving grace for observations of microquasar-ISM interactions may well be the fact that they vary on observable time scales: As argued above, the periods over which signs of this interaction should be visible is short, but because the sources are strongly variable, we may observe them during this short timescale if we can trigger observations based on the observed variability of the accreting source. In the case of HMXBs, we also know that the sources themselves must be young (in many cases significantly younger than $\Delta t_{\mathrm{Salpeter}}$).

This is best demonstrated by the successful detection of moving hot spots in the sources XTE J1550-564 (Corbel et al. 2002), H1743-322 (Corbel et al. 2005), and possibly XTE J1752-223 (Ratti et al. 2012), as well as the arcsecond scale extended X-ray emission found along the jet axis in Circinus X-1 (Sell et al. 2010). Especially in the case of XTE J1550, it is clear that the observed radio and X-ray knots moving away from the black hole are the signatures of jet ejections with the ISM: they are observed to decelerate, as would be expected for a momentum-driven propagation of jet material.

The fact that the X-ray hot spots first appeared several years after the ejection supports the notion that XTE J1550 in particular, and microquasars in general are surrounded by undetectable radio lobes (Heinz 2002), through which jet ejections can travel essentially unhindered (as AGN jets can traverse radio lobes without losses), decelerating and re-brightening only once they encounter the ISM surrounding the lobe (Heinz 2002; Steiner and McClintock 2012). Thus, observations of hot spots (ejections) can indirectly be used as diagnostics of otherwise invisible microquasar radio lobes.

4.3 Terminal Shocks and Shells

While moving hot spots are powerful probes of the energetics of individual XRB outbursts, they do not provide the same long term average measure of jet power that X-ray cavities and shocks provide of the power of AGN jets in clusters.

The two benchmark sources for which evidence of a long term jet-inflated stationary shock exists are Circinus X-1 and Cygnus X-1.

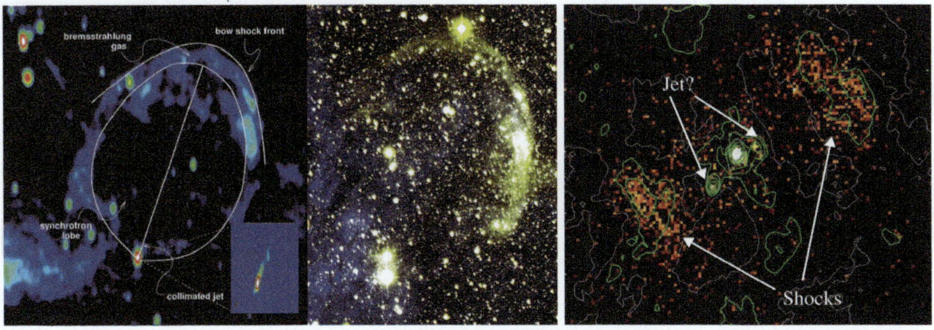

Fig. 7 *Left:* Westerbork radio observation of the shocked shell around Cygnus X-1 (Gallo et al. 2005). The image size is 16.4 × 17.5 arcminutes, which corresponds to 9 × 9.7 pc at a distance of 1.9 kpc (Reid et al. 2011); *Center:* Hα and [OIII] image of the Cygnus X-1 nebula (see also (Russell et al. 2007)) showing the shock-enhanced [OIII]/Hα ratio of the shell. The image size is 9 × 14 arcminutes, which corresponds to 5 × 7.7 parsecs at a distance of 2.1 kpc; *Right: Chandra* X-ray image of the stationary terminal shock of the Circinus X-1 jet (Sell et al. 2010). The image size is 2.1 × 1.8 arcminutes, which corresponds to a physical scale of 5.5 × 4.2 parsecs at a distance of 8 kpc (Jonker et al. 2007)

Cygnus X-1 presents a thermally emitting shell in radio and optical with a radius of about 3 parsec, shown in Fig. 7 (Gallo et al. 2005), generally aligned with the direction of the VLBA jet (Stirling et al. 2001). This shell has been interpreted as the swept up ISM around the invisible synchrotron lobe blown by the jet, consistent with the expectations of low surface brightness of microquasar lobes. In this sense, Cygnus X-1 displays a radio cavity akin to the X-ray cavities found in clusters.

Based on limits of the shock velocity from the observed [OIII] emission, Russell et al. (2007) derive an estimated jet power of $P_{\rm CygX-1} \sim 4 - 14 \times 10^{36}\,{\rm ergs\,s^{-1}}$. It should be kept in mind, however, that this is roughly of the same order of magnitude as the power of the wind expected from the O9.7Iab companion star of the black hole in Cygnus X-1, which may well be contributing to the inflation of this cavity.

Unlike in AGN, the shells around microquasars may be radiative, given the higher densities and lower temperatures of the ISM. The dynamics of the shock may therefore be significantly more complicated (e.g., the pre-ionization of the ISM by the shock's radiation field will affect the shock jump, and the temperature evolution behind the shock).

In the case of Circinus X-1, deep *Chandra* observations reveal two clearly stationary X-ray shocks along the jet direction, about half way across the radio nebula, as shown in Fig. 7 (Heinz et al. 2007; Sell et al. 2010).

The emission from these shocks is consistent with pure powerlaw emission with a spectral index of $\alpha = 1$, and co-spatial with excess radio synchrotron emission within the nebula. The combined radio and X-ray spectra point towards a simple cooled injection synchrotron spectrum, with a spectral break of $\Delta\alpha = 0.5$, indicating that the age of the plasma inside the shocks is of order $\Delta t \sim 1600$ yrs. This provides a very robust lower limit on the average jet power of $\langle P_{\rm CirX-1}\rangle \geq 3 \times 10^{35}\,{\rm ergs\,s^{-1}}$.

4.4 Bow Shock Nebulae and Trails

One of the most important differences between microquasars and AGN jets is that the central engine of microquasars is not situated at the center of its host galaxy. This implies that XRBs are not at rest with respect to the ISM, evidenced by the substantial velocity dispersion of

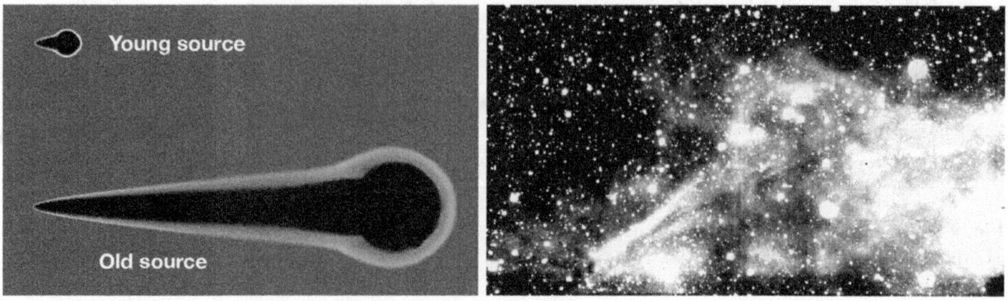

Fig. 8 *Left:* Hydrodynamic simulation of a microquasar bow shock nebula early (*top*) and late (*bottom*) in it evolution; *Right: VLT* observation of the Hα bow shock nebula around SAX J1712.6-3739 (Yoon et al. 2011)

LMXBs (Heinz et al. 2008). For example, XTE J1118+480 is moving at an estimated 145 km/s relative to the local standard of rest (Mirabel et al. 2001).

Jets moving this rapidly through the background medium will be affected by the ram pressure of the ISM. This makes them the microquasar equivalent to bent double and narrow angle tail radio sources. According to Yoon et al. (2011), the observational signatures we should expect from this interaction are (a) a bow shock nebula similar to those observed ahead of pulsar winds, visible in free-free emission and traditional shock tracers such as Hα and [OIII], (b) a trail of radio plasma at low surface brightness, (c) a terminal bubble inflated by the back flow of plasma along the trail, surrounded by a shell of shocked ISM (see Fig. 8).

As in the case of bent double radio sources, observations of microquasar bow shock nebulae offer additional diagnostic power: The length of the trail is simply $l = v_{\rm XRB}\Delta t$, where $v_{\rm XRB}$ is the microquasar velocity relative to the ISM and Δt the age of the radio source. Because the terminal bubble is inflated by a constant fraction of the total energy flowing back along the trail, it simply follows the expansion law described in Eq. (5), proportional to $R \propto t^{3/5}$. The ratio of bubble radius to trail length must decrease with time. As is the case in bent doubles, the radius of curvature of the jet (if observable) is set by the ratio of jet thrust to ISM ram pressure following Eq. (14).

The only candidate XRB bowshow nebula to date is SAX J1712.6-3739, shown in Fig. 8. This nebula shows the characteristic Mach cone in Hα, and deep VLT observations hint at the existence of a terminal bubble with properties consistent with those predicted in Yoon et al. (2011). More extensive surveys of LMXBs, preferentially in the Galactic plane, where densities are sufficiently high to produce a detectable bow shock, will be necessary to quantify how prevalent these nebulae are and to develop the into a similarly quantitative tool as bent double radio galaxies have become.

References

S.W. Allen, R.J.H. Dunn, A.C. Fabian, G.B. Taylor, C.S. Reynolds, The relation between accretion rate and jet power in X-ray luminous elliptical galaxies. Mon. Not. R. Astron. Soc. **372**, 21–30 (2006). doi:10.1111/j.1365-2966.2006.10778.x

E.L. Blanton, S.W. Randall, T.E. Clarke, C.L. Sarazin, B.R. McNamara, E.M. Douglass, M. McDonald, A very deep Chandra observation of A2052: bubbles, shocks, and sloshing. Astrophys. J. **737**, 99 (2011). doi:10.1088/0004-637X/737/2/99

J. Castor, R. McCray, R. Weaver Astrophys. J. Lett. **200**, 107 (1975)

S. Corbel, R.F. Fender, A.K. Tzioumis, J.A. Tomsick, J.A. Orosz, J.M. Miller, R. Wijnands, P. Kaaret Science **298**, 196–199 (2002)

S. Corbel, P. Kaaret, R.P. Fender, A.K. Tzioumis, J.A. Tomsick, J.A. Orosz, Discovery of X-ray jets in the microquasar H1743-322. Astrophys. J. **632**, 504–513 (2005). doi:10.1086/432499

T. Di Matteo, S.W. Allen, A.C. Fabian, A.S. Wilson, A.J. Young, Accretion onto the supermassive black hole in M87. Astrophys. J. **582**, 133–140 (2003)

G.M. Dubner, M. Holdaway, W.M. Goss, I.F. Mirabel, A high-resolution radio study of the W50-SS 433 system and the surrounding medium. Astron. J. **116**, 1842–1855 (1998)

T.A. Enßlin, M. Brüggen, On the formation of cluster radio relics. Mon. Not. R. Astron. Soc. **331**, 1011–1019 (2002)

T.A. Enßlin, S. Heinz, Radio and X-ray detectability of buoyant radio plasma bubbles in clusters of galaxies. Astron. Astrophys. **384**, 27–30 (2002). doi:10.1051/0004-6361:20020207

A.C. Fabian, J.S. Sanders, S. Ettori, G.B. Taylor, S.W. Allen, C.S. Crawford, K. Iwasawa, R.M. Johnstone, P.M. Ogle, Chandra imaging of the complex X-ray core of the Perseus cluster. Mon. Not. R. Astron. Soc. **318**, 65–68 (2000)

A.C. Fabian, A. Celotti, K.M. Blundell, N.E. Kassim, R.A. Perley Mon. Not. R. Astron. Soc. **331**, 369–375 (2002)

A.C. Fabian, J.S. Sanders, C.S. Crawford, C.J. Conselice, J.S. Gallagher, R.F.G. Wyse, The relationship between the optical Hα filaments and the X-ray emission in the core of the Perseus cluster. Mon. Not. R. Astron. Soc. **344**, 48–52 (2003)

A.C. Fabian, J.S. Sanders, G.B. Taylor, S.W. Allen, C.S. Crawford, R.M. Johnstone, K. Iwasawa, A very deep Chandra observation of the Perseus cluster: shocks, ripples and conduction. Mon. Not. R. Astron. Soc. **366**, 417–428 (2006). doi:10.1111/j.1365-2966.2005.09896.x

H. Falcke, N.M. Nagar, A.S. Wilson, J.S. Ulvestad, Radio sources in low-luminosity active galactic nuclei. II. Very long baseline interferometry detections of compact radio cores and jets in a sample of LINERs. Astrophys. J. **542**, 197–200 (2000)

B.L. Fanaroff, J.M. Riley, The morphology of extragalactic radio sources of high and low luminosity. Mon. Not. R. Astron. Soc. **167**, 31–36 (1974)

R.P. Fender, Powerful jets from black hole X-ray binaries in low/hard X-ray states. Mon. Not. R. Astron. Soc. **322**, 31–42 (2001)

W.R. Forman, E. Churazov, M. Markevitch, P. Nulsen, A. Vikhlinin, M.C. Begelman, H. Boringer, J. Eilek, S. Heinz, R. Kraft, F. Owen Astrophys. J. (2007). arXiv:astro-ph/0604583. doi:10.1086/519480

E. Freeland, R.F. Cardoso, E. Wilcots, Bent-double radio sources as probes of intergalactic gas. Astrophys. J. **685**, 858–862 (2008). doi:10.1086/591443

E. Gallo, R. Fender, C. Kaiser, D. Russell, R. Morganti, T. Oosterloo, S. Heinz, A dark jet dominates the power output of the stellar black hole cygnus X-1. Nature **436**, 819–821 (2005). doi:10.1038/nature03879

S. Heinz, Radio lobe dynamics and the environment of microquasars. Astron. Astrophys. **388**, 40–43 (2002)

S. Heinz, R.A. Sunyaev, The non-linear dependence of flux on black hole mass and accretion rate in core-dominated jets. Mon. Not. R. Astron. Soc. **343**, 59–64 (2003)

S. Heinz, M. Brüggen, A. Young, E. Levesque, The answer is blowing in the wind: simulating the interaction of jets with dynamic cluster atmospheres. Mon. Not. R. Astron. Soc. **373**, 65–69 (2006). doi:10.1111/j.1745-3933.2006.00243.x

S. Heinz, N.S. Schulz, W.N. Brandt, D.K. Galloway, Evidence of a parsec-scale X-ray jet from the accreting neutron star circinus X-1. Astrophys. J. Lett. **663**, 93–96 (2007). doi:10.1086/519950

S. Heinz, H.J. Grimm, R.A. Sunyaev, R.P. Fender, Blazing trails: microquasars as head-tail sources and the seeding of magnetized plasma into the ISM. Astrophys. J. **686**, 1145–1154 (2008). doi:10.1086/591435

L.C. Ho, Nuclear activity in nearby galaxies. Annu. Rev. Astron. Astrophys. **46**, 475–539 (2008). doi:10.1146/annurev.astro.45.051806.110546

L.C. Ho, J.S. Ulvestad, Radio continuum survey of an optically selected sample of nearby Seyfert galaxies. Astrophys. J. Suppl. Ser. **133**, 77–118 (2001)

J. Holt, C.N. Tadhunter, R. Morganti, Fast outflows in compact radio sources: evidence for AGN-induced feedback in the early stages of radio source evolution. Mon. Not. R. Astron. Soc. **387**, 639–659 (2008). doi:10.1111/j.1365-2966.2008.13089.x

P.G. Jonker, G. Nelemans, C.G. Bassa, Detection of the radial velocity curve of the B5-A0 supergiant companion star of cir X-1? Mon. Not. R. Astron. Soc. **374**, 999–1005 (2007). doi:10.1111/j.1365-2966.2006.11210.x

H. Kaneda, M. Tashiro, Y. Ikebe, Y. Ishisaki, H. Kubo, K. Makshima, T. Ohashi, Y. Saito, H. Tabara, T. Takahashi, Detection of inverse-compton X-rays from lobes of the radio galaxy fornax A. Astrophys. J. Lett. **453**, 13 (1995). doi:10.1086/309742

M. Kunert-Bajraszewska, M.P. Gawroński, A. Labiano, A. Siemiginowska, A survey of low-luminosity compact sources and its implication for the evolution of radio-loud active galactic nuclei—I. Radio data. Mon. Not. R. Astron. Soc. **408**, 2261–2278 (2010). doi:10.1111/j.1365-2966.2010.17271.x

R.A. Laing, A.H. Bridle, Relativistic models and the jet velocity field in the radio galaxy 3C 31. Mon. Not. R. Astron. Soc. **336**, 328–352 (2002). doi:10.1046/j.1365-8711.2002.05756.x

B.R. McNamara, P.E.J. Nulsen, Heating hot atmospheres with active galactic nuclei. Annu. Rev. Astron. Astrophys. **45**, 117–175 (2007). doi:10.1146/annurev.astro.45.051806.110625

B.R. McNamara, M.W. Wise, P.E.J. Nulsen, L.P. David, C.L. Carilli, C.L. Sarazin, C.P. O'Dea, J. Houck, M. Donahue, S. Baum, M. Voit, R.W. O'Connell, A. Koekemoer, Discovery of ghost cavities in the X-ray atmosphere of Abell 2597. Astrophys. J. Lett. **562**, 149 (2001)

B.R. McNamara, P.E.J. Nulsen, M.W. Wise, D.A. Rafferty, C. Carilli, C.L. Sarazin, E.L. Blanton, The heating of gas in a galaxy cluster by X-ray cavities and large-scale shock fronts. Nature **433**, 45–47 (2005)

B.R. McNamara, M. Rohanizadegan, P.E.J. Nulsen, Are radio active galactic nuclei powered by accretion or black hole spin? Astrophys. J. **727**, 39 (2011). doi:10.1088/0004-637X/727/1/39

A. Merloni, S. Heinz, A synthesis model for AGN evolution: supermassive black holes growth and feedback modes. Mon. Not. R. Astron. Soc. **388**, 1011–1030 (2008). doi:10.1111/j.1365-2966.2008.13472.x

A. Merloni, S. Heinz Mon. Not. R. Astron. Soc. (2007). doi:10.1111/j.1365-2966.2007.12253.x

I.F. Mirabel, V. Dhawan, R.P. Mignani, I. Rodrigues, F. Guglielmetti, A high-velocity black hole on a galactic-halo orbit in the solar neighbourhood. Nature **413**, 139–141 (2001)

B.J. Morsony, S. Heinz, M. Brüggen, M. Ruszkowski, Swimming against the current: simulations of central AGN evolution in dynamic galaxy clusters. Mon. Not. R. Astron. Soc. **407**, 1277–1289 (2010). doi:10.1111/j.1365-2966.2010.17059.x

B.J. Morsony, J.J. Miller, S. Heinz, E. Freeland, E. Wilcots, M. Brüggen, M. Ruszkowski, Simulations of bent-double radio sources in galaxy groups. Mon. Not. R. Astron. Soc. **431**, 781–792 (2013). doi:10.1093/mnras/stt210

P.E.J. Nulsen, D.C. Hambrick, B.R. McNamara, D. Rafferty, L. Birzan, M.W. Wise, L.P. David, The powerful outburst in Hercules A. Astrophys. J. Lett. **625**, 9–12 (2005b). doi:10.1086/430945

P.E.J. Nulsen, B.R. McNamara, M.W. Wise, L.P. David, The cluster-scale AGN outburst in hydra a. Astrophys. J. **628**, 629–636 (2005a). doi:10.1086/430845

C.P. O'Dea, The compact steep-spectrum and gigahertz peaked-spectrum radio sources. Publ. Astron. Soc. Pac. **110**, 493–532 (1998). doi:10.1086/316162

C.P. Odea, F.N. Owen, Multifrequency VLA observations of the prototypical narrow-angle tail radio source NGC 1265. Astrophys. J. **301**, 841–859 (1986). doi:10.1086/163948

F.N. Owen, J.A. Eilek, N.E. Kassim, M87 at 90 centimeters: a different picture. Astrophys. J. **543**, 611–619 (2000)

R.F. Penna, J.C. McKinney, R. Narayan, A. Tchekhovskoy, R. Shafee, J.E. McClintock, Simulations of magnetized discs around black holes: effects of black hole spin, disc thickness and magnetic field geometry. Mon. Not. R. Astron. Soc. **408**, 752–782 (2010). doi:10.1111/j.1365-2966.2010.17170.x

R.A. Perley, J.W. Dreher, J.J. Cowan, The jet and filaments in cygnus a. Astrophys. J. Lett. **285**, 35–38 (1984). doi:10.1086/184360

D.A. Rafferty, B.R. McNamara, P.E.J. Nulsen, M.W. Wise, The feedback-regulated growth of black holes and bulges through gas accretion and starbursts in cluster central dominant galaxies. Astrophys. J. **652**, 216–231 (2006). doi:10.1086/507672

E.M. Ratti, P.G. Jonker, J.C.A. Miller-Jones, M.A.P. Torres, J. Homan, S. Markoff, J.A. Tomsick, P. Kaaret, R. Wijnands, E. Gallo, F. Özel, D.T.H. Steeghs, R.P. Fender, The black hole candidate XTE J1752-223 towards and in quiescence: optical and simultaneous X-ray-radio observations. Mon. Not. R. Astron. Soc. **423**, 2656–2667 (2012). doi:10.1111/j.1365-2966.2012.21071.x

M.J. Reid, J.E. McClintock, R. Narayan, L. Gou, R.A. Remillard, J.A. Orosz, The trigonometric parallax of cygnus X-1. Astrophys. J. **742**, 83 (2011). doi:10.1088/0004-637X/742/2/83

C.S. Reynolds, S. Heinz, M.C. Begelman, Shocks and sonic booms in the intracluster medium: X-ray shells and radio galaxy activity. Astrophys. J. Lett. **549**, 179–182 (2001)

D.M. Russell, R.P. Fender, E. Gallo, C.R. Kaiser, The jet-powered optical nebula of cygnus X-1. Mon. Not. R. Astron. Soc. **376**, 1341–1349 (2007). doi:10.1111/j.1365-2966.2007.11539.x

M. Ruszkowski, M. Brüggen, M.C. Begelman, Cluster heating by viscous dissipation of sound waves. Astrophys. J. **611**, 158–163 (2004). doi:10.1086/422158

P.A.G. Scheuer, Morphology and power of radio sources, in *IAU Symp. 97: Extragalactic Radio Sources*, ed. by D.S. Heeschen, C.M. Wade, 1982, pp. 163–165

P.H. Sell, S. Heinz, D.E. Calvelo, V. Tudose, P. Soleri, R.P. Fender, P.G. Jonker, N.S. Schulz, W.N. Brandt, M.A. Nowak, R. Wijnands, M. van der Klis, P. Casella, Parsec-scale bipolar X-ray shocks produced by powerful jets from the neutron star circinus X-1. Astrophys. J. Lett. **719**, 194–198 (2010). doi:10.1088/2041-8205/719/2/L194

K.C. Steenbrugge, K.M. Blundell, Multiwavelength study of CygnusA—I. Precession and slow jet speeds from radio observations. Mon. Not. R. Astron. Soc. **388**, 1457–1464 (2008). doi:10.1111/j.1365-2966.2007.12665.x

J.F. Steiner, J.E. McClintock, Modeling the jet kinematics of the black hole microquasar XTE J1550-564: a constraint on spin-orbit alignment. Astrophys. J. **745**, 136 (2012). doi:10.1088/0004-637X/745/2/136

A.M. Stirling, R.E. Spencer, C.J. de la Force, M.A. Garrett, R.P. Fender, R.N. Ogley, A relativistic jet from cygnus X-1 in the low/hard X-ray state. Mon. Not. R. Astron. Soc. **327**, 1273–1278 (2001)

A. Tchekhovskoy, R. Narayan, J.C. McKinney, Efficient generation of jets from magnetically arrested accretion on a rapidly spinning black hole. Mon. Not. R. Astron. Soc. **418**, 79–83 (2011). doi:10.1111/j.1745-3933.2011.01147.x

V. Tudose, R.P. Fender, C.R. Kaiser, A.K. Tzioumis, M. van der Klis, R.E. Spencer, The large-scale jet-powered radio nebula of circinus X-1. Mon. Not. R. Astron. Soc. **372**, 417–424 (2006). doi:10.1111/j.1365-2966.2006.10873.x

J.C. Vernaleo, C.S. Reynolds, AGN feedback and cooling flows: problems with simple hydrodynamic models. Astrophys. J. **645**, 83–94 (2006). doi:10.1086/504029

D. Yoon, B. Morsony, S. Heinz, K. Wiersema, R.P. Fender, D.M. Russell, R. Sunyaev, Jet trails and mach cones: the interaction of microquasars with the interstellar medium. Astrophys. J. **742**, 25 (2011). doi:10.1088/0004-637X/742/1/25

DOI 10.1007/978-1-4939-2227-7_21
Reprinted from *Space Science Reviews* Journal, DOI 10.1007/s11214-013-0018-2

The Supermassive Black Hole—Galaxy Connection

Andrew King

Received: 18 February 2013 / Accepted: 12 August 2013 / Published online: 19 September 2013

Abstract The observed scaling relations imply that supermassive black holes (SMBH) and their host galaxies evolve together. Near-Eddington winds from the SMBH accretion discs explain many aspects of this connection. The wind Eddington factor $\dot{m}$ should be in the range ~ 1–30. A factor $\dot{m} \sim 1$ give black hole winds with velocities $v \sim 0.1c$, observable in X-rays, just as seen in the most extreme ultrafast outflows (UFOs). Higher Eddington factors predict slower and less ionized winds, observable in the UV, as in BAL QSOs.

In all cases the wind must shock against the host interstellar gas and it is plausible that these shocks should cool efficiently. There is detailed observational evidence for this in some UFOs. The wind sweeps up the interstellar gas into a thin shell and propels it outwards. For SMBH masses below a certain critical (M–σ) value, all these outflows eventually stall and fall back, as the Eddington thrust of the wind is too weak to drive the gas to large radii. But once the SMBH mass reaches the critical M–σ value the global character of the outflow changes completely. The wind shock is no longer efficiently cooled, and the resulting thermal expansion drives the interstellar gas far from the black hole, which is unlikely to grow significantly further. Simple estimates of the maximum stellar bulge mass M_b allowed by self-limited star formation show that the SMBH mass is typically about $10^{-3} M_b$ at this point, in line with observation.

The expansion-driven outflow reaches speeds $v_{\rm out} \simeq 1200\ {\rm km\,s^{-1}}$ and drives rates $\dot{M}_{\rm out} \sim 4000\ {\rm M_\odot\,yr^{-1}}$ in cool (molecular) gas, giving a typical outflow mechanical energy $L_{\rm mech} \simeq 0.05 L_{\rm Edd}$, where $L_{\rm Edd}$ is the Eddington luminosity of the central SMBH. This is again in line with observation. These massive outflows may be what makes galaxies become red and dead, and can have several other potentially observable effects. In particular they have the right properties to enrich the intergalactic gas with metals.

Our current picture of SMBH-galaxy coevolution is still incomplete, as there is no predictive theory of how the hole accretes gas from its surroundings. Recent progress in understanding how large-scale discs of gas can partially cancel angular momentum and promote dynamical infall offers a possible way forward.

A. King (✉)
Department of Physics & Astronomy, University of Leicester, Leicester LE1 7RH, UK
e-mail: ark@astro.le.ac.uk

Keywords Black holes · Galaxies · Feedback

1 Introduction

It is now generally accepted that the centre of all but the smallest galaxies contains a supermassive black hole (SMBH). More recently, astronomers have shown that the mass M of the hole correlates strongly with properties of the host galaxy. Typically the hole mass M is a small constant fraction of the stellar bulge mass M_b, i.e.

$$M \sim 10^{-3} M_b \tag{1}$$

(Häring and Rix 2004), and most strikingly, there appears to be a fairly tight relation of the form

$$M \simeq 2 \times 10^8 \ \mathrm{M}_{\odot} \sigma_{200}^4 \tag{2}$$

between the SMBH mass and the velocity dispersion $\sigma = 200\sigma_{200}$ km s^{-1} of the host galaxy's central bulge (Ferrarese and Merritt 2000; Gebhardt et al. 2000). The precise slope here is debated, with some observers favouring slightly larger values than 4. Selection effects may mean that Eq. (2) may represent a maximum SMBH mass for a given velocity dispersion σ (Batcheldor 2010).

There is only one likely physical origin for these correlations. In most respects (direct gravitational effect, total mass, etc) the black hole is a very minor component of the host galaxy. But in growing its mass by accretion, it releases energy $\eta M c^2 \sim 2 \times 10^{61} M_8$ erg (where $\eta \simeq 0.1$ is the accretion efficiency and $M_8 = M/10^8 \ \mathrm{M}_{\odot}$), far larger than the binding energy $\sim M_b \sigma^2 \sim 8 \times 10^{58} M_8 \sigma_{200}^2$ erg of a host bulge of mass $M_b \sim 10^3 M$ and velocity dispersion $\sigma = 200\sigma_{200}$ km s^{-1}. The huge disparity here shows that the host cannot ignore the presence of the hole, even though it is utterly insignificant in all other ways.

So an explanation of the SMBH-galaxy connection must involve a way of communicating some of its accretion energy to the host galaxy. Evidently this communication must be fairly inefficient, as otherwise we would have the opposite problem that the hole's accretion energy could disrupt the host entirely. Assembling a galaxy with a black hole in its centre is rather similar to building a house with a bomb in the basement. We will see that at a certain point the hole always has a decisive effect on its host.

2 Black Hole Winds

A credible observational candidate for the mechanism linking the SMBH accretion energy and the host galaxy emerged very shortly after the M–σ relation was established. X-ray absorption lines (particularly hydrogen- and helium-like iron) corresponding to outflow at velocities $v \sim 0.1c$ were seen in a number of Seyfert galaxies (e.g. Pounds et al. 2003; Reeves et al. 2003 (see the chapter by Pounds for a full discussion). Outflows like this are now detected in a large fraction of nearby AGN (Tombesi et al. 2010, 2011) and have been nicknamed UFOs ('ultrafast outflows'). This implies that they must have large solid angles $4\pi b$, with $b \sim 1$, i.e. they are quasispherical. The P Cygni profile of the X-ray iron line in PG1211+143 is direct evidence of this.

It is natural to imagine that such fast, near-spherical outflows from bright accreting objects such as active galactic nuclei should be connected with the Eddington limit, where radiation pressure balances gravity. Estimates using the observed velocities and ionization states of the wind gas quickly showed that this was very likely. On quite general grounds, SMBH mass growth must be closely associated with accretion at or near this limit. The Soltan relation (Soltan 1982) shows that the largest SMBH gained most of their mass by luminous accretion, i.e. during AGN phases. Since the fraction of AGN among all galaxies is small, this strongly suggests that when SMBH grow, they are likely to do so as fast as possible. If they are fed gas from their surroundings in a galaxy bulge with velocity dispersion σ the maximum possible accretion rate is dynamical, i.e.

$$\dot{M}_{\rm dyn} \simeq \frac{f_g \sigma^3}{G}, \tag{3}$$

where f_g is the gas fraction. The dynamical rate applies when a gas mass which was previously self-supporting against gravity is destabilized and falls freely on to the black hole. Equation (3) describes the case where gas is initially in rough virial equilibrium in the bulge of a galaxy with velocity dispersion σ and gas mass fraction f_g. We will see below that $\dot{M}_{\rm dyn}$ is a large overestimate of the rate $\dot{M}$ at which the black hole actually accretes. Parameterizing, we find

$$\dot{M}_{\rm dyn} \simeq 2.8 \times 10^2 \sigma_{200}^3 \ {\rm M}_\odot \ {\rm yr}^{-1} \tag{4}$$

where I have taken $f_g = 0.16$, which is the cosmological baryon fraction of all matter. We compare this with

$$\dot{M}_{\rm Edd} = \frac{L_{\rm Edd}}{\eta c^2} = \frac{4\pi G M}{\kappa \eta c} \tag{5}$$

where $L_{\rm Edd}$ is the Eddington luminosity and κ is the electron scattering opacity. We evaluate this for $\eta = 0.1$ and black hole masses M lying close to the observed M–σ relation (2) to find

$$\dot{M}_{\rm Edd} \simeq 4.4 \, \sigma_{200}^4 \ {\rm M}_\odot \ {\rm yr}^{-1} \tag{6}$$

and thus an Eddington ratio

$$\dot{m} < \frac{\dot{M}_{\rm dyn}}{\dot{M}_{\rm Edd}} \simeq \frac{64}{\sigma_{200}} \simeq \frac{54}{M_8^{1/4}}. \tag{7}$$

This ratio is significantly overestimated, since the rate $\dot{M}_{\rm dyn}$ assumes that the infalling gas has lost *all* its angular momentum. Retention of even a tiny fraction of it instead forces the gas to orbit the black hole and form an accretion disc, which is a bottleneck for accretion: gas moves inwards through a disc on the viscous timescale

$$t_{\rm visc} = \frac{1}{\alpha}\left(\frac{R}{H}\right)^2 \left(\frac{R^3}{GM}\right)^{1/2} \tag{8}$$

Here $\alpha \sim 0.1$ is the Shakura-Sunyaev viscosity parameter, while the disc aspect ratio H/R is almost constant with radius, and close to 10^{-3} for an AGN accretion disc (e.g. Collin-Souffrin and Dumont 1990). We note that $t_{\rm visc}$ approaches a Hubble time even for disc radii of only 1 pc.

With this in mind we can evaluate the significance of the ratio (7). The mass M_8 lies between ~ 0.1 and ~ 10 for most black holes in AGN, and as we have seen $\dot{M}_{\rm dyn}$ is a generous upper limit to $\dot{M}$. So relatively modest values $\dot{m} \sim 1$ of the Eddington ratio are likely in SMBH growth episodes.

This suggests that we should consider the properties of quasispherical winds from SMBH accreting with modest Eddington ratios $\dot{m} = \dot{M}/\dot{M}_{\rm Edd} \sim 1$. It is well known that winds of this type have electron scattering optical depth $\tau \sim 1$ measured from infinity inwards to a distance of order the Schwarzschild radius $R_s = 2GM/c^2$ (e.g. King and Pounds 2003). This means that on average every emitted photon scatters about once before escaping to infinity, which in turn suggests that the total wind momentum must be of order the photon momentum, i.e.

$$\dot{M}_{\rm w} v \simeq \frac{L_{\rm Edd}}{c}, \tag{9}$$

as is for example also found for the winds of hot stars. This is the basic result which emerges from the classic analysis by Shakura and Sunyaev (1973) of disc accretion at such mass inflow rates, and corresponds to the Model A wind solutions discussed in the chapter by Ohsuga and Mineshige (2013). Using (5) gives the wind velocity

$$v \simeq \frac{\eta}{\dot{m}} c \sim 0.1c. \tag{10}$$

This agrees with the fact that winds are always found to have a terminal velocity typical of the escape velocity from the radius at which they are launched, here $\sim 100R_s$. As expected, using this value of v in the mass conservation equation

$$\dot{M}_{\rm w} = 4\pi b R^2 v \rho(R), \tag{11}$$

where $\rho(R)$ is the mass density, self-consistently shows that the optical depth τ of the wind is ~ 1 (cf. King and Pounds 2003, Eq. (4)). From (10) we get the wind mechanical luminosity as

$$\dot{M}_{\rm w} \frac{v^2}{2} \simeq \frac{L_{\rm Edd}}{c} \frac{v}{2} \simeq \frac{\eta}{2} L_{\rm Edd} \simeq 0.05 L_{\rm Edd}. \tag{12}$$

The chapter by Ohsuga and Mineshige shows in detail that winds of this type (Model A) are a natural outcome of super-Eddington accretion In particular Model A winds are predicted (cf. Table 2 of Ohsuga and Mineshige 2013) to have mechanical luminosities $\sim 0.1 L_{\rm Edd}$, in agreement with Eq. (12).

Since the wind moves with speed $\sim 0.1c$, it can persist long after the AGN is observed to have become sub-Eddington. The duration of the lag is $\sim 10R/c$, where R is the radial extent of the wind (i.e. the shock radius, as we shall see below). For R larger than ~ 3 pc this lag is at least a century, and far longer lags are possible, as we shall see. This may be the reason why AGN showing other signs of super-Eddington phenomena (e.g. narrow-line Seyfert 2 galaxies) are nevertheless seen to have sub-Eddington luminosities (e.g. NGC 4051: Denney et al. 2009).

We can use (10), (11) to estimate the ionization parameter

$$\xi = \frac{L_i}{NR^2} \tag{13}$$

of the wind. Here $L_i = l_i L_{\rm Edd}$ is the ionizing luminosity, with $l_i < 1$ a dimensionless parameter specified by the quasar spectrum, and $N = \rho/\mu m_p$ is the number density. This gives

$$\xi = 3 \times 10^4 \eta_{0.1}^2 l_2 \dot{m}^{-2}, \tag{14}$$

where $l_2 = l_i/10^{-2}$, and $\eta_{0.1} = \eta/0.1$.

Equation (14) shows that the wind momentum and mass rates determine its ionization parameter: for a given quasar spectrum, the predominant ionization state is such that the threshold photon energy defining L_i, and the corresponding ionization parameter ξ, together satisfy (14). This requires high excitation: a low threshold photon energy (say in the infrared) would imply a large value of l_2, but the high value of ξ then given by (14) would require the presence of very highly ionized species, physically incompatible with such low excitation.

For suitably chosen continuum spectra, it is possible to envisage a range of solutions of (14), and it is even possible that a given spectrum may allow more than one solution, the result being specified by initial conditions. For a typical quasar spectrum, an obvious self-consistent solution of (14) is $l_2 \simeq 1$, $\dot{m} \simeq 1$, $\xi \simeq 3 \times 10^4$. This describes the case where the quasar is currently radiating at the Eddington limit. However as remarked after Eq. (11), we can also have situations where the quasar's luminosity has dropped after an Eddington episode, but the wind is still flowing, with $\dot{m} \simeq 1$. In this case the ionizing luminosity $10^{-2} l_2 L_{\rm Edd}$ in (14) takes a lower value, giving a lower value of ξ. For example a quasar of luminosity $0.3 L_{\rm Edd}$ would have $\xi \sim 10^4$. This corresponds to a photon energy threshold appropriate for heliumlike or hydrogenlike iron (i.e. $h\nu_{\rm threshold} \sim 9$ keV). We conclude that *Eddington winds from AGN are likely to have velocities* $\sim 0.1c$, *and show the presence of helium- or hydrogenlike iron.* Zubovas et al. (2011) show that this is likely to hold even for AGN which are significantly *sub*-Eddington.

These results suggest that it is no coincidence that UFOs have wind velocities $v \sim 0.1c$, and further that they are all found by identifying blueshifted resonance lines of Fe XXV, XXVI in absorption (Tombesi et al. 2010, 2011). More generally, we can see from (10), (14) that a larger Eddington factor $\dot{m}$ is likely to produce slower winds which are less ionized, perhaps suggesting that systems with $\dot{m} > 1$ appear as BAL QSOs. Zubovas et al. (2011) investigate this idea in detail and tentatively confirm it.

3 The Wind Shock

The black hole wind must have a significant effect on its host galaxy, impacting directly on the host interstellar medium (ISM). The basic pattern of the interaction is similar to that of a stellar wind hitting the interstellar medium. The wind (here from the vicinity of the SMBH) is slowed in an inner (reverse) shock, where the temperature approaches $\sim 10^{11}$ K if ions and electrons reach equipartition (but see the discussion below). The shocked wind gas sweeps up the host ISM at a contact discontinuity further out, driving an outer (forward) shock into this gas (see Fig. 1).

The dynamics of this interaction differ markedly depending on whether or not the shocked wind gas is cooled rapidly compared to its flow time. If the gas is strongly cooled ('momentum-driven flow'), most of the preshock kinetic energy is lost to radiation, and the postshock gas transmits essentially only its ram pressure (9) to the host ISM.

In the opposite limit of inefficient cooling, the postshock gas retains most of its original energy, and uses the resulting mechanical luminosity $\simeq 0.05 L_{\rm Edd}$ (cf. Eq. (12)) to expand adiabatically into the ISM. This 'energy-driven flow' is much more violent, allowing the black hole accretion energy to drive the interstellar gas away. We shall see in Sects. 4 and 6 below that black hole outflows are momentum-driven for SMBH masses M below the critical M–σ mass, but change to energy-driven if the SMBH mass grows even slightly above it.

The decision between momentum and energy driving is clearcut. It is easy to show that ordinary atomic two-body cooling processes have no significant effect in cooling the wind

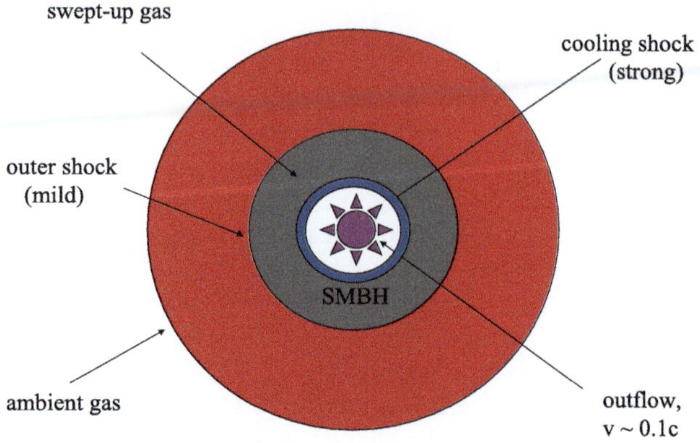

Fig. 1 Schematic view of the shock pattern resulting from the impact of an Eddington wind on the interstellar gas of the host galaxy. A supermassive black hole (SMBH) accreting at just above the Eddington rate drives a fast wind (velocity $u = v \sim \eta c \sim 0.1c$), whose ionization state makes it observable in X-ray absorption lines. The outflow collides with the ambient gas in the host galaxy and is slowed in a strong shock. The inverse Compton effect from the quasar's radiation field rapidly cools the shocked gas, removing its thermal energy and strongly compressing and slowing it over a very short radial extent. This gas may be observable in an inverse Compton continuum and lower-excitation emission lines associated with lower velocities. The cooled gas exerts the preshock ram pressure on the galaxy's interstellar gas and sweeps it up into a dense shell ('snowplough'). This shell's motion drives a milder outward shock into the ambient interstellar medium. This shock ultimately stalls unless the SMBH mass has reached the value M_σ satisfying the M–σ relation. (From King 2010)

shock. But the radiation field of the Eddington-accreting supermassive black hole has a characteristic temperature of only $\sim 10^7$ K, far lower than the wind shock. For shocks close to the SMBH, this radiation field is sufficiently intense that the inverse Compton effect (cf. Ciotti and Ostriker 1997) cools the electrons of the postshock wind gas more rapidly than their flow time $\sim R/\sigma$ (see below), and we are in the momentum-driven regime. For shocks at larger radii R the radiation energy density, and so the cooling rate, decrease as R^{-2}, until for R greater than a critical cooling radius

$$R_C \sim 500 M_8^{1/2} \sigma_{200} \text{ pc} \tag{15}$$

(King 2003, 2005; Zubovas and King 2012b) the cooling time becomes longer than the flow time, and we enter the energy-driven regime. In the language of the Introduction, we see that the momentum-driven regime corresponds to the relatively stable case of inefficient coupling of energy from SMBH accretion to the host galaxy, while when energy-driving sets in, the 'bomb' goes off, and something drastic happens to the host.

The inverse Compton effect is so efficient in cooling the post-shock electrons that we should question the assumption we made above, that the post-shock wind gas achieves equipartition between ions (which have all the initial postshock energy) and electrons (which gain energy from the ions in Coulomb collisions and perhaps by other processes before losing it to radiation). Faucher-Giguère and Quataert (2012) show that with Coulomb energy transfer alone, there is a significant parameter space where the gas does not reach equipartition. The resulting lower electron temperature reduces the ability of the shocked gas to radiate away its energy, and thus decreases the cooling radius R_C below the estimate (15). This might be detectable in future detailed observations, but probably does not greatly change the qualitative picture described here unless the estimate of R_C becomes comparable with the

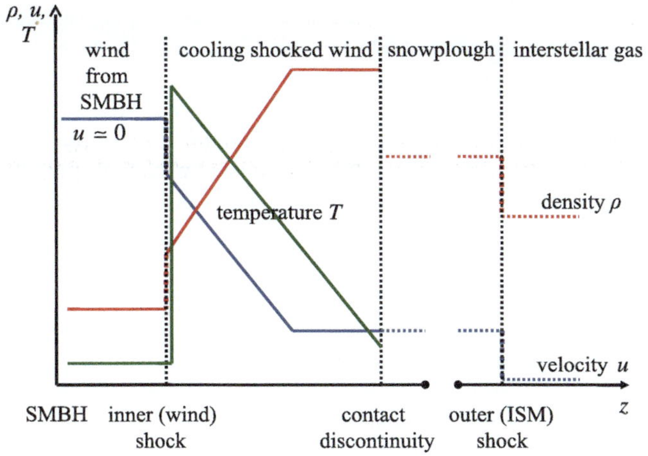

Fig. 2 Impact of a wind from an SMBH accreting at a super-Eddington rate on the interstellar gas of the host galaxy: schematic view of the radial dependence of the gas density ρ, velocity u and temperature T. At the inner shock, the gas temperature rises strongly, while the wind density and velocity respectively increase (decrease) by factors ~ 4. Immediately outside this (adiabatic) shock, the strong Compton cooling effect of the quasar radiation severely reduces the temperature, and slows and compresses the wind gas still further. This cooling region is very narrow compared with the shock radius (see Fig. 1), and may be observable through the and inverse Compton continuum and lower-excitation emission lines. The shocked wind sweeps up the host ISM as a 'snowplough'. This is more extended than the cooling region (cf. Fig. 1), and itself drives an outer shock into the ambient ISM of the host. Linestyles: *red, solid*: wind gas density ρ; *red, dotted*: ISM gas density ρ; *blue, solid*, wind gas velocity u; *blue, dotted*, ISM gas velocity u; *green, solid*, wind gas temperature T. The *vertical dashed lines* denote the three discontinuities, inner shock, contact discontinuity and outer shock. (From King 2010)

black hole radius of influence

$$R_{\rm inf} \simeq \frac{GM}{\sigma^2} \simeq 8\frac{M_8}{\sigma_{200}^2}\ {\rm pc}, \tag{16}$$

which requires extreme parameters. Faucher-Giguère and Quataert (2012) remark that their arguments do not rule out substantial momentum-conserving phases. In particular there may be microinstabilities coupling electron and ion temperatures more rapidly that Coulomb collisions.

It is plausible then that the observed UFO winds produce momentum-driving through strongly cooling shocks close to the SMBH. All but a small fraction of the mechanical luminosity (12) is harmlessly radiated as an inverse Compton continuum with characteristic photon energy ~ 1 keV as the postshock electrons lose energy to the AGN radiation field. Technically such cooling shocks are called 'isothermal', in that the gas temperature rapidly returns to its preshock value. To conserve momentum the gas is also strongly slowed and compressed as it cools. Figure 2 shows a schematic picture of the whole region between the two shocks.

This picture suggests that the postshock velocity of the X-ray emitting gas should correlate with its temperature (or roughly, ionization) while Compton cooling is dominant. Once this cooling has compressed the gas sufficiently, two-body processes such as free-free and bound-free emission must begin to dominate, since they go as the square of the density, and their cooling times decrease with temperature also (Pounds and King 2013). The chapter by Pounds shows that there is direct observational evidence for both of these effects. So some observed UFOs appear to correspond to Eddington winds coupling to momentum-driven outflows. This means that the wind shocks are close enough to the AGN that cooling dominates. Estimates from observation (cf. the chapter by Pounds) again support this—the shock observed in NGC 4051 (Pounds and Vaughan 2011) is at an estimated distance from the SMBH of ~ 0.1 pc, well within the cooling radius R_C (15). Further, the frequent detection of very slow systems ('warm absorbers') at distance less than a few parsecs from the black hole is indirect support for the idea of shocks occurring well within the cooling radius.

4 Dynamics: Origin of the M–σ Relation

We now have a simple picture of how quasispherical Eddington winds from supermassive black holes interact with their host galaxies. At least at the beginning of the interaction, the shock are close enough to the SMBH that inverse Compton cooling from the AGN radiation field puts the flow in the momentum-driven regime. This means that the region between the initial (reverse) shock first slowing the wind, and the contact discontinuity where it impacts and sweeps up the host ISM, is geometrically very narrow. The outer (forward) shock accelerating the ISM is also strongly cooled, so that the 'snowplough' region of swept-up ISM is also geometrically narrow. So for many purposes we can idealize the whole region between the inner and outer shocks as a single outward-moving gas shell whose mass grows as it sweeps up more and more of the host ISM.

As a further simplification, we assume that the matter in the central bulge of the host galaxy is distributed roughly as an isothermal sphere with velocity dispersion σ, so that the mass within radius R is

$$M(R) = \frac{2\sigma^2 R}{G} \tag{17}$$

With a roughly constant gas fraction f_g, the mass of the narrow swept-up shell at radius R is then

$$M_g(R) = \frac{2 f_g \sigma^2 R}{G} \tag{18}$$

The simple spherical geometry now gives the equation of motion of the shell as

$$\frac{\mathrm{d}}{\mathrm{d}t}\big[M_g(R)\dot{R}\big] + \frac{G M_g(R)[M + M(R)]}{R^2} = \frac{L_{\mathrm{Edd}}}{c}, \tag{19}$$

where M is the SMBH mass. Using (17), (18) and the definition of L_{Edd} (Eq. (5)) we can simplify this to

$$\frac{\mathrm{d}}{\mathrm{d}t}(R\dot{R}) + \frac{GM}{R} = -2\sigma^2\left(1 - \frac{M}{M_\sigma}\right), \tag{20}$$

where

$$M_\sigma = \frac{f_g \kappa}{\pi G^2}\sigma^4. \tag{21}$$

Multiplying through by $R\dot{R}$ allows us to find the first integral

$$R^2\dot{R}^2 = -2GMR - 2\sigma^2\left[1 - \frac{M}{M_\sigma}\right]R^2 + \text{constant} \tag{22}$$

For large R we have

$$\dot{R}^2 \simeq -2\sigma^2\left[1 - \frac{M}{M_\sigma}\right] \tag{23}$$

which cannot be satisfied for $M < M_\sigma$. If the SMBH mass is below M_σ, outflows cannot reach large radius because the Eddington thrust of the black hole wind is too small lift the weight of the interstellar gas against the galaxy bulge potential. All attempts by the SMBH to remove the gas from its surroundings fail: in practice any gas shell it drives outwards ultimately becomes too massive, and so tends to fall back (and presumably fragment).

To evaluate M_σ we need to specify some average gas fraction f_g. If we want to consider protogalaxies forming at high redshift the obvious value to adopt the value $f_g = \Omega_{\text{baryon}}/\Omega_{\text{matter}} \simeq 0.16$ (Spergel et al. 2003). Clearly there can be deviations from this value for galaxies forming later: galaxies may have gained abundant gas ($f_g \to 1$) or have been largely swept clear ($f_g \to 0$). In the latter case we would predict very low black hole masses which are presumably unobservable. With the gas fraction f_g fixed at the cosmological value $f_c = 0.16$, the expression

$$M_\sigma = \frac{f_g \kappa}{\pi G^2}\sigma^4 \simeq 3.1 \times 10^8\ \mathrm{M}_\odot \sigma_{200}^4 \tag{24}$$

is remarkably close to the observed relation (2), even though it now contains no free parameter.

The derivation here adopts the simplest possible description of a galaxy spheroid as an isothermal sphere (cf. Eq. (17)), and one might wonder what happens if the galaxy is more complex than this. For any choice of cumulative mass $M(R)$ it is clear that we can find a first integral of the equation of motion (19) simply by multiplying through by $M(R)\dot{R}$. Then one can again ask under what conditions the swept-up momentum-driven shell can reach large radii. McQuillin and McLaughlin (2012) do this for three more halo distributions (Hernquist 1990; Navarro et al. 1996, 1997; Dehnen et al. 2006). They show that relations very similar to (24) hold in each case, where σ is replaced by the peak circular speed of the halo in each case. Observationally this implies relations like (24) between SMBH mass and asymptotic circular speed in galaxy spheroids. So we can conclude that the outflow physics we have discussed so far quite generally picks out a relation like (24), remarkably close to that actually observed.

5 What Happens when $M = M_\sigma$?

The last statement is encouraging, but we still have to explain precisely what happens when the black hole reaches the critical M_σ mass. The M–σ relation is central to any theory of how supermassive black holes grow in the centres of galaxies, so a number of different approaches appear in the literature. All of these implicitly appeal to the idea that at the critical M_σ mass the SMBH drives away the gas which it might otherwise accrete, and so terminates its own mass growth. Many of these attempted explanations assume that the physics of the wind interaction is characterized purely either by energy-driving or momentum-driving throughout. In fact neither assumption gives an M–σ relation in agreement in observation.

The earliest derivations (Silk and Rees 1998; Haehnelt et al. 1998) of a critical SMBH mass implicitly assume energy-driving, i.e. that the wind shocks do not cool. We shall see in Sect. 9 that these all lead to critical masses significantly lower (typically by a factor $\sim \sigma/c \sim 10^{-3}$) than allowed by observations.

The assumption of pure momentum-driving leads to more subtle difficulties. Fabian (1999) does not consider shock cooling, but does assume that the outflow momentum alone defines the critical SMBH mass. Specifically, this paper explicitly assumes sub-Eddington accretion, and a slow black hole wind with speed v_w (later taken to be $0.1c$) and mechanical luminosity a fixed fraction a of $L_{\rm Edd}$. Equating the wind thrust with the weight of the overlying gas in an isothermal potential produces a critical mass $M_{\rm crit} = (v_w/2ac)M_\sigma$. This at first sight looks promising, since one would find $M_{\rm crit} = M_\sigma$ if it could be arranged that $v_w \simeq 2ac$. But by definition $a = \dot{M}_w v_w^2/2L_{\rm Edd}$ and $L_{\rm Edd} = \eta \dot{M}_{\rm Edd}c^2$, so we have

$$M_{\rm crit} = \frac{v_w}{2ac}M_\sigma = \frac{\eta c}{v_w}\frac{\dot{M}_{\rm Edd}}{\dot{M}_w}M_\sigma \simeq \frac{\dot{M}_{\rm Edd}}{\dot{M}_w}M_\sigma M_\sigma \gg M_\sigma \tag{25}$$

if accretion is significantly sub-Eddington ($\dot{M}_w < \dot{M} < \dot{M}_{\rm Edd}$) as assumed at the outset. In other words, pure momentum driving produces a viable M–σ relation if and only if the wind outflow momentum is very close to Eddington: (9) is the crucial relation, and assuming any thrust weaker than Eddington means that the black hole grows to a larger mass.

Murray et al. (2005) consider a wider problem, namely the stability of a galaxy under momentum deposition by direct radiation pressure on dust grains. The pressure comes both from stars (via radiation pressure and supernovae) and from a central accreting SMBH. The paper identifies a maximum Eddington-like luminosity $L_M \simeq 4(f_g c/G)\sigma^4$ which allows a galaxy to retain its gas, and argues that the central black hole will clear the central regions of the galaxy and be unable to accrete if its accretion luminosity exceeds this value. Assuming that the AGN accretion luminosity is close to the Eddington value $L_{\rm Edd}$, this produces an expression similar to (24). However it is clear that reaching the mass (24) is a necessary condition for driving away gas and halting SMBH mass growth, but not sufficient. More momentum (and so more SMBH accretion and mass growth) is needed to achieve this. McQuillin and McLaughlin (2012) show that if momentum-driving alone is to push gas far from the SMBH a mass $\gtrsim 3M_\sigma$ is needed (see also Ishibashi and Fabian 2012). Debuhr et al. (2011) find a similar result numerically. So the true critical mass implied by the calculation of Murray et al. (2005) is significantly higher than M_σ, and thus noticeably larger than the observed value.

Debuhr et al. (2011) attempt to overcome this by assuming that the dust has optical depth $\tau > 1$, which reduces the critical mass by a factor $1/\tau$. However there are further difficulties here: the optical depth of a swept-up shell varies as $M(R)/4\pi R^2 \propto 1/R$, and so drops sharply as the shell expands, until the pressure force is too weak to drive it outwards and it stalls. (Ishibashi and Fabian 2012 indeed appeal to this property to suggest that star formation in massive galaxies proceeds from inside to outside as shells of dusty gas are driven out and then stall. They obtain realistic radii provided that the dust opacity is very large, i.e. $\kappa_{\rm dust} \simeq 1000\kappa$.) Silk and Nusser (2010) argue that strong cooling will always enforce momentum-driving by the central SMBH, and also note that this cannot remove cannot remove the gas from its vicinity. They then appeal to star formation to add extra momentum and so produce the M–σ relation. However in ruling out energy-driving by the SMBH they appear to have considered only the cooling of the ambient interstellar medium. This is irrelevant to the question of energy or momentum-driving. As we have seen, it is the cooling of the shocked wind gas which decides this, and the analysis of Sect. 3 shows that the wind shocks are Compton cooled inside R_C and not cooled outside it.

This discussion shows that ending SMBH mass growth cleanly at the correct M–σ mass is slightly more complex. Pure energy-driving is too strong, and produces a critical mass much smaller than observed, while simplistic momentum-driving would require a critical mass which is about a factor 3 larger than observed. In fact *the real significance of the M–σ relation is that at this point the shocked outflow undergoes a global change from momentum to energy-driving.* We saw above that for $M < M_\sigma$ the wind shocks are efficiently Compton cooled, enforcing momentum-driving. Under these conditions the Eddington thrust of the SMBH cannot push the ambient host gas far away. All this changes radically once the SMBH begins to exceed this critical mass. We can see from (22), (23) that the SMBH momentum thrust is able to lift the swept-up shell of interstellar gas to larger radii. Importantly, this means that even for very small increases ($O(R_{\rm inf}/R_C) \sim 10^{-2}$) of M above M_σ, a shell can now reach radii $> R_C$, causing the outflow to become energy-driven. So at this point the shocked outflow can now use all of its energy to drive gas away as the shocked wind expands into the interstellar medium of the host. It is this property that causes a global change in the relation between the black hole and its host at the critical mass M_σ. For masses below this value, momentum-driven shells push against the interstellar gas, and make little progress—they are confined within R_C, so within scales of ~ 500 pc, and repeatedly stall and fall back. But once the SMBH mass exceeds the critical value things change radically—the full mechanical luminosity (12) is now used to drive gas away. Thus a significant fraction $\eta/2 = 0.05$ of the gravitational binding energy release of the SMBH now expels gas: the bomb goes off.

6 Clearing out the Galaxy

The global change from momentum- to energy-driving signalled by the black hole mass reaching M_σ has profound implications.

The geometry of the outflow changes completely (see Fig. 3). The shocked wind region is no longer narrow, but large and expanding. It is easy to show that the wind shock initially moves inwards to the cooling radius, where it remains fixed, at least initially. If the energy-driven flow persists for long enough, the shock eventually moves outwards to the point where the wind ram pressure just balances the gas pressure of the shock-heated wind (Faucher-Giguère and Quataert 2012). The contact discontinuity is always at far larger radii, and pushes into the shocked interstellar gas, sweeping it up into a shell. Since the shocked wind region has expanded strongly under its internal gas pressure, any pressure gradient within it is very small, so we can regard it as having constant total pressure P. This is larger than the ram pressure ρv^2 appearing in (19). The equation of motion for the inner radius R of the swept-up gas shell (the contact discontinuity) becomes

$$\frac{\mathrm{d}}{\mathrm{d}t}\big[M_g(R)\dot{R}\big] + \frac{GM_g(R)M(R)}{R^2} = 4\pi R^2 P \tag{26}$$

(in the second term on the lhs we neglect the contribution $GM_g M/R^2$ of the black hole gravity, as the shell is far outside the SMBH sphere of influence radius

$$R_{\rm inf} \simeq \frac{GM}{\sigma^2} \ll R_C. \tag{27}$$

To determine the problem we add the energy equation, which constrains P by specifying the rate that energy is fed into the shocked gas, minus the rate of $P\mathrm{d}V$ working on the ambient gas and against gravity:

$$\frac{\mathrm{d}}{\mathrm{d}t}\left[\frac{4\pi R^3}{3}\cdot\frac{3}{2}P\right] = \frac{\eta}{2}L_{\rm Edd} - P\frac{\mathrm{d}}{\mathrm{d}t}\left[\frac{4\pi}{3}R^3\right] - 4f_g\frac{\sigma^4}{G}\dot{R} \tag{28}$$

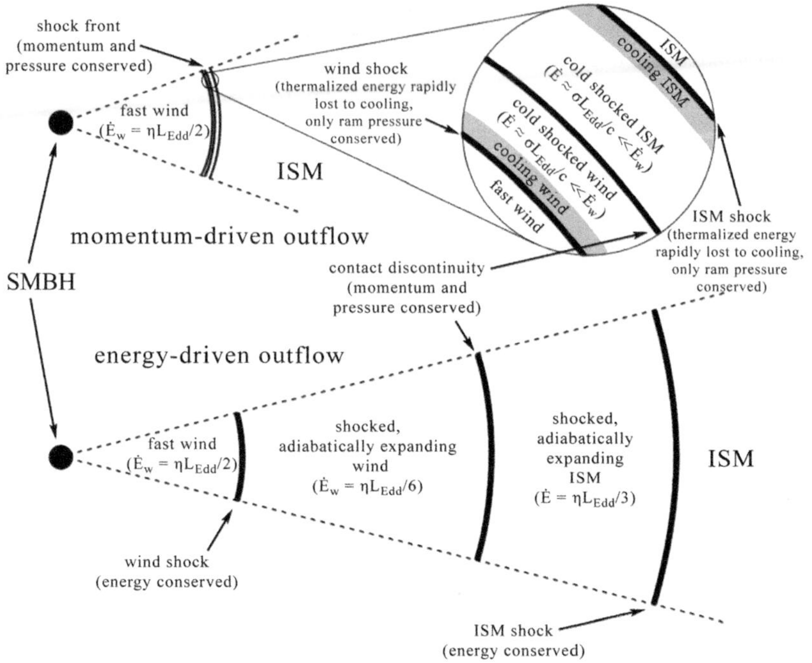

Fig. 3 Schematic picture of momentum-driven (*top*) and energy-driven (*bottom*) outflows. In both cases a fast wind (velocity $\sim 0.1c$) impacts the interstellar gas of the host galaxy, producing an inner reverse shock slowing the wind, and an outer forward shock accelerating the swept-up gas. In the momentum-driven case, the shocks are very narrow and rapidly cool to become effectively isothermal. Only the ram pressure is communicated to the outflow, leading to very low kinetic energy $\sim (\sigma/c)L_{\rm Edd}$. In an energy-driven outflow, the shocked regions are much wider and do not cool. They expand adiabatically, communicating most of the kinetic energy of the wind to the outflow (in simple cases divided in a ratio of about 1:2 between the shocked wind and the swept-up gas). The outflow radial momentum flux is therefore greater than that of the wind. Momentum-driven conditions hold for shocks confined to within ~ 1 kpc of the AGN, and establish the M–σ relation (24) (King 2003, 2005). Once the supermassive black hole mass attains the critical M–σ value, the shocks move further from the AGN and the outflow becomes energy-driven. This produces the observed large-scale molecular outflows which probably sweep the galaxy clear of gas. (From Zubovas and King 2012a)

Here we have assumed a specific heat ratio $\gamma = 5/3$, used (12) to specify energy input from the outflow and (17) to simplify the gravity term $GM(R)M(R)/R^2$. (Note that in the momentum-driven case the energy equation did not appear, as by definition it was implicitly assumed to account for the removal of all the wind energy other than that associated with the ram pressure.)

Now using (26) to eliminate P from (28), and replacing the gravity terms as before, we get

$$\frac{\eta}{2}L_{\rm Edd} = \dot{R}\frac{\mathrm{d}}{\mathrm{d}t}\big[M(R)\dot{R}\big] + 8f_g\frac{\sigma^4}{G}\dot{R} + \frac{\mathrm{d}}{\mathrm{d}t}\left\{\frac{R}{2}\dot{R}\frac{\mathrm{d}}{\mathrm{d}t}\big[M(R)\dot{R}\big] + 2f_g\frac{\sigma^4}{G}R\right\} \tag{29}$$

leading to

$$\frac{\eta}{2}L_{\rm Edd} = \frac{2f_g\sigma^2}{G}\left\{\frac{1}{2}R^2\dddot{R} + 3R\dot{R}\ddot{R} + \frac{3}{2}\dot{R}^3\right\} + 10f_g\frac{\sigma^4}{G}\dot{R}, \tag{30}$$

which in the energy-driven regime replaces Eq. (19) in describing the motion of the interface between wind and interstellar gas.

Using $M = M_\sigma$ from (21) in $L_{\rm Edd}$, this equation has a solution $R = v_e t$ with

$$2\eta c = 3\frac{v_e^3}{\sigma^2} + 10 v_e \tag{31}$$

The assumption $v_e \ll \sigma$ leads to a contradiction ($v_e \simeq 0.02c \gg \sigma$), hence this equation has the approximate solution

$$v_e \simeq \left[\frac{2\eta\sigma^2 c}{3}\right]^{1/3} \simeq 925\sigma_{200}^{2/3}\ \mathrm{km\,s^{-1}} \tag{32}$$

This solution is an attractor. At radii R larger than the cooling radius R_C, Compton cooling is ineffective, and the extra gas pressure accelerates the previously momentum-driven gas shell to this new higher velocity. We see this in the numerical solutions of Fig. 4, where all solutions quickly converge to it, regardless of initial conditions. This figure also confirms that once the driving by the AGN switches off, the shell decelerates as predicted by the analytic solution of (30) with $L_{\rm Edd} = 0$ found by King et al. (2011):

$$\dot{R}^2 = 3\left(v_e^2 + \frac{10}{3}\sigma^2\right)\left(\frac{1}{x^2} - \frac{2}{3x^3}\right) - \frac{10}{3}\sigma^2 \tag{33}$$

where $x = R/R_0 \geq 1$.

The solutions (32), (33) describe the motion of the contact discontinuity where the shocked wind encounters swept-up interstellar gas (see Fig. 3). This interface is strongly Rayleigh-Taylor (RT) unstable, because the shocked wind gas is highly expanded and so has much lower mass density than the swept-up interstellar gas outside it. The RT instability is difficult to handle numerically. Deductions concerning the average spatial scale $R(t)$, velocity and energy of the outflow are likely to be believable, but we should be very cautious in accepting any results which depend strongly on the detailed nature of the interface between the shocked wind and the swept-up interstellar medium. An example where this may be relevant is the puzzling nature of the observed outflows: despite their high velocities ($\sim 1000\ \mathrm{km\,s^{-1}}$) much of the outflowing gas is in molecular form. In line with this, interstellar gas entering the forward shock is efficiently cooled by two-body radiation processes. A preliminary analysis suggests that the interstellar gas is likely to have a multiphase structure, with most of the mass able to cool all the way from the forward shock temperature $\sim 10^7$ K back to low temperatures, ending in largely molecular form, despite now being entrained in an outflow with the velocity of the forward shock. However cooling is affected by the topology of the gas flow and the total area of interfaces between different gas phases, so a full numerical calculation showing this explicitly is currently impossible.

Molecular outflows like this are seen in recent observations. Feruglio et al. (2010), Rupke and Veilleux (2011), and Sturm et al. (2011) use molecular spectral line observations to reveal fast ($v_{\rm out} \sim 1000\ \mathrm{km\,s^{-1}}$) large-scale (kpc) and massive ($\dot{M}_{\rm out} \sim 1000\ \mathrm{M_\odot\,yr^{-1}}$) outflows in the nearby quasar Mrk 231. Other galaxies show indications of similar phenomena, e.g. Lonsdale et al. 2003; Tacconi et al. 2002; Veilleux et al. 2009; Riffel and Storchi-Bergmann (2011a, 2011b) and Sturm et al. (2011). These appear to show quasar feedback transforming young, star-forming galaxies into red and dead spheroids. All three groups reach this conclusion for Mrk231 by noting that the mass outflow rate $\dot{M}_{\rm out}$ and the kinetic energy rate $\dot{E}_{\rm out} = \dot{M}_{\rm out} v_{\rm out}^2/2$ of the outflow are too large to be driven by star formation,

Fig. 4 Evolution of an energy-driven shock pattern for the case $\sigma = 200\ \mathrm{km\,s^{-1}}$, $f_g = 10^{-2}$ computed numerically from the full equation (30). *Top*: radius vs time, *middle*: velocity vs time, *bottom*: velocity vs radius. The curves refer to different initial conditions: *black solid*—$R_0 = 10$ pc, $v_0 = 400$ km/s; *blue dashed*—$R_0 = 100$ pc, $v_0 = 1000$ km/s; *red dot-dashed*—$R_0 = 50$ pc, $v_0 = 200$ km/s. All these solutions converge to the attractor (32). The *vertical dashed line* marks the time $t = 10^6$ yr when (for this case) the quasar driving is switched off. All solutions then follow the analytic solution (33). A case where the quasar remains on for a Salpeter times $\sim 4 \times 10^7$ yr would sweep the galaxy clear of gas. (From King et al. 2011)

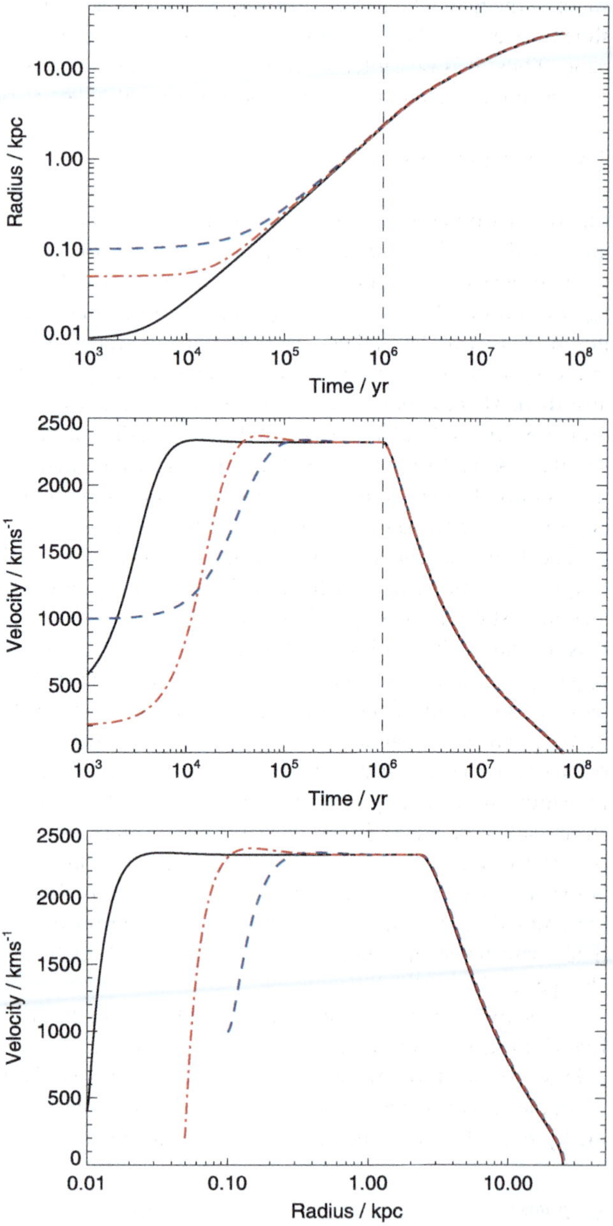

but as we shall see, comparable with those predicted for AGN feedback. The kinetic energy rate is a few percent of the likely SMBH Eddington luminosity, just as expected for an energy-driven outflow powered by a black hole wind (cf. Eq. (12)).

To work out the predicted mass rate of an energy-driven outflow we need to know how fast the outer shock moves and entrains new molecular material. This shock must run ahead of the contact discontinuity into the ambient interstellar medium in such a way that the velocity jump across it is a factor $(\gamma + 1)/(\gamma - 1)$ (where γ is the specific heat ratio). This fixes its velocity as

$$v_{\rm out} = \frac{\gamma+1}{2}\dot{R} \simeq 1230\sigma_{200}^{2/3}\left(\frac{lf_c}{f_g}\right)^{1/3} \ {\rm km\,s^{-1}} \tag{34}$$

(where we have used $\gamma = 5/3$ in the last form, the factor $l \simeq 1$ allows for a small deviation from the precise Eddington luminosity, and $f_c \simeq 0.16$ is the cosmological value of f_g). This corresponds to a shock temperature of order 10^7 K for the forward shock into the interstellar medium (as opposed to $\sim 10^{10-11}$ K for the wind shock). Since the outer shock and the contact discontinuity are very close together when energy-driven flow starts (see Fig. 3) this means that the outer shock is at

$$R_{\rm out}(t) = \frac{\gamma+1}{2}R(t) = \frac{\gamma+1}{2}v_e t. \tag{35}$$

The outflow rate of shocked interstellar gas is then just the rate at which the outer shock runs into the ambient gas, i.e.

$$\dot{M}_{\rm out} = \frac{{\rm d}M(R_{\rm out})}{{\rm d}t} = \frac{(\gamma+1)f_g\sigma^2}{G}\dot{R}. \tag{36}$$

Assuming $M = M_\sigma$, the wind outflow rate is

$$\dot{M}_{\rm w} \equiv \dot{m}\dot{M}_{\rm Edd} = \frac{4f_c\dot{m}\sigma^4}{\eta c G}. \tag{37}$$

We define the mass-loading factor for the outflow as the ratio of the mass flow rate in the shocked ISM to that in the wind:

$$f_{\rm L} \equiv \frac{\dot{M}_{\rm out}}{\dot{M}_{\rm w}} = \frac{\eta(\gamma+1)}{4\dot{m}}\frac{f_g}{f_c}\frac{\dot{R}c}{\sigma^2}. \tag{38}$$

Then the mass outflow rate is

$$\dot{M}_{\rm out} = f_{\rm L}\dot{M}_{\rm w} = \frac{\eta(\gamma+1)}{4}\frac{f_g}{f_c}\frac{\dot{R}c}{\sigma^2}\dot{M}_{\rm Edd}. \tag{39}$$

If the AGN is still radiating at a luminosity close to Eddington, we have $\dot{R} = v_e$, and using (32) gives

$$f_{\rm L} = \left(\frac{2\eta c}{3\sigma}\right)^{4/3}\left(\frac{f_g}{f_c}\right)^{2/3}\frac{l^{1/3}}{\dot{m}} \simeq 460\sigma_{200}^{-2/3}\frac{l^{1/3}}{\dot{m}}, \tag{40}$$

and

$$\dot{M}_{\rm out} \simeq 4060\sigma_{200}^{10/3}l^{1/3}\ {\rm M_\odot\,yr^{-1}} \tag{41}$$

for typical parameters, $f_{\rm g} = f_{\rm c}$ and $\gamma = 5/3$. Once the central quasar is no longer active, the mass outflow rate declines as $\dot{R}/v_e$ times this expression, with $\dot{R}$ given by (33). Clearly if the outflow persists for long enough it will sweep away all the galaxy's gas, leaving it red and dead. A detailed analysis by Zubovas and King (2012b) shows that the required time is of order the Salpeter timescale of the SMBH, i.e. this must grow slightly above the M–σ value to clear out the galaxy. This analysis also shows that the final energy-driven outflow can put metal-enriched gas far outside the galaxy, suggesting that supermassive black holes may be the main driver of this process.

We note that the crucial element here is the sudden and global change in the character of black hole outflows once the critical M–σ mass is reached. The outflows are comparatively weak and on small scales before this point, but become far more energetic after it, probably sweeping the interstellar gas out of the galaxy in a single event. The need for this abrupt change probably explains the difficulty in modelling the process of metal enrichment in cosmological simulations.

Equations (34), (41) give the approximate equality

$$\frac{1}{2}\dot{M}_{\rm w}v^2 \simeq \frac{1}{2}\dot{M}_{\rm out}v_{\rm out}^2, \tag{42}$$

i.e. most of the wind kinetic energy ultimately goes into the mechanical energy of the outflow, as expected for energy driving. (The jump conditions across the contact discontinuity show that if the quasar is still active, the shocked wind retains 1/3 of the total incident wind kinetic energy $\dot{M}_{\rm w}v^2/2$, giving 2/3 to the swept-up gas.)

Equation (42) means that the swept-up gas must have a momentum rate greater than the Eddington value $L_{\rm Edd}/c$, since we can rewrite it as

$$\frac{\dot{P}_{\rm w}^2}{2\dot{M}_{\rm w}} \simeq \frac{\dot{P}_{\rm out}^2}{2\dot{M}_{\rm out}}, \tag{43}$$

where the $\dot{P}$'s are the momentum fluxes. With $\dot{P}_{\rm w} = L_{\rm Edd}/c$, we have

$$\dot{P}_{\rm out} = \dot{P}_{\rm w}\left(\frac{\dot{M}_{\rm out}}{\dot{M}_{\rm w}}\right)^{1/2} = \frac{L_{\rm Edd}}{c}f_{\rm L}^{1/2} \sim 20\sigma_{200}^{-1/3}l^{1/6}\frac{L_{\rm Edd}}{c} \tag{44}$$

where $f_{\rm L}$ is the mass loading factor of the outflow. The factor $f_{\rm L}^{1/2} \sim 20$ is the reason why observations consistently show $\dot{M}_{\rm out}v_{\rm out} > L_{\rm Edd}/c$.

The predictions of this section agree well with observations of molecular outflows in galaxies (see Tables 1 and 2 of Zubovas and King 2012a). The energy-driven outflows from SMBH which have just reached their M–σ masses appear capable of sweeping galaxies clear of gas and making them red and dead. A robust observational signature of this is the expected mechanical luminosity

$$L_{\rm mech} \sim \frac{\eta}{2}L_{\rm Edd} \simeq 0.05L_{\rm Edd}, \tag{45}$$

which does appear to characterize most observed outflows of this type.

It is difficult to go further than these simple estimates. In particular, the Rayleigh-Taylor instability of the interface between the shocked wind gas and the interstellar medium seems likely to be a significant obstacle for numerical treatments.

7 The SMBH Scaling Relations

At the beginning of this chapter we noted the proportionality between M and M_b—see (1). Since M_b is the stellar mass of the bulge we need some idea of how efficient star formation can be. We follow a simple argument from Power et al. (2011) quantifying the effect of star formation on the virialized gas in a dark matter halo. Massive stars dominate the mass and momentum outflow from star clusters through supernovae and stellar winds (e.g. Leitherer et al. 1992). For a Salpeter initial mass function (IMF) and solar metallicity stars the contributions peak at stellar masses of 12 $M_\odot$ for supernovae, and about 50 $M_\odot$ for momentum outflow via winds. Integrated over time, these two feedback mechanisms contribute roughly equally to the momentum outflow from massive clusters. The characteristic main sequence age for stars dominating momentum feedback is thus $t_{\rm ms} \simeq 10^7$ years. This is shorter than both the dynamical time in the dark matter halo and the Salpeter time $t_{\rm S} \simeq 4 \times 10^7$ years. This means that the rate of stellar momentum feedback is fixed by the rate at which new massive stars form and replace the ones which are dying (see Fig. 12 in Leitherer et al. 1992). A single star produces wind momentum at the rate $\sim L/c$, where L is its luminosity, so in this regime star formation injects momentum at the global rate

$$\dot{p}_* \simeq \epsilon_* c \dot{M}_* \tag{46}$$

where $\epsilon_* \simeq 2 \times 10^{-3}$ (essentially the product of the nuclear efficiency and the Schönberg-Chandrasekhar mass fraction). Hence the total momentum produced by a stellar mass M_* is simply

$$p_* \simeq \epsilon_* c M_* \tag{47}$$

Feedback must inhibit star formation when gas which is not in stars acquires typical velocities σ. If M_0 was the gas mass when star formation began (presumably close to the virial value) this happens when the total momentum injected into the bulge gas reaches

$$p_{\rm inh} \simeq (M_0 - M_*)\sigma. \tag{48}$$

The derivation of the $M_{\rm BH} - \sigma$ relation in Sect. 4 shows that SMBH feedback is confined to a small region around it until its mass reaches M_σ. So only stellar feedback is potentially able to inhibit star formation in most of the galaxy bulge until this point, and the total stellar mass M_b cannot exceed a value $M_{\rm max}$ such that $p_* = p_{\rm inh}$. Thus from (47) we find

$$\epsilon_* c M_{\rm max} \simeq (M_0 - M_{\rm max})\sigma \tag{49}$$

or

$$M_b < M_{\rm max} \simeq \frac{M_0 \sigma}{\epsilon_* c + \sigma} \simeq M_0 \frac{\sigma}{\epsilon_* c}. \tag{50}$$

since $\epsilon_* c > \sigma$ in all cases. Now M_0 itself is limited: it cannot exceed a value $f_g M_V$, where f_g is the gas fraction and

$$M_V \simeq \frac{\sqrt{2}}{10} \frac{\sigma^3}{G} t_H \tag{51}$$

is the virial mass corresponding to velocity dispersion σ (one can think of this as the result of accretion at the dynamical rate (3) for a Hubble time t_H). So we find the estimate

$$M_b \sim \frac{0.14 f_g t_H \sigma^4}{\epsilon_* c G}. \tag{52}$$

Now comparing with (21) we get

$$M \simeq M_\sigma \sim \frac{1.8 \kappa \epsilon_* c}{\pi G t_H} M_b \sim 10^{-3} M_b, \tag{53}$$

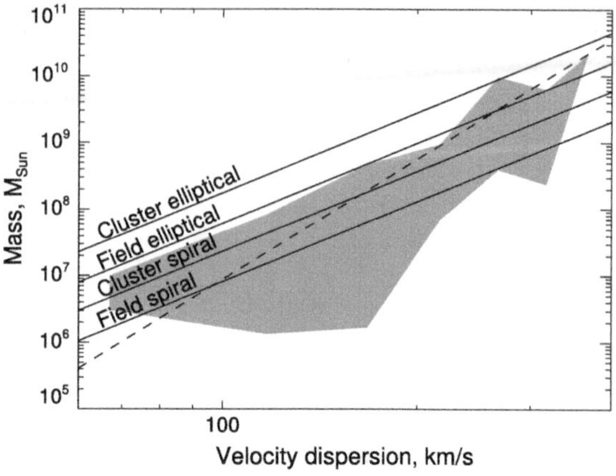

Fig. 5 The four (in reality three, as cluster spirals are rare) M–σ relations (*solid lines*) and their combined effect on observational fits (*dashed line*). All *solid lines* have slopes $M \propto \sigma^4$ and the *dashed line* has $M \propto \sigma^6$. The *grey area* is the approximate locus of data points in Fig. 3 of McConnell et al. (2011). (From Zubovas and King 2012b)

in agreement with observation (cf. Eq. (1)).

This argument is very simplified in that it does not take account of whether the galaxy is a spiral or an elliptical. This can potentially affect the M–σ relation because in particular in an elliptical, the much larger bulge mass means that energy-driving by the central SMBH wind must continue for longer in order to expel the remaining gas. Zubovas and King (2012b) find that energy-driving, and therefore SMBH mass growth, must continue for about 2 Salpeter times after the hole reaches M_σ to achieve this. So the final SMBH mass in an elliptical is

$$M_{\rm final} \sim \exp 2M_\sigma \sim 7.5M_\sigma \tag{54}$$

Similarly, environment can affect M, M_b to some degree. The bulges of cluster ellipticals can gain gas—especially Brightest Cluster Galaxies (BCGs), which are near the centre of the cluster potential, and some BCGs are known to contain unusually massive SMBH (McConnell et al. 2011). There appear to be essentially *three* parallel but offset M–σ relations for spirals, field and cluster ellipticals (see Fig. 5). In principal there is also a relation for cluster spirals, although such galaxies are rare. As we can see from the Figure, the spread in offsets may mean that an observed sample tends to produce a slightly higher slope than 4. All three types of galaxies obey the M–M_b relation (1) within the errors, as growth of the SMBH above M_σ correlates with higher M_b.

Finally, with a suitable mass-to-light ratio $M/L \sim 3$, Eq. (52) reproduces the observed Faber-Jackson relation between the stellar luminosity of ellipticals and their velocity dispersion

$$L \sim 2 \times 10^{10}\,{\rm L}_\odot \sigma_{200}^2, \tag{55}$$

suggesting that this relation is a result of (stellar) momentum driving limiting the star formation rate in a spherical system, as first shown by Murray et al. (2005). A similar argument (McLaughlin et al. 2006) shows that nucleated galaxies (i.e those whose central regions are dominated by nuclear star clusters, with no detectable sign of the presence of a supermassive black hole) there is an offset M–σ relation between the mass of the cluster and the velocity dispersion of the form

$$M_c \simeq 20M_\sigma = 6 \times 10^9 \sigma_{200}^4\ {\rm M}_\odot, \tag{56}$$

where typically the galaxy is small, with $\sigma \lesssim 120\ {\rm km\,s^{-1}}$. The factor ~ 20 offset in cluster mass for a given σ appears because momentum-driving by the stellar winds of an ensemble of cluster stars is about 20 times less efficient per unit mass than from a black hole accreting at the Eddington rate.

8 Energy Driving in General

The simple analytic approach adopted here for studying the SMBH-galaxy connection allows us to connect to results derived using different methods.

Some early attempts (e.g. Silk and Rees 1998, Haehnelt et al. 1998) to derive the M–σ relation invoked Eddington winds, but implicitly assumed energy-driving. With hindsight we can now see that even without knowing that Compton cooling always enforces momentum-driving instead, this approach cannot work. From the equation of motion (30) for an energy-driven outflow we can see that gas is always driven out at constant speed for *any* SMBH mass, however small: setting $R = v_e t$ and using the definition of $L_{\rm Edd}$ (cf. Eq. (5)) gives the outflow velocity v_e from

$$M = \frac{3 f_g \kappa \sigma^2 v_e^3}{\pi G^2 c \eta}. \tag{57}$$

This simply expresses the fact that the adiabatic shock pushes the interstellar gas away as a hot atmosphere for *any* SMBH mass. If one assumes in addition that only a value $v_e \sim \sigma$ is physically consistent (e.g. with the need for the gas to acquire the escape velocity, and so continue moving out once the central accretion and hence energy-driving turns off) this gives a mass

$$M = \frac{3 f_g \kappa}{\pi G^2 c \eta}\sigma^5 = \frac{3\sigma}{\eta c} M_\sigma = 0.02 M_\sigma = 6 \times 10^6\ {\rm M_\odot} \sigma_{200}^5 \tag{58}$$

which is clearly too small in comparison with observations. Silk and Rees (1998) actually give the expression

$$M \sim \frac{\kappa \sigma^5}{G^2 c} = \frac{\pi \sigma}{f_g c} M_\sigma, \quad \left(\text{which evaluates as } 4 \times 10^6\ {\rm M_\odot} \sigma_{200}^5\right) \tag{59}$$

Cosmological simulations often produce an empirical M–σ relation as part of much larger structure formation calculations. Limits on numerical resolution inevitably require a much more broad-brush approach. Accretion on to the hole is generally represented by the Bondi formula or ad hoc parameterizations based on it. This takes no account of the gas's specific angular momentum and the fact that the hole must in reality accrete from a disc (cf. Sect. 2 above, and Sect. 9). The formula also assumes that the black hole determines the accretion rate on to the hole at large radii, whereas in reality this assumption cannot be correct outside the sphere of influence $R_{\rm inf}$. The effect of the SMBH on its surroundings is usually modelled by injecting energy into the numerical 'galaxy'. The amount of energy is then iterated until the right relation appears. This empirical approach (e.g. Di Matteo et al. 2005, 2008; Hopkins and Hernquist 2006; Booth and Schaye 2009) seems to require energy to be injected at a rate of $0.05 L_{\rm Edd}$ to give the correct relation. This is precisely the mechanical luminosity (12) of a black hole wind with the Eddington momentum $\dot{M}_{\rm out} v = L_{\rm Edd}/c$. If this energy were all used to drive an outflow it would fix the black hole mass at the Silk-Rees mass (58) or (59) above) which is too small compared with observations. The fact that the simulations instead actually iterate to something like the correct M–σ value (24) suggests that they somehow arrange that the injected energy only couples to the surroundings at the very inefficient rate which occurs in momentum driving. In practice there must be some effective cooling. The physics ensuring this in reality operates at lengthscales far below the resolution of any conceivable cosmological simulation, so the effective cooling must be implicit in some of the 'sub-grid' physics which all such simulations have to assume. As we saw in Sects. 4–6, the real significance of the M–σ relation is that momentum-driving of outflows gives way to much more violent energy-driving. It seems unlikely that the simulations can capture this.

9 The Feeding Problem

We have seen that quasispherical near-Eddington winds from supermassive black holes naturally predict a number of effects in good agreement with observation, suggesting that they are a fundamental ingredient in understanding how SMBH and galaxies interact. Yet we are still some way from a fully deterministic theory of supermassive black holes and their hosts. The reason is clear: we do not yet have a convincing picture of how the holes are able to accrete large amounts of gas from their surroundings. It is remarkable that we are able to make considerable progress despite this—this is largely because SMBH generally grow their masses at rates close to Eddington.

The existence of the M–σ relation highlights the problem. By definition, this relation says that the hole knows a lot about the structure of its host galaxy, in particular at distances *greater* than its radius of influence

$$R_{\rm inf} \sim \frac{GM}{\sigma^2} \sim 20 M_8 \sigma_{200}^{-2}\ {\rm pc} \tag{60}$$

Yet because all gas has angular momentum, we also know that the hole accretes mass only through a standard thin accretion disc, whose viscous timescale approaches a Hubble time t_H at disc radii of a few tenths of a parsec, far smaller than $R_{\rm inf}$ (e.g. King and Pringle 2006, 2007). In other words, the hole knows about σ, but accretes from a structure that does not.

This problem reveals itself in various ways. For example, we argued above that there are slightly differing M–σ relations for spirals and field and cluster ellipticals, because the differing amounts of energy needed to clear the galaxy of gas in each case require the SMBH to accrete specific amounts beyond the 'bare' M–σ relation (24) in each case. But what actually tells the hole to accrete just the right amount in each case? There is apparently an inescapable coupling between large scales (the swept-up gas) and very small ones (the residual gas accreting on to the SMBH).

Gas falling inside $R_{\rm inf}$ must have lost almost all its original specific angular momentum—gas far from the hole does not have velocities aimed with minute accuracy at the SMBH. So the connection between large and small scales that we need must be connected to this process. Attempts to solve the angular momentum problem generally appeal to ways of transporting it outwards. A very similar problem is central to accretion disc theory, where the process invoked to transport angular momentum is called viscosity. Despite considerable progress we still do not know precisely how this works, and again it means that accretion disc theory too is fundamentally incomplete.

But the net total angular momentum of the gas in the central regions of a galaxy is likely to be close to zero. So a more promising way of promoting rapid gas infall might invoke *cancellation* of the angular momenta of different gas flows. This possibility is implicit in the idea that central accretion on to SMBH might be *chaotic* (King and Pringle 2006, 2007; King et al. 2008), i.e. have no preferred direction, which is attractive on other grounds also.

There are several ways this cancellation might occur: for example the outflows associated with feedback might cause counterrotating gas streams to collide. Or repeated collisions with momentum-driven shell would cause gas orbits to become very eccentric, so that these orbits would cross and collide very close to the black hole. Recently another possibility has emerged which uses the behaviour of accretion discs which are inclined with respect to a symmetry axis of the effective potential of the accretor. A simple example is a disc whose plane is not orthogonal to the spin of an accretion black hole. The gas orbits in the disc then precess about the spin axis by the Lense-Thirring effect, Viscous dissipation means that the hole-disc system always tries to reach axisymmetry, usually by propagating a warp in the disc which travels out and gradually co- or counter-aligns the whole disc. However the existence of the warp affects the way in which viscous stresses behave: if the viscous process transporting angular momentum is locally close to isotropy, a sufficiently inclined warp actually tends to *weaken* the stresses trying to keep the disc flow together (Ogilvie 1999). The result is that the disc can break into distinct planes, with only tenuous and rapid gas flows between them (cf. Lodato and Price 2010).

The most spectacular outcome of this process arises if the disc plane is inclined by an angle of more than 45° (and less than 135°) to the spin (Nixon et al. 2012). For once the disc breaks, the almost-disconnected inner and outer disc planes precess independently at rather different rates (the precession period varies as [disc radius]^3). At these large inclinations, this means that after about one-half of a precession period of the inner disc, its gas motions are partially opposed to those of the outer disc. Viscous spreading of the discs inevitably causes these opposed gas streams to cancel part of their specific angular momenta, and so fall inwards to smaller radii (see Fig. 6). The disc is torn up by the effect of the precession and the nonlinear evolution of viscosity in warps. The angular momentum cancellation and consequent infall occur because the gas draws on the angular momentum reservoir represented by the black hole spin to produce partially opposed gas flows. One might at first think that since the cancellation only affects the component of angular momentum orthogonal to the axis of precession, and then only by factors of order unity, that this would not have a large effect. However the specific angular momentum of a near-Keplerian ring goes as $j \propto R^{1/2}$, and the viscous timescale for viscous transport (which would otherwise specify the local accretion rate) goes as

$$t_{\rm visc} = \frac{1}{\alpha}\left(\frac{R}{H}\right)^2 \left(\frac{R^3}{GM}\right)^{1/2} \propto j^3 \qquad (61)$$

for a typical α-disc, whose aspect ratio H/R is almost constant with radius. So in a disc inclined at an angle of 75°, cancellation of angular momentum between precessing rings would reduce j by a factor 4, the circularization radius by a factor 16, and the viscous timescale by a factor 64.

This calculation treats the specific case of the Lense-Thirring effect on a small-scale accretion disc. However breaking and tearing are quite generic features of all accretion flows in a nonspherical potentials. The effects are much stronger if the accreting object is an SMBH binary (as might result from a galaxy merger): the effective potential is rather similar to that for LT precession but much stronger. (It differs qualitatively in that—unlike for the LT effect—prograde gas orbits exchange angular momentum with the accretor through resonances, while retrograde ones do not. Figures 8, 9 shows the results of a recent calculation of tearing in this case (Nixon et al. 2013).

On a still larger scale, the effective potential of the centre of a galaxy bulge almost certainly has similar properties. It seems very likely that the gas infall rate may be drastically increased in this case also. If so, this may offer a clue to how SMBH feeding works.

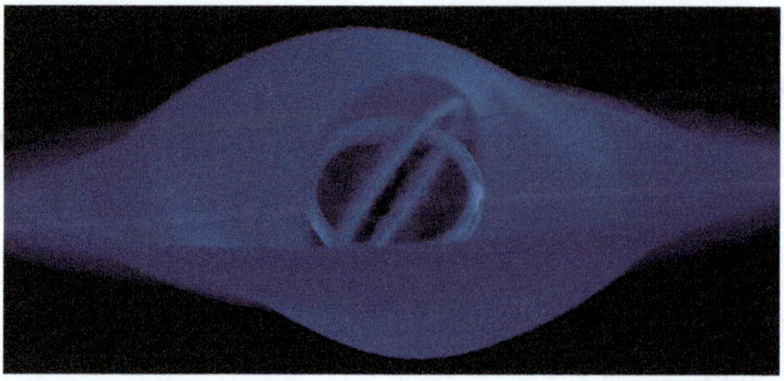

Fig. 6 Disc tearing. Gas in an accretion disc inclined to the spin of a central black hole breaks into rings which precess independently and eventually become counterrotating. This allows angular momentum to cancel and promotes rapid (dynamical) infall (see Fig. 7). Similar effects on much larger scales can occur if the gravitational potential is not spherically symmetric and gas orbits are suitably inclined. This may allow rapid infall from large scale to small and so alleviate the feeding problem discussed here. (From Nixon et al. 2012)

Fig. 7 Effect of disc tearing: accretion rates with time for simulations at *small* (*black*) and *large* (*red*) disc inclination to the black hole spin. The accretion rate is in arbitrary units. The time unit is chosen so that the dynamical time at the inner edge of the discs is ~ 350. The plot shows the mass flow rate through $R = 50$ Rg and not the final accretion rate on to the black hole. (From Nixon et al. 2012)

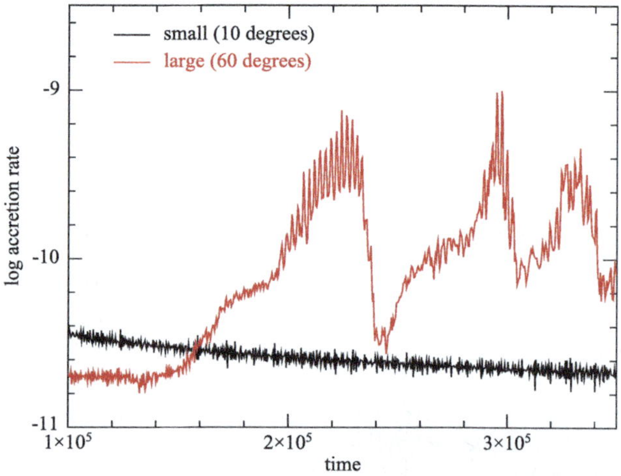

10 Discussion

We have shown that simple quasispherical near-Eddington winds expelled from supermassive black hole accretion discs explain many aspects of the SMBH-galaxy connection. First, the wind Eddington factors should be in the range ~ 1–30. Eddington factors ~ 1 predict black hole winds with velocities $v \sim 0.1c$, observable in X-rays, just as seen in UFOs. For higher Eddington factors this picture predicts slower winds, observable in the UV, as plausibly seen in BAL QSOs. The picture predicts that the wind shocks against the host ISM should cool efficiently, and in several UFO systems there is clear and detailed evidence for this. For SMBH masses below a precisely-defined value M_σ all these outflows stall and fall back. Rapid motions of large baryonic masses like this are likely to be very effective in influencing the distribution of dark matter associated with galaxies, for example in removing dark matter cusps (cf. Pontzen and Governato 2012; Garrison-Kimmel et al. 2013).

Once the SMBH mass reaches M_σ, its thrust can expel most of the gas from its vicinity and is unlikely to grow significantly further. This agrees with observations of the M–σ relation. Simple estimates of the maximum stellar bulge mass M_b also show that the SMBH mass is typically about $10^{-3} M_b$, in line with observation.

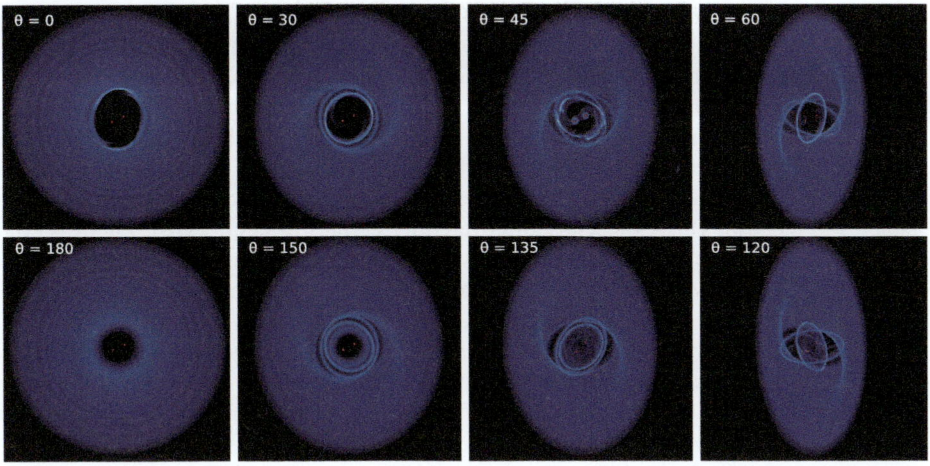

Fig. 8 Tearing of misaligned gas discs around an SMBH binary. Column density plots of the simulations after $\simeq 80$ orbits of the binary, viewed along its axis. The *top row* shows discs prograde with respect to the binary; from *left* to *right*, discs inclined at angles $\theta = 0, 30, 45, 60$ degrees to the binary plane. The *bottom row* shows the corresponding retrograde discs with $\theta = 180, 150, 135, 120$ degrees. Although these snapshots are taken after the same number of binary orbits, this is not the same number of disc precession times, since this depends on the disc tilt. The maximum precession is achieved at $\theta = 45°$ and $135°$. (From Nixon et al. 2013)

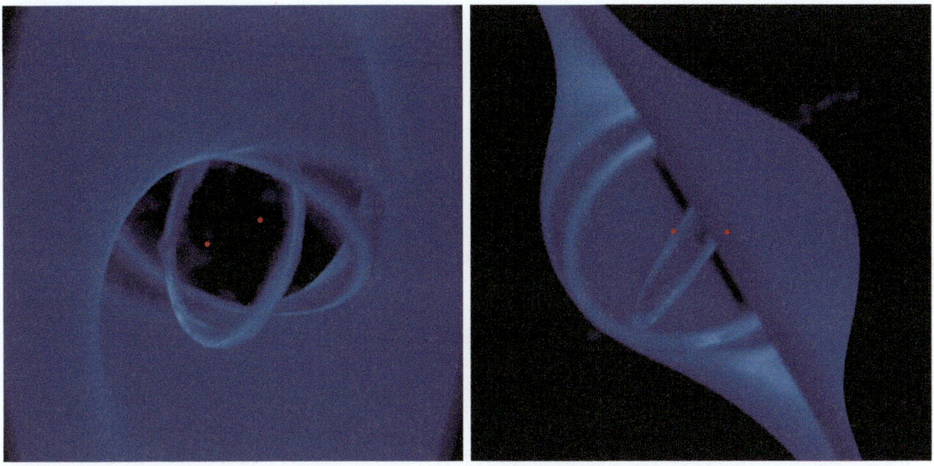

Fig. 9 Tearing of gas discs around an SMBH binary. The 3D structure of the $\theta = 60°$ snapshot from Fig. 8 (From Nixon et al. 2013)

As the SMBH tries to grow above the M–σ mass, the theoretical analysis shows that the wind shock is no longer efficiently cooled, and sweeps up the surrounding interstellar gas much more rapidly. The SMBH outflow becomes energetic and mass-loaded, with speeds $v_{\rm out} \simeq 1200\ {\rm km\,s^{-1}}$ and cool (largely molecular) mass outflow rates $\dot{M}_{\rm out} \sim 4000\ {\rm M_\odot\,yr^{-1}}$, well beyond what starbursts can reasonably drive. Molecular line observations of several galaxies reveal outflows with properties like these. This may be what makes galaxies become red and dead. The same process accounts for the slight differences in the normalization of the M–σ relation between spirals, field ellipticals and cluster ellipticals, particularly BCGs. The outflows may have other important properties in addition to those discussed here. In particular they could be the main agency enriching the intergalactic gas with metals.

In addition to this list, there are a number of other predicted effects which are observable or potentially so. In the energetic molecular outflows discussed in the last Section the wind shock presumably accelerates cosmic ray particles, and gamma rays result when these hit the colder ISM. The outflows are therefore directly comparable with the gamma-ray emitting bubbles in our own Galaxy recently discovered by *Fermi* (Su et al. 2010). Zubovas et al. (2011) interpret these as relics of the Milky Way's last quasar outburst about 6 Myr ago by noting that the greater density of the Galactic plane must pinch a quasi-spherical quasar outflow into a bipolar shape. The gamma-ray emission from the molecular outflow galaxies should be intrinsically stronger than in the Milky Way, but the long integration time required to detect the Galactic bubbles may mean that the outflows are undetectable with current gamma-ray instruments. Perhaps more promisingly, these cosmic ray electrons cool and emit synchrotron radiation in the radio band. This radiation may be observable, so it would be interesting to check whether there are kpc or sub-kpc scale radio bubbles associated with the outflows. A related effect of energy-driven outflows in a galactic bulge is that although the greater pressure of a galaxy disc pinches them into bipolar shapes, the resulting overpressure each side of the disc plane may stimulate starbursts in the disc (Zubovas et al. 2011).

All this suggests that the current theoretical picture of how supermassive black hole influence their host galaxies is promising. But it still lacks the vital ingredient of a proper theory of how the holes actually gain gas. Recent work suggesting how gas may lose specific angular momentum by interacting with a central reservoir may offer a way forward.

Acknowledgements I thank ISSI for its hospitality, and Ken Pounds, Jim Pringle, Kastytis Zubovas, Chris Nixon, Sergei Nayakshin, and Chris Power for comments on this chapter, and for many fruitful interactions on these subjects over the years.

References

D. Batcheldor, Astrophys. J. **711**, L108 (2010)
C.M. Booth, J. Schaye, Mon. Not. R. Astron. Soc. **398**, 53 (2009)
L. Ciotti, J.P. Ostriker, Astrophys. J. **487**, L105 (1997)
S. Collin-Souffrin, A.M. Dumont, Astron. Astrophys. **229**, 292 (1990)
J. Debuhr, E. Quataert, C.-P. Ma, Mon. Not. R. Astron. Soc. **412**, 1341 (2011)
W. Dehnen, D.E. McLaughlin, J. Sachania, Mon. Not. R. Astron. Soc. **369**, 1688 (2006)
K.D. Denney et al., Astrophys. J. **702**, 1353 (2009)
T. Di Matteo, V. Springel, L. Hernquist, Nature **433**, 604 (2005)
T. Di Matteo, J. Colberg, V. Springel, L. Hernquist, D. Sijacki, Astrophys. J. **676**, 33 (2008)
A.C. Fabian, Mon. Not. R. Astron. Soc. **308**, L39 (1999)
C.-A. Faucher-Giguère, E. Quataert, Mon. Not. R. Astron. Soc. **425**, 605 (2012)
L. Ferrarese, D. Merritt, Astrophys. J. **539**, L9 (2000)
C. Feruglio, R. Maiolino, E. Piconcelli et al., Astron. Astrophys. **518**, L155 (2010)
S. Garrison-Kimmel, M. Rocha, M. Boylan-Kolchin, J. Bullock, J. Lally, arXiv:1301.3137 (2013)
K. Gebhardt et al., Astrophys. J. **539**, L13 (2000)
M.G. Haehnelt, P. Natarajan, M.J. Rees, Mon. Not. R. Astron. Soc. **300**, 817 (1998)
N. Häring, H.-W. Rix, Astrophys. J. **604**, L89 (2004)
L. Hernquist, Astrophys. J. **356**, 359 (1990)
P.F. Hopkins, L. Hernquist, Astrophys. J. Suppl. Ser. **166**, 1 (2006)
W. Ishibashi, A.C. Fabian, Mon. Not. R. Astron. Soc. **427**, 2998 (2012)
A.R. King, Astrophys. J. **596**, L27 (2003)
A.R. King, Astrophys. J. **635**, L121 (2005)
A.R. King, Mon. Not. R. Astron. Soc. **402**, 1516 (2010)

A.R. King, K.A. Pounds, Mon. Not. R. Astron. Soc. **345**, 657 (2003)
A.R. King, J.E. Pringle, Mon. Not. R. Astron. Soc. **373**, L90 (2006)
A.R. King, J.E. Pringle, Mon. Not. R. Astron. Soc. **377**, L25 (2007)
A.R. King, J.E. Pringle, J.A. Hofmann, Mon. Not. R. Astron. Soc. **385**, 1621 (2008)
A.R. King, K. Zubovas, C. Power, Mon. Not. R. Astron. Soc. **415**, L6 (2011)
C. Leitherer, C. Robert, L. Drissen, Astrophys. J. **401**, 596 (1992)
G. Lodato, D.J. Price, Mon. Not. R. Astron. Soc. **405**, 1212 (2010)
C.J. Lonsdale, C.J. Lonsdale, H.E. Smith, P.J. Diamond, Astrophys. J. **592**, 804 (2003)
N.J. McConnell, C.-P. Ma, K. Gebhardt, S.A. Wright, J.D. Murphy, T.R. Lauer, J.R. Graham, D.O. Richstone, Nature **480**, 215 (2011)
D.E. McLaughlin, A.R. King, S. Nayakshin, Astrophys. J. **650**, L37 (2006)
R.C. McQuillin, D.E. McLaughlin, Mon. Not. R. Astron. Soc. **423**, 2162 (2012)
N. Murray, E. Quataert, T.A. Thompson, Astrophys. J. **618**, 569 (2005)
J.F. Navarro, C.S. Frenk, S.D.M. White, Astrophys. J. **462**, 563 (1996)
J.F. Navarro, C.S. Frenk, S.D.M. White, Astrophys. J. **490**, 493 (1997)
C. Nixon, A. King, D. Price, J. Frank, Astrophys. J. **757**, L24 (2012)
C. Nixon, A. King, D. Price, Mon. Not. R. Astron. Soc. **434**, 1946 (2013)
G.I. Ogilvie, Mon. Not. R. Astron. Soc. **304**, 557 (1999)
K. Ohsuga, S. Mineshige, Outflow launching mechanisms. Space Sci. Rev. (2013). doi:10.1007/s11214-013-0017-3
A. Pontzen, F. Governato, Mon. Not. R. Astron. Soc. **421**, 3464 (2012)
K.A. Pounds, A.R. King, Mon. Not. R. Astron. Soc. **433**, 1369 (2013)
K.A. Pounds, S. Vaughan, Mon. Not. R. Astron. Soc. **413**, 1251 (2011)
K.A. Pounds, J.N. Reeves, A.R. King, K.L. Page, P.T. O'Brien, M.J.L. Turner, Mon. Not. R. Astron. Soc. **345**, 705 (2003)
C. Power, K. Zubovas, S. Nayakshin, A.R. King, Mon. Not. R. Astron. Soc. **413**, L110 (2011)
J.N. Reeves, P.T. O'Brien, M.J. Ward, Astrophys. J. **593**, L65 (2003)
R.A. Riffel, T. Storchi-Bergmann, Mon. Not. R. Astron. Soc. **411**, 469 (2011a)
R.A. Riffel, T. Storchi-Bergmann, Mon. Not. R. Astron. Soc. **417**, 2752 (2011b)
D.S.N. Rupke, S. Veilleux, Astrophys. J. **729**, L27+ (2011)
N.I. Shakura, R.A. Sunyaev, Astron. Astrophys. **24**, 337 (1973)
J. Silk, A. Nusser, Astrophys. J. **725**, 556 (2010)
J. Silk, M.J. Rees, Astron. Astrophys. **331**, L1 (1998)
A. Soltan, Mon. Not. R. Astron. Soc. **200**, 115 (1982)
D.N. Spergel et al., Astrophys. J. Suppl. Ser. **148**, 175 (2003)
E. Sturm, E. González-Alfonso, S. Veilleux et al., Astrophys. J. **733**, L16+ (2011)
M. Su, T.R. Slatyer, D.P. Finkbeiner, Astrophys. J. **724**, 1044 (2010)
L.J. Tacconi, R. Genzel, D. Lutz, D. Rigopoulou, A.J. Baker, C. Iserlohe, M. Tecza, Astrophys. J. **580**, 73 (2002)
F. Tombesi, M. Cappi, J.N. Reeves, G.G.C. Palumbo, T. Yaqoob, V. Braito, M. Dadina, Astron. Astrophys. **521**, A57+ (2010). arXiv:1006.2858
F. Tombesi, M. Cappi, J.N. Reeves, G.G.C. Palumbo, V. Braito, M. Dadina, Astrophys. J. **742**, 44 (2011)
S. Veilleux, D.S.N. Rupke, D.-C. Kim et al., Astrophys. J. Suppl. Ser. **182**, 628 (2009)
K. Zubovas, A.R. King, Astrophys. J. **745**, L34 (2012a)
K. Zubovas, A.R. King, Mon. Not. R. Astron. Soc. **426**, 2751 (2012b)
K. Zubovas, A.R. King, S. Nayakshin, Mon. Not. R. Astron. Soc. **415**, L21 (2011)

DOI 10.1007/978-1-4939-2227-7_22
Reprinted from *Space Science Reviews* Journal, DOI 10.1007/s11214-014-0072-4

Multi-Wavelength Variability

Accretion and Ejection at the Fastest Timescales

Phil Uttley · Piergiorgio Casella

Received: 25 March 2014 / Accepted: 9 July 2014 / Published online: 5 August 2014

Abstract Multiwavelength variability data, combined with spectral-timing analysis techniques, provides information about the causal relationship between different physical components in accreting black holes. Using fast-timing data and long-term monitoring, we can probe the behaviour of the same components across the black hole mass scale. In this chapter we review the observational status of multiwavelength variability in accreting black holes, from black hole X-ray binaries to AGN, and consider the implications for models of accretion and ejection, primarily considering the evidence for accretion disc and jet variability in these systems. We end with a consideration of future prospects in this quickly-developing field.

Keywords Multi-wavelength · Variability · Black holes

1 Introduction

A wealth of observational results reveal how time-averaged spectra from variable objects, effectively 'wash away' a large amount of relevant information contained in the variability. The emission from the different components of the accretion flow/outflow is now known to change rapidly over a wide range of timescales, as short as minutes in the lowest-mass AGN and milliseconds in the black-hole X-ray binaries (BHXRBs). Resolving such variability down to the shortest possible time-scales is crucial in order to fully disentangle the physical processes causing the emission.

The different emitting components are in fact necessarily inter-connected via the inflowing/outflowing matter itself, at least some fraction of which is likely to be transiting from

P. Uttley
Anton Pannekoek Institute, University of Amsterdam, Science Park 904, 1098 XH Amsterdam, The Netherlands
e-mail: p.uttley@uva.nl

P. Casella (✉)
Osservatorio Astronomico di Roma, INAF, Via Frascati 33, 00040, Monteporzio Catone, Italy
e-mail: piergiorgio.casella@oa-roma.inaf.it

one component to the other. As different physical components emit at different wavelengths, multi-wavelength studies can offer the unique opportunity to observe the mass flow through the different components of the accretion inflow (disc, hot flow/corona) and outflow (jet, wind).

This is especially true for those spectral components whose energy budget is somewhat connected to their internal variability, as is the case for the jet, and those whose origin is still unclear, as is the case of the X-ray power-law component. Different works describe the latter in terms of inverse-Compton emission from hot inflowing electrons (Poutanen, this issue) or from the base of an outflowing, collimated jet (Pe'er, this issue). Despite the many physical differences between these models, the degeneracy between them is still large and currently unresolved. Multi-wavelength variability studies may prove to be key here, as the component(s) emitting the photons in X-rays (the energy range where most of the variability is observed and traditionally studied, see Sect. 2.1 and Belloni and Stella, this issue), could very well (and probably does) emit at longer wavelengths as well. Thus, simultaneous observations over the full energy range of the emission, i.e. both at optical and X-ray wavelengths, might allow us to solve this conundrum.

Furthermore, we can use correlated multi-wavelength variability to search for lags between bands and also study the time-lags within a waveband (e.g. between soft and hard X-rays) to study the *causal* relationship between the different components in the spectrum. Huge observational progress has been made in this area in recent years, both in the measurement of correlations between X-ray and optical/IR emission, in both AGN and X-ray binaries, as well as the detailed study of time-lags within the X-ray band, which has also led to the discovery of X-ray reverberation (see also Reynolds, this issue). We will review this observational progress in this chapter, and also give a flavour of the models to explain the observed behaviour, which are still at an early stage but already show significant promise as a means to understand the physical structure of the emitting regions close to black holes. We also point the reader to the other relevant chapters in this issue (e.g. Pe'er, Poutanen) for a more detailed look at some of the models and associated issues.

1.1 Stellar-Mass vs. Supermassive BHs: Complementary Information

The study of the physics of the accretion onto BHs cannot overlook the issue of the similarities and differences between accretion on to stellar-mass BHs in XRBs and supermassive BHs in AGN. On one hand, it is reasonable to assume that the physics at play is substantially the same. On the other hand, some differences are expected, associated with the different energy densities at work. In fact, the very existence of BHs over such a broad range of masses can be seen as an outstanding opportunity to tackle these studies from two different, complementary perspectives. In AGN, since their black holes are $\sim 10^{5-8}$ times more massive than those in BHXRBs, the intrinsic luminosities (at the same fraction of the Eddington luminosity) are very high, and the characteristic timescales are very long compared to those in BHXRBs, from minutes for the shortest dynamical times, to years or longer for viscous time-scales. Thus, at least for the nearest AGN, the number of photons per characteristic timescale is typically several orders of magnitude higher than in stellar-mass BHs, allowing us a high-precision look at the detailed physics of individual variations on the fastest variability time-scales.

On the other hand, stellar-mass BHs have much shorter characteristic timescales (viscous, dynamical or thermal), so that data can be collected over large numbers of cycles to give an excellent picture of their average variability properties, which is extremely useful when the variability processes are themselves stochastic, as they are for accreting black holes.

We can thus probe—with high precision in just a few hours—the signals associated with viscous evolution of the accretion flow which would be observed over only a few cycles in 10 years of AGN monitoring. Crucially also, in BHXRBs we are able to observe the long-term evolution of the accretion flow through different spectral states, which may well be realised in AGN on much longer (and cosmologically relevant) time-scales (see also Koerding; Fender & Gallo, this issue). Using multiwavelength techniques on observations of the same BHXRB in different states can thus allow us to unlock the changes in the inner structure of accreting black holes and connect them to long-term behaviour such as jet and wind formation: again, a topic of key relevance for the study of supermassive black holes and their feedback into the environment.

1.2 Multi-Wavelength Variability Tools and Techniques

In this article, we will review the state of the art of multi-wavelength variability of accreting BHs. In particular, we will focus on the most recent results obtained with two relatively new methods/strategies as well as a tried-and-trusted (but previously difficult to do) old strategy. On the one hand, the use of novel timing tools, like the covariance spectrum (Wilkinson and Uttley 2009) and the detailed study of lag vs. energy (e.g. Uttley et al. 2011; Kara et al. 2013), together with the large-area and soft-response of satellites like *XMM-Newton*, are allowing us for the first time to identify and study the disc variability driving the variability of the hard X-ray component as well as measuring the signature of reverberation of the variable power-law off of the disc, thus broadening our interpretative approach. These methods simply extend the traditional Fourier cross-spectral techniques to measure correlations and lags (see Nowak et al. 1999 for a useful review) to look at the causal relationship between much finer energy bins relative to a broad reference band, chosen to optimise signal-to-noise.

At the same time, high time resolution observations have started—in the early 2000s—to become available at wavelengths different than the "traditional" X-rays, after the first isolated results obtained with proportional counters a few decades ago (Motch et al. 1983). Dedicated fast optical photometers with good quantum efficiency (e.g. Kanbach et al. 2003; Dhillon et al. 2007) as well as fast modes implemented in existing near-infrared detectors, allow us to tackle old questions from a new perspective. The fastest timescales of the accretion onto stellar-mass BHs are now accessible for studying in the synchrotron emission from both the very same electron population responsible for the X-ray Comptonized emission, and—at longer wavelengths—the electron population outflowing in the relativistic jet.

Finally, the previous decade saw significant improvements in the availability of more traditional long-term monitoring data. Before the launch of the *Rossi X-ray Timing Explorer* (*RXTE*), high-quality X-ray monitoring was near-impossible to achieve, since previous observatories were not set up for the flexible scheduling and fast-slewing to targets which allow efficient monitoring. After the launch of *RXTE*, X-ray monitoring to study the detailed long-term variability of AGN (as well as catching X-ray transients in different stages of their outbursts) became much easier. At the same time, optical monitoring campaigns became easier to obtain, thanks to the work of dedicated groups with access to institutional telescope time, as well as the development of a new generation of smaller—sometimes robotic—telescopes (e.g. the Liverpool Telescope and the SMARTS telescopes), which are ideal for carrying out photometric monitoring if AGN. Finally, the launch of *Swift* satellite has allowed combined X-ray and optical/UV monitoring from space, adding another string to the multiwavelength-monitoring bow.

2 X-ray Power-Law Variability

The X-ray power-law continuum component is the best-studied with spectral-timing techniques, but arguably still the least understood, given the uncertainties in the physical origin of this component. In this section, we will discuss what is known about the spectral-timing behaviour of the X-ray power-law and then describe the attempts to model and understand this behaviour. We shall focus throughout this chapter on the broadband noise variability, rather than quasi-periodic behaviour, since more substantial progress has been made in understanding the spectral-timing behaviour of the broadband noise, and clearer connections can be made between BHXRBs and AGN, which only show one clear example of a QPO to date (Gierliński et al. 2008). However, we note here that the QPO variability seems to be primarily associated with the power-law component and may be produced by a distinct physical mechanism from that producing the broadband noise, as discussed by Belloni & Stella (this issue).

2.1 The Most Variable Spectral Component?

The X-ray band shows some of the most rapid, largest-amplitude variability of all wavebands from accreting objects, including both black hole X-ray binaries and AGN, although the variability time-scales themselves need to be corrected for the black hole mass (see Koerding, this issue). For many years, it has been realised that the power-law emission dominates this variability.

In AGN, it is clear that the X-ray power-law is the dominant emission component in the X-ray band, since with the exception of a few low-mass, high accretion rate AGN, the disc should emit primarily in the optical/UV. Various spectral variability techniques, such as time-resolved spectroscopy (Markowitz et al. 2003), simple flux-flux plots (Taylor et al. 2003; Ponti et al. 2006) as well as rms-spectra (Vaughan and Fabian 2004; Papadakis et al. 2007; Arévalo et al. 2008b), show that flux variability in AGN is dominated by changes in normalisation of the power-law component, sometimes with no corresponding change in power-law spectral shape. There is some evidence for variable absorption driving some of the observed X-ray variability in AGN, most notably on short, sub-day time-scales in the 'eclipses' seen in the now-famous case of NGC 1365 (e.g. Risaliti et al. 2005), but also on longer time-scales (weeks to months) in other AGN (e.g. Lamer et al. 2003; Rivers et al. 2011). These absorption variations are generally interpreted in terms of the observer's line of sight passing through the so-called Broad Line Region (BLR) which produces the broad optical emission lines seen in type 1 AGN, or the inner edge of the dusty torus just beyond the BLR. The BLR can act as a variable partial covering screen or more appropriately a 'mist', since the BLR clouds may span a range of size scales and may have a smaller physical size than the central X-ray emitting region.

There are also claims that the larger-amplitude X-ray variability and 'harder-when-fainter' X-ray spectral changes observed in a large number of AGN are driven at least partly by absorption by a combination of variable partial covering and reflection, from material as close as 100 gravitational radii (light-minutes-hours) from the black hole (Legg et al. 2012; Tatum et al. 2013). This possibility is arguably most convincing in sources which show the most extreme spectral variability (Miller et al. 2009; Lobban et al. 2011), although this is hotly debated even in these cases, with much of the evidence centering on the interpretation of the X-ray time delays seen in AGN (e.g. see Miller et al. 2010; Zoghbi et al. 2011 and Reynolds, this issue). However, given the ubiquity of AGN variability even where there is little spectral evidence for variable absorption, the similarities between AGN and BHXRB variability properties and the correlations between the optical and X-ray continuum variability, which we outline below, it seems likely that the variability in most AGN is primarily intrinsic and not driven by absorption. We will therefore assume that this is the case in the remainder of this section.

In BHXRBs, the X-ray band contains emission from both the disc blackbody and the power-law, although in the hard states the disc emission can only be studied with CCD detectors that extend spectral coverage to energies <2 keV (Miller et al. 2006; Wilkinson and Uttley 2009). For this reason, the study of spectral-timing in BHXRBs, pioneered by satellites with proportional-counter detectors such *Ginga* and later *RXTE*, has necessarily been biased towards studying the power-law in the hard states (e.g. Nowak et al. 1999; Revnivtsev et al. 2001). In the softer states on the other hand, both the power-law and disc components can be studied and there it is clear that the power-law dominates the variability and disc emission seems to be rather stable. For example, rms-spectra of the soft state of the high mass BHXRB Cyg X-1 show clearly that the power-law varies but not the disc emission (Churazov et al. 2001). This effect could help to explain why the broadband fractional rms of BHXRBs decreases as they get softer (e.g. Belloni et al. 2005): the weakly-varying/constant disc component becomes stronger and this dilutes the intrinsic power-law variability amplitude.

2.2 Time-Lags

The first Fourier cross-spectral study of the time lags in accreting black holes (Miyamoto et al. 1988) showed that the lags in hard state BHXRBs were hard (i.e. variations of hard photons lag those in softer photons) and also dependent on Fourier frequency, i.e. variability time-scale. Specifically, the phase lag increases weakly with frequency, approximately as $\phi \propto \nu^{0.3}$, or equivalently, *time lag* scales as $\tau \propto \nu^{-0.7}$ (Fig. 1, top-left panel). Typical hard-state hard lags are around 1 per cent of the sampled variability time-scale (e.g. between 3–4 keV and 8–14 keV, Nowak et al. 1999), so the effect is relatively small, but despite that, the time lags at the lowest well-sampled ($\sim$mHz) frequencies are in excess of thousands of the light-crossing time for one gravitational radius. The lag between two energies scales roughly linearly with the logarithmic energy separation (Nowak et al. 1999), although the best measurements to date of the lag-energy dependence of the power-law component indicate a more complex shape than a simple log-linear law, notably with 'wiggles' seen around the energy of the iron Kα line (Kotov et al. 2001, and see Fig. 1, right panels).

The above-noted properties are well-known, but another less-often noted but important feature of the hard state lags is that, although they follow an approximate power-law dependence on Fourier-frequency when measured over several decades in frequency, the best measurements indicate that they show a step-like structure, with the time-lags appearing to be relatively constant over broad ranges in frequency before 'stepping' to a lower lag value. As noted by Nowak (2000); Kotov et al. (2001), the steps appear to correspond to the frequencies where the dominant component in the power spectrum switches from one broad Lorentzian to another (see Fig. 1, left panels). In other words, each variability component in the power-spectrum appears to have its own, roughly constant, time-lag, with the time-lag increasing as the time-scale of the component increases.

The hard-state lags are also seen to evolve with the changes in the power spectrum which correlate with X-ray spectral changes. The most systematic work so far has been done for Cyg X-1, with Pottschmidt et al. (2000, 2003) showing that the lag in a fixed frequency band increases as the spectral index steepens (and the frequencies of the broad Lorentzian power spectral components correspondingly increase), with large lags seen during the intermediate

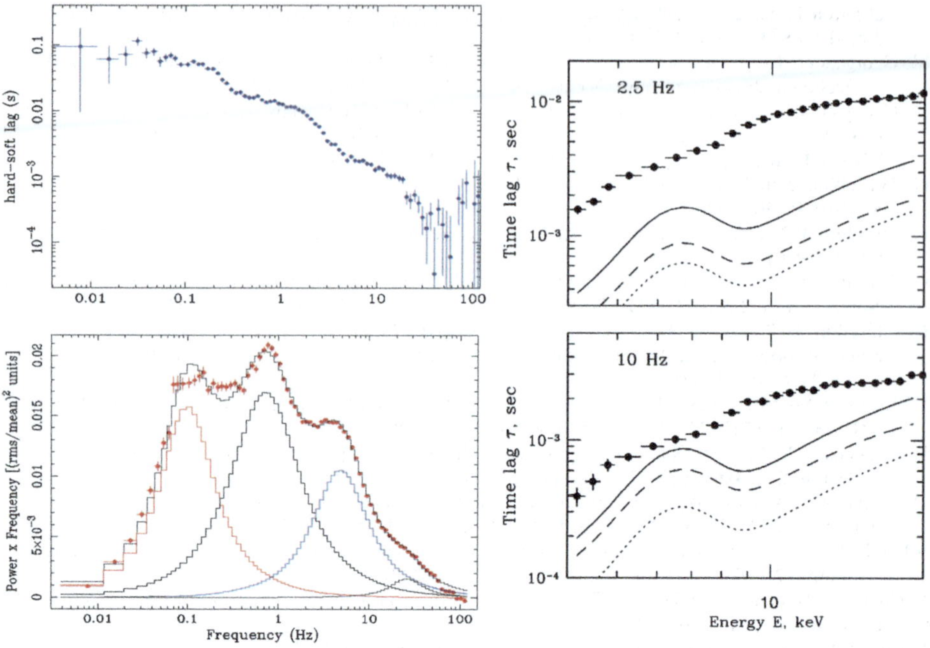

Fig. 1 *Left*: The *top panel* shows the 8–13 keV vs. 2–4 keV time-lag dependence on frequency for Cyg X-1 in the hard state (observation from December 1996). Although the general trend follows $\tau \propto \nu^{-0.7}$, the 'stepped' nature of the lags is very clear. The steps appear to correspond to the transitions between the broad Lorentzian components which approximately fit the hard state power spectrum, shown below. *Right*: Lag (with respect to variations in the 2.8–3.7 keV band) versus energy, measured for variations at 10 Hz and 2.5 Hz for the hard state of Cyg X-1. The figure, taken from Kotov et al. (2001), also shows the expectations of simple reflection models for the lags. Although the lags do not look like they are caused by simple reflection, they also clearly do not follow the simple log-linear law expected from Compton upscattering, and show interesting 'wiggles' around the Fe K features

states and so-called 'failed state transitions'. Interestingly, the soft state of Cyg X-1, where variability is dominated by the power-law spectral component, shows smaller lags, with an amplitude and frequency-dependence similar to the hard state (although without the clear 'steps'). The lags in the soft states of transient low mass BHXRB sources are not well-studied to date, due to their much lower variability amplitudes than in Cyg X-1 (fractional rms $\sim$1 % or less versus >10 % in Cyg X-1), and their power-law spectral components are correspondingly much weaker relative to the disc emission than in Cyg X-1.

Studying the lags in AGN to equivalent (mass-scaled) frequencies as sampled in BHXRBs is not yet possible, since the intermittent sampling that was used to monitor long-time-scale AGN variability is not sufficient to measure lags which are only a small fraction of the observed variability time-scales. However, days-long, intense (quasi)continuous observations can probe the lags on shorter time-scales (equivalent to $\sim$1 second or less in BHXRBs). Fourier-frequency-dependent time-lags in AGN X-ray variability were first observed in data from *RXTE* observations of the Seyfert galaxy NGC 7469 and later in *XMM-Newton* and *ASCA* data for several other AGN (e.g. Vaughan et al. 2003; McHardy et al. 2004; Arévalo et al. 2006).

These results indicated a similar picture to that seen in hard state BHXRBs (and Cyg X-1 in the soft state), with hard time-lags decreasing inversely with variability time-scale (consistent with $\tau \propto \nu^{-0.7}$ but with large uncertainty) and following an approximately log-linear scaling with energy. Ark 564, the only AGN known which shows a doubly-broken, or more likely multi-Lorentzian, power-spectrum,[1] added to the similarity with BHXRBs when combined short and mid-time-scale data revealed a sharp drop to a possible negative (i.e. soft) lag at a frequency consistent with the change-over in power-spectral components (McHardy et al. 2007). Subsequently, following the first highly significant detection in 1H0707-495 (Fabian et al. 2009), soft high-frequency lags have been confirmed in *XMM-Newton* data from many other AGN (e.g. De Marco et al. 2013), and these are now considered to be associated with reverberation from relatively close to the black hole, which we will not discuss further here (but see Reynolds, this issue). The origin of the hard lags observed in AGN at lower Fourier frequencies is still debated, however.

2.3 Models for Longer Time-Scale Lags: Light-Travel Times, or Something Slower?

When measuring time-lags it is implicit that there is a correlation (or anti-correlation for large phase lags) between the different energy bands, i.e. the variability is *spectrally coherent* to some degree. A correlation implies either a direct causal connection between the mechanisms producing the variations in photons in each band, or a causal link of both bands to some underlying variable process. The lag contains causal constraints on these physical situations and the amplitude of the lag is strongly determined by the speed with which signals produce variations in each energy band. The fastest speed which can be considered is light-speed and many models for lags invoke light-travel times as the primary origin of the lag.

Early proposals for the physical origin of lags in BHXRBs focussed on Compton upscattering, where lags are imparted by the travel times of photons in the hot upscattering plasma which produces the power-law emission (Payne 1980; Kazanas and Hua 1999). Higher energy photons have undergone more scatterings and hence spend more time in the scattering medium, lagging behind the lower-energy photons which escape sooner. Since the logarithmic change in energy scales with the number of scatterings (as does the time spent in the upscattering region), this model predicts a log-linear energy dependence in the lags, approximately what is observed. However, the size of the spherical upscattering region would need to be extremely large ($\sim$a light-second across) to explain the large lags seen at low-frequencies, so these models were subsequently discounted on energetics grounds (Nowak et al. 1999). Such an extended hot ($\sim$100 keV) plasma could not be maintained at such scales.

A more plausible Compton upscattering model for lags was proposed by Reig et al. (2003) and subsequently developed by Giannios et al. (2004); Kylafis et al. (2008). The basic mechanism (light-travel times associated with multiple scatterings) is the same as for spherical scattering regions, but the energetics problem is avoided by positing that the upscattering takes place in the jet, which is inferred to extend to large scales, based on the flat-spectrum radio emission seen in the hard state. Seed photons can be trapped in the jet by being preferentially upscattered in the forward direction, so that they can upscatter several times before leaving the jet, thus imparting the energy-dependent lags. One difficulty with this model is that one should expect a significant fraction of the X-ray continuum emission to be produced hundreds of gravitational radii from the disc, so that the illumination pattern should be relatively flat compared to the disc-like emissivity or steeper profiles inferred from spectral fitting of the Fe Kα line (e.g. see Reynolds, this issue).

[1] Other AGNs show power spectra consistent with singly-broken or bending power-laws, similar to those seen in BHXRB soft states, e.g. McHardy et al. (2004).

The fundamental difficulty with models which invoke light-travel time to explain the low-frequency hard lags in BHXRBs is that the inferred sizes of the emitting regions are too large to be physically plausible and/or consistent with other aspects of the data. To address this problem, other types of slower signal were considered, most likely associated with the accretion flow itself. One possibility is the cylindrical propagation of waves or other fluctuations through a corona or hot flow with a temperature that increases with decreasing radius, to produce hard lags (Nowak et al. 1999). A likely scenario seems to be that the lags are associated with the propagation of mass accretion fluctuations (Kotov et al. 2001; Arévalo et al. 2006), since the same scenario can explain many other timing properties, including the linear rms-flux relation observed in accreting sources (Uttley and McHardy 2001; Uttley et al. 2005). Thus the large lags could be associated with the relatively slow, viscous propagation speeds of signals, meaning that the signals themselves may originate from relatively small regions, within 100 R_G of the black hole.

Although the propagating accretion fluctuation model can qualitatively explain the large lags observed at low frequencies in BHXRBs, the situation is less clear for AGN. Arguably the AGN lags, being of similar relative magnitude and time-scale dependence as the BHXRBs, could plausibly share the same origin as the BHXRB lags. However, even if that is true, AGN lag measurements do not yet extend to long enough time-scales to effectively rule out light-travel time origins for the lags. Thus the origin of the hard lags in AGN is currently a matter of debate, with light-travel time associated with reflection being the major alternative to a slower signal propagation effect (e.g. see Zoghbi et al. 2011; Legg et al. 2012). It is worth noting that evidence is currently emerging that the reflection signatures expected in the lag vs. energy spectrum are only seen in the high-frequency lags, not at lower frequencies (Kara et al. 2013), and furthermore, there is evidence that even AGN with apparently weak reflection also show hard lags (Walton et al. 2013). Thus, it seems that the low-frequency hard lags do not show any evidence that they are associated with reflection. The same is certainly true for BHXRBs, where the evidence against reflection as the origin of hard low-frequency lags is even more compelling (Cassatella et al. 2012b).

3 Disc Blackbody Variability

3.1 AGN

The optical continuum in AGN has been known to be significantly variable almost since the discovery of these enigmatic objects. This variability can be clearly associated with the AGN accretion disc, which is the most likely physical origin of the 'big blue bump' component responsible for the optical continuum (Koratkar and Blaes 1999). Studies of the optical variability in different continuum filters showed that variations from the *R* band through to the UV are very well correlated, with only short *red* lags observed, so that longer wavelengths lag shorter wavelengths by up to a day (e.g. Wanders et al. 1997; Sergeev et al. 2005). This behaviour posed a puzzle, because the emission from different wavebands should originate primarily at different radii in a multi-temperature accretion disc. For intrinsic disc fluctuations to propagate between these regions (either on a viscous or thermal time-scale) should take much longer than the observed sub-day lag times. Furthermore, one might expect that fluctuations should propagate from the outside in, with accretion variations propagating inwards with the accretion flow, from cooler to hotter parts of the disc. This effect should produce blue lags rather than the observed red lags.

The likely solution to the puzzle of short optical lags in AGN is to suppose that the optical variability is driven by heating by the central, variable X-ray source (Krolik et al. 1991).

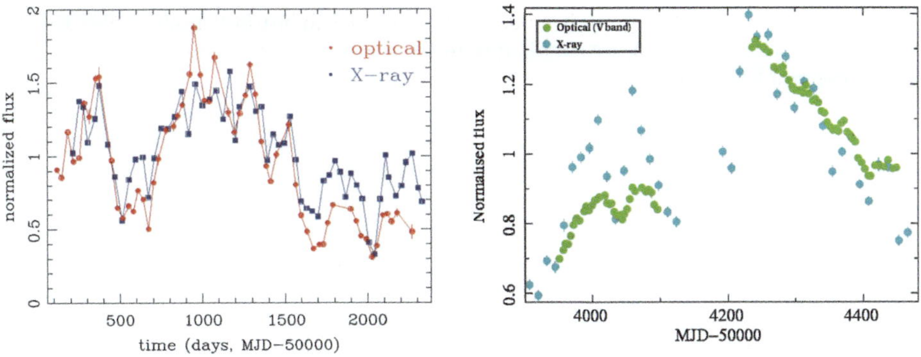

Fig. 2 *Left*: Correlated optical continuum (5100 Å) and 2–10 keV X-ray light curves of the Seyfert galaxy NGC 5548 (adapted from Uttley et al. 2003). The optical light curve has been corrected for the constant host galaxy optical light. The larger optical variability amplitude implies that long-term optical variations are due to intrinsic disc variability and not driven by X-ray heating of the disc. This result is strengthened when considering the luminosities in the optical/UV and X-ray components. *Right*: Correlated optical and X-ray variability in the quasar MR 2251-178 (adapted from Arévalo et al. 2008a). The large long-term optical variation also implies intrinsic disc variability. However, the picture is complicated by the small but still correlated optical variations on short time-scales. These short-term correlated variations with short lags (<10 days) can be explained if X-ray heating drives the short-term optical variability from the disc

The lags could then be simply interpreted as being due to the differential light-travel time between the emitting regions, with redder bands being emitted in the lower-temperature parts of the disc further away from the central X-ray source, hence short, red lags are observed. The detailed lag versus wavelength dependence (both amplitude and functional form) is consistent with this picture, assuming a standard disc temperature profile (Cackett et al. 2007).

The success of the X-ray heating model for explaining AGN optical lags led to an assumption that all AGN optical variability could be explained in this way. However, a further breakthrough was made once the long-term *RXTE* X-ray monitoring programmes used to study Seyfert galaxy power spectra were combined with simultaneous optical monitoring. After an early mixed picture for several AGN (Nandra et al. 2000; Peterson et al. 2000; Maoz et al. 2002), it became clear that most Seyfert galaxies show a significant correlation between X-ray and optical variations (see Fig. 2, left panel). On short time-scales of days–weeks, the optical variations have significantly smaller amplitude than the corresponding X-ray variations. However, on longer time-scales of months–years, larger amplitude fractional rms variations are seen in the optical compared to X-rays (Uttley et al. 2003; Arévalo et al. 2008b, 2009; Breedt et al. 2009, 2010). This behaviour is difficult to explain if X-ray heating dominates the variability, since some component of the optical light should be associated with intrinsic disc heating, which if it were constant would dilute the fractional amplitude of variations to a lower value than seen in the driving X-ray emission. The natural interpretation is that a significant component of *long term* optical variability is intrinsic to the disc. The fact that this component is also correlated with long-term X-ray variations further suggests that instabilities in the accretion disc may ultimately drive the X-ray variability on those time-scales. Perhaps the strongest argument that large-amplitude AGN optical variability is primarily due to intrinsic disc variability, made by Gaskell (2008), is the fact that the SEDs of many Seyfert galaxies are dominated by the big blue bump, which can be an order of magnitude more luminous than the total luminosity of the X-ray power-law (assuming a standard cut-off energy of $\sim$100 keV). Combined with the large amplitude of optical variability seen on long time-scales, it seems very unlikely on simple energetics grounds that X-ray heating can be responsible for most of the optical variability.[2]

These problems notwithstanding, we still need to explain the short optical lags, and these are best considered by noting that they occur in response to *short-time-scale (days–weeks) X-ray variability*. The response of the disc to rapid X-ray variability dominates the location of the central peak of the cross-correlation function (CCF) which is used to define the lags, rather than the overall CCF shape which is also affected by long time-scale variability. Putting things together, AGN optical variability may be best explained by a composite picture, where long-time-scale, large amplitude variations are produced by intrinsic variations in the accretion disc (perhaps in mass accretion rate), while short-time-scale smaller amplitude variations (which dominate the optical lag measurements) are driven by X-ray heating (see Fig. 2, right panel). The fact that the long-term X-ray variations are also correlated with long-term optical variations could be explained if the X-ray power-law is also driven by the same mass-accretion variations that drive the optical variability. If much of the optical variability is driven by mass-accretion variations propagating through the disc, why do we not see the long, blue lags that we expect from such a model? Arévalo et al. (2008a) address this question by showing how the X-ray heating effect pushes the peak of the CCF to short red lags, while the broad base of the CCF is pushed towards long blue lags by the long-term intrinsic disc variability. Hence the CCF is asymmetric but still peaks sharply and close to zero, as observed. The data obtained so far are not yet capable of detecting the asymmetry on long time-scales however.

3.2 BHXRBs

As noted above, in the softer states which show SEDs increasingly dominated by disc emission, variability is weak and when significant variability is seen it seems to be dominated by the power-law emission. These results led to the scenario proposed by Churazov et al. (2001) where the standard, optically thick disc is intrinsically stable and variability seen in the power-law component (including in the hard states) is produced by an unstable, hot, optically thin flow which occurs just inside the disc truncation radius. The outer and inner radii of this hot flow were suggested as the origins of the low and high-frequency breaks (or equivalently, broad Lorentzians) seen in hard state power spectra.

One outstanding—and in hindsight, crucial—issue with this picture was that it was based on timing results from proportional counter instruments, notably on *RXTE*, which are not sensitive to photons below 2 keV. Therefore, the actual variability of the cool ($kT < 0.5$ keV) discs in the hard state could not be studied. This problem was solved relatively recently, with the first X-ray timing studies of the hard state which extend to soft X-rays, pioneered with the EPIC-pn instrument on board *XMM-Newton*. These studies made use of advanced spectral-timing techniques, namely the covariance and lag-energy spectra, to show that, not only is the disc substantially variable in the hard state (in fact it can be more variable than the power-law) on time-scales of a second or longer (Wilkinson and Uttley 2009), but the disc variations *precede* power-law variations on these time-scales (Uttley et al. 2011). Variations of the power-law lag the correlated variations of the disc by a tenth of a second or even longer (see Fig. 3).

The hard state disc variability observed by *XMM-Newton* has a number of important implications for the origin of the variability as well as the interpretation of spectral-timing data.

[2]Note that significant X-ray beaming towards the disc can be ruled out by the observed X-ray reflection component which is not stronger than expected, in most cases.

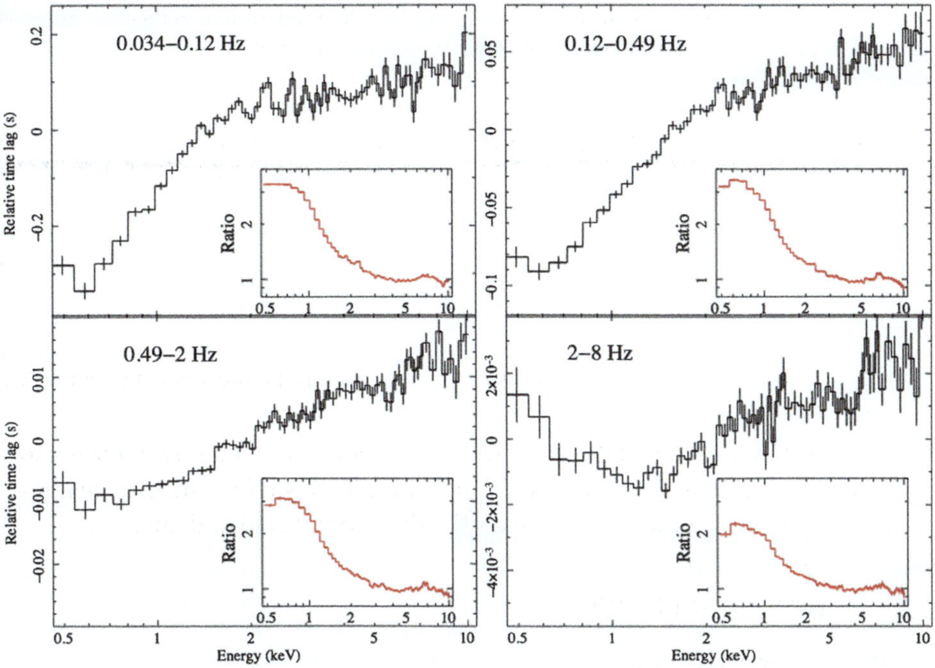

Fig. 3 Comparison of lag vs. energy for four different Fourier frequency ranges of the X-ray variations of the hard state BHXRB GX 339-4. The *insets* show the covariance spectra plotted as a ratio to a Galactic-absorbed power-law, clearly showing as a soft excess the variable disc emission first revealed by Wilkinson and Uttley (2009). The disc photons clearly precede the power-law photons in the lag-energy spectra, by up to tenths of a second at the lowest frequencies. As the disc variability becomes weaker above 1 Hz, the disc photon lead changes to a short lag, which can be easily explained by the increasing dominance of X-ray heating and associated thermal reverberation lags on these short time-scales. Figure taken from Uttley et al. (2011)

Firstly, the obvious implication is that in the hard state, variability (at least on time-scales of seconds or longer) is driven by intrinsic accretion variations in the disc, just as long-term variability appears to be driven by intrinsic disc variations in AGN. Secondly, the observed lag of the power-law with respect to the disc sets strong and probably fatal constraints on models where the power-law lags are due to Compton scattering, e.g. in the jet. This is because the disc is almost certainly the source of seed photons, and in these models the large power-law to disc lags would then imply a distance of the scattering region from the disc of tenths of a light-second, i.e. thousands of gravitational radii. If the X-ray power-law is located such a large distance from the disc, it would surely produce very different reflection spectra than observed. Furthermore, even if the power-law lags are not due to light-travel times in an upscattering region, the upscattering region cannot itself subtend much of a solid angle as seen from the disc (e.g. the disc cannot be embedded in a corona), or the large lags of the power-law relative to the disc would be washed out since disc photons would be rapidly upscattered to power-law photons. The implication is that any upscattering region that produces the power-law must be compact and central, as seen from the disc.

A further result from the hard state disc variability studies is that on short time-scales, <1 s, the direction of disc lags appears to switch so that disc photons start to lag behind power-law variations, with lags of around 1 to 2 ms (see Fig. 3). The natural interpretation of this result is that we are seeing the same X-ray heating of the disc by the power-law as inferred on short time-scales in AGN. The implied distances between the disc and the power-law emitting region are only a few tens of gravitational radii or even less. Again, the power-law emitting region appears to be compact, and if it is related to the jet, it must be the base of the jet.

3.3 Towards a Unified Picture

Discs appear to be inherently unstable in BHXRB hard states, but they are clearly stable in soft states. Does the stabilisation begin within the hard state (perhaps associated with the increase in luminosity prior to the state transition), or does it occur around the state transition? The harder (>2 keV) X-ray variability amplitude also starts to drop continuously during the hard state and then across the state transition, with only a temporary drop when type B QPOs are observed (Belloni et al. 2005). Since disc variations drive the harder X-ray variability (at least the lower-frequency component, Uttley et al. 2011), it seems likely that the stabilisation of disc variability is not strictly linked to the state transition itself but occurs continuously with the overall spectral softening of the source which begins in the hard state, perhaps linked to rises in overall accretion rate that trigger the state transition. Recent work looking at disc variability in *XMM-Newton* data from hard states of SWIFT J1753.5-0127 strongly supports this picture, showing that the disc is less variable in a softer, higher-luminosity hard state than in a harder, lower-luminosity harder state (Cassatella et al. 2012a). Data from the *Swift* observatory, which enabled monitoring of the soft and hard X-ray variability of SWIFT J1753.5-0127 throughout a mini-outburst, also supports this picture (Kalamkar et al. 2013).

The work on disc variability in BHXRBs raises a further question over the very similar patterns seen in AGN variability. The Seyfert AGN studied with optical/X-ray monitoring mostly show disc-dominated SEDs, i.e. they appear like soft states and not hard states, and yet their discs are clearly variable. Why do soft-state-like AGN show disc variability while soft state BHXRBs don't? Perhaps the instability is somehow linked to the radiation pressure instability in discs, since AGN discs should be radiation-pressure dominated while BHXRB discs are more likely to be gas-pressure dominated. If this is the case, it remains puzzling that soft-state BHXRB discs—which should be *more* radiation-pressure dominated—appear to be more stable than hard state discs.

A further question remains over the origin of the higher-frequency variability, seen on sub-second time-scales in BHXRBs and on time-scales of days or less in AGN, where reprocessing of higher-energy photons seems to dominate the disc variability. The simplest interpretation is that this variability is intrinsic to the corona, hot inner flow, or base of the jet—whichever region produces the power-law emission. However, this need not be the case: mass accretion variations at higher Fourier frequencies in the disc could be filtered by the more extended disc emissivity profile to suppress the variability of the direct disc emission at these frequencies (Uttley et al. 2011; Cassatella et al. 2012a). It remains an important open question as to whether or not the disc is responsible for driving all the non-jet continuum variability, or only that on longer time-scales.

4 Synchrotron Variability

Synchrotron emission from jets has been known as a key emission process in AGN almost since their original discovery in the radio band. A wealth of observational efforts over the last couple of decades, have shown that jets are almost ubiquitous in accreting stellar mass

BHs as well. In BHXRBs, jets are always present throughout the hard state, while their emission is observed to be highly quenched in the soft state (Russell et al. 2011). As we have discussed in the previous Sections, the accretion flow is highly variable, especially in the hard state, where the properties of such variability are also observed to tightly correlate with the jet radio emission (Migliari et al. 2005). Thus, it is perhaps natural to ask if and how such variability is transferred into the jet. The observed jet quenching in the soft state, in which very little variability is observed in the flow, unavoidably broadens the question to the role played by the variability in triggering, powering and influencing the jet itself.

In this section, we shall review the observational evidences for jet variability, with an eye on the ongoing theoretical efforts and on the future perspectives.

4.1 AGN: Jet Variability, a Difficult Task

The study of jet variability in AGN has been, historically, mostly limited to the highly relativistically-beamed sources such as blazars. The variability observed in the beamed objects is almost certainly telling us about variability processes produced in the jet, such as shocks which together with the beaming effects can explain the relatively rapid variations seen, especially at very high energies. Since our main focus is on the multiwavelength connections between distinct physical components, we will not cover here the literature on blazar variability which is entirely a jet phenomenon and would require a dedicated Chapter by itself. However, we will see later how models used to explain blazar variability, in conjunction with accretion-driven variations, can yield useful new ideas for the mechanism that drives the jet emission and variability in BHXRBs.

In non-blazar AGN we have an opportunity to study separately the variations from the jet and other continuum components by studying correlations between variations in other wavebands and the radio emission from the jet. However, due to the relatively faint nature of non-beamed objects, radio monitoring has been difficult to arrange due to the limited sensitivity of most radio telescopes: this is a particular problem for the weak radio-quiet objects, leading to only a small amount of literature in this area (e.g. Mundell et al. 2009). None-the-less, combined X-ray and radio monitoring campaigns lasting several years were made possible with the observing flexibility of *RXTE*. Bell et al. (2011) reported tentative evidence for an X-ray/radio correlation in the low-luminosity AGN (and possible hard-state analogue) NGC 7213, with 8.4 GHz radio lagging the X-rays by a few weeks (and 4.8 GHz radio lagging by two weeks longer still), assuming that the correlation is real. Such a lag is consistent with a relatively compact jet in this AGN, which has a black hole mass around $10^8\ M_\odot$. Although NGC 7213 showed significant radio variability and some evidence for an X-ray/radio correlation, the lower mass and higher accretion rate radio-quiet Narrow Line Seyfert 1 AGN, NGC 4051 shows only marginal evidence for radio variability with $\sim$25 % fractional rms, with no clear evidence for a correlation with X-rays which are highly variable (Jones et al. 2011). However, NGC 4051 does show perhaps the clearest evidence yet for a highly-collimated jet in a radio-quiet AGN, in the form of VLBI images showing a pair of unresolved steep-spectrum radio 'hotspots' equidistant from and aligned with the weak, central flat-spectrum radio core (Giroletti and Panessa 2009; Jones et al. 2011). These findings are important because they strongly suggest that weak jets are present in a type of AGN which is considered to be an analogue of soft state BHXRBs (e.g. see Koerding, this issue).

4.2 BHXRBs: OIR Reveals Non-thermal Variability

The data on radio/X-ray correlations (see Körding, this issue) have provided strong evidence for a direct connection between the accretion flow and jets. Past attempts to correlate rapid radio and X-ray variability have confirmed that correlations can be found over different timescales, but that there is little variability in the radio band on timescales shorter than several minutes (e.g. Gleissner et al. 2004; Nipoti et al. 2005; Wilms et al. 2007). This is perhaps not surprising, as the observed radio emission typically comes from at least 10^{14} cm from the black hole, a light travel time of tens of minutes away from the central engine. This is because in a conical jet, the radio emission from the parts of the jet closest to the black hole is absorbed, and the synchrotron self-absorption frequency varies as a function of height along the jet (see Pe'er, this issue). Nonetheless, one of the more interesting potential ways to understand the link between the accretion disc and the jet is to probe the variability of the jet and the variability of the disc together, to see how they are coupled. As jets in BHXRBs are known to emit from radio to at least near-infrared (sometimes optical) wavelengths, it is reasonable to look for jet variability at these shorter wavelengths.

As a figure of merit, it is expected that the spatial scale from which the near-infrared emission should come will be approximately 10^4 times smaller than that from which the radio emission comes (i.e. the ratio of wavelengths) giving a light travel time of tens of milliseconds across the IR emitting region. Even with a mildly sub-luminal jet speed, this allows for the likelihood of detecting jet variability on timescales in the 0.1–10 second range, in which most X-ray binaries show most of their variability, without smearing due to light travel-time effects or propagation times along the jet to the region which emits in the IR.

The first hints for non-thermal variable emission at optical wavelengths came from high-speed simultaneous optical/X-ray photometry of the black-hole candidate XTE J1118+480. This source showed a complex, asymmetric cross-correlation function, consisting of a sharp peak corresponding to an optical lag of a few tenths of a second (as it might be expected from reprocessing from the disc), but also an unusual negative value of the cross-correlation function for X-rays lagging behind the optical emission (Kanbach et al. 2001 and see Fig. 4, left panel). A further complication came with the observations of two other sources with the same technique (Gandhi et al. 2008; Durant et al. 2008; Gandhi et al. 2010), which also showed very complex optical variability, albeit extremely different from each other. This left several open questions about the actual physical differences between them, and the correct interpretation of the optical/X-ray correlated variability.

A possible physical explanation for this unusual behaviour came from the 'common reservoir model' (Malzac et al. 2004), in which part of the optical emission comes from the relativistic jet while the X-ray emission comes from a corona above the accretion disc. Both the jet and the corona stochastically tap the same magnetic energy reservoir, which is also filled stochastically. The fraction of the power that goes into the jet is higher than the fraction that goes into the corona. The correlation is then due to the filling of the reservoir. The time lag of the correlation is assumed to be the time scale on which energy is dissipated in the jet. The anti-correlation is due to the fact that when the jet happens to be tapping more than its usual fraction of the reservoir's energy, the energy is drained sufficiently that the X-ray emission will be suppressed. This model was able to reproduce rather accurately the optical/X-ray CCF of XTE J1118+480, albeit with a large number of free parameters.

An alternative explanation came from Veledina et al. (2011, 2013), who suggested that the optical emission is produced by synchrotron emission in a magnetised, extended hot accretion flow (for a detailed review of this model, see Poutanen, this issue). A possible caveat to this otherwise promising model is the need for a very large (hundreds of gravitational radii) outer radius of the hot inflow, which is not compatible with what other observations seem to suggest about the location of the X-ray power-law emission and the disc truncation radius (e.g. Uttley et al. 2011). Such large radii however are needed if one wants to obtain significant near-infrared emission from the inflow, whose emission would otherwise be rapidly quenched at wavelengths longer than ~ultraviolet-optical.

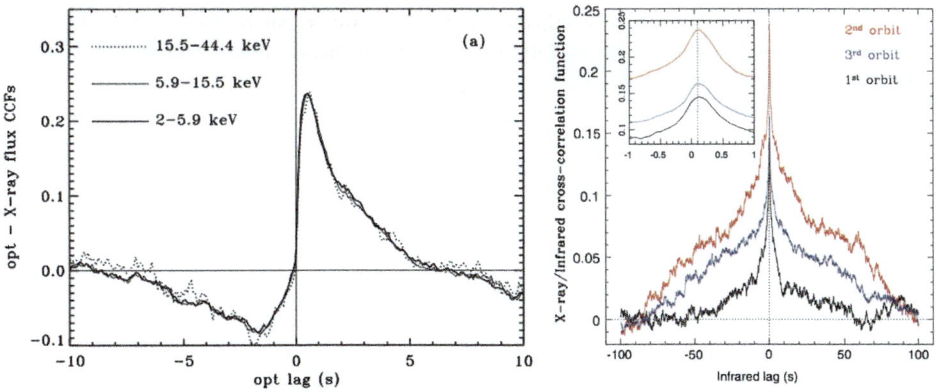

Fig. 4 *Left*: Optical/X-ray cross-correlation functions of XTE J1118+480. Positive values of the lag correspond to optical lagging X-rays. Note the complex asymmetric structure, although rapid optical variations which contribute to the sharp peak are clearly lagging the X-rays by a fraction of a second. Figure taken from Malzac et al. (2004). *Right*: IR/X-ray cross-correlation functions of GX 339-4. The shape is much more symmetric than that seen in the optical for XTE J1118+480, although a significant IR lag of ~0.1 s is still seen, lending strong support to the jet model for the IR variability

Leaving aside the details of each individual model, the scenarios trying to explain the optical variable emission can be divided into two categories, depending on whether such emission is interpreted as coming from the inflow or from the outflow. The two hypotheses differ substantially in the predictions at longer wavelengths: as already mentioned, the inflow is expected to contribute less and less toward longer wavelengths, while the emission from the jet is expected to remain roughly constant—or even to increase (if optically thin)—at wavelengths longer than optical.

These ambiguities were eventually solved by the first fast IR photometry of a BHXRB (Casella et al. 2010). The IR variability appeared to be strongly correlated with the X-rays, with a very small (~0.1 seconds) time delay, and a nearly symmetric CCF (see Fig. 4, right panel). Simple calculations and brightness-temperature arguments ruled out a disc-reprocessing origin for this IR variability. Similarly, the inflow origin could be easily ruled out, as in that case an anti-correlation would be expected, let alone the fact that a synchrotron self-Compton, one-zone scenario would require the infrared emission to lead, not to lag the X-rays. Furthermore, and perhaps conclusively, unrealistically large values of the outer hot inflow radius (i.e., of the disc truncation radius) would be needed in order to obtain sufficient infrared emission to explain the data.

These IR data represented the first unambiguous evidence for sub-second variability in an X-ray binary jet. By making a number of reasonable model-dependent assumptions, the measured time delay between X-rays and infrared photons could be used to estimate either a lower limit of $\Gamma > 2$ for the bulk Lorentz factor of the jet (interpreting the delay in terms of travel time), or a magnetic field intensity of $\sim 10^4$ G at the base of the jet itself (interpreting the delay in terms of cooling time). Furthermore, the Fourier power spectrum of the IR variability showed a clear cutoff at ~1 Hz, which was not visible in the X-ray variability. Interpreting this as a signature of the size of the IR-emitting region in the jet, it was possible to obtain an estimate of $\sim 10^{10}$ cm, consistent with that expected for a mildly relativistic jet in these sources.

These quantitative estimates remain as of today largely model dependent, as they will need larger datasets, e.g. by monitoring a single outburst through its accretion rate evolution, in order to suppress the systematic uncertainties caused by the underlying assumptions. Nevertheless, the qualitative yet robust conclusion that the so-called "steady jet" is all but steady remains, challenging the current long-standing view for a steady radio-flat compact jet in the hard state. Perhaps more importantly, the discovered jet variability represents a new promising tool to track matter (and/or internal shocks) through the jet and provides a new tool to measure the geometry and the physics of the jet itself.

4.3 A Variable Compact Jet

A confirmation of the actual complexity of such a variable jet in BHXRBs came from observations at longer wavelengths. Gandhi et al. (2010) reported on four-band WISE simultaneous observations of the BHXRB GX 339-4 over the 3–22 μm range, revealing large spectral variability of the jet emission on timescales as short as 11 seconds. The 24-hour long observations showed dramatic variations of the slope and normalisation of the spectrum, implying that the spectral break associated with the transition from self-absorbed to optically thin jet synchrotron radiation was varying across the full observed wavelength range. This result suggests that either the magnetic field intensity and/or the size of the acceleration zone above the jet base was being modulated by factors of $\sim$10 on relatively short timescales.

More recently, Corbel et al. (2013) reported on a possibly even more complex jet spectral variation from the same source, as revealed by the Herschel telescope at 70 and 160 μm during the outburst decay, with the far-IR flux lying largely above a simple extrapolation of the radio to near-IR spectrum. This result clearly confirmed that the jet spectral emission is far from being as simple and steady as thought in the past (see also Fender & Gallo and Pe'er, this issue).

A question remains open, as to *how* the variability is transferred from the X-ray emitting regions (either a hot inflow or the base of the jet itself) into and along the jet, i.e. in the form of either density or speed fluctuations, or a combination of both.

The answer to this question holds a rather large potential as a tool to investigate jet physics. In fact, internal shocks arising from bulk Lorentz factor fluctuations linked to the variable inflow were used recently as a key ingredient by two jet models. Based on a model by Spada et al. (2001) for blazar variability (itself based on models for Gamma Ray Burst jet emission), Jamil et al. (2010) first tried to reproduce a $\sim$flat radio-to-infrared jet spectrum in BHXRBs by assuming a white-noise distribution of shell bulk Lorentz factor values. The differential kinetic energy between two colliding shells is assumed to be dissipated into the jet, providing the electron re-heating needed to at least partially balance the energy losses (radiative and/or adiabatic; for a detailed discussion of this issue see Pe'er, this issue). More recently, Malzac (2013) generalized this scenario, testing different power-spectral shapes for the variation of the bulk Lorentz factor, namely letting the power-law index of the power spectrum vary from -1 to 2. These works have shown how dissipation in internal shocks driven by flicker noise fluctuations can indeed balance the energy losses, reproducing the observed jet spectra, including the flux amplitude and the location of the self-absorption break frequency. This scenario clearly also predicts strong multiwavelength variability which, assuming some ad-hoc relation between the observed X-ray variability and the speed fluctuations in the jet, can indeed reproduce some of the observed properties (Malzac, private communication).

These exciting new developments in modelling the jet spectra and variability properties raise further questions about the role of the accretion disc, because as we have discussed in Section 3.2, it appears likely that much of the variation in the X-ray power-law which is correlated with jet variability is actually driven by fluctuations in the blackbody-emitting disc. Thus, an unstable thin disc may be required to explain the variable flat-spectrum jet emission observed in the hard state. We can speculate further that the absence of significant compact jet emission in the soft state may be linked to the absence of variability from the accretion disc in the soft state. That is, without a variable disc there are no variations in bulk Lorentz factor and hence no shocks and consequent reheating to produce emission along the jet. The jet would retain its bulk kinetic energy however, which may lead to observable signatures in the soft states, but further investigation is needed to test this possibility.

5 The Future

A number of multi-wavelength campaigns have shown that variability observed over a broad range of wavelengths and from all physical components is a key ingredient for understanding accretion/ejection physics. Furthermore, advanced spectral-timing techniques are revealing the causal links between the physical components with the X-ray band. The observational picture is becoming more and more clear, with strong evidence being discovered of how the accretion fluctuations propagate through the accretion disc, possibly into the hot inflow and then into and along the jet. Future developments will be driven primarily by advances in technology and new instrumentation, together with advances in application of the relevant techniques and modelling.

5.1 Advances in Modelling

Already a basic phenomenological picture is emerging, with the pieces of the puzzle falling into place. But the physical models are still substantially behind the data. Part of the problem is that we are not yet sure what causes the variability, although it appears very likely to be produced in the accretion flow. However, given the clear phenomenological picture which has emerged in recent years, where we can start to separate out the disc and power-law variations and their associated lags, there is some hope that even simple models for fluctuations propagating through the flow will be useful to fit the data, and that the data will itself put constraints on the models. There is already a strong indication that variations must originate at relatively small radii to produce the observed lags via viscous propagation, and also that the viscosity parameter α may need to be fairly high to explain the observed lags (Arévalo and Uttley 2006; Uttley et al. 2011). Another recent breakthrough has been the discovery of X-ray reverberation lags (see also Reynolds, this issue). These lags are more easily modelled since we are dealing with light-travel time effects, and the observed lags and variability time-scales are too short to be significantly effected by the more complex and uncertain effects of propagation in the flow. Already, the Fe Kα reverberation signatures are being modelled using raytraced, general relativistic transfer functions (Cackett et al. 2013) and there is justifiable optimism that this approach will allow us to confidently map the inner structure of accreting compact objects for the first time. The picture that emerges from reverberation studies can naturally fed back into the other models, as it will reveal the geometry of the disc and power-law components which are crucial parts of the picture.

The approaches to modelling data outlined above are more analytical and 'top-down' in terms of the application of the physics. For example, a source of variability must be assumed to be present in the accretion flow, and possibly injected into the jet, its amplitude and power-spectrum being specified as variables in order to fit the data. In contrast, a 'bottom-up' approach would allow the variability to self-consistently emerge from the physics of the accretion and ejection process. The current best hope for this emergent picture comes from magneto-hydrodynamic (MHD) simulations of accretion flows and jets, which have made substantial advances in the past decade, due to improvements in the techniques along with significant growth in computing power (e.g. see O'Neill et al. 2011; McKinney et al. 2012 for examples). These simulations are now beginning to examine the variability that is a natural outcome of, e.g. magnetic instabilities in these flows, and we expect substantial advances in this area over the coming decade, which could allow the first fully-self-consistent modelling of variable accretion and ejection. As a note of caution however, the coupling of these effects to radiation processes is a challenging task and may present a significant obstacle to modelling the high-accretion rate flows which we observe. Therefore it is likely that the analytical approach will be the primary driver for understanding the observations for some time to come.

5.2 Advances in Observational Capabilities

Over the next decade we expect significant advances in the quality of multiwavelength data on both long and short time-scales due to a number of important developments. Firstly, on time-scales of days or longer, (near)-simultaneous high-cadence monitoring of sources will be possible in both the optical and radio, due to the development of synoptic and monitoring survey telescopes in the optical (culminating with the *Large Synoptic Survey Telescope* (*LSST*|), at the end of the decade), together with *Square Kilometre Array* (*SKA*) pathfinders (e.g. *MeerKAT* and *ASKAP*) and ultimately the *SKA* itself, which all have significant wide-field monitoring capabilities. The situation in X-rays is less clear however. Historically, the development of X-ray All-Sky Monitors has lagged behind that of pointed X-ray telescopes, with the monitoring instruments being seen as secondary 'service' instruments to identify new (and relatively bright) X-ray transient sources. Their sensitivity for monitoring fainter sources, e.g. AGN, is low. To match the quality of monitoring data that will be provided in the optical and radio in the future, dedicated all-sky or wide-field X-ray monitors need to be developed, perhaps even as separate observatories in their own right. Possible candidates include the Wide-Field Monitor instrument (WFM) on board the proposed *Large Observatory for Timing* (*LOFT*), or a sensitive all-sky observatory based on 'Lobster-Eye' optics, but at the time of writing, nothing like this is guaranteed to be flown at the same time as the optical and radio observatories are taking data. If this remains the case, this would be a very unfortunate missed opportunity: the strong connection between X-ray and optical variability shows that X-ray information is crucial to understanding the observed optical behaviour in accreting compact objects (and almost certainly also the jet behaviour in the radio).

Fortunately, the next decade will see significant advances in observations of rapid variability in all wavebands, which will provide revolutionary new data for the study of accretion and ejection XRBs in particular. Because of the relatively extended scale of radio emission from XRB jets, we do not expect to see extremely rapid variability in the radio band, although we will have the observational capability to detect it and we should be prepared for surprises. The most rapid jet variability is expected in the optical/IR and this will be readily detected by a new generation of fast CCD cameras (e.g. HIPERCAM, the successor to ULTRACAM), whose development has been spurred by the study of exoplanetary transits around compact stars, but will naturally lend themselves to fast timing studies of XRBs. Especially important is the parallel growth in the availability of 4–8 m class telescopes, on which to mount such instruments, since photon count rates are the key to sensitivity to rapid variability. Many of these new observatories are queue-scheduled (e.g. the *South-African Large Telescope*, *SALT*) or robotically operated, allowing rapid follow-up of transients. At

Fig. 5 The dependence of signal-to-noise on effective area, for the detection of a 1 R_g/c Fe Kα lag at high-frequencies (100 Hz in BHXRBs, scaled by black hole mass in AGN). A typical range of fluxes is assumed for each source class

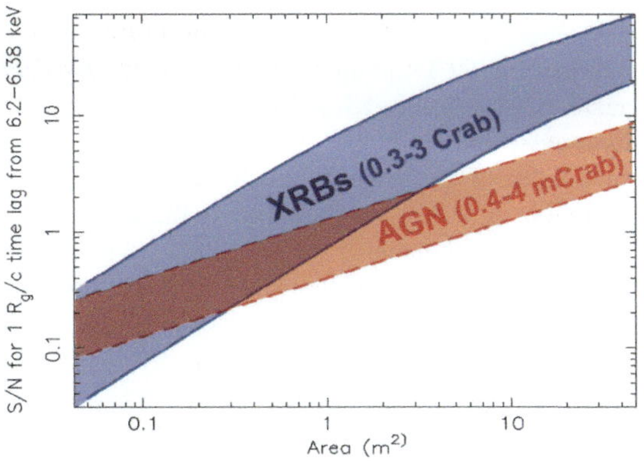

the same time, an important advance in optical/IR detector technology is the development of Microwave Kinetic Inductance Detectors (MKIDs) which are able to measure the energies of individual optical/IR photons without using filters or gratings, obtaining similar spectral resolutions to those possible with X-ray CCD instruments (e.g. the ARCONS camera, Mazin et al. 2013). The deployment of MKIDs on large telescopes will thus allow detailed spectral-timing analysis of rapid variability within in the optical-IR range, similar to the developments that have been pioneered in the X-ray band over the past few years.

In the X-ray band, we anticipate continued operation of *XMM-Newton*, with its fast EPIC-pn timing capability, into the next decade, allowing the continued application of X-ray spectral-timing techniques to new transients, as well as simultaneous observations with the new fast optical/IR instrumentation. For simultaneous X-ray/optical/IR spectral-timing, a limiting factor will be the X-ray count rates. The main limitations of the EPIC-pn instrument when observing bright sources in its fast timing mode are the effects of pileup and deadtime, as well as strong telemetry constraints, meaning the EPIC-pn is effectively limited to observing sources up to $\sim$0.1 Crab, typically corresponding to BHXRBs in the hard state. The upcoming *ASTROSAT* mission will provide *RXTE*-like collecting area and timing capability, together with a larger collecting area that *RXTE* at hard energies (which may prove interesting to study any X-ray power-law component of the jet). Another important development will be the deployment of the *Neutron star Interior Composition Explorer* (*NICER*) telescope on the International Space Station in 2016. Although this instrument's core mission is to study X-ray pulsars, its very fast timing-capability coupled with a soft response and CCD-like spectral resolution will allow the detailed spectral-timing studies of hard-state X-ray binaries pioneered by *XMM-Newton* to be extended to the more luminous states, where key questions of jet and wind-formation can be studied.

In the longer term, the full potential of X-ray spectral-timing can only be realised at the high count rates facilitated by large-area detectors. This point is illustrated in Fig. 5, which shows the signal-to-noise ratio expected for the detection of high-frequency time-lags around the Fe Kα line, as a function of detector effective area at that energy. It is important to note that the sensitivity of lag measurements increases more rapidly for XRBs than for AGN, such that for effective areas above $\sim$1 m^2, XRB measurements of lags (e.g. for reverberation studies) become significantly better than those for AGN. The reason for this is that spectral-timing involves the correlation of two light curves, leading to a term in the error which is due to the multiplication of Poisson noise errors from both light curves. This term dominates the lag-error in the regime where there are few photons per characteristic variability time-scale (i.e. the situation for XRBs), leading to a linear scaling of S/N

with count rate (and hence, area) in this regime. Pushing detector areas to 10 m^2—while maintaining CCD-quality spectral resolution—could completely revolutionise the study of accretion/ejection in X-ray binaries by allowing high-resolution reverberation maps of their innermost regions to be made. The development of large-area silicon-drift detectors puts such detector areas within reach for launch in the 2020s, e.g. the proposed *LOFT* mission (Feroci et al. 2012). Furthermore, in the soft X-ray band the *ATHENA* mission, which is now confirmed for launch in the late 2020s, will allow significant advances in studying lags of disc blackbody components and soft X-ray reverberation in AGN (Nandra et al. 2013).

References

P. Arévalo, P. Uttley, Investigating a fluctuating-accretion model for the spectral-timing properties of accreting black hole systems. Mon. Not. R. Astron. Soc. **367**, 801–814 (2006). doi:10.1111/j.1365-2966.2006.09989.x

P. Arévalo, I.E. Papadakis, P. Uttley, I.M. McHardy, W. Brinkmann, Spectral-timing evidence for a very high state in the narrow-line Seyfert 1 Ark 564. Mon. Not. R. Astron. Soc. **372**, 401–409 (2006). doi:10.1111/j.1365-2966.2006.10871.x

P. Arévalo, P. Uttley, S. Kaspi, E. Breedt, P. Lira, I.M. McHardy, Correlated X-ray/optical variability in the quasar MR2251-178. Mon. Not. R. Astron. Soc. **389**, 1479–1488 (2008a). doi:10.1111/j.1365-2966.2008.13719.x

P. Arévalo, I.M. McHardy, A. Markowitz, I.E. Papadakis, T.J. Turner, L. Miller, J. Reeves, Fourier-resolved energy spectra of the Narrow-Line Seyfert 1 Mkn 766. Mon. Not. R. Astron. Soc. **387**, 279–288 (2008b). doi:10.1111/j.1365-2966.2008.13216.x

P. Arévalo, P. Uttley, P. Lira, E. Breedt, I.M. McHardy, E. Churazov, Correlation and time delays of the X-ray and optical emission of the Seyfert Galaxy NGC 3783. Mon. Not. R. Astron. Soc. **397**, 2004–2014 (2009). doi:10.1111/j.1365-2966.2009.15110.x

M.E. Bell, T. Tzioumis, P. Uttley, R.P. Fender, P. Arévalo, E. Breedt, I. McHardy, D.E. Calvelo, O. Jamil, E. Körding, X-ray and radio variability in the low-luminosity active galactic nucleus NGC 7213. Mon. Not. R. Astron. Soc. **411**, 402–410 (2011). doi:10.1111/j.1365-2966.2010.17692.x

T. Belloni, J. Homan, P. Casella, M. van der Klis, E. Nespoli, W.H.G. Lewin, J.M. Miller, M. Méndez, The evolution of the timing properties of the black-hole transient GX 339-4 during its 2002/2003 outburst. Astron. Astrophys. **440**, 207–222 (2005). doi:10.1051/0004-6361:20042457

E. Breedt, P. Arévalo, I.M. McHardy, P. Uttley, S.G. Sergeev, T. Minezaki, Y. Yoshii, C.M. Gaskell, E.M. Cackett, K. Horne, S. Koshida, Long-term optical and X-ray variability of the Seyfert galaxy Markarian 79. Mon. Not. R. Astron. Soc. **394**, 427–437 (2009). doi:10.1111/j.1365-2966.2008.14302.x

E. Breedt, I.M. McHardy, P. Arévalo, P. Uttley, S.G. Sergeev, T. Minezaki, Y. Yoshii, Y. Sakata, P. Lira, N.G. Chesnok, Twelve years of X-ray and optical variability in the Seyfert galaxy NGC 4051. Mon. Not. R. Astron. Soc. **403**, 605–619 (2010). doi:10.1111/j.1365-2966.2009.16146.x

E.M. Cackett, K. Horne, H. Winkler, Testing thermal reprocessing in active galactic nuclei accretion discs. Mon. Not. R. Astron. Soc. **380**, 669–682 (2007). doi:10.1111/j.1365-2966.2007.12098.x

E.M. Cackett, A. Zoghbi, C. Reynolds, A.C. Fabian, E. Kara, P. Uttley, D.R. Wilkins, Modelling the broad Fe K\alpha\ reverberation in the AGN NGC 4151. ArXiv e-prints (2013)

P. Casella, T.J. Maccarone, K. O'Brien, R.P. Fender, D.M. Russell, M. van der Klis, A. Pe'Er, D. Maitra, D. Altamirano, T. Belloni, G. Kanbach, M. Klein-Wolt, E. Mason, P. Soleri, A. Stefanescu, K. Wiersema, R. Wijnands, Fast infrared variability from a relativistic jet in GX 339-4. Mon. Not. R. Astron. Soc. **404**, 21–25 (2010). doi:10.1111/j.1745-3933.2010.00826.x

P. Cassatella, P. Uttley, T.J. Maccarone, Accretion flow diagnostics with X-ray spectral timing: the hard state of SWIFT J1753.5-0127. Mon. Not. R. Astron. Soc. **427**, 2985–2997 (2012a). doi:10.1111/j.1365-2966.2012.22021.x

P. Cassatella, P. Uttley, J. Wilms, J. Poutanen, Joint spectral-timing modelling of the hard lags in GX 339-4: constraints on reflection models. Mon. Not. R. Astron. Soc. **422**, 2407–2416 (2012b). doi:10.1111/j.1365-2966.2012.20792.x

E. Churazov, M. Gilfanov, M. Revnivtsev, Soft state of Cygnus X-1: stable disc and unstable corona. Mon. Not. R. Astron. Soc. **321**, 759–766 (2001). doi:10.1046/j.1365-8711.2001.04056.x

S. Corbel, H. Aussel, J.W. Broderick, P. Chanial, M. Coriat, A.J. Maury, M.M. Buxton, J.A. Tomsick, A.K. Tzioumis, S. Markoff, J. Rodriguez, C.D. Bailyn, C. Brocksopp, R.P. Fender, P.O. Petrucci, M. Cadolle-Bel, D. Calvelo, L. Harvey-Smith, Formation of the compact jets in the black hole GX 339-4. Mon. Not. R. Astron. Soc. **431**, 107–111 (2013). doi:10.1093/mnrasl/slt018

B. De Marco, G. Ponti, M. Cappi, M. Dadina, P. Uttley, E.M. Cackett, A.C. Fabian, G. Miniutti, Discovery of a relation between black hole mass and soft X-ray time lags in active galactic nuclei. Mon. Not. R. Astron. Soc. (2013). doi:10.1093/mnras/stt339

V.S. Dhillon, T.R. Marsh, M.J. Stevenson, D.C. Atkinson, P. Kerry, P.T. Peacocke, A.J.A. Vick, S.M. Beard, D.J. Ives, D.W. Lunney, S.A. McLay, C.J. Tierney, J. Kelly, S.P. Littlefair, R. Nicholson, R. Pashley, E.T. Harlaftis, K. O'Brien, ULTRACAM: an ultrafast, triple-beam CCD camera for high-speed astrophysics. Mon. Not. R. Astron. Soc. **378**, 825–840 (2007). doi:10.1111/j.1365-2966.2007.11881.x

M. Durant, P. Gandhi, T. Shahbaz, A.P. Fabian, J. Miller, V.S. Dhillon, T.R. Marsh, SWIFT J1753.5-0127: a surprising optical/X-ray cross-correlation function. Astrophys. J. Lett. **682**, 45–48 (2008). doi:10.1086/590906

A.C. Fabian, A. Zoghbi, R.R. Ross, P. Uttley, L.C. Gallo, W.N. Brandt, A.J. Blustin, T. Boller, M.D. Caballero-Garcia, J. Larsson, J.M. Miller, G. Miniutti, G. Ponti, R.C. Reis, C.S. Reynolds, Y. Tanaka, A.J. Young, Broad line emission from iron K- and L-shell transitions in the active galaxy 1H0707-495. Nature **459**, 540–542 (2009). doi:10.1038/nature08007

M. Feroci, L. Stella, M. van der Klis, T.J.-L. Courvoisier, M. Hernanz, R. Hudec, A. Santangelo, D. Walton, A. Zdziarski, D. Barret, T. Belloni, J. Braga, S. Brandt, C. Budtz-Jørgensen, S. Campana, J.-W. den Herder, J. Huovelin, G.L. Israel, M. Pohl, P. Ray, A. Vacchi, S. Zane, A. Argan, P. Attinà, G. Bertuccio, E. Bozzo, R. Campana, D. Chakrabarty, E. Costa, A. De Rosa, E. Del Monte, S. Di Cosimo, I. Donnarumma, Y. Evangelista, D. Haas, P. Jonker, S. Korpela, C. Labanti, P. Malcovati, R. Mignani, F. Muleri, M. Rapisarda, A. Rashevsky, N. Rea, A. Rubini, C. Tenzer, C. Wilson-Hodge, B. Winter, K. Wood, G. Zampa, N. Zampa, M.A. Abramowicz, M.A. Alpar, D. Altamirano, J.M. Alvarez, L. Amati, C. Amoros, L.A. Antonelli, R. Artigue, P. Azzarello, M. Bachetti, G. Baldazzi, M. Barbera, C. Barbieri, S. Basa, A. Baykal, R. Belmont, L. Boirin, V. Bonvicini, L. Burderi, M. Bursa, C. Cabanac, E. Cackett, G.A. Caliandro, P. Casella, S. Chaty, J. Chenevez, M.J. Coe, A. Collura, A. Corongiu, S. Covino, G. Cusumano, F. D'Amico, S. Dall'Osso, D. De Martino, G. De Paris, G. Di Persio, T. Di Salvo, C. Done, M. Dovčiak, A. Drago, U. Ertan, S. Fabiani, M. Falanga, R. Fender, P. Ferrando, D. Della Monica Ferreira, G. Fraser, F. Frontera, F. Fuschino, J.L. Galvez, P. Gandhi, P. Giommi, O. Godet, E. Göğüş, A. Goldwurm, D. Götz, M. Grassi, P. Guttridge, P. Hakala, G. Henri, W. Hermsen, J. Horak, A. Hornstrup, J.J.M. in't Zand, J. Isern, E. Kalemci, G. Kanbach, V. Karas, D. Kataria, T. Kennedy, D. Klochkov, W. Kluźniak, K. Kokkotas, I. Kreykenbohm, J. Krolik, L. Kuiper, I. Kuvvetli, N. Kylafis, J.M. Lattimer, F. Lazzarotto, D. Leahy, F. Lebrun, D. Lin, N. Lund, T. Maccarone, J. Malzac, M. Marisaldi, A. Martindale, M. Mastropietro, J. McClintock, I. McHardy, M. Mendez, S. Mereghetti, M.C. Miller, T. Mineo, E. Morelli, S. Morsink, C. Motch, S. Motta, T. Muñoz-Darias, G. Naletto, V. Neustroev, J. Nevalainen, J.F. Olive, M. Orio, M. Orlandini, P. Orleanski, F. Ozel, L. Pacciani, S. Paltani, I. Papadakis, A. Papitto, A. Patruno, A. Pellizzoni, V. Petráček, J. Petri, P.O. Petrucci, B. Phlips, L. Picolli, A. Possenti, D. Psaltis, D. Rambaud, P. Reig, R. Remillard, J. Rodriguez, P. Romano, M. Romanova, T. Schanz, C. Schmid, A. Segreto, A. Shearer, A. Smith, P.J. Smith, P. Soffitta, N. Stergioulas, M. Stolarski, Z. Stuchlik, A. Tiengo, D. Torres, G. Török, R. Turolla, P. Uttley, S. Vaughan, S. Vercellone, R. Waters, A. Watts, R. Wawrzaszek, N. Webb, J. Wilms, L. Zampieri, A. Zezas, J. Ziolkowski, The large observatory for X-ray timing (LOFT). Exp. Astron. **34**, 415–444 (2012). doi:10.1007/s10686-011-9237-2

P. Gandhi, K. Makishima, M. Durant, A.C. Fabian, V.S. Dhillon, T.R. Marsh, J.M. Miller, T. Shahbaz, H.C. Spruit, Rapid optical and X-ray timing observations of GX 339-4: flux correlations at the onset of a low/hard state. Mon. Not. R. Astron. Soc. **390**, 29–33 (2008). doi:10.1111/j.1745-3933.2008.00529.x

P. Gandhi, V.S. Dhillon, M. Durant, A.C. Fabian, A. Kubota, K. Makishima, J. Malzac, T.R. Marsh, J.M. Miller, T. Shahbaz, H.C. Spruit, P. Casella, Rapid optical and X-ray timing observations of GX339-4: multicomponent optical variability in the low/hard state. Mon. Not. R. Astron. Soc. **407**, 2166–2192 (2010). doi:10.1111/j.1365-2966.2010.17083.x

C.M. Gaskell, Accretion disks and the nature and origin of AGN continuum variability, in *Revista Mexicana de Astronomia y Astrofisica Conference Series*, vol. 32 (2008), pp. 1–11

D. Giannios, N.D. Kylafis, D. Psaltis, Spectra and time variability of Galactic black-hole X-ray sources in the low/hard state. Astron. Astrophys. **425**, 163–169 (2004). doi:10.1051/0004-6361:20041002

M. Gierliński, M. Middleton, M. Ward, C. Done, A periodicity of $\sim$1 hour in X-ray emission from the active galaxy RE J1034+396. Nature **455**, 369–371 (2008). doi:10.1038/nature07277

M. Giroletti, F. Panessa, The faintest Seyfert radio cores revealed by VLBI. Astrophys. J. Lett. **706**, 260–264 (2009). doi:10.1088/0004-637X/706/2/L260

T. Gleissner, J. Wilms, G.G. Pooley, M.A. Nowak, K. Pottschmidt, S. Markoff, S. Heinz, M. Klein-Wolt, R.P. Fender, R. Staubert, Long term variability of Cygnus X-1. III. Radio-X-ray correlations. Astron. Astrophys. **425**, 1061–1068 (2004). doi:10.1051/0004-6361:20040280

O. Jamil, R.P. Fender, C.R. Kaiser, iShocks: X-ray binary jets with an internal shocks model. Mon. Not. R. Astron. Soc. **401**, 394–404 (2010). doi:10.1111/j.1365-2966.2009.15652.x

S. Jones, I. McHardy, D. Moss, N. Seymour, E. Breedt, P. Uttley, E. Körding, V. Tudose, Radio and X-ray variability in the Seyfert galaxy NGC 4051. Mon. Not. R. Astron. Soc. **412**, 2641–2652 (2011). doi:10.1111/j.1365-2966.2010.18105.x

M. Kalamkar, M. van der Klis, P. Uttley, D. Altamirano, R. Wijnands, Swift X-ray telescope timing observations of the black hole binary SWIFT J1753.5-0127: Disk-diluted fluctuations in the outburst peak. Astrophys. J. **766**, 89 (2013). doi:10.1088/0004-637X/766/2/89

G. Kanbach, C. Straubmeier, H.C. Spruit, T. Belloni, Correlated fast X-ray and optical variability in the black-hole candidate XTE J1118+480. Nature **414**, 180–182 (2001)

G. Kanbach, S. Kellner, F.Z. Schrey, H. Steinle, C. Straubmeier, H.C. Spruit, Design and results of the fast timing photo-polarimeter optima, in *Society of Photo-Optical Instrumentation Engineers (SPIE) Conference Series*, ed. by M. Iye, A.F.M. Moorwood. Society of Photo-Optical Instrumentation Engineers (SPIE) Conference Series, vol. 4841 (2003), pp. 82–93. doi:10.1117/12.460323

E. Kara, A.C. Fabian, E.M. Cackett, P. Uttley, D.R. Wilkins, A. Zoghbi, Discovery of high-frequency iron K lags in Ark 564 and Mrk 335. Mon. Not. R. Astron. Soc. **434**, 1129–1137 (2013). doi:10.1093/mnras/stt1055

D. Kazanas, X.-M. Hua, Modeling the time variability of accreting compact sources. Astrophys. J. **519**, 750–761 (1999). doi:10.1086/307379

A. Koratkar, O. Blaes, The ultraviolet and optical continuum emission in active galactic nuclei: the status of accretion disks. Publ. Astron. Soc. Pac. **111**, 1–30 (1999). doi:10.1086/316294

O. Kotov, E. Churazov, M. Gilfanov, On the X-ray time-lags in the black hole candidates. Mon. Not. R. Astron. Soc. **327**, 799–807 (2001). doi:10.1046/j.1365-8711.2001.04769.x

J.H. Krolik, K. Horne, T.R. Kallman, M.A. Malkan, R.A. Edelson, G.A. Kriss, Ultraviolet variability of NGC 5548—Dynamics of the continuum production region and geometry of the broad-line region. Astrophys. J. **371**, 541–562 (1991). doi:10.1086/169918

N.D. Kylafis, I.E. Papadakis, P. Reig, D. Giannios, G.G. Pooley, A jet model for Galactic black-hole X-ray sources: some constraining correlations. Astron. Astrophys. **489**, 481–487 (2008). doi:10.1051/0004-6361:20079159

G. Lamer, P. Uttley, I.M. McHardy, An absorption event in the X-ray light curve of NGC 3227. Mon. Not. R. Astron. Soc. **342**, 41–45 (2003). doi:10.1046/j.1365-8711.2003.06759.x

E. Legg, L. Miller, T.J. Turner, M. Giustini, J.N. Reeves, S.B. Kraemer, Direct measurement of the X-ray time-delay transfer function in active galactic nuclei. Astrophys. J. **760**, 73 (2012). doi:10.1088/0004-637X/760/1/73

A.P. Lobban, J.N. Reeves, L. Miller, T.J. Turner, V. Braito, S.B. Kraemer, D.M. Crenshaw, Contemporaneous Chandra HETG and Suzaku X-ray observations of NGC 4051. Mon. Not. R. Astron. Soc. **414**, 1965–1986 (2011). doi:10.1111/j.1365-2966.2011.18513.x

J. Malzac, Internal shocks at the origin of the flat spectral energy distribution of compact jets. Mon. Not. R. Astron. Soc. **429**, 20–24 (2013). doi:10.1093/mnrasl/sls017

J. Malzac, A. Merloni, A.C. Fabian, Jet-disc coupling through a common energy reservoir in the black hole XTE J1118+480. Mon. Not. R. Astron. Soc. **351**, 253–264 (2004). doi:10.1111/j.1365-2966.2004.07772.x

D. Maoz, A. Markowitz, R. Edelson, K. Nandra, X-Ray versus optical variations in the Seyfert 1 nucleus NGC 3516: a puzzling disconnectedness. Astron. J. **124**, 1988–1994 (2002). doi:10.1086/342937

A. Markowitz, R. Edelson, S. Vaughan, Long-term X-ray spectral variability in Seyfert 1 galaxies. Astrophys. J. **598**, 935–955 (2003). doi:10.1086/379103

B.A. Mazin, S.R. Meeker, M.J. Strader, P. Szypryt, D. Marsden, J.C. van Eyken, G.E. Duggan, A.B. Walter, G. Ulbricht, M. Johnson, B. Bumble, K. O'Brien, C. Stoughton, ARCONS: a 2024 pixel optical through near-IR cryogenic imaging spectrophotometer. Publ. Astron. Soc. Pac. **125**, 1348–1361 (2013). doi:10.1086/674013

I.M. McHardy, I.E. Papadakis, P. Uttley, M.J. Page, K.O. Mason, Combined long and short time-scale X-ray variability of NGC 4051 with RXTE and XMM-Newton. Mon. Not. R. Astron. Soc. **348**, 783–801 (2004). doi:10.1111/j.1365-2966.2004.07376.x

I.M. McHardy, P. Arévalo, P. Uttley, I.E. Papadakis, D.P. Summons, W. Brinkmann, M.J. Page, Discovery of multiple Lorentzian components in the X-ray timing properties of the Narrow Line Seyfert 1 Ark 564. Mon. Not. R. Astron. Soc. **382**, 985–994 (2007). doi:10.1111/j.1365-2966.2007.12411.x

J.C. McKinney, A. Tchekhovskoy, R.D. Blandford, General relativistic magnetohydrodynamic simulations of magnetically choked accretion flows around black holes. Mon. Not. R. Astron. Soc. **423**, 3083–3117 (2012). doi:10.1111/j.1365-2966.2012.21074.x

S. Migliari, R.P. Fender, M. van der Klis, Correlation between radio luminosity and X-ray timing frequencies in neutron star and black hole X-ray binaries. Mon. Not. R. Astron. Soc. **363**, 112–120 (2005). doi:10.1111/j.1365-2966.2005.09412.x

J.M. Miller, J. Homan, D. Steeghs, M. Rupen, R.W. Hunstead, R. Wijnands, P.A. Charles, A.C. Fabian, A long, hard look at the low/hard state in accreting black holes. Astrophys. J. **653**, 525–535 (2006). doi:10.1086/508644
L. Miller, T.J. Turner, J.N. Reeves, The absorption-dominated model for the X-ray spectra of type I active galaxies: MCG-6-30-15. Mon. Not. R. Astron. Soc. **399**, 69–73 (2009). doi:10.1111/j.1745-3933.2009.00726.x
L. Miller, T.J. Turner, J.N. Reeves, V. Braito, X-ray reverberation in 1H0707-495 revisited. Mon. Not. R. Astron. Soc. **408**, 1928–1935 (2010). doi:10.1111/j.1365-2966.2010.17261.x
S. Miyamoto, S. Kitamoto, K. Mitsuda, T. Dotani, Delayed hard X-rays from Cygnus X-1. Nature **336**, 450–452 (1988). doi:10.1038/336450a0
C. Motch, M.J. Ricketts, C.G. Page, S.A. Ilovaisky, C. Chevalier, Simultaneous X-ray/optical observations of GX339-4 during the May 1981 optically bright state. Astron. Astrophys. **119**, 171–176 (1983)
C.G. Mundell, P. Ferruit, N. Nagar, A.S. Wilson, Radio variability in Seyfert nuclei. Astrophys. J. **703**, 802–815 (2009). doi:10.1088/0004-637X/703/1/802
K. Nandra, T. Le, I.M. George, R.A. Edelson, R.F. Mushotzky, B.M. Peterson, T.J. Turner, The origin of the X-ray and ultraviolet emission in NGC 7469. Astrophys. J. **544**, 734–746 (2000). doi:10.1086/317237
K. Nandra, D. Barret, X. Barcons, A. Fabian, J.-W. den Herder, L. Piro, M. Watson, C. Adami, J. Aird, J.M. Afonso et al., The Hot and Energetic Universe: A White Paper presenting the science theme motivating the Athena+ mission. ArXiv e-prints (2013)
C. Nipoti, K.M. Blundell, J. Binney, Radio-loud flares from microquasars and radio-loudness of quasars. Mon. Not. R. Astron. Soc. **361**, 633–637 (2005). doi:10.1111/j.1365-2966.2005.09194.x
M.A. Nowak, Are there three peaks in the power spectra of GX 339-4 and Cyg X-1? Mon. Not. R. Astron. Soc. **318**, 361–367 (2000). doi:10.1046/j.1365-8711.2000.03668.x
M.A. Nowak, J. Wilms, J.B. Dove, Low-luminosity states of the black hole candidate GX 339-4. II. Timing analysis. Astrophys. J. **517**, 355–366 (1999). doi:10.1086/307189
S.M. O'Neill, C.S. Reynolds, M.C. Miller, K.A. Sorathia, Low-frequency oscillations in global simulations of black hole accretion. Astrophys. J. **736**, 107 (2011). doi:10.1088/0004-637X/736/2/107
I.E. Papadakis, Z. Ioannou, D. Kazanas, Fourier-resolved spectroscopy of active galactic nuclei using XMM-Newton data. I. The 3–10 keV band results. Astrophys. J. **661**, 38–51 (2007). doi:10.1086/513307
D.G. Payne, Time-dependent comptonization—X-ray reverberations. Astrophys. J. **237**, 951–963 (1980). doi:10.1086/157941
B.M. Peterson, I.M. McHardy, B.J. Wilkes, P. Berlind, R. Bertram, M. Calkins, S.J. Collier, J.P. Huchra, S. Mathur, I. Papadakis, J. Peters, R.W. Pogge, P. Romano, S. Tokarz, P. Uttley, M. Vestergaard, R.M. Wagner, X-ray and optical variability in NGC 4051 and the nature of narrow-line Seyfert 1 galaxies. Astrophys. J. **542**, 161–174 (2000). doi:10.1086/309518
G. Ponti, G. Miniutti, M. Cappi, L. Maraschi, A.C. Fabian, K. Iwasawa, XMM-Newton study of the complex and variable spectrum of NGC 4051. Mon. Not. R. Astron. Soc. **368**, 903–916 (2006). doi:10.1111/j.1365-2966.2006.10165.x
K. Pottschmidt, J. Wilms, M.A. Nowak, W.A. Heindl, D.M. Smith, R. Staubert, Temporal evolution of X-ray lags in Cygnus X-1. Astron. Astrophys. **357**, 17–20 (2000)
K. Pottschmidt, J. Wilms, M.A. Nowak, G.G. Pooley, T. Gleissner, W.A. Heindl, D.M. Smith, R. Remillard, R. Staubert, Long term variability of Cygnus X-1. I. X-ray spectral-temporal correlations in the hard state. Astron. Astrophys. **407**, 1039–1058 (2003). doi:10.1051/0004-6361:20030906
P. Reig, N.D. Kylafis, D. Giannios, Energy and time-lag spectra of galactic black-hole X-ray sources in the low/hard state. Astron. Astrophys. **403**, 15–18 (2003). doi:10.1051/0004-6361:20030449
M. Revnivtsev, M. Gilfanov, E. Churazov, Reflection and noise in the low spectral state of GX 339-4. Astron. Astrophys. **380**, 520–525 (2001). doi:10.1051/0004-6361:20011413
G. Risaliti, M. Elvis, G. Fabbiano, A. Baldi, A. Zezas, Rapid compton-thick/compton-thin transitions in the Seyfert 2 galaxy NGC 1365. Astrophys. J. Lett. **623**, 93–96 (2005). doi:10.1086/430252
E. Rivers, A. Markowitz, R. Rothschild, An occultation event in Centaurus A and the Clumpy Torus Model. Astrophys. J. Lett. **742**, 29 (2011). doi:10.1088/2041-8205/742/2/L29
D.M. Russell, J.C.A. Miller-Jones, T.J. Maccarone, Y.J. Yang, R.P. Fender, F. Lewis, Testing the jet quenching paradigm with an ultradeep observation of a steadily soft state black hole. Astrophys. J. Lett. **739**, 19 (2011). doi:10.1088/2041-8205/739/1/L19
S.G. Sergeev, V.T. Doroshenko, Y.V. Golubinskiy, N.I. Merkulova, E.A. Sergeeva, Lag-luminosity relationship for interband lags between variations in B, V, R, and I bands in active galactic nuclei. Astrophys. J. **622**, 129–135 (2005). doi:10.1086/427820
M. Spada, G. Ghisellini, D. Lazzati, A. Celotti, Internal shocks in the jets of radio-loud quasars. Mon. Not. R. Astron. Soc. **325**, 1559–1570 (2001). doi:10.1046/j.1365-8711.2001.04557.x
M.M. Tatum, T.J. Turner, L. Miller, J.N. Reeves, The global implications of the hard X-ray excess in Type 1 active galactic nuclei. Astrophys. J. **762**, 80 (2013). doi:10.1088/0004-637X/762/2/80

R.D. Taylor, P. Uttley, I.M. McHardy, The nature of X-ray spectral variability in Seyfert galaxies. Mon. Not. R. Astron. Soc. **342**, 31–35 (2003). doi:10.1046/j.1365-8711.2003.06742.x

P. Uttley, I.M. McHardy, The flux-dependent amplitude of broadband noise variability in X-ray binaries and active galaxies. Mon. Not. R. Astron. Soc. **323**, 26–30 (2001). doi:10.1046/j.1365-8711.2001.04496.x

P. Uttley, I.M. McHardy, S. Vaughan, Non-linear X-ray variability in X-ray binaries and active galaxies. Mon. Not. R. Astron. Soc. **359**, 345–362 (2005). doi:10.1111/j.1365-2966.2005.08886.x

P. Uttley, R. Edelson, I.M. McHardy, B.M. Peterson, A. Markowitz, Correlated long-term optical and X-ray variations in NGC 5548. Astrophys. J. Lett. **584**, 53–56 (2003). doi:10.1086/373887

P. Uttley, T. Wilkinson, P. Cassatella, J. Wilms, K. Pottschmidt, M. Hanke, M. Böck, The causal connection between disc and power-law variability in hard state black hole X-ray binaries. Mon. Not. R. Astron. Soc. **414**, 60–64 (2011). doi:10.1111/j.1745-3933.2011.01056.x

S. Vaughan, A.C. Fabian, A long hard look at MCG-6-30-15 with XMM-Newton—II. Detailed EPIC analysis and modelling. Mon. Not. R. Astron. Soc. **348**, 1415–1438 (2004). doi:10.1111/j.1365-2966.2004.07456.x

S. Vaughan, A.C. Fabian, K. Nandra, X-ray continuum variability of MCG-6-30-15. Mon. Not. R. Astron. Soc. **339**, 1237–1255 (2003). doi:10.1046/j.1365-8711.2003.06285.x

A. Veledina, J. Poutanen, I. Vurm, A Synchrotron self-compton-disk reprocessing model for optical/X-ray correlation in black hole X-ray binaries. Astrophys. J. Lett. **737**, 17 (2011). doi:10.1088/2041-8205/737/1/L17

A. Veledina, J. Poutanen, I. Vurm, Hot accretion flow in black hole binaries: a link connecting X-rays to the infrared. Mon. Not. R. Astron. Soc. **430**, 3196–3212 (2013). doi:10.1093/mnras/stt124

D.J. Walton, A. Zoghbi, E.M. Cackett, P. Uttley, F.A. Harrison, A.C. Fabian, E. Kara, J.M. Miller, R.C. Reis, C.S. Reynolds, Hard X-ray lags in active galactic nuclei: testing the distant reverberation hypothesis with NGC 6814. Astrophys. J. Lett. **777**, 23 (2013). doi:10.1088/2041-8205/777/2/L23

I. Wanders, B.M. Peterson, D. Alloin, T.R. Ayres, J. Clavel, D.M. Crenshaw, K. Horne, G.A. Kriss, J.H. Krolik, M.A. Malkan, H. Netzer, P.T. O'Brien, G.A. Reichert, P.M. Rodriguez-Pascual, W. Wamsteker, T. Alexander, K.S.J. Anderson, E. Benitez, N.G. Bochkarev, A.N. Burenkov, F.-Z. Cheng, S.J. Collier, A. Comastri, M. Dietrich, D. Dultzin-Hacyan, B.R. Espey, A.V. Filippenko, C.M. Gaskell, I.M. George, M.R. Goad, L.C. Ho, S. Kaspi, W. Kollatschny, K.T. Korista, A. Laor, G.M. MacAlpine, M. Mignoli, S.L. Morris, K. Nandra, S. Penton, R.W. Pogge, R.L. Ptak, J.M. Rodriguez-Espinoza, M. Santos-Lleo, A.I. Shapovalova, J.M. Shull, S.A. Snedden, L.S. Sparke, G.M. Stirpe, W.-H. Sun, T.J. Turner, M.-H. Ulrich, T.-G. Wang, C. Wei, W.F. Welsh, S.-J. Xue, Z.-L. Zou, Steps toward determination of the size and structure of the broad-line region in active galactic nuclei. XI. Intensive monitoring of the ultraviolet spectrum of NGC 7469. Astrophys. J. Suppl. **113**, 69 (1997). doi:10.1086/313054

T. Wilkinson, P. Uttley, Accretion disc variability in the hard state of black hole X-ray binaries. Mon. Not. R. Astron. Soc. **397**, 666–676 (2009). doi:10.1111/j.1365-2966.2009.15008.x

J. Wilms, K. Pottschmidt, G.G. Pooley, S. Markoff, M.A. Nowak, I. Kreykenbohm, R.E. Rothschild, Correlated radio-X-ray variability of galactic black holes: a radio-X-ray flare in Cygnus X-1. Astrophys. J. Lett. **663**, 97–100 (2007). doi:10.1086/520508

A. Zoghbi, P. Uttley, A.C. Fabian, Understanding reverberation lags in 1H0707-495. Mon. Not. R. Astron. Soc. **412**, 59–64 (2011). doi:10.1111/j.1365-2966.2010.17883.x

DOI 10.1007/978-1-4939-2227-7_23
Reprinted from *Space Science Reviews* Journal, DOI 10.1007/s11214-013-0026-2

Black Hole Studies: Overview and Outlook

Thomas J. Maccarone

Received: 21 February 2013 / Accepted: 28 September 2013 / Published online: 19 November 2013

Abstract In this article, I will attempt to give an overview of the motivations for studying black holes and of the current major problems in the field. I will also give some perspectives on what can be done in the future, focusing on instrumentation which has already been approved. This chapter will necessarily be more speculative than the other chapters in this volume.

Keywords Accretion · Accretion disks · Galaxies: active · X-Rays: binaries

1 Introduction

The first rapid expansion in our knowledge about accretion onto compact objects took place in the 1970s with the first satellite observations from Ariel V and Uhuru. The basic foundations of accretion theory were laid out then, as well as the start of the development of detailed phenomenology. On multiple occasions in recent years, major re-discoveries have been made of phenomena first found in that era. The launch of the Rossi X-ray Timing Explorer began a new era of rapid growth in our understanding of accretion, which has been further augmented by the Chandra and XMM-Newton Observatories. At the same time, multi-wavelength campaigns from the radio through the X-rays and γ-rays, once common only for blazar research, have become far more common. In many cases, these campaigns have made excellent use of small optical telescopes, while in more recent years, they have benefitted from the largest optical telescopes on Earth, and from the upgraded Jansky Very Large Array. The chapters in this book should neatly outline the progress made in the past decade or so, in the context of what has been known for quite some time. In this chapter, I will attempt to lay out some key questions remaining in accretion phenomenology and theory, the key motivations for addressing these questions, and discuss the landscape of future facilities likely to exist, and what should be possible with these facilities. In particular, I will attempt to demonstrate the complimentarity of observations of X-ray binaries and of active

T.J. Maccarone (✉)
Department of Physics, Texas Tech University, Box 41051, Lubbock, TX, 79404-1051, USA
e-mail: thomas.maccarone@ttu.edu

galactic nuclei as two different means for getting at the basic physics of accretion by black holes. Finally, I aim to make clear that several upcoming projects will have profound implications for our understanding of black hole accretion, even if, in some cases the motivations of those projects are to study quite different phenomena.

2 Key Problems in Black Hole Astrophysics

The key problems in black hole astrophysics remain more or less as they have been for quite some time: what are the key system parameters (e.g. masses and spins); what is the accretion geometry and on what does it depend; what mechanism supplies the mass and kinetic energy to relativistic jets, what are the jets made of (i.e. protons, pairs, Poynting flux, or some combination of the three), and how does jet production depend on system parameters?

One can think about the key goals of black hole research from a few points of view—that of the physicist and that of the astronomer. I distinguish these two classes of scientist not on the basis of the methodology they use nor the objects they study, but based on whether their primary motivation is to understand the Universe and the objects within it, or to use the Universe as a laboratory for understanding physics that cannot be well-tested in conventional laboratories.

From the physicist's point of view, black holes represent one of the few classes of systems in the Universe in which "strong field" general relativity can be studied. Neutron stars also present these opportunities, but for the case of neutron stars, the relativistic effects will often be entangled with poorly understood nuclear physics, such that the physics one is likely to test with neutron stars is more likely to be physics of the strong interaction.[1] Supermassive black holes, through their relativistic jets, are generally thought to be the sites from which the highest energy cosmic rays are launched—and these cosmic rays can be $\sim 10^8$ times more energetic than the highest energy particles that are currently being produced at the Large Hadron Collider. Finally, merging black holes are a likely source of strong gravitational radiation. There are two important implications of this—that they might merely be detectable would be exciting as arguably the most stringent test of general relativity to date; but furthermore, if general relativity is the correct description of gravity, then, in principle, a well-measured black hole merger can be used as a "standard siren" for cosmology, giving a clean distance estimate without the systematics that plague electromagnetic distance indicators (see e.g. Schutz 1999). Braneworld cosmologies can be tested using the mere presence of stellar mass black holes, since in these theories, Hawking radiation can work much more quickly than in standard $3 + 1$ dimensional general relativity—for example, the presence of a stellar mass black hole in a globular cluster, presuming that the black hole formed at the same time as the rest of the cluster's stars, yields a constraint on the size scale of extra dimensions in the prominent Randall-Sundrum 2 model that is about ten times as stringent as that from torsion pendulum measurements (Gnedin et al. 2009).

From the astronomer's point of view, accretion physics itself is interesting. Some aspects of accretion physics learned from black hole accretion disks may be applicable to cataclysmic variables, accreting neutron stars, and protostars. In terms of implications for understanding "live" stars, black holes represent the remnants of massive stars. Both the

[1]On the other hand, it should also be noted that the maximum mass of a neutron star is far more sensitive to whether general relativity properly describes gravity than is anything measureable from black hole accretion disks (see e.g. Psaltis 2008).

mass distribution and the spin distribution of black holes yield important information about the supernovae that produced the black holes (and possibly about the late stages of stellar evolution of massive stars). A concern will remain that the measurement techniques currently available to get at black hole masses and spins work only for accreting objects, and are predominantly applied in low mass X-ray binaries. Outside globular clusters, the black holes in soft X-ray transients are produced predominantly (perhaps exclusively) through common envelope evolution, so that the mass distribution of black holes we can measure may not represent the typical population seen in wide binaries and single stars. Most theoretical work on the topic suggests that black holes in wide binaries or seen as single stars should typically be more massive than those in close binaries (see e.g. Fryer et al. 2012).

Supermassive black holes tie in with other areas of astronomy in quite a different manner than do stellar mass black holes. The close correlation between masses of galactic bulges and supermassive black holes can be taken as evidence that the growth of the two components is intimately intertwined, perhaps due to winds from black holes accreting near the Eddington rate (see e.g. King 2003). Watson et al. (2011) argue that the tight relationship between the radius of an AGN's broadline region and its luminosity makes AGN good standard candles—something long hoped for, given that AGN are so bright they can be seen to much larger distances than can normal galaxies. Jets from AGN are also widely argued to be the energy source that stops cooling flows from developing in galaxy clusters—or at least stops them from being as extreme as they might be (see e.g. Di Matteo et al. 2005). Feedback processes—mostly due to radiative feedback—from lower mass black holes have been proposed to be important in other contexts, reionizing dwarf galaxies (Justham and Schawinski 2012) and perhaps the whole high redshift universe (Mirabel et al. 2011).

On quite a different note, active galactic nuclei also present one of the few means to probe the cosmic infrared background free from systematics due to zodiacal and Galactic foreground light—one can look at the spectra of gamma-ray emitting blazars, and their evolution with redshift, and determine how much light is being absorbed by photon-photon pair production (see e.g. Krawczynski et al. 2000; Ackermann et al. 2012). At the same time, the fact that AGN jets can often produce substantial gamma-ray fluxes means that attempts to make indirect detections of dark matter must taken into account black hole activity. How important this is depends considerably on whether a putative dark matter particle will produce a continuum spectrum or a line spectrum through decay and/or annihilation.

Finally, the popular understanding of what X-rays do—pass through solids—can be taken advantage of. Many star formation indicators, for example, are strongly affected by the presence of dust. X-rays, coming mostly from X-ray binaries, may thus be useful star formation indicators, since hard X-rays are largely unaffected by dust (e.g. Grimm et al. 2003; Symeonidis et al. 2011).

3 The Complimentarity of Active Galactic Nuclei and X-Ray Binaries

Much of the physics of accretion should be common to active galactic nuclei and X-ray binaries. Anything which is purely general relativity/gravity—like many (but not all) of the key timescales for accretion—should be "scale-free", while other properties such as the disk temperatures and the corresponding atomic physics properties of the accretion disks may show significant qualitative behavioral changes as a function of black hole mass.

In some sense, then, the fundamental differences due to mass are mostly differences in the type of variability which is easily probed. The rapid variaiblity of accreting systems is best studied in active galactic nuclei; assuming that the characteristic timescale of variability will be linearly proportional to the black hole mass, then a nearby Seyfert galaxy, with a black hole of $10^7 M_\odot$ at a distance of 10 Mpc emitting near the Eddington luminosity will have a count rate per light crossing time of the Schwarzschild radius which is about 10^4 times as high as a stellar mass black hole of $10 M_\odot$ at 1 kpc emitting at its Eddington limit—provided that the background count rate is negligible compared to the source count rate, and that timeseries can be made that are long enough for stationarity to develop, the effects of having a greater mass win out over the effects of being more nearby.

On the other hand, long timescale variability can be studied much more easily in X-ray binaries. In the approximation of an α-disk, the viscous timescale is $3 \times 10^5 \alpha^{-4/5} \dot{M}_{16}^{-3/10} m_1^{1/4} R_{10}^{5/4}$ seconds (Eq. 5.69 of Frank et al. 2002), where α is the dimensionless viscosity parameter, $\dot{M}_{16}$ is the mass transfer rate in units of 10^{16} g/s, m_1 is the accretor mass in units of solar masses, and R_{10} is the radius from which the mass transfer is taking place, in units of cm. For a 10 $M_\odot$ black hole accreting at Eddingon, with $\alpha = 0.1$ assumed then the viscous timescale from 3 R_{SCH} will be about 75 seconds. For a $10^6 M_\odot$ black hole accreting at the Eddington limit, the viscous timescale from $3R_{SCH}$ will be about 40 years—thus fundamental change in the accretion rate should not be expected in active galactic nuclei over the career duration of the typical astronomer, while stellar mass black holes can show dramatic variability, with even the viscous timescale from the outer edge of typical accretion disks being of order a week to a month.

In fact, X-ray binaries with black hole accretors routinely show variability in X-ray luminosity by a factor of 10^6. In doing so, these systems present ideal places to study the effects of changing accretion rate on other properties of the system, with some key examples being the discoveries of spectral state phenomenology in both black hole and neutron star X-ray binaries (Tananbaum et al. 1972; van der Klis 1994), the discovery of the correlation between spectral state and whether a strong jet is present (Tananbaum et al. 1972; Fender et al. 1999). X-ray transients essentially serve as single-object controlled experiments—the mass and spin of the black hole cannot change on short timescales; changes that are seen must almost certainly be due to changes in accretion rate. Gleaning such information from studies of active galactic nuclei requires large, carefully selected samples of objects with excellent supporting observations (e.g. reverberation mapping or M–σ correlation estimates of the black hole masses) to estimate the system parameters. AGN thus serve as the best probes of "weather" in accretion flows, while X-ray binaries serve as the best probes of "climate".

System parameters represent another case where AGN and XRBs are complimentary. For example, the range of masses of black holes in active galactic nuclei has historically been taken to span a factor of 1000, from 10^6 to $10^9 M_\odot$. Recent work has found evidence for black holes both of higher masses at the high end of the range (e.g. van den Bosch et al. 2012) and lower masses at the low end of the range (e.g. Filippenko and Ho 2003). Stellar mass black holes less than $3M_\odot$ would be hard to distinguish from neutron stars,[2] while stellar mass black holes heavier than 300 $M_\odot$ would exceed the masses of the most massive stars in the Galaxy. In practice, the observed range of masses for stellar mass black holes spans only a factor of about 2, with theoretical models requiring a series of special, but not altogether unrealistic, circumstances to produce black holes more than about 30 $M_\odot$ (e.g. Belczynski et al. 2010). While stellar mass black holes span a much smaller range of masses than do supermassive black holes, they present the attractive feature that many of them have been measured precisely (the intrinsic scatter in the M–σ relation is about a factor of 2—Gültekin et al. 2009—while the uncertainties on the masses of stellar mass black holes are

[2] Although it is worth noting that some reasonable attempts have been made to identify surface effects from neutron stars which are prominent enough that absence of them would be evidence of absence of a surface.

typically about 25 %—McClintock and Remillard 2006), and there is considerable hope to measure more of them precisely in the near future. Some supermassive black hole masses have been measured precisely, mostly using megamasers (e.g. Kuo et al. 2011), and while there are prospects to provide more such measurements, megamasers are found in only a small fraction of active galactic nuclei, and furthermore, the megamasers tend to be found disproportionately in AGN with obscurred central engines (e.g. Braatz et al. 1997).

Spins represent a similar situation. The spins of supermassive black holes are far more likely to evolve over time than those of stellar mass black holes; supermassive black holes are generally thought to form from seeds which may be stellar mass black holes, but almost certainly must be a relatively small fraction of their final masses (Soltan 1982), and their spins may be changed by major or minor mergers as well; stellar mass black holes are generally not thought to accrete more than 10 % of their final masses, and they are unlikely to undergo mergers while still remaining in mass-transferring binary systems. The spins of stellar mass black holes thus are telling us something about stellar evolution and/or supernova explosions, while the spins of supermassive black holes are probably telling us something about the processes by which they grow. Spin measurements also represent a key difference between stellar mass and supermassive black holes. For the stellar mass black holes, there are two methods which are being used in earnest to estimate spins—fitting the continuum from the accretion disk, and fitting the reflection spectra. It may also turn out that the quasi-periodic oscillations seen from these systems give independent mechanisms for estimating the spins. Radiation pressure is far more important in AGN accretion disks than in X-ray binary disks, and it has turned out to be a much more difficult problem to develop continuum spin-fitting models for AGN than for X-ray binaries (see Done et al. 2012, for example, for a discussion of the state of the art for detailed AGN disk continuum modelling).

Certain spectral components vary in peak wavelength as a functions of both mass, and accretion rate. The accretion disk itself will have a temperature which scales as $M^{-1/4}$, so that bright X-ray binary accretion disks peak in the soft X-rays, while the accretion disks from bright active galactic nuclei peak in the ultraviolet or the bluer optical bands. At the same time, the frequency at which conical jets become optically thin to synchrotron self absorption scales roughly as $M^{-1/3}$ (Heinz and Sunyaev 2003), and the jets from X-ray binaries tend to peak in the infrared or optical bands.

Again, this leads to certain things being easier for X-ray binaries and certain other things. For X-ray binaries, the bolometric luminosities tend to be strongly dominated by the spectral region from 1–200 keV, except in the cases of sources at very low fractions of the Eddington limit. Other interesting parts of the accretion flow may overlap in wavelength—e.g. the infrared-through-ultraviolet emission in some systems may be produced with substantial contributions from thermal reprocessing of X-rays into optical photons in the outer accretion disk; jet emission (e.g. Casella et al. 2010); and synchrotron emission the inner optically thin accretion flow (e.g. Veledina et al. 2012)—indeed, this topic probably represented the most spirited debate of the ISSI workshop. Separating these apart will probably require a combination of timing and polarization measurements.

For active galactic nuclei, there are often peaks in the broadband spectrum in the mid-to-far infrared corresponding to dust reprocessing, in the blue-through-ultraviolet corresponding to the thermal accretion disk, and in the hard X-rays corresponding to an optically thin "corona." These peaks tend to be well-separated in most AGN, so one can usually be fairly certain about from which component the light is coming; the chief difficulties with broadband AGN spectroscopy are that one often needs to cover a very wide range of wavelengths to develop a good picture of the nature of an object, and that the extreme ultraviolet band is often a very important part of the spectrum, and, of course, tends to be the band in which observations are most difficult.

Studies of relativistic jets, too, are best done in a complimentary manner between active galactic nuclei and X-ray binaries. The first note made of jets from active galactic nuclei was made in 1918, but studies of the phenomenon did not take place in earnest until the discovery of radio emission much later. Radio emission was used to identify the optical counterpart to Cygnus X-1, and some work went on throughout the 1970's and 1980's investigating the nature of radio emission from these objects (including a reference to Cygnus X-3 as a "nano-quasar"—Phinney 1982—more than a decade before the usage of the term microquasar became commonplace), but a key discovery in the 1990's of superluminal motion from GRS 1915 + 105 motivating an explosion in the prominence of studies of X-ray binaries' radio emission (Mirabel and Rodríguez 1994).

As with resolution of jets, essentially any attempt to resolve the black hole event horizon will be far easier with supermassive black holes than with stellar mass ones—most especially the Galactic Center, and the black hole in M87 (see e.g. Heino Falcke's article in this volume for a detailed discussion of this possibility). While fluxes scale with d^{-2}, angular sizes scale only with d^{-1}, so the effects of having larger black holes win out over the effects of being further away for the purposes of making size scale measurements.

4 Some Key Challenges

There are a few key challenges in understanding black holes and the physics of accretion onto them that are widely understood to be important problems to solve. There are also some which are just as important, but which are the topic of less attention. In some cases, the solutions to the problem are clear, and just require hard work and telescope time (in some cases on telescopes which have not yet been built). In other cases, the solutions are less clear, and require real breakthroughs.

4.1 Challenge I: Systematic Errors in Our Measurements

Two main classes of systematic errors in measurements can be considered. One is the case of systematic errors in the data (or more specifically in the calibration of the data)—and this is a serious issue in X-ray astronomy that has traditionally received surprisingly little attention. Calibration in X-ray astronomy is based primarily on rocket flight measurements of the Crab made in the late 1960's. Given that the supernova which produced the Crab pulsar and the associated nebula took place about 1000 years ago, and these rocket flights were made about 45 years ago, the implicit assumption in much of X-ray astronomy that the Crab is a constant source should be questioned. The further discovery of flaring activity from the Crab in recent years strengthens the case for variability from the Crab. It is generally taken as a given that the calibration of X-ray spectrum will not be accurate to much better than a few percent,[3] and the broad wings in relativistic iron lines are typically not more than about 5 % of their local continuum flux—it is surprising that relatively little discussion has been made of the effects of systematic errors on black hole spin measurements—in fact, many refereed journal articles claim black hole spin measurements without any consideration of systematic errors in the fitting process.

[3]With the Rossi X-ray Timing Explorer, for many sources, including the Crab, acceptable fits in χ^2 terms can be found with relatively simple spectral models, with the systematic errors added to the data being no more than 1 %. This is suggestive of the calibration being good to about 1 %, but still runs into questions about how well the Crab spectrum is really known.

The other case of systematic errors in measurements comes from the use of models which make simplifications not justified empirically. To give an example, let us look at the methodology typically used for iron line fitting measurements of black hole spins. Ordinarily, for example, the emissivity profile of the line is assumed to follow a power law, and the ionization parameter is assumed to be constant both in time and in space.[4] The assumption that the emissivity profile is constant in time is clearly a dangerous one, since black hole X-ray binaries are often observed to vary by large factors on short timescales when strong reflection components are seen, and it is clear that at least the flux of the line varies in response to the continuum changes (e.g. Miller and Homan 2005). The assumption that the ionization parameter of the disk is constant spatially implicitly assumes a specific relationship between the disk's density profile and its emissivity profile.

The agreement between most of the continuum fitting measurements and iron line measurements for the small sample of objects for which both techniques have been applied is discussed in Chris Reynolds' article in this volume, and may be taken to provide some support for the idea that neither method is too highly flawed; on the other hand, with only 6 objects having been studied using both techniques, and two objects having statistically inconsistent spins, and two objects having 90 % confidence intervals spanning more than 0.4 in the value of the dimensionless spin parameter, the case that the spin estimates from the two techniques agree with one another because both techniques are accurate, rather than for reasons of random coincidence is not particularly strong.

4.2 Challenge II: Things for Which We Have No Measurement Techniques, or Measurement Techniques that Do Not Meet Our Current Desires

One of the goals of studies of accretion is to estimate the parameters of black holes, and to probe general relativity. The masses of stellar mass black holes can be measured with reasonably good precision, but the masses of supermassive black holes come predominantly from techniques like the M–σ relation or reverberation mapping. Gültekin et al. (2009) find that the M–σ relation typically has a factor of about 2 scatter, and similar scatter is found for reverberation mapping measurements (Peterson et al. 2004).

For black hole spins, to give a key example, an even more severe challenge presents itself. The techniques currently being employed to make estimates of black hole spins require geometrically thin, optically thick, Keplerian accretion disks that extend in to the innermost stable circular orbit around their host black holes. A general consensus has developed that the hard states of accreting black holes do not fit into this picture—there is a "hole" in the inner accretion disk.[5] Furthermore, it seems likely that at high accretion rates (i.e. above 0.3 times the Eddington luminosity), the accretion disks are not entirely Keplerian (e.g. Gu 2012), instead being partially pressure-supported.

A possible solution may exist—timing data may help provide black hole spin estimates (see Tomaso Belloni's article in this volume for a thorough discussion of timing measurements in accreting black holes). In particular, the Lense-Thirring precession (or some other mechanism) may provide a "clock" that produces low frequency quasi-periodic oscillations

[4] Although in some cases, variation in the ionization parameter is taken into account to allow, e.g., for narrow emission lines far from the black hole (e.g. Reynolds et al. 2012; Steiner et al. 2012).

[5] Some groups do debate whether this change happens exactly at the transition to the hard state, or at a somewhat lower fraction of the Eddington luminosity (e.g. Rykoff et al. 2007), but not whether this effect happens at all.

that are often seen in X-ray binaries. It might be possible, then to calibrate the spin-mass-accretion rate-QPO frequency relations in spectral states where the spins can be measured reliably, and then to estimate the spins in systems that never enter soft states from just the black hole masses, accretion rates and QPO frequencies. In general, timing measurements are "cleaner" than spectroscopic measurements, in the sense that the calibration uncertainties are usually negligible, so a good model, or even a good phenomenological relationship, that allows spins to be extracted from timing measurements would be of incredible benefit to black hole astrophysics. Methods for estimating black hole spins from timing measurements are clearly less developed than those for measuring black hole spins from spectra, but further exploration should be done to develop them, as they have the potential for being more robust than spectra.

5 How Bright is the Future for Black Hole Research?

Several recent papers have given overviews of the current and near-future landscape of space missions available (Ubertini and Gehrels 2012; Watson et al. 2011). To be sure at the present time, it appears that X-ray astronomy is exiting its current "golden age". At the present time, several missions are flying which provide dramatic improvements in some fundamentally important region of parameter space relative to missions that flew before them. The X-ray missions which are currently funded for launch are predominantly niche missions which provide valuable combinations of instruments that should allow substantial progress to be made, but which may not have the same potential for impact as more major observatories. Some proposed missions do exist with that potential—e.g. Athena (Barcons et al. 2012) would provide a factor of three improvement in effective area over XMM-Newton, along with wider field imaging, and better spectral resolution[6]. LOFT (Feroci et al. 2012) would provide about 20 times the collecting area of RXTE and about 5 times its spectral resolution.

There was spirited debate at the ISSI meeting over whether serious progress could be made in understanding accretion even if neither of these missions is selected. Certain topics which require high resolution spectroscopy of faint objects (e.g. a better understanding of disk winds from AGN) require something like Athena for major progress to be made. At the same time, many major results in recent years have come from well-orchestrated multi-wavelength campaigns using combinations of small ground-based telescopes, old radio telescopes, and medium-budget X-ray missions like Swift and RXTE. There is no reason to imagine that discoveries of that nature will stop if no new major X-ray mission is launched, and there is an array of projects being developed which work at longer wavelengths than X-rays which have excellent potential for being used to learn about black holes.

The author's own position skews toward the viewpoint that the next decade will be an excellent era to study black holes for scientists willing to make good use of the instruments that will be available, and so the focus of the remainder of this article will be on the opportunities that will clearly exist rather than being aimed at promoting new missions that might take place in the more distant future. A clear point of concern is that there needs to be some kind of imaging X-ray observatory in the sky to make the best use of other facilities (e.g. JWST when it launches), and at the present time, there is no mission even under consideration for funding which would have angular resolution within a factor of 5 of what Chandra can deliver.

[6]The improvement for the new concept Athena+ would be a factor of about 5 in effective area over XMM-Newton

5.1 Mid-sized and Small X-Ray Missions

While the future is uncertain for a mission that would provide capabilities in a similar vein to those of Chandra and XMM, albeit with a major increase in collecting area, the future is quite bright for smaller X-ray missions. Each of these will provide substantial new capabilities in a small niche, and the niches taken as a whole cover a good deal of what one would want in terms of advances in understanding accretion physics. Of particular note are the soon-to-be-launched ASTROSAT and recently launched NuStar. The advantages from NuStar—orders of magnitude higher sensitivity above 10 keV than has ever been achieved—are straightforward, so I will focus on the advantages from ASTROSAT.

Spectroscopic measurements of black hole spins currently dominate the literature. At the same time, the highest frequency quasi-periodic oscillations from stellar mass black hole accretion disks are faster than the Keplerian periods for these objects at the innermost stable circular orbit—a range of models exist in which the QPO frequencies give strong information about the spin of the black hole (e.g. Abramowicz and Kluźniak 2001; Rezzolla et al. 2003). At the same time, in nearly all models, the QPO frequency is inversely proportional to the mass of the black hole for a given spin period. ASTROSAT is due to launch soon, and while its collecting area is often presented as being the same as that of RXTE, in fact, at the highest energies, above 20 keV, where the high frequency QPOs from black holes are strongest (e.g. Strohmayer 2001), the collecting area of ASTROSAT is 3–5 times higher than that of the Proportional Counter Array on RXTE. ASTROSAT should be more efficient at finding high frequency QPOs from black hole X-ray binaries in outburst than was RXTE. Additionally, in recent years, an explanation for the discrepancies between ellipsoidal modulation inclination angles found by different groups—namely poor subtraction of the accretion light due to strong variability of the accretion light (Cantrell et al. 2010); furthermore, with this explanation has come a solution to the problem—collection of sufficiently large data sets with a sufficiently large number of filters that the accretion light can be estimated in a largely model independent manner. One can envision, in the near future, a large enough sample of QPO measurements for black holes whose masses are well-estimated that one could look for an empirical relation between other spin estimates for the black holes and the deviations from a purely linear relationship between mass and frequency. Even in the absence of a consensus QPO model, then, one would have hope of being able to use the QPOs to help test the reliability of the spectroscopic spin measurement techniques.

5.2 LSST and the Other Optical Transient and Variability Projects

One of the highest priorities in the recent US-based decadal survey was the Large Synoptic Survey Telescope. The chief contribution of the LSST will be to provide coverage of the entire night sky (from Chile) to a magnitude of 24.5 in the optical roughly every three days. For active galactic nuclei, it is clear that this will be a boon. It will not be necessary to obtain continuum measurements, but rather only to obtain spectroscopic measurements, in order to do standard reverberation mapping—one will be able to take for granted that photometric support data will be available, reducing the need either to switch instruments or to obtain data from multiple telescopes.

Additionally, it will be possible to compute reliable power spectra of a large sample of active galactic nuclei with LSST. It may, then, be possible to make reliable estimates of the masses of the central black holes of many systems without standard reverberation mapping techniques, if a reliable calibration can be made of a relation between a characteristic variability timescale and the black hole mass (or, perhaps, some combination of the mass and accretion rate—see e.g. Kelly et al. 2013). Such a relation has been found in the X-rays already (McHardy et al. 2006), and suffers in its applicability mainly due to the fact that the measurements needed use telescope time very intensively, and sometimes require campaigns of more than a year. Neither of these would be problematic for the LSST, since the observations needed would be ones that would be made anyway.

For X-ray binaries, LSST will act, effectively, as an all-sky monitor. Except in the most obscurred locations of the Galaxy, an all-sky monitor that reaches $i = 24.5$ will be more sensitive than existing X-ray all-sky monitors are for soft X-ray transient events. It has already been seen that regular optical monitoring, even with 1-m class telescopes, has detected the outbursts of X-ray binaries before they showed up with the RXTE all-sky monitor (Jain et al. 2001). Additionally, unlike X-ray all-sky monitors, LSST will give ellipsoidal modulations for free for interesting objects; in many cases, the orbital periods of new X-ray binaries will be known before they go into outburst.

Finally, detecting these ellipsoidal modulations, in combination with eROSITA measurements of the X-ray fluxes will often identify X-ray binaries which have not gone into outburst. eROSITA is expected to reach a flux limit of about 10^{-14} ergs/sec/cm^2 from 0.5–10 keV—this corresponds to a luminosity level of 8×10^{31} ergs/sec at a distance of 8 kpc—thus most neutron star X-ray binaries in quiescence should be detectable with eROSITA, and the longer period black hole X-ray binaries should be detectable as well. The opportunity will thus be opened to estimate masses for a much larger sample of black holes than has currently been studied. Additional new transients in outburst may be discovered in real time by other projects (e.g. PanStarrs and the Palomar Transient Factory; and retroactively by DASCH).

5.3 Gaia and X-Ray Binaries

The Gaia satellite is scheduled for launch in October 2013. It should provide full, accurate astrometric information for all stars brighter than 15th magnitude within about 5 kpc, and quite good astrometric information for a much larger number of stars. A large part of the error budget for black hole spins measured using the continuum-fitting measurement technique comes from the distance uncertainties to the sources. Mass estimates are also affected, albeit more mildly, by distance uncertainties. Full astrometric information plus masses plus spins will give interesting information about black hole formation—e.g. it will become possible to look for correlations between any natal kicks applied to black holes with the spins of the black holes. A second contribution Gaia will be able to make to the study of black holes is to look for black holes in wide binary systems through both astrometric wobble (for long period binaries) and radial velocity variation (for shorter period binaries). These systems may probe a mass distribution for black holes that is more typical of the mass spectrum for black holes produced over the lifetime of the Galaxy than is the mass distribution that is easily observed in black hole X-ray binaries.

5.4 The Next Generation of Large Optical Facilities and Detectors

Over the next decade, we should see the launch of the James Webb Space Telescope, which will provide a collecting area about 5 times as large as the Hubble Space Telescope, with an optimization in the infrared. We should also see the opening of at least one of the 30-meter class ground-based optical telescopes. These large telescopes should provide much better opportunities to do fast infrared timing, and to do optical follow-up to measure masses of black holes, especially those in crowded regions or those embedded deeply in the Galactic

Plane. The Wide Field Infrared Survey Telescope should also be launched within that time, and notably, its exoplanet microlensing survey lines up well with the Chandra Galactic Bulge Survey (Jonker et al. 2011), so some of the best ellipsoidal modulation curves of X-ray binaries should be generated by WFIRST, and these should be for some of the first X-ray binaries selected from their quiescent emission, rather than their outburst emission.

5.5 New Radio Facilities

The new class of radio facilities currently being detected, and planned for the future, will also provide a wealth of information about about accretion and about source parameters than we currently have. For stellar mass systems, much higher duty cycle monitoring will be possible for looking at radio/X-ray correlations than is currently being done. The Southern Hemisphere locations of the Square Kilometer Array pathfinders (ASKAP and MeerKAT) will allow for the bulk of X-ray binaries, which tend to be in the Galactic Plane and the Galactic Bulge, to be observed. The present situation, with the most powerful radio telescopes being located in the Northern Hemisphere, has, in some cases, hampered efforts to study X-ray binaries in the radio. Furthermore, even with MeerKAT, it should be possible to follow some of the brightest extragalactic X-ray binaries, and as the Square Kilometer Array is built, even more parameter space for studies of extragalactic X-ray binaries will be opened up.

For active galactic nuclei radio monitoring may take a step forward. Because the variability timescales of AGN are long, especially in the radio, the VLA has not represented an ideal instrument. A large fraction of active galactic nuclei, especially those with low mass black holes (i.e. those which vary most quickly), are located in regions of diffuse radio emission due to star formation. The array configuration changes of the VLA mean that different configurations detect different amounts of the diffuse emission, so that it can be difficult to separate genuine source variability from changes in the amount of detected diffuse emission. Having the large collecting areas, relatively long baselines, and consistent array configurations of the next generation of radio telescopes will allow ideal radio monitoring campaigns to be done.

Additionally, one can think more about what the Square Kilometer Array may due for black hole mass measurements. At the present time, black hole mass measurements with similar precision to those routinely made for stellar mass black holes are available for only a small number of supermassive black holes—the Galactic Center and those measured from maser proper motions. The Square Kilometer Array should be able to detect pulsars out to distances of a few tens of Mpc. Timing pulsars as they orbit around the Galactic Center black hole has long been suggested as a good technique for measuring its mass, and maybe even its spin. The chief problem in detecting such pulsars in the Galactic Center is that the column density of ionized gas in that direction is too large; for nearby galaxies which are not viewed edge-on, this should not be a problem. The question will remain whether there are enough bright pulsars in a given galaxy to make good measurements of black hole masses (particularly if natal kicks remove the pulsars from the central regions of the galaxies)—but there remains great potential from this technique.

6 Some Encouraging Global Socioeconomic Signs

In the current economic downturn, with science budgets being cut over a lot of the Western world, a tendency has come to think that the best times for science are in the past. One can, of course, attempt to keep a sense of perspective and expect the economy to rebound, and the science budgets to rise with it. Perhaps more importantly, it is worth looking at who is funding many of the new facilities that we *are* discussing, and how revolutionary we expect them to be. Some future X-ray/soft gamma-ray missions that we do expect to be launched include SVOM (which is mostly a Chinese mission); ASTROSAT (which is mostly an Indian mission with British and Canadian participation as well); and MIRAX (a Brazilian mission). The Square Kilometer Array includes a major South African contribution that goes far beyond merely hosting the observatory, and the Large Millimeter Telescope is predominantly a Mexican venture. The nations we now often call "developing" are doing just that—cuts in funding of science in the world's wealthiest nations are being offset, at least in part, by increases in funding from nations with emerging economies. Furthermore, these nations are not just bringing money to the table, but they are also rapidly increasing the number of talented scientists, as they begin to increase the rate at which they are developing the talent within their countries. At this ISSI workshop, the participants were all based in the United States, Europe and Japan. It is very unlikely that an ISSI meeting on this topic 10 years from now would have the same geographic distribution of participants.

7 Conclusions

In conclusion, there are a lot of opportunities to do cutting edge science on black hole accretion in the upcoming years. To be sure, a Great Observatory class NASA mission or an L-class ESA mission would be a tremendous boon to black hole accretion research; at the same time, facilities at longer and shorter wavelengths should be tremendously valuable, as should the continued development of theoretical work.

Acknowledgements I wish to thank the organizers and the participants of the workshop for an interesting program.

References

M.A. Abramowicz, W. Kluźniak, Astronomy & Astrophysics **374**, L19 (2001)
M. Ackermann, M. Ajello, A. Allafort et al., Science **338**, 1190 (2012)
X. Barcons, D. Barret, A. Derouchelle et al. (2012). arXiv:1207.2745
K. Belczynski, T. Bulik, C.L. Fryer et al., Astrophysical Journal **714**, 1217 (2010)
J.A. Braatz, A.S. Wilson, C. Henkel, Astrophys. J. Suppl. **110**, 321 (1997)
A. Cantrell et al., Astrophys. J. **710**, 1127 (2010)
P. Casella, T.J. Maccarone, K. O'Brien et al., Monthly Notices of the Royal Astronomical Society **404**, L21 (2010)
T. Di Matteo, V. Springel, L. Hernquist, Nature **433**, 604 (2005)
C. Done, S.W. Davis, C. Jin, O. Blaes, M. Ward, Monthly Notices of the Royal Astronomical Society **420**, 1848 (2012)
R. Fender, S. Corbel, T. Tzioumis et al., Astrophysical Journal Letters **519**, L165 (1999)
M. Feroci, L. Stella, M. van der Klis et al., Experimental Astronomy **34**, 415 (2012)
A.V. Filippenko, L.C. Ho, Astrophysical Journal Letters **588**, L13 (2003)
J. Frank, A. King, D.J. Raine, in *Accretion Power in Astrophysics*, ed. by J. Frank, A. King, D. Raine, p. 398 (Cambridge University Press, Cambridge, 2002). ISBN 0521620538
C.L. Fryer, K. Belczynski, G. Wiktorowicz et al., Astrophysical Journal **749**, 91 (2012)
O.Y. Gnedin, T.J. Maccarone, D. Psaltis, S.E. Zepf, Astrophysical Journal Letters **705**, L168 (2009)
H.-J. Grimm, M. Gilfanov, R. Sunyaev, Monthly Notices of the Royal Astronomical Society **339**, 793 (2003)
W.-M. Gu, Astrophysical Journal **753**, 118 (2012)
K. Gültekin, D.O. Richstone, K. Gebhardt et al., Astrophysical Journal **698**, 198 (2009)
S. Heinz, R.A. Sunyaev, Monthly Notices of the Royal Astronomical Society **343**, L59 (2003)

R.K. Jain, C.D. Bailyn, J.A. Orosz, J.E. McClintock, R.A. Remillard, Astrophysical Journal Letters **554**, L181 (2001)
P.G. Jonker, C.G. Bassa, G. Nelemans et al., Astrophysical Journal Supplement **194**, 18 (2011)
S. Justham, K. Schawinski, Monthly Notices of the Royal Astronomical Society **423**, 1641 (2012)
B.C. Kelly, T. Treu, M. Malkan, A. Pancoast, J.-H. Woo (2013). arXiv:1307.5253
A. King, Astrophysical Journal Letters **596**, L27 (2003)
H. Krawczynski, P.S. Coppi, T. Maccarone, F.A. Aharonian, Astronomy & Astrophysics **353**, 97 (2000)
C.Y. Kuo, J.A. Braatz, J.J. Condon et al., Astrophys. J. **727**, 20 (2011)
J.E. McClintock, R.A. Remillard, *Compact Stellar X-ray Sources* (2006), p. 157
I.M. McHardy, E. Koerding, C. Knigge, P. Uttley, R.P. Fender, Nature **444**, 730 (2006)
J.M. Miller, J. Homan, Astrophysical Journal Letters **618**, L107 (2005)
I.F. Mirabel, L.F. Rodríguez, Nature **371**, 46 (1994)
I.F. Mirabel, M. Dijkstra, P. Laurent, A. Loeb, J.R. Pritchard, Astronomy & Astrophysics **528**, A149 (2011)
B.M. Peterson, L. Ferrarese, K.M. Gilbert et al., Astrophysical Journal **613**, 682 (2004)
E.S. Phinney, Monthly Notices of the Royal Astronomical Society **198**, 1109 (1982)
P. Psaltis, Astrophys. J. **688**, 1282 (2008)
C.S. Reynolds, L.W. Brenneman, A.M. Lohfink et al., Astrophys. J. **755**, 88 (2012)
L. Rezzolla, S. Yoshida, T.J. Maccarone, O. Zanotti, Monthly Notices of the Royal Astronomical Society **344**, L37 (2003)
E.S. Rykoff, J.M. Miller, D. Steeghs, M.A.P. Torres, Astrophysical Journal **666**, 1129 (2007)
B.F. Schutz, Classical and Quantum Gravity **16**, A131 (1999)
A. Soltan, Monthly Notices of the Royal Astronomical Society **200**, 115 (1982)
J.F. Steiner, R.C. Reis, A.C. Fabian et al., Monthly Notices of the Royal Astronomical Society **427**, 2552 (2012)
T.E. Strohmayer, Astrophysical Journal Letters **554**, L169 (2001)
M. Symeonidis, A. Georgakakis, N. Seymour et al., Monthly Notices of the Royal Astronomical Society **417**, 2239 (2011)
H. Tananbaum, H. Gursky, E. Kellogg, R. Giacconi, C. Jones, Astrophysical Journal Letters **177**, L5 (1972)
P. Ubertini, N. Gehrels, A Space Astronomy Global Road Map for the Next Decades, in *3th COSPAR Scientific Assmebly*, vol. 39 (2012)
M. van der Klis, Astrophysical Journal Supplement **92**, 511 (1994)
R.C.E. van den Bosch, K. Gebhardt, K. Gültekin et al., Nature **491**, 729 (2012)
A. Veledina, J. Poutanen, I. Vurm (2012). arXiv:1210.0236 [astro-pH.HE]
D. Watson, K.D. Denney, M. Vestergaard, T.M. Davis, Astrophysical Journal Letters **740**, L49 (2011)

GPSR Compliance

The European Union's (EU) General Product Safety Regulation (GPSR) is a set of rules that requires consumer products to be safe and our obligations to ensure this.

If you have any concerns about our products, you can contact us on ProductSafety@springernature.com

In case Publisher is established outside the EU, the EU authorized representative is:

Springer Nature Customer Service Center GmbH
Europaplatz 3
69115 Heidelberg, Germany

Batch number: 10837402

Printed by Printforce, the Netherlands